IRRIGATION ENGINEERING

For

Degree, Diploma and A.I.M.E. (India) Students

In SI UNITS

Gurcharan Singh
Joint Director (Retd.)
Directorate of Technical Education
Rajasthan, Jodhpur

STANDARD BOOK HOUSE

unit of: **RAJSONS PUBLICATIONS PVT. LTD.**

1705-A, Nai Sarak, PB.No. 1074, Delhi-110006 Ph.: +91-(011)-23265506
Show Room: 4262/3, First Lane, G-Floor, Gali Punjabian, Ansari Road, Darya Ganj, New Delhi-110002 Ph.: +91-(011)43551085 Tel Fax : +91-(011)43551185, Fax: +91-(011)-23250212

E-mail: sbh10@hotmail.com www.standardbookhouse.com

Published by:

RAJINDER KUMAR JAIN

Standard Book House

Unit of: Rajsons Publications Pvt. Ltd.

1705-A, Nai Sarak, Delhi - 110006

Post Box: 1074

Ph.: +91-(011)-23265506 Fax: +91-(011)-23250212

Showroom:

4262/3. First Lane, G-Floor, Gali Punjabian,

Ansari Road, Darya Ganj.

New Delhi-110002

Ph.: +91-(011)-43551085, +91-(011)-43551185

E-mail: sbhl0@ hotmail.com

Web: www.standardbookhouse.com

Sixth Edition : 2021

Price: **Rs. 220.00**

ISBN: 978-81-89401-32-0

Typeset by:

C.S.M.S. Computers, Delhi.

Printed by:

Radha Press, Delh-110031

Preface to the Second Edition

The book "Irrigation Engineering" has completely been revised, enlarged and updated. Latest Indian standards pertaining to the design of various irrigation structures have been incorporated. Lot of latest literature has been added in almost all the chapters so that the present book may fulfill the latest need of both degree and diploma students of Civil Engineering. The author appreciates the effort put in by the publishers in bringing out the book in such a nice form. Suggestions from the readers for the improvement of the forthcoming editios of the book will be most gratefully acknowledged.

July 2010 GURCHARAN SINGH

Preface to the First Edition

The subject "Irrigation Engineering" has assumed importance since last 30 to 40 years. Continued increase in population, particular in developing countries, at a very fast rate has caused scarcity of food. The real answer to food problem, is increased production of food articles; which is possible only by artificial irrigation of fields.

India has a very large potential for irrigation, because area and water resources both are abundantly available. Abundance of area for irrigation arid availability of lot of water resources are probably the reasons that most of the early irrigation practices and theories were developed in India.

There is lot of variations in rainfall in different regions of India. Some of the areas have very little rainfall insufficient to grow any crop. Other areas have sufficient rainfall but its distribution is not as required by the crops. Scanty rainfall and erratic distribution both necessitate artificial irrigation.

The purpose of this book is to present the subject in most concise form. Simplicity of language is the main feature of the book. The book is completely in MKS units and covers the syllabus of all the Indian Universities, State Technical Boards, and A.M.I.E. (India) examinations. The book should be equally useful to practising Engineers as reference book. Examples of almost all the important irrigation works have been solved and then illustrated in neat drawing charts. Khosla's Charts, Lacey's and Garret diagrams all are in MKS units.

Every effort was made to eliminate printing errors. I would appreciate if printing errors are brought to my notice and Suggestions to bring about improvements in the book are most welcome.

I am thankful to all my friends who have rendered great help by their valuable suggestions. In last I am thankful to Shri R.K. Jain, Prop. Standard Book House, without whose efforts this venture would not have reached the readers.

July 1980 **GURCHARAN SINGH**

Contents

1

Introduction and Necessity of Irrigation

1.1 INTRODUCTION

Water is the greatest natural gift to the mankind. All the human beings and plant life thrive on the earth only because of water. All the early civilisations flourished along great rivers, because man had realised in the very beginning that water is essential for life. Today, out of total water consumed for human needs in the world, about 80 per cent is used for raising the crops. The use of water for raising the crops is known as *irrigation*. Irrigation has assumed prominence throughout the world as it is the only way by which food production may be augmented for the ever increasing population of the world. Crops are raised by rains also in most parts of the world, but uncertainty of rains, and amount of rains are uncontrollable features. There may not be any rains or there may be insufficient rain when it is most required by the crop. This aspect affects the yield of the crop. Hence the only way of supplying required amount of water to the crop, when it is most required, is by way of irrigation.

The sources of water may be divided under two heads—namely—*surface sources* and *ground sources*. Streams, lakes etc., are the surface sources whereas wells, infiltration gallories, and springs are examples of ground sources. Harnessing these sources and delivering water to the fields for raising crops is known as irrigation and is the subject matter of this book.

At present it is estimated that only about 15% of the agriculturable land of the world is receiving assured irrigation facilities whereas present sources of water are enough to provide irrigation to about 50% of the irrigable area of the world.

Hence it can be said that there is a lot of scope to bring about improvements in irrigation methods. To supplement the food requirements of fast increasing population, new technologies like sea water disalienation and even weather modification may have to be evolved.

1.2 HYDROLOGICAL CYCLE

It is a natural system by which moisture keeps on circulating from ocean to atmosphere and again to ocean. Water from oceans enters atmosphere due to evaporation and forms clouds. The water carried by clouds again returns to ocean in form of precipitation of rainfall. Precipitation water again joins ocean by way of percolation through ground and by way of flow through rivers and streams. The part of water absorbed by the soil is used by the plants or vegetation, for their growth. Part of soaked water keeps on moving downwards under the influence of gravity and joins the ground water reservoir. The water from ground water reservoir is obtained in form of wells. Hence we see that different sources of water keep on replenishing each other by way of the elements of the hydrologic cycle i.e. precipitation, evaporation, stream flow and ground water conditions. The elements of the hydrological cycle should be measured accurately to assess the capability of water resources and its proportion which can be used for irrigation and other human requirements.

1.3 IRRIGATION

Artificial application of water to the soil for raising crops is known as irrigation. It is a science which pertains to the planning, designing, controlling and maintenance of irrigation works so that available water resources may be used in the best possible manner. Irrigation works involve construction of dams, reservoirs, canals, headworks, cross-drainage works, river training works etc. Methods of application of water to crops, measures to prevent water logging and generation of hydroelectric power also come under the purview of irrigation.

1.4 WHY IRRIGATION IS NECESSARY

Irrigation of crops becomes necessary under the following circumstances.

1. When rainfall is very small. In such a case rainfall is not sufficient to mature the crops. Water has to be supplied to the crops artificially. In such a case irrigation works are constructed at far places where water for irrigation is available. This water is then conveyed to the deficient areas through a system of canals and used for irrigation purposes. Rajasthan Canal is an example of such an irrigation system. This canal carries surplus waters of Punjab rivers to and zones of Rajasthan where rainfall is very scanty.

2. When rainfall is not distributed over the crop period. There are certain areas where overall annual rainfall statistics speak of sufficient rainfall for growing crops. But rainfall not being uniformly distributed for the crop period, it is not possible to mature the crops by rainfall alone. Rainfall may be sufficient or even excessive, during early periods of the crop, and later on, there may be no rains or very little rains to cause the proper maturity of the crop. The ultimate result of such a condition is that either crops may die altogether or their yield is affected. In such areas excess water during excessive rainfalls is collected and used for irrigation purposes when rainfall is very scanty. In India rainfall is very small during winters. Most of the irrigation projects in India are based on this concept. Excessive rain water during heavy rains is kept stored in form of reservoirs by constructing dams and weirs. This stored water is not used during monsoons but used during winter for growing second crop i.e. Rabi crop.

3. For growing cash and other commercial crops. Irrigation also becomes necessary where it is proposed to grow cash crops and other commercial crops, which generally require very large amount of water, and that too at fixed intervals. Overall rainfall data may be appealing not to instal any irrigation system, but requirements of water of cash crops at fixed intervals which cannot be assured by rainfall alone may warrant the necessity of installation of an irrigation system.

1.5 FUNCTIONS OF IRRIGATION WATER

Water when applied to fields performs the following functions:

1. It supplies moisture to the soil which is essential for the life of bacteria favourable to the plant growth.
2. It lowers the temperature of the soil during summer and thus provides more favourable conditions for plant growth.
3. It saves the plants from the harmful effects of frost during intensive colds.
4. It is essential for the chemical action within the plant leading to its growth.
5. It supplies moisture or water to the soil at fixed interval so as to replenish the moisture content of the soil.
6. Tillage of the fields becomes easy as irrigation water renders the fields soft.
7. It also reduces the concentration of the harmful salts in the soil. Harmful salts get washed out or diluted by irrigation water.
8. It helps dissolve chemical manures and renders them effective for plant growth. Thus water acts as nutrient carrier.
9. It helps in bringing up ground water table. Higher ground water table is very helpful in the installation of tube wells. Tube wells relieve lot of irrigation pressure from canals and canal water thus saved, may be used for developing new areas.
10. Irrigation water is essential for producing nourishing food for the plants.

1.6 ADVANTAGES OR BENEFITS OF IRRIGATION

Earlier, irrigation projects were sanctioned only if after deducting the maintenance charges, they would ensure return of minimum prescribed rate of interest on the capital outlay. Sometimes unproductive schemes were also sanctioned. Nowadays, irrigation projects are sanctioned on the basis of benefit cost ratio. This ratio should be generally more than 1.5. Direct and indirect benefits from irrigation project have been summarised as follows :

1. Food production is increased. This helps in the solution of food problem of any country.
2. Production of other crops also increases.
3. Due to assured controlled supply of irrigation water it is possible to grow cash crops like cotton, sugarcane, tobacco etc.
4. Famine fear is eliminated for ever. In a particular year if there are no rains at all the yield of the crops is bound to suffer. But suffering will not be very accute as production of the crops may not be as it should be, but some production will be there. It may not be possible to export production that year, but whatever, has been produced may be sufficient for local needs.
5. Irrigation projects are so designed that they bring some revenue to the state. This revenue adds to the income of the state, and state can undertake certain welfare measures with it.
6. It helps the country in the attainment of self-sufficiency in food production. This results in saving the lot of foreign exchange.
7. The people of the region become prosperous. This is because yield of the crops and value of land increases due to irrigation facilities.
8. Irrigation canals and reservoirs may be used as sources for domestic and industrial water supply schemes.
9. Irrigation canals provide facilities for cattle watering, boating, bathing and other recreations.
10. Large deep canals with very low velocity of flow can be used for inland navigation.
11. Major river valley projects are generally planned as multipurpose schemes. They usually provide hydro-electric power along with irrigation. Falls in the canals are also used to generate additional electricity which may be used for industrialisation of rural areas.
12. Almost all the irrigation canals are provided with inspection road on one of their banks. These roads may be used as a means of communication with the interior far flunged areas where there are no *pucca* roads.
13. Due to percolation of water, large strips of land along both the banks of the canals remain damp. Trees may be planted in these strips. These

trees grow without any watering. These trees yield lot of timber and add in the timber wealth of the country. Rows of trees also add to the aesthetic view of the country.

14. Irrigation raises the ground water table in the area, due to constant seepage and percolation of water from canals. So long as this water table remains well below the root zone of the crops, it is beneficial. Tube well irrigation can be easily installed and the water obtained from it can be used for augmenting the irrigation or for raising the cash crops.
15. Fish industry can be developed along the banks of the reservoirs.
16. The people in irrigated areas are more civilized being more educated. Their living habits, behaviour, and way of talking are more civilized.
17. Being more civilized and educated, the people of the area become more aware of their rights and responsibilities for the nation.
18. Because of plantation along the canal banks, chances of rains in the area are improved. Plantation also prevents soil erosion.

1.7 DISADVANTAGES OR ILL-EFFECTS OF IRRIGATION

It is the law of the nature that where something causes advantages, it will definitely cause certain disadvantages also. Following may be the ill-effects of excess irrigation:

1. Due to careless irrigation or excess irrigation, ponds and depressions in the are a get filled up with water. This may lead to creation of breeding places for mosquitoes.
2. Over irrigation may cause excessive rise in the ground water table and thus may be the cause of water logging and salt efflorescence of the area.
3. The atmosphere is generally damp, which is not good for health.
4. Presence of canals affect the drainage conditions of the area. For this lot of cross-drainage works have to be constructed, which are very costly.
5. Bridges have to be constructed on canals to pass railway lines and roads.
6. Irrigation projects are generally big projects and require lot of money and skill in their construction.
7. Deep canals may prove dangerous for people who do not know swimming.
8. The people in irrigated areas are prosperous and consider their own importance more than others. The people even kill each other on petty disputes of land and irrigation.

1.8 TYPES OF IRRIGATION SYSTEMS

For irrigation purposes both ground and surface water resources are utilized. Irrigation systems may broadly be divided under two heads:

1. Flow irrigation or gravity irrigation, and
2. Lift irrigation or pumped irrigation.

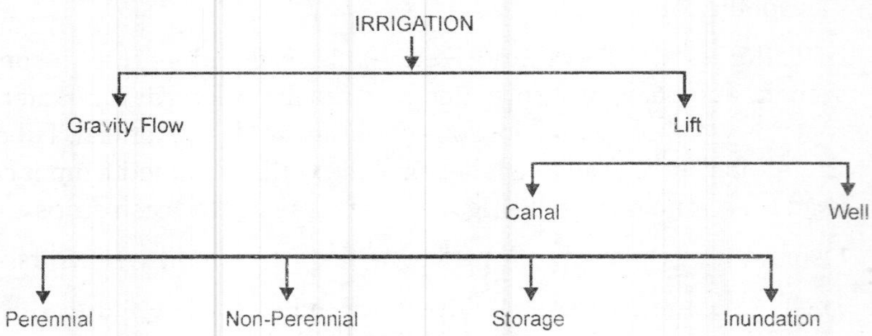

Fig. 1.1. *Systems of irrigation.*

1. Flow Irrigation. In this system of irrigation the water is conveyed to the fields by gravity only. In this system irrigation water to be used for irrigation is supplied at such an elevation that it reaches the fields under gravity through a system of canals and connected works. Flow irrigation may further be sub-divided into following two classes.

(i) Perennial irrigation, and

(ii) Flood or inundation irrigation.

In case of perennial irrigation water is supplied to the fields throughout the crop period at the controlled rate. If water is available for growing two crops then water is supplied to the fields throughout the year. This system of irrigation thus requires some storage works in form of dam, barrage or storage weir so as to store the excess flood water. The stored water is released when regular flow in the stream falls short of the requirements for crops.

Inundation or flood irrigation is also sometimes known as tidal irrigation. In this case the area is not irrigated by controlled irrigation water but inundated during monsoons when rivers are in spate. There is no control over the water. Fields once inundated retain moisture for considerable time and bring the crops sown in them to maturity by no more watering or by one or two minor waterings. Natural rains may even eliminate possibilities of minor waterings also.

Depending upon the source from which the canals draw their water flow, irrigation can be further sub-divided into two types as follows.

(a) Direct irrigation, and

(b) Storage irrigation.

(*a*) *Direct irrigation.* In this system an obstruction in form of low weir or barrage is constructed across the river to raise the water level to such an elevation that the flow is diverted to the off taking canal. This system is also known as *run-of-the river scheme.* It is also sometimes named as *river canal irrigation.* In this case there is practically no storage of water. If discharge in the river is more than the off taking canal capacity, the excess water is allowed to flow down the river. If discharge in the river is less than the capacity of the off taking canal, the canal can be run with full discharge only for a day or two as there is no storage to augment the supplies for longer times. Hence canal will thereafter run with reduced discharge which is available from the river. Sarda Canal System and Ganga Canal System of U.P., are the examples of direct irrigation schemes. Direct irrigation projects are generally of smaller magnitude as there is no rigid control over the supplies.

(*b*) *Storage irrigation system.* In this case a solid barrier such as dam or a storage weir is constructed across the river and excess flood water is stored behind it. A large storge reservoir or an artificial lake is formed behind the dam. The stored water is used to run the canals full of water when discharge in the river falls short of the canal capacity. This system involves construction of a dam and as such proves to be costly. To make them economical, these schemes are usually planned as multipurpose schemes. Generally more than one canal is taken off and also hydro-electric-power is generated. Bhakra Dam Scheme of Punjab and Ramganga Scheme of U.P. are the examples of storage schemes. In such schemes, since lot of storage is developed behind the darn, therefore considerable land property gets submerged upstream of the dam.

Direct irrigation system is adopted when river is perennial and carries enough discharge during lean months, to run the canal full. Storage scheme is adopted when river either is not perennial or flows with insufficient discharge during lean months.

2. Lift Irrigation or Pumped Irrigation. Lift irrigation may be classified as :

(i) Life irrigation from surface water, and

(ii) Lift irrigation from ground water.

In case of lift irrigation from surface water, lift canals are taken either from river directly or from parent gravity flow canal. In this case water is lifted with the help of big pumps and thrown into the lift canal. The water in the lift canal flows under gravity. When lift canal, after some run of length again comes in cutting, the water may again be lifted. In lift canals water may be lifted in stages depending upon the total lift required. Lifted water flows under gravity in the canal.

In the case of lift irrigation from ground water, water is lifted from deep tube wells. Tube well is bored in the field where its water is to be used. Tube well is fitted with pump which is run either with the help of electric motor or diesel engine. This method is the most economical method of utilizing ground water resources.

1.9 DEVELOPMENT OF IRRIGATION IN INDIA

History reveals that India, China and Egypt knew the science of irrigation from very primitive stage. In the early days of civilization irrigation works were not given any significance as population of the world at that time was not much. In the early days crops were grown on rain water only. But as the time passed on, population also went on growing and food requirements increased. It was not possible to fulfil the food requirements from crops gown by rains only and thus man realized the importance of the rivers to increase the food output. In early stages of irrigation, banks of rivers were used to be cut and nearby areas or fields used to be inundated with river water. In doing so sometimes cut in the river bank used to become too large and this used to submerge large areas under floods, thus causing lot of damage. Although some irrigation works were constructed earlier also but irrigation started taking shape from Mughal period in India. Muslim rulers took keen interest in constructing canals for irrigation purposes. These canals were primarily inundation canals. There were no regulation works on these canals for exercising control on them. Later on most of these canals were converted to modern perennial canals. Inundation canals are still in existence in Sindh province of Pakistan. Sultan Firoz Tughlaq constructed a number of canals from Yamuna and Sutlej rivers in fourteenth century. Some of these canals were later remodelled by Akbar in sixteenth century. Shah Jahan also constructed a number of new canals and remodelled some old ones.

Foundation of modern irrigation in India was laid during British period. In the early days of their rule they laid more emphasis on improvement of existing irrigation works. Western Yamuna Canal, Eastern Yamuna Canal, and Cauvery Delta systems were remodelled and put to use for modern irrigation. Great Ganga Canal, Eastern Yamuna Canal and aqueduct over Solani river were the major works done by Sir Proby Cantley in North India. Sir Arthur Cotton worked in Southern India and was responsible for the development of Godavari and Krishna Deltas.

After independence most of the important and productive irrigation works of Punjab and Sindh went to Pakistan and India had to face very acute problem of food shortage. This fact made Indian Government aware about the importance of irrigation. In first five year plan 27.2 per cent of the total outlay was set aside for the development of new potential of irrigation and power. In second and third five year plans this outlay was 19 per cent and 22 per cent respectively. Lists of some of the important irrigation works constructed before and after independence have been given in Tables 1.1 and 1.2.

Table 1.1. *Pre-independence Irrigation Works*

S. No.	*Name of the project*	*Year of completion of the project*	*Irrigation potential in hectare × 1000*
1.	Western Yamuna Canal	1817	413
2.	Great Ganga Canal	1856	655
3.	Lower Ganga Canal	1880	506
4.	Sirhind Canal	1884	935
5.	Cauvery Delta Systems	1889	434
6.	Godavari Delta System	1890	497
7.	Krishna Delta System	1898	405
8.	Sarda Canal	1930	525

Table 1.2. *Post-independence Irrigation Works*

S. No.	*Name of the project*	*Area irrigated in million hectares*	*Total cost of irrigation works in crores of Rs.*
1.	Damodar Valley	0.51	34.7
2.	Magi Project	0.30	42.8
3.	Kosi Project	0.56	24.8
4.	Chambal Project	0.57	54.9
5.	Nagarjuna Sagar	0.83	91.1
6.	Hirakud Project	0.87	93.3
7.	Gandak Project	1.35	49.6
8.	Rajasthan Canal	1.41	206.0
9.	Bhakra Nangal	1.46	101.9

Most of the major irrigation projects in India are multipurpose projects. They also generate hydro-electric power along with irrigation.

1.10 IRRIGATION POTENTIAL IN INDIA

India's total geographical area is about 328 million hectares, out of which about 200 million hectares is cultivable. At present about 50 per cent of the geographical area i.e. 164 million hectares, is under cultivation.

We know India is country of villages. About 75 per cent of its vast population lives in villages. The main occupation of this 75 per cent population is agriculture. Hence agriculture is the main industry of India and it will continue to be so in future for considerable times to come. India has very large water resources, in form of ground water, large rivers, and lakes etc. India also has very large area of thirsty tracts which if provided with irrigation facilities can yield very large quantities of food and other crops. We see that India has water resources and the land to use them and thus there is very large scope of development of irrigation and power potential. Availability of water and land are probably the root causes that India has the pride of having some of the earliest irrigation works.

Surface water resources for India have been estimated at about 178 million hectare metres out of which only 35 per cent i.e. about 56 million hectare metres can be used for irrigation. The ground water resources have been estimated at about 22 million hectare metres.

By 1970–71, India had total area under irrigation about 39 million hectares. It has been estimated that with full utilization of available water resources only 82 million hectares will receive irrigation water. In other words only 50 per cent of the cultivable area (164 million hectares) can ultimately be irrigated.

India started its planned development of irrigation potential with the initiation of its first five year plan in 1951. At the end of fourth five year plan i.e., 1974, only 25 per cent of the total cropped area i.e. 164 million hectares had come under irrigation, and remaining 75 per cent of cropped area was still dependent upon rainfall alone. In fifth five year plan 10 million hectares additional area also has come under irrigation. At this rate it is expected that India will attain its full irrigation potential of 82 million hectares in about coming 20 years.

Food production can be increased by following two ways:

(i) By bringing more of cultivable area under irrigation.

(ii) By increasing the production per hectare.

Our water resources can irrigate upto 82 million hectares only. If after achieving this potential of irrigation still we fall short in food, then we have to depend entirely on increase in production per hectare. Per hectare production can be increased by following measures:

(i) With assured irrigation.

(ii) Adopting multiple cropping.

(iii) Judicious use of irrigation water.

QUESTIONS

1.1 (a) What are the sources of water that can be used for irrigation purposes ?

(b) Explain hydrological cycle of water.

1.2 Define irrigation. Why irrigation is necessary ?

1.3 Enumerate the purposes or functions of water as regards irrigation.

1.4 Discuss the benefits or advantages and disadvantages or ill effects of irrigation.

1.5 What are different systems of irrigation as regard to flow of water ? Discuss each system in details giving systems of their working.

1.6 Write an essay on development in irrigation in India. Also give a list of major irrigation projects that have come up in post-independence period.

1.7 Write an essay on irrigation potential in India.

2

Soil-Water-Plant Relationship

2.1 PHYSICAL PROPERTIES OF SOILS

Soil is a medium which provides support, nutrients and oxygen to the plants. Soils keep water stored and supply slowly to plants to maintain their growth. Soils also provide the necessary porocity which is essential for the growth of the plant. Large depths of soil do not carry any importance for an Agronomist. He is mainly concerned with the upper crest of the soil which is primarily used for growing crops. This upper crest retains water and furnishes it for plant growth. The water retained in the upper crest is also termed as *the belt of soil water*. The depth of the belt of soil water depends upon, the type of soil and vegetation. The depth of this belt may extend from a metre to several metres below the surface.

Upper layer of soil contains, air, minerals, free carbonates besides moisture. This layer may also contain remains of plants and animals at various stages of decomposition. Bacteria, fungi, protozoa, worms and insects etc. are the forms of plant and animal life.

All the soils are the resultant of the disintegration of rocks. The resulting soils may have the same mineral composition as that of parent rock or different because of the changes brought about by water, carbon dioxide and organic materials. The particle size of the soil may vary from very fine clay particle to as large as boulder size.

Physical properties as given above affect the fertility of the soil and growth of the plants. The supply of food to the plant is also related to the physical properties of the soil. The most important physical properties of the soil which markedly affect the growth of the plant are *soil texture* and *soil structure*. According to grain size, the soils may be classified as clay, sand and silt. U.S. Department of agriculture classified various soils with the help of Triangular Texture Diagram shown in Fig. 2.1. All the soils classified by this method contain varying percentages of clay, sand and silt. If percentages of clay, sand and silt for a particular soil are given it can be easily classified using this diagram.

Soil structure refers to the arrangement of the soil particles. Soil structure depends upon shape, composition and electrical properties of soil particles. There are so many other parameters which also effect the soil structure.

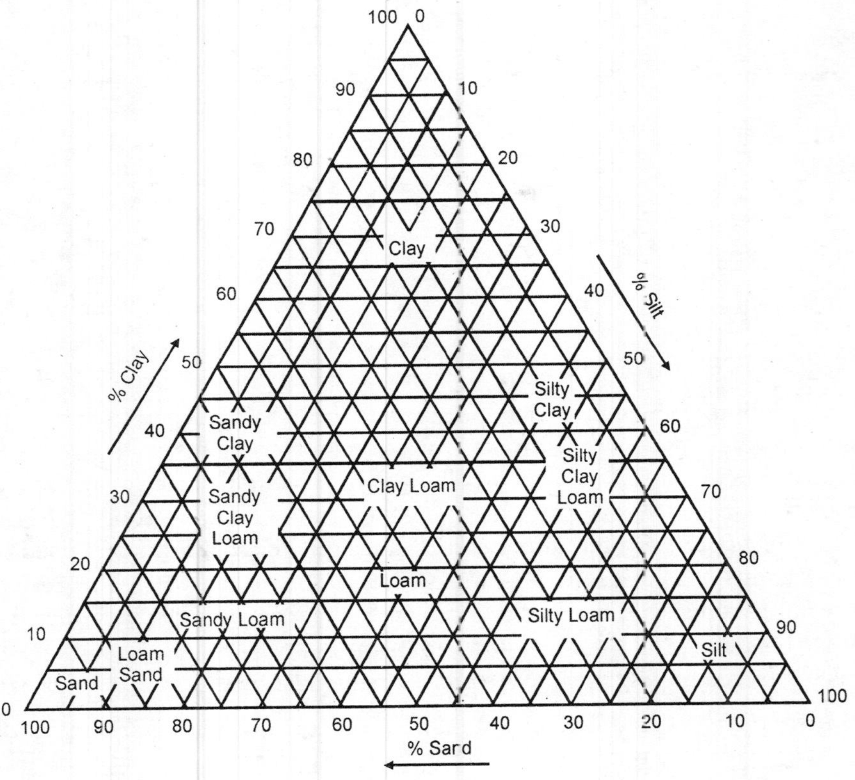

Fig. 2.1. *Triangular texture diagram.*

2.1.1 Classification of Soils

The soils may be classified in following ways.

1. Classification Based on Age of Formation of Soil

According to this classification soils may be youthful soil, Mature soil and Senile soils, youthful soils are fully pervious, whereas mature soils have low permeability. The senile soils become very hard and have very low permeability. Senile soil normally gives very little or no productivity.

2. Classification Based upon Geological Process of Formation

Following soils are described under this classification:

(*i*) *Alluvial soils.* These soils are formed by deposition of water borne materials. It is a very fertile type of soil for crops.

(*ii*) *Residual soils.* These soils are the resultant of disintegration of rocks under various natural actions.

(*iii*) *Volcanic ash.* These soils are formed by deposition of Volcanic Ash from volcanic eruptions.

(*iv*) *Glacial soils.* These soils are formed by transportation and deposition by glaciers.

(*v*) *Eolian soils.* They are the resultant of deposition by wind action.

(*vi*) *Colluvial soils.* They are formed as a result of deposition by rainwash below foot hills.

(*vii*) *Aggradation soils.* They are the accumulation soils.

(*viii*) *Degradation soils.* These are the continuously zooning out soils.

Pan and Clay Pan. There are impervious hard layers formed due to compaction or cementation by silica, iron oxide or by calcium chloride etc.

3. Classification Based on Salt Content

(*i*) *Ped-o-cal.* These are the soils rich in calcium carborate.

(*ii*) *Ped-al-fir.* These soils rich in aluminium or iron salts come in this class.

(*iii*) *Humus.* These are the soils are rich in organic salts.

Soils may also be classified on regional basis and particle size basis. Partical size basis classification is mainly useful for engineering structures point of view.

2.2 TYPES OF INDIAN SOILS

From agricultural point of view, the Indian soils may be classified into following categories.

1. Red Soils. Red soils are light textured porous and friable soils. These soils are the residual soils left at the surface as a result of decay of the underlying parent rocks. They conceal the parent rocks under themselves. Most red soils are sandy loam or sandy clay in texture. They are red in colour and contain very low lime content. These soils are usually deficient in nutritional matters like nitrogen, phosphorus, lime and organic matter. But they are very responsive to green manure, chemical fertilizer, animal dung manure, and irrigation. Such soils are found in southern and central parts of India. They are found in Tamil Nadu, Karnatak, South-east Maharshtra, Central Andhra Pradesh, Western Orissa, and Southern Madhya Pradesh.

2. Laterite Soils. These soils are also residual soils, often open and porous. They also lack nutrition elements like nitrogen, phosphorus, lime and potassium.

They are also red in colour. These soils may be cut and used as building stones. These soils are found in Andhra Pradesh, Madhya Pradesh, Orissa, Assam, Kerala and Southern Maharashtra.

3. Black Soils. These soils are mostly clays. They develop deep cracks during drying. They are black in colour. They are sometimes called black cotton soils as they are considered very good soils for growing cotton. They are quite fertile if adequate rainfall is available. These soils respond very well to the application of chemical fertilizers, green manure and animal dung manure. These soils are located in Maharashtra, M.P., Karnatak, A.P. and Tamil Nadu. It is very difficult to build, building foundation in them.

4. Alluvial Soils. These are very productive soils. They are formed by deposition under water. These soils are found in Indo-Gangetic Plains. They also respond to phosphorus and nitrogen fertilizers. They absorb fair amount of rain water and as such act very good source for ground water reservoir. These soil are suitable for any crop.

5. Sandy or Desert Soils. This soil occurs usually in low rainfall areas. It is almost pure sand and is found in Gujarat, Punjab and Rajasthan. The soil is free draining. It has high pH value and low organic content. This soil is quite productive if irrigation facilities are available. This soils is easily blown by strong winds and dust storms. This soil is very good for melon, barley, wheat, bajra, til etc.

6. Clayey Soils. Such soils are also known as heavy soils. They are not easily drainable, otherwise they are rich with nutritions. These soils are found in U.P., Punjab and parts of Haryana. They are very fertile soils.

7. Saline or Alkaline Soils. These soils can be used for growing crops only when they are adequately drained. Otherwise salts go on accumulating and reach such a concentration that no crops can be grown in them. Such soils are located in semi-arid zones of Bihar, U.P., Gujarat and Punjab.

8. Peat and Mashy Soil. This soil is found in low lying areas along sea coast. This soil is formed by partly decayed plants and other organic matter. This soil may produce very good crops of rice when properly drained and fertilized. It contains organic matter varying from 25 per cent to as such as 90 per cent. The organic matter is mostly in form of decayed vegetation of marshy areas.

9. Humus or Forest Soil. This soil is mostly found in areas having thick forests. This soil consists of organic matter and forest dead vegetation. This soil is very fertile.

The soils may also be classified as follows depending upon their composition:

1. Matiar soil.
2. Matiar Domat.
3. Domat.
4. Baluar Domat.

5. Baluar soil.
6. Humus soil.
7. Usar soil.

Matiar soil and Matiar Domat both come under the category of heavy soil. Matiar soil consists of larger content of clay than Matiar Domat. The clay content in both these soils varies from 20 per cent to 40 per cent. Both these soils do not readily absorb water but once they absorb, they do not allow it to be drained off. These soils can be improved by adding phosphatic and other fertilizers. Addition of fertilizers improves the drainability and ventilation of the soil. Crops which require moisture continuously like sugarcane and rice, are considered best for such soils. Since these soils do not allow absorbed water to be easily drained, they maintain continuous supply of moisture to the crops.

Clay content, in domat soil is never more than 20 per cent. This soil has good drainage and ventilation capacity. This soil is good for any crop. Wheat, barley, sugarcane, maize, peas and potato are grown very successfully in it.

Baluar Domat has relatively larger percentage of sand. Clay content is about 10 per cent. Good irrigation and adequate fertilizers are essential for growing good crops in such soils. Groundnut, cereals, mustard, spiked millet and foddar crops are the usual crops grown in it.

Baluar soils contain about 5 per cent clay and the remaining part is all sand. This soil cannot retain water and thus dries out very soon. Productivity of this soil can be improved by adding green manure and clay. Melon, groundnut, spiked millet are the suitable crops for such soils.

Humus soils are found in forests which contain lot of decaying organic matter and vegetation. These soils are good for crops. Such soils are found in hilly tracts having thick forests.

Usar soils contain salts which are harmful for the crops. Concentration of salts is such that it is impossible to grow any crops in them.

2.3 SUITABILITY OF SOIL FOR IRRIGATION

Particle size, compactness, position of water table, depth of soil, and presence of organic matter in the soil, are the usual aspects which influence the depth of available water in the root zone of the crops. Heavy soils like matiar and domat, can retain water for considerable time and as such are considered suitable for growing crops that require larger amount of irrigation water. Sugarcane, rice, wheat etc. are such crops which can be grown in matiar soils. Baluar or sandy soils cannot retain water for considerable time. Such soils are considered suitable for crops which do not require much irrigation water. Baluar soils being easily drainable, require frequent watering whereas matiar soils require watering at considerable interval because they can retain water. Suitable soil conditions for irrigation may be summarised as follows :

1. The soil should be moisture retentive type so that it may feed the crop for considerable time.
2. The soil should not suffer from erosion, so that fertile soils remain intact in heavy rains and winds.
3. The soil should be adequately porous so that proper aeration of root zone of the crops may take place.
4. The soil should possess sufficient nutrients suitable for proper growth of the crops.
5. The soil should not possess undue concentration of harmful salts. The soil should be free from impurities which may favour the development of parasitic organisms for the crops.
6. The soil should be easily tillable so that it may be easily prepared for sowing new crops.

2.4 IRRIGATION WATER

Functions of irrigation water have been explained in Chapter 1 (Sec. 1.5). Water and nutrients are the two most essential requirements of the crop. Both these requirements persist right from the germination of seeds to the maturity of the crop.

Normally any water is suitable for irrigation. But water may be declared unsatisfactory in the following circumstances :

1. If water is charged with toxic chemicals which are injurious to plants or to persons using plants as food.
2. If water contains such chemicals which react with the soil and form unhealthy environment for the plant life.
3. If water is charged with injurious bacteria.

Presence of salts of sodium, calcium, magnesium and pottassium, in excess amount, are harmful for crops. But if present in small amount some of them prove even beneficial. Presence of boron content in water is of great importance for some crops. Sugarcane can withstand large concentration of boron in irrigation water; but sensitive crops like groundnut, beans etc. cannot tolerate its excess amount. Boron concentration of more than 2 ppm is harmful to most of the crops. Three standards of irrigation water with respect to presence of total salts, sodium, boron and electrical conductivity have been given here. Class I water is goof for any type of crop. Class II water is injurious to sensitive crops only but good for other crops. Class III water is considered unsuitable for most of the crops. The salt content is usually expressed in parts per million (ppm) or milligram per litre (mg/l), milli equivalent per litre ($ME\varphi/l$). The electrical conductivity is expressed in micromhos per centimetre ($EC \times 10^6$).

Table 2.1

Class of water	*Total salt content in ppm*	*Sodium %*	*Boron content in ppm*	*Electrical conductivity $EC \times 10^6$*
I	0 – 700	60	0.0 – 0.5	0 – 1000
II	700 – 2000	60 – 75	0.5 – 2.0	1000 – 3000
III	Above 2000	Over 75	Over 2.0	Over 3000

Suitability of water for irrigation is also dependent upon the pH value and total dissolved solids. Excess salts present in water affect the plant growth and thus crop yield. If total dissolved solids in water are upto 400 ppm all the waters are considered fit for irrigation purposes. If total dissolved solids exceed 1200 ppm, water is considered unsuitable for irrigation. Waters having total dissolved solids upto 600, 800 and 1000 ppm will be suitable for irrigation if their pH values do not exceed 9.0, 8.5, and 8.0 respectively.

Salinity Concentration of Soil Solution (C_s)

The salinity concentration of soil solution is also one of the measures of assessing the suitability of irrigation water. It can be found by following formula.

$$C_s = \frac{CQ}{Q-(C_u - P_{\text{eff}})}$$

where C = Concentration of salt in irrigation water

Q = Total quantity of water applied to the soil

C_u = Consumptive use of water

P_{eff} = Useful rainfall

Table 2.2. *Classification of Irrigation Water Based on Salt*

Type of water	*Suitability for irrigation*
1. C_1 water has low salinity. It has electrical conductivity between 100 and 250 micromhos/cm at 25°C	Suitable for all types of crops and all kinds of soils. Permissible under normal irrigation practices except in soil of extremely low permeability
2. Medium salinity water C_2. Its electrical conductivity lies between 250 to 750 micromhos/cm at 25°C	This kind of water can be used if moderate amount of leaching occurs. Normally salt tolerant plants can be grown without much salinity control
3. High salinity water (C_3). Its electrical conductivity lies between 750 and 2250 micro-mhos/cm at 25°C	unsuitable for soil with restricted drainage only high-salt tolerant plants can be grown
4. Very high salinity water (C_4). Its electrical conductivity is more than 2250 micro-mhos/cm at 25°C	Unsuitable for irrigation

Based upon the salt concentration or electrical conductivity, US Department of Agriculture has classified water in four categories C_1, C_2, C_3 and C_4. C_1 water has low conductivity, C_2 medium conductivity, C_3 high conductivity and C_4 very high conductivity. For this classification refer Table 2.2.

Effects of Sodium Concentration in Water

Higher sodium percentage in irrigation water after some time gives rise to a soil having a large percentage of replaceable sodium in the colloid farm. Such a soil is often known as black alkali. The percentage of sodium is found from the following equation.

$$\text{Sodium \%} = \frac{100\,\text{Na}}{\text{Ca+ Mg+ Na+ K}}$$

The soil grains breakdown with higher percentage of sodium in irrigation water. This renders soil less permeable and its tilth becomes easy. Prolonged use of irrigation water with 85 per cent or higher percentage of sodium may render even sandy soils impermeable. The soil becomes plastic and sticky when wet and forms clods and crust on drying.

Sodium Absorption Ratio (SAR) is factor which classifies the irrigation waters based on sodium concentration. SAR is found out using following equation

$$\text{SAR} = \frac{\text{Na}}{\sqrt{\dfrac{\text{Ca + Mg}}{2}}}$$

Based upon SAR, the water may be of four categories namely S_1, S_2, S_3 and S_4. S_1 category of water has low sodium content; S_2 medium sodium content, S_3 higher sodium content, and S_4 very high sodium content. The characteristics of four types of irrigation water have been given in Table 2.3.

Table 2.3. *SAR, Classification of Water*

Type of water	*Suitability for irrigation*
1. Low sodium water (S_1) SAR range 0–10	Suitable for all types of crops and all types of soils, except for those crops which are highly sensitive to sodium
2. Medium sodium water (S_2) SAR range 10–18	Suitable for coarse textured or organic soil with good permeability. Relatively unsuitable in fine textured soils
3. High sodium water (S_3) SAR range 18–16	Harmful for almost all types of soils. Requires good drainage high leaching, gypsum addition
4. Very high sodium water (S_4) SAR range above 26	Unsuitable for irrigation

2.5 PRESENCE OF WATER IN SOIL

Water present in the soil may be classified under three heads:

1. Hygroscopic water,
2. Capillary water, and
3. Gravitational water.

1. Hygroscopic Water. When a soil sample after having been completely dried in oven is put in open atmosphere, it absorbs some amount of water from it. The amount of water so absorbed by the oven dried soil sample is termed as *hygroscopic water*. This water is not capable of any movement by the action of gravity force or capillary force. Amount of this water in soil can vary only if there is change in the moisture content in atmosphere. This water is not available for plant growth.

2. Capillary Water. This water is that water content in soil excess of hygroscopic water, which exists in the pore space of the soil, due to molecular attraction. This

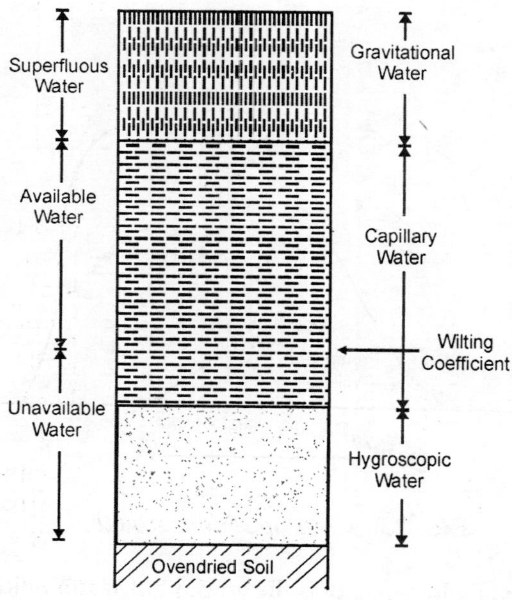

Fig. 2.2. *Various forms of water.*

water is held in form of a thin continuous film around the soil particles due to forces of surface tension. It is this water which remains available for the plant growth between successive waterings to the crops. See Fig. 2.2.

3. Gravitation Water. It is that part in excess of hygroscopic and capillary water, which will drain out of the soil under the action of gravity. This water passes down and joins the ground water table. This water is also known as superfluous water. It is also not available for the growth of plant as it readily gets drained whenever it finds conditions favourable for it.

Water present in the soil may be classified as *unavailable, available* and *superfluous*. This classification is based on the availability of the soil water for plant growth. See Fig. 2.2.

2.6 SOME IMPORTANT TERMS PERTAINING TO SOIL WATER

1. Saturation Capacity of Saturation Point. It is the amount of water required to fill all the pore spaces among soil particles by replacing all the air present in pore spaces. It is the maximum moisture holding capacity of the soil. It is also sometimes known as total capacity. It is upper limit of possible moisture content. When porosity of the soil is known, the saturation capacity of the soil can be expressed as equivalent cm of water per metre of soil depth. If porosity of the soil is say 40 per cent by volume. The moisture in each meter of saturated soil is equivalent to a depth of 40 cm on the field surface. It is also known as saturation point.

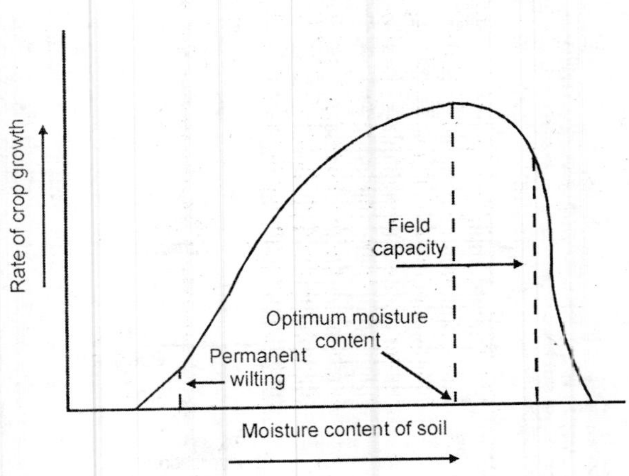

Fig. 2.3. *Crop optimum growth.*

2. Field Capacity. Field capacity is the amount of water held in the soil after free drainage has removed most of the gravity water. The concept of field capacity of soil is extremely important in arriving at the amount of water available in the soil for plant use. Most of the gravitational water gets drained out from the soil before it can be used by the plants.

3. Wilting Point. It is that maximum water content in the soil at which plants can no longer extract water sufficient to maintain their growth. The wilting may be permanent or temporary. If plants are unable to extract sufficient water to meet their growth requirements, the water content present in soil is termed as *permanent wilting point* or *wilting coefficient*. This moisture content lies at the lower end of the available water. A plant is considered wilted permanently when it cannot recover even after being placed in a saturated atmosphere. On a hot windy day, sometimes plants wilt at noon but again recover in the cooler portion of the day in the evening. Such wilting is known as temporary wilting.

Wilting point is closely indicated by the moisture retained against a tension of 15 atmospheres. Permanent wilting percentage can be estimated approximately by dividing the field capacity by a factor varying from 2.0 to 2.4 depending upon the amount of silt present in the soil. It is seen that the wilting coefficient for most of the soils is about 150 per cent of the hygroscopic water.

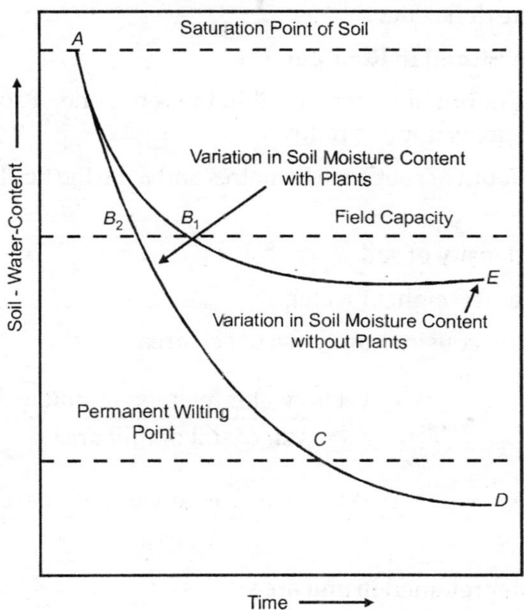

Fig. 2.4

4. Available Moisture. It is the difference in water content of the soil between field capacity and permanent wilting coefficient. It is the actual available moisture of water to the plant to maintain its growth.

5. Readily Available Moisture. It is that portion of available water that can be easily extracted by plant. Its amount is about 75% of the available moisture. For all the above terms, see Fig. 2.2.

6. Moisture Equivalent. Moisture equivalent is used as a single factor to which the properties of the soils can be related within reasonable limits. The amount of moisture equivalent very nearly equals the field capacity for a medium textured soil.

Moisture equivalent = field capacity
= 1.8 to 2.0 permanent wilting point
= 2.7 to 3.0 hygroscopic coefficient.

Moisture equivalent is an artificial property of moisture of the soil. It is used as an index of the natural properties. It is expressed as the percentage of moisture retained in a small wet soil sample 1 cm deep, when subjected to a centrifugal force of 1000 times as great as gravity and for a duration of 30 minutes.

Soil-moisture Deficiency

The amount of water required to bring the soil moisture content of the soil to its field capacity is known as the soil-moisture deficiency. Sometimes it is also referred as field-moisture deficiency.

Depth of Water Stored in Root Zone

The estimate of depth of water stored in the roof zone of soil containing water upto field capacity is done as follows.

Let d be the depth of root zone in metres and F_o be the field capacity expressed as ratio.

Let λ = density of soil

λ_w = unit weight of water

For calculation consider unit area of soil area.

Thus
$$F_o = \frac{\text{Wt. of water retained in unit area}}{\text{Wt. of soil of unit area}}$$

$$= \frac{\text{Wt. of water retained in unit area}}{\lambda \times 1 \times d}$$

$\therefore$ Wt. of water retained in unit area

$$= F_o \lambda \times 1 \times d$$

$$\text{Depth of water stored} = \frac{F_o \lambda d}{\lambda_w} \text{ metres}$$

This depth of water will be available for consumptive use or evapo-transpiration.

$$\text{Available moisture depth} = \frac{\lambda d}{\lambda_w} \text{(Field capacity-wilting coefficient)}$$

Example 2.1. *Loamy soil has field capacity of 23% and wilting coefficient of 11%. The dry unit weight of soil 1.5 g/cm³. If the depth of root zone is 600 mm, find out the storage capacity of the soil. Irrigation water is applied when M.C. of soil falls to 13%. If the water application efficiency is 70%, find out the depth required to be applied is the field.*

Solution

Maximum storage capacity = Available moisture

$$= \frac{\lambda d}{\lambda_w} \text{(Field capacity – wilting coefficient)}$$

$$= \frac{1.5 \times 0.6}{1} (0.23 - 0.11) \text{ metres}$$

$$= 0.108 \text{ m} = 108 \text{ mm.}$$

Depth of irrigation water

$$= \frac{\lambda d}{\lambda_w} \text{(Field capacity – optimum M.C.)}$$

$$= \frac{1.5 \times 0.6}{1} (0.23 - 0.13)$$

$$= 0.09 \text{ m} = 90 \text{ mm.}$$

Field irrigation requirement

$$= \frac{90}{0.70} = 128.6 \text{ mm.}$$

2.7 SOIL WATER AND PLANT GROWTH

Permeability of soil, soil structure, texture of soil are the most effective parameters which control the downward flow of water through the soil. When irrigation water is applied to the field its diffusion into the soil starts immediately. Rate of diffusion of water depends largely upon the permeability of the soil. In clayey soils diffusion of applied water is very slow but in sandy and sandy loam it is quite fast. The irrigation water which is in excess than required to wet the root zone will ultimately join the water table and thus goes waste. From this it follows that deep watering is actually not required and only shallow waterings will be quite enough to wet the root zone. In this way lot of precious irrigation water can be saved from going waste and water thus saved an be used for bringing more areas under irrigation.

Evaporation from the surface of irrigated soil, starts immediately after watering. After some time surface soil dries up and process of evaporation slows down. The moisture in soil left thereafter, is used by the plants for transpiration and metabolism.

Utilisation of soil moisture by plants steadily goes on decreasing as soil suction increases. When rate of transpiration becomes more than the rate of absorption, initial wilting conditions set in. With the depletion of moisture from the soil, equal to one atmosphere suction, temporary wilting starts occurring. When depletion reaches 15 atmospheres suction, permanent wilting takes place. At one atmosphere suction, the photosynthetic activity of plants is affected and the relative growth rate tends to slow down. This is the point when irrigation water must be applied to the field to revive the growth of plants.

Availability of water to plants is widely dependent upon the texture of the soil. Besides this shallow water table, soil stratification, depth of soil etc. are the other parameters which influence the availability of water to plants. Root depth and rot spread or proliferation are the most important factors which influence the extraction of moisture from soils. Shallow rooted crops require frequent watering whereas deep rooted crops require watering at considerable interval. Deep rooted crops derive moisture from large depths and as such do not require frequent watering. Shallow rooted crops require frequent watering as soil from the top surface and its vicinity dries very quickly.

The moisture lying between the range of field capacity and permanent wilting point is the available moisture for plant growth. The knowledge of this range is very essential so as to understand the behaviour of soil water in relation to plant growth.

The water needs of different crops, during their various stages of growth, are different. Figure 2.3 shows the water requirements of wheat during different stages of growth. The soil should be such so as to meet the varying demand of moisture for the plant. For most judicious use of irrigation water, irrigation schedules should be so planned that only just sufficient water is applied to the crop at right time.

Depth and Frequency of Irrigation Watering

As already explained earlier the available moisture is between wilting point in the lower side and field capacity on the upper side. About 75 per cent of the available moisture is easily extracted by the plants and thus is known as *readily available moisture*.

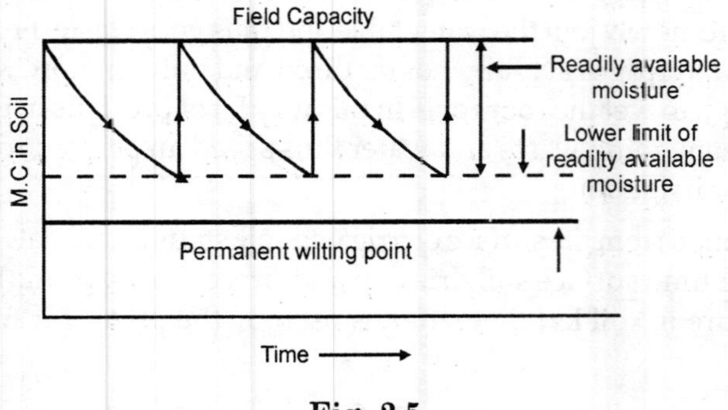

Fig. 2.5

Hence in order to maintain proper yield of the crop, the moisture content in the soil should be between the *field capacity* and the optimum water content upto which the soil moisture may be allowed to be depleted in the root zone. Various moisture levels in the soil have been shown in Fig. 2.5.

Readily available moisture is shown in the Fig. 2.6 which lies between field capacity and optimum moisture content.

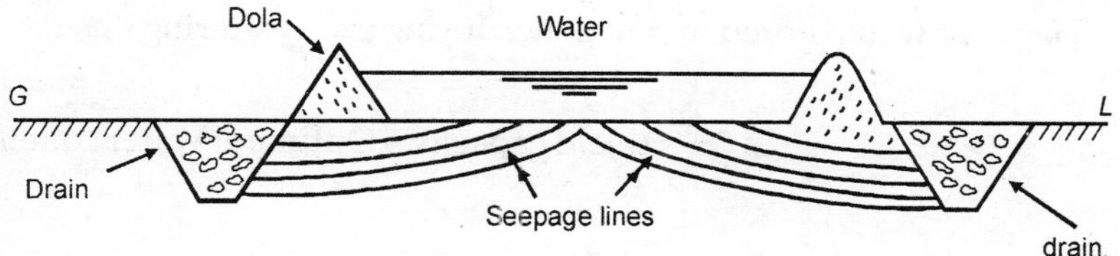

Fig. 2.6. *Underground drainage of soil.*

The amount of water supplied during watering should be such that the water content is equal to the field capacity. Now water will gradually start being utilised consumptively by plants and the soil moisture starts falling. When the water content in the soil reaches the optimum value, fresh dose of irrigation water has to be applied so that water content again gets raised to the field capacity of the soil.

The frequency of irrigation water depends upon the amount of available water contained in the root zone of the soil and the rate of consumptive use.

If d *is* the root zone depth is metres, F_c the field capacity, M_o the optimum moisture content and d_w the depth of each watering, the relation between all the form elements is as follows

$$d_w = \frac{\lambda}{\lambda}(F_c - M_o) \text{ metres}$$

where λ = unit weight of soil

λ_w = unit weight of water

Both F_c and M_o are to be expressed as ratio. If C_u is the daily consumptive use rate, the frequency of watering f is given by

$$f = \frac{d_w}{C_u} \text{ days.}$$

Example 2.2. *Field capacity of soil 25 per cent, permanent wilting point 13 per cent, Density of soil 1.5 g/cm³, effective depth of root zone 700 min and daily consumptive use rate for the crop is 10 mm. If the soil is clay loam what should be the frequency of watering so that yield of the crop is not effected.*

Solution

Available Moisture = field capacity – permanent wilting point

= 25 – 13 = 12%

Let readily available moisture be 75 per cent of the available moisture

$\therefore$ Readily available moisture = 12 × 0.75 = 9%

Optimum moisture = 25 – 9 = 16%

Hence when irrigation watering is done, moisture is raised from 16% to 25%

$\therefore$ Depth of water stored in root zone, during each watering

$$= \frac{\lambda d}{\lambda_w} \text{ (Field capacity – Optimum water content)}$$

$$= \frac{1.5 \times 0.7}{1} \text{ } (0.25 - 0.16) \text{ m}$$

$$= \frac{1.5 \times 0.7}{1} \times 0.09$$

$$= 0.0945 \text{ m } = 94.5 \text{ mm}$$

$$\text{Watering frequency} = \frac{\text{Depth of water available for evapo–transpiration}}{\text{Daily consumptive rate}}$$

$$= \frac{94.5}{10} = 9.45 \text{ days i.e. 10 days.}$$

2.8 HARMFUL SALTS IN SOILS

If amount of harmful salts becomes excessive in the soil no crop can be grown over it. Sodium carbonate, sodium chloride, sodium sulphate etc. are some of the harmful salts. But of these, sodium carbonate which is also known as white salt, is considered to be the most harmful. If percentage of harmful salts reaches 0.15 per cent it affects the production and if this percentage reaches 0.25 per cent it renders the land completely barren. Other salts are also present in the soil but unless their percentage crosses the limit they do not affect the growth of the crops.

2.9 EFFECT OF pH VALUE

The pH value of the soil suitable for crops should lie between 6.0 and 8.5. If pH value exceeds 10.5 the land becomes completely unsuitable for crops.

2.10 METHODS OF IMPROVING FERTILITY OF LAND

If deficiencies of any land are made good it becomes suitable for growing crops. If land is too clayey it can be improved by adding sand to it. Similarly sandy soils can be improved by adding suitable amount of clay to it. Soils can be improved as regards its fertility by adding:

(a) Organic manure;

(b) Chemical manure;

(c) Green manure.

Cowdung manure is organic manure. It is available in village in large amount, as every villager maintains animals.

All the manures manufactured by large companies are chemical manures. Super phosphate, urea, pottassium sulphate, are the forms of chemical manures.

Green manure is developed by sowing some crop in the field. When crop becomes about 20–30 cm high it is ploughed in the field which after decomposition gives nutrients to the crops to be gown over it.

Land Reclamation. When, due to over irrigation or any other reason, lot of harmful salts get deposited on top layers of soil, it becomes unproductive. If amount of harmful salts exceeds certain limit the land may not be in position to growth any crop. Such lands have to be reclaimed. Such land is known as kallar land. Following are the methods of land reclamation.

1. Leaching method.
2. Scraping the saltish crust.
3. Surface drainage.
4. By adding some chemical so as to neutralize the harmful salts.
5. By sowing some special crop which would neutralize the harmful salts.

Out of these methods leaching method is very common. In this method the land is always kept filed with water. The salts get dissolved in water and percolate deep in the land. Thus concentration of salts at the surface is reduced. Deep ditches are also constructed along both the sides of the field. Percolating water escapes into these ditches. The ditches carry this water along with dissolved salts away from the fields.

In scraping method the top layer of soil is scraped completely and thus concentration of harmful salts is reduced.

In surface drainage method, water is filled in the field. The water is agitated by moving animals in the field. The salts get dissolved in water. The water is lastly drained out of fields by means of drainage ditches constructed along the fields.

Preparation of Land for Irrigation. Whenever a new area of land is brought under irrigation the following steps are taken so that irrigation water could profitably be applied to this new area.

1. The area is cleared from jungles, bushes, wild trees and grass etc. The deep roots of the trees and bushes should be extracted.
2. The land should thereafter be cleared and levelled. Higher patches should be scraped and filled into the depressions. The process of levelling the land is very very important as unless this is done, irrigation water will

fill only the depressions and duty of water will be very slow.

3. A small regular falling gradient should be provided to the field.
4. The land should be properly ploughed.
5. The land should be divided into small suitable sized plots with the help of small earthen levees according to the method of irrigation to be adopted.
6. Field channels should be made at proper interval to facilitate proper distribution of water to the entire field.
7. If there is possibility of water logging proper drainage measure should be taken.

QUESTIONS

2.1 What are the physical properties of the soils that are important from irrigation point of view ?

2.2 Enumerate the properties of various soils briefly. Why alluvial soils are more productive ? What is the significance of clay content and sand content in the soil ?

2.3 (a) How do you weigh the suitability of any particular soil for any particular crop ?

(b) What should be the salt content, sodium content and boron content in the water suitable for irrigation ?

2.4 By drawing a neat sketch show various forms of soil moistures that can be present in the soil. Explain the terms, superfluous water, available water, and unavailable water.

2.5 Define the terms (i) wilting point, (ii) field capacity, (iii) saturation point, (iv) moisture equivalent.

2.6 Describe in details the relation between soil moisture and plant growth. Show this relationship diagramatically also.

3

Delta, Duty, Base Period and Consumptive Use of Water

3.1 WATER REQUIREMENTS OF CROPS

The subject of water requirements for crops is of direct significance wherever irrigation is practised. The areas may be divided into three categories namely – arid region, semi-arid region and humid region. Growing of crops in arid regions is almost impossible without irrigation as rainfall in this region is very rare. In semi-arid regions irrigation may be optional but crops like sugar cane, vegetables, rice etc. requiring more of water can only be grown by irrigation. In humid regions irrigation plays a protective role against possible failure or deficiency of rainfall during the crop season. In all the region, it is only irrigation which can ensure optimum yield of the crops.

The water requirement of a particular crop does not remain uniform in different areas. It varies according to variation in climate, rainfall, and type of soil. India is a very large country having lot of variation in its different areas as regards rainfall and type of soil. Hence water requirement of a particular crop in a particular region cannot be considered applicable for whole of the country.

3.2 OPTIMUM USE OF WATER

The quantity of water supplied to a particular crop during its growth period, which results in the maximum yield of the crop is known as *optimum water requirement*. This water requirement includes water supplied by irrigation and by precipitation both. In order to get the maximum yield with least amount of irrigation water, knowledge of optimum requirement of water for the crop is essential. Supply of irrigation water, both more, and less than the optimum requirement, affects yield adversely. If less than optimum amount of irrigation water is supplied, the plant has to expend extra energy to get moisture from the soil which would otherwise have been utilised for the growth of the plant. This affects the yield. On the other hand if more than optimum requirement of water is supplied it would expel the air from soil pores and fill them with water. This will prevent free circulation of fresh air which is essential for the metabolic activity to prepare food for the plant. This also affects yield. Relation between yield and depth of irrigation water for the crops follows the pattern shown in Fig. 3.2. On the same ground the relation between rate of growth of the plant and moisture content of soil, can be drawn as shown in Fig. 3.1. At moisture content below

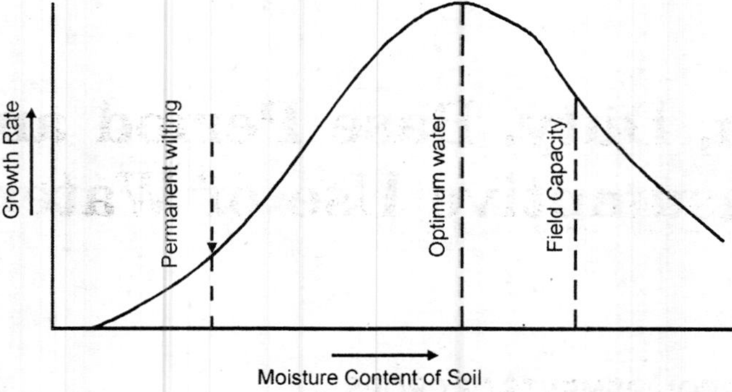

Fig. 3.1. *Relation between M.C. and growth rate.*

permanent wilting point practically no growth of the plant is possible as moisture is not available from soil for this purpose. As the moisture content in the soil is slowly increased, the rate of growth of crop also correspondingly increases as shown in Fig. 3.1. This phenomenon continues till optimum moisture content is reached. After this any increase in moisture content would affect the growth

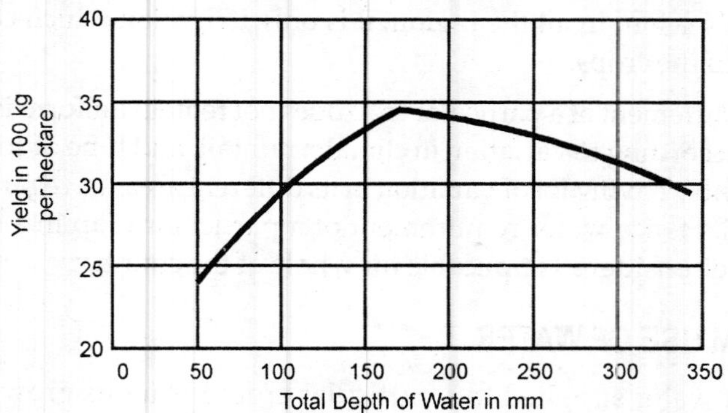

Fig. 3.2. *Relation between yield and water depth.*

adversely. Thus in order to achieve maximum yield of the crop growth of the plants should be optimum. Irrigation schedule should be so planned and regulated that crops receive (Fig. 3.1) only optimum amount of moisture in each watering and total depth of water supplied is as per the requirements of optimum yield, see Fig. 3.2.

3.3 FACTORS AFFECTING WATER REQUIREMENTS OF CROPS

Following are the possible factors that affect the water requirements of crops.

1. Texture and structure of the soil and its moisture storage capacity.
2. Position of ground water table.
3. Slope of the ground.
4. Drainage conditions.
5. Climatic conditions like rainfall temperature, wind movement, and relative humidity.
6. The system of irrigation adopted.
7. Intensity of irrigation.
8. Type and amount of manure applied to fields.

3.4 DUTY, DELTA AND BASE PERIOD

1. Duty. Duty represents the relationship between the quantity of water and the area irrigated by it of some particular crop the water being supplied for the entire period of the growth of that crop. It can also be stated as the irrigating capacity of water for specific crop. The area irrigated by unit quantity of water flowing for full growth period of some specified crop is known as the *duty of water* for that particular crop. If 5 cumecs of water is required for maturing 10000 hectares of say wheat right from sowing to harvesting, the duty of irrigation water for wheat will be $\frac{10000}{5} = 2000$ hectares/cumec. Duty of water for different crops is different.

2. Base Period. It is the time, usually in days, for which a crop occupies a field to attain its full maturity. This time is counted from the day when irrigation water is first issued to field for preparing it for sowing the crop to last watering before crop is harvested.

3. Delta. It is the total depth of water, required to be supplied during the entire period of the crop. The depth of the water is denoted by Δ (Delta). For example, if a crop requires say 7 waterings at an interval of 15 days and depth of each watering being say 75 cm, then Delta or depth of water for the crop will be $7 \times 7.5 = 52.5$ cm. In this case base period of the crop will be $15 \times 7 = 105$ days. Of the total depth of water part may be supplied by precipitation and part by irrigation.

4. Crop Period. The time in days from the instant of sowing the crop to that of harvesting, is known as crop period. Crop period is very nearly equal to the base period of the crop. The reason being that base period is counted from the watering given to field for preparing field for sowing. The watering given to field for preparing for sowing is known as 'Paleo' Watering. The time from paleo water to instant of sowing the seeds may be 3–4 days depending upon the climatic conditions and crop being sown. So we see that base period starts 3–4 days earlier than crop period. Now let us consider last watering before harvesting. Crop is harvested three four days after the last watering. Base period of crop ceases at last watering but crop period finishes only at harvesting. Hence base period of crop starts 3–4 days before crop period and ends 3–4 days before the crop period.

The duty of water can be expressed in the following four ways :

1. Area in number of hectares (acres) which a unit discharge (one cumec) running for base period can irrigate.
2. By total depth of water in metres.
3. Number of hectares of crop matured by one million cubic metres of stored water.
4. Number of hectare-metres expended per hectare irrigated area.

In U.P. and other northern states of this country, duty of canal water is usually expressed as the area of crop is thousand hectares that can be matured with a flow of one cumec of water flowing for base or crop period. This unit of duty is considered convenient for design of channels or canals.

3.5 DUTY OF CANAL WATER AT VARIOUS PLACES OF MEASUREMENT

Duty of canal water is always expressed as so many hectares of area matured by one cumec of water flowing for base period. But still this definition is not complete. In duty figures it is necessary to state.

1. Base period and,
2. The place of measurement of duty.

If one cumec water can bring to maturity say 2000 hectares of wheat and base period of wheat is say 120 days, then we write this duty figure as 2000 hectare/ cumec at the field for a base period of 120 days.

Canals are taken from rivers which may be quite far off from the area which they are going to irrigate. River water is diverted to the canal with the help of weir, barrage or dam. This diverted water reaches the fields for irrigation through a net work of canals and distributories. Some of the flowing canal water is lost due evaporation and percolation in the way before it reaches the fields for irrigation. The water lost in the flow is known as conveyance losses. Conveyance losses do not remain constant but vary with the length of the canal. It is because of variation in the conveyance losses that duty of canal water also undergoes change. Duty of canal water is maximum at the field channel and goes on decreasing as we proceed towards the direction against the flow of water. Duty is minimum at the diversion head works of the main canal. This fact can be explained by considering a net work of canals shown in Fig. 3.3.

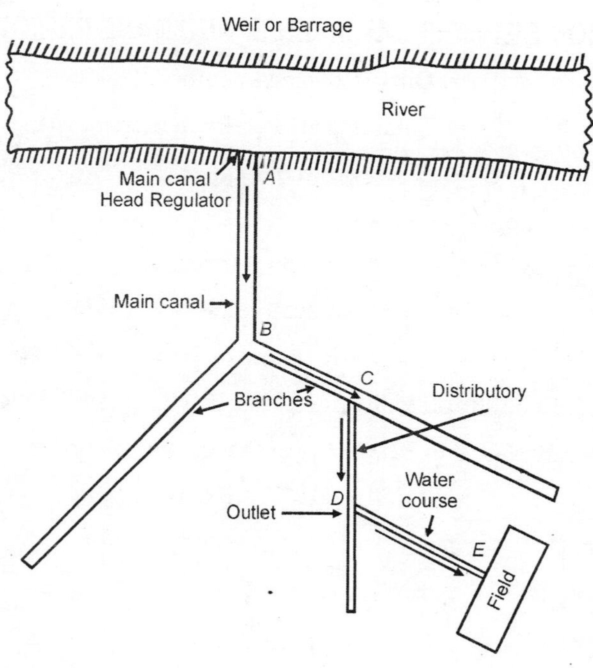

Fig. 3.3

Let *A, B, C, D,* be off taking points of main canal, branch distributory, and water course (outlet) respectively. Let *E* be the point on the water course from where water is flowing to the fields. Let water reaching point *E* be one cumec. In order to make one cumec water reach point *E*, some extra water has to be supplied from outlet point *D* as some water is lost during flow from *D* to *E*. Let conveyance losses between *D* and *E* points be say 0.1 cumec. Thus to make available 1 cumec water at point *E*, 1.1 cumec water will have to be left from *D*. Similarly let conveyance losses between *A* and *B*, *B* and *C*, and *C* and *D* each again be 0.1 cumec. Then to make one cumec water available at *E* 1.1 cumec will have to be left from *D*, 1.2 cumec from *C*, 1.3 cumec from *B* and 1.4 cumec from *A*. If one cumec water at *E* is irrigating 2000 hectare in base period then duties of water at point *E*, *D, C, B* and *A* respectively will be $\frac{2000}{1}, \frac{2000}{1.1}, \frac{2000}{1.2}, \frac{2000}{1.3}$, and $\frac{2000}{1.4}$, i.e. 2000, 1818.2, 1666.67, 1538.5, 1428.6 hectares. These duty figures clearly reveal that duty is maximum at field and goes on decreasing towards upstream side, and it is minimum at main canal head works.

The duty at the head of main canal is known as gross quantity at head of branch canal as lateral quantity, at outlet as outlet factor and lastly at the head of land being irrigated as net quantity.

3.6 RELATION BETWEEN DUTY, DELTA AND BASE PERIOD

Let D = Duty in hectares/cumec.

Δ = Total depth of water in metres.

B = Base period in days.

Take a field area of D hectares.

Water required to fill this area upto Δ metre depth

$$= \Delta \times D \text{ hectare-metres}$$

$$= \Delta \times D \times 10^4 \text{ cubic metres.} \quad (1)$$

This water is to be supplied by one cumec discharge in B days (base period).

Hence water available in B days by discharge of one cumec

$$= 1 \times 24 \times 60 \times 60 \times B \text{ cubic-metres} \quad (2)$$

Equating (1) and (2), we get

$$\Delta \times D \times 10^4 = B \times 24 \times 60 \times 60$$

$$D = \frac{B \times 24 \times 60 \times 60}{\Delta \times 10^4}$$

$$= 8.64 \frac{B}{\Delta} \text{ hectares.}$$

3.7 FACTORS AFFECTING DUTY

The duty of water depends upon following factors.

1. Soil characteristics through which canal runs. If canal is unlined and soil through which it flows is coarse grained, seepage and percolation losses will be too much and duty of water will be reduced. If on the other hand canal is either lined or runs through fine grained soil, the losses will be considerably small and duty of water will be more.

2. Soil characteristic of fields. If soil of the field is deep coarse grained, percolation losses will be more. If, however, a hard pan is present at a depth of 1 or 2 m, the percolation losses will be less and duty will be more.

3. Undulated fields. If fields which are to be irrigated are quite undulated, the duty of water will be less. Lower portions are filled with more water whereas higher portions get less depth of water.

4. Climatic conditions. Temperature, wind, rainfall and humidity are the climatic conditions which affect the duty of water. High winds, high temperature, less rainfall and less humidity conditions, require more of irrigation water and duty is thus reduced. If rainfall is taken into consideration while calculating duty figures, it is known as *Duty inclusive of rainfall.* If rainfall is not considered it is called *Duty exclusive of rainfall.*

5. Base period of the crop. The crops having longer base period, require waterings for longer time and thus duty will be less and vice-versa.

6. Depth of water or Delta. The crops which require lesser overall depth (Delta) of water have more of duty. Delta is different for different crops. Hence duty of water depends upon type of crop.

7. Method of application of irrigation water. Method of application of irrigation water affects duty figures. Flood irrigation has lesser duty than Furrow irrigation. Sub-irrigation methods give highest duty.

8. Preparation of fields. If fields are prepared by deep ploughs, the water retention capacity of the unsaturated zone of soil gets increased. When such soils are irrigated, although they require more of water for irrigation, but the number of waterings are reduced. This also helps in increasing the duty, to some extent.

9. Position of fields in relation to canal. If fields are situated very near to the canal supplying water for irrigation, conveyance losses beyond outlet will be small and duty of water will be more.

10. Method of assessment. Assessment of irrigation if done on volumetric basis, the duty of water is bound to increase. If assessment is done on irrigated area basis, the farmers do not use water carefully and thus duty becomes low.

11. Longitudinal slope to fields. If fields are given a little longitudinal downward slope towards the farther end, duty of water is increased.

12. Awareness of farmers. If farmers are made to understand the importance of irrigation water, they use it more judiciously in their fields without allowing water to go waste. It helps in increasing the duty of water. Farmers maintain their water courses free from grass and other weed growths and do not allow water to go waste anywhere.

13. If water courses are lined and levels of water courses for each field are kept as per requirements, the duty of water increases considerably. In lined water courses having regular longitudinal slope, all the water coming out of outlet rushes to fields with more velocity and does not remain stationary anywhere and hence gives increased duty figures.

3.8 METHODS OF IMPROVING DUTY

The factors affecting the duty have been explained in previous article of this chapter. If these factors are somehow made less effective the duty of water can be considerably increased. Methods of improving duty have been enumerated as follows :

1. By selecting such method of applying water which would cause least wastage by way of seepage, percolation, and evaporation.
2. Land should be properly ploughed and levelled so that a thin sheet of water may evenly be spread over it.
3. The fields should be frequently ploughed. This measure reduces loss of moisture specially when ground water is within capillary reach of the ground surface.
4. Fields should be divided into small *kiaries*. Smaller the size of *kiaries* more will be the duty of water.
5. Area to be irrigated should be located as near the canal as possible. This reduces coveyance losses and duty is increased. In other words idle length of the canal should be minimum.
6. Volumetric method of assessment should be adopted.
7. Only that much land should be given to the farmers as they can carefully look after.
8. Farmers should be given training for judicious use of water.
9. All the canals, water courses should be lined.
10. The alignment of canal either in sandy soil or in fissured rocks should be avoided.
11. Severe punishments should be awarded for those who get irrigation water by unlawful means. Deliberate canal cuts should be prevented by resorting to effective patrolling of canals by the staff.
12. Water tax should be increased. This measure will force farmers to use water more economically.
13. Research stations should be established in adequate number and research in regard to moisture conservation, and economical use of water should be continuously carried out.

3.9 DEFINITIONS OF SOME MORE IMPORTANT TERMS

Some of the common terms usually used in regard to the watering of the fields have been explained here.

1. Outlet Factor. This term has already been explained earlier. It is nothing but duty of canal water at outlet.

2. Paleo Watering. This is the watering applied to the fields before sowing the crops. This watering is given so as to develop sufficient moisture in the unsaturated zone of the soil so that seeds may germinate easily.

3. Kor Watering. After sowing, when crops have grown a few centimetres this watering is given. It is the first watering after sowing of the crop. Depth of kor watering is always more than the subsequent watering to be given to the crops in its growth or base period.

4. Kor Depth. The depth of water applied during Kor watering is known as *Kor depth.*

5. Kor Period. It is that initial part of the base period for a crop, in which Kor watering has to be necessarily given to the crop, otherwise its yield may be affected. While designing the capacity of the canal, Kor watering is always taken into account. Maximum canal discharge is required during Kor period as Kor watering has to be given in limited Kor period of the crop. Kor watering requires maximum discharge in limited time.

6. Time Factor. It is the ratio of the number of days the canal has actually run to the number of days or irrigation period. For example, if irrigation period is 15 days and canal has run only for 10 days during it, the time factor would be $\frac{10}{15}$ i.e., $\frac{2}{3}$. When we call days of irrigation, the day also includes the night.

7. Capacity Factor. It can be defined as the ratio of the mean supply discharge to the full supply or capacity discharge of the canal.

8. Cumec-day. It is the quantity of water supplied by one cumec discharge flowing for one-day. It is equal to 8.64 hectare metre.

9. Root Zone Depth. It is that maximum depth of the soil upto which roots of the crop extend and derive water from the soil.

10. Full Supply Co-efficient. It is obtained by dividing area estimated to be irrigated during base period by the design full supply discharge of the canal at its head during peak demand. It is also sometimes referred as "duty on capacity".

11. Nominal Duty. In some areas farmers have to submit an application to the irrigation authorities, stating their water demands in advance of every crop season. They are then issued permits authorising to use water. Farmers have to pay full amount for the water irrespective of the area they actually irrigate. The nominal duty is the ratio of the area for which permits have been issued to the mean supply for the base period.

12. Open discharge. It is the ratio of the number of cumec-days to the number of days the canal has actually been used for irrigation.

13. Gross-Command Area (G.C.A.). It is the total area over which water can flow under gravity. Generally in a canal system the whole area is divided into a number of watersheds and drainage lines. The canal is located on the watershed so that water may flow from it on both the sides due to gravity. The irrigation is carried out only upto drainage lines. Beyond drainage line another canal located on other watershed will be irrigating. Thus a particular canal can do irrigation only upto drainage lines lying on both the sides of the canal. Thus Gross command area (G.C.A.) is the total area lying between drainage boundaries which can be irrigated by the canal system.

14. Culturable Command Area (C.C.A.). It is obtained by deducting unculturable areas like Abadi area, ponds, barren land, reserved forest lands etc. from gross command area (G.C.A.). It is only that area over which cultivation is possible.

15. Incommand Areas. In culturable command areas, there may be such areas where cultivation is possible but their levels are such that they cannot be supplied water under gravity. Such areas are declared incommand areas and no water is allotted to them. Such areas may be irrigated by tube wells or may be left as pastures or may be brought under dry cropping (Barani cropping). With passage of time the incommand areas may become low due to shifting of soil by storms. Such areas may later be declared command areas and provided with irrigation water.

16. Culturable Cultivated Area. It is the area in which crops are grown at a particular crop season or time.

17. Culturable Uncultivated Area. It is that area where irrigation is possible but crops are not grown in a particular crop season. This is the area which is given rest in a particular season. In a canal system of irrigation, water is not supplied for full culturable land in a crop season. This is done because if whole of the culturable area has come under one season there will be no area left for the crops of next season. Hence the area left for the crops of next season is known as culturable uncultivated area. Following are its advantages.

(i) Land gets rest and acquires fertility again.

(ii) Temporary pastures for animals are provided.

(iii) Chances of water-logging the area are reduced.

18. Water Allowance. It is the authorised discharge of water in cumecs per thousand hectares of culturable command area.

3.10 METHODS OF DETERMINING DUTY OF WATER

The duty of water can be found out by any of the following three methods.

1. Inductive method.
2. Critical growth period basis and
3. Consumptive use of water basis.

All the three methods have been explained one by one.

1. Duty by Inductive Method. This method is based mostly on past experiments and experiences. Various states have developed standard duty tables depending upon the methods of irrigation adopted by them. These tables are prepared usually not by considering chemical properties of soils and climatic conditions. Methods adopted by various states to determine discharge in the channels to meet the crop requirements vary. Following are the methods mostly adopted:

(a) In this method the percentages of areas under different crops are worked out and water allowance for the entire C.C.A. for a particular channel fixed. Example of this method may be 0.17 cumec per 1000 hectares. This method of fixing the duty is practiced in Punjab, Haryana, and Rajasthan.

(b) In the method *outlet discharge factors* for different crops of the region are standardised. Based on outlet discharge factor, and crops grown in each season, discharges required for Kharif and Rabi seasons are separately worked out. The channel is designed for the maximum of the two discharges.

(c) In this method, based on the standard duties of various crops and area under cultivation in each crop, month-wise required discharges are worked out. The channel is designed to meet the maximum requirements of discharge, required in a month.

Example 3.1. *Determine the maximum discharge at the outlet of a distributary having following cropped areas and duties of water.*

Kind of crop	*Area in hectares*	*Duty in thousand hec/cumec*
1. Rice	*3000*	*0.75*
2. Sugarcane	*2000*	*0.60*
3. Rabi crops	*10000*	*1.75*
4. Other Kharif crops	*4000*	*1.50*

Solution. Discharges required in cumec for maturing different crops are as follows.

Kind of crop	*Area in hectares*	*Duty in 1000 hectares/cumec*	*Discharge required in cumecs*
Rice	3000	0.75	4.0
Sugarcane	2000	0.60	3.33
Rabi crops	10000	1.75	5.71
Other Kharif crops	4000	1.50	2.67

The discharges required during different seasons are as follows :

Crop	*Hot weather*	*Kharif*	*Rabi*
Rice	—	4.0	
Sugarcane	3.33	3.33	3.33
Rabi crops	—	—	5.71
Other Kharif crops		2.67	—
Total	3.33	10.0	9.04

It is seen from this table that maximum discharge of 10 cumecs is required during Kharif. Hence distributary should be designed for this discharge.

2. Critical Growth Period Basis. This method is based on the concept that crops require different quantities of water at various stages of their growth. The water requirements of all the crops are usually maximum during initial growth and then at flowering period. It is minimum during maturing period. Out of maximum demand periods, Kor period generally requires the maximum discharge. This is because Kor watering itself requires more water, and more so this watering has to be supplied to the crop during the limited Kor period. If crop plants do not receive adequate water during Kor period, yield of the crop is affected significantly. Depth of Kor watering U.P., Punjab and Haryana at outlet head, is usually taken as 13.3 cm, 16.50 cm and 19.10 cm respectively for wheat, sugarcane and rice. The Kor periods for wheat, sugarcane and rice are specified as follows :

Rainfall	*Rice*	*Sugarcane*	*Rabi*
Below 100 cm	$2\frac{1}{2}$ weeks	4 weeks	8 weeks
Between 100 and 115 cm	$2\frac{1}{2}$ weeks	6 weeks	6 weeks
Above 115 cm	$2\frac{1}{2}$ weeks	6 weeks	4 weeks

If any channel is capable to satisfy the needs of water during Kor period, it will satisfy the needs of all other periods of the crops.

Example 3.2. *Areas to be irrigated in Kharif and Rabi are respectively 2000 and 3000 hectares. Kor depths for Kharif and Rabi crops are 19 cm and 13.5 cm and Kor periods are respectively $2\frac{1}{2}$ weeks and 4 weeks. Find out the outlet discharge.*

Solution. Outlet discharge for Kharif

$$= \frac{8.64 \times B}{\Delta} \quad (B = 2\tfrac{1}{2} \text{ weeks})$$

$$= \frac{8.64 \times 17.5}{0.19} = 795.8 \text{ hectares}$$

$= 0.7958$ thousand hectares.

Outlet discharge for Rabi

$$= \frac{8.64 \times 28}{0.135} \; (B = 4 \text{ weeks})$$

$= 1792$ hectares

$= 1.792$ thousand hectares.

Required discharge for Kharif

$$= \frac{2000}{0.7958 \times 1000}$$

$= 2.51$ cumecs.

Required discharge for Rabi

$$= \frac{3000}{1.792 \times 1000}$$

$= 1.674$ cumecs.

Hence channel should be designed for an outlet discharge of 2.51 cumecs.

3. Consumptive Use of Water. Consumptive use of water is also known as evapo-transpiration use of water by the crop. It is the depth of water consumed by the crop by evaporation and transpiration during the crop period. This depth of water also includes the water consumed by accompanying weed growth in the field. Water or moisture supplied by dew or rainfall and which subsequently evaporates away without entering the plant system is also considered as part of the consumptive use. Thus consumptive use of water is the sum total of the quantity of water applied to the field by irrigation plus the quantity of water supplied by rainfall minus the quantity of water removed as surface run off and plus or minus the quantity of water absorbed by the root zone soil moisture neglecting the deep percolation losses.

The consumptive use expresses the sum total of the volume of water in evaporation and transpiration and can be expressed as hectare-metres or hectar-cm. for the growth of the plant. Once the consumptive use of water is known it is very easy to fix the duty of water.

Consumptive use of water by the crop is primarily governed by meteorological factors. As a consequence, the water requirements of different crops are found out largely by the length of their growth periods and seasonal changes in climate. The nature of soil does not affect the consumptive use of water by the crop but affects only the frequency and depth of irrigation. Shallow soils require relatively more number of waterings of lesser depth, whereas deep soils require lesser number of waterings but of larger depths. But in both the cases, the amount of water required by the crops remains the same provided the climatic conditions are identical. Evapo-transpiration rate curve for potato crop has been shown in

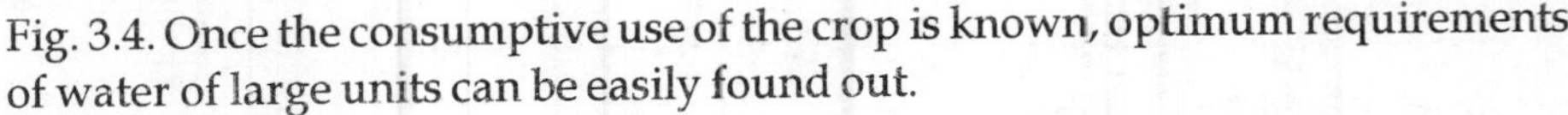

Fig. 3.4. Once the consumptive use of the crop is known, optimum requirements of water of large units can be easily found out.

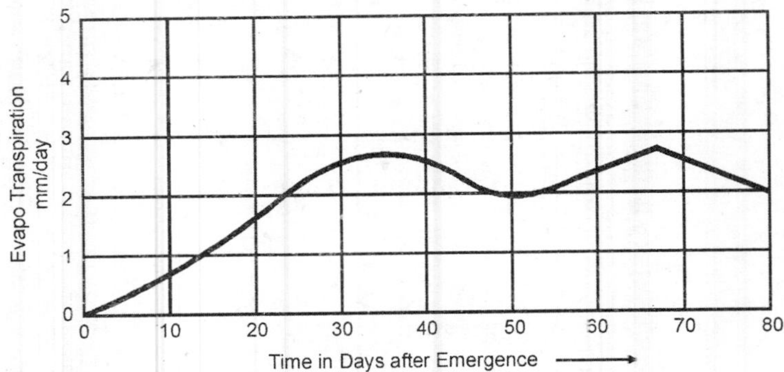

Fig. 3.4. *Evapo-transpiration of water with time.*

Evaporation. It is transfer of water from liquid state to vapour state. The rate of evaporation from the surface is proportional to the difference between the vapour pressure at the surface and the vapour pressure in the air over lying. When irrigation water is supplied to the field by flooding method, large amount of water is lost by direct evaporation from the surface of the soil without going into the life cycle of the plants.

Transpiration. It is the process by which plants dissipate water from the surface of leaves, stalks and trunks, during the period of their growth.

Transpiration Ratio. It is ratio of the weight of water transpired by the plant during its growth period to the weight of dry matter produced by the plant. When we talk of the dry matter produced by the plant, matter in roots is excluded from it. The average values of transpiration ratio for rice vary between 650 and 700 and that for wheat between 550 and 575.

3.11 FACTORS THAT AFFECT CONSUMPTIVE USE OF WATER

Following factors affect consumptive use of water:

1. Evaporation, which is dependent upon humidity,
2. Mean monthly temperature,
3. Cropping pattern,
4. Growing season of the crop,
5. Monthly precipitation in the area,
6. Wind velocity,
7. Methods of irrigation and irrigation practices,
8. Soil and topography of the region,
9. Depth of water applied for irrigation.

3.12 CONSUMPTIVE USE OF WATER FOR SOME CROPS

Consumptive use of water of some principal crops of U.P. as per the experiments conducted by Irrigation Research Institute, U.P. are as follows.

Crop	*Consumptive use of water*	*Duration in Weeks*
Rice	22.5 cm	3
Wheat	7.5 cm to 10 cm	4
Sugarcane	10 cm	3

Methods of Direct Measurement of Consumptive Use

The following are the five methods which can be used to measure the consumptive use of water.

1. Inflow and outflow studies
2. Tank and lysimeter method
3. Soil moisture studies
4. Field experimental plots
5. Integration method.

Each method has been explained here in brief.

1. Inflow and outflow studies. This method is useful for comparatively larger areas. It gives annual consumptive use. The value of annual consumptive use may be found using following equation,

$$C_u = (I + P) + (G_s - G_e) - R.$$

where C_u = Valley consumptive use in hectare-metre

I = Total inflow during 12 months of the year

P = Yearly precipitation on valley floor

G_s = Ground storage at the beginning of the year

G_c = Ground storage at the end of the year

R = Yearly outflow

All these volumes have to be measured in hectare-metre.

2. Tank and Lysimeter method. 3 m deep and 10 m square tanks are set flush with the ground level. It is good if tanks of larger area are adopted as they provide better resemblance to root development. In order to obtain satisfactory proper growth of the plant constant moisture conditions are maintained within the tank. The quantity of water required to maintain constant moisture conditions within the tank is the *consumptive use of water.* In Lysimeters the bottom of the tank is made pervious. Some of the water applied in the tank is used as consumptive use and the remaining part of water drains through the pervious bottom and gets collected in a pan. The difference between water applied and the water collected in the pan is known as consumptive use of water.

3. Soil-moisture studies. This method is found suitable where soil is fairly uniform and ground water is deep enough that it does not affect the soil moisture within the root zone of the soil. Soil moisture measurements are done before and after each irrigation. The quantity of water extracted per day from the soil is determined for each period. A curve is plotted between rate of use and time. This curve gives seasonal use.

4. Field experimental plots. Irrigation water is applied to the field experimental plot in such a way that there is neither run-off nor deep percolation. Yield obtained from different fields are plotted against the total water used. The curve so obtained forms the basis for determination of the consumptive use it is seen that for every type of crops, the yield increases rapidly with an increase of water used to a certain point, and then starts decreasing with further increase in water. The break point in the curve reflects amount of water which is consumptive use of water. This method is better than tank and Lysimeter method.

5. Integration method. In this method the total area is divided into following areas:

1. Areas under irrigation for different crops
2. Areas under natural vegetation
3. Area under water surface
4. Area under bare land.

This method consists summation of the products of the above mentioned different areas with their respective unit consumptive uses. The summation of the results is known as annual consumptive use for the whole area. It is measured in hectare-metre units.

Methods of Determination of Consumptive Use of Water by Use of Equations.

Following three methods are commonly used

1. Penman-method
2. Blaney-Criddle method
3. Hargreaves class A pan evaporation method.

1. Penman method. Consumptive use of water is determined by the following equation,

$$E_T = \frac{\Delta H + 0.27\, E_a}{\Delta - 0.27}$$

Values of H and E_a are given by the following equations,

$$H = R_A(1-r)\left(0.18 - 0.55\frac{n}{N}\right) - \sigma T_a^4\left(0.56 - 0.092\sqrt{e_d}\right) \times \left(0.10 + 0.90\frac{n}{N}\right)$$

$$E_a = 0.35\,(e_a - e_d)\,(1 + 0.0098\,U_2)$$

where

H = Daily heat budget at surface is mm H_2O/day

R_A = Men monthly extra terrestrial radiation is mm H_2O/day (Table 2.3)

r = Reflection coefficient of surface

n = Actual duration of bright surface

N = Maximum possible duration of bright

σ = Boltzmann constant

σT_a^4 = mm H_2O / day (Table 2.4)

e_d = Saturation vapour pressure at mean dew point mm Hg (actual vapour pressure in air)

E_a = Evaporation is mm H_2O/day

e_a = Vapour pressure at mean air temperature

U_2 = Mean wind speed at 2 m above the ground (mile/day)

U_1 = Measured wind speed is mile/day at height h is ft

Δ = Slope of saturated vapour pressure curve of air at absolute temperature T_a is °F (mm Hg/°F)

E_T = Evapo-Transpiration is mm H_2O/day.

The above given equations are quite complex. In order to facilitate calculation work several factors are determined as follows.

R_A can be determined from Table 3.1,

σT_a^4 can be read from Table 3.2.

The saturation vapour pressure (e_a) can be found from Fig. 3.4 given here. Δ can be read from Fig. 3.5. Tables 3.1 and 3.2 as well as Fig 3.4 and 3.5 were

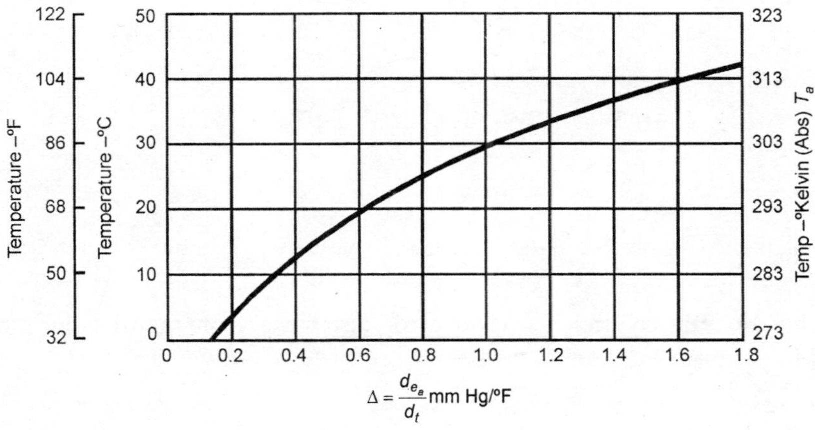

Fig. 3.5

developed by Mr. Criddle. He suggested a Tabular form for the systematic solution of Panman equation.

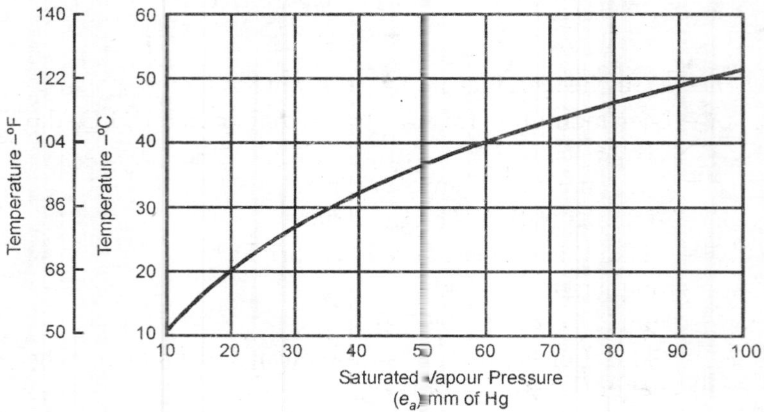

Fig. 3.6

2. Blaney–Criddle method. Blaney–Criddle gave following equation which expresses the consumptive use of water in terms of temperature and day time hours.

$$C_u = kf\text{(inches)}$$

where f = Monthly consumptive use factor $= \dfrac{t \times p}{100}$

k = Crop factor to be determined for each crop. Its value depends upon certain environmental conditions.

t = Mean temperature is °F

p = % of day time hours of the year, occurring during the period.

When expressed is metric units the above equation reduces to

$$C_u = k\frac{p}{40}[1.8t + 32] = kf.$$

where t = Temperature is °C

C_u = Monthly consumptive use in cm.

$$f = \frac{p}{40}(1.8t + 32)$$

If C_u is the seasonal consumptive use, its value is given by

$$C_u = k\varepsilon f$$

If however the coefficient k (crop factor) also varies from month to month, the total consumptive use or seasonal consumptive use is given by

$$C_u = \varepsilon kf$$

The value of p is normally taken from sun-shine Table 3.4.

3. Hargreaves class A pan evaporation method. Consumptive use or Evapo-Transpiration of water is given by following equation

$$\phi C_u \text{ or } E_t = k\,E_p.$$

where E_p = Class A pan evaporation, which can be found by Christiansen equation

E_p = $0.459\, R \times C_t \times C_w \times C_h \times C_s \times C_c$ (cm)

R = Extra-Terrestrial radiation is cm to be found from the Tables.

C_t = Coefficient for temperature, to be found from equation.

$C_t = 0.393 + 0.02796\, T_c + 0.0001189\, T_c^2$

where T_c is the mean temperature in °C

C_w = Coefficient of wind velocity given by equation

$C_w = 0.708 + 0.0034\, W - 0.0000038\, W^2$

where W is the mean wind velocity is km/day at 0.5 m above ground surface

C_h = Coefficient of relative humidity which is given by equation.

$C_h = 1.250 - 0.0087\, H + 0.75 \times 10^4\, H^2 - 0.83 \times 10^{-8}\, H^4$

where H is the mean % relative humidity at noon or average relative humidity for 11 and 18 hours.

C_s = Coefficient for % of possible sunshine and is given by

$C_s = 0.542 + 0.0085 - 0.78 \times 10^{-4}\, S^2 + 0.62 \times 10^{-6}\, S^3$

where S is the mean sun-shine in percent

C_e = Coefficient of elevation, given by

$C_e = 0.97 + 0.00984\, E$

where E is the elevation is 100 m

k = Consumptive use coefficient which is different for different crops. For each crop it depends upon many environmental factors. It is found experimentally. Table 3.3 gives the value of k for some Indian crops.

Irrigation Efficiency

In Chapter 1 under Sec. 1.10 irrigation potential of Indian resources have been explained. There, it has been stated that existing water resources of the country are adequate to irrigate only 50% of the cultivable area. This aspect makes it imperative for Indian engineers to devise ways and means to conserve water and make its judicious and economical use in irrigation.

Studies conducted on canal irrigation projects located in Northern India reveal that 15–20% of the water supplied from diversion head works is lost in mass and branch canals only, 6 to 8% losses occur is major and minor distributions, losses in field channels amount from 20 to 22%, so we see that 41% to 50% water supplied from diversion head works is lost only in conveyance before it reaches the fields for irrigation but of remaining about 50%, water supplied to fields, about 22% is further lost due to surface evaporation, deep percolation, and irregular distribution. Thus the effective percentage of water that is utilised by the crops is the form of evapo-Transpiration or consumptive use, is only 28-29% of the total water released in the main canal from diversion head works.

Due to permeability of soils, Topography and climatic conditions, it is impossible to achieve 100% utilisation of water for the growth of crops. But the above said losses in conveyance and other means suggest ample scope of improvement in the efficiency of irrigation. With correct management and by adopting such measures that would cause least losses, reasonably high efficiency can definitely be achieved.

Efficient use of irrigation water is expected from each user as well as the planners. Even under the best method of irrigation not all the water applied during irrigation is stored in the root zone. The efficiency of irrigation water is the ratio of water output to the water input and is expressed as percentage. The following use the various types of irrigation efficiencies

1. Water conveyance efficiency
2. Water application efficiency
3. Water use efficiency
4. Water storage efficiency
5. Water distribution efficiency
6. Consumptive use efficiency and
7. Project efficiency.

1. Water conveyance efficiency. This terms indicates the ratio of irrigation water that is available at the outlets to the fields, to irrigation water supplied from the diversion point into the main canal

$$\eta_c = \frac{W_f}{W_r} \times 100$$

where η_c = Water conveyance efficiency

W_f = Water delivered to the farm or plot

W_r = Water diverted from the river or reservoir into the main canal.

2. Water application efficiency. It is the ratio of the quantity of water stored into the root zone of the crops to the quantity of water delivered to the field.

It is expressed as follows

$$\eta_a = \frac{W_s}{W_f} \times 100$$

where η_a = Water application efficiency

W_s = Water stored in the root zone during the irrigation

W_f = Water delivered to the farm or plot.

Table 3.1. *Mid-Monthly Intensity of Solar-Radiation (R_a) on a horizontal surface in mm of water-evaporation per day (after Criddle)*

Month	*Southern Hemisphere*									*Northern Hemisphere*									
	90	*80*	*70*	*60*	*50*	*40*	*30*	*20*	*10*	*0*	*10*	*20*	*30*	*40*	*50*	*60*	*70*	*80*	*90*
January	17.6	17.3	16.5	16.6	17.1	17.3	17.3	16.8	15.8	14.5	12.8	10.8	8.5	6.0	3.6	1.30	–	–	–
February	10.7	10.5	11.2	12.7	14.1	15.2	15.8	16.0	15.7	15.0	13.9	12.3	10.5	8.3	5.9	3.5	1.10	–	–
March	1.9	3.6	6.1	8.4	10.5	12.2	13.6	14.6	15.1	15.2	14.8	13.9	12.7	11.0	9.1	6.8	4.3	1.8	–
April	–	–	1.9	4.3	6.6	8.8	10.8	12.5	13.8	14.7	15.2	15.2	14.8	13.9	12.7	11.1	9.1	7.8	7.9
May	–	–	0.1	1.9	4.1	6.4	8.7	10.7	12.4	13.9	15.0	15.7	16.0	15.9	15.4	14.6	13.6	14.6	14.2
June	–	–	–	0.8	2.8	5.1	7.4	9.6	11.6	13.4	14.8	15.8	16.5	16.7	16.7	16.5	17.0	17.8	18.1
July	–	–	–	1.2	3.3	5.6	7.8	10.0	11.9	13.5	14.8	15.7	16.2	16.3	16.1	15.7	15.8	16.5	16.8
August	–	–	0.8	2.9	5.2	7.5	9.6	11.5	13.0	11.2	15.0	15.3	15.3	14.8	13.9	12.7	11.4	10.6	11.2
September	–	3.3	3.8	5.2	8.5	10.5	12.1	13.5	14.4	14.9	14.9	14.4	13.5	12.2	10.5	8.5	6.8	4.0	2.6
October	7.0	7.1	8.8	10.7	12.5	13.8	14.8	15.3	15.3	15.0	14.1	12.9	11.3	9.3	7.1	4.7	2.4	0.2	–
November	15.3	15.0	14.5	15.2	16.0	16.5	16.7	16.4	15.7	14.6	13.1	11.2	9.1	6.7	4.3	1.9	0.1	–	–
December	19.3	18.9	18.1	17.5	17.8	17.8	17.6	16.9	15.8	14.3	12.4	10.3	7.9	5.5	3.0	0.9	–	–	–

Table 3.2. *Values of σT_a^4 for various temperatures (After Criddle)*

Temperature in °kelvin (Abs)	σT_a^4 *mm* H_2O*/day*	*Temperature in °F*	σT_a^4 *(mm* H_2O*/day)*
270	10.73	35	11.48
275	11.51	40	11.96
280	12.40	45	12.55
285	13.20	50	12.94
290	14.26	55	13.45
295	15.30	60	13.95
300	16.34	65	14.52
305	17.46	70	15.10
310	18.60	75	15.65
315	19.85	80	16.25
320	21.15	85	16.85
325	22.50	90	17.46
—	—	95	18.10
—	—	100	18.80

Table 3.3. *Values of k for some Indian crops (To be used in equation $E_f = kE_p$)*

Percentage of crop growing season	*Wheat (Ludhiana)*	*Maize (Ludhiana)*	*Wheat (Poona)*	*Cotton (Poona)*
0	0.14	0.40	0.30	0.22
5	0.17	0.42	0.40	0.22
10	0.23	0.47	0.51	0.23
15	0.33	0.54	0.62	0.24
20	0.45	0.63	0.73	0.26
25	0.60	0.75	0.84	0.35
30	0.72	0.85	0.92	0.58
35	0.81	0.96	0.96	0.80
40	0.88	1.04	1.10	0.95
45	0.90	1.07	1.10	1.03
50	0.91	1.09	1.10	1.08
55	0.90	1.10	0.91	1.08
60	0.89	1.11	0.80	1.07
65	0-86	1.10	0.65	1.05
70	0.83	1.07	0.51	1.00
75	0.80	1.04	0.40	0.93
80	0.76	1.00	0.30	0.85
85	0.71	0.97	0.20	0.73
90	0.65	0.89	0.12	0.62
95	0.58	0.81	0.10	0.56
100	0.51	0.70	0.10	0.40
Seasonal value of *k*	0.61	0.86	0.61	0.68

The loss of irrigation water during water application is by I surface run off R_f from the farm and II deep percolation D_f below the farm root-zone soil.

Hence $$W_f = W_s + R_f + D_f$$

or $$W_s = W_f - (R_f + D_f)$$

Hence $$\eta_a = \frac{W_f - (R_f + D_f)}{W_f} \times 100$$

Following are the factors which responsible for low water application efficiency.

1. Irrigation land surfaces
2. Irrigation streams being either very small or excessively large
3. Shallow soils underlain by gravels of light permeability
4. Wrong irrigation methods
5. Field preparation not being proper
6. Compact impervious soil
7. Steep slopes of land surfaces
8. Excessive single application of water
9. Long irrigation runs
10. Careless attendance during water application to the fields.

In a well designed surface irrigation system water application efficiency should at least be about 60% and that for sprinkler irrigation system about 75%. This efficiency focuses the attention of the suitability of the method of application of water to the crops.

3. Water use efficiency. Water use efficiency is the ratio of water beneficially used including leaching water, to the quantity of water delivered. It is expressed as follows.

$$\eta_u = \frac{W_u}{W_d} \times 100$$

where η_u = water use efficiency

W_u = water used beneficially or consumptively

W_d = water delivered.

4. Water storage efficiency

$$\eta_s = \frac{W_s}{W_\eta} \times 100$$

where η_s = Water storage efficiency

Table 3.4. *Sunshine percent day light hours*

Latitude in Degrees	January	February	March	April	May	June	July	August	September	October	November	December
						North						
0	8.5	7.66	8.49	8.21	8.50	8.22	8.50	8.49	8.21	8.50	8.22	8.50
5	8.32	7.57	8.47	8.29	8.65	8.41	8.67	8.60	8.23	8.42	8.07	8.30
10	8.13	7.47	8.45	8.37	8.81	8.60	8.86	8.71	8.25	8.34	7.91	8.10
15	7.94	7.36	8.43	8.44	8.98	8.80	9.05	8.83	8.28	8.26	7.75	7.88
20	7.74	7.25	8.41	8.52	9.15	9.00	9.25	8.96	8.30	8.18	7.58	7.66
25	7.53	7.14	8.39	8.61	9.33	9.23	9.45	9.09	8.32	8.09	7.40	7.42
30	7.30	7.03	8.38	8.72	9.53	9.49	9.67	9.22	8.33	7.99	7.19	7.15
32	7.20	6.97	8.37	8.76	9.62	9.59	9.77	9.27	8.34	7.95	7.11	7.05
34	7.10	6.91	8.36	8.80	9.72	9.70	9.88	9.33	8.36	7.90	7.02	6.92
36	6.99	6.85	8.35	8.85	9.82	9.82	9.99	9.40	8.37	7.85	6.92	6.79
38	6.87	6.79	8.34	8.90	9.92	9.95	10.10	9.47	8.38	7.80	6.82	6.66
40	6.76	6.72	8.33	8.95	10.02	10.08	10.22	9.54	8.39	7.75	6.72	7.52
42	6.63	6.65	8.31	9.00	10.14	10.22	10.35	9.62	8.40	7.69	6.62	6.67
44	6.49	6.58	8.30	9.06	10.26	10.38	10.49	9.70	8.41	7.63	6.49	6.21
46	6.34	6.50	8.29	9.12	10.39	10.54	10.64	9.79	8.42	7.57	6.36	6.04
48	6.17	6.41	8.27	9.18	10.53	10.71	10.80	9.89	8.44	7.51	6.23	5.86
50	5.98	6.30	8.24	9.24	10.68	10.91	10.99	10.00	8.49	7.45	6.10	5.65

Latitude in Degree	January	February	March	April	May	June	July	August	September	October	November	December
						South						
0	8.50	7.66	8.49	8.21	8.50	8.22	8.50	8.49	8.21	8.50	8.22	8.50
5	8.68	7.76	8.51	8.15	8.34	8.05	8.33	8.38	8.19	8.56	8.37	8.68
10	8.86	7.78	8.53	8.09	8.18	7.85	8.14	8.27	8.17	8.62	8.53	8.88
15	9.05	7.98	8.55	8.02	8.02	7.65	7.95	8.15	8.15	8.68	8.70	9.10
20	9.24	8.09	8.57	7.94	7.85	7.43	7.76	8.03	8.13	8.76	8.87	9.33
25	9.49	8.21	8.60	7.84	7.66	7.20	7.54	7.90	8.11	9.36	9.04	9.58
30	9.70	8.33	8.62	7.73	7.45	6.96	7.31	7.76	8.07	8.97	9.24	9.85

W_s = Water stored in the root zone during irrigation

W_h = Water required in the root zone prior to irrigations

= (Field capacity – Available moisture)

This efficiency gives an insight to how completely the required water has been stored in the root zone during irrigation.

5. Water distribution efficiency. Water distribution efficiency is determined from the following equation

$$\eta_d = 100\left[1-\frac{y}{d}\right]$$

where η_d = Water distribution efficiency

y = Average numerical deviation in depth of water stored from average depth stored during irrigation

d = Average depth of water stored during irrigation.

This efficiency evaluates the degree to which water is uniformly distributed throughout the root zone. The more uniformly the water is distributed, the better will be the crop response. This efficiency also provides a measure for comparing various methods or systems of water application.

6. Consumptive use efficiency. It is defined by following ratio

$$\eta_{cu} = \frac{W_{cu}}{W_d}\times 100$$

where η_{cu} = consumptive use efficiency

W_{cu} = normal consumptive use of water

W_d = net amount of water depleted from root zone soil.

This efficiency evaluates the loss of water by deep percolation and by excessive surface evaporation after an irrigation.

7. Project efficiency (η_p)

$$\eta_p = \frac{W_{cu}}{W_r}\times 100$$

η_p = Project efficiency

W_{cu} = Amount of water used in form of consumptive use by the crop

W_r = Amount of water supplied from diversion head works

We see that no efficiency is 100%. This is due to losses of irrigation water by many ways. If some how these losses are reduced, the project efficiency can be considerably increased. These losses can be minimised by liming of canals and water courses, by designing deep narrow canals, by preparing the fields properly and by adopting so many other measures by which loss of irrigation water is reduced.

Irrigation Requirements of Crops

The following terms are used to find out the irrigation requirements of any crop during its base period.

1. Effective rainfall (R_c). That part of precipitation falling during the growing period of a crop which is available is form of consumptive use of the crop, is known as *effective rainfall.*

2. Consumptive irrigation requirement (CIR). It is the amount of irrigation water required to meet the consumptive needs of the crop during its full growth.

$$CIR = C_u - R_e$$

where C_u = Consumptive use of water

3. Net irrigation requirement (NIR). It is defined as the amount of irrigation water required at the field to meet the consumptive needs of water as well as other needs such as leaching etc. Mathematically

$$NIR = C_u - R_e + \text{water lost is deep percolation for the purpose of leaching etc.}$$

4. Field irrigation requirement (FIR). It is the amount of water required to meet "net irrigation requirements" plus the water lost in percolation in the field water courses, field channels and in field application of water. If η_a is the water application efficiency, FIR can be expressed as follows.

$$FIR = \frac{NIR}{\eta_a}$$

5. Gross irrigation requirements (GIR). It is the sum of water required to satisfy the field irrigation requirement and the water lost as conveyance losses in distribution upto the field. If η_c, is the water conveyance efficiency then GIR is expressed as follows:

$$GIR = \frac{NIR}{\eta_c}$$

In order to find the FIR or GIR of any crop it is essential to know the monthly or periodical pan evaporation data and also the knowledge of pan evaporation co-efficient *k*. Table 3.5 shows the specimen calculations for wheat at a certain location.

$$\eta_a = 0.68, \quad \eta_c = 0.80$$

The volumes of Table 3.5 are self explanatory.

Table 3.5. *Determination of irrigation requirements of wheat period of grow being 10th Nov to 19th march i.e. 30 days*

Internal	*No of days upto to mid point of interval*	*% of growing reasons =* $col(2) \times \frac{100}{130}$	E_P *(cm)*	*K*	$C = K\,E_P$ *(cm)*	*Re (cm)*	$NIR = C_u - R_e$ *(cm)*	$FIR = \frac{MIR}{n_c}$ *(cm)*	$GIR = \frac{FIR}{n_c}$ *(cm)*
1	*2*	*3*	*4*	*5*	*6*	*7*	*8*	*9*	*10*
Nov 10 – 30	11	8.5	15.8	0.21	3.32	–	3.32	4.88	6.10
Dec 1 – 31	37	28.5	13.1	0.68	8.91	0.71	8.20	12.50	15.60
Jan 1 – 31	68	52.0	12.8	0.90	11.52	0.52	11.00	16.20	20.20
Feb 1 – 28	97	74.5	15.0	0.80	12.00	–	12.00	17.65	22.60
March 1 – 19	121	93.0	16.2	0.60	9.72	–	9.72	14.60	18.25
					$\Sigma = 45.47$		$\Sigma = 44.24$	$\Sigma = 65.83$	$\Sigma = 82.75$

Example 3.3. *Using Blaney-Criddle equation and a crop factor k = 0.72 for certain crop whose details have been given in the Table below, determine the following*

(i) Consumptive use

(ii) Consumptive irrigation requirements

(iii) Field irrigation requirement

The water application efficiency is 0.7 and latitude of the place is 25°N.

Month	*Monthly temp (A_V) °C*	*Monthly % of day time hours of the year*	*Useful rainfall in cm*
(1)	*(2)*	*(3)*	*(4)*
October	*25°C*	*8.09*	–
November	*20°C*	*7.40*	–
December	*16°C*	*7.42*	*1.5*
January	*14°C*	*7.53*	*0.9*
February	*15°C*	*7.14*	–

Solution. The data of column 3 of this Table has been taken from Table 3.4 corresponding to a latitude of 25° N.

The monthly temperature and useful rainfall are taken as the average values of the last 10 years.

The consumptive use is completed from Blaney–Criddle equation.

$$C_u = k\,\varepsilon f$$

where $$f = \frac{p}{40}(1.8t + 32)$$

Computation is done in Tabular form as follows

Month	*t°C*	*p%*	*f*
(1)	(2)	(3)	(4)
October	25	8.09	15.57
November	20	7.40	12.58
December	16	7.42	11.28
January	14	7.53	10.77
February	15	7.14	10.53
			Σf= 60.73

$$C_u = k\,\varepsilon f = 0.72 \times 60.73 = 43.73 \text{ cm}$$

$$R_e = 1.5 + 0.9 = 2.4 \text{ cm}$$

$$\text{CIR} = C_e - R_e = 43.73 - 2.40 = 41.33 \text{ cm}$$

NIR = CIR (Since no water is used for deep percolation)

$$\text{FIR} = \frac{\text{NIR}}{\eta_a} = \frac{41.33}{0.7} = \mathbf{59.04\ cm}$$

3.13 DUTY OF WELL WATER

Duty of well water is expressed as for tank irrigation. Discharge of the well can be easily found out and then area of crop matured by this discharge, can be expressed as so many hectares per cumec. Duty of well water is usually more than the duty of water obtained from tanks or irrigation channels. Following are the reasons for this :

1. Well is usually located in the middle of the area it has to irrigate. This aspect reduces the length of the channels and thus evaporation losses are very much reduced during conveyance of water. Moreover channels are very small and are usually lined which further eliminate the loss of water by seepage.
2. The water is given to the field only when it is most required.
3. Field Kiaries are usually very small in relation to tank or canal irrigation. Water spreads on the Kiary in a very short time, allowing minimum loss by percolation.
4. Wells are generally one's individual property. The farmer can run it as and when he desires. If there are rains wells are not required to be run. On the other hand canals will continue to supply water irrespective of there being rain or not.

3.14 BENEFITS OF DUTY FIGURES

1. Duty figures of any project can be compared with duty figures of similar other projects and inference is drawn whether our project is being managed well or there is scope of any improvement.

2. Duty figures also help in fixing the discharge capacity of new canals. If the crops that will be sown in the new area after installation of irrigation canal are known and also area available that require irrigation is known, the total amount of water for which canal should be designed can be easily estimated using duty figures for the crops proposed to be sown. By adding some conveyance losses, the overall capacity of the canal can be fixed.

3. Approximate amount of revenue that will be realised from the proposed area can be estimated.

Example 3.4. *Duty of water for a particular crop is 1000 hectare per cumec. The base period of the crop is 110 days. Find out the depth of water.*

Solution. Depth of water

$$\Delta = \frac{8.64 \times B}{D} \text{ m.}$$

$$= \frac{8.64 \times 100}{1000} \text{ m.}$$

$$= 0.95 \text{ m.}$$

$$= \mathbf{95\ cm.}$$

Example 3.5. *The total available gross command area of an irrigation channel is 100000 hectares. If culturable command area is 80% of the G.C.A. and intensities of irrigation for Rabi and Kharif crops are respectively 60% and 35% find out the discharge in the canal at its head. The duty of Rabi and Kharif crops at the head of the canal are 1500 and 700 hectares per cumec respectively.*

Solution. Culturable command are C.C.A.

$$= 100000 \times 0.80$$

$$= 80000 \text{ hectares}$$

Area under Kharif crop

$$= 80000 \times \frac{35}{100} = 28000 \text{ hectares.}$$

Area under Rabi crop

$$= 80000 \times \frac{60}{100} = 48000 \text{ hectares.}$$

The ratio of area irrigated in Rabi crop and Kharif crop is known as crop ratio.

$$\text{Crop ratio} = \frac{48000}{28000} = \frac{12}{7}$$

Discharge required at the head of the canal for Kharif crop

$$= 28000 \times \frac{1}{700} = 40 \text{ cumecs}$$

Discharge required at the head of the canal or Rabi crop

$$= \frac{48000}{1500} = 32 \text{ cumecs.}$$

Discharge at the head of the canal as per normal demand of the crops

$$= 40 \text{ cumecs.}$$

Normally 20 to 25% additional discharge is required for the Kor period. Hence canal will hot be designed for 40 cumecs but for

$$40 \times \frac{125}{100} = 50 \text{ cumecs.}$$

Example 3.6. *A village has 2000 hectares of C.C.A. out of which 20% area is under the cultivation of perennial crop i.e. sugarcane, 50% area is under wheat cultivation whose duty at the head of the outlet is 2000 hectare/cumec. Duty of sugarcane is 700 hectare/cumec. If demand of water during Kor period gets increased by 20% of the average demand find out the discharge for which village water course has to be designed.*

Solution. Area under sugarcane

$$= 2000 \times 0.2$$

$$= 400 \text{ hectares.}$$

Area under wheat $= 2000 \times 0.50 = 1000$ hectares.

Discharge for sugarcane

$$= \frac{400}{700} = 0.57 \text{ cumecs.}$$

Discharge for wheat

$$= \frac{1000}{2000} = 0.50 \text{ cumecs.}$$

Sugarcane is perennial crop which requires irrigation water for whole of the year. Hence water course will have to carry the combined discharge simultaneously for both the crops.

Combined discharge for both the crops

$$= 0.57 + 0.5 = 1.07 \text{ cumecs.}$$

Maximum discharge for which water course should be designed

$$1.07 \times \frac{120}{100}$$

i.e., $$1.284 \text{ cumecs.}$$

Example 3.7. *Different crops, their base period, area under irrigation of each crop and duty of water at head of the canal for different crops are given as follows. If whole of this area is under the command of one canal, find out the discharge required in the canal. Value of time factor may be taken 3/4.*

Crops	*Base period*	*Area in hectares*	*Duty at head in hectares/cumec*
1. *Sugarcane*	300	800	600
2. *Over lap of sugarcane in summer*	30	150	600
3. *Wheat*	120	100	1700
4. *Spiked millet or Bajra*	100	500	2200
5. *Hot weather vegetables*	110	350	700

Solution. Required discharge for different crops

1. For sugarcane

$$= \frac{800}{600} = 1.33 \text{ cumecs.}$$

2. For overlapping sugarcane

$$= \frac{150}{600} = 0.25 \text{ cumecs.}$$

3. For wheat $= \dfrac{700}{1700} = 0.41$ cumecs.

4. For spiked millet

$$= \frac{500}{2200} = 0.227 \text{ cumecs.}$$

5. For hot weather vegetables

$$= \frac{350}{700} = 0.50 \text{ cumecs.}$$

Base period of sugarcane is 300 days. Hence it will require water in Rabi, Rain and hot weather also.

Discharge required during Rabi

$$= 1.33 + 0.41 = 1.74 \text{ cumecs.}$$

Discharge required during Rainy season

$$= 1.33 + 0.227 = 1.557 \text{ cumecs.}$$

Discharge required during hot weather

$$= 1.33 + 0.25 + 0.50 = 2.08 \text{ cumecs.}$$

Overlap discharge for the sugarcane is required only during hot season. Hence maximum demand of water is 2.08 cumecs, which is required during hot weather.

$$\text{Time factor} = \frac{\text{Number of days for which canal has actually run}}{\text{Number of days for which canal should run}}$$

$$= \frac{3}{4}.$$

Hence full discharge required at head of the canal considering time factor

$= 2.08 \times \dfrac{4}{3} = 2.76$ cumecs.

If say 20% extra demand is expected during Kor period then discharge in the canal should be

$$2.76 \times 1.2 = 3.312 \text{ say } 3.31 \text{ cumec.}$$

Hence canal should be designed for a maximum discharge of 3.31 cumecs.

Example 3.8. *Intensity of irrigation, duties and base periods of various crops under a canal system are given in the Table below. If reservoir losses are 10% and conveyance loss in canal system 20%. Find out the reservoir capacity.*

Crop	*Base period*	*Duty at field in hectare/cumec*	*Area under each crop in hectares*
Wheat	120	1800	5400
Sugarcane	330	800	6000
Cotton	210	1200	3000
Rice	130	800	3000
Vegetables	100	700	1500

Solution

1. Wheat. Discharge $= \dfrac{5400}{1800} = 3$ cumec

Volume of water in 120 days

$= 3 \times 120 = 360$ cumec-day

2. Sugarcane. Discharge

$= \dfrac{6000}{800} = 7.5$ cumec.

Volume of water $= 7.5 \times 330 = 2475$ cumec-day.

3. Cotton. Discharge

$= \dfrac{3000}{1200} = 2.5$ cumec.

Volume of water $= 2.5 \times 210 = 525$ cumec-day.

4. Rice. Discharge

$= \dfrac{3000}{800} = 3.75$ cumec.

Volume of water $= 3.75 \times 130$

$= 487.5$ cumec-day.

5. Vegetables. Discharge

$$= \frac{1500}{700} = 2.143 \text{ cumec.}$$

Volume of water $= 2.143 \times 100 = 214.3$ cumec-day.

Total volume of water required for all the crops

$= 360 + 2475 + 525 + 487.5 + 214.3$

$= 4061.8$, say 4062 cumec-day.

One cumec-day $= 1 \times 24 \times 60 \times 60 \text{ m}^3$

One hectare-metre $= 1 \times 10^4 \text{ m}^3$.

$$\therefore \text{ One cumec day} = \frac{1 \times 24 \times 60 \times 60}{1 \times 10^4} = 8.64 \text{ hectare-metre.}$$

Total volume of water

$= 4062 \times 8.64$ hectare-metres

$= 35095.68$ hectare-metres

$= 35096$ hectare-metres (say)

20% water lost in conveyance.

Hence water volume for canal

$$= 35096 \times \frac{100}{80}$$

$= 43870$ hectare-metre.

10% water is lost in reservoir.

Hence reservoir capacity

$$= 43870 \times \frac{100}{90}$$

$= 48744.4$ hectare-metres.

$= 48750$ hectare-metres (say)

Example 3.8. *From the following data, determine the field capacity of the soil.*

Depth of root zone $= 1.5$ *m.*

Present water content $= 5\%$.

Dry density of soil $= 1.6 \text{ gm/cm}^3$.

Water applied to the soil $= 7000 \text{ m}^3$

Water loss due to evaporation = 10%.

Area of the plot = one hectare = 10^4 m^2.

Solution. Total water supplied = 7000 m^3

Loss of water = 10%

∴ Water absorbed by soil

$$= 7000 \times \frac{90}{100} = 6300 \text{ m}^3$$

$$= 6300 \times 10^6 \text{ gms.}$$

Total dry weight of the soil

$$= 10000 \times 1.5 \times 1.6 \times 10^6 \text{ gms.}$$

∴ Percentage of water added

$$= \frac{6300 \times 10^6 \times 100}{10000 \times 1.5 \times 1.6 \times 10^6}$$

$$= 26.25\%.$$

Hence new water content = 5% + 26.25%

= 31.25%

Hence field capacity is 31.25%

QUESTIONS

3.1 Write a note on water requirements of crops. How use of water for a particular crop can be optimized ?

3.2 Enumerate various factors that affect the water requirements of the crops.

3.3 Define Duly, Delta, and Base period. Establish relationship among them.

3.4 In how many ways the duty of water can be expressed ? Why duty of water varies at various take off points on a canal system ? Discuss with a suitable example.

3.5 (a) Explain the points that affect duty of water.

(b) Discuss the measures by which duty of water can be improved.

3.6 Define the following terms :

(i) Kor period (ii) Outlet factor

(iii) Kor watering (iv) Paleo watering

(v) Time factor (vi) Capacity factor

(vii) G.C.A and (viii) C.C.A.

3.7 What are various methods of determining duty of water ? Explain inductive method with the help of suitable numerical example.

3.8 What do you understand by the term consumptive use of water ? Why this method of evaluating duty of water is considered best ?

3.9 A field channel has a C.C.A of 4000 hectares. The intensity of irrigation for gram is 30% and for wheat is 50%. Gram has a Kor period of 18 days whereas wheat 15 days. Kor depths for gram and wheat are respectively 12 cm and 15 cm. Calculate the discharge of the field channel.

(**Ans.** 3.248 cumecs)

3.10 The discharge available from a tube well is 180 m^3/hour. If tube well is worked for 4800 hours in a year, find out the C.C.A. that this tube well can command. The intensity of irrigation is 50% and the average depth of Rabi and Kharif crops is 48 cm.

(**Ans.** 360 hectares)

4

Methods of Irrigation

4.1 EFFICIENCIES OF IRRIGATION

In Chapter 1 under the Sec. 1.10, irrigation potential of Indian resources have been explained. There, it had been stated that existing water resources in the country are adequate to irrigate only 50 per cent of the cultivable area. This aspect makes it imperative for Indian Engineers to devise ways and means, to conserve water and make its judicious and economical use in irrigation.

Studies conducted on canal irrigation projects located in northern India reveal that 15–20% of the water supplied from diversion head works is lost in main and branch canals only 6–8% losses occur in major and minor distributries. Losses in field channels amount from 20–22%. So we see that 41% to 50% water supplied from diversion head works is lost only in conveyance before it reaches the fields for irrigation. Out of remaining about 50% water supplied to fields, about 22% is further lost due to surface evaporation, deep percolation and irregular distribution. The effective percentage of water that is utilised by the crops in the form of evapotranspiration or consumptive use, is only 28–29% of the total water released in the main canal from diversion head works. From this study, it is clear that only 28–29% of the water supplied from the diversion head works is used by the crops and remaining 71 to 72% water is lost by one way or the other.

Due to permeability of soils, topography, and climatic conditions, it is impossible to achieve 100% utilisation of water for the growth of crops. But there is ample scope to improve the efficiency of irrigation. With correct management, and by adopting such measures that would cause least losses, reasonably high efficiency can definitely be achieved. Some important terms applicable to the efficiency of irrigation water are given as follows.

1. Project Efficiency

$$E_i = \frac{W_e}{W_r}$$

where E_i = Project efficiency

W_e = Amount of irrigation water used in form of evapo-transpiration by the crop

W_r = Amount of irrigation water supplied at the diversion point.

Conveyance losses occur right from the diversion head works of the main canal to the entry points to the fields. These losses are caused by evaporation and seepage. If somehow these losses are reduced to minimum project efficiency can be considerably increased. These losses can be minimised by lining of canals and water courses, by designing deep and narrow canals and adopting so many other measures.

2. Conveyance Efficiency. This term indicates the ratio of irrigation water that is available at the outlets, to irrigation water supplied from the diversion point into the main canal. If can be expressed mathematically as follows :

$$E_c = \frac{W_d}{W_r}$$

E_c = Conveyance efficiency

W_d = Amount of irrigation water available at the outlets to the fields

W_r = Amount of irrigation water supplied from the diversion point.

3. Application Efficiency. It is the ratio of irrigation water used by the crops as consumptive use, to irrigation water supplied at the outlet points of the fields. It can be written as follows :

$$E_a = \frac{W_e}{W_d}$$

where E_a = Water application efficiency

W_e = Amount of water used by crops as consumptive use

W_d = Amount of irrigation water available at the outlets to the fields.

In our country water application efficiency is very low and it requires some serious thinking on the part of Engineers to improve it, by devising suitable measures.

4.2 BOARD CLASSIFICATION OF METHODS OF IRRIGATION AND THEIR EFFICIENCIES

Methods of irrigation cab be broadly classified under three headings as follows:

1. Surface irrigation methods.
2. Spray or sprinkler irrigation.
3. Sub-soil methods of irrigation.

In surface irrigation methods, irrigation water is spread in form of thin sheet on the surface of the fields to be irrigated. The application efficiency of surface methods is low as lot of irrigation water is lost by evaporation and deep percolation. If fields are prepared in suitable lengths and slope, and surface roughness and soil infilteration ratio are some how scientifically designed and controlled, the application efficiency of these methods can be increased as much as 50 to 60 per cent.

In the case of spray or sprinkler irrigation, water is supplied to crops in forms of rains. Irrigation efficiency upto 80% is possible to be achieved by this method.

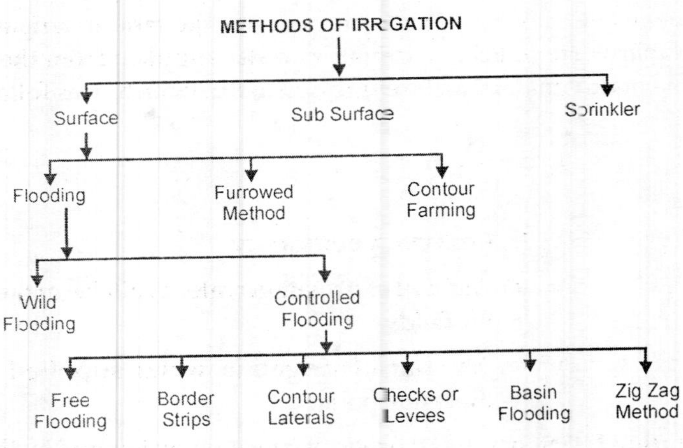

Fig. 4.1. *Methods of irrigation.*

In sub-soil method of irrigation water is not applied at the surface of the fields but at the root zone of the crops. It is possible to achieve even higher than 80% irrigation efficiency by this method of irrigation.

4.3 METHODS OF SURFACE IRRIGATION

Following are the methods of surface irrigation:

1. Wild or uncontrolled flooding.
2. Free flooding.
3. Contour laterals.
4. Border strip method.

5. Check flooding method.
6. Basin flooding method.
7. Zigzag method.
8. Furrowed method.

All these methods have been explained one by one.

1. Wild or Uncontrolled Flooding. In this method water is applied to rather unprepared fields without exercising any control on the direction of spread of water. No levees are generally made to guide the spread of water. This method of irrigation is adopted in inundation irrigation wherein water from streams during floods is taken and allowed to spread on vast areas lying adjacent to the stream. This is a very wasteful method of irrigation and can be used only where water is available abundantly at nominal or not cost.

2. Free Flooding. This method is a somewhat improved method than wild or uncontrolled flooding method. This method is mainly adopted where fields have not been properly prepared and field soil is so hard that it will allow very small amount of water to be absorbed by top soil before flooding water gets evaporated. In this method, fields are divided into large plots and water is

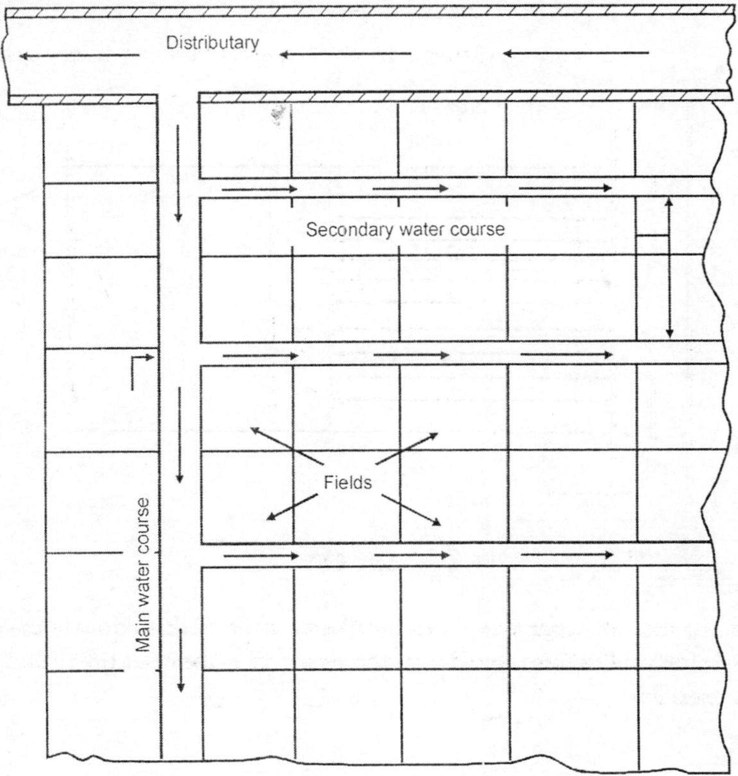

Fig. 4.2. *Free flooding.*

allowed to flood or spread on whole of the field so as to give time to the soil to absorb sufficient amount of water. The fields are provided with permanent supply water course. This method is very much prevalent in most parts of U.P. where soil is of very heavy type.

4.3.1 Analysis of Time to Cover a Strip Area by Irrigation Water

Depending upon the rate of infiltration of water into the soil the size of the outlet supplying water to the fields is decided. Total area in the command of an outlet also is a deciding factor in fixing the size of the outlet. When a large stream of water is applied to unit area of the soil of low infiltration rate, excessive run off occurs. Conversely when a small stream is applied to a unit area of soil of high infiltration rate, excessive depth of water lost by deep percolation. The relation between size of water stream and time rate of water application over a given area of land can be established easily as follows.

Let A = Area of land covered at any time t in hectares

I = Rate of infiltration in m/hr

q = Quantity of water in ha-m/hr

t = Time in hours

y = Average depth of sheet of flowing water in metres

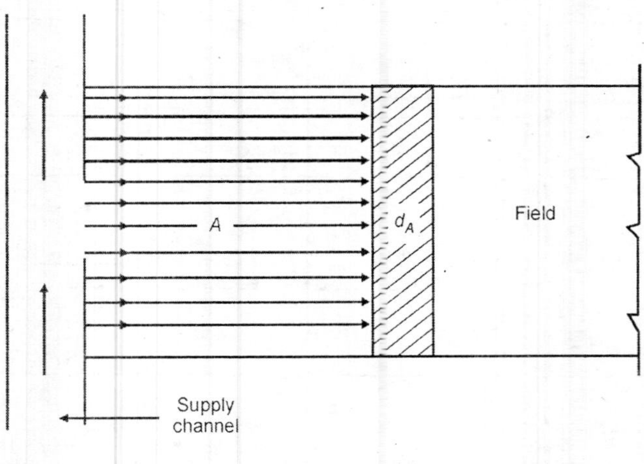

Fig. 4.3

Total quantity of water flowing in small time interval dt is equal to the quantity of water infiltrated in this time dt over the area A plus the quantity of surface flow over the area dA

Thus

$$qdt = ydA + IA\,dt$$

or

$$(q - IA)\,dt = ydA$$

$$dt = \frac{ydA}{q - IA} \quad (1)$$

Considering I and y as constants and integrating we get

$$t = 2.303\frac{y}{I}\log_{10}\frac{q}{q - IA} \quad (2)$$

where $t = 0 \;\; A = 0$ and hence constant of integration is zero

In Eq. (2) we have taken I as constant, actually it decreases as the soil gets saturated. Rewrite Eq. (2) as follows

$$\log_{10}\frac{q}{q - IA} = \frac{I}{2.303\,y} = x \text{ (say)}$$

$$\therefore \quad 10^x = \frac{q}{q - IA}$$

from which

$$A = \frac{(10^x - 1)q}{10^x I} = \frac{q}{I} \quad \text{(Approx.)} \quad (3)$$

Equation (3) gives the maximum area that can be irrigated with a stream of discharge q.

Example 4.1. *0.04 cumecs of water is being discharged by an outlet at the head of a field of area 0.2 hectare. If the average depth of flow is expected to be 80 mm and average infiltration rate for the soil is 50 mm/hr determine the time required to cover this area.*

Solution

$$\begin{aligned} q &= 0.04 \text{ cumecs} \\ &= 0.04 \times 3600 \text{ m}^3/\text{hour} \\ &= 0.0144 \times \text{hectare-metre/hour} \\ y &= 80 \text{ min} = 0.08 \text{ m} \\ I &= 50 \text{ mm/hr} = 0.05 \text{ m/hr} \\ A &= 0.2 \text{ hectare.} \end{aligned}$$

Use equation

$$t = 2.303\frac{y}{I}\log_{10}\left(\frac{q}{q - IA}\right)$$

$$= \frac{2.303 \times 0.08}{0.05}\log_{10}\left(\frac{0.0144}{0.0144 - 0.2 \times 0.05}\right)$$

$$= 1.90 \text{ hrs}$$

$$\text{Maximum area that can be irrigated} = \frac{q}{I}$$

$$= \frac{0.0144}{0.05} = 0.288 \text{ hectare.}$$

3. Contour Laterals. This method may be used for irrigating steeper terrains. In this method a number of small contour laterals (drains) are constructed. The spacing of these ditches depends upon the grade of field between adjacent ditches. Spacing is less for steep slope and more for flat terrains. This is also not a very good method of irrigation.

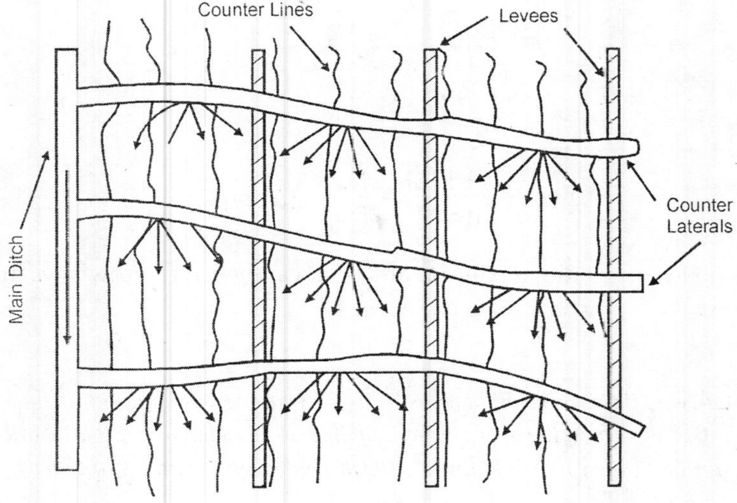

Fig. 4.4. *Contour laterals method.*

4. Border Strip Method. This method is very much adopted in canal irrigation in Northern India. In this, the farm is divided into series of strips. The strips are 10 to 20 m wide and 100 m to 300 m long. In sandy soils, the size of the strip is usually 8 to 10 m wide and about 50 m long. In case of sandy soils, strips are made of shorter lengths as otherwise head reaches of strips absorb excessive water and thus affect duty of water. The method has very scientifically laid system of water courses. The water from the water courses is led to the strips and water moves in each strip in form of a thin film, uniformly spread along the width. As soon as one strip gets filled, water flowing into it, is stopped. The water is then opened to the adjoining strip. Size of the water course may vary from 0.015 to 0.3 cumec and slope of the strip in the direction of flow of water may be 0.5 to 2% depending upon the porosity of the soil. The whole of the form is divided into long strips with the help of low or small earthen levees. See Fig. 4.6.

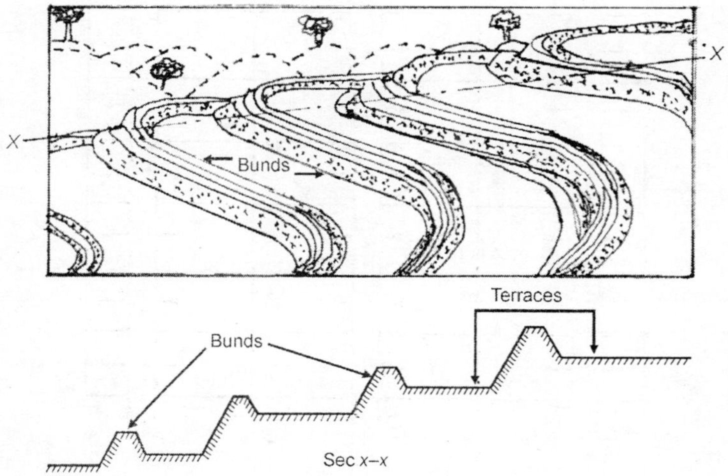

Fig. 4.5. *Isometric veiw of contour laterls.*

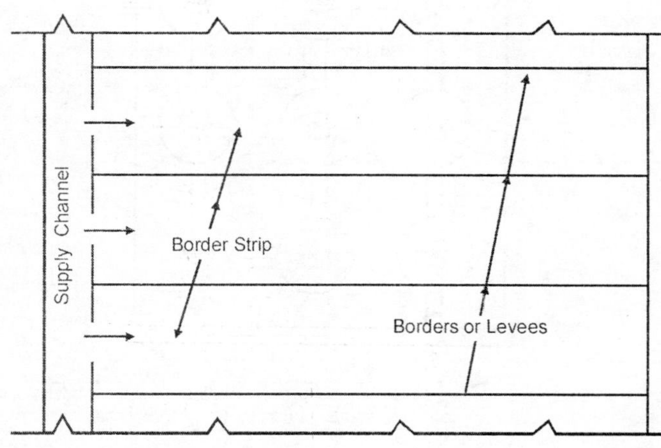

Fig. 4.6. *Border strip method.*

5. Check Flooding Method. This method consists of a comparatively large stream which discharges water into a well prepared and levelled plot. The plot is surrounded by checks or levees about 30 cm in height. The levees are broad based at the bottom so that water filled in the plot may not escape. This method is found suitable for very permeable soils, as plot gets filled up very soon without allowing much water to percolate. It is equally suitable for heavy soils also. The water filled in the field can remain held up in the levees and allows heavy hard soil sufficient time to absorb sufficient water.

In the levelled farm, the plots or kiarsies are generally rectangular. But if the ground has substantial slope the checks or levees may follow the contours at

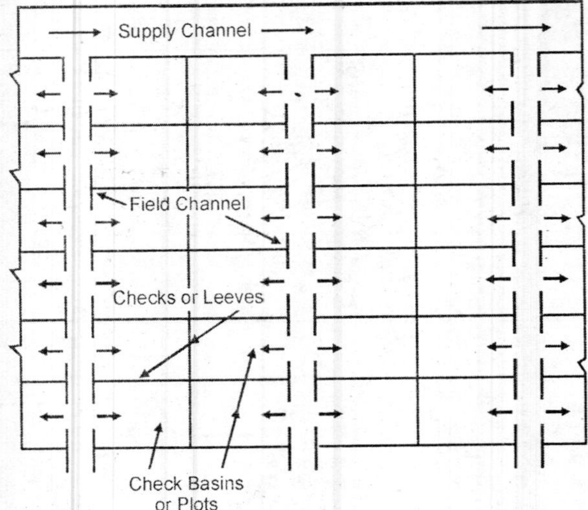

Fig. 4.7. *Check irrigation.*

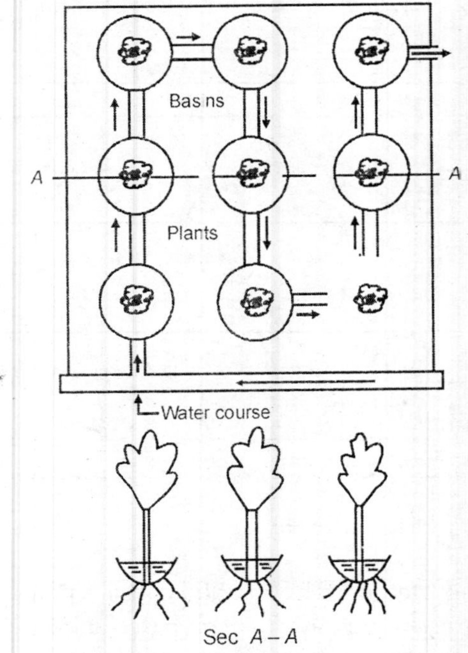

Fig. 4.8. *Basin irrigation.*

vertical intervals of 5 to 12 cm. Contour levees may be connected with cross-levees at convenient places. In this method plots having areas varying from 0.2 to 0.8 hectares have been found to give best results.

6. Basin Flooding. It is also a sort of deep check flooding method. It is mostly adopted for the irrigation of orchards. The basins are formed for each individual plant or for a group of plants. All these basins remain connected with each other. The water from supply ditch flows the basins. Sometimes portable pipes or large hoses may be used in place of ditches.

7. Zig-zag Method of Flooding. In this method a large plots is subdivided into a number of small size plots and all the plots remain connected with one another as shown in Fig. 4.9.

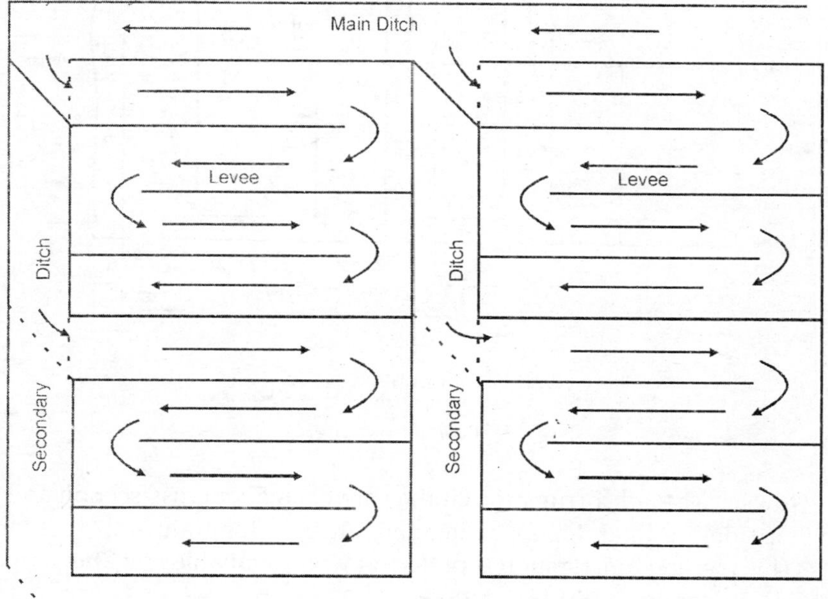

Fig. 4.9. *Zig-zag irrigation.*

Water is led into the first part which after filling takes a circuitous route and floods the second, third and so on, small plots. This method is not much practiced as number of levees hamper farming operations with modern farm machinery.

8. Furrowed method. In this method of irrigation the fields is divided into ridges and furrows. Duty of canal water is very much increased by this method as only part of the surface of the fields has to be flooded. Crops are sown on ridges and watering is done in furrows. This method of irrigation is adopted for certain crops which give better yield in this way. All the crops which give their yield from under the soil are usually grown by this method. Length of furrow or ditch varies from about 3 m for gardens to as much as 500 m for field crops. The most common length of the ditch varies from 100 m to 200 m. The general slope of the furrows varies from 1 per cent to 3 per cent. This method of irrigation is not suitable for very sandy soils, as in that case losses by percolation become excessive. Spacing of the adjacent furrows depends upon the type of soil and also the crop to be sown in it. Depth of furrows varies from 20 cm to 30 cm.

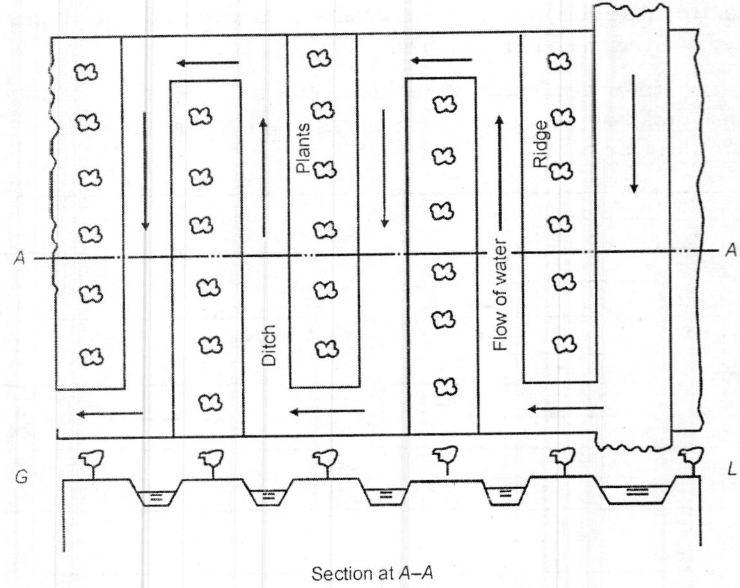

Fig. 4.10. *Furrow method.*

Furrows are called corrugations if they are of smaller cross-section and run with more longitudinal slope. Corrugations are used for grain and other forage crops. This method of irrigation is preferred when available irrigation streams are small and topography of land uneven.

Spray of Sprinkler Irrigation. In this method of irrigation water is sprayed on the surface of the soil. This method of irrigation is preferred under the following conditions.

(i) If soil is too porous and surface irrigation over it involves lot of wastage of water.

(ii) Where topography of the soil is undulated and levelling of land for surface irrigation is not possible.

(iii) Land is steeply sloping, and easily erodible.

(iv) Availability of water is too small to be used as surface irrigation.

The method consists of a system of pipes comprising main pipes and lateral pipes. The pipe system is generally portable which can be easily dismantled at the time of preparation of the fields for sowing and reassembled after sowing. Laterals are so spaced and nozzle system is so fixed that spray water is uniformly sprinkled over the entire fields. Mainly three types of sprinklers are used. They are fixed nozzle pipe, perforated pipe and rotating sprinklers. Figures 4.11 and 4.12 show fixed nozzle pipe system and perforated pipe systems. Fixed nozzle type sprinklers are outdated not used any more. Perforated pipe type sprinklers are useful for orchards and nurseries. The pressure being low the water may be supplied from an overhead tank instead of pumps.

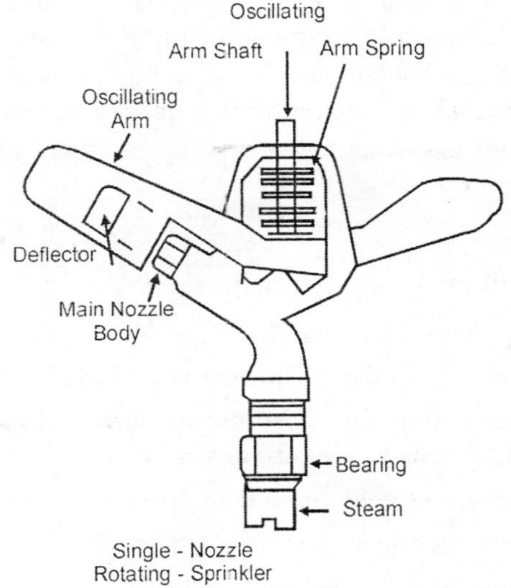

Fig. 4.11

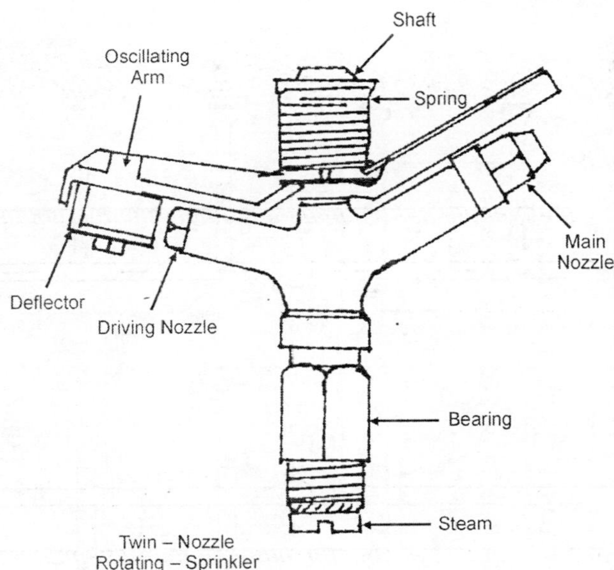

Fig. 4.12

Rotating sprinklers is the latest system of sprinkler irrigation. The rotating sprinklers consists of one or two nozzles mounted on a body which is rotated slowly about vertical axis by the action of a deflecting vane connected to it. The jet of water issuing from one of the nozzles impinges on the vane and thrusts it aside. The motion of the vane is controlled by light spring which tends to bring it back to original position, but the return of the vane is stopped by a stop on the body which also rotates through a small angle by the impulse. The vane again intercepts the jet and the cycle is repeated. Single nozzle Rotating sprinkler and Twin-nozzle rotating sprinklers are shown in Figs. 4.11 and 4.12. The irrigation water is pumped into the mains with the help of a pump. Pumped water gets distributed on land inform of spray through nozzles fitted on laterals. This system has following advantages.

Advantages

(i) Water is uniformly applied on the field.

(ii) Soil erosion does not take place.

(iii) Deep or shallow watering can be given as per the demand of the crop.

(iv) Land can be prepared easily as clay puddling does not take place at the surface as happens in the case of surface irrigation.

(v) Crops can be effectively saved from frost.

(vi) Surface run off is eliminated.

(vii) Intensity of irrigation is more.

(viii) Small streams of water cab be efficiently used.

(ix) Fertiliser application can be better controlled.

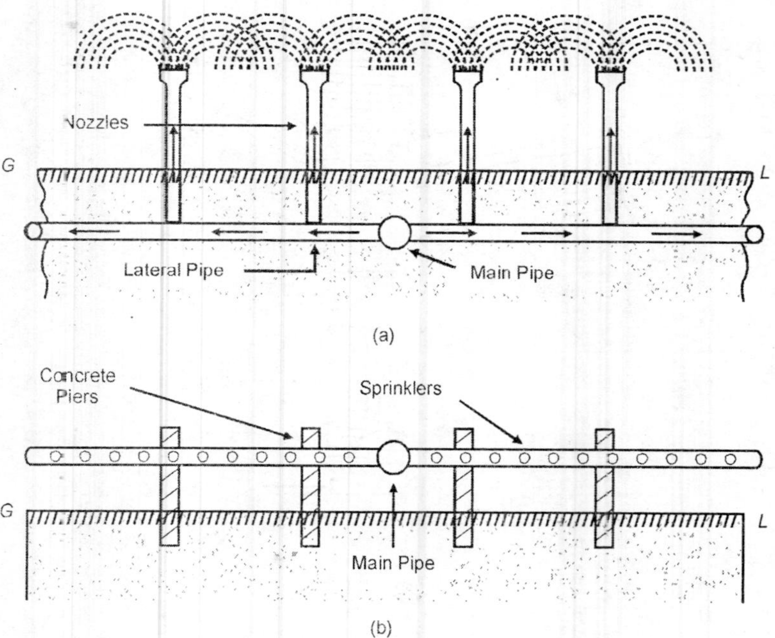

Fig. 4.13. *Spray of sprinkler irrigation.*

Disadvantages

(i) Involves heavy initial investment.

(ii) Pipe system has to be assembled and dismantled frequently.

(iii) Requires power for running the pumps.

(iv) Heavy clayey soils cannot be irrigated.

(v) Even strong winds affect sprinkling.

(vi) Pipes may be frequently choked and thus disturb the sprinkling operation.

(vii) Water used should be clean and free from sand to avoid choking trouble.

Sub-surface Irrigation. This system is also sometimes named as drip or trickier irrigation. Water is not applied at the surface of the soil but only at the root zone of the crops with the help of embedded system of porous pipes. The system consists of main line, submains, and laterals. Lateral pipes are porous pipes whereas mains and submains are used to distribute water to the laterals. The laterals allow irrigation water to drip slowly at almost zero pressure, and keep the soil around the pipes, which is also root zone of plants, constantly wet. This method eliminates loss of water by evaporation. This method is most expensive, but is best to provide irrigation in arid zones where soils generally have poor structure and high salt content.

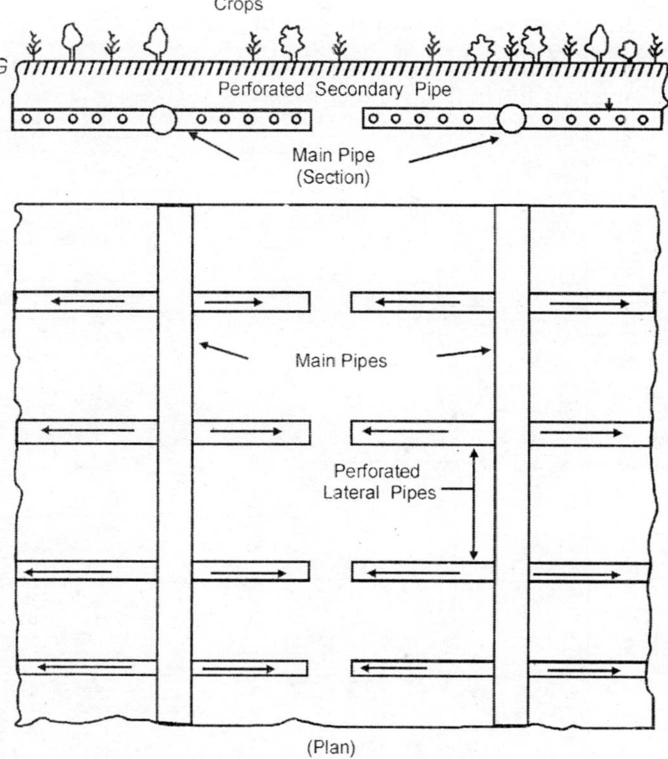

Fig. 4.14. *Sub-surface irrigation.*

This method of irrigation has been very successful in Israel. It has practically developed the whole of Arava Desert, with very little amount of water. Conditions similar to Arava Desert of Israel, exist in West Rajasthan and Gujarat and theexperiences of Israeli experts can be requisitioned to develop these areas. Experiments have also confirmed that the yield of crops like tomato, pepper, musk melon, cucumber exceeds far more than that obtained by sprinkler or surface methods of irrigation.

QUESTIONS

4.1 Explain the terms :

(i) Efficiency of irrigation. (ii) Project efficiency.

(iii) Conveyance efficiency. (iv) Application efficiency.

4.2 Enlist the various method of surface irrigation. Briefly describe following methods :

(i) Basin flooding method. (ii) Furrowed method.

(iii) Free flooding method. (iv) Border strip method.

4.3 Under what conditions sprinkler method of irrigation is preferred. Write down advantages and disadvantages of this method.

4.4 Describe sub-soil method of irrigation in details. Discuss the applicability of this method for Indian conditions.

5

Principal Crops of India and Methods of Assessment

5.1 CLASSIFICATION OF CROPS

Based upon the use and nature, the crops can be classified as given in Table 5.1.

Table 5.1

Classification of crops	*Examples of such crops*
1. Cereals	Wheat, Rice, Barley, Oats etc.
2. Pulses	Moong, Arhar, Gram, Peas, Soyabeen etc.
3. Oil seeds	Groundnut, Linseed, Mustard, Castor etc.
4. Millets	Maize, Bajra etc.
5. Vegetables	Potato, Peas, Tomato, Onion, Cabbage etc.
6. Fruit crops or garden crops	Mango, Apple, Pear etc.
7. Plantation crops	Tea, Coffee, Sugarcane, Coconut etc.
8. Folders	Barseem, Senji etc.
9. Spices	Ginger, Pepper, Corridar.
10. Narcotics	Poppy, Hashis, Tabasco.
11. Fibre crops	Hemp, Sisal, Jute, Cotton.

5.2 WET AND DRY CROPS

Based upon the uses of irrigation water, the crops can be classified as *wet crops* and *dry crops*.

Wet Crops. Wet crops are those crops which require irrigation water.

Dry Crops. These are such crops which do not require irrigation water.

All the crops require moisture to maintain their growth. Some crops require less amount of water while others more quantity of water. Any crop may be dry crop or wet crop. It all depends upon the rainfall and water requirements of the crops If rainfall is causal then only such crops should be sown as dry crops which require comparatively less amount of water. If rainfall is quite enough for ordinary cereal crops they can be sown as dry crops. The same crops which can be grown as dry crop in moderate rainfall region can be grown as wet crop in areas of scanty rainfall. The crops which require lot of water and that too for quite a long base period, can generally be grown by irrigation only. Dry crops are generally those crops which require very little water and that water is supplied by natural rainfall. Yield of wet crop is more than dry crop.

5.3 PRINCIPAL CROP SEASONS

Agriculturally the year may be divided into two principal seasons namely Rabi and Kharif.

Rabi Season. In North India this season is also known as *Hari* season. Wheat, Barley, Gram, Mustard, Peas etc. are the main Rabi crops. Sowing of these crops is done in autumn and harvesting in springs. Usual sowing time of these crops starts from middle of October and ends at the end of November. Harvesting time of these crops starts from 15th March and ends by 15th April. Sowing and harvesting times may vary slightly depending upon the crop and local climatic conditions.

Kharif Season. This reason in also known as *Sawni* season. Cotton, Millets, Paddy, Groundnut etc. are the main Kharif crops. Kharif crops are sown from 15 March and onwards say up to beginning of May. Harvesting of these crops starts from middle of October and continues up to the end of November. Harvesting and sowing times, again, may vary slightly depending upon the type of crop and local climatic conditions.

Sometimes short duration Rabi and Kharif crops are harvested and fields remain vacant for say one or two months, before next main crop of the season is sown in it. In such circumstances some short duration crops such as fodder and vegetables may be sown in the vacant fields and harvested before next main crop of the season is sown in the same field. Such short duration crops which are grown during the intervening periods between two main crops of the successive seasons are known as zaid crops. The zaid crops which are grown after the main crop of Rabi season are known as *Zaid Rabi* crops. Similarly zaid crops which are grown after the main crop of Kharif season are known as *Zaid Kharif* crops.

Table 5.2 gives the relevant data about various crops of Northern India.

Table 5.2

Name of the crop	*Sowing time*	*Harvesting time*	*Seeds per hectare in kg*	*Average yield in quintals cm*	*Average total water depth in cm*
KHARIF CROPS					
1. Transplanted rice	June-July	Oct.-Nov.	80 kg	15	125-150
2. High yielding rice	July	November	90 kg	40	125-150
3. Maize	June	Sept-Oct.	14 kg	15	45
4. Hybrid maize	June	Sept-Oct.	18 kg	28	45
5. Bajra or spiked millets	July-Aug.	October	2 kg	6	30
6. Great millets (Juar)	July	October	10 kg	20-30	30
7. Arhar	July-Aug.	March	20 kg	20	30
8. Urad or Moong	July	November	15 kg	8	30
9. Cotton	May-June	Nov.-Jan.	15	12	45
10. Til	Aug.	November	15	3.5	35
11. Groundnut	May	Nov.Dec.	30	20	45
RABI CROPS					
1. (a) Wheat	Oct-Nov.	April	70-80	15	37.5
(b) High yielding	Oct.	April	80-100	50	45
2. Gram	Sept-Oct.	March	40-50	20	30
3. Barley	October	March-April.	90-130	20	30
4. Mustard	October	Feb.-March	1.5 kg	12	45
5. Peas	Oct-.-Nov.	March	30-40	12	55
6. Potatoes	Sept.-Oct,	Feb.	400-900	200	80
7. Linseed	Oct.-Nov.	March	10	10	45
8. Tabacco	Oct,-Feb.	Feb.-March	—	20	60
9. Vegetables	Aug.-Nov.	Oct.-April		—	45
10. Sugarcane	Feb.-Mar.	Dec.-March	5t cut pieces	50 (Gur)	90

5.4 BRIEF DESCRIPTION OF MAIN CROPS

Sowing time, harvesting time, seed requirements, average yield and average water depth requirements, of most of the crops have been given in Table 5.2. Still some important crops require special mention which has been given here.

1. Wheat. It is the main food cereal crop. It forms main food for millions of people all over the world. In India Indo-Gangetic plain is the main region where this crop is predominantly grown. It grows best in well draining soils like clayey loams and loam. This crop is sometimes sown by mixing with barley, gram peas or linseed. But in case of high yielding varieties it should be sown alone. Sowing of this crop starts from mid-October and continues up to mid-November. Harvesting is done from mid-March to mid-April. Common high yielding varieties of Wheat are Sharbati, Sonalika, Kalyan Sona, Sonara, C – 273, C – 281 etc. This crop requires total water depth of 37.5 cm which is provided in three to four waterings each of 7–10 cm depth.

2. Paddy or Rice. Like wheat, it is also a main food crop of millions of people. It is grown predominantly in Central, Eastern and Southern, parts of the country. However, it is grown practically in all parts of our country. The most suitable land is clayey loams to heavy clays. High temperature and high humidity provide excellent conditions for its cultivation. There are varieties that can be grown throughout the year. Early variety of rice is sown in May to July and harvested in September–October. The late or transplanted variety is sown in July and harvested in November, Sabarmati, Padma, IR – 8, and Taichoon are some of the high yielding varieties of rice. Water depth for ordinary or early variety rice is 90 cm and that for high yielding or transplanted variety 125–150 cm. Ammonium sulphate is the best chemical fertilizer for it.

3. Sugarcane. It is the chief source of sugar and is considered one of the cash crops. Since it occupies the field from February to December, practically for full year, it is also known as perennial crop. It covers both the seasons of crop, i.e. Rabi and Kharif. The soil which suits best for the growth of this crop is loamy and clayey loamy soils. This crop is mainly grown in Gangetic plains. When grown in heavier soils like clayey loams, proper drainage of the fields is very important. Being long duration crop it requires thorough preparation of fields and adequate amount of manures. Ammonium sulphate chemical fertilizer is mostly used. COS – 416, CD – 313, 395, 839 are the common varieties whereas CO – 356, 393, 1148, 1158 and COS – 245 are the high yielding varieties. It is mostly sown in Haryana, UP, Rajasthan and parts of Bihar, M.P., and Maharashtra.

4. Cotton. It is also a long duration crop of Kharif season. It is a fibrous crop which is used mainly in textile industry and Khadi Gramodyog. It occupies fields for about 8 months from April to December. It is also a deep rooted crop which drives its food from large depths of say 0.5 m or so. Average depth of water required for its growth in about 45 cm which is provided in form of 6–7 waterings each of depth of about 7.5 cm. It is also a cash crop.

5. Gram. It is a Rabi crop. It requires very little water for its growth. It can be grown in form of dry crop if one or two good rains take place in the season. It is also a very important crop. It is mostly grown in sandy or loamy soil.

6. Maize. It is a coarse food cereal. It grows on good well drained fertile loamy soils. It can be grown both in tropical as well as in cold regions. It requires more water in the early stage of its growth. It requires green and compost manures in field so as to give good yield. Ganga hybrid – 1, Ganga hybrid – 101, Deccan hybrid and Ranjit hybrid are the main high yielding varieties of maize.

7. Potato. It is one of the subsidiary food crops. It is grown best in moist cold regions. It is sown either in form of cut pieces having 2 to 3 eyes or in form of whole tubers. This crop is sown at the ridges of a furrowed field. It gives its yield from below the surface level of the field. It may give yield of about 500 quintals per hectare. Great scott and Military special are its early varieties. Kufri Kundan, Fulva and Darjeeling red, are its high yielding varieties.

5.5 SOME IMPORTANT TERMS AS APPLIED TO CROPS

There are some important terms as applied to crops :

1. Dry crops. These crops grown without irrigation.

2. Wet crops. These crops grown with irrigation water.

3. Crop ratio. It is the ratio of the areas irrigated in Rabi and Kharif seasons.

$$\text{Crop ratio} = \frac{\text{Area irrigated during Rabi season}}{\text{Area irrigated during Kharif season}}$$

Water requirements of Kharif crops is much more than those of rabi crops. The crop ratio in two seasons is kept such that water requirements of crops sown in the area in both the seasons require almost uniform discharge from the canal throughout the year.

4. Overlap Allowance. Sometimes, it so happens that crop of previous crop season has not yet been harvested and the crop of coming season has also been sown. Under such circumstances irrigation water will have to be provided for some time to both the crops. The time period in days for which crop of previous season overlaps the new sown crop is known as overlap allowance. Since during overlap period, both the crops require water, additional supplies will have to be supplied to the canal to cater the needs of both the crops.

5. Deep Rooted Crops. Such crops whose roots lead to comparatively larger depths, are known as *deep rooted crops.* Cotton, Arhar, Hemp etc. are the examples of deep rooted crops.

6. Shallow Rooted Crops. The crops whose roots remain in the upper layer of the soil are termed as shallow rooted crops. Wheat, Barley are the examples of such crops.

7. Leguminous Crops. These are such crops which increase the fertility of the soil by being in the field. These crops gather Nitrogen in their root zone and this nitrogen is utilised by the subsequent crop in that field in form of fertilizer.

8. Garden Crops. These are fruit crops. They require watering throughout the year.

9. Root Zone Depth. It is that depth of the soil upto which roots of crops may lead.

10. Mixed Crop. When two crops are sown in the same field simultaneously the resulting crop is known as mixed crop. Combination of wheat and gram, cotton and til, cotton and moong, are the examples of mixed crops.

11. Double Crop. When two crops of short duration are grown in the same field in the same season one after the other, they are known as double crop.

12. Rotation of Crops. Different crops have different depths of their root-zones, and as such draw out nutrients from different depths of soil, during the process of their growth. There are also certain crops like gram, Pea, Barseem etc., also known as Leguminous crops, which replenish the nitrogen content of the soil through their roots. If the same crop is sown in the same field repeatedly, the yield of the crop is almost certain to fall. This happens because soil does not get time to replenish the already depleted fertility of the soil of previous season. By practicing rotation of the crops in the same field, the fertility of the field can be maintained at its usual level without leaving fields empty. Following points should be considered in rotation of the crops.

1. In rotation of the crops deep rooted and shallow rooted crops should be sown alternately. By doing so the crop nutrients from the soil would be used uniformly from different depths of the soil. When shallow rooted crop is in the field, nutrients from top soil only would be used and soil lying in deeper layer will get time and rest to replenish their nutrient content. Reverse would be the case when deep rooted crop is in the field.

2. If leguminous crop is introduced in rotation, the fertility of the soil can be increased instead of maintaining at constant level. Such crops increase nitrogen content in the soil.

Following are the advantages of rotation of the crops:

1. Nutrients from different depths of soil are used more balancedly.
2. Fertility of the field can be improved by introducing leguminous crops in rotation.
3. The fields are not required to be left vacant after each crop.
4. Crop disease in the fields can be checked.

Following may be the useful rotations of crops:

1. Cotton—Spiked millet—Gram.
2. Cotton—Wheat or Gram.
3. Wheat—Great millet—Gram.
4. Rice—Grain.

5.6 FERTILITY OF SOIL

A soil is said to be fertile when it contains all the nutrients which are essential for growing a healthy crop. Presence of nitrogen, organic materials, and soluble compounds in a soil are the usual fertilizing agents. Physical properties of soil, adequate amount of water and drainability of the soil are such elements which help in making fair utilisation of the nutrients. Fertility of the soil can be maintained by following measures.

1. By spreading all the farm wastes in the field. On disintegration they supply fertilizing agents to the soil.
2. Preparing land by scientific methods.
3. Practicing rotation of crops, and further by introducing leguminous crop in the rotation.
4. By applying suitable chemical fertilizer.
5. By green manure.
6. By adopting measures which prevent soil erosion.
7. By allowing storm dust to be entrapped in freshly ploughed fields.
8. By allowing silt of water deposit in fields.
9. By applying composite manure to the fields from time to time.

5.7 ASSESSMENT OF IRRIGATION WATER

Irrigation water supplied to the farmers through canals, or reservoir is at the expense of the government. Farmers are charged suitably by the government for having used this facility. The fixation of such charges is known as assessment of irrigation water. Following are the reasons for which farmers are charged :

1. To recover the cost of the project due to which this facility has been possible.
2. To collect revenue so that it may augment the state resources.
3. To collect the maintenance cost of the project.
4. To put some restrictions on farmers against careless and uneconomical use of water.

5.8 METHODS OF ASSESSMENT

The following are the possible methods by which farmers may be charged.

1. Volumetric basis
2. Irrigated area basis
3. Crop rate basis
4. Seasonal basis
5. Composite rate basis
6. Permanent assessment basis.

1. Volumetric Basis. This method is based on actual volume of water supplied to the farmers. Theoretically this method seems to be most logic method of assessment, but there are practical difficulties. Water meters will have to be fitted on each outlet and permanent staff will have to be deployed to maintain these meters. However this method can be used in tube well irrigation.

2. Irrigated Area Basis. In this method total area of different crops, which has been matured by the water is charged. The record of area of different crops is maintained by Patwaries of Revenue Department. After harvesting of the crops each framer is asked to pay levies according to the total area irrigated by him irrespective of the crops sown by him.

3. Crop Rate Basis. This method is almost similar to irrigated area basis method. The only difference is that areas under each crop are charged at different rates and not at uniform rate as in case of irrigated area basis. Each crop has its own market value. The crops which give more returns and consume more water are charged at enhanced rates. For example sugarcane gives more returns than any other crop and as such charged heavily. Almost all the canal irrigated lands in India are levied according to this method of assessment.

4. Seasonal Basis. This method is not used in India. In this method uniform charges are charged from all the farmers of a particular region depending upon the crop they grow.

5. Composite Rate Basis. This is actually combination of land revenue and levies for water charges. Since some states collect both these revenues together, this method of assessment is named composite rate method.

6. Permanent Assessment Basis. This method is also known as betterment levy basis. In this method canal system is not a regular feature of supplying irrigation water to the farmers. Farmers may be having their own systems of irrigation but during drought period when wells and tanks go dry, water is supplied through this system of canal to protect the crops of the farmers. For making this protective facility available farmers are charged at fixed rates on yearly basis. During droughts, farmers are authorised to use irrigation water from these canals. These canals are fed from some perennial river, lake or artificial reservoir.

5.9 INTERNAL DISTRIBUTION OF WATER

Internal distribution of water amongst the farmers is carried out on *Wara bandi basis*. *Wara bandi* is enforced by Executive Engineer of Irrigation Department. Under this system whole of the area under the charge of a particular outlet is to be supplied water once in a week. Each farmer is allotted fixed hours of irrigation water in a fixed day of the week. Hours of irrigation to each farmer, are allotted in direct proportion to the irrigable area held by him. Any farmer flouting the *wara bandi* or forcibly taking water of any other farmer, is suitably dealt in court of law.

5.10 METHODS OF TAKING UNAUTHORISED IRRIGATION WATER FROM CANAL

Following are the some methods of taking unauthorised irrigation water from canal :

1. By making deliberate cuts in canals and flooding the adjoining area. This mischief is mostly done by those farmers whose land lies in the vicinity of the canal alignment.

2. Tempering the outlets.
3. Fixing wooden planks touching the free sprout of water emanating from the outlet.
4. By adopting to syphons.

All these mischiefs should be challenged in court of law and guilty persons should be severely punished.

QUESTIONS

5.1 (a) Give the detailed classification of various crops.

(b) What are the principal crop seasons ? Explain each season in details.

5.2 What are various methods of assessment of irrigation water ? Briefly discuss each method.

5.3 Write short notes on :

(i) Rotation of crops.

(ii) Fertility of soil.

5.4 Explain the terms, double crop, mixed crop, leguminous crop, overlap allowance, dry crop, wet crop.

❑❑❑

6

Surface Hydrology and Discharge Measurement

6.1 HYDROLOGY AND HYDROLOGICAL CYCLE

Hydrology. It is a science which concerns with the occurrence, distribution and movement of water. It may also be defined as the science that deals with the depletion and replenishment of water sources. When we say movement of water, it includes movement of water in atmosphere, on the surface of the earth, and below the surface of the earth. In atmosphere, water is present in form of vapours. On surface of earth it is in form of rivers, lakes, ponds and snow etc. Below the earth surface it is present in form of moist soils. The water occurring below the surface of the earth is known as *ground water*. Ground water occupies the voids of the soil. In saturated condition, all the pores of soil are completely filled with water.

Hydrological Cycle. Except for the deep ground water, the total water supply of the earth remains in constant circulation between earth and atmosphere. Transference or circulation of water from earth to atmosphere and then back to earth is known as hydrological cycle. Water is evaporated to atmosphere from oceans, rivers, lakes, ponds etc. This water again falls over the earth in form of precipitation. The part of precipitation again evaporates back to atmosphere, part is soaked by the ground, and remaining part is either held in form of lakes and ponds or flows in form of rivers. Rivers again join the oceans and evaporation of water again occurs from them. The water that has percolated in to the ground either joins oceans or is brought to the surface by means of wells, infiltration gallerries etc. So circulation of water from earth to atmosphere and back to earth is a continuous process. It is this circulation of water or hydrological cycle, which keeps on using and then replenishing the water resources. Precipitation, evaporation, stream flow and ground water conditions, are known as essential elements of hydrological cycle. By measuring the elements of

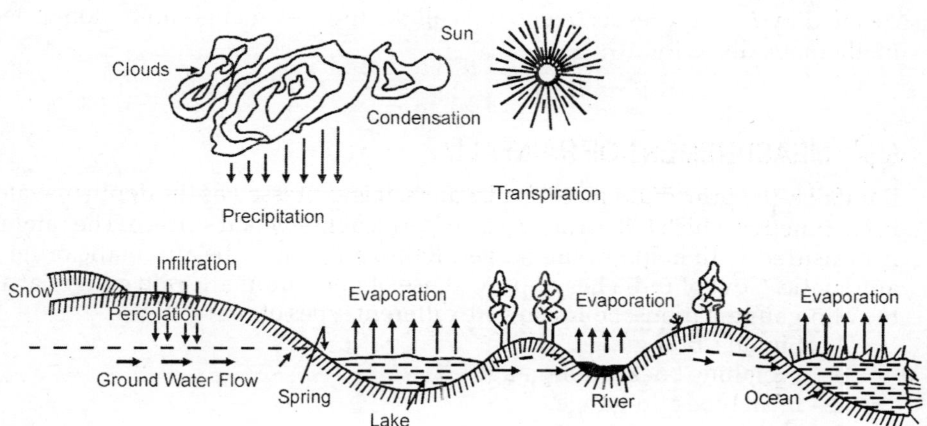

Fig. 6.1. *Hydrological cycle.*

hydrological cycle it is possible to assess accurately the potential of water resources and their proportion that can be used for irrigation and other purposes. A diagrammatic view of hydrological cycle has been shown in Fig. 6.1.

6.2 PRECIPITATION, RUN-OFF, EVAPORATION AND TRANSPIRATION

The fall of moisture or water over the surface of the earth from atmosphere is known as *precipitation*. Precipitation includes rainfall, snow, hail, sleet and frozen water in any form.

Evaporation. It is loss of water from surfaces of oceans, rivers, lakes, and moist soils. The water lost by evaporation is in form of vapours and is held in form of clouds.

Transpiration. It is the process by which water is lost from the leaves of the plants. Sometimes combined loss of water by evaporation and transpiration is referred as total evaporation of water.

Run-off. Unevaporated portion of precipitation is known as run-off. A part of precipitation water that falls on the surface gets evaporated and transpirated and remaining part is called run-off. Run-off water ultimately runs to the ocean through surface and sub-surface streams of water. The run-off may be classified as *surface run-off, sub-surface run -off* and *ground water flow.*

Water, flowing on the land is known as *surface run-off.* This water first reaches the streams and river, and then ultimately joins sea.

The portion of precipitation that infiltrates into the surface soil runs as *sub-surface run-off* and joins streams and rivers.

The portion of precipitation that percolates down and joins ground water reservoir is known as *ground water flow*. Ground water flow ultimately joins the ocean.

If precipitation is denoted by P and total evaporation and various run-offs are denoted by E and R respectively, then all the three elements can be connected mathematically as follows:

$$P = R + E.$$

6.3 MEASUREMENT OF RAINFALL

Rainfall is the principal source of all waters. It is expressed as the depth of water in centimetres which falls on a pucca, impermeable levelled surface. The rainfall is measured with help of rain-gauges. Rain-gauges may be automatic or non-automatic. Govt of India has approved use of non-automatic rain gauges at all the rain-gauge stations. Following are different types of rain-gauges.

1. Simon's rain-gauge.
3. Weighing bucket rain-gauge.
2. Float type rain-gauge.
4. Tipping bucket rain-gauge.

1. Simon's Rain-gauge. A typical Simon's rain gauge is shown in Fig. 6.2. It is also known as non-recording type of rain-gauge, as it does not record the rate of

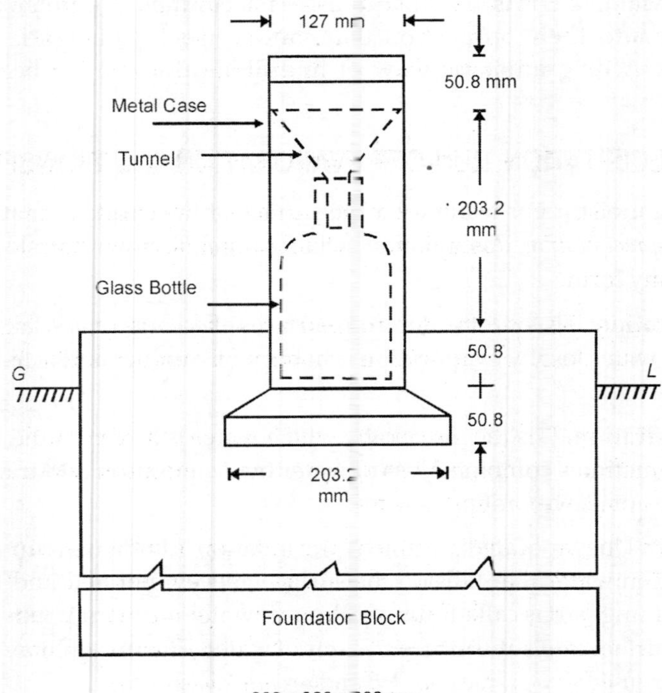

Fig. 6.2. *Simon's rain gauge.*

rainfall at any moment but only collects rain water. It consists of a funnel fixed at the top of a receiving bottle. The receiving bottle is about 8 to 10 cm in diameter and is encased in the metal casing. This bottle is fixed in the ground slightly below the ground level. The rain water enters the bottle through the funnel and gets collected in the bottle. The bottle can collect about 10 cm to 12 cm of rain. Rainfall is recorded daily at 8 A.M. by a man deputed for this purpose. If the rain fall is too much and is likely to exceed the capacity of the bottle, then two or three intermediate readings are taken and their sum is recorded as the total rainfall during past 24 hours. Receiving bottle is graduated to read up to the accuracy of 0.2 mm rainfall. The rain-gauge station should be located on fairly levelled open ground and away from such objects which may affect the fall of rain water into the receiving bottle of the rain-gauge. (See Fig. 6.2.)

2. Float Type Rain-gauge. It is a type of automatic rain-gauge. The instrument consists of a funnel and a rectangular container. The container consists of a float which remains connected to a pointer. The pointer moves on a recording drum which is kept moving at such a rate that drum completes exactly one revolution in 24 hours. When level of rain water collected in the container fluctuates, the fluctuation is transmitted to the pointer which records it on the graph paper wrapped on the revolving drum. Thus with the help of this rain gauge besides recording the full rainfall, the rate of raining at any moment can also be recorded.

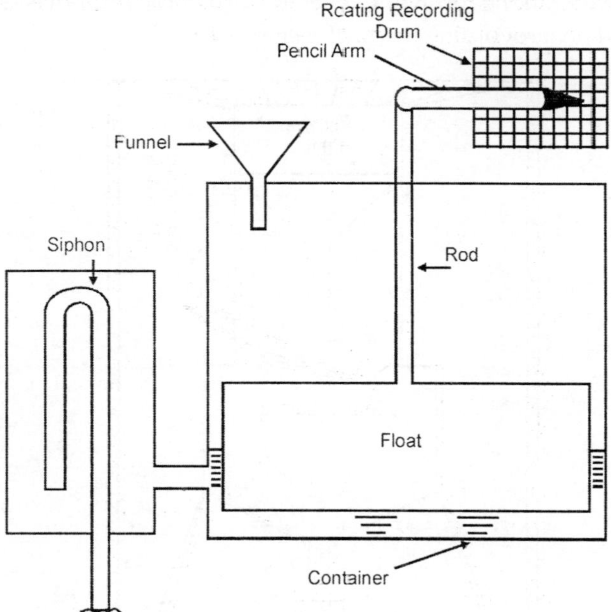

Fig. 6.3. *Float type rain-gauge.*

When container gets filled up with rain water and float just touches the top of the container, the water from the container is emptied with the help of siphon automatically. (See Fig. 6.3).

3. Weighing Bucket Rain-gauge. This rain-gauge consists of a receiver bucket which remains supported over a spring or lever balance. A pointer remains

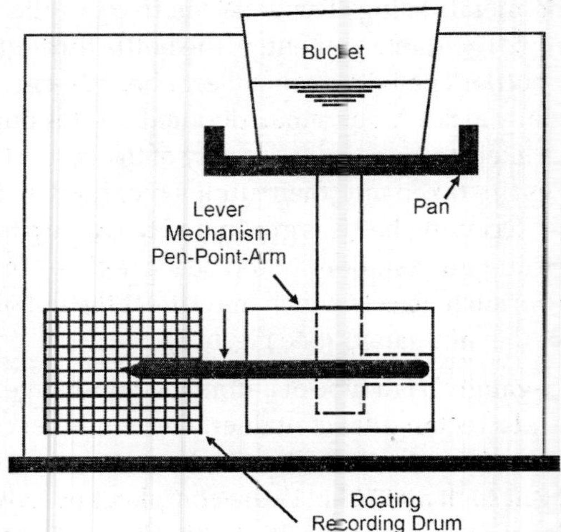

Fig. 6.4. *Weighing bucket type rain-gauge.*

connected with the balance which records the rainfall on the basis of weight of collected rain water in the receiver bucket. Record of the rainfall is recorded on a graph wrapped on a revolving drum. (See Fig. 6.4.)

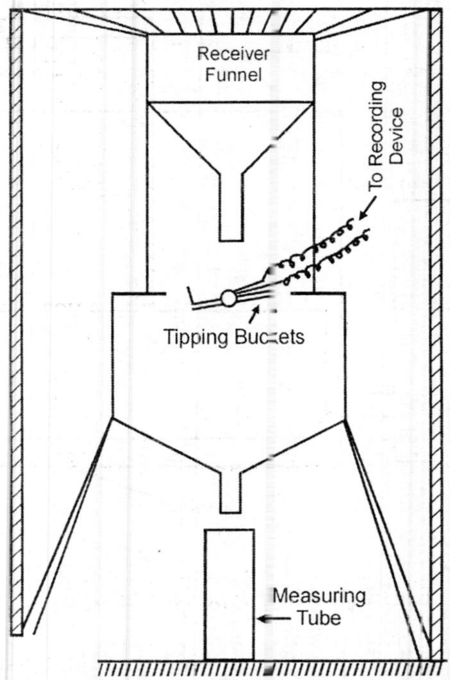

Fig. 6.5. *Tipping bucket rain-gauge.*

4. Tipping Bucket Rain-gauge. It consists of a pair of buckets which remain pivoted under the funnel of the rain-gauge. The buckets come under the funnel one by one. One bucket receives 0.25 mm of rainfall. Once a bucket is filled with rain water, it tips and discharges its contents into the reservoir and second bucket immediately comes under the funnel. Tipping of the buckets is recorded on the revolving drum with the help of a pencil and rainfall is estimated from it. (See Fig. 6.5.)

In all the self recording rain gauges, there is a pencil point in contact with the graph paper, which records the cumulative rainfall in the form of a graph. This graph is a plot between time and the cumulative rainfall and is known as mass curve.

The slope of this curve gives intensity of rainfall for any given period. Since the gauges reflect the cumulative rainfall, they are sometimes known as *integrated rain gauges* also.

6.4 AVERAGE ANNUAL RAINFALL AND INDEX OF WETNESS

The amount of rain collected by a rain gauge in 24 hours is known as daily rainfall and the amount collected in one year is known as *annual rainfall*. This annual rainfall at a given station should be recorded over a number of years say 35 to 40 years or so. In India this rainfall cycle period is taken about 35 years. The mean of the annual rainfall over a period of 35 years or so it, therefore, known as *average annual rainfall* at the given station.

When we talk of the rainfall of a given place we generally refer to the average annual rainfall of that place. Thus when we speak of rainfall figures of a particular place, it means that this figure has been averaged over a long period of about 35 years. This is known as normal rainfall. But in any given year the rain may not be equal to this amount. It may be less than this average value or may exceed it. The ratio of the actual rainfall in a given particular year at a given place, to the normal rainfall of that place is known as *index of wetness*.

$$\text{Index of wetness} = \frac{\text{Actual rainfall in a given years at a given place}}{\text{Normal rainfall of that place}}$$

Index of wetness thus gives some idea of the wetness of the year in relation to normal rainfall. 50 per cent index of wetness means a rainfall deficiency of 50 per cent. A deficiency of about 30–45 per cent is known as large deficiency, 45–60 per cent as serious deficiency and more than 60 per cent as disastrous deficiency.

The year in which the rainfall is less than the average annual rainfall, is called a bad or a sub-normal year and if it is more than the average it is known as a good year. If the rainfall in a particular year is approximately equal to the annual average value, then it is known as a normal year or an average year.

6.5 COMPUTATION OF AVERAGE OR MEAN RAINFALL OVER DRAINAGE BASIN

A given drainage basin is divided into number of parts and rain-gauge stations are uniformly established over the entire basin. The density of rain-gauges in the basin depends upon how accurate we want to estimate the rainfall. The following methods are generally used to work out the mean rainfall of an area of basin.

1. Arithmetical mean method,
2. Thiessen's polygon method, and
3. Isohyetal method.

1. Arithmetical Mean Method. It is one of the simplest methods. It consists of averaging all the amounts of rainfalls that have been recorded at various rain gauge stations in the drainage basin.

$$P = \frac{P_1 + P_2 + P_3 + \ldots\ldots + P_n}{n}$$

where P = The mean precipitation on the basin

$P_1, P_2, P_3 \ldots\ldots P_n$ = Precipitations or rainfall values measured at n different rain gauge stations.

If the stations are uniformly distributed over the area and the rate of rainfall also does not differ much at various stations, this method gives quite satisfactory results. But if there are many topographic differences and the rainfall rates also vary considerably, then this method may give misleading results.

2. Thiessen's Polygon Method. This method is also known as weighted mean method and is considered more accurate. In this method adjacent rain gauge stations (Fig. 6.6) say A, B, C, D are joined by straight lines. Draw perpendicular bisectors on all these lines. Lines joining A, B, C, D rain gauge stations, and the bisector lines form a net work known as Thiessen's net work

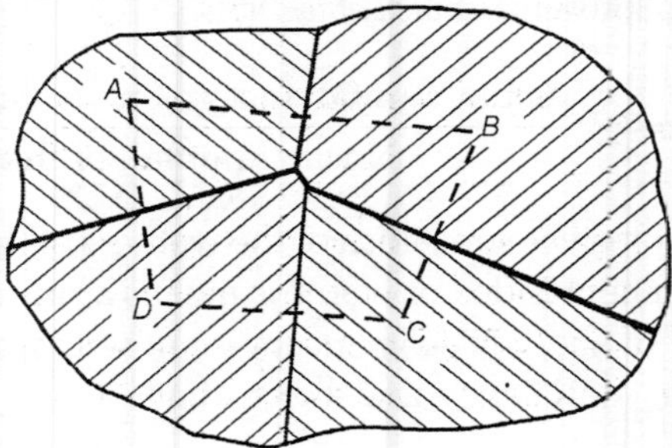

Fig. 6.6. *Thiessen's polygon method.*

of polygons and each polygon contains one rain gauge station. It is assumed that the area enclosed in a particular polygon is nearer to the rainfall station located in that polygon. The rainfall reading of a particular rain gauge represents the rainfall uniformly distributed over the area of the polygon of which this rain gauge is in charge. Average precipitation or rainfall can be computed as follows. See Fig. 6.6.

$$P = \frac{A_1 P_1 + A_2 P_2 + \ldots\ldots + A_n P_n}{A_1 + A_2 + \ldots\ldots + A_n}$$

where $A_1, A_2, A_3 \ldots A_n$ = areas of polygons

$P_1, P_2, P_3 \ldots P_n$ = rainfall in respective polygonal areas

P = average precipitation

n = number of polygons or rain gauge stations.

3. Isohyetal Method. Isohyets are the contours of equal rainfall. They are drawn on the map by using common sense after the rainfall at each station is plotted. The area between the adjacent isohyets is either estimated on the graph paper or measured by a planimeter. Let them be $A_1, A_2, A_3 \ldots\ldots\ldots A_n$ and let the average precipitation for these areas be $P_1, P_2, P_3 \ldots\ldots P_n$. The mean precipitation P on the basin is given by

$$P = \frac{P_1 A_1 + A_2 P_2 + A_3 P_3 \ldots\ldots + P_n A_n}{A_1 + A_2 + A_3 \ldots\ldots + A_n}$$

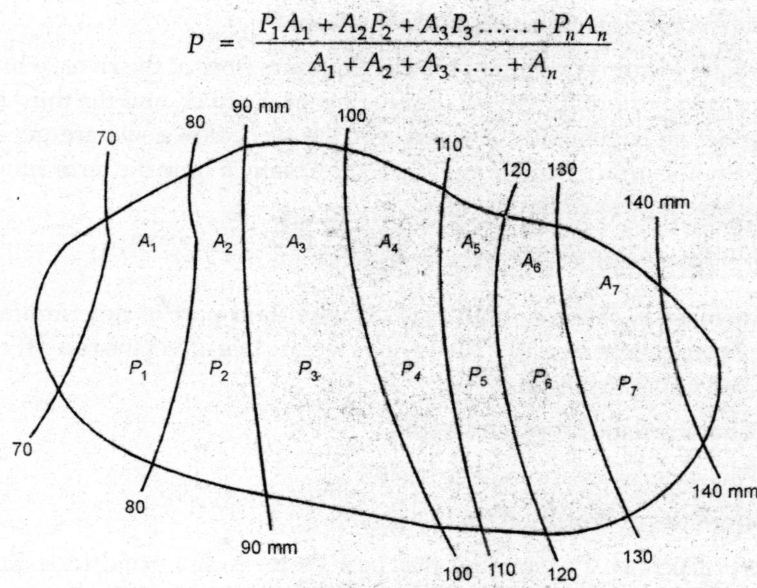

Fig. 6.7. *Isohyetal method.*

6.6 RUN-OFF

It is defined as the portion of the precipitation that somehow makes its way towards rivers or oceans etc. as surface or sub-surface flow. The discharge flowing in a river is the run-off from the basins drained by that river. The terms discharge, stream flow, and run off all refer to the same thing.

When rain falls, a part of it is intercepted by vegetation. Some of it is stored in depressions on the ground surface and is known as *depression storage,* which later infiltrates or evaporates. Some of the precipitation is absorbed by the soil, the amount of which depends upon the soil, moisture conditions existing at the time of precipitation. If the rain continues further, water starts infiltrating to the water table and if rate of rainfall or the rate at which the water is reaching the ground exceeds the infiltration rate, then this excess water starts collecting on the surface, as surface detention and this water flows overland and joins the streams, rivers, lakes, oceans, etc. This flow is known as *surface runoff.* The water which perculates without joining the water table and joins the stream, as sub-surface flow, is known as *sub-surface storm flow* and is considered as a part of surface run-off. On the other hand, the water that percolates to the ground water table and later after a long time joins the river or stream, is known as *ground water flow.* The run off thus actually consists of three parts :

1. Surface run off,
2. Ground water flow, and
3. Direct precipitation over the river stream.

The second factor is important for the minimum flow of the river, while the first factor is important for the maximum flow of the river, and the third factor being negligible is generally ignored. For the peak flows, we are generally concerned with surface run-off and, therefore, many a time the term run-off is exclusively used for surface run-off.

6.7 FACTORS AFFECTING THE RUN-OFF

The characteristics of the rainfall play an important part in determining the amount of consequent run-off'. The various factors that affect the run-off can be summarised under two heads.

1. Characteristics of precipitation,
2. Characteristics of drainage basin.

1. Characteristics of Precipitation

(a) Type of precipitation. Precipitation may be in the form of rain or drizzle. Run-off pattern or the hydrograph of run off is considerably governed by this factor. If precipitation occurs in the form of heavy rain, it will immediately produce bulk of run off (Peak flow of short duration). If precipitation is in the form of a drizzle it will produce run off at a slow and steady rate.

(b) Rain intensity. Rain intensity has a lot of effect on the run off. If the intensity of rain increases, the run off increases rapidly. For example, if the intensity is increased four times, the run off may increase nine times or so. For example let there be a rain in progress sufficient to make the infiltration capacity constant,

say 0.5 cm/hr. Now if the intensity of rain is 0.8 cm/hr the run off (strictly speaking excess rain) will occur at the rate of 0.3 cm/hr. Now let the intensity of rain be increased to a value say 12 cm/hr (four times), the resulting run off rate will be equal to 2.7 cm/hr (nine times). Thus an intense rain of the type shown in Fig. 6.8 (a) will definitely produce much more run off than a uniform rain of the type shown in Fig. 6.8 (b) provided the infiltration capacity remains the same throughout the storm period.

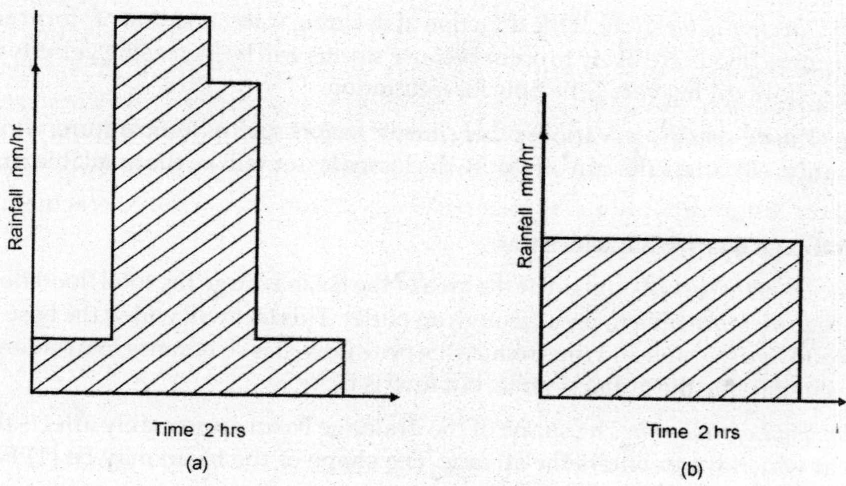

Fig. 6.8. (*a*) *Intense rainfall* (*b*) *Slow rainfall.*

Although the total amount of rain in Fig. 6.8 (a) and (b) is same, still rain (a) will produce higher amounts of run off while rain (b) is likely to produce much less run off.

(c) *Rainfall duration.* It is important because infiltration capacity goes on reducing with longer duration of rainfall till it becomes constant. If the infiltration is less, the surface run-off will be more. Thus, in some cases, a longer duration rain may produce considerable run-off even when its intensity is mild. Further if there is a rain extended over large periods of time, the water-table may rise quite high. The water table may reach the ground level and in such case there will be no filtration and floods may become even more serious.

(d) *Rainfall distribution.* Run-off from a basin is very much dependent upon the rainfall distribution. The rain may fall either on the whole basin or on a small part of it. For small drainage basins the peak flows are generally the result of intense rains falling over small areas. On the other hand for large drainage basins, the peak flows are the result of storms of lesser intensity but covering large areas. The rainfall distribution is generally expressed by the *distribution coefficient* which for a given storm can be obtained by dividing the maximum rainfall at the point by the mean rainfall on the basin.

(*e*) *Soil moisture deficiency.* The run off also depends upon the soil moisture present at the time of the rainfall. If a rainfall occurs after a long dry spell of time, the soil is dry and it can absorb large amounts of water. In such a condition even intense rain may fail to produce any appreciable run off. On the other hand if there are persistent rains, the soil will be already wet and infiltration will be very small. In such conditions even small rains may cause considerable floods.

(*f*) *Direction of the storm.* If the direction of draining water and that of storm are same, more floods are likely to occur because storms will be increasing velocity of surface flow, giving very little time for infiltration.

(*g*) *Climatic conditions.* Various other climatic factors such as temperature, wind, humidity, etc. affect the run off. More the losses lesser will be the available run-off.

2. Drainage Basin Characteristics

(*a*) *Size of the basin.* If spread or the area of the basin is large the total flood flow will require more time to pass through an outlet. This fact will widen the base of the flood hydrograph and thus reduce the peak flow. It is so because total volume of water passing the outlet is same, but time is more.

(*b*) *Shape of the basin.* The shape of the drainage basin appreciably affects the rate at which water enters the stream. The shape of the basin may be (1) Fan

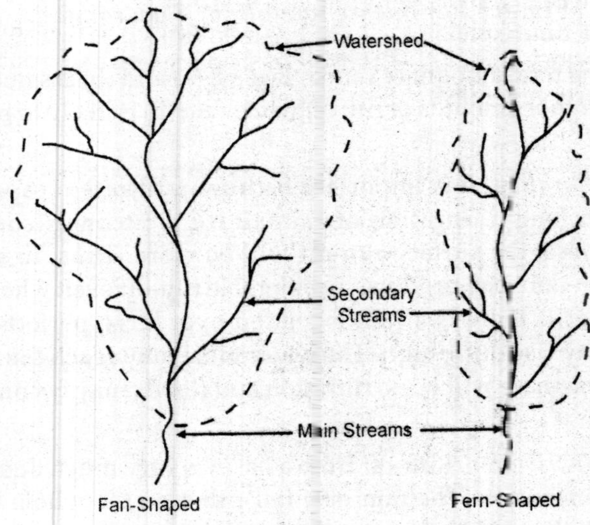

Fig. 6.9. *Type of catchments.*

shaped and (2) Fern leaf shaped. Fan shaped basins give greater run off than Fern-leaf basins. This is because the tributaries in the case of Fan shaped basin are nearly of same size and length and lead their discharges at the outlet point practically at the same time. In the case of Fern-leaf catchments flood discharge is less as water from the tributaries lying near the outlet point passes out of outlet point, before water from upper areas reaches the outlet point.

(c) *Slope of the catchment.* If catchment area has more slope, discharge available in form of run-off will be more.

(d) *Nature of surface of basin.* If surface of the catchment is very permeable to large depth, run-off available will be small. Such soil conditions absorb more of rain water.

(e) *Topography of the basin.* The catchment having more rugged surface, more of vegetation gives less amount of run off. If artificial storages are available in the catchment, available run-off is again decreased.

6.8 SOME TERMS IN MOST COMMON USE IN REGARD TO RUN-OFF

1. Hydrograph. It is a curve or plot of discharge versus time at any section of a river. It represents the flow characteristic of the river.

2. Period of Surface Run-off. It is the time taken by the surface run off to pass the given section of the river, after the surface run-off makes its first appearance at the section.

3. Period of Rise. It is the time taken by the surface run-off to reach its maximum value from the time of its beginning.

4. Time of Concentration. It is the time required by the water to reach the outlet point from the most remote point of the drainage area. When a storm has been in progress for a time equal to the time of concentration, it is assumed that all the part of the catchment start contributing to the discharge at the outlet. This time of concentration is generally denoted by T_c.

5. Time of Over Land Flow. The excess rainfall finds its way over land to the river stream and appears as surface run off, but only after some delay. In other words, there exists a lag between the time when excess rainfall occurs and the time when it appears at run-off at the outlet of the drainage basin. This lag is the time for which water has flown through the basin before entering into a defined drain and is known as *time of overland flow* or *concentration time of overland flow*. This delay varies throughout the basin and throughout the storm period. This time of overland flow is a variable factor depending upon the slope, the type of surface of the ground, the length of path of flow and so many other factors.

6.9 COMPUTATION OF RUN-OFF

The Run-off available from a basin can be computed daily, weekly, monthly or yearly. Following are the methods which can be used for finding out the run-off:

1. Using empirical formulae and tables
2. Infiltration characteristics method
3. Rational method, and

4. **Unit hydrograph method.**

All these methods have been explained separately.

6.10 RUN-OFF BY USING EMPIRICAL FORMULAE AND TABLES

1. Run-off Coefficient Method. The volume of run-off can be directly computed approximately by using an equation of the form

$$R = KP$$

where R is run-off, P is rainfall or precipitation and K, a constant depending upon the surface of the drainage area.

Truly speaking this equation cannot be rational, because the run-off not only depends upon the precipitation but also upon the recharge of the basin. This equation gives more and more reliable results as the imperviousness of the drainage area increases and the value of K tends to approach unity. This formula is used in the computation of run off from small areas, especially for urban areas where the percentage of imperviousness of the area is quite high. This method should be avoided for rural areas and for major storms. The values of K for different conditions may be taken as given in Table 6.1.

Table 6.1. *Values of Run-off Coefficient (K)*

Typed of area	*Value of K*		
	Flat area 0 – 5% slope	*Rolling area 5 to 10% slope*	*Hilly area 10 to 30% slope*
1. Cultivated areas			
(a) Open sandy loam	0.30	0.40	0.52
(b) Clay and silt loam	0.50	0.60	0.72
(c) Tight clay	0.60	0.70	0.82
2. Urban areas			
30% area impervious (paved)	0.40	0.50	–
50% area paved	0.55	0.65	–
70% area paved	0.65	0.80	–
3. Pastures			
Open sandy loam	0.10	0.16	0.22
Clay and silt loam	0.30	0.36	0.42
Tight clay	0.40	0.55	0.60
4. Forested or wood land			
Open sandy loam	0.10	0.25	0.30
Clay and silt loam	0.30	0.35	0.50
Tight clay	0.40	0.50	0.60

Mr. T.G. Barlow carried out his studies of catchments in U.P. He suggested coefficients of run-off for different catchments as follows. He classified different catchments in five class as A, B, C, D and E.

Table 6.2. *Coefficient of Run-offs as given by T.G. Barlow for Average Monsoon*

Class of catchment	*Catchment characteristic*	*K*
A	Flat, cultivated and black cotton soils	0.10
B	Flat, partly cultivated various soils	0.15
C	Average soils	0.20
D	Hilly and plains with little cultivation	0.35
E	Very hilly and steep with hardly any cultivation	0.45

These coefficient are to be further multiplied by coefficient given below to convert them to suit the different conditions of rain fall. These coefficient also had been given by Mr. T.G. Barlow.

Table 6.3

Condition of rainfall	*Coefficients for different classes of catchments*				
	A	B	C	D	E
1. Continuous downpour	1.50	1.50	1.60	1.70	1.80
2. Average or varying rainfall, no continuous downpour	1.0	1.0	1.0	1.0	1.0
3. Light rain, no heavy downpour	0.70	0.80	0.80	0.80	0.80

2. Strange's Tables and Curves. Mr. W.L. Strange carried out his experiments in South India. He classified the catchments into three categories, namely, good, average, and had. He also had taken initial surface conditions for each catchment as dry, damp and wet. Daily run-off for dry, damp and wet conditions for different intensities of rainfall is given in Table 6.4.

Table 6.4. *Yield and Daily Run-off according to Strange*

Daily rainfall in mm	*Run-off coefficient and yield when original state of catchments*					
	Dry		*Damp*		*Wet*	
	K	*Yield in mm*	K	*Yield in mm*	K	*Yield in mm*
6.25 ($2\frac{1}{2}$")	—	—	—	—	0.08	0.5
12.5 (5")	—	—	0.06	0.75	0.12	1.5
18.75 ($7\frac{1}{2}$")	—	—	0.08	1.50	0.16	3.0
25.0 (10")	0.03	0.75	0.11	2.75	0.18	4.5
31.25 ($12\frac{1}{2}$")	0.05	1.56	0.14	4.37	0.22	6.88

Contd.

Table 6.4. *Contd.*

37.50 (15")	0.06	2.25	0.16	6.00	0.25	9.37
43.75 ($17\frac{1}{2}$")	0.08	3.50	0.19	8.31	0.30	13.10
50.0 (20")	0.10	5.0	0.22	11.0	0.34	17.0
62.5 (25")	0.15	9.37	0.29	18.15	0.43	26.9
75.0 (30")	0.20	15.0	0.37	27.75	0.55	41.25
100.00 (40")	0.30	30.0	0.50	50.00	0.70	70.0

Mr. Strange also gave curves relating rainfall and run-off and conditions of the basin both on daily and yearly basis. These curves have been reproduced in Fig. 6.10.

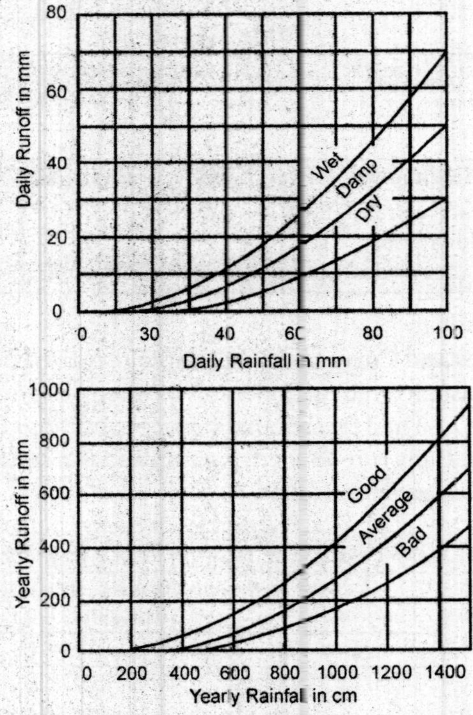

Fig. 6.10. *Strange's curves.*

3. Empirical Formulae. Run-off can be approximately computed by using empirical formulae. Some of the formulae in most common use have been given as follows :

(i) Lacey's formulae
$$R = \frac{P}{1+\frac{305F}{PS}}$$

(ii) Khosla's formulae $R = P - \frac{T-32}{3.74}$

(iii) Inglis formulae for ghat areas

$$R = 0.85\,P - 30.5$$

(iv) Inglis formulae for non-ghat areas

$$R = \frac{(P-17.8)}{254}P$$

In all the empirical formulae given here R and P are respectively run-off and precipitation and both have been considered in centimetres. T is mean temperature in 0° F

In Lacey's formulae S represents catchment factor and F, monsoon duration factor. The values of S for catchments A, B, C, D and E as classified by Mr. T.G. Barlow are respectively 0.25, 0.60, 1.0, 1.70 and 3.45. The value of F is taken as 0.45, 1.0 and 1.5 for very short, standard, and very long duration rainfalls respectively.

6.11 RUN-OFF BY USING INFILTRATION CHARACTERISTICS

The process, whereby water enters the surface strata of the soil and thus moves downward towards the water-table is known as *infiltration*. In fact when water falls on the soil, a small part of it is first of all absorbed by the top thin layer of soil so as to replenish the soil moisture deficiency. After this any excess water moves downward where it is trapped in the voids and becomes *ground water*.

The amount of stored ground water mainly depends upon the number of voids present in the soil. The number of voids further depend upon the size, shape, arrangement, and degree of compaction of the soil. Hence different soils will have different number of voids and hence different capacities to absorb water. The maximum rate at which a soil in any given condition is capable of absorbing water, is called its *infiltration capacity*.

It is evident that rain water will enter the soil at full capacity rate only during the periods when rainfall rate exceeds the infiltration capacity. When the rainfall intensity is less than the infiltration capacity, the prevailing infiltration rate is approximately equal to the rainfall rate. Hence the actual prevailing infiltration rate may be equal to or less than the infiltration capacity. This actual prevailing rate at which the water is entering the given soil at any given time is known as *infiltration rate*.

If the rainfall intensity exceeds the infiltration capacity the difference is called as the rainfall excess rate. This excess water is first of all accumulated on the ground as surface detention and then flows over land into the streams. The water below the water-table is known as the *ground water* and the water above the water-table is known as *soil moisture*.

When a graph is drawn between infitration capacity and duration of rainfall in hours we get a curve known as infiltration capacity curve. Infiltration capacity is very high at the beginning of a rainfall. As the duration of rain goes on prolonging the infiltration capacity also goes on decreasing. After a certain period (of the order of 1 to 3 hours) the infiltration capacity tends to become constant.

When infiltration capacity curve (I.C.C.) is superimposed on the rainfall intensity pattern, the resultant amount will represent run-off. Hatched area in Fig. 6.11 represents the rate of run-off. This method of computing run-off can be used very easily if the rainfall rate never falls below the infiltration capacity rate. But natural rains are sometimes below and sometimes above the prevailing infiltration capacity and thus distort the capacity-time curve. It is generally assumed that the infiltration capacity at any time is determined by the mass infiltration which has occurred upto that time. Thus if a rain begins at low rate and the rainfall during the first hour is two-third of the infiltration the capacity rate at the end of the hour will be taken as the capacity that would have prevailed at $\frac{2}{3}$ hrs and not at one hour.

This method of computing run-off is used for small catchments having uniform characteristics. The rate of run-off and volume of run-off can be determined by deducting infiltration from the rain-fall.

6.12 USE OF INFILTRATION INDICES

The infiltration capacity curve as shown in Fig. 6.11 cannot be used for computing

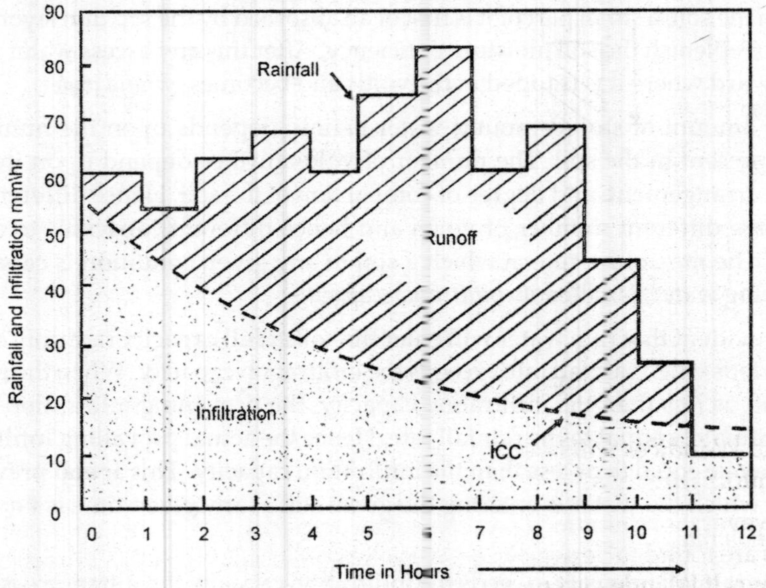

Fig. 6.11. *Infiltration rate and run-off.*

run-off from large basins. It is, because, in large basins the infiltration capacity as well as rainfall rate vary from point to point. Moreover sub-surface flow (interflow) will also be substantial. Since this water-flow is a part of infiltration, it will not normally be included in the run-off compute by using infiltration capacity curve determined on a small test plot. Run-off volumes for large areas are computed using *infiltration indices. W* and φ are the two commonly used indices.

W-index in the average infiltration rate or the infiltration capacity averaged over the whole storm period and is given as follows :

$$W\text{-index} = \frac{P-R}{T_r}$$

where P = Total precipitation or rainfall.

R = Total run-off.

T_r = Duration of rainfall in hours.

φ-index may be defined as the average rate of loss of precipitation such that the volume of rainfall in excess of that rate will be equal to the volume of direct run-off φ-index can also be stated as the rate of rainfall above which the rainfall volume equals the run-off volume. φ-index can be represented as shown in Fig. 6.12.

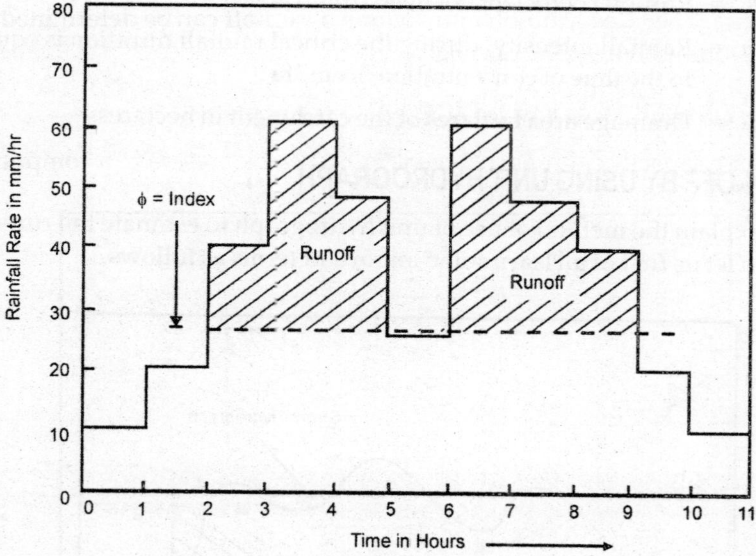

Fig. 6.12. *φ-index and run-off.*

φ and *W*-indices will be equal for a uniform rainfall but may not be equal for non-uniform rainfalls.

However for rains which are reasonably uniform or for heavy rains, these two indices are found to be nearly equal. The run-off coefficient *k* can be determined as follows if W-index is known,

$$k = \frac{p - W\text{-index}}{p}$$

where p = rate of rainfall or rainfall intensity.

6.13 RATIONAL METHOD OF ESTIMATING RUN-OFF

This method is a very useful method for evaluating the peak rate of run-off. This method is based on the fact that if a rainfall is applied to an impervious surface at a constant rate, the resulting run-off from the surface would finally reach a rate equal to the rate of rainfall. In the beginning only a certain amount of water will reach the outlet, but after sometime, the water will start reaching the out let from the entire area and in this case the run-off rate would become equal to the rainfall rate. The time required to reach this equilibrium condition is known as time of concentration and the peak rate of run-off would be equal to the rate of rainfall. This is the basis of the rational method. The peak rate of run-off can be estimated using the following formula :

$$R_p = \left(\frac{1}{36}\right) kpA$$

where R_p = Peak rate of run-off in cumecs.

k = Run-off coefficient obtained from Table 6.1.

p = Rainfall intensity, during the critical rainfall duration is equal to the time of concentration in cm/hr.

A = Drainage area hectares of the catchment in hectares.

6.14 RUN-OFF BY USING UNIT HYDROGRAPH

Before we explain the method of use of unit hydrograph to estimate the run-off from a basin let us first of all learn some important terms as follows.

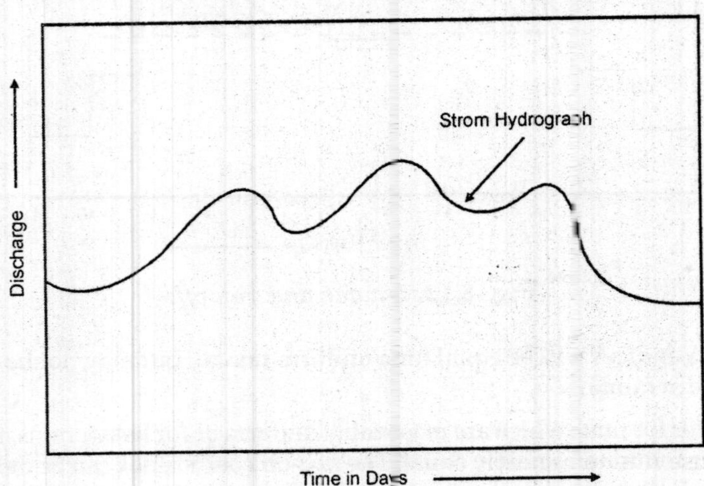

Fig. 6.13. *Storm hydrograph.*

1. Hydrograph. It is graphical relation between discharge or flow, against time at a particular point of a stream or river. Hydrograph represents the time distribution of total run-off at the point of measurement. As volume of run-off, discharge or flow, is obtained by multiplying discharge with time, the area under the hydrographs gives the volume of flow during that period.

The hydrographs have three types of flows :

(*i*) *Surface run-off or* water flowing in the stream or river.

(*ii*) *Sub-surface storm flow* i.e. infiltrated water in the top layers of soil. This water reaches the streams within short time. It is also known as inter-flow or influent stream.

(*iii*) *Ground water flow* or water contributed as underground flow from the ground water reservoir fed by infiltration from previous rainfall and also some by the storm in question.

At the start of any hydrograph, there is contribution to the run-off from the ground water reservoir accumulated in the soil during previous rainfall. Due to fresh rainfall when water level in the streams rises, they contribute water to the ground water. The contribution of ground water to surface flow may be represented by the dotted line in Fig. 6.14.

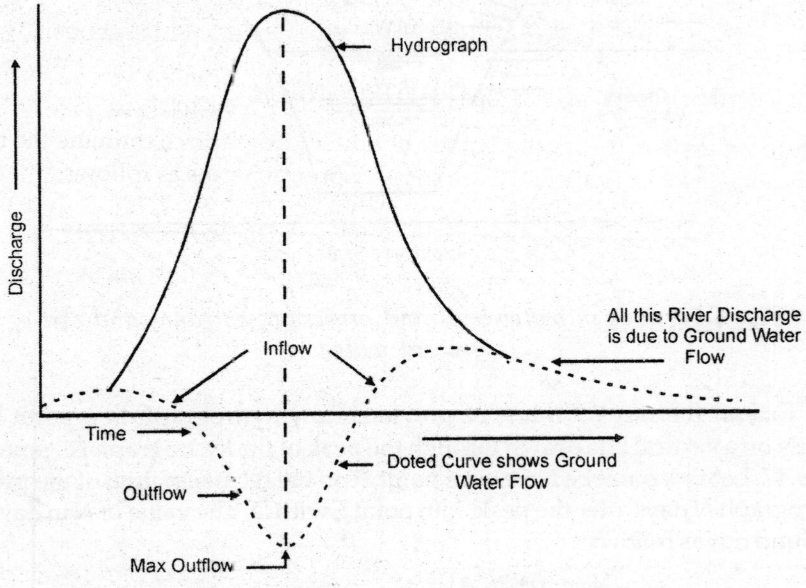

Fig. 6.14. *Ground water inflow and outflow.*

Because it is not possible to determine, the actual shape of dotted line curve and also since in any major flood rise, the ground water table contribution is a small percentage of the total flow, it is sufficient to assume this line as straight line. The surface run off is above this line whereas ground water flow lies below this line.

6.15 COMPONENT PARTS OF A STORM HYDROGRAPH

Figure 6.17 shows a hydrograph for any isolated duration of rainfall. A is the point from which hydrograph starts rising. The hydrograph continues to rise at a very steep rate till peak point B is reached. After this flood discharge starts receding. AB limb of the hydrograph is called rising limb and BD limbs as the receding limb. On limb BD, there is a point C known as point of inflection. It has already been stated in this article, that the hydrographs have three types of flows, *over land flow* (surface runoff), *Interflow* (influent streams or subsurface flow) and *ground water flow*. Overland flow and interflow are generally grouped together and this combined flow is known as *Direct run-off.* During floods the streams contribute ground water to the soil but during low water flows, streams derive most of its water from ground water. See Figs 6.14, 6.15 and 6.16. Direct run-off and ground water run-off in a particular hydrograph can be separated by the following methods :

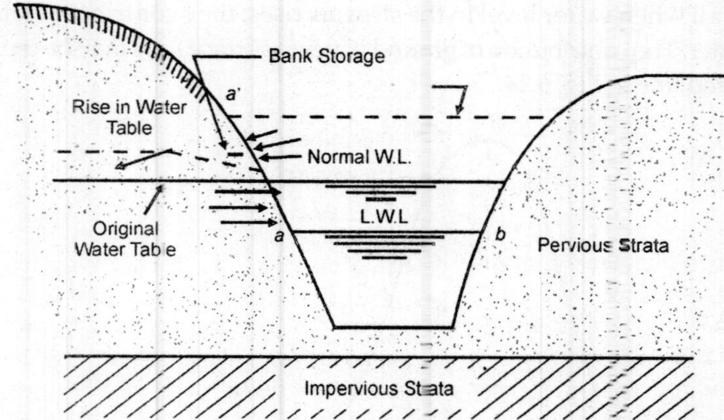

Fig. 6.15. *Rise in water level and process of entrance and exit of ground water.*

1. Extend the recession line of previous flow hydrograph to a point lying exactly on a vertical line drawn through the peak of the hydrograph i.e. point B of Fig. 6.17. Let this point be E. Select a point D on the recession limb of the present hydrograph N days after the peak. Join point E with D. The value of N in days can be found out as follows:

$$N = 0.826\ A^{1/5}$$

$$A = \text{catchment area in sq. km.}$$

Thus AED (Fig. 6.17) is the dividing line between *ground water* and *Direct run-off.* The area $ABCDEA$ represents the volume of direct run-off and area below line AED represents ground water.

2. Refer Fig. 6.17. The base flow is obtained by simply drawing a line AD tangential to both the limbs of the hydrograph at its lower portion. This is the most approximate method and is useful only for preliminary estimates.

Since a base flow forms a very insignificant part of total run-off during high floods after a constant value of base flow is assumed.

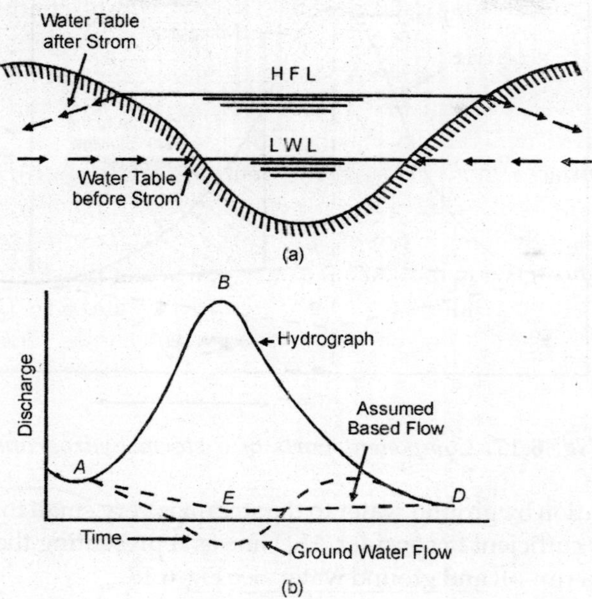

Fig. 6.16. *(a) Flow water to ground and then to stream (b) Assumed base flow and actual ground water flow in case of a hydrograph.*

2. Valley Storage. As more and more water enters the stream, the discharge and level of water goes on increasing. The difference between the water levels of stream before and after the floods is known as valley storage. It is that amount of water which is temporarily stored in the stream. This storage is not stand still, but moves as a sheet of water.

3. Channel Storage. Unless and until flood water remains within the banks of the stream, the valley storage is known as channel storage. The flow in the stream beyond point *C* i.e., the point of inflection on the recession limb of a hydrograph and above the ground water line is almost entirely due to channel storage. The point. indicates the position when surface run-off has almost finished and channel discharge is almost entirely due to ground water flow.

4. Depletion Curve or Base Flow. Look to Fig. 6.16 (a) and (b). When level of water in the stream is lower than the floods, there is flow from ground water to the stream. This ground water is the storage retained by the soil from previous floods. On the other hand when water level in the stream rises because of flood water there is a flow from stream to the ground water. After floods, as soon as water level in the stream recedes, the ground water again starts contributing flow to the streams. The line *AED* denotes the ground water line when there were no floods. It is this line which denotes estimated or assumed base flow or depletion curve. Actual contribution of ground water is shown by the dotted line Fig. 6.16 (b).

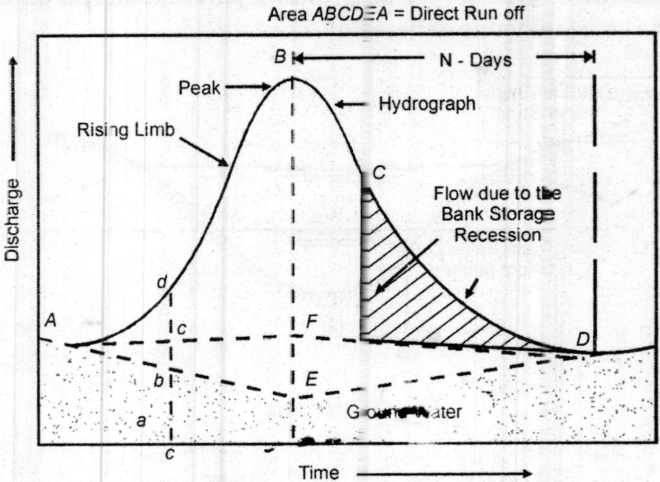

Fig. 6.17. *Component parts of a storm hydrograph.*

Since contribution by ground water to the stream is very small in relation to the total flow it is sufficient to consider *AED* line as representing the dividing line between direct run-off and ground water. See Fig. 6.18.

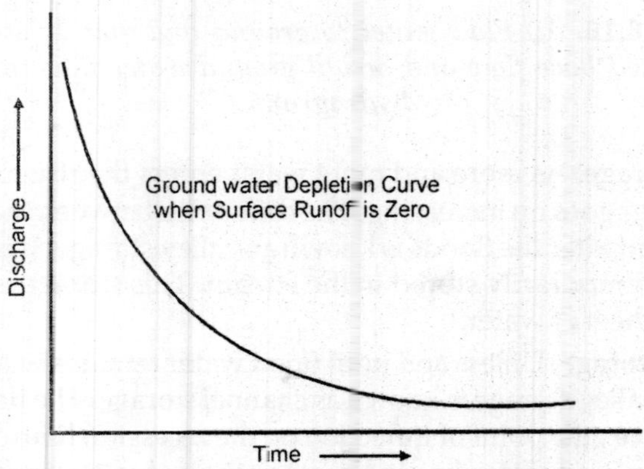

Fig. 6.18. *Ground water depletion.*

6.16 CALCULATION OF DIRECT RUN-OFF FROM STORM HYDROGRAPH

From storm hydrograph measure ordinates at fixed time intervals. These ordinates represent discharge or flow (Q). Also measure the ordinates of base flow, at the same time interval as adopted earlier. By subtracting ordinates of base flow from corresponding ordinates of total discharge Find out the net ordinates which represent direct run-off in that time interval. Add up all the ordinates of direct run-off which is nothing but total volume of direct run-off. By dividing total volume of direct run-off by area of the basin, direct run-off available from the catchment is obtained. The method can very well be understood from the following illustrative example.

Date and time	*Ordinates of hydrograph (Q) in cumecs*	*Base flow in cumecs*	*Direct run-off R*
1-5-79	—	—	—
4 A.M.	8	8	0
6 A.M.	20	6	14
8 A.M.	25	6	19
10 A.M.	32	5	27
12 noon	50	5	45
2 P.M.	70	4	66
4 P.M.	30	5	25
6 P.M.	20	7	13
8 P.M.	8	8	0

$\Sigma R = 209$

Direct run-off in depth of water

$$= \frac{\text{Total volume of direct run-off}}{\text{Area of the basin}}$$

$$= \left[\frac{\Sigma E \times t \times 60 \times 60}{A \times 10^6}\right] \times 100$$

$$= \frac{0.36\ (\Sigma R) \times t}{A}\ \text{cm}$$

where t = time interval in hours between successive ordinates

A = Area of catchment in km^2

Q = Discharge in cumecs.

ΣR = Sum of discharge ordinates in cumecs or direct run-off is cumecs.

In above example $t = 2$ hrs. and let A is 40 km^2.

$$\therefore \quad \text{Direct run-off} = \frac{0.36 \times 209 \times 2}{40} = 3.762\ \text{cm} = 37.62\ \text{mm}$$

6.17 UNIT HYDROGRAPH

The basis of unit hydrograph is that if two identical rainfalls occur on a drainage basin having identical conditions prior to the rains, the hydrograph of run-off from the two storms would be expected to be same.

In actual conditions two identical storms are rare to occur. Most of the practical storms vary in duration, amount, and distribution of rainfall.

A *unit hydrograph* is a hydrograph of a rainfall of a specified duration and areal pattern, resulting in a run-off of 1 cm. In other words, the total volume of water contained in this run-off hydrograph will be equal to 1 cm depth of water on the catchment. The specified duration is such that the hydrographs of any other storms of like durations are assumed to have the same shape but with ordinates of flow in proportion to the run-off volumes. This specified duration is known as unit duration and the storm of this specified duration is called unit storm. Various storms of varying intensities and having different volumes of run-off, but a particular *unit duration,* will have the same shape of their run-off hydrographs. Different unit hydrographs for different unit duration can be prepared. A particular unit hydrograph when once prepared for a particular unit duration can be utilised for evaluating the run-off hydrographs of other storms of the like durations.

The number of unit hydrographs for a given basin is theoretically infinite, since there may be one for every possible rainfall duration, and for every possible distribution pattern in the basin. Practically only a certain limited number of hydrographs can be used for a given basin. It is a common practice to ignore variations in rainfall distribution within a basin. This is reasonable for small catchments, but for large catchments, this assumption is not valid because these variations are usually too large to be ignored. Some catchments are so large that a storm covers only a small portion of the basin. Unit hydrograph cannot give precise results and should not be used for areas larger than 8000 sq km.

The number of unit durations is also reduced to keep only a limited number of unit hydrographs for a basin.

6.18 ASSUMPTIONS IN THE THEORY OF UNIT HYDROGRAPH

It is essential to learn about the assumptions considered in the theory of unit hydrographs :

1. The rainfall is uniformly distributed within the specified duration.
2. The rainfall is uniformly distributed in the whole of the catchment area.
3. The base or time duration of the hydrograph of direct run-off due to an effective rainfall of unit duration is constant.
4. The ordinates of direct run-off of common base time are directly proportional to the total amount of direct run-off represented by each hydrograph
5. For a given basis the hydrograph of run-off due to a given period of rainfall reflects all the combined physical characteristics for the catchment.

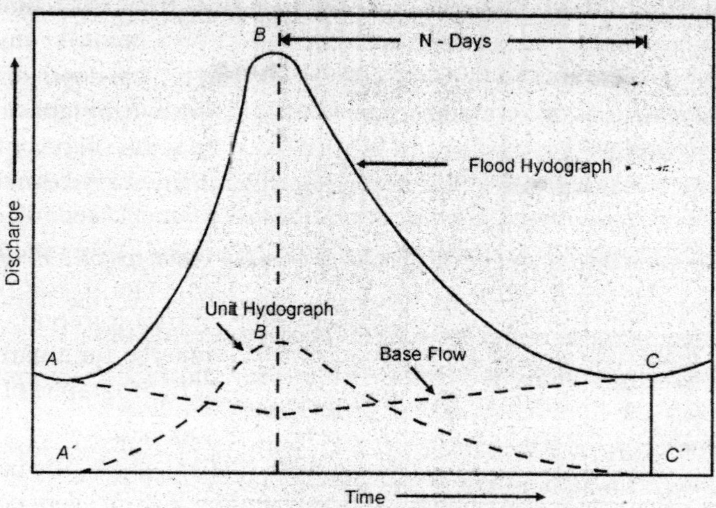

Fig. 6.19. *Flood hydrographs and unit hydrograph.*

6.19 CONSTRUCTION OF UNIT HYDROGRAPH

The unit hydrograph method is used for determination of the maximum flood discharge of a stream and also for developing a flood hydrograph corresponding to any anticipated rainfall. For this a unit hydrograph of suitable unit duration is derived from an observed hydrograph of the drainage basis under consideration. Using the unit hydrograph, the flood hydrograph corresponding to any rainfall of the same unit duration can be obtained. Use following steps to derive a unit hydrograph.

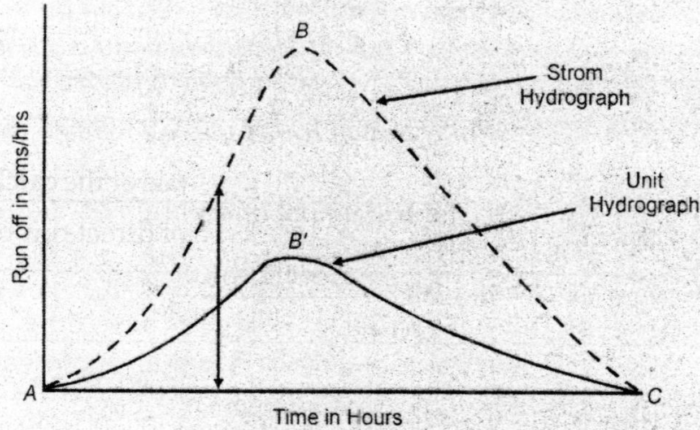

Fig. 6.20. *Construction of unit to hydrograph from flood hydrograph.*

1. Choose a hydrograph resulting from an isolated, intense, short-duration rainfall of nearly uniform distribution over the entire drainage basis as well as within its duration.

2. To obtain direct run-off deduct the base flow from the total run off represented by the hydrograph. For base flow separation any of the methods explained earlier may be used.
3. Draw a series of ordinates at some uniform time interval for the entire hydrograph.
4. From these ordinates of the hydrograph obtain the ordinates of direct run-off by deducting the corresponding ordinates of baseflow.
5. Compute the volume of direct run off which is equal to $(\Sigma R \times t \times 3600)$ m^3.

   where ΣR is the sum of the direct run-off ordinates in cumecs and t is the time interval in hours between the successive ordinates.
6. Divide the volume of direct run-off by the area of the drainage basis to obtain the equivalent depth of direct run-off as follows.

$$\text{Direct run-off depth} = \left(\frac{\Sigma R \times t \times 3600}{A \times 10^6}\right) \text{m}$$

$$= \frac{0.30\, \Sigma R \times t}{A} \text{cm.}$$

7. By dividing each of the ordinates of direct run-off by the depth of direct run-off, obtain the ordinates of the unit hydrograph. By joining the tops of these ordinates by a smooth curve a unit hydrography is obtained. The unit duration of the obtained unit hydrograph is same as that of effective rain-fall corresponding to which the resulting hydrograph has been used for driving the unit-hydrograph.

Example 6.1. *Date, time, total discharge in cumecs and base flow in cumecs are given as follows. Readings were taken at an interval of 4 hours. Find out the unit hydrograph ordinates. Also find the excess depth of run-off. Area of the basin is 30 km^2.*

Date	1.5.09						2.5.09		
Time	4 A.M.	8 A.M.	12 Noon	4 P.M.	8 P.M.	12 Night	4 A.M.	8 A.M.	12 Noon
Q in cumecs	6	10	28	60	110	90	68	9	6
Base flow cumecs	6	5.5	4.5	3.5	2.5	2.5	3.0	5.0	6.0

Solution

Direct run-off in cm

$$V = \frac{\text{Total volume of direct run-off}}{\text{Area of the basin}}$$

$$= \left[\frac{348.5 \times 4 \times 60 \times 60}{30 \times 10^6}\right] \text{m}$$

$$= \left[\frac{348.5 \times 4 \times 60 \times 60}{30 \times 10^6}\right] 100 \text{ cm}$$

$$= 16.728 \text{ cm}$$

Date (1)	*Time* (2)	*Q in cumecs* (3)	*Base flow in cumecs* (4)	*Ordinates of direct run-off (cumecs)* (5)	*Unit hydrograph ordinates* (6)
	4 A.M.	6	6	0	0
	8 A.M.	10	5.5	4.5	0.269
	12 Noon	28	4.5	23.5	1.405
1.5.09	4 P.M.	60	3.5	56.05	3.378
	8 P.M.	110	2.5	107.5	6.426
	12 Night	90	2.5	87.5	5.230
	4 A.M.	68	3.0	65.0	3.886
2.5.09	8 A.M.	9	5.0	4.0	0.239
	12 Noon	6	6.0	0	0.0

$\Sigma R = 348.5$

ordinates of unit hydrograph are calculated by dividing all the ordinates of direct run-off of column (5) by 16.728 cm.

Hence excess rainfall is 16.728 cm.

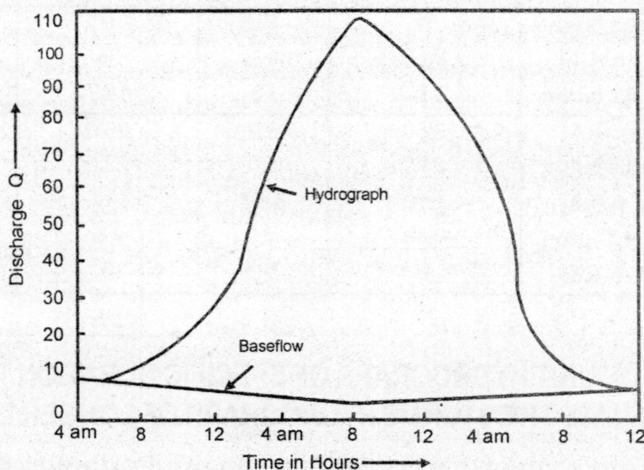

Fig. 6.21. *Hydrograph*

Storm hydrograph and base flow are shown in Fig. 6.21. By adopting values of Col. 6 the unit hydrograph can be developed on the same storm hydrograph.

6.20 TO DEVELOP STORM HYDROGRAPH FROM UNIT HYDROGRAPH

The unit hydrograph of a specified unit duration can be easily used to develop the hydrograph of storms of the same duration. The total volume of run-off (in cm) resulting from a storm when divided by 1 cm will give the multiplying factor i.e. the factor by which the ordinates of the unit hydrograph must be multiplied in order to obtain the ordinates of storm hydrograph of the same duration. The following example illustrates the method by which storm hydrograph is developed from unit hydrograph of the same duration.

Example 6.2. *Date, time, ordinates of unit hydrograph and base flow are given as follows. Determine the ordinates of direct run-off and total discharge ordinates. Rainfall excess is 10 cm. Time of duration is 6 hrs.*

Date	*5.5.09*				*6.5.09*			
Time	*12 Night*	*6 A.M.*	*12 Noon*	*6 P.M.*	*12 Night*	*6 A.M.*	*12 Noon*	*6 P.M.*
Unit hydrograph in cumecs	*0*	*0.36*	*1.67*	*2.50*	*270*	*1.90*	*0.46*	*0*
Base flow in cumecs	*5.0*	*4.0*	*3.5*	*3.0*	*3.0*	*3.50*	*4.0*	*5.0*

Solution

Date (1)	*Time* (2)	*Ordinates of unit hydrograph (cumecs)* (3)	*Base flow (cumecs)* (4)	*Ordinates of direct run-off = (3) ← n* (5)	*Total discharge ordinate* (6)
5.5.09	12 Night	0	5.0	0	5.0
	6 A.M.	0.36	4.0	3.6	7.6
	12 Noon	1.67	3.5	16.7	20.2
6.5.09	6 P.M.	2.50	3.0	25.0	28.0
	12 Night	2.70	3.0	27.0	30.0
	6 A.M.	1.90	3.50	19.0	22.50
	12 Noon	0.46	4.0	4.6	8.6
	6 P.M.	0	5.0	0	5.0

6.21 USE OF UNIT HYDROGRAPH OF SPECIFIC DURATION TO EVALUATE THE STORMS HYDROGRAPH OF LONGER DURATION

The unit hydrograph of a specified duration can be used to evaluate the hydrograph of storms of longer durations. If a unit hydrograph of say 3 hours duration is given, it can be used to evaluate the hydrograph of 6 hours, 9 hours to 12 hours duration. This is accomplished by dividing the storm in to a number of storm hydrographs, each of 3 hours duration. Ordinates of different hydrographs are computed separately and then added to get the final storm hydrograph. But care should be taken to see that the hydrograph of each subsequent part of the storm will start at a lag time of each part i.e. 3 hours as shown in Fig. 6.22.

In Fig. 6.22 a 3 hours unit hydrograph is available and it is required to compute the flood hydrograph for a rainfall lasting 9 hours. With variable intensities of rainfall. The rainfall intensity rates are n_1 cm/3 hr for the first of 3 hours, n_2 cm/3 hr for the second 3 hours and n_3 cm/3 hr for the last 3 hours. The storm is divided into three parts and flood hydrograph of each point are found separately and added to get the final flood hydrograph as shown in Fig. 6.22.

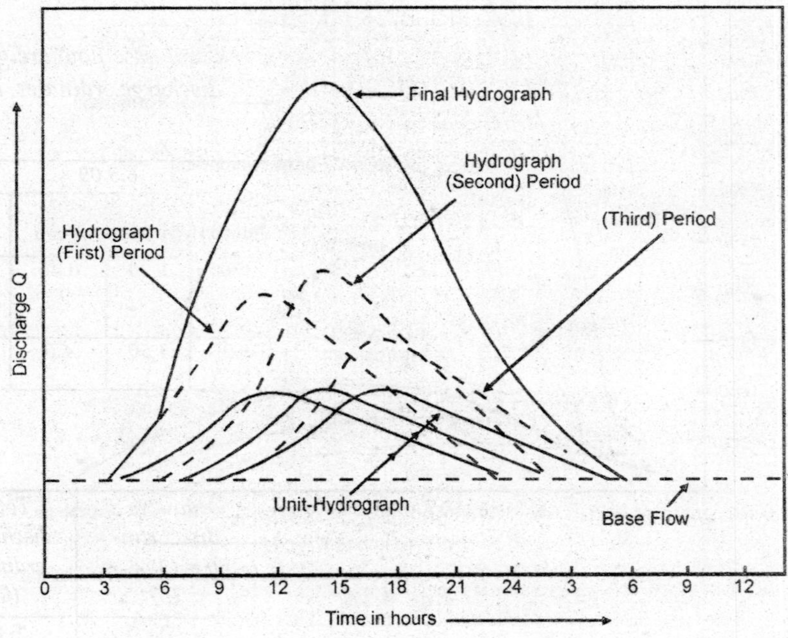

Fig. 6.22

Example 6.3. *Illustrates the development of storm hydrograph from a 2 hours storm with 3 6.4 and 5.4 cm intensities of rainfall during subsequent intervals of 2 hours. Find the ordinates of a storm hydrograph resulting from a 2 hours storm with rainfall 3.0, 6.4 and 5.4 cm during subsequent intervals of 2 hours. The ordinates of unit hydrograph of 2 hours are given as follows.*

Hours Ordinates of unit hydrograph in cumecs	2 A.M 0	4 A.M. 100	6 A.M. 250	8 A.M. 300	10 A.M. 500	12 Noon 450
	2 P.M. 400	4 P.M. 300	6 P.M. 250	8 P.M. 150	10 P.M. 50	12 Night 0

Assume initial loss of 6 mm, infiltration index as 2 mm/hr and Base Flow 10 cumecs.

Solution. 1. Rainfall excess during first 2 hours

$$= 30 - 2 \times 2 - 6 = 20 \text{ mm} = 2 \text{ cm}$$

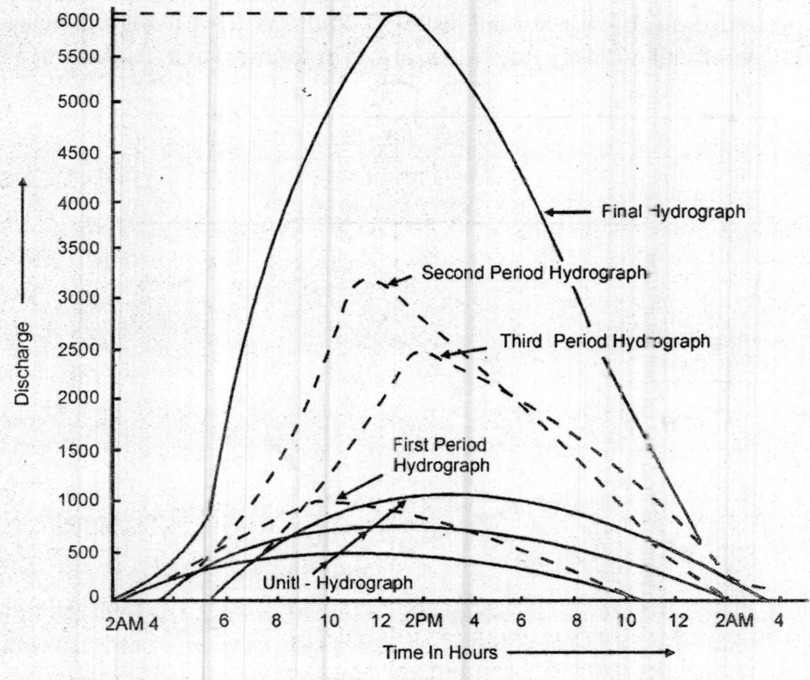

Fig. 6.23

2. Rainfall excess during second 2 hours

$$= 64 - 2 \times 2 = 60 \text{ mm} = 6.0 \text{ cm}.$$

3. Rainfall excess during the last 2 hours

$$= 54 - 2 \times 2 = 50 \text{ min} = 5 \text{ cm}.$$

Rainfall excess as ratio of unit rainfall of 1 cm during subsequent 2 hours intervals are 2 cm, 6 cm and 5 cm. The example has been solved in tabular form as follows. The computations of run-off due to 2 cm rainfall excess will start from 2 A.M. The computations of run-off due 6 cm and 5 cm rainfall excesses will start at 4 A.M. and 6 A.M. respectively. See Fig. 6.23.

Example 6.4. *In a typical 4 hours storm producing 5 cm run-off a catchment, the flows in stream are as follows*

Time in hours	0	2	4	6	8	12	16	20
Flow in	0.0	1.22	4.05	6.75	5.67	3.375	1.35	0.0

Time in hours	Ordinates 2 hrs unit hydrograph (cumecs)	Rainfall excess in cm/-2 hrs	Surface run-off from rainfall excess during successive unit periods (cumecs)				Base flow in (cumecs)	Total discharge in (cumecs)
			2 cm	6 cm	5 cm	Total		
(1)	(2)	(3)	(4)	(5)	(6)	(7)	(8)	(9)
2 A.M.		2.0	0			0	10	10
4 A.M.	100	60	200	0		200	10	210
6 A.M.	250	5.0	500	600	0	1100	10	1110
8 A.M.	300		600	1500	500	2600	10	2610
10 A.M.	500		100	1800	1250	4050	10	4060
12 Noon	450		900	3000	1500	5400	10	5410
2 P.M.	400		800	2700	2500	6000	10	6010
4 P.M.	300		600	2400	2250	5250	10	5260
P.M.	250		500	1800	2000	4300	10	4310
8 P.M.	150		300	1500	1500	3300	10	3310
10 P.M.	50		100	900	1250	2250	10	2260
12 Night	0		0	300	750	1050	10	1060
2 A.M.				0	250	250	10	260
4 A.M.					0	0	10	10

Estimate the peak flow and the time of its occurrence in a flood created by an 8 hour storm which produces 2.5 cm of run-off during the first 4 hours and 3.75 cm of run-off during the next 4 hours. Assume base flow as negligible.

Solution. First of all the hydrograph of a run-off is obtained from the given storm of 4 hours duration and is plotted by using the various given values of time and flow.

Since this hydrograph contains 5 cm of run-off volume, we must multiply its ordinates by $\frac{1}{5}$, in order to obtain unit hydrograph of 4 hours duration. Both the hydrographs are shown in Fig. 6.24.

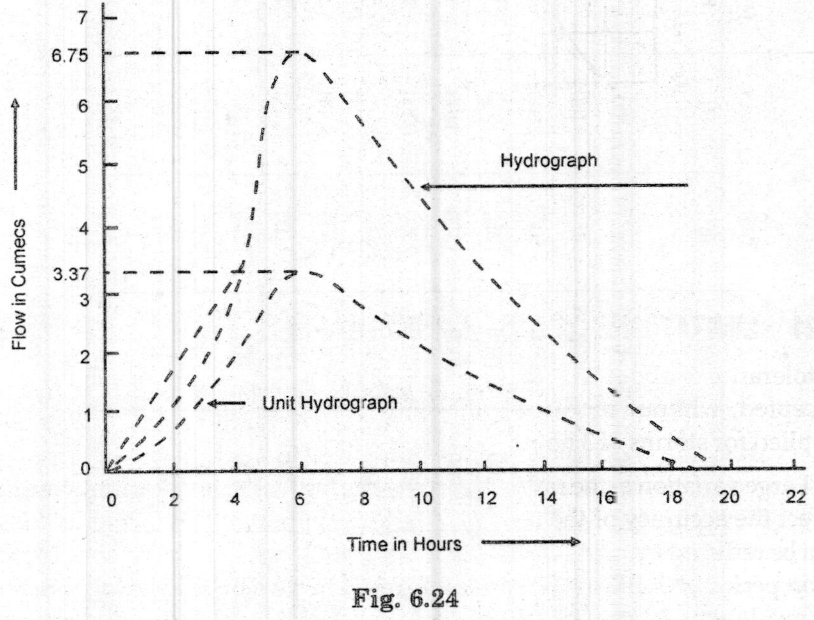

Fig. 6.24

The storm of 8 hours duration has been divided into two parts each of 4 hours duration. During the first part 2.5 cm of run-off is produced and during the second part 3.75 cm of run-offs is produced. The hydrograph for the first part is the same as the unit hydrograph. Multiplying factor is $\frac{2.5}{1} = 2.5$.

The hydrograph for the second part can be obtained by multiplying its ordinates by $\frac{3.75}{1} = 3.75$. But this hydrograph of the second part must be plotted 4 hours after the first part hydrograph as shown in Fig. 6.25. These two graphs are added together so as to obtain the final total hydrograph of 8 hours duration rain. This is also shown in Fig. 6.25. The peak value is found to be 7.312 cumecs and it occurs at a time 10 hours from the start of the rain.

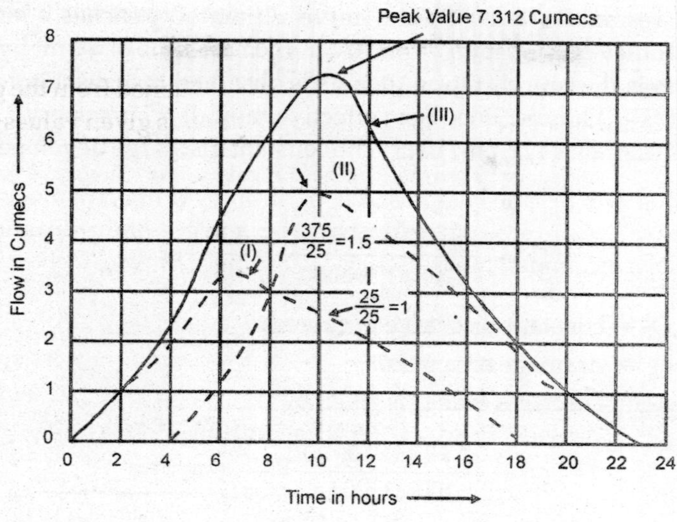

Fig. 6.25

6.21 LIMITATIONS OF UNIT HYDROGRAPHS

A tolerance of about 25% of the adopted unit hydrograph duration is generally accepted, without serious errors. Thus a 4 hours unit hydrograph might be applied for storms of 3 hours to 5 hours effective duration.

Large variation in the rainfall intensity during the unit period may considerably affect the accuracy of the unit hydrograph approach. Errors due to this reason can be reduced by using unit hydrographs for relatively short time periods. The short period of unit hydrographs can be used to develop the hydrograph resulting from a long rain of varying intensity. Experience has shown that the best unit period is about one fourth of the basin lag i.e. the time from the centre of mass of rainfall to the peak of the hydrograph. In general it has been found that the period of unit storm should always be less than the period of rise (probably half or so).

Sometimes the unit hydrographs having a similar range in durations are averaged to obtain an average unit hydrograph. An average unit hydrograph is preferred to a single storm unit hydrograph since the averaging tends to minimise the errors in data.

6.22 S-HYDROGRAPH (OR S-CURVE)

S-Hydrograph is S-shaped curve. It is a hydrograph of direct surface discharge resulting from successive storms each producing 1 cm effective rainfall in unit duration. This hydrograph is derived by summing up the ordinates of a series of unit hydrographs of same unit duration spaced at intervals of the unit duration. It is sometime also called summation curve. It is a continuously rising curve and resembles letter S. The curve ultimately reaches a stage when discharge achieves a constant value. If T is the base time of a unit hydrograph and t_r hours is the unit duration then its is observed that discharge becomes constant at $(T-t_r)$ hours. Since 1 cm effective rainfall occurs every t_r hours the intensity of rainfall is $1/t_r$ cm/hour. The constant discharge developed in given by

$$Q = \frac{A(1000\times1000)\times1}{100\times t_r\times3600} = \frac{2.778}{t_r}A \qquad (1)$$

where Q = constant discharge in cumees

A = drainage area in km^2

If A is taken in hectares then Q is given by

$$Q = \frac{A\times100\times100\times1}{100\times3600\,t_r} = \frac{A}{36\,t_r}$$

Construction of S-hydrograph has been illustrated by an example given here in Fig. 6.26 shows this Example. 7 unit hydrographs of 4 hours interval and all the successive unit hydrograph beginning at a lag of 4 hours. The example has been given in Table 6.5.

Table 6.5. *Derivation of S-Hydrograph from a given hour unit hydrograph.*

Time in hours	*Ordinates of 4-hr unit hydrography (cumecs)*	*Ordinates of successive 4-hr unit hydrographs each lagging by 4 hrs unit cumecs*						*Ordinates of S-hydrograph (cumecs)*
1	2	3	4	5	6	7	8	9
0	0	—	—	—	—	—	—	0.0
4	20	0.0	—	—	—	—	—	20.0
8	49.0	20.0	—	—	—	—	—	69.0
12	35.0	49.0	20	0	—	—	—	104.0
16	22.0	35.0	49	20	0	—	—	126.0
20	12.0	22.0	35	49	20	—	—	138.0
24	6.0	12.0	22	35	49	20	0	144.0
28	0	6.0	12	22	35	49	20	144.0

The area of the catchment, is 207.4 sq km. Hence constant discharge as per equation

$$Q = \frac{2.778\times207.4}{4} = 144.04 \text{ cumecs.}$$

It is same is found out by Table 6.5. Constant discharge is reached at 28 – 4 = 24 hours.

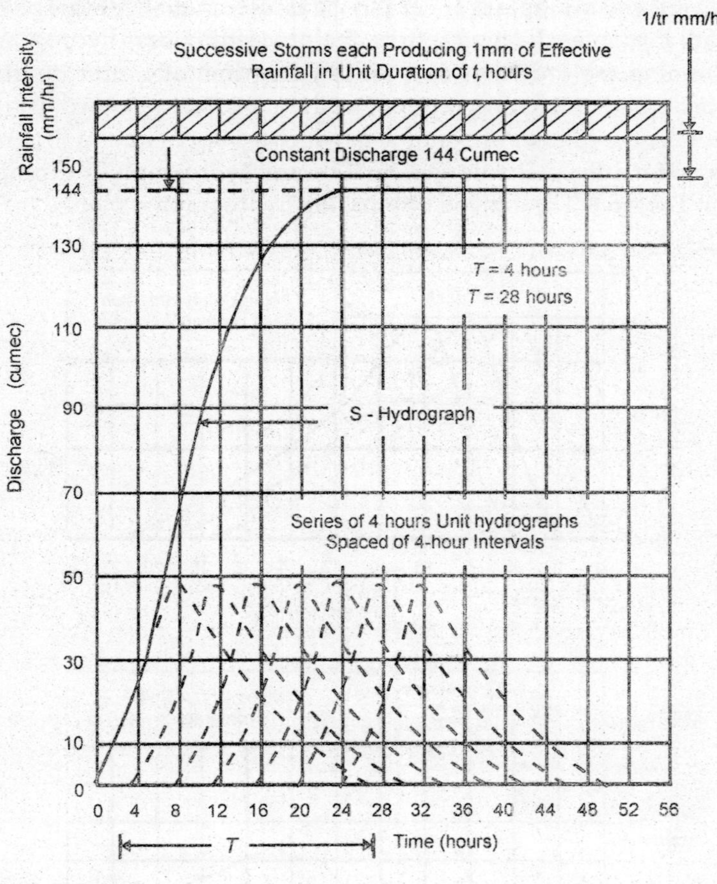

Fig. 6.26

Derivation of unit hydrograph of different unit duration from a unit hydrograph of given unit duration.

If a unit hydrograph of desired unit duration is not available, it can be derived by converting an available unit hydrograph of known unit duration. If the unit duration of the available unit hydrograph is smaller than the unit duration of the desired unit hydrograph and it is an integral multiple then by the principle of super-position the unit hydrograph of desired unit duration may be obtained. However a more general method using S-hydrograph may be adopted to derive a unit hydrograph of unit duration either longer or shorter than that of the available unit hydrograph. In this method the unit duration of the desired unit hydrograph need not be an integral multiple of the unit duration of the available unit hydrograph. Both these methods have been described here.

(a) *Construction of longer period unit hydrograph from a given unit hydrograph of shorter unit period.* It is desired to construct a unit hydrograph of unit period or duration t'_r from available unit hydrograph of unit duration t_r such that $t'_r > t_r$. Also t'_r is an integral multiple of t_r. For this n number of unit hydrograph of t_r unit duration are drawn each lagging from the proceeding unit hydrograpy by t_r hours. Summing the ordinates of n unit hydrograph of t_r unit duration, the ordinates of desired unit hydrograph of unit duration t'_r are obtained. The method is illustrated by the following example in which a 6 hours unit hydrograph has been obtained for a given 2 hours unit duration unit hydrograph. The computation are given in Table 6.6. The derived 6 hours unit hydrograph is plotted in Fig. 6.27.

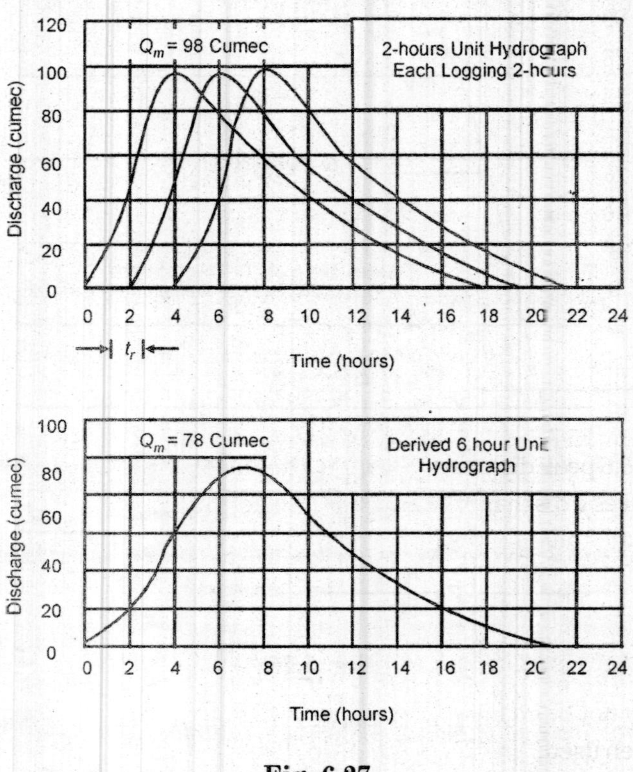

Fig. 6.27

Figure 6.27 (a) shows these 2 hours unit hydrographs each lagging by 2-hours. It can be observed that the peak discharge Q'_m of the derived unit hydrograph is less than the peak discharge Q of the given unit hydrograph. The peak of the derived unit hydrograpy occurs later than that of the given unit hydrograph and the derived unit hydrograph has a longer base time by $(n - 1)\, t_r$, hours than that of the given unit hydrograph. This is imperative because increase in unit duration results in a lower intensity of the rainfall for a longer time which results the discharge to be extended over a longer duration.

Table 6.6

Time in hrs	*Ordinates of 2 hrs unit hydrograph*	*Ordinates of 2 hrs unit hydrograph lagged by 2 hrs in (cumecs)*	*Ordinates of 2 hr unit hydrograph lagged 4 hrs by in (cumecs)*	*Total of* Col 2 + 3 + 4 *cumecs*	*Ordinates of 6 hrs unit hydrograph (cumecs)* $\left(\frac{\text{Col.5}}{3}\right)$
(1)	*(2)*	*(3)*	*(4)*	*(5)*	*(6)*
0	0	—	—	0	0
2	40	0	—	40	13.33
4	98	40	0	138	46.0
6	78	98	40	216	72.0
8	54	78	98	230	76.67
10	38	54	78	170	56.67
12	26	38	54	118	39.33
14	16	26	38	80	26.67
16	8	16	26	50	16.67
18	0	8	16	24	8.0
20	—	0	8	8	2.67
22	—	—	0	0	0

From Table 6.6 peak discharge occurs at 7 hours when peak of the derived unit hydrograph gives a discharge of about 78 cumecs.

(*b*) *Construction of shorter or longer unit duration unit hydrograph from a given unit hydrograph.*

Let a unit hydrograph of unit duration t'_r hours be required to be constructed from a unit hydrograph of unit duration t_r hours. t'_r may be more or less than t_r and also t'_r may not be an integral multiple of t_r. For this purpose S-hydrograph method has been used.

As already explained earlier, from the given unit hydrograph of t_r unit duration construct S-hydrograph. Figure 6.28 shows S-hydrograph so obtained. This hydrograph represents a constant effective rainfall of intensity $\frac{1}{t_r}$ cm/hour.

A lagged S-hydrograph is also constructed by shifting the original S-hydrograph by t'_r hours i.e. equal to the unit duration of the desired unit hydrograph. This lagged S-hydrograph is also shown in Fig. 6.28.

The area enclosed between the original and the lagged S-hydrographs represents the total run off due to a rainfall of intensity $\frac{1}{t_r}$ cm/hour in a duration of t'_r hours. In other words it represents a total run-off of $\frac{t'_r}{t_r}$ cm in t'_r hours. The

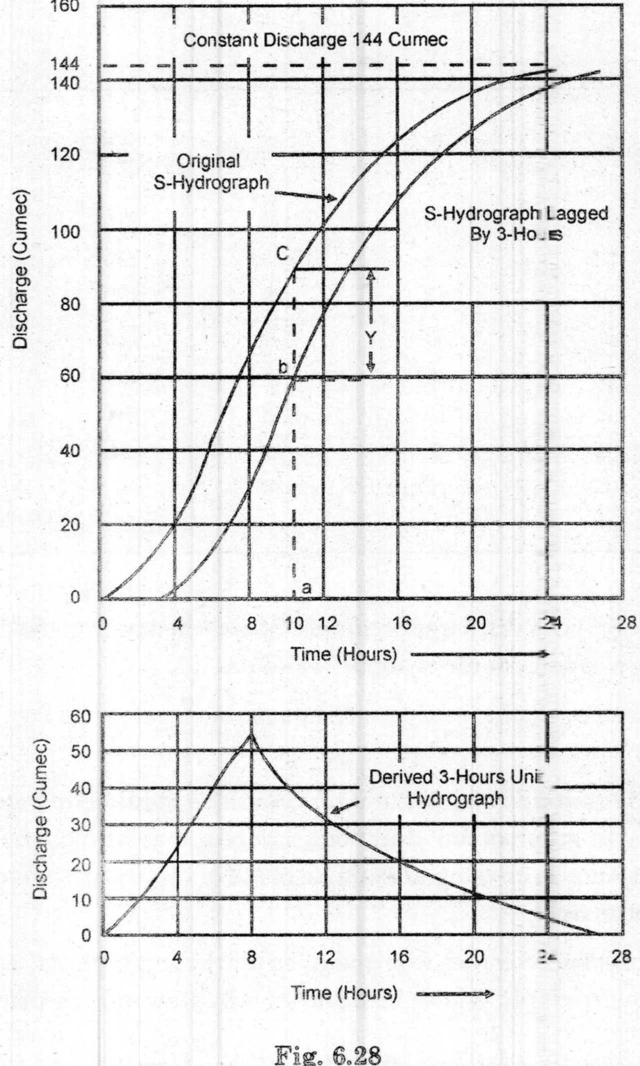

Fig. 6.28

difference between the ordinates of the original and the lagged S-hydrographs i.e. Δy represents the rate of run-off at any instant of time. It can also be said that the values of Δy at different instants of time represent the ordinates of a hydrography having a unit duration of t'_r hours and a total run-off of t'_r/t_r cm. The ordinates of a t'_r hour unit hydrograph may be obtained by the principle of linearity as

$$\frac{\Delta y}{(t'_r / t_r)} = \frac{\Delta y \times t_r}{t'_r}$$

This method has been explained by an example in which a 3 hours unit hydrograph has been derived from a 4-hour unit hydrograph. The S-hydrograph derived from a 4-hour unit hydrograph given in in Table 6.5 here has been adopted to derive a 3-hour unit hydrograph. The complete method is given in Table 6.7. The ordinates of derived 3-hour unit hydrograph are plotted for the data computed in column 5 of Table 6.7. A smooth curve is obtained as shown in Fig. 6.28 (b). Generally the unit hydrograph so obtained may not be a smooth curve. Hence smoothening has to be done. The values of the ordinates along the smoothened curve are given in column (6) of Table 6.7. However as a check the sum of the ordinate in column (5) and (6) of Table 6.7 must be the same.

Table 6.7. *Derivation of 3-hr unit hydrograph from 4-hr unit hydrograph by S-hydrograph method.*

Time in hrs	*Ordinates of S-hydrograph (cumecs) from Table 6.5*	*Ordinates of tagged S-hydrograph cumecs*	Δy *cumecs*	*Ordinates of 3-hr unit hydrograph (cumec)* $= \Delta y \times \frac{4}{3}$	*Ordinates of 3-hr unit hydrograph (smoothened) (cumecs)*
(1)	(2)	(3)	(4)	(5)	(6)
0	0.0		0	0	0
4	20.0	4.0	16.0	21.30	23.0
8	69.0	30.0	39.0	52.0	52.0
12	104.0	79.0	25.0	33.3	320
16	126.0	110.0	16.0	21.3	21.0
20	138.0	128.0	10.0	13.3	13.5
24	144 0	140.0	4.0	5.3	4.0
27	144.0	144.0	0	0	1.0
28	144.0	144.0	0	0	0
		ΣR		146.5	146.5

16.23 DERIVATION OF UNIT HYDROGRAPH FROM COMPLEX STORMS

Generally a rainfall never develops a single peaked hydrograph because in most cases rate of rainfall during rainfall keeps varying. Hence generally during rainfall multiple peaked hydrograph are developed. As such a unit hydrograph may be required to be derived from hydrographs of complex storms.

If the storms are sufficiently separated so that the individual storm results in a well defined peak it is possible to separate the hydrographs resulting from each of the storms. In such cases the unit hydrograph may be derived from each of these hydrographs by method explained earlier in this chapter. If it is not possible to separate the hydrograph resulting from complex storm then the following method may be adopted to derive the unit hydrograph.

To illustrate let us consider a complex storm consisting of three successive storms of respective rainfalls of R_1, R_2 and R_3 each of t_r duration.

Let $Q_1, Q_2, Q_3 \ldots\ldots Q_4$ be the ordinates of the direct run-off hydrograph due to complex storm.

In this case we have considered three storms and hence there will be three unit hydrographs each operating at a time lag equal to the duration of the storm i.e. t_r, let the ordinates of the unit hydrograph be $U_1, U_2, U_3, \ldots U_m$ at successive times. The run-off hydrograph due to complex storm is developed by superposing the successive storm hydrographs. It means that at any instant of time the direct run-off ordinate of the hydrograph due to complex storm is equal to the sum of the ordinates of hydrographs due to each of the storms. Also by the principle of linearity the ordinates of each storm hydrograph may be obtained as the effective rainfall of the storm times corresponding ordinate of the unit hydrograph.

Thus for the given complex storm hydrograph shown in Fig. 6.29.

First ordinate $Q_1 = R_1 U_1$

Since Q_1 and R_1 are known, U_1 can be determined.

Second ordinate $Q_2 = R_1 U_2$

Since Q_2 and R_2 are known, U_2 can be determined.

Third ordinate $Q_3 = R_1 U_3 + R_2 U_1$

Since all the values except U_3 are known the same can be computed.

Fourth ordinate $Q_4 = R_1 U_4 + R_2 U_2$

From this U_4 can be determined. Fifth ordinate $Q_5 = R_1 U_5 + R_2 U_3 + R_3 U_1$

From this U_5 can be determined $Q_N = R_1 U_M + R_2 U_{n-2} + R_3 U_{n-4} + \ldots\ldots$

It should be noted that n is less than m.

Thus by repeating the same procedure the ordinates $U_1, U_2, U_3 \ldots\ldots U_m$ of the unit hydrograph can be found out following inferences may be drawn from the complex storm hydrograph shown in Fig. 6.29.

1. *A*, *B*, and *C* curves represent the respective recession limbs of the three storm hydrographs.

2. The ordinates of curve *A* are equal to $R_1 \times$ the successive ordinates of the first unit hydrograph.

3. The ordinates of curve *B* are equal to $R_2 \times$ successive ordinates of second unit hydrograph + successive ordinates of curve *A*.

4. The ordinates of curve *C* are equal to (R_3 × successive ordinates of third unit hydrograph + successive ordinates of curves *A* + *B*).

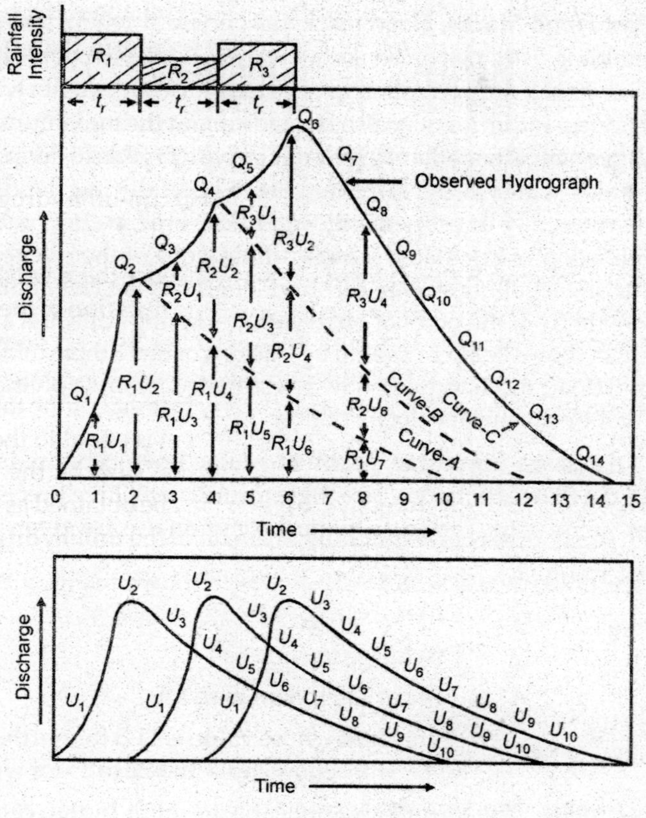

Fig. 6.29

6.24 FLOODS AND THEIR ESTIMATION

A flood may be defined as an overflow coming from some river or from some other body of water. Rivers may get flooded either due to an excessive rainfall or due to melting of snow or due to some obstruction in the flow. Whenever water overflows the banks of the river, it is said to be flooded. The floods my be rated as *ordinary, standard project* and *maximum probable floods.* The floods may also be called peak flow. Magnitude of floods or peak flow, can be estimated by following methods :

1. From past flood marks.
2. By flood discharge formulae.
3. By flood frequency studies.
4. Unit hydrographs.

Run-off analysis by unit hydrographs has already been explained in this chapter and thus may be referred. The remaining three methods are being explained here one by one.

1. From Past Flood Marks. Flood marks are generally left by the floods on old trees or old buildings, situated on banks of the river. Local old people from villages located at river banks may be asked to reveal as to what extent river generally rises during floods. Inquiries may also be made about the maximum level which river might have ever attained during floods in past 35 years. Having known the level up to which water in the river may have risen during floods, the cross-section of the river may be plotted and water flow area, wetted perimeter, and hydraulic mean depth, calculated. Longitudinal slope of the river can be found out by conducting levelling survey for short lengths of the river both up stream and down stream from the point of observation. Having collected all these data mean velocity of flow can be computed by Chezy's or any other suitable formula. The maximum flood discharge can now be estimated by multiplying water flow area of the river by the mean velocity of flow.

2. Flood Discharge Formulae. Many formulae have been in use for rough estimation of the flood discharge. No single formula will have precise results for all the places. Some of the formulae in most common use are given as follows :

(i) *Dicken's formula*

$$Q = CA^{3/4}$$

where Q = Discharge in cumecs

A = Area of catchment in sq km

C = A constant whose value is 11.5 for northern India, 14 to 19.5 for central India and 22 to 25 for western India.

(ii) *Ryve's formula.* This formula is applicable to south Indian catchments

$$Q = CA^{2/3}$$

where Q = Discharge in cumecs

A = Area of catchment in sq km

C = Constant, whose value is taken as 6.8 for areas within 24 km from the sea coast, 8.8 for areas within 24–160 km from the coast and 10.1 for limited area near hills.

(iii) *Inglis formula*

$$Q = \frac{123A}{\sqrt{A+10.4}} \text{ or } 123\sqrt{A}$$

where Q = cumecs

A = area of basin in sq. km

(iv) *Nawab Jung Bahadur formula.* This formula was developed for catchments of Hyderabad Deccan

$$Q = CA^{\left(0.92 - \frac{\log A}{14}\right)}$$

Q and A are respectively in cumecs and sq km and value of C is taken 48 to 60.

(v) W.P. Creager's formula

$$Q = 46\ C\ (A^{0.894}\ A^{-0.048})\ \text{cumecs}$$

$$C = 30 \text{ to } 100$$

$$A = \text{Area of catchment in sq km.}$$

(vi) Jarvis formula

$$Q = C\sqrt{A}$$

Q and A are same as stated in other formula ; value of C is taken from 1.77 to 177.

(vii) Fanning's formula

$$Q = CA^{5/6}$$

Q and A are same as above, C is taken 2.54.

3. By Flood Frequency Studies. In this method the prediction for the future floods is made based upon the pat records. For the success of any of the probability method sufficient past records must be available.

Flood Frequency. This term denotes the possibility of flood being equalled or exceeded. A 20% frequency (F) means that there are 20% chances of flood being equalled or exceeded.

Recurrence Interval. It is the time interval after which a similar flood can be expected. If recurrence interval is denoted by T_r and frequency of flood by F, then both these terms can be connected as follows

$$T_r = \frac{100}{F} \text{ or } F = \frac{100}{T_r}.$$

If data of annual floods for N number of years is given, then they should be arranged in decreasing order of their magnitude. The highest flood should be placed at the top and lowest flood at the bottom of the list. The Serial Number of highest flood will be one and that of lowest flood N. If Serial Number of a particular flood in the order is say m, its recurrence interval can be found by any of the following three methods.

(i) California method

$$T_r = \frac{N}{m}$$

(ii) Hazen's method

$$T_r = \frac{2N}{2m-1}.$$

(iii) Gumbel's method

$$T_r = \frac{N}{m+C-1}$$

where C = Gumbel's correction which depends upon $\frac{m}{N}$ ratio. The value of C corresponding to $\frac{m}{N}$ ratios is given as follows :

$\frac{m}{N}$	1	0.9	0.8	0.7	0.6	0.5	0.4	0.3	0.2	0.1	0.08	0.04
C	1	0.95	0.88	0.845	0.78	0.73	0.66	0.59	0.52	0.4	0.38	0.28

In order to find any design flood of desired frequency a graph is plotted between Discharge versus frequency or recurrence interval. This plot is made on a special paper known as probability paper. Unknown floods of any recurrence interval can be found out from this plot.

Example 6.5. *The following Table gives flood data for 12 years, recorded at a site of a certain river.*

Year	*Discharge in cumecs*	*Year*	*Discharge in cumecs*
1999	2400	2005	3000
2000	6000	2005	3600
2001	1800	2006	4200
2002	3300	2007	8400
2003	7200	2008	4800
2004	6600	2009	5400

Find the recurrence interval for the flood of various magnitudes by the following methods :

(i) California method

(ii) Hazen's method

(iii) Gumbel's method.

Solution. The recurrence interval has been calculated by the three methods as given in Table. The value of *c* for Gumbel's method has been taken from the Table of previous : $N = 12$.

Order No. (m)	Peak yearly discharge in cumecs in arranged descending order	Recurrence interval		
		California method $T_r = \frac{N}{m}$	Hazen's method $T_r = \frac{2N}{2m-1}$	Gumbel's method $T_r = \frac{N}{m+C-1}$
1	8400.	12	24	32.43
2	7200	6	8	8.16
3	6600	4	4 8	4.70
4	6000	3	3.43	3.32
5	5400	24	2.67	2.57
6	4800	2.0	2.20	2.09
7	4200	1.7	1.84	1.77
8	3600	1.5	1.60	1.667
9	3300	1.33	1.40	1.354
10	3000	1.20	1.30	1.212
11	2400	1.10	1.14	1.095
12	1800	1.0	1.04	1.00

Example 6.6. *Values of precipitations at 6 gauge stations are 10.5, 5.5, 7.5, 12.6, 8.4 and 13.5 mm What is average precipitation?*

Solution

Station	Precipitation in mm
1	10.5 min
2	5.5 mm
3	7.5 mm
4	12.6 min
5	8.4 mm
6	13.5 mm
	58.0 mm

Average precipitation

$$= \frac{58}{6} = 9.67 \text{ mm.}$$

Example 6.7. *There are 5 rain gauge stations A, B, C, D, E and Areas in charge of each rain gauge and precipitations recorded by gauge stations are given as follows. Determine the average precipitation by Thiessen polygon method :*

Gauge Station	*A*	*B*	*C*	*D*	*E*
Area as per Thiessen polygon in km²	*68*	*72*	*56*	*44*	*50*
Rainfall in mm	*25.0*	*29.0*	*18.0*	*22.0*	*15.0*

Solution

Rain gauge Station	*Area as per Thiessen Polygon (A)*	*Precipitation in mm (P)*	*A × P*
A	68	25	1700
B	72	29	2088
C	56	18	1008
D	44	22	968
E	50	15	750
	290 km²		6514

Average precipitation

$$= \frac{\Sigma AP}{\Sigma A} = \frac{6514}{290} = 22.462 \text{ mm.}$$

Example 6.8. *A catchment area has right isohyets of value 7, 8, 9, 10, 11, 12, 13, 14 cm. Areas enclosed between successive isohyets are respectively 20 70, 121, 79, 50, 44, 26 square kilometres. Determine the average value of pricipitation in the area. See Fig. 6.7.*

Solution

Isohyet (cm)	*Area enclosed between successive isohyets in sq km*	*Average precipitation*	*Area × average precipitation*
7	20	7.5	150.0
8	70	8.5	595.0
9	121	9.5	1149.5
10	79	10.5	829.5
11	50	11.5	575.0
12	44	12.5	550.0
13	26	13.5	351.0
14	410		4200.00

Average precipitation

$$\frac{4200}{410} = 10.244 \text{ cm.}$$

QUESTIONS

6.1 Explain the terms Hydrology, Hydrological cycle, Precipitation, Run off, ground water, surface run-off.

6.2 Explain the working of different types of rain gauges with the help of neat sketches.

6.3 Explain various methods of computation of mean rainfall. Also give the limitations of each method.

6.4 What are various factors that affect the run off ?

6.5 Define the following terms :

(i) Hydrograph, (ii) Surface run-off,

(iii) Time of concentration, (iv) Time of overland Flow,

(v) Period of rise.

6.6 What are the methods of computing runoff from a catchment area ? Give various formulae stating clearly the area for which each is applicable.

6.7 Explain various methods of determining flood discharge in a stream. Give various flood discharge formulae applicable to Indian catchments.

6.8 What is a hydrograph ? Draw a single peaked hydrograph and name its various parts.

6.9 What do you understand by the term unit hydrograph ? Explain the method of constructing unit hydrograph from a storm hydrograph by taking practical example.

6.10 Explain the infiltration characteristics of a soil. What do you understand by the term infiltration indices ? How infiltration indices are used to compute the coefficient of run-off ?

6.11 Explain the terms : Channel storage, Depletion curve or Rase Flow. Direct Run-off, Interflow (influent stream or subsurface flow), ground water flow.

6.12 Given below are the observed flows from a storm of 6 hrs duration on a stream with a drainage area of 632 sq km.

Date	*3.5.09*				*4.5.09*			
Time	12 Night	6 A.M.	12 Noon	6 P.M.	12 Night	6 A.M.	12 Noon	6 P.M.
Flow in cumecs	34	225	500	400	300	225	180	34

Assuming constant base flow 20 cumecs derive and plot a 6 hour unit hydrograph. How many cm of rainfall excess does the above storm hydrograph represent ?

6.13 Find the ordinates of a storm hydrograph resulting from a 3 hour storm with rainfall of 4 cm, 9 cm and 6cm during subsequent 3 hour intervals. The ordinates of unit hydrograph are given in the following Table.

Hours	03	06	09	12	15	18	21	24	03	06	09	12
Ordinates of unit hydrograph in cumecs	0	50	175	400	600	750	700	550	300	250	20	0

Assume an interval loss of 10 mm, infiltration index of 5 mm/hr, and base flow of 10 cumecs.

Ground Water Hydrology

7.1 GROUND WATER HYDROLOGY

It is a science of the occurrence, distribution and movement of water below the surface of the earth.

7.2 GROUND WATER RESERVOIR

The water absorbed by the soil which percolates downwards ultimately joins the ground water reservoir. It is that water which is readily available all the time. It can be availed by digging wells. Ground water reservoir is a big natural storage of water. It is estimated that the total ground water potential is as large as to be equivalent to the capacity of oceans.

The main source of ground water is rainfall, or to be very precise, precipitation. The part of rainfall that infiltrates into the ground percolates downwards unless checked by impervious layer. Since impervious layer does not allow infiltrated water to go further down, it gets stored in the soil above the impervious layer forming *ground water reservoir*. This water remains held up in the voids of the soil. The layers of soil in which water is held up in soil pores are known as water bearing formations. They act as storage reservoirs for ground water and also as conduits for transmission of water.

If the surface of the earth is cut deep, we find existence of water at different depths in different forms. As regards the existence of water at different depths, the earth's crust can be divided into various zones namely:

(i) Zone or rock fracture and

(ii) Zone of rock flowage.

The depth of the zone of rock flowage is not known accurately. Its presence is estimated to lie kilometres below the surface of the earth This zone does not contain interstices as stresses here are beyond the elastic limits and the rock remains more or less in a state of plastic flow. Water present in this zone is known as internal water and a hydraulic engineer has nothing to do with this water.

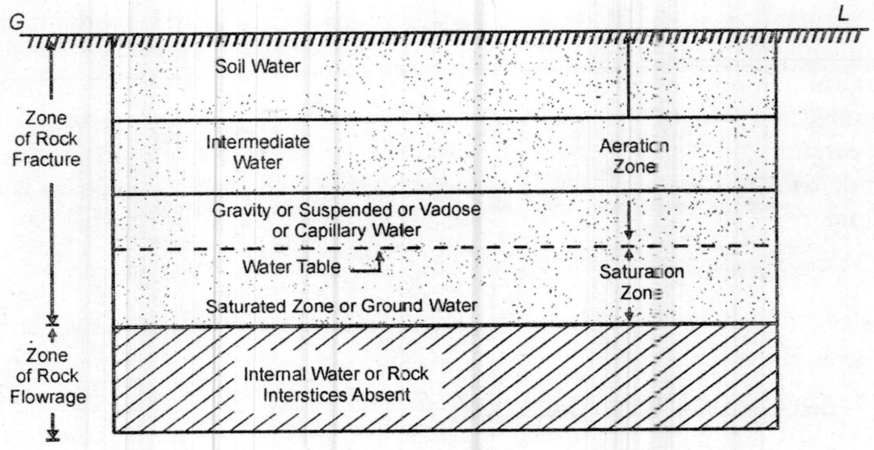

Fig. 7.1. *Section of earth's crust.*

Zone of rock fracture lies above the zone of rock flowage. Interstices do exist in this zone and water is stored in the voids. The amount of water present in the voids is dependent upon porosity of the zone.

The zone of rock fracture can be further sub-divided into two zones, one is the *zone of saturation* and the other is the *zone of aeration.* Zone of saturation is below the water table and the zone of aeration above the water table. In the zone of saturation, water exists within the interstices or pores and is known as *ground water.* This is the most important zone for a ground water hydraulic Engineer, because he has to tap out this water. Water in this zone is under hydrostatic pressure. The upper level of zone of saturation is called Water Table. The space above the Water Table and below the surface of the earth is known as the zone of aeration. In this zone, water exists by molecular attraction. The gravity water moves through this zone. The water in this zone is not at hydrostatic pressure. The thickness of this zone varies from almost zero in marshy and low lying areas to about 300 m or so in arid regions.

The zone of aeration can be further divided intro three classes depending upon number of voids present. The capillary fringe is the belt overlying the zone of saturation and it does contain some pore water and is thus a continuation to the zone of saturation. The upper depth of the soil at the surface which is penetrated by the roots of vegetation is known as *soil zone*. The remaining thickness lying above capillary fringe and below the soil zone is known as intermediate zone. Intermediate zone is also called pellicular zone. See Fig. 7.1.

7.3 DRAINAGE OF GROUND WATER

This term is used for extracting water from below the water table, with the help of wells, springs, infiltration galleries etc. The ground water is drained either under a natural process like springs, or it may be drained artificially by constructing wells etc. The water so drained may be used to fulfil the domestic, municipal, industrial, and irrigation requirements. The ground water is a very large and important National Resource and has been heavily exploited in our country for different uses. Although this source has been heavily tapped, still research in this direction has been nominal. There is need of more research work in this field so that a fuller and wiser use of this resource may be made in future.

7.4 SOME IMPORTANT TERMS

Following are some important terms which are frequently used in the exploitation of ground water reservoir.

1. Aquifer. The permeable formations which permit appreciable water to move through them under ordinary field conditions are known as *aquifers*. These are geological formations made of gravel or sand, which possess lot of ground water.

2. Aquicludes. These are such formations which contain water in them but are not capable of transmitting. Clay formations are the examples of such formations.

3. Aquifuge. These are such formations which neither contain any water nor allow any transmission of water through them. Rock formations are the examples of such formations.

4. Porosity. Porosity is defined as the percentage of the voids present in a given volume of aggregate. Mathematically it can be expressed as

$$\text{Porosity} = \frac{\text{Total volume of voids}}{\text{Total volume of the soil}}$$

$$= \frac{V_v}{V} \times 100$$

The porosity of rock and unconsolidated material may vary considerably. Generally porosity does not exceed 40% except in very poorly compacted materials. For uniform loose sand and dense sand it is 45% and 35% respectively. For clays, it may vary from 20% to as much as 75%. Porosity as worked out for soils having exactly spherical particles is 47.6% and for soils having rhombohedral packing of soil particles as 26%. See Fig. 7.2.

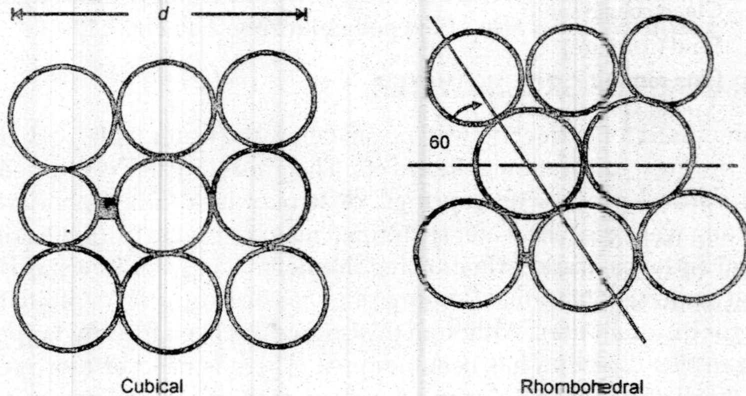

Fig. 7.2. *Voids is cubical and Rhombohedral grain mass.*

5. **Specific Yield.** The amount of ground water that can be extracted by gravity drainage from a saturated water bearing material is known as the yield. When it is expressed as ratio of the volume of the total material drained, then it is known as *specific yield.*

$$\therefore \quad \text{Specific yield} = \frac{\text{Volume of water obtained by gravity drainage}}{\text{Total volume of the material drained}} \times 100$$

6. **Specific Retention.** The quantity of water retained by the material against the pull of the gravity is termed as *specific retention* or *field capacity*.

Specific yield or field capacity

$$= \frac{\text{Volume of water held against gravity drainage}}{\text{Total volume of the material drained}} \times 100$$

It is evident that the sum of the specific yield and specific retention is equal to the porosity of the material.

7. **Water Table.** The upper surface of the zone of saturation lying under the ground level is known as *Water Table.*

8. **Permeability of the Soil.** The permeability of the soil is measured by the coefficient of permeability usually denoted by k. It is defined as the rate of flow of water through unit cross-sectional area under unit hydraulic gradient and a temp. of 60° F. Coefficient of permeability of soil is usually expressed in cm/sec. Values of permeability coefficients for some usually found soils are given as follows :

Type of soil	*Coefficient of permeability in cm/sec*
Clean gravel	One and more
Clean coarse sand	1.0 – 0.01
Sand (mixture)	0.01 – 0.005
Fine sand	0.05 – 0.001
Silty sand	0.002 – 0.0001
Silt	0.0005 – 0.00001
Clay	0.000001 and smaller

9. Transmissibility of Soil. Transmissibility of the soil is measured by the coefficient of transmissibility, which is denoted usually by T. Coefficient of transmissibility is defined as the rate of flow of water through a vertical strip of the aquifer of unit width and full depth under a unit hydraulic gradient and at a temp of 60° F.

Actually permeability and transmissibility both represent same thing so far as their physical significance is concerned. There is only a mathematical difference between the two. Both these terms represent the capability of a formation to pass water through it. The capability of the entire soil of full width and depth is represented by permeability while that of the soil of full depth and unit width is known as transmissibility (see Fig. 7.3).

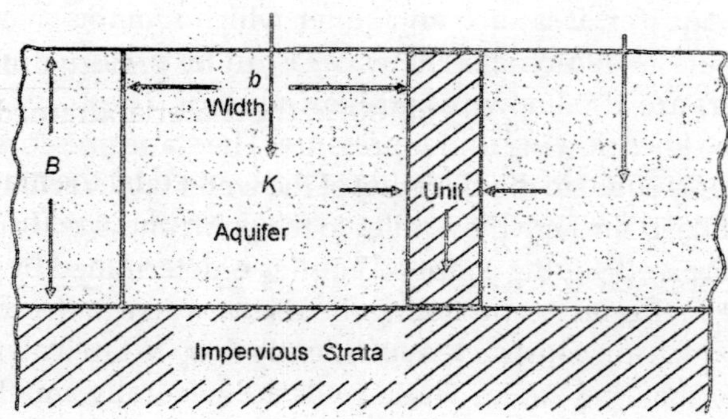

Fig. 7.3. *Permeability and transmissibility of water.*

Coefficient of permeability k and coefficient of transmissibility (T) can be related by following equation.

$$T = Bk$$

where B = Thickness of the aquifer. (Fig. 7.3)

7.5 TYPES OF AQUIFERS

Aquifers are mainly of two types.

1. Unconfined aquifer.
2. Confined aquifer.

1. Unconfined Aquifer. The topmost water bearing strata having no confining impermeable over-burden is known as *unconfined aquifer*. This aquifer is also known as *non-artesian aquifer* or *water table aquifer*. Water table in such an aquifer varies in undulating form depending upon the storage of water within it. The gravity wells are constructed to tap water from unconfined aquifer only. The water rise in such wells is equal to water rise in a piezometer connected to the water table.

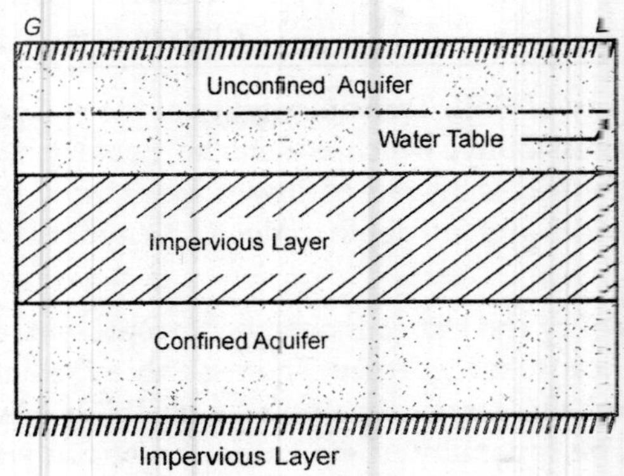

Fig. 7.4. *Confined and unconfined aquifers.*

2. Confined Aquifer. It is such an aquifer which remains overladen by an impermeable strata or aquiclude. The water is under pressure in this aquifer. Confined aquifers are also known as artesian aquifers. They can be considered analogous to pipe lines flowing under pressure. The static pressure at a point in a confined aquifer is equal to the elevation of the water table recharging the area minus the loss of head through the aquifer to the point under consideration. The rise of water in the well, taping confined aquifer is determined by connecting a piezometer tube to this layer and noting the rise of water in the tube. Artesian wells are those which are constructed sufficiently deep to tap water from 2nd or even lower water bearing strata. These wells are generally small in size with diameters less than 30 cm

When pressure of water in a certain water bearing formation is so high that water rises above the ground level automatically such a well is known as *flowing well*. If, however, the water level in such a well remain below the ground level, but is above the local water table it is known as the *artesian well*.

The aquifers can be classified as confined or unconfined depending upon the absence or presence of water table within them. Confined aquifers do not have nay Water Table as water in them is always under pressure. Unconfined aquifers are such formations which have equal pressure at all the points on its surface and pressure is equal to hydrostatic pressure. Therefore in an unconfined well, water will rise up to the water table level, but in confined well water will rise up to the piezometric head. The line joining the various pierometric heads in various tightly cased wells tapping the aquifer is known as piezometric surface.

Perched aquifer. It is a special condition of aquifer which may exist in unconfined aquifer. If an impervious stratum of small area extent lying within zone of saturation is found to support a body of saturated soil above it, then this body of

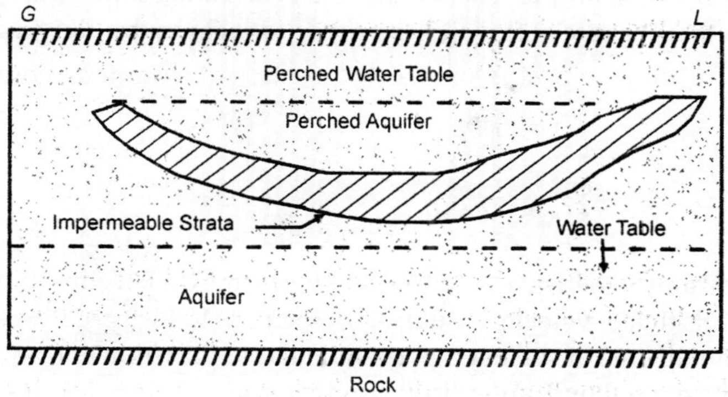

Fig. 7.5. *Perched water table.*

saturated soil is known as *perched aquifer.* The top surface of water held in the perched aquifer is known as *perched water table.* See Fig. 7.5.

7.6 COEFFICIENT OF STORAGE

Water discharged from an aquifer or recharged into an aquifer represents the change in its storage volume. It is very easy to determine this change for unconfined aquifers. Rise or fall of water table in given time is determined experimentally. Multiplying this by average specific capacity during the time, the change of the storage volume can be found out.

In the case of confined aquifers, change in pressure produces only small change in storage volume. The hydrostatic pressure within an aquifer partially supports the over burden and partially the solid structure of the aquifer. When hydrostatic pressure is reduced by pumping water from an quifer, the load on it is increased and aquifer gets consequently compressed and thus forces some water out of it. In addition, lowering of the pressure causes expansion and subsequent release of water. The water yielding capacity of an artesian acquifer can be expressed by its storage coefficient, generally denoted by A.

The storage coefficient (A) for a confined aquifer is equal to the volume of the water released from the aquifer of unit cross-sectional area and full height, when piezometric surface depresses by unity. In general, storage co-efficient is defined as the volume of water than an aquifer releases or stores per unit surface area of the aquifer per unit change in the head normal to that surface.

7.7 WELL HYDRAULICS

The velocity of flow and consequent discharge through a soil depends upon various factors, such as the type of soil, the arrangement of the grains, viscosity of water etc. The various physical factors decide as to whether the flow will be laminar or turbulent. In hydraulics, the flow in pipes is generally turbulent but in soils the flow is usually laminar. We know that in laminar range of flow the velocity (V) of flow is proportional to hydraulic gradient (I). This fact was confirmed by Mr. Darcy and hence later on it was called Darcy's law of flow. This law states that the rate of flow (Q) is proportional to hydraulic gradient.

$$Q \propto I$$

$$= KIA$$

$$\frac{Q}{A} = V = KI$$

where K = Darcy's coefficient of permeability, I being dimensionless the dimension or unit of K is that of velocity (usually cm/sec.) V = Velocity of flow and A = Total cross-sectional area of soil mass perpendicular to direction of flow. Hence these days whole of well hydraulics is dependent upon Darcy's law of flow through soils. This law is applicable for Laminar flows only.

7.8 STEADY RADIAL FLOW TO A WELL

The analysis of flow towards well through soil was first studied by Mr. Dupuit in 1863. The method of analysis as adopted by Mr. Dupuit was later modified by Mr. Thiem in 1906. Flow or discharge formulae, for both the analysis, for confined and unconfined formations, have been derived as follows.

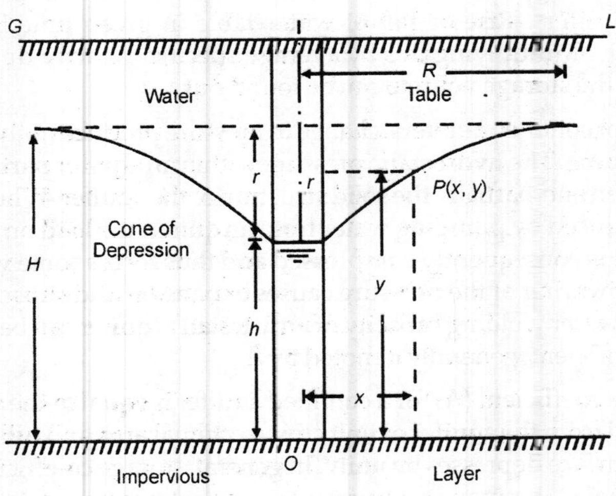

Fig. 7.6. *Well hydraulics is unconfined aquifer.*

Dupuit's theory. Level of water in the well which is not flowing, is same as that of water table. When water is pumped from the well, its level is depressed and water from surrounding soil starts entering the well. As soon as water starts entering the well, a depression in water table surrounding the well is created.

This depression is called the *draw down curve* or *the cone of depression*. Draw down at any point away from well is the vertical distance by which a water table is lowered from the normal water table.

Discharge from a well in unconfined aquifer.

Let
Q = Discharge from the well.
R = Influence radius
r = Radius of well
H = Depth of water table above the impervious layer
s = Draw down at the centre of the well
h = Depth of water in the well after pumping above impervious layer. (Fig. 7.6)

Consider centre of the well at the bottom as the origin. Let P be a point on the draw down curve and xy its co-ordinates. From Darcy's law

$$Q = kA_x i_x$$

where i_n = Hydraulic gradient at point P which is also represented by $\frac{dy}{dx}$

A_x = Area of saturated part of the aquifer at point P
$= 2\pi x \times y = 2\pi xy$

Substituting these values of A_x and I_x in Darcy's formula we get

$$Q = k2\pi xy\frac{dy}{dx}$$

$$Q\frac{dy}{dx} = 2\pi ky\ dy$$

Integrating between limits R, r for x, and H, h for y we get,

$$\int_r^R Q\frac{dy}{x} = 2\pi k \int_h^H y\,dy$$

$$Q[\log^e x]_r^R = 2\pi k\left[\frac{y^2}{2}\right]_h^H$$

From which $$Q = \frac{\pi k(H^2 - h^2)}{\log_e \frac{R}{r}} \qquad (1)$$

$$= \frac{1.36k(H^2 - h^2)}{\log_{10} \frac{R}{r}} \quad (2)$$

The value of R is selected by experience. The values of R may be taken varying from 100 to 300 m. Selection of R is done on approximate basis only, as discharge does not change by appreciable amount if R is changed within quite wide ranges. Value of R can also be estimated by Sichardt expression given below :

$$R = 3000 s \sqrt{k} \quad (3)$$

where R and s are in metres and k is in m/sec.

If draw down observations are taken from two observation wells located at radial distances of r_1 and r_2 and if the depths of water in the wells are respectively h_1 and h_2, then expression (1) given above can be expressed as follows :

$$Q = \frac{\pi k(h_2^2 - h_1^2)}{\log_e \frac{r_1}{r_2}}$$

$$= \frac{1.36\, k(h_2^2 - h_1^2)}{\log_{10} \frac{r_2}{r_1}} \quad 3(a)$$

Draw down $s = H - h.$

$$H = s + h$$

or $H + h = s + 2h$

Write again Eq. (1)

$$Q = \frac{\pi k(H^2 - h^2)}{\log_e \frac{R}{r}}$$

$$= \frac{\pi k(H - h)(H + h)}{\log_e \frac{R}{r}}$$

$$= \frac{\pi k s(s + 2h)}{\log_e \frac{R}{r}}$$

(Since $H - h = s$, $H + h = s + 2h$)

If $h = L$ the length of the strainer then

$$Q = \frac{\pi k s(s+2h)}{\log_e \frac{R}{r}} = \frac{\pi k s(s+2L)}{\log_e \frac{R}{r}}$$

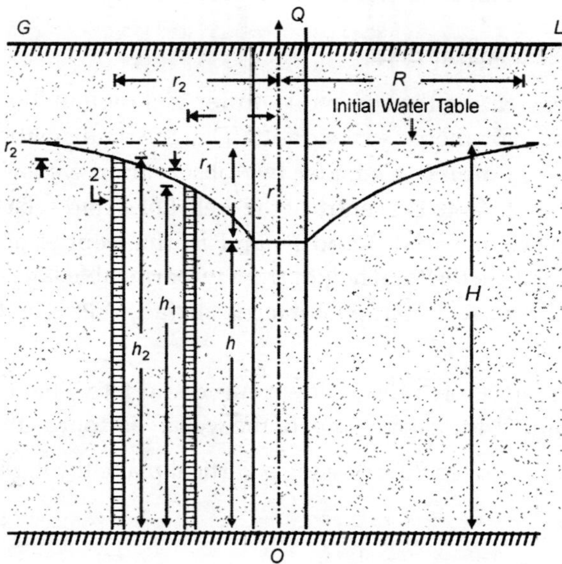

Fig. 7.7. *Unconfined aquifer well.*

$$= \frac{k 2\pi s\left(L + \frac{s}{2}\right)}{\log_e \frac{R}{r}}$$

$$= \frac{2.72\, ks\left(L + \frac{s}{2}\right)}{\log_{10} \frac{R}{r}} \qquad (4)$$

7.9 ASSUMPTIONS OF DUPUIT'S THEORY

Following are the some assumptions of Dupuit's Theory:

1. Aquifer is homogeneous, isotropic and of infinite aerial extent.
2. The velocity of flow is r oportional to tangent of hydraulic gradient and not sine.

3. The flow is horizontal and uniform throughout the vertical section.
4. Entire thickness of aquifer is contributing water to the well.
5. Coefficient of transmissibility remains constant at all places and all the time.
6. Flow is laminar and Darcy's law is applicable.
7. Ground water conditions remain constant all the time.

7.10 CONFINED AQUIFERS

A well fully penetrating the confined aquifer is shown in Fig. 7.8. Let *AB* line represent initial piezometric surface and *CD* in the final position of water level in the well after pumping. Consider two observation wells 1 and 2 at radial distances of r_1 and r_2 from the centre of the well. Let s_1 and s_2 be the draw downs at observation wells 1 and 2 and h_1 and h_2 the depths of water in them respectively. Let *H* be the total height from impermeable layer lying below the confined aquifer up to initial piezometric surface. Let *b* be the thickness of confined aquifer.

Consider a point *P* on a draw down curve and let *xy* be its coordinates by considering point *O*, the centre point of the well at bottom as the origin.

Apply Darcy's law for flow passing a vertical plane through point *P*.

$$Q = kA_x i_x$$

$$A_x = 2\pi xb$$

= Cross-sectional area of flow measured alround the well at point *P*

$$i_x = \frac{dy}{dx} = \text{Hydraulic gradient at } P$$

$$\therefore \quad Q = k \times 2\pi xb \frac{dy}{dx}$$

$$Q\frac{dy}{dx} = 2\pi kb\, dy.$$

Integrating between the limits [*R*, *r*] for *x* and *H*, *h* for *y* we get

$$Q\int_r^R \frac{dx}{x} = 2\pi kb \int_h^H dy$$

$$Q[\log_e x]_r^R = 2\pi kb[y]_h^H$$

By further simplifying

$$Q = \frac{2\pi kb(H-h)}{\log_e \frac{R}{r}} \quad (5)$$

or

$$Q = \frac{2.72\, bk(H-h)}{\log_{10} \frac{R}{r}} \quad (6)$$

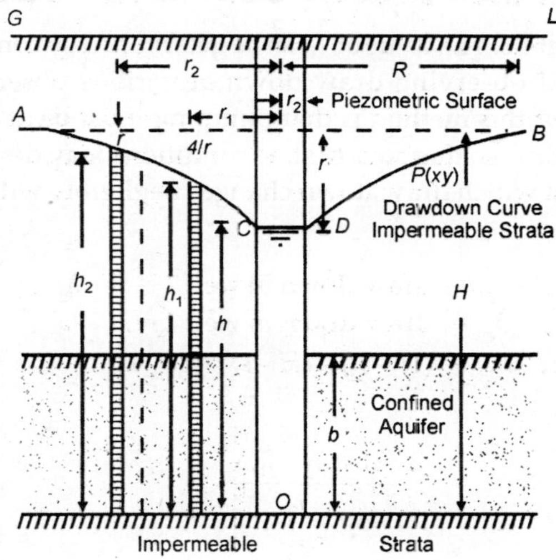

Fig. 7.8. *Confined aquifer well.*

But $H - h = s$

$$\therefore \quad Q = \frac{2\pi bks}{\log_e \frac{R}{r}} = \frac{2.72\, bks}{\log_{10} \frac{R}{r}} \quad (7)$$

While considering the two observation wells 1 and 2 at radial distances of r_1 and r_2 the discharge Eq. (6) takes the shape as follows :

$$Q = \frac{2.72\; bk(h_2 - h_1)}{\log_{10} \frac{r_2}{r_1}} \quad (8)$$

If h_x is the depth at any radial distance r_x the discharge Eq. (6) can be written as

$$Q = \frac{2\pi kb(h_x - h)}{\log_e \frac{r_x}{r}} \quad (9)$$

Equating Eqs (5) and (9), we get

$$h_x - h = (H-h)\frac{\log_e \frac{r_x}{r}}{\log_e \frac{R}{r}} \qquad (10)$$

Equation (10) shows that head varies linearly with the logarithm of the distance regardless of the rate of discharge.

7.11 DETERMINATION OF COEFFICIENT OF TRANSMISSIBILITY (*T*)

(a) Confined Aquifers. The value of *T* can be determined by conducting prolonged pumping test and observing draw down at various observation wells. But essential feature of this method is that pumping must be at a uniform rate for sufficiently long time so that steady state conditions may develop. Steady state condition is one at which draw down changes negligibly with time.

Refer Fig. 7.8

$$s_1 = \text{draw down in well 1} = H - h_1$$

$$s_2 = \text{draw down in well 2} = H - h_2$$

$$h_2 - h_1 = (H - s_2) - (H - s_1) = s_1 - s_2$$

Use Eq. (8)

$$Q = \frac{2.72\,bk(h_2 - h_1)}{\log_{10}\frac{r_2}{r_1}}$$

$$= \frac{2.72\,T(h_2 - h_1)}{\log_{10}\frac{r_2}{r_1}} \qquad (\because T = kb)$$

$$= \frac{2.72\,T(s_1 - s_2)}{\log_{10}\frac{r_2}{r_1}}$$

$$\therefore \quad T = \frac{Q}{2.72(s_1 - s_2)} \times \log_{10}\frac{r_2}{r_1} \qquad (11)$$

If we choose $r_2 = 10\,r_1$, then $\log_{10}\frac{r_2}{r_1} = 1$

and $$T = \frac{Q}{2.72(s_1 - s_2)}$$

$$= \frac{Q}{2.72\,\Delta_s} \tag{12}$$

where Δ_s = difference in draw downs at the two wells so selected that $r_2 = 10\, r_1$

In this method, draw down observations $s_1, s_2, s_3 \ldots\ldots s_x$ are taken in observation wells located at $r_1, r_2, r_3 \ldots\ldots r_x$ distance from the centre of the main well. Now a graph is plotted by taking draw downs along ordinate and $\log_{10} r_x$ as abscissa. The plot will be a straight line as shown in Fig. 7.9. From this plot the value of Δ_s can be obtained for one log cycle of distance. After substituting this value in Eq. (12) value of T can be determined.

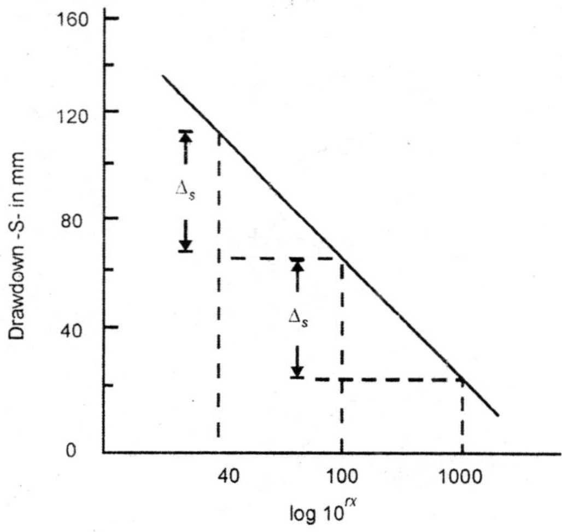

Fig. 7.9

(b) Unconfined Aquifers. Refer Fig. 7.7

$$h_1 = H - s_1 \text{ and } h_2 = H - s_2$$

$$\therefore \quad h_2^2 - h_1^2 = (H - s_2)^2 - (H - s_1)^2$$

$$= H^2 - 2Hs_2 + s_2^2 - H_2 + 2Hs_2 - s_1^2$$

$$= (2Hs_1 - s_1^2) - (2Hs_2 - s_2^2)$$

$$= 2H\left\{\left(s_1 - \frac{s_1^2}{2H}\right) - \left(s_2 - \frac{s_2^2}{2H}\right)\right\}$$

But $\quad s_1 - \dfrac{s_1^2}{2H} = s_1'$

and $\quad s_2 - \dfrac{s_2^2}{2H} = s_2'$

$$(h_2^2 - h_1^2) = 2H\,(s_1' - s_2') \tag{13}$$

s_1' and s_2' are the modified values of draw down.

Substituting these values in Eq. 3 (a)

$$Q = \frac{1.36\,k(h_2^2 - h_1^2)}{\log_{10}\dfrac{r_2}{r_1}}$$

$$= \frac{1.36\,k\;2h(s_1' - s_2')}{\log_{10}\dfrac{r_2}{r_1}}, \qquad (\because\; h_2^2 - h_1^2 = 2h(s_1' - s_2'))$$

$$= \frac{2.72\,kH(s_1' - s_2')}{\log_{10}\dfrac{r_2}{r_1}}$$

$$= \frac{2.72\;T(s_1' - s_2')}{\log_{10}\dfrac{r_2}{r_1}} \qquad (\because T = kH)$$

$$T = \frac{Q}{2.72}\;\frac{\log_{10}\dfrac{r_2}{r_1}}{(s_1' - s_2')}$$

$$= \frac{Q}{2.72}\;\frac{\log_{10}\dfrac{r_2}{r_1}}{\Delta s'}$$

If $\quad r_2 = 10\,r_1$ then $\log_{10}\dfrac{r_2}{r_1} = 1$

$$\therefore \qquad T = \frac{Q}{2.72\;\Delta s'} \tag{14}$$

If we compare Eqs. (12) and (14) we find them identical except that $\Delta s'$ in (14) is based on modified values of draw down whereas it is based on actual values of draw down in case of Eq. (12).

In this case also a graph is drawn similar to one drawn in case of confined aquifer. But here values of draw downs taken along ordinate are modified values of draw downs. See Fig. 7.10.

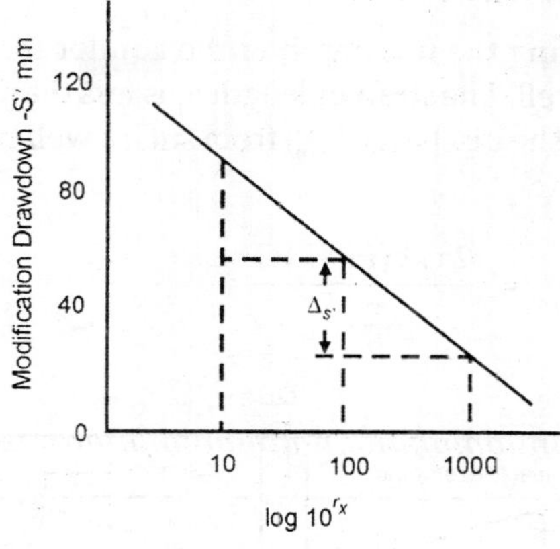

Fig. 7.10

7.12 FULLY PENETRATING ARTESIAN GRAVITY WELL

Due to high rate of pumping, sometimes, level of water in a confined aquifer depresses below the top of the confined aquifer as shown in Fig. 7.11. In such a

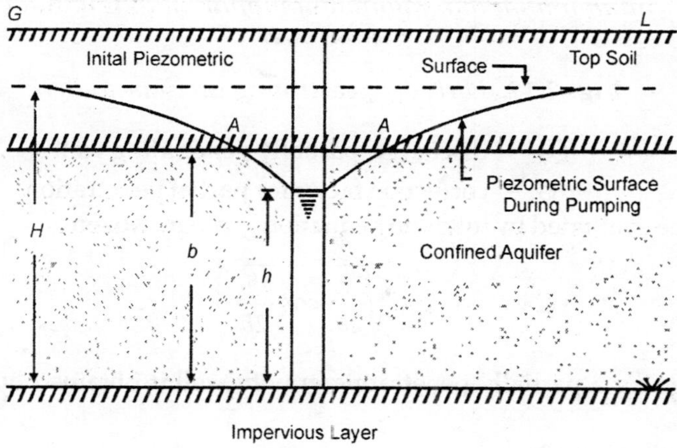

Fig. 7.11. *Fully penetrating artesian gravity well.*

case the flow pattern near the well, upto the point (A) where draw down curve cuts the top of confined aquifer behaves as a gravity well. Beyond point A the flow from the well is artesian flow. This type of well is known as a *combined artesian-gravity well.* The flow from such a well can be found out using the following formula given by Muskat

$$Q = \frac{\pi k(2bH - b^2 - h^2)}{\log_e \frac{R}{r}} \qquad (15)$$

7.13 PARTIALLY PENETRATING ARTESIAN WELL

If well is not penetrating the full depth of the aquifer it is known as partially penetrating artesian well. The strainer length b_1 is less than the aquifer thickness b_2, shown in Fig. 7.12. The discharge (Q_p) from such a well can be computed from the following equation :

$$Q_p = \frac{2\pi kb(H-h)G}{\log_e \frac{R}{r}} \quad (16)$$

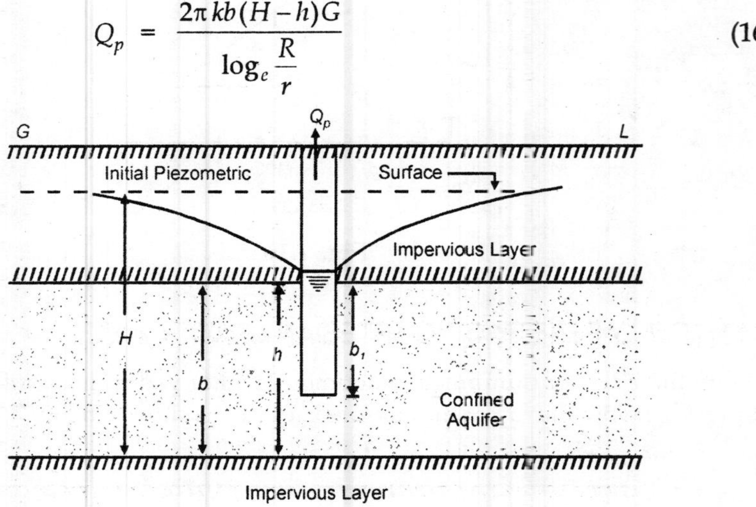

Fig. 7.12. *Partially penetrating artesian well.*

Q_p = discharge partially penetrating well
G = correction factor for partial penetration

which can be evaluated by following equation given by Kozeny.

$$G = \frac{b_1}{b}\left(1+7\sqrt{\frac{r}{2b_1}}\cos\frac{\pi b_1}{2b}\right) \quad (17)$$

Discharge for a partially penetrating gravity well is obtained by following formula :

$$Q_p = \frac{\pi k(d_1^2 - h_w^2)}{\log_e \frac{R}{r}} \times G \quad (18)$$

$$G = \left(1+7\sqrt{\frac{r}{2d_1}}\cos\frac{\pi d_1}{2d}\right) \quad (19)$$

where d_1 = Actual penetration depth below the water table.
d = Total depth aquifer below the water table
h_w = Depth of water in the well.

7.14 SPHERICAL FLOW IN A WELL

Such a condition of (low in the well happens when well just penetrates the top surface of the confined aquifer as shown in Fig. 7.13. Since flow towards the well is purely spherical, formula 16 does not apply. The discharge from such a well can be found out from equation

$$Q_s = 2\pi\, kr\,(H - h) \tag{20}$$

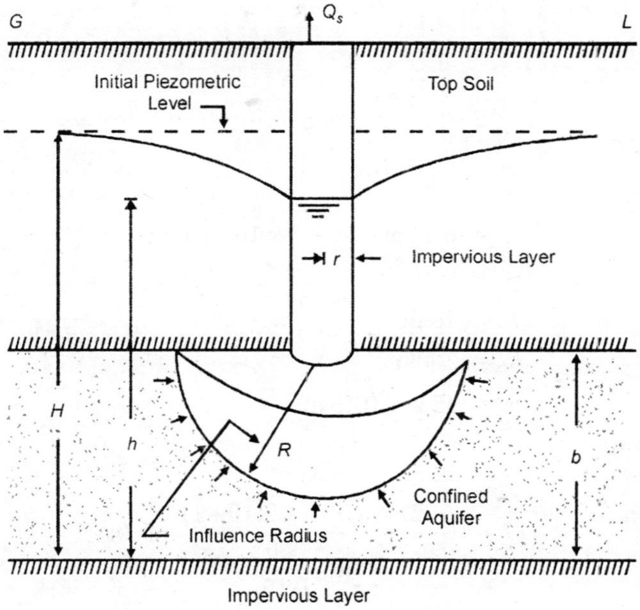

Fig. 7.13. *Spherical flow in a well.*

For the case of simple radial flow in a fully penetrating well, the discharge Q is given by

$$Q = \frac{2\pi\, kb(H-h)}{\log_e \dfrac{R}{r}}$$

$$\therefore \quad \frac{Q_s}{Q} = \frac{2\pi\, kr(H-h)}{2\pi\, kb(H-h)} \times \log_e \frac{R}{r}$$

$$= \frac{r \log_e \dfrac{R}{r}}{b} = 2.3\frac{r}{b} \, \log_{10} \frac{R}{r} \tag{21}$$

When we use Eq. (21) in actual practical problems we come to know that spherical flow in a well is very much less than the radial flow.

Example 7.1. *A tube well is installed penetrating full thickness of an unconfined water logged layer. Find the discharge of the tube well from following data:*

Diameter of tube well	= 30 m
Drawn down	= 2 m
Effective length of strainer pipe	= 10 m
Coefficient of permeability	= 0.10 cm/sec
Radius at zero draw down	= 300 m

Solution

$$Q = \frac{2.72\,kS\left(L+\dfrac{S}{2}\right)}{\log_{10}\dfrac{R}{r}}$$

$$k = 0.1 \text{ cm/sec} = 1\times10^{-3} \text{ cm/sec}$$

$$s = 2 \text{ m}$$

$$L = 10 \text{ m}$$

$$R = 300 \text{ m}$$

$$r = 15 \text{ cm}, 0.15 \text{ m}.$$

Putting values

$$Q = \frac{2.72\times1\times10^{-3}\times2(10+1)}{\log_{10}\dfrac{300}{0.15}} \text{ m}^2/\text{sec}$$

$$= 18.1\times10^{-3} \text{ m}^3/\text{sec}$$

$$= \mathbf{18.1 \text{ lit/sec.}}$$

Example 7.2. *Design a tube well in a confined aquifer for following data :*

Required discharge	Q = 160 lit/sec
Thickness of confined layer	= 30 m
Radius of influence	= 300 m
Permeability coefficient	= 120 m/day
Draw down	= 5 m.

Solution

$$Q = 160 \text{ lit/sec} = 0.16 \text{ cumecs}$$

$$b = 30 \text{ m}$$

$$s = 5 \text{ m}$$

$$k = 120 \text{ m/day} = \frac{120}{60\times60\times24} = \frac{2}{60\times24} \text{ m/sec}$$

$$R = 300 \text{ m}, \quad r = ?$$

$$Q = \frac{2.72\, bk\, S}{\log_{10} \frac{R}{r}}$$

Substituting the values

$$0.16 = \frac{2.72 \times 30 \times 5 \times 2}{\log_{10} \frac{R}{r} \times 60 \times 24}$$

$$\therefore \quad \log_{10} \frac{R}{r} = \frac{2.72 \times 30 \times 5 \times 2}{60 \times 24 \times 0.16} = 3.54$$

$$\frac{R}{r} = \text{Antilog } 3.54 = 3467$$

$$r = \frac{300}{3467} = 0.087 \text{ m} = 8.60 \text{ cm}$$

Diameter of well $= \mathbf{2 \times 8.60 = 17.20 \text{ cm.}}$

7.15 INTERFERENCE AMONG WELLS

When two wells are constructed so close to each other that their individual draw down curves intersect within their radius of influence, their individual discharges are affected. Figure 7.14 shows two similar wells located at distance B apart and B is less than the influence radius R.

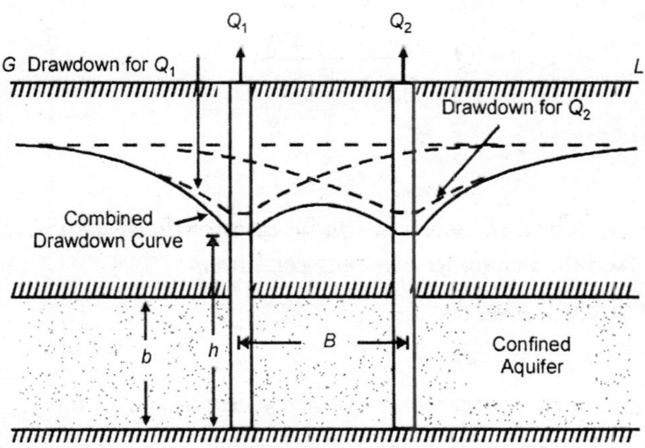

Fig. 7.14

Discharges of both the wells for confined and unconfined aquifers are computed from following formulae.

1. Confined Aquifer

$$Q_1 = Q_2 = \frac{2\pi\ kb(H-h)}{\log_e \frac{R^2}{rB}}$$

If there were only one well, then its discharge under similar conditions of draw down would have been as follow :

$$Q = \frac{2\pi\ kb(H-h)}{\log_e \frac{R}{r}}$$

Since $R > B$ $\therefore$ $\frac{R^2}{rB}$ will definitely be more than $\frac{R}{r}$.

Hence $Q > Q_1$.

Thus discharge of each well has decreased due to interference.

If three wells are located at the vertices of an equilateral triangle of sides B, when B is less than R, then all the wells have same characteristics, and their discharges will be same as given by following formula :

$$Q_1 = Q_2 = Q_3 = \frac{2\pi\ kb(H-h)}{\log_e \frac{R^3}{rB^2}}$$

2. Unconfined Aquifer

$$Q_1 = Q_2 = \frac{\pi k(H^2 - h^2)}{\log_e \left(\frac{R}{r} \times B\right)}.$$

Example 7.3. *A tube well penetrates fully an unconfined aquifer. Determine the discharge from the tube well under following conditions :*

Diameter of the well = 40 cm

Draw down = 4 m.

Effective length of the strainer under the above draw down = 10 m

Coefficient of permeability of aquifer = 0.05 cm/sec.

Radius of influence of circle = 300 m.

Solution

$$Q = \frac{2.7\,kS\left(L+\dfrac{s}{2}\right)}{\log_{10}\dfrac{R}{r}}$$

$$k = 0.05 \text{ cm/sec} = \frac{0.05}{100} \text{ m/sec} = 5 \times 10^{-4} \text{ m/sec}$$

$$s = \text{draw down} = 4 \text{ m}$$

$$L = 10 \text{ m}, R = 300 \text{ m}, r = 0.2 \text{ m}$$

$$Q = \frac{2.72 \times 5 \times 10^{-4} \times 4(10+2)}{\log_e \dfrac{300}{0.20}} \text{ m}^3/\text{sec}$$

$$= \frac{2.72 \times 5 \times 4 \times 12}{10000 \times 3.1760}$$

$$= 0.02055 \text{ m}^3/\text{sec}$$

$$= \mathbf{20.55 \text{ lit/sec.}}$$

Example 7.4. *A well penetrates fully an 8 m thick water bearing stratum of medium sand having coefficient of permeability of 0.006 m/sec. The well radius is 15 cm and to be worked under a draw down of 3 m at the well face. Calculate the discharge from the well. What will be % increase in the discharge if the radius of the well is doubled ? Take R = 300 m in each case.*

Solution. Use equation

$$Q = \frac{2.72\,bk\,S}{\log_{10}\dfrac{R}{r}}$$

$$b = 8 \text{ m}, k = 0.006 \text{ m/sec}, S = 3 \text{ m}, R = 300 \text{ m}$$

$$Q = \frac{2.72 \times 8 \times 0.006 \times 3}{\log_{10}\dfrac{300}{r}} = \frac{0.3917}{\log\dfrac{300}{r}}$$

I. When $r = 15$ cm = 0.15 m.

$$Q = \frac{0.3917}{\log_{10}\dfrac{300}{0.15}} = \frac{0.3917}{3.301} = 0.119 \text{ cumecs}$$

II. When $r = 30$ cm = 0.30 m

$$Q = \frac{0.3917}{\log_{10}\frac{300}{0.30}} = \frac{0.3917}{3.000} = 0.1305 \text{ cumecs}$$

$$\% \text{ increase} = \frac{0.1305 - 0.1190}{0.119} = 9.7\%$$

Hence we see that size of the well does not have much bearing on the discharge.

Example 7.5. *While conducting a prolonged pumping test following observations were noted:*

Well diameter = *30 cm*

Discharge from the well = *300 m³/hr*

R.L. of water in the well before commencement of pumping = 140.5 m

R.L. of water at constant pumping = 130.5 m

R.L. of impervious layer = 110.5 m

R.L. of water in observations well = 138.5 m

Radial distance of observation well = 50 m

Determine :

I. *The field permeability of unconfined aquifer.*

II. *Error in field permeability if there is no observation well and radius of influence is assumed 300 m.*

III. *Actual radius of influence based on the observations of test well.*

Solution. I. Since aquifer is unconfined, use formula

$$Q = \frac{1.36\, k(h_2^2 - h_1^2)}{\log_{10}\frac{r_2}{r_1}}$$

Use suffix 2 for observation well and 1 for face of the well

$$r_1 = 15 \text{ cm} = 0.15 \text{ in}, \; r_2 = 50 \text{ m}$$

$$h_1 = 130.5 - 110.5 = 20 \text{ m}$$

$$h_2 = 138.5 - 110.5 = 28 \text{ m}$$

$$h_2 + h_1 = 28 + 20 = 48 \text{ m}$$

$$h_2 - h_1 = 28 - 20 = 8 \text{ m}$$

$$Q = 300 \text{ m}^3/\text{hr}$$

$$300 = \frac{1.36 \times k \times 48 \times 8}{\log_{10} \frac{0.50}{0.15}}$$

$$k = \frac{300 \times \log_{10} 333.3}{1.36 \times 48 \times 8} \text{m/hour}$$

$$= \frac{300 \times 2.5228}{1.36 \times 48 \times 8}$$

$$= 1.45 \text{ m/hr}$$

$$= 34.8 \text{ m/day.}$$

II. If k is found out on the basis of assumed radius of influence

$$Q = \frac{1.36\ k(H^2 - h^2)}{\log_{10} \frac{R}{r}}$$

$$\frac{R}{r} = \frac{300}{0.15} = 2000$$

$$H = 140.5 - 110.5 = 30 \text{ m}$$

$$h = 130.5 - 110.5 = 20 \text{ m}$$

$$H + h = 30 + 210 = 50$$

$$H - h = 30 - 20 = 10.$$

Substituting the values

$$300 = \frac{1.36 \times k \times 50 \times 10}{\log_{10} 2000}$$

$$k = \frac{300 \times \log_{10} 2000}{1.36 \times 50 \times 10}$$

$$= \frac{300 \times 3.3010}{1.36 \times 50 \times 10} = 1.456 \text{ m/hr}$$

$$= 35 \text{ m/day}$$

$$\% \text{ error} = \frac{(35 - 34.8)100}{34.8} = \mathbf{0.6\%.}$$

III. *Computation of actual radius of influence*

$$Q = \frac{1.36\,k(H^2 - h^2)}{\log_{10}\dfrac{R}{r}}$$

$$k = 1.45 \text{ m/hr.}$$

$$H + h = 50 \text{ m.}$$

$$H - h = 10 \text{ m.}$$

$$Q = 300 \text{ m}^3\text{/hr}$$

$$\therefore \quad 300 = \frac{1.36 \times 1.45 \times 50 \times 10}{\log_{10}\dfrac{R}{r}}$$

$$\log_{10}\frac{R}{r} = \frac{1.36 \times 1.45 \times 50 \times 10}{300} = 3.287$$

$$\frac{R}{r} = 1936$$

$$R = 1936 \times r = 1936 \times 0.15$$

$$= \mathbf{290.4 \text{ m.}}$$

Example 7.6. *There are two tube wells spaced at 50 m distance. Both the tube wells are of 20 cm diameter and penetrate a confined aquifer for its full depth of 10 m. If radius of in influence for each well is 200 m and coefficient of permeability 30 m/day, calculate:*

(i) The discharge if only one well is discharging under a depression head of 4 m.

(ii) The per cent decrease in the discharge when both the wells are discharging simultaneously under the depression head of 4 m.

Solution. (i) Only one well is discharging

$$Q = \frac{2.72 \times bks}{\log_{10}\dfrac{R}{r}}$$

$$b = 10 \text{ m}, \ k = 30 \text{ m/day}$$

$$\log_{10}\frac{R}{r} = \log_{10}\frac{200}{0.1} = 3.301$$

$$Q = \frac{2.72 \times 10 \times 30 \times 4}{3.301} = \text{m}^3\text{/day}$$

$= 990 \text{ m/day}$

$= \mathbf{41.25 \ m^3/hr.}$

(ii) When both the wells are working simultaneously

$$Q_1 = Q_2 = \frac{2\pi kb(H-h)}{\log_e \frac{R^2}{rB}}$$

$$= \frac{2.72\, kbs}{\log_{10} \frac{R^2}{rB}}$$

$$= \frac{2.72 \times 10 \times 30 \times 4}{\log_{10} \frac{200 \times 200}{0.1 \times 50}}$$

$$= \frac{2.72 \times 10 \times 30 \times 4}{3.9031}$$

$= 836.25 \text{ m}^4\text{/day.}$

$= \mathbf{34.84 \ m^3/hr}$

% decrease in the discharge

$$= \left(\frac{41.25 - 34.84}{41.25}\right) \times 100$$

$= \mathbf{15.54\%.}$

Example 7.7. *A tube well commands 32 hectares of land under cotton. When the well is discharging at constant rate, the water level in the well drops to 110 m. The level of high land to be irrigated is 122 m. The duty of cotton is 800 hectare/cumec on the field. Taking the pump efficiency as 75%, calculate the input H.P. of the pump.*

Solution. Discharge required $= \frac{32}{800} = 0.04$ cumec.

Weight of water to be lifted per second

$= 0.04 \times 1000 = 40$ kg

Minimum static lift of the pump

$= 122 - 110 = 12$ m

Consider 0.2 m head required for the water to flow to the field from the well, total lift against which pump has to work

$= 12 + 0.2 = 12.2$ m

Work done by pump in lifting water

$$= 12.2 \times 40 = 488 \text{ kg-m/sec}$$

$$1 \text{ H.P.} = 75 \text{ kg-m/sec}$$

Output H.P. of pump

$$= \frac{488}{75} = 6.5$$

$$\text{Input H.P. of pump} = \frac{6.5}{0.7} = \mathbf{9.3 \text{ H.P.}}$$

7.16 YIELD OF OPEN WELL

Open wells are large diameter dug wells. They only penetrate into the top unconfined aquifer. The hydraulics of a lined open well is different from that of a tube-well. The following are the two methods most commonly used for determining the yield of the open well.

1. Constant level pumping method.
2. Recuperation test method.

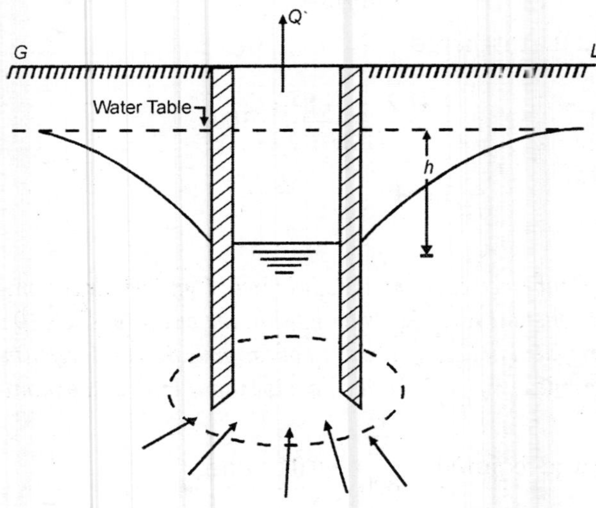

Fig. 7.15. *Flow in the open well.*

1. Constant Level Pumping Method. In this method a pump having suitable regulation arrangement is the most essential feature.

The pump is run and water level is depressed in the well by amount say h. This depression in level of water is known as depression head. After depressing the level of water to certain depression head, the speed of the pump is so adjusted that thereafter level of water in the well becomes stationary. At this stage whatever water enters the well is immediately pumped out. The rate at which amount of water is being pumped out at this stage is the rate at which water can be availed from this well. This rate of pumping when expressed as per hour, is known as the yield of the well per hour.

The formula for discharge in cumecs from an open well with impervious lining can be derived by Darcy's law as follows :

$$q = kiA = kA\frac{h}{L} = \frac{k}{L}Ah$$

$$= cAh \qquad (1)$$

$$\left(\therefore \frac{k}{L} = c \text{ constant}\right)$$

This formula can be directly found out as follows

$$Q = Av$$

$$= Ach \qquad (\because v \propto h) \qquad (1a)$$

In both these formulae

Q = Discharge in cumecs.

A = Area of flow in m^2 at the base of the well. It is taken $\frac{4}{2}$ of sectional area of the well at the base because of cavity formation.

v = Mean velocity of percolating water in the well in m/sec.

h = Depression head in metres.

C = A constant whose value depends upon the soil formation around the well. Its unit is in m/sec under unit head.

The value of A is increased by factor $\frac{4}{3}$ to make allowance for cavity formation at the base of the well.

From expression (1) above it is obvious that discharge in the well increases with increase in percolation head. However the percolation head cannot be increased beyond a certain critical value as otherwise the percolation velocity will increase to such an extent that soil particles start being disturbed and dislodged. The depression head causing the critical velocity of flow is known as the critical depression head. Open wells are generally worked at $\frac{1}{3}$ of the critical head and such a head is known as working head. Thus yield obtained corresponding to the critical depression head is known as *maximum* or *critical yield*. The yield corresponding to working head is called the *maximum safe yield*. The maximum safe yield of a well can be obtained by pumping test.

2. Recuperative Test Method. Though the constant level pumping test method is more accurate in assessing the safe yield of an open well, it has some practical difficulties. In this method pumping rate has to be so regulated and adjusted that constant level is maintained in the well, which is not very easy. Recuperation test method is the most practical method and hence mostly adopted. In this, water

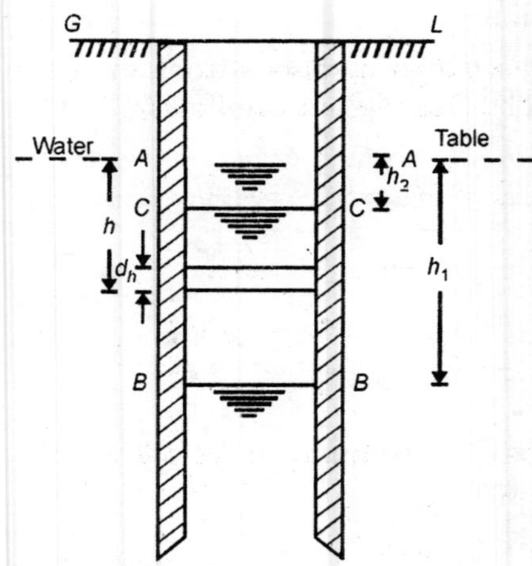

Fig. 7.16. *Recuperetive test in an open well.*

level is depressed to certain level by pumping and then pumping is stopped. The time taken by water to recuperate to the normal level is noted from stoppage of pump. After noting this data, the rate of discharge from a well can be calculated as follows:

Consider Fig. 7.16 an open well.

Let AA = Level of water before start of pumping.

BB = Level of water in the well when pumps are stopped.

h_1 = Depression head i.e. difference in levels of AA and BB planes.

CC = Level of water in the well T time after the stoppage of pumps.

h_2 = Available depression head after lapse of T time after the pumps were stopped.

h = Depression at any time t from the stoppage of pumps.

dh = Rise in water level in small time dt.

In time t, from stoppage of pumps, the water level recuperates by $(h_1 - h)$ metres.

In small fraction time dt, let head recuperates by a value dh metres.

Volume of water (dV) entered in the well during time dt is

$$A \times dh$$

i.e. $$dV = Adh \quad (1)$$

A = cross-sectional area of the well at its bottom. If Q is the rate of discharge at time t under depression head h, the volume of water entering the well in time dt is given by

$$dV = Qdt$$

But $\quad Q \propto h$ or $Q = kh$ (2)

$$dV = kh\,dt \quad (3)$$

k is a constant which is dependent upon the type of soil at the base of the well, through which water is entering the well.

From Eqs (1) and (3), we get

$$khd_t = -Adh \quad (4)$$

Negative sign shows that with increase in t, h decreases.

Integrating Eq. (4) between limits of 0 to T and h_1 to h_2 we get

$$khdt = -Adh$$

$$\frac{k}{A}\int_0^T dt = \frac{k}{A}\int dt$$

$$\frac{k}{A}[T] = [\log_e h]_{h_2}^{h_1}$$

$$\frac{k}{A} = \frac{1}{T}\log_e \frac{h_1}{h_2}$$

$$= \frac{2.303}{T}\log_{10}\frac{h_1}{h_2} \quad (5)$$

Thus knowing the values of h_1, h_2, T from a recuperative test, the value of $\frac{k}{A}$ is found out. $\frac{k}{A}$ is known as the *specific yield* or *specific capacity* of an open well in cubic metre/hour/m^2 of area through which water percolates under one metre depression head. After knowing $\frac{k}{A}$, the discharge Q from a well under a constant head H can be calculated as follows :

$$Q = kh \quad (6)$$

$$Q = \frac{k}{A}AH$$

$$= \frac{2.303}{T}\left(\log_{10}\frac{h_1}{h_2}\right)AH \text{ m}^3/\text{hr} \quad (7)$$

In this expression T is in hours. If it is converted into seconds, then Q will be in cumecs. In expression (7) if H is the maximum head, the corresponding Q will be maximum yield. If H is average depression head, Q will be average yield. If value of $\frac{k}{A}$ is not determined by recuperation test, the value $\frac{k}{A}$ for clay, fine sand and course sand, may be taken as 0.25, 0.50 and 1.00 respectively as per Mr Marrot.

Example 7.8. *In a recuperation test following data was collected :*

Diameter of the open well	$= 3\ m$
Initial head of percolation	$= 4\ m$
Final depression head	$= 2.5\ m$
Time of recuperation	$= 2\ hrs$

Determine the specific capacity of the well and yield under a head of 3 m.

Solution. Use formula $\frac{k}{A} = 2.303\,T \log_{10} \frac{h_1}{h_2}$

or

$$k = 2.303 \frac{A}{T} \log_{10} \frac{h_1}{h_2}$$

$$A = \frac{\pi}{4} d^2 = \frac{\pi}{4} \times (3)^2 = 2.25\pi \text{ m}^2$$

$$T = 2 \text{ hrs}$$

$$h_1 = 4 \text{ m}, \quad h_2 = 2.5 \text{ m}$$

$$k = 2.303 \times \frac{2.25\pi}{2} \log_{10} \frac{4}{2.5}$$

$$= 13.02 \text{ m}^3/\text{hrs/metre head.}$$

$$Q = kH$$

$$= 13.03 \times 3 = 39.06 \text{ m}^3/\text{hrs.}$$

7.17 EFFICIENCY OF A WELL

We know that discharge from a well is approximately proportional to the draw down. The discharge per unit draw down is called specific capacity of the well. This specific capacity will be different for different well designs. For finding the best draw down discharge conditions for a well, the well may be operated under varying draw down conditions and then a graph may be plotted between discharge and draw down as shown in Fig. 7.17. The curve obtained is a straight line upto a certain stage of draw down, beyond which the draw down increases stage of draw down, beyond which the draw down increases disproportionately to the yield. This places an optimum and efficient limit to the draw down which may be allowed to be created in a well. This is generally found to be 70% of the maximum draw down which can be created in a well.

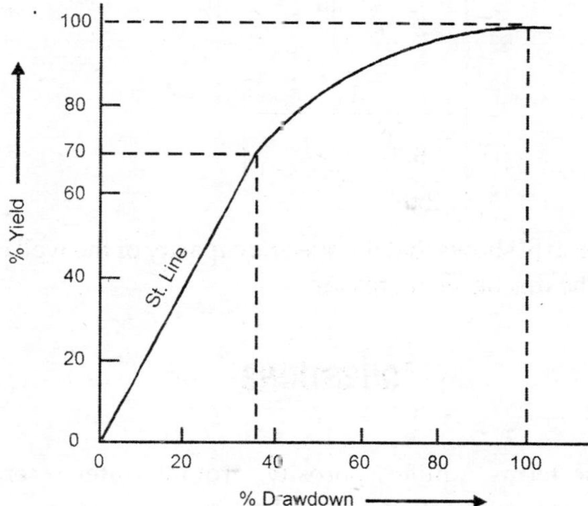

Fig. 7.17. *Efficiency of a well.*

Well Losses. When water is being pumped out of a well, the total draw down developed includes not only that of a logarithmic draw down curve at the well face, but also draw down caused by the flow of water through well screen and axial movement within the well. The draw down caused by later effect is known as *well loss*. The magnitude of this well loss may be taken equal to CQ^n as the flow in the vicinity of well face is **turbulent.**

Total draw down thus can be obtained by adding the two draw downs.

$$\text{Total draw down} = \frac{Q\log_e \frac{R}{r}}{2\pi bk} + CQ^n$$

$$CQ^n = \text{well loss}$$

$$\frac{Q\log_e \frac{R}{r}}{2\pi bk} = \text{Aquifer loss}$$

7.18 SPECIFIC CAPACITY OF AN ARTESIAN WELL

It is the yield of the well per unit draw down. It is considered as a measure of the effectiveness of the well.

$$\text{Specific capacity} = \frac{\text{Discharge of a well}}{\text{Drawn down}}$$

$$= \left[\frac{Q}{\dfrac{Q\log_e \dfrac{R}{r}}{2\pi bk} + CQ^n} \right]$$

$$= \left[\frac{1}{\dfrac{\log_e \dfrac{R}{r}}{2\pi bk} + CQ^{(n-1)}} \right]$$

This equation clearly shows that the specific capacity of the well is not constant but decreases as the discharge increases.

QUESTIONS

7.1 Explain the terms Aquifer, porosity, ground water reservoir, zone of saturation, zone of aeration, water table.

7.2 (a) Draw section of earth and show presence of water at different depths from the surface.

(b) What do you understand by the terms permeability and transmissibility of the soil ? What is the difference between the two terms ?

7.3 What are different types of aquifers ? Write down the characteristic of each in details.

7.4 Derive an expression for the discharge obtained from a tube well penetrating fully an unconfined aquifer.

7.5 Derive an expression for the discharge obtained from a tube well fully penetrating a confined aquifer.

7.6 Explain Dupuit's theory. What arc the assumptions of the theory ?

7.7 Derive expressions for determining the coefficients of transmissibility for both confined and unconfined aquifers.

7.8 Explain the methods of finding out the yield of an open well. Explain recuperation test method in details.

7.9 Write short notes on:

(i) Well losses.

(ii) Specific capacity.

(iii) Well efficiency.

(iv) Interference of wells when located very near to each other.

7.10 Derive expressions for discharge for following conditions:

(i) Fully penetrating artesian gravity well.

(ii) partially penetrating artesian wells.

(iii) Spherical flow in a well.

❑❑❑

8

Well Irrigation

8.1 WELL IRRIGATION

When water obtained from wells is used for irrigation purpose it is known as well irrigation. Well irrigation essentially requires construction of wells; hydrological characteristics of ground water have been discussed in Chapter 7. In this chapter following things in regard to wells will be discussed:

1. Classification of wells and their application.
2. Construction of wells.
3. Various types of strainers used in tube wells.
4. Methods of lifting water from wells.

8.2 CLASSIFICATION OF WELLS

The wells may be classified under two heads.

1. Open wells or dug well.
2. Tube wells.

1. Open Wells or Dug Wells. The ancient wells were all open wells. They are constructed by digging large diameter holes in the earth. Although there have been some improvement in methods of digging the wells, in some areas old conventional methods are still prevalent. These wells are initially used to get water for drinking purposes, but later on their use for irrigation purpose was also appreciated. Open wells are generally open masonry wells having diameters varying from 2 m to as much as 7 m. Their depth seldom exceeds 20 m. Their discharge is about 18 m^3/hr or about 0.005 cumecs depending upon the depth of water table and nature of soil comprising the aquifer. The walls of the open well may be built of precast concrete, blocks, brick, or stone masonry. The wall is known as staining wall or lining wall of the well. The thickness of lining or staining wall varies between 38 cm and 75 cm depending upon the depth of the well.

The yield of an open well is limited because such wells can be dug only to a limited depth. The ground water storage is also limited in the top most aquifer. Moreover open wells cannot be worked at velocity of percolation more than the critical velocity. The velocity more than the critical, causes dislocation of soil grains and consequently develops hollows behind the well lining. Hence, limited velocity of percolation, limited depth of well and limited ground water storage in the top most aquifers, put limit on the yield of the open well.

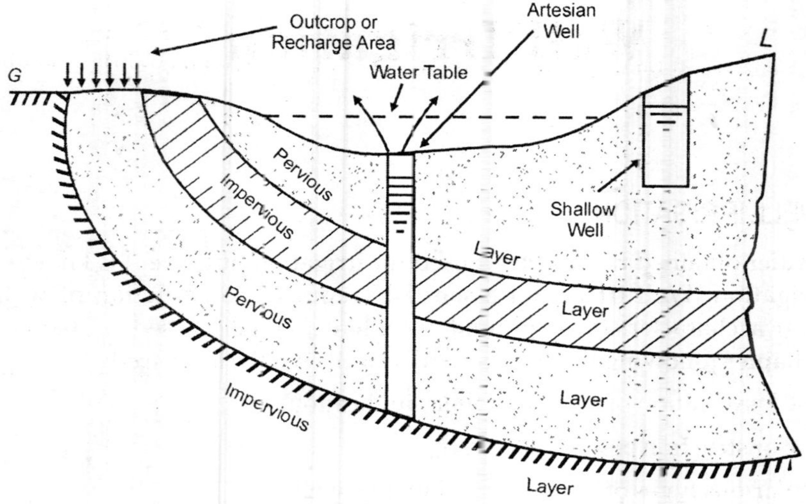

Fig. 8.1. *Open wells.*

However yield of the open well can be increased, if open well is constructed in top aquifer but in addition 8 to 10 cm diameter hole is bored to puncture the lower layers of water bearing strata.

Open wells may further be classified into two types.

1. Shallow wells and

2. Deep wells.

1. Shallow Well. It is such an open well which gets its supplies from upper most aquifer. If such wells penetrate for small depth under water table, they may even go dry in summer. This is because water from the ground reservoir will soon be drawn out and as soon as water table goes below the bottom of the well it will go dry. Hence in order to get water from such wells for more time or for the year round they should be carried deep into the ground water.

Quality of water obtained from shallow wells may not be good for drinking but it is quite good for irrigation purposes. In fact shallow wells can hardly provide substantial irrigation as availability of water is very limited.

2. Deep Well. The deep wells get their supplies from water bearing straws lying below an impervious stratum. The theory of deep wells is based on the percolation of water into the under lying aquifers from out-crops.

The out-crop is the exposed area of the pervious layer or aquifer, lying below the impervious layer. It is this area from where rain water enters the pervious layers lying below the impervious layer. After getting entry into the layer from out-crop water starts seeping along the slope of the aquifer and reaches the well site. During travel from out crop to well site, water gets thoroughly purified. But during this process certain salts may get dissolved in water and make it hard. Deep wells may get supplies from more than one confined aquifer.

Deep wells are not necessarily deeper than shallow wells. It is not the depth of the well but the criteria of getting their supplies. Shallow wells get their water from top most layer but deep wells get their supplies from confined aquifers lying below the impervious layers. A 'shallow well' might have more depth than a 'deep well'.

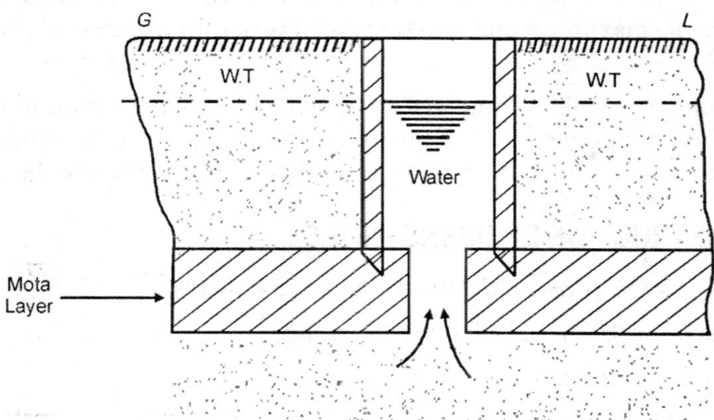

Fig. 8.2. *Mota layer well.*

The impervious layer on which deep wells generally rest and draw their supplies from pervious formations lying below the impervious layer, through a hole is known as Mota layer. The term Mota layer is also sometimes known as 'Matbarwa' or 'Magasan' layer. This layer refers to a formation of clay, cemented sand, Kankar or other hard materials which are often found lying a few metres below the water table in the sub soil. These names are not used for hard material layers lying above the water table. The main advantage of 'mota' layer is that it gives structural support to the open well resting on its surface. It is useful for unlined and partly lined wells and is indispensible for a heavy masonry well which would not remain stable without Mota layer support under steady use.

The Mota layer is found throughout the Indo-Gangetic plain. The Mota layer may be present as continuous layer or may be in form of local formation. It has different thicknesses at different places.

8.3 CAVITY FORMATION IN OPEN WELLS

We have already discussed in Chapter 7 that when water is pumped out of the well there is fall in the level of water. This fall in the level of water from the static water table conditions causes water held in the surrounding soil to flow towards the well. As more and more water is drawn from the well the level of water goes on depressing causing increase in depression head. Increased depression head leads to increase in the velocity of percolating water. Velocity of percolating water is maximum at the face of the well but goes on decreasing as distance from the well increases. This is because depression head goes on decreasing as we move away from the well. At certain rate of withdrawal of water from the well, the velocity increases to such an extent that it starts lifting up of the soil particles. As more and more soil particles are lifted, a hollow is formed in the bottom of the well, resulting in increased effective area so that ultimately the velocity falls below the critical value and then no further sand goes out of the well. This forms a cavity below the bottom of the well. The formation of such a cavity below shallow well is dangerous as it may lead to the subsidence of the well lining. The maximum rate of taking out of water from such wells is therefore limited.

In case of deep wells resting on a mota layer the cavity formation below the bored hole is not dangerous. A large cavity may deliberately be formed so as to increase the area of percolation and thus higher yields can be obtained.

8.4 CONSTRUCTION OF OPENS WELLS

Open wells may be constructed in following three ways:

1. Kacha wells or wells having no lining,
2. Wells having pervious lining and
3. Lined wells.

1. Kacha Wells. These wells are of small depths and can be constructed in boulder or hard soils. They can be constructed at places where water table is very near to the surface of the ground. They can be only temporary source of irrigation water. They may collapse easily and thus sometimes prove very dangerous.

2. Wells having Pervious Lining. In this, the lining of the well is done in dry bricks or stones without using any mortar. The seeping water can enter the well through the open joints of dry masonry. Thus inflow in this case is radial and not axial. Such wells are generally plugged at the bottom by means of concrete. If bottom is not plugged, then water will be entering the well both, from bottom as axial flow and sides as radial flow. When such wells are constructed in fine sandy soil, the pervious lining is generally surrounded by gravel to act as vertical filter. This vertical filter prevents ingress of sand along with seeping water.

3. Lined Wells. In this case wells are lined by stone or brick masonry or by cement concrete. For constructing such a well first of all a pit is excavated with the help of hand tools. The diameter of the pit is kept slightly larger in diameter than the finished diameter of the well. After the pit has been dug a kerb is laid at its bottom and masonry is raised over it. The masonry is raised above ground. Now earth is excavated and bailed out of well and masonry is simultaneously raised over lining wall projecting above the ground level. As the earth is excavated from below the kerb, the kerb along with masonry above, sinks in the hollow thus formed. This process of sinking and raising masonry above the projected lining wall is continued till well reaches the water table. After this excavation is done with the help of *Jhams* which is nothing but a self closing bucket which is tied to a rope and worked up and down over a pulley. When Jham is thrown into the well, it dislodges some soil and as the Jham is pulled up the soil cuttings get retained but the water oozes out. Sinking is continued till mota layer is reached. After this a small diameter hole is bored at the centre of the well to puncture the below lying confined aquifer.

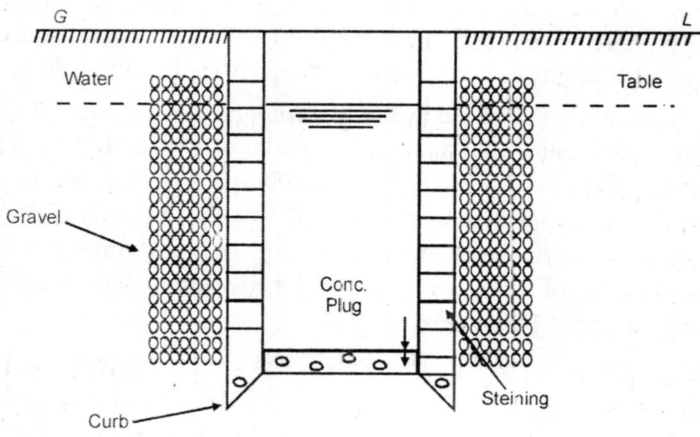

Fig. 8.3. *Open well with pervious lining.*

If mota layer is not available then bottom of the well is filled with gravel, boulders etc. Gravel filled at the bottom acts as a filter and prevents sand particles from being lifted along with seeping water. Flow in this well is from the bottom only.

Sometimes, to utilize the water available in top most aquifer, small weep holes are left in the pucca lining of the well. In that case, along with water from bottom of the well, water from the sides can be admitted in the well. Yield test and recuperative tests for the open wells have already been explained in Chapter 7.

8.5 TUBE WELL

This well is a type of deep well, tapping ground water from one or several confined aquifers lying under the surface of the earth. As its name suggests, it is a tube inserted in the ground. The parts of the tube which come in level with confined aquifers are slotted and those which come in level with impervious layers are hollow pipes not having any slots. The parts of the pipe which are hollow and do not have any slots at this surface are known as blind pies. The parts of pipe that have slots at their surface are known as strainer pipes or perforated pipes. Thus a tube well consists of strainer pipes as well as blind pipes. There are several types of strainers or perforated pipes which have been explained later in this chapter. The discharge of tube well is several times more than open well. The reasons of greater discharge are the following:

1. Tube wells can tap several water bearing stratum.
2. Water can seep into the tube well with several times more velocity than the critical velocity of flow in the case of open well. This becomes possible because strainer pipes allow only water to pass through and no sand particles.
3. There is no danger of well being subsided.
4. Flow in the well being radial and that too from several water bearing layers, the cross-sectional area contributing water to well becomes quite large and thus quite large discharge can be obtained from the tube well.

Tube wells may be constructed by the government or by the individual farmers. The average yield from a tube well is of the order of 40 to 50 lit/sec. Tube wells have been installed to give yield as high as 200 to 250 lit/sec. The depth of the tube wells may vary from 20 m to as high as 500 m. The diameter of the bore hole is kept 60 cm for top 60 m depth. Below this depth of hole diameter is reduced to 56 cm. The diameter of strainer is 25 cm and draw down 10 m. Each deep tube well can irrigate about 15 hectares of land.

Besides deep tube wells, shallow tube wells are also constructed by small farmers. The depths of such tube wells vary from 15 m to 30 m or so. Such tube wells may give about 15 lit/sec yield, if located at proper place. Each shallow tube well can irrigate 5 to 8 hectares of land.

Except in desert areas of Rajasthan, porous sand and gravel layers charged with ground water are extensively found in India. The entire area from Himalayas to Vindhya Chal, coastal area, Narmada Valley etc. consist of deep alluvial soils. Tube wells can be easily installed in such soils and are very useful for irrigation. It is in this context that the tube wells are assuming greater and greater importance in these regions in India.

It is very difficult to construct a tube well in rocky region. It is actually not required in such areas as land for irrigation is generally not available.

8.6 TYPES OF TUBE WELLS

The tube wells may be further classified into following three types:

1. Strainer wells

2. Cavity wells
3. Slotted wells.

All the three types of tube wells have been discussed one by one.

1. Strainer Type Tube Well. It is the most commonly used type of tube well in our country. It is so common that whenever we talk of a tube well, it is automatically understood as referring to strainer type tube well. This tube well consists of blind pipes and strainer pipes or screen pipes. This tube well is generally unsuitable for very fine sandy strata, because in that case screen openings are likely to get choked easily.

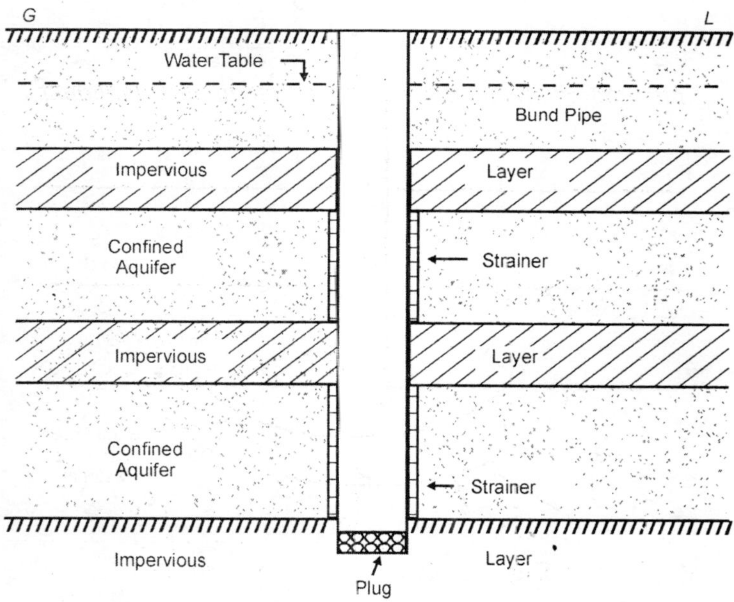

Fig. 8.4. *Tube well (Strainer type).*

The boring of the hole is generally carried out by using casing pipe of about 5 to 10 cm larger than the diameter of the well pipe. Thus a 15 cm diameter well, will require a bore hole of 20 to 25 cm diameter. After boring the hole, the well pipe assembly consisting of blind and strainer pipes is lowered into the bore hole. The lengths of blind and strainer pipes are so adjusted that the blind pipes about against impervious layers and strainer pipes against the water bearing aquifers as shown in Fig. 8.4.

A short blind pipe should be provided at the bottom so as to permit settlement of any sand particles that have passed through the strainer. The well is generally plugged at bottom by cement concrete.

2. Cavity Type Tube Well. This is such a tube well which does not require any strainer. It draws its supplies from a cavity developed in the water bearing layer at the bottom end of the well. In strainer type tube wells flow of water through strainers is horizontal and radial. But in this well water entres well from bottom of the well without passing through any screen and flow of water is normal to the spherical cavity formed at the bottom. The essential requirements for a cavity well to function efficiently is to have confined aquifer of good specific yield and the aquifer should have a stronger impervious material layer. The impervious layer is punctured and a cavity is developed by pumping, by a sand pump. In the initial stage of pumping, fine sand comes out with water and consequently a cavity is formed. As the spherical surface area of the cavity increases outwards, the radial critical velocity decreases and sand particles stop entering the well. See Fig. 8.5.

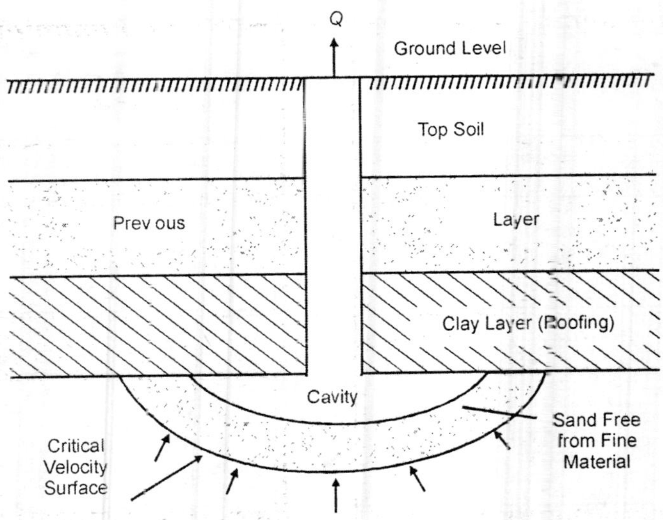

Fig. 8.5. *Cavity type tube well.*

The principle behind this well is same as that of a deep well. The only difference is that in case of open deep well only first aquifer lying below mota layer is punctured whereas in case of strainer tube well any aquifer lying at any position can be tapped.

3. Slotted Type Tube Well. This well is made at such places where sufficient depth of water bearing stratum is not available even at a depth of 80 m to 100 m and also when suitable strong roofing layer is not available for the construction of a cavity well.

This well consists of a slotted wrought iron tube penetrating a highly pervious confined aquifer. The size of the slots may be 25 mm × 3 mm and spaced at 10 to 12 cm centre to centre. In order to prevent the entry of fine sand particles into the pipe, the pipe is surrounded by a mixture of gravel and bajri. This mixture is called shrouding.

First of all a casing pipe 40 cm diameter is lowered and aquifer is penetrated for a depth of about 5 m. The slotted pipe, sometimes also known as education pipe of 15 cm diameter is then lowered. Gravel is then poured from top in the annular space left between inside of the casing pipe and outside of education pipe. The gravel is filled for 3 to 4 m higher than the slotted pipe. Now the casing pipe is withdrawn 5 cm at a time and well is developed with the help of compressed air pumped into the education pipe. Finally when casing pipe is fully withdrawn, the annular space between the casing pipe and the education pipe is suitably plugged. By developing the well with the help of compressed air, the sand surrounding the gravel is freed from finer particles and the chances of

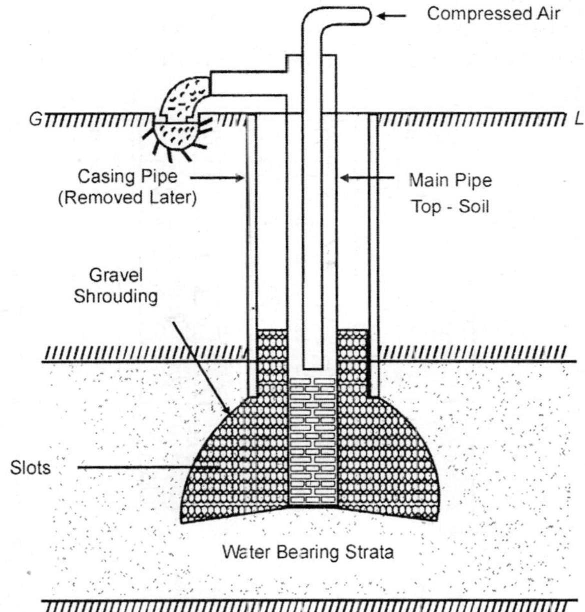

Fig. 8.6. *Slotted type tube well.*

filter getting choked are very much reduced. The difference between slotted tube well and strainer tube well are the following.

(i) Slotted well can tap only one water bearing strata whereas strainer well can tap many.

(ii) Slots of slotted tube well are protected by gravel shrouding so as to prevent entry of sand particles in the well. In case of strainer tube wells strainer is used to prevent sand particles entering the well. This well is not much in use and hence not very important.

8.7 TYPES OF STRAINERS

The strainer pipe is a perforated pipe which is provided with such an arrangement that only water can be admitted inside the pipe. Various patented types of strainers are available in the market.

Some of the strainer are :

1. Cook's strainer
2. Tej strainer
3. Ashford strainer
4. Leggett strainer
5. Phoenix strainer
6. Layne and Bowler strainer
7. Brownlie strainer
8. Esbee strainer
9. Mesh strainer.

1. Cook's Strainer. It is made from solid drawn brass tube and is very costly. It consists of wedge shaped slots at its surface. Slots are wider at the inside and finer at the outer surface. The size of the slots depends upon the coarseness of the sand in which it is to be used. It is not used much in India. The size of the slots may vary from 0.1 mm to 0.4 mm.

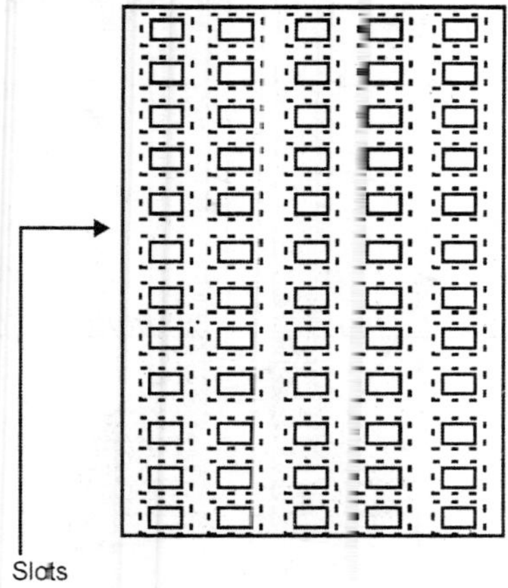

Fig. 8.7. *Cook's strainer.*

2. Tej Strainer. It is just like Cook's strainer. The strainer tube is made by folding and bending the brass sheet. Slots are cuts in the sheet in required gauge before sheet is bent to form a tube. It is very much manufactured in India. Its diameter is 7.5 cm and above and length 2.5 m. Vertical joint of the sheet is closed by brazing. If larger than 2.5 in length is required more than one lengths of strainer are joined together by screwed collar joint.

3. Ashford Strainer. It is also a brass tube having round holes in it. A steel wire is wrapped over the brass tube and then a wire mesh is soldered to the wound wire. The wire is wound to maintain a clear space between strainer and wire mesh. The wire mesh is finally protected and strengthened by a wire net around it.

4. Leggett Strainer. It is a very costly strainer. It is not used in India. It consists of a usual strainer which is also fitted with a special strainer cleaning device. The device can be operated from the ground. Whenever strainer is clogged this device is operated and strainer is cleaned without any difficulty.

5. Phoenix Strainer. It consists of a cadmium plated mild steel tube. Cadmium plating keeps the strainer free from choking and corrosion. Its slots are also made from inside the tube with the help of a special machine.

6. Layne and Bowler Strainer. It also consists of a slotted or perforated steel or wrought iron pipe. A wedge shaped steel wire is wound around the tube at suitable pitch.

7. Brownlie Strainer. Steel plates are bent in form of polygonal convolutes. Suitable perforation holes are made in convolutes. The convolute shaped perforated tube is enclosed in a wire mesh of copper wires. It is considered one of the best strainers as wire mesh remains *slightly* away from the perforated tube.

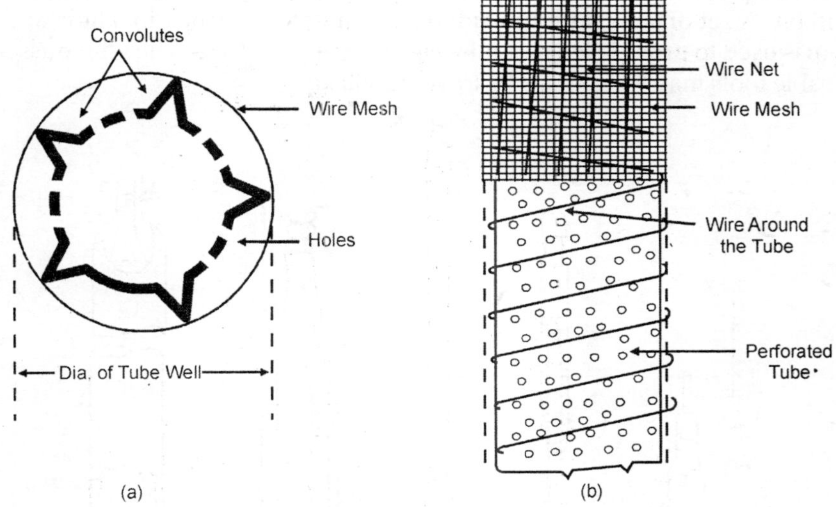

Fig. 8.8. *(a) Brownlie strainer (b) Ashford strainer.*

8. Esbee Strainer. It consists of a steel frame made of iron rods and wound by coir. The coir acts as a screen which allows water to pass but prevents sand. Its ends are so rivetted to screwed pipe that they can be easily joined to the next pipe.

9. Mesh Strainer. It consists of a metal plate in which slots are punched. The punched strip is not removed. It is mostly made from iron plate which is wound in form of a pipe after making slots in it. It is claimed that they do not choke.

8.8 METHODS OF DRILLING TUBE WELLS

Wells for domestic purposes which require small quantity of water are bored with auger turned by hand. Smaller wells in unconsolidated formations may be constructed by cutting action of a jet of water. None of these two methods is however applicable to deep tube wells. Deep and high yield wells are constructed by drilling. There are various techniques of boring the well hole. All the techniques have their own merits and demerits, depending upon the formation in which drilling is to be done. Most commonly used drilling methods have been described as follows.

1. Cable tools method or percussion drilling method.

2. Hydraulic rotary method.
3. Reverse rotary or jetting method.

1. Cable Tool Method or Percussion Drilling Method. This method of drilling of the well hole is based on the use of striking force, (hammering and cutting) of the drilling bit, attached to the lower end of the cable. The drilling bit is alternately lifted and dropped in the hole to break the formation. In this method a standard well drilling rig consisting of a mast, a multiline hoist, a walking beam, and an engine are all mounted on a truck in assembled form so that equipment may be easily transported.

A string of cable tools, consists of a rope socket, a set of jars, a drill stem and a drill bit. A set of jars is used to aid in loosening tools struck in a hole and drill stem is used to provide length and weight to the cable tools. The whole assembly of cable tools may weight 100 kg to over 2000 kg.

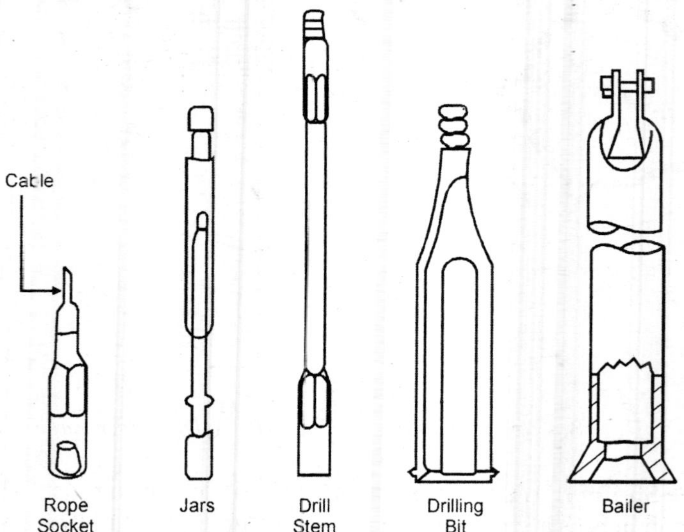

Fig. 8.9. *Percussion drilling tools.*

A pit is dug at the site where well is to be bored. A casing pipe along with drive shoe is erected vertically in the pit and the string of drilling tools is inserted in the first length of the casing pipe. The string of tools is alternately lifted and dropped with the help of engine. After penetrating for 1 m to 1.5 m the string of tools along with bit is taken out of the hole and a bailer is inserted to remove the cuttings from the hole. The bailer is a pipe fitted with a valve at the bottom. When it is dropped in the hole, the valve at the bottom opens and cut material enters the bailer. When bailer is lifted up, valve automatically closes and cut material remains held in the bailer. Bailer may be lifted and dropped several times so that it may be filled with the cut material. The bailer is then pulled out emptied at the surface. The length of the bailer varies from 3 m to 12 m, depending upon its diameter.

When cut material is fully bailed out of well hole, the string of tools is again inserted and penetration carried out further. The string of tools is slightly kept rotating so that a round hole is drilled. Water is also put in the drilled hole to form the paste with the cut material so that it may be easily bailed out. Secondly water also softens the formation which is being cut by the drill bit. When one length of casing pipe is sunk, subsequent length is joined either by threading or welding. This method is used for drilling through consolidated rock materials.

In soft and fissured rock formations, drilling is carried out with the help of a tripod and manual labour. In that case truck mounted with mast, multiline, walking beams, engine, string of tools etc. are not required. Tripod erected over the proposed well hole acts as a mast and manual labour provides the force for lifting and dropping the cutting equipment. In this case 'sludger' is used in place of string of tools. Sludger is more or less similar to bailer. It is made of steel pipe 2 to 4 m long fitted with a hard steel cutting shoe at the bottom. A flap valve is also fitted near its lower end which allows cut material to enter the sludger pipe and then does not allow it to fall down, when sludger is lifted out of drilled hole. Sludger is inserted in the hole to be drilled and is lifted and dropped by manual labour through a rope which passes over a pulley fixed to the tripod. A platform

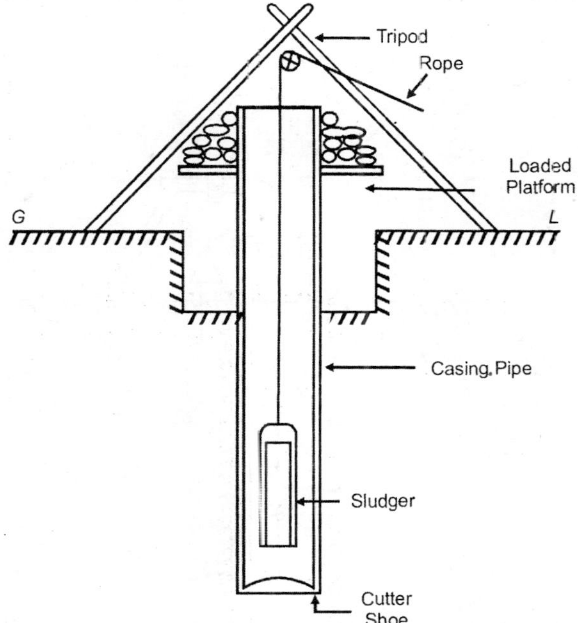

Fig. 8.10. *Percussion method by manual labour.*

is attached at the upper end of the casing pipe and weight is placed on it to help drive the casing pipe slowly. When sludger is full with the cut material, it is taken out and emptied. This process is continued till boring upto the desired level is achieved. If some rock formation is encountered embedded in soft ground, it can be cut using a string of drilling tool instead of sludger.

In both the cases mentioned above, the record of material found at different depth is maintained. The positions of strainer pipes and blind pipes are decided based on this record. This record reveals what type of formation exists at what depth. Strainer pipes are positioned against aquifers and blind pipes against aquicludes.

2. Hydraulic Rotary Method. This method is used for drilling large bores in unconsolidated formations. This is a very fast method of drilling well holes. This method consists of a drilling bit attached at the bottom end of a string of hollow pipes. Drilling mud or Bentonite slurry, is continuously pumped down through the string of hollow pipe and released through a nozzle in the drill bit. This released drill mud is then carried up through the annular space between the drill pipe and the drilled hole, along with cut and loosened material. In this case no casing pipe is required since the drilling mud forms a clay lining and supports the walls of the drilled hole. The rising drilling mud carrying cut material is taken to the settling basin where cut material settles. The mud is recirculated to the hole. In order to maintain required consistency, clay and water are added to the circulating mud from time to time. Complete record of settled materials is maintained and type of formations at different depths predicted on its basis.

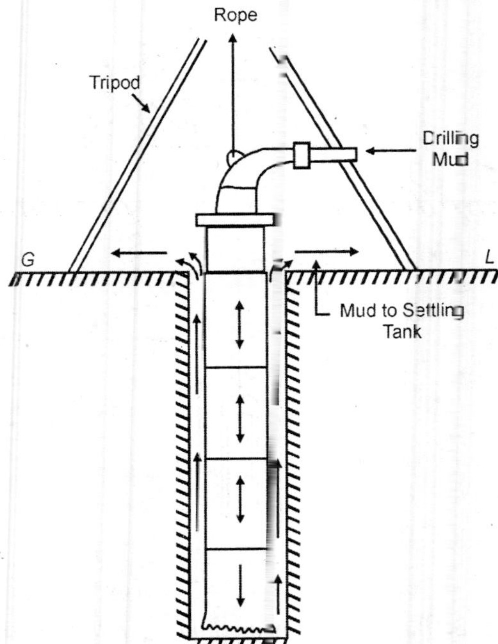

Fig. 8.11. *Hydraulic rotary method.*

After reaching the desired level, the drilling equipment is taken out and well pipes comprising strainer as well as blind pipes are lowered at appropriate levels. After fitting the well pipes the well hole walls which are coated by clay or Bentonite are to be washed. Washing of the walls is done by lowering drill pipe and drill bit in the well again and forcing water containing calgon (sodium-hexa-meta-phosphate) in the bore.

In this case of drilling, drill bit is the most important equipment. It has hollow shanks and one or more centrally located holes for jetting the mud into the bottom of the hole. There is a drill rod made of heavy pipe. At one end of drill rod drill bit is attached and at other end it remains attached to a square rod known as kelly. A rotating table rotates the drill rod and slides downwards as the hole deepens. Rotating table remains closely fitted to the kelly.

3. Reverse Rotary Method. This method is also known as jetting method. It is a modified form of hydraulic rotary method. It is also useful for making large wells in unconsolidated formations. The tools required for this method are hollow drill, a drill pipe, and water swivel. The cuttings are removed by a suction pipe of a large capacity centrifugal pump. The walls of the hole during drilling remain supported by hydrostatic pressure acting against the film of fine-grained material deposited on the walls by drilling water. Clay or bentonite are not necessary to be added to the drilling mud.

The mixture of water and cuttings is passed through settling tank and after settlement the effluent is again recirculated. Casing and cleaning of the walls is done in the same way as in hydraulic rotary method.

8.9 COMPLETION OF WELL

During boring of the well hole by any method it should be ensured that bored hole remains straight and vertical.

If well hole is located in consolidated formation, water enters the well hole directly and no casing is required as all the surroundings are quite stable. But if formations are unconsolidated a casing is necessary to support the outside material and also to help in freely admitting the water into the well. The casing should either contain perforations or its lower part may be replaced by a screen or strainer so that water may enter the well.

Sometimes in order to increase the effective diameter of the well, the screen or strainer is surrounded by a layer of gravel. Gravel layer keeps fine material out of the well. Such a gravel packed well will have a greater specific capacity than one of the same diameter not surrounded by gravel. The thickness of the gravel may vary with type of formation and method of drilling. However a minimum thickness of 15 cm is generally used.

While drilling a new tube well it should be ensured that it would not affect the discharge of other nearby existing tube wells. The distance of the proposed well from existing wells should not normally be less than 1.5 km. The diameter of the well pipe may be reduced theoretically, as we go deep down the well; but it is not recommended. A single sized or at the most two sized tube may be used in actual practice. The velocity of flow through the tube is generally limited between the range of 1.5 m/sec to 4.5 m/sec. Once the design discharge of the tube well and the velocity of flow are known, the diameters of the pie can be easily found out.

The nearest available size of the pipe in the market may then be adopted.

Normal life of a tube well in Northern India is about 20 years.

8.10 WELL SHROUDING

It is a process by which fine particles of soil coming along the seeping water into the well, are prevented from entering the well. It is accomplished by interposing a coarse grained material such as gravel between strainer pipe and the aquifer soil. This process is essential in sandy and unconsolidated formations of aquifer and also for slotted tube wells.

8.11 WELL DEVELOPMENT

It is a process by which the fine particles are removed from aquifer formation surrounding the strainer pipe of the well. Following favourable effects are obtained from well development :

(i) Specific capacity of the well increases.

(ii) Flow of sand into the well is prevented.

(iii) Life of the well is increased.

The well can be developed by following methods :

(i) by pumping,

(ii) by surging,

(iii) by compressed air,

(iv) by back washing,

(v) by dry ice.

(i) Pumping. In this method, pumping of water is done non-continuous, irregularly and at variable rate. By this fine material surrounding the well gets agitated and carried out of the well.

(ii) Surging. A bailer or hollow surge block is moved up and down in the well briskly. By this finer particles are agitated and enter the well from where they are pumped out. Sometimes calgon is also added to well water so as to act as dispersing agent for fine particles.

(iii) Compressed Air. In this method compressed air is continuously injected which develops an air lift pump system in the well which carries sand particles with water and are pumped out. After developing with compressed air, surging is also done to fully develop the well.

(iv) Back Washing. In this method water is forced in the reverse direction by means of compressed air pressure. All the sand and clay material which is truck around the strainer is agitated and removed.

(v) Dry Ice Method. It is a chemical method in which hydrochloric acid and solid sodium dioxide, known as dry ice is used. First of all HCl is poured into the well and well is capped. Now compressed air is forced into the well which forces the HCl solution into the soil formation. Now cap is removed and blocks of solid sodium dioxide (dry ice) are put Into the well. Due to sublimation CO_2 gas is released and this gas develops a very high pressure in the well. When the pressure is released muddy water is automatically thrown out of the well. This method is not used in India.

8.12 WELL TROUBLES

Failure of the well or reduction in the yield is not uncommon. Following may be the reason and remedies.

1. Reduction of discharge may be due to a draft greater than the percolation of water. Lowering the pumps may increase the yield for the time being, but permanent remedy is to reduce pumpage.

2. The casing pipe or screen pipe may have collapsed, partially or completely, blocking the flow of water. This trouble necessitates replacement of the casing or screen pipes. If the material in which the well is drilled is unconsolidated it may have to be abandoned.

3. Casing pipe may have corroded or leaked. Due to this, water may escape out into the ground or contaminated ground water may enter the well. The remedy lies in withdrawing and replacing the casing pipe. The life of the casing pipe depends upon the character of water handled by it. If water being handled is highly corrosive, C.I. or concrete casing may be used.

4. Screens cause trouble by corrosion or incrustation. There is little that can be done about corrosion, but some remedies are available against incrustation. All forms of incrustation are remedied by acid treatment.

8.13 METHODS OF LIFTING WATER

Water is to be lifted from wells before it is used for irrigation. Sometimes water may have to be lifted from canals, streams, and rain water filled in low lying areas to use it for irrigation. The methods of lifting water vary depending on the source, depth, quantity to be lifted and availability of power. The following are the various methods of lifting water:

1. Rope, pulley and bucket method
2. Dhenkli or lever or let method
3. Mote, charas or leather bag method
4. Nar, Rahat or Persian wheel
5. Basket method
6. Doon method
7. Archimedian screw method
8. Pump method.

Each of the above methods have been briefly discussed as follows.

1. Rope-pulley and Bucket Method. This method consists of a pulley over which a rope is passed. A bucket is tied to the end of the rope hanging in the well and other end of the rope is held by a man standing at the surface. The bucket is lowered into the well, and pulled out with help of rope and pulley after being filled with water. This method is suitable only when well is not deep and quantity of water to be lifted is small.

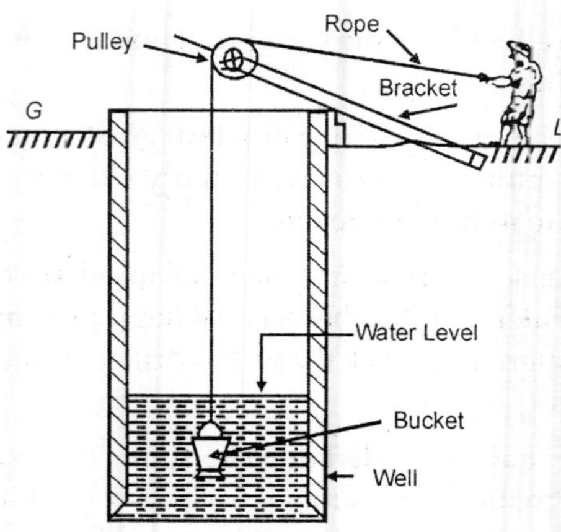

Fig. 8.12. *Bucket method.*

2. Dhenkli or Lever Method. This method consists of a lever-rod rocking over the top of a vertical upright post. A suitable counter weight is tied on one side of the lever and a bucket is suspended from the other end into the well. Length of the

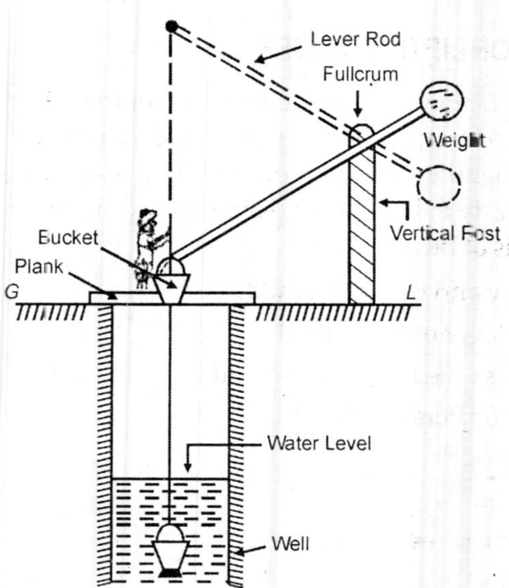

Fig. 8.13. *Dhenkli or lever method.*

lever rod on the side of the bucket is longer than that on the side of counter-weight. The bucket when filled is lifted a little to set the lever action. The bucket full of water moves up automatically due to moment of the counter weight. When bucket reaches the ground it is used for irrigation and empty bucket is again sent into the well. But this time lever will have to be pulled against the moment of counter weight so that bucket may reach water and get filled with it.

3. Mote, Charas or Leather Bag Method. This method has a pulley fitted on the well on two uprights. A large leather bag also called Mote or charas is tied at one end of the rope. The other end is used to lift filled leather bag of water. In this

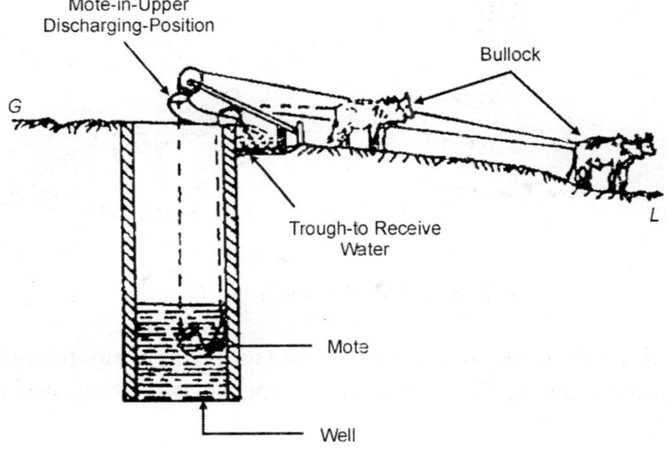

Fig. 8.14. *Mote or Charas method.*

case lifting force is applied by a pair of bullocks. This method is used when water level in the well is quite deep and irrigation is to be provided to a little larger area. About 1.5 hectare land can be irrigated by this method. In this case bullocks move on straight inclined path. See Fig. 8.14.

4. Nar, Rahat or Persian Wheel. This method was very common in Northern India. Still it can be noticed in U.P., Haryana and Punjab. Persian wheel consists of an arrangement of wheels and gears run by bullocks. An endless chain of several buckets is put on a big wheel. The lower end of chain of buckets remains under well water. As the wheel is moved by bullocks through a system of gears, the buckets filled with water come to the top and discharge water into a trough from where water flows to the fields. In this case pair of bullocks moves on circular path. By this method 3–5 hectare of land can be irrigated. See Fig. 8.15.

5. Basket Method. This method is used for lifting rain water from low lying areas to the near by fields. By this method water can be lifted for 60 to 80 cm only. The basket can be worked by one man or two men. The basket has ropes. It is filled and discharged by swing motion.

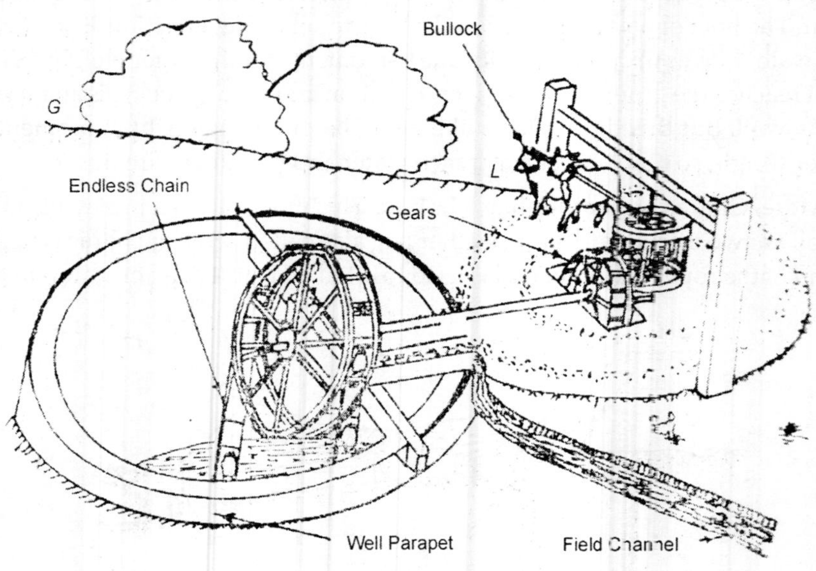

Fig. 8.15. *Persian wheel.*

6. Doons. Like Basket method it is also used for lifting water for small heights of say 1 m. Doon is a wooden or metallic channel closed at one end and open at the other. It is supported on a wooden bully which acts as its fulcrum. A doon in action is shown in Fig. 8.16.

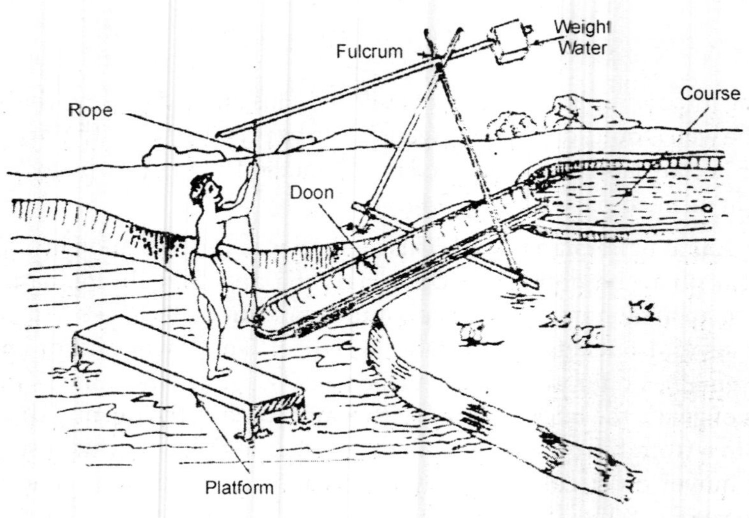

Fig. 8.16. Doons.

This method is very much prevalent in Bengal.

7. Archimedian Screw. It consists of a cylinder in which a screw is fitted. The cylinder is kept inclined with its lower end immersed in water to be lifted. Upper

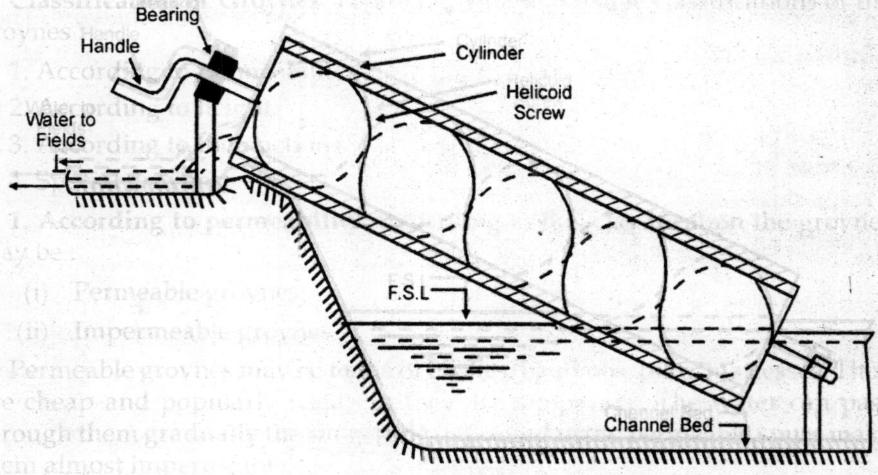

Fig. 8.17. *Archimedian screw.*

end is kept abutting the field to be irrigated. When screw in the cylinder is rotated, it lifts water as water enclosed in screw cannot go back. In this case also lift of water seldom exceeds 1 m. Inclination of the cylinder is about 30° with the horizontal.

8. Pump. There are several types of pumps which can be used for lifting water. Pumps may be divided into (1) constant displacement and (2) variable displacement pumps.

Constant displacement pumps deliver the same volume of water against any head within which they can operate.

Variable displacement pumps deliver water in volume varying inversely with the head. The use of constant displacement pumps is confined to places where discharge is low. Variable displacement pumps which are mostly used for pumping water may be divided into the following classes.

1. Centrifugal pump
2. Bore hole type pump
3. Jet pump
4. Air lift pump.

1. Centrigugal Pump. This pump lifts water by creating the required pressure with the help of centrifugal action. The maximum suction head under which the pumps can practically work effectively is about 6 m to 8 m. Hence this pump can be used only at places, where the fluctuations in water table plus the depression head is limited to about 8 m. For larger values the bore hole type pump is to be used.

A section of a tube well using centrifugal pump is shown in Fig. 8.18.

In this arrangement a sump well is made so as to place the pump at the required level. The pump is to be set slightly above the water table so as to avoid the submergence of the pump. The minimum water level should not be lower than the pump level by more than 6 to 8 m, otherwise pump will not work. The

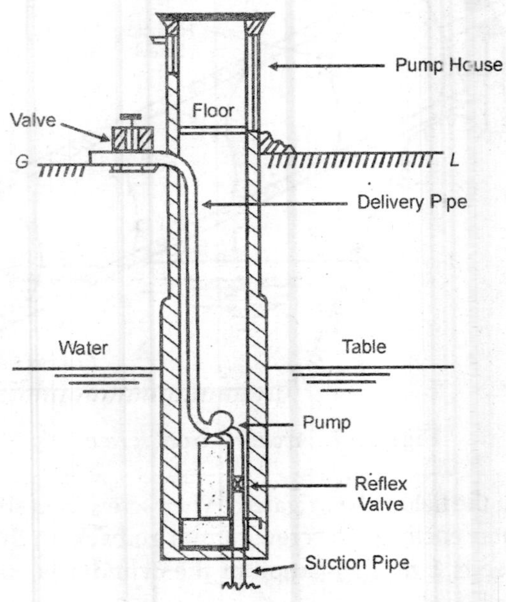

Fig. 8.18. *Centrifugal pump.*

sump well is generally of a large size and deep enough and has to be plugged at the bottom with concrete. This arrangement is quite costly and the depression head plus the fluctuations in water table is also limited. This arrangement has become out of date these days. The bore hole type pump is generally used in the modern times.

2. Bore Hole Type Pump. Such pumps are of very compact designs and can be lowered in the casing pipe itself. The top 20 to 30 m of the bore hole and casing pipe is generally kept larger than the remaining bore, so as to accommodate the pump in the casing itself. The pump is driven by a direct coupled electric motor of a vertical shaft type, and is placed at the top of the line shaft at ground level. Thus construction of pump well is completely eliminated. This is a better alternative to the centrifugal pump.

3. Jet Pump. It is a combination of a centrifugal pump and an ejector pump. The pump is a single stage at the top of the well and the ejector is located at the suction screen in the well. A part of the water discharged by the pump flows down through the ejector, where it helps to improve flow into the pump and up the discharge tube. This pump is suitable for lifts more than 8 m and has a capacity of upto 200 lit/min. These pumps are inefficient and not used much.

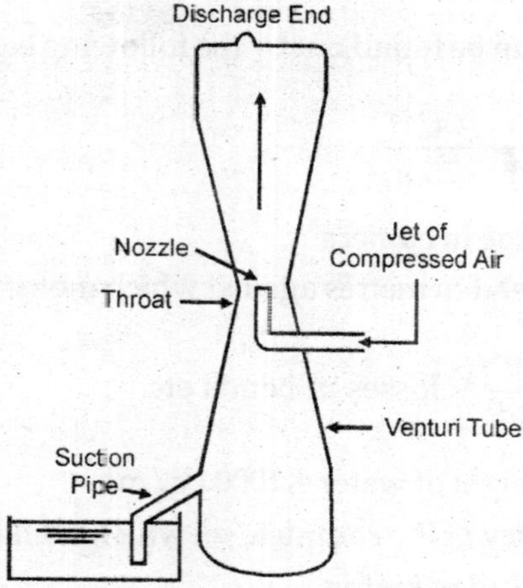

Fig. 8.19. *Jet pump.*

4. Air Lift Pumps. Compressed air is released into the discharge pipe lowered into the well. When air is admitted into the well, the water recedes from the level of static head to the bottom of the discharge pipe. The displaced column of liquid

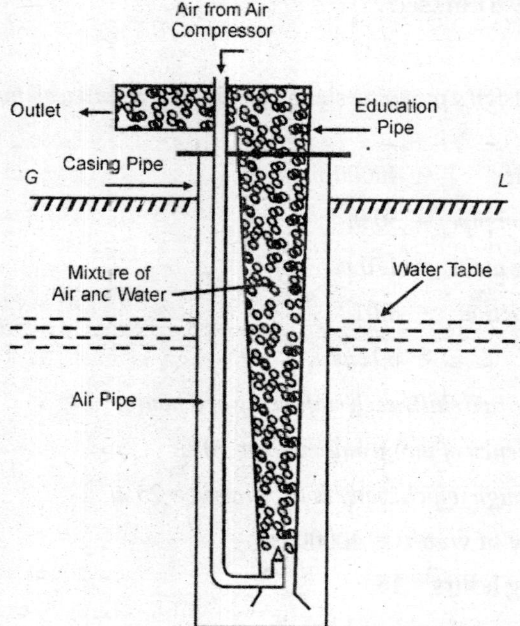

Fig. 8.20. *Air lift pump.*

rises up the discharge pipe. Air bubbles mix with water and this reduces the specific gravity of the column of water sufficiently to lift it to the surface. It is rarely used for pumping purposes.

8.14 HORSE POWER OF PUMP MOTOR

Horse power of motor can be found out by the following equation

$$\text{H.P.} = \frac{wQH}{75\,n}$$

where Q = Discharge in cumecs.

H = Total head in metres against which motor has to work.

$$= h + \frac{4\,flv^2}{2\,gd} + \text{losses in bends etc.}$$

w = Unit weight of water = 1000 kg/m³

n = Efficiency of the complete set which is taken about 60 to 65%.

h = Static head in metres

f = Friction coefficient

l = Length of the pipe in metres

v = Velocity of flow in m/sec

d = diameter of the pipe in metres

g = 9.81 m/sec².

Example 8.1. *Design a pumping station to lift water from a well to the field. Following is the data.*

Quantity of water = 40000 m³

Length of suction pipe = 30 m

Length of rising main = 170 m

Coefficient of friction = 0.01

Pipe diameter = 60 cm

Pumps work for two shifts each shift being of 8 hours.

Combined efficiency of pump and motor = 70%.

Static head through which water is to be raised = 25 m.

Solution. Quantity of water = 40000 m³

Total pumping hours = 16

Per hour required pumping capacity

$$= \frac{40000}{16} = 2500 \text{ m}^3$$

Discharge in cumecs

$$= \frac{2500}{60 \times 60} = 0.7 \text{ m}^3$$

$$wQ = 1000 \times 0.7 = 700 \text{ kg/sec}$$

$$h = 25 \text{ m}$$

$$hf = \frac{4 flv^2}{2gd} = \frac{flQ^2}{3d^5} = \frac{0.01 \times (30 + 170) \times (0.7)^2}{3 \times (0.60)^5}$$

$$= 4.34 \text{ m}$$

$$H = 25 + 4.34 = 29.34 \text{ m}$$

$$\text{B.H.P.} = \frac{wQH}{75 \times n} = \frac{700 \times 29.37}{75 \times 0.70} = 391.2$$

8.15 OTHER SOURCES OF GROUND WATER

Besides different types of wells following may also sometimes becomes source of water for irrigation.

1. Infiltration galleries
2. Springs
3. Karez.

1. Infiltration Galleries. They are horizontal or nearly horizontal tunnels which are constructed through water bearing strata along the river banks to collect the inflow of ground water from river. They are also known as horizontal wells. The galleries are made of bricks, or stones. To collect more of water in the gallery, several porous drains may be laid in direction perpendicular to the

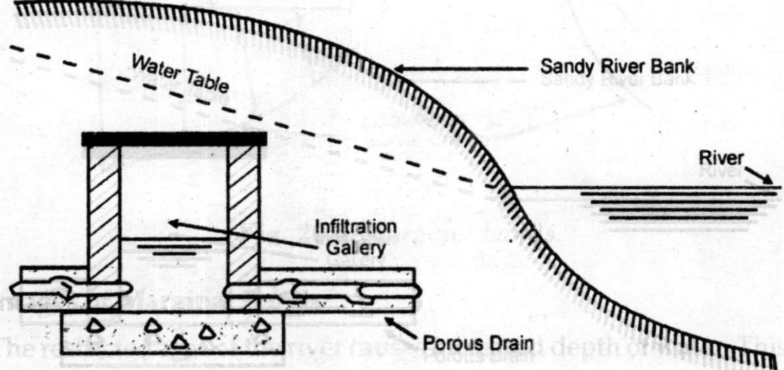

Fig. 8.21. *Infiltration galleries.*

direction of the gallery. The galleries are laid at some longitudinal slope so that the water collected in the gallery is led out of gallery under gravity flow. Galleries are provided with man holes for inspection and maintenance. Infiltration galleries are not used for getting water for irrigation purposes.

2. Springs. When ground water automatically comes to the surface and starts flowing it is called a spring. Springs are found in hills. They also supply very little quantity of water and as such are not adopted for irrigation purposes. Springs may be classed into following three types:

(a) Artesian springs

(b) Gravity or shallow springs

(c) Surface springs.

In the case of artesian springs underground water comes to the surface under hydraulic pressure. This spring is formed when an aquifer lying under an impervious layer gets punctured or some how fissured.

Gravity spring develops when ground water table gets exposed due to ups and downs in the surface.

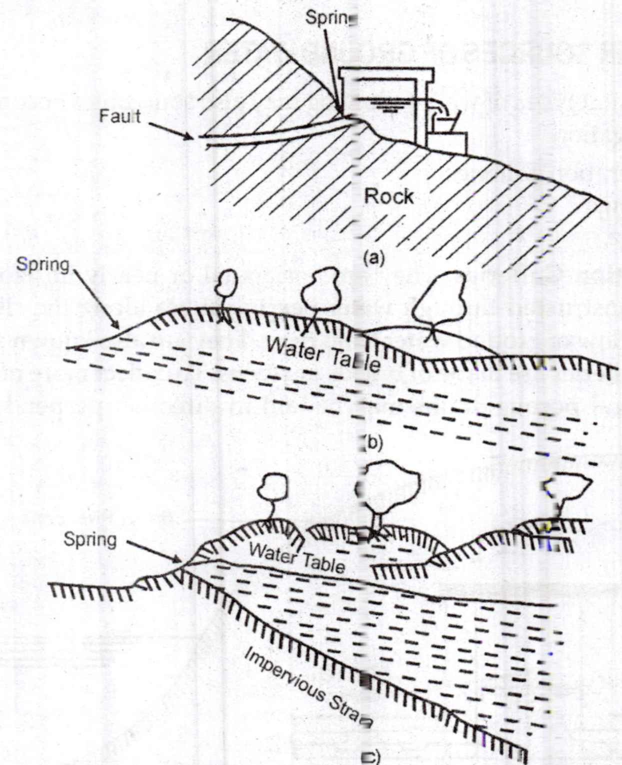

Fig. 8.22. *(a) Artesian spring (b) and (c) surface springs.*

Surface spring is developed when an impervious stratum which is supporting the ground water reservoirs is out cropped. Formation of various springs is given in Fig. 8.22.

3. Karez. In it a tunnel laid along the hill side to tap water from under ground springs. The tunnel has some longitudinal slope. As it lies in hard stratum it is

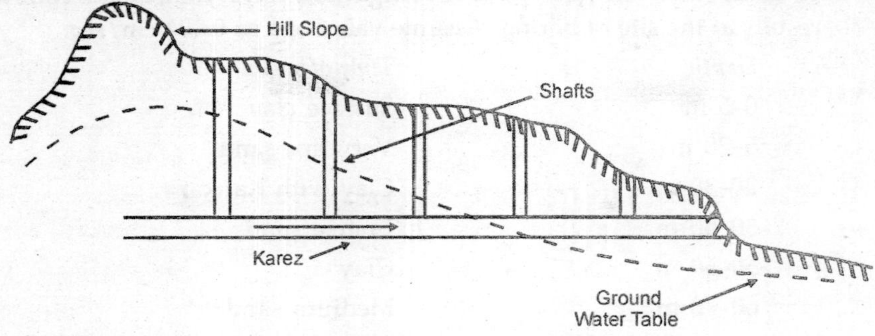

Fig. 8.22. *Karez.*

usually not lined. Water from Karez is led into an open channel at its down stream end and used as deemed suitable. Such tunnels are existing in Baluchistan and North West frontier in Pakistan.

QUESTIONS

8.1 Give various classifications of wells. What is basic difference between shallow well and deep well ?

8.2 Describe the method of construction of an open well. Under what circumstances the open wells require lining ?

8.3 How deep wells are classified ? Describe the construction of a cavity tube well and slotted tube well.

8.4 Describe in details, the construction of a strainer tube well.

8.5 Describe various types of strainers, used in the construction of a strainer tube well.

8.6 Describe various methods of drilling hole for the installation of tube well.

8.7 Explain the terms :

1. Shrouding of the well
2. Development of the well
3. Well troubles
4. Out crop
5. Calgon

8.8 Describe various methods commonly used for lifting water. Explain each method with the help of neat sketch.

8.9 Explain the following terms

1. Spring
2. Karez
3. Infiltration galleries.

8.10 Design a tube well to deliver 0.0417 cumec of water at a depression head of 5 m. The average water level is 10 m below the ground in October and

15 m in July. The geological investigations have yielded the following results at the site of boring. Assume value of k as 0.004 cm/sec.

Depth	*Type of strata*
0–5 m	Surface clay
5–20 m	Very fine sand
20–30 m	Clay with Kankar
30–50 m	Coarse sand
50–60 m	Clay
60–70 m	Medium sand
Below 70 m	Clay with sand stone.

(Ans. Pipe dia 14.6 cm.
Velocity of flow 2.36 m/sec.
Length of strainer 29.2 m)

❑❑❑

9

Reservoir Planning

9.1 INTRODUCTION

Dams and weirs are some of the barriers which when constructed across the rivers and streams, cause accumulation of water behind them. The water thus accumulated in form of an artificial lake is known as **reservoir.** In broad sense, any collected water in form of a pool or lake, may be termed as reservoir. Dams and reservoirs are the most important elements of any multipurpose river basin development.

River water may be used for generation of hydroelectric power, irrigation, and water supply source for some town or city. All these uses require a constant or almost constant supply of water. But we know that discharge in rivers and streams remains fluctuating. During rains, the water in the river may be more than the requirements and during dry period the discharge in the river may become less than the minimum requirements. Thus to store the excess water, flowing during floods, dams are constructed across the rivers. This stored water is used to augment the supplies of the river during dry weather. By this arrangement it is possible to generate hydroelectric power, provide irrigation facilities and make available drinking water for whole of the year.

Actually dams and reservoirs are complementary to each other. Reservoir can be developed only by constructing dam, conversely when dam is constructed reservoir is bound to develop. Both dams and reservoirs, being very important elements of any multipurpose irrigation project, have to be very carefully planned, designed and operated. This involves proper selection of site, for dam and reservoir, fixation of capacity of reservoir, safe yield of the storage and other connected works.

9.2 PURPOSES OF STORAGE

The storage of water may be done for following purposes.

1. To act as source of water for any public water supply scheme.
2. To augment the irrigation supplies when discharge in the river is smaller than the demand.
3. To maintain some minimum level of water for generation of hydroelectric power during lean months of discharge.
4. To increase the depth of water to facilitate navigation.
5. For reducing the flood havoc downstream. It is achieved as flood waters are temporarily held up in the reservoir.
6. To render water comparatively silt and debris free so that it may not cause any damage in hydroelectric generation equipment or pumping equipment if water is to be pumped for some public water supply scheme or any other purpose.
7. Growing useful aquatic life
8. Recreation.

9.3 CLASSIFICATION OF RESERVOIRS

Depending upon the purposes served, the reservoirs may be classified into following categories.

1. Storage or conservation reservoirs
2. Flood control reservoirs
3. Distribution reservoirs
4. Multipurpose reservoirs.

1. Storage or Conservation Reservoirs. These reservoirs are primarily used to maintain minimum supplies of water for irrigation, hydroelectric generation, domestic and industrial water supply schemes, etc. during lean months of discharge in the rivers. We know that discharge in the rivers remains changing day to day, and season to season. During high floods the excess water in the river goes waste while in dry months it may not be sufficient to meet the minimum needs. The storage reservoir is constructed to store the excess water of floods and released gradually as and when required.

2. Flood Control Reservoirs. This reservoir is also called *flood mitigation reservoir*. The main purpose of this reservoir is to temporarily store the flood water and release slowly at a safe rate after the floods, so that it may not cause any damage on the down stream side. So this reservoir may be said a flood prevention reservoir also. This reservoir requires provision of large spillways and sluice-ways so that excess stored flood water is rapidly released downstream, but only at the safe rate. Figure 9.1 shows hydrographs at certain point on the river. *ABC* is the natural hydrograph at the dam site having a maximum flood discharge Q_1. After construction of the dam the hydrograph gets modified because of the development of a reservoir at the back of the dam. *AB'DC* is the modified hydrograph showing maximum flood discharge as Q_2. Thus, because of reservoir the flood discharge has reduced from Q_1 to Q_2. Shaded are *ABDB'A* represents the storage to be provided in the reservoir. The area *DEC* represents the excess volume being released from the reservoir after the floods have receded. Flood control reservoirs may be further classified into two categories.

(i) Storage reservoir of Detention basins.

(ii) Retarding basins or retarding reservoirs.

(i) Storage reservoirs or Detention basins. The reservoirs whose spillways and sluice outlets are fitted with gates and valves, are known as storage reservoirs or detention basins. Gated spillways and gated sluiceways provide more flexibility in operation. They help in exercising better control on the reservoir and thus reservoir water can be used more wisely and usefully. This reservoir is costly as it involves cost of gates and valves. Detention basins or storage reservoirs are preferred on large rivers, as better controls on flood waters can be exercised and water can be released from the reservoir at better controlled rate so as not to cause any damage on the down stream side.

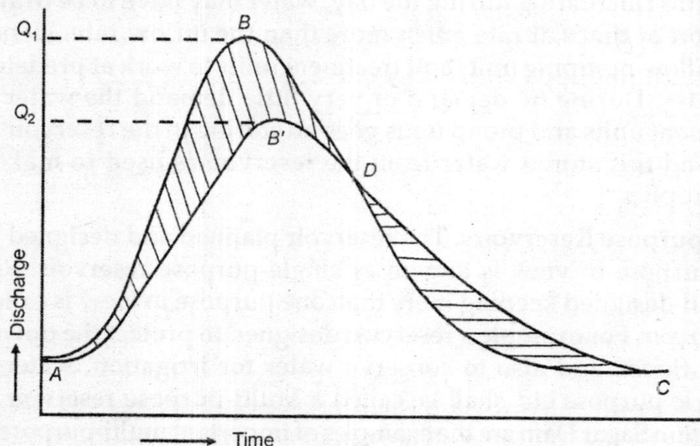

Fig. 9.1. *Original hydrograph ABC modified hydrograph ABC.*

(ii) Retarding basins or retarding reservoirs. In these reservoirs there are no gates at spillways and sluice outlets. In this case sluice and spillway's joint maximum discharging capacity is at the most equal to the maximum safe carrying capacity of the channel down stream. As floods occur, the reservoir first of all gets filled up, upto normal level. At this time sluice outlets are discharging out water from the reservoir. As level in the reservoir increases further, discharge through sluice ways also increases. At certain level of reservoir, water also starts escaping through

ungated spillways. As the level of water further rises the discharge over spill ways also increases. At some particular level a balance will be struck between the inflow and outflow of the reservoir. At this time level of water in the reservoir will become stable. This condition happens only when flood inflow in the reservoir is equal to the outflow from the reservoir.

Now when floods occurring recede, inflow in the reservoir will also decrease but the outflow is at the same maximum rate. Hence flood water which had accumulated in the reservoir will now be slowly flowing out of the reservoir. This reservoir has the following advantages over detention basin.

(a) Gates are not required to be provided at sluice ways and spillway crests.

(b) Since there are no gates at the spillways, chances of human error in opening the gates during floods cannot take place.

(c) Since water from the reservoir is driven out in few days after floods, the land during maximum floods remains submerged only temporarily. This submerged land can be used for growing very good crops. But no habitation on this land should be allowed.

Retarding basins are preferred on small rivers.

3. Distribution Reservoir. It is a small capacity reservoir which is mainly constructed to meet the water supply requirements of a particular city. It is made of masonry or cement concrete and may be covered from the top. This reservoir is filled by treated water at some constant rate. Since demand of water remains fluctuating during the day, water may have to be drawn from this reservoir at times at rate much more than the inflow rate. Hence these reservoirs allow pumping units and treatment units to work at predetermined constant rates. During no demand or very little demand the water coming from treatment units and pump units goes on storing in the reservoir. During peak demand this stored water from the reservoir is used to make up the required supplies.

4. Multipurpose Reservoirs. The reservoir planned and designed keeping only one purpose in view is known as single-purpose reservoir. Reservoir planned and designed keeping more than one purpose in view is called *Multi purpose reservoir*. For example a reservoir designed to protect the downstream areas from floods and also to conserve water for irrigation, water-supply, hydroelectric purpose etc. shall be called a Multi-purpose reservoir. Bhakra Dam, Nagarjun Sagar Dam are the examples of important multi-purpose projects of India.

9.4 INVESTIGATIONS FOR RESERVOIR PLANNING

For any reservoir planning, various investigations, which are required to be carried out, are the following :

1. Engineering surveys
2. Hydrological investigations
3. Geological investigations.

1. Engineering Surveys. The area of proposed reservoir is extensively surveyed and contour plan is prepared. Contour plan is very useful for fixing the height of the dam, evaluating the area that will be submerged ; and fixing the positions of spill ways, sluice ways and other river training works, required for the development of a reservoir. *Area-elevation curve* and *storage elevation curves* can be drawn from the contour plan which help in fixing the maximum operating level of the resevoir and thus help in fixing the height of the dam.

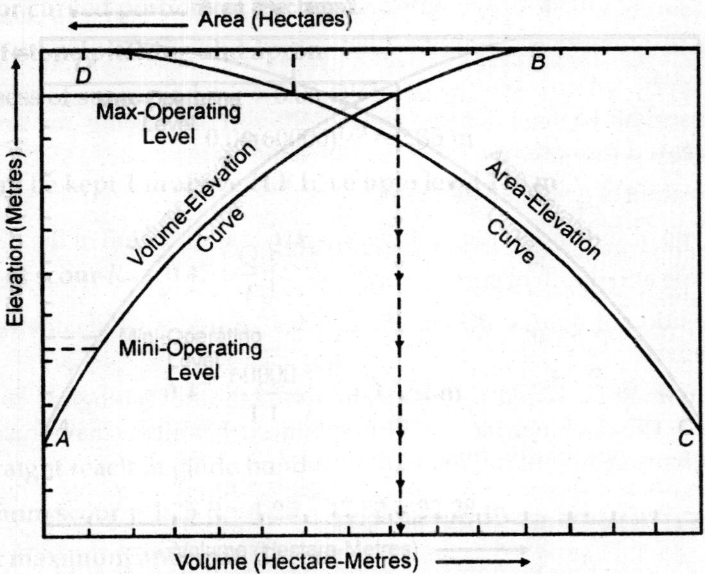

Fig. 9.2. *Area-elevation and Volume-elevation curves.*

Area elevation curve. A contour plan of the proposed site of a reservoir is prepared. Considering from the bottom let A_1, A_2, A_3......A_n be the areas of successive contours. These areas can be easily determined with the help of planimeter. The value of area A_1 is minimum at the base of the basin. As elevation of the contours increases the area under them also increases.

A curve as *CD* between area elevation is shown in Fig. 9.2.

Storage elevation curve. The volume of storage corresponding to different contours can be calculated by using either prismoidal formula or trapezoidal formula. We have seen that area of the contours goes on increasing as we proceed from bottom of the reservoir towards top. Consequently volume of storage also goes on increasing as we proceed from bottom of the reservoir towards top. The volumes of storage for different contours may be calculated and curve between storage and elevation may be plotted as shown by curve *AB* in Fig. 9.2.

2. Hydrological Investigations. It is very important aspect of reservoir planning. Capacity of reservoir, its potential for irrigation, power generation etc. all depend upon the availability of water from the reservoir. Capacity of spillway, storage capacity, height of dam, etc. all depend upon the amount of available water. Hence rainfall records, catchment characteristics, spread of catchment, climatic characteristics etc. will all have to be considered before any accurate assessment of the available run off is made.

3. Geological Investigations. In all the major Civil Engineering projects, geological advice is considered as very essential. It costs very little but relieves the Engineer from the worry of possible presence of fissured rocks or other geological faults which may cost too much when they come to notice later. Following information is sought while carrying out the geological carrying investigations.

(i) Reservoir basin should be water tight without having any embedded fissured formations.

(ii) Position of ground water table.

(iii) Type of soil and its suitability in regard to foundation of the dam.

(iv) Type and depth of surface soils.

(v) Position of quarry sites for materials required for the construction of dam etc.

Trial pits may be bored in the catchment area and geological conditions investigated. The catchment should be free from such faults as may cause serious leakages when catchment is filled with water.

9.5 SELECTION OF RESERVOIR SITE

Before finally selecting the reservoir site following factors should be seriously considered.

1. Catchment area should have such geological conditions that percolation and absorption losses are minimum.

2. Available run-off should be maximum.

3. The site should be free from fissured rocks. This will avoid possibilities of leakage when reservoir is full to capacity.

4. The reservoir site must have adequate capacity.

5. The reservoir basin should have a deep narrow opening in the valley so that the length of the dam may be kept minimum.

6. Heavily silt laden tributaries should not lead their discharge to the reservoir.

7. Suitable site for dam should be available. It will be an ideal site if dam is constructed at the narrow and shallow part of the river which lies down stream of the deep river. It is very important point as cost of dam is often a controlling factor in selection of the reservoirs site.

8. Site should be such that deep reservoir is formed. Deep reservoir would store more of water and expose minimum area at the surface for evaporation.

9. If earthen dam is proposed to be constructed, then separate suitable site for spillway works should be available.

10. Reservoir site should be well connected by rail and road.

11. Materials for the construction of dam should be available nearby.

12. The soil formation at reservoir site should be free from harmful salts.

13. If reservoir water is to be used for irrigation, the dam site should be near the area proposed to be irrigated. This would reduce the length of the canal system and consequently the cost of the project.

14. Reservoir should not submerge habited area or areas of fertile lands or gardens.

15. River banks should be hard, strong and high so that cost on river training works is minimum.

9.6 SOME IMPORTANT TERMS

1. Normal Pool Level. This is the maximum level of water in the reservoir to which the water will rise during ordinary operating conditions. This level corresponds either to the crest level of the spillway or top level of the spillway gates. This level is also known as maximum conservation level.

2. Minimum Pool Level. The lowest elevation of water which has to be maintained in the reservoir under normal operating conditions is called minimum pool level. The elevation is guided either by the lowest outlet in the dam or by minimum head required for efficient functioning of turbines.

3. Maximum Pool Level. During floods water starts flowing over the spillways. But discharge over spill-ways increases with further rise in reservoir level. The maximum level attained by water during worst design flood is known as the *maximum pool level.* This level of water is essentially above the *Normal pool level.*

9.7 STORAGE ZONES OF A RESERVOIR

1. Dead Storage. The volume of water stored below the minimum pool level of the reservoir is known as *dead storage.* This storage is not of such use in the operation of the reservoirs.

2. Useful Storage. The volume of water stored in a reservoir between minimum pool level and normal pool level is known as useful storage. The useful storage may be further classified into conservation storage and flood mitigation storage in a multi-purpose reservoir.

3. Surcharge Storage. The volume of water stored between normal pool level and the maximum pool level is known as *surcharge storage.* This storage is an uncontrolled storage. It exists only till floods are in progress and cannot be retained for later use.

4. Bank Storage. When reservoir is full of water some amount of water seeps in the permeable banks of the reservoir. This seeped water comes out as soon as the reservoir level gets depleted. This amount of seeped water which becomes available after the reservoir is depleted is known as Bank storage. Amount of bank storage may amount to several per cent of the reservoir storage depending upon the geological formations of the banks. This storage increases the capacity of the reservoir above that indicated by the elevation-volume curve as shown in Fig. 9.2.

5. Valley Storage. Some amount of water is stored by the stream channel even before a dam is constructed. This storage is known as *valley storage*. Amount of valley storage is variable as it depends upon the rate of flow in the reservoir. After

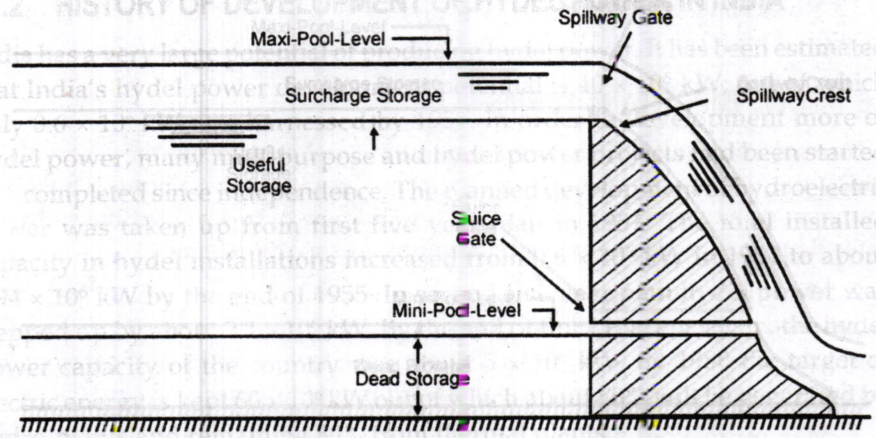

Fig. 9.3. *Various storages.*

the construction of dam the valley storage increases. The net increase in the storage is equal to the storage capacity of the reservoir and less than natural valley storage. The valley storage thus reduces the effective storage capacity of the reservoir. This storage is not of much consequence in conservation reservoirs, but the available storage for flood mitigation is reduced as follows :

Effective storage for flood mitigation

= Useful storage + Surcharge storage – Valley storage.

9.8 VARIOUS YIELDS OF RESERVOIR

1. Yield of the Reservoir. It is the amount of water that can be supplied by the reservoir in a specified interval of time. The specified time interval may vary from a day for a small distribution reservoir to a month or year for large conservation reservoirs. If we say that three million cubic metres of water can be supplied from a reservoir in a year then its yield is 3000000 m^3/year. The yield of the reservoir is dependent upon the inflow and thus varies from time to time.

2. Safe Yield. It is also known as firm yield. It is the maximum quantity of water that can be supplied from the reservoir with full guarantee during the worst dry period.

3. Design Yield. The critical period for a reservoir is generally considered, when natural flow in the reservoir is minimum. There is possibility that sometimes the minimum natural flow in the reservoir may even fall short of guaranteed yield. Hence a lower value than the guaranteed yield or safe yield may be taken for design purpose. This yield whose value is smaller than the safe or firm yield is known as *design yield.* The value of *design yield* for a reservoir to be used for water supply is taken less than the safe yield. In the case of reservoirs used for irrigation purpose the design yield may be taken slightly more than the safe yield as crops can tolerate some deficiency of water during exceptionally dry season.

4. A Secondary Yield. The quantity of water available in excess of safe yield is known as *secondary yield.* This yield is available during period of high inflows. This secondary yield of the reservoir can be used either to generate extra hydro-electric power or for irrigation of extra lands.

5. Average Yield. The arithmetic average of the safe yield and the secondary yield considered for a number of years is known as *average yield.*

The storage capacity of the reservoir and its yield are very much interdependent. The water is stored in the reservoir to fulfill the safe yield requirements. If capacity of the reservoir is more it can certainly provide more water and hence yield is more. The reservoirs are designed to meet a specific demand of water. The capacity of the reservoir and the yield are governed by the following storage equation :

Inflow – outflow = Increase in storage.

9.9 MASS INFLOW CURVE

Mass inflow is a curve which represents the cumulative flow in a reservoir at any particular instance. It is a plot between cumulative inflow in the reservoir with time. It can be prepared with the help of a hydrograph of the river for the dam site for a large number of years. We know hydrograph is a plot or curve between discharge versus time.

Figure 9.4 (a) shows a hydrograph of a river for the dam site for a number of years and Fig. 9.4 (b) a corresponding mass curve prepared from the hydrograph of Fig. 9.4 (a). Let starting year of the hydrograph is 1998. The total quantity of water that has flown through the river from 1998 to a time t_1 (say 2001) has been represented by the hatched area. In mass curve, the corresponding ordinate at time t_1 (ordinate *AB*) will represent the total volume of water indicated by the hatched area of hydrograph curve. Similarly the ordinates of the mass curve corresponding to other years can be determined from the hydrograph curve and plotted. When all such points are joined free hand we get the mass curve corresponding to the hydrograph. It is evident that a mass curve will continuously rise as it is the plot of accumulated inflow versus time. Periods of no inflow would be indicated by horizontal lines on the mass curve. The slope of the mass curve at any time gives the rate of inflow at that time. Mass inflow curve is sometimes known as Ripple diagram also.

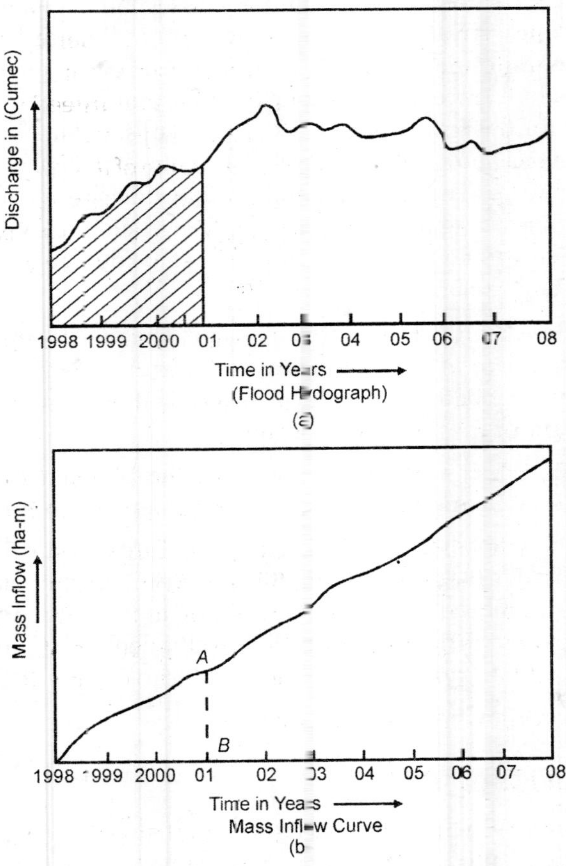

Fig. 9.4. *Flood hydrograph and mass inflow curve.*

9.10 DEMAND CURVE

It is a plot between accumulated demand of water from reservoir with time. If demand of water is at constant rate then demand curve becomes a straight line. The slope of the demand curve represents the rate of demand. If demand is variable the demand curve does not remain a straight line but assumes curved shape. See Fig. 9.5 given on next page.

9.11 DETERMINING RESERVOIR CAPACITY FOR GIVEN YIELD

The mass inflow curve and the demand curve are used to determine the required reservoir capacity. From the flood hydrograph of several years prepare a mass inflow curve as described earlier. Draw demand curve near the mass inflow curve keeping scale of both as the same. Let A_1, A_2, A_3......etc. be the peak points in the mass inflow curve. Draw A_1B_1, A_2B_2, A_3B_3 etc. lines parallel to demand line

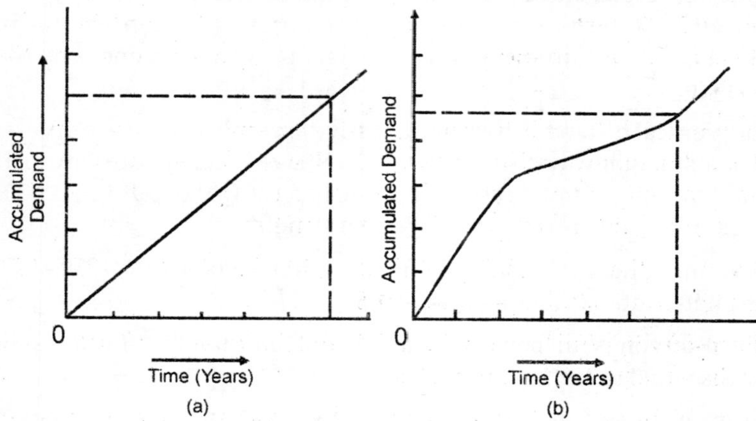

Fig. 9.5 *Demand curves (a) Constant rate (b) Variable rate.*

at points A_1, A_2, A_3 etc., respectively. It is evident that demand lines A_1B_1, A_2B_2, A_3B_3 etc., all represent rate of withdrawal or demand from the reservoir. The intersection of demand line A_1B_1, A_2B_2, A_3B_3 etc., with mass inflow curve at points B_1, B_2, B_3 represent reservoir to be full. The vertical intercepts between demand lines A_1B_1, A_2B_2, A_3B_3 etc., from mass inflow curve represent water

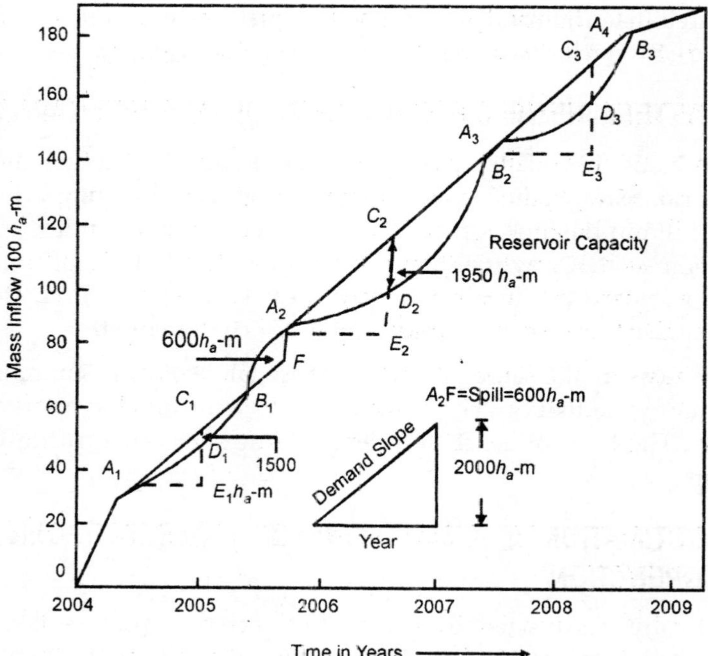

Fig. 9.6. *Determining reservoir capacity.*

required to be stored in the reservoir. C_1D_1, C_2D_2, C_3D_3 etc., are the vertical intercepts between demand curve and mass inflow curve. The largest of these intercepts represents the required reservoir capacity. Figure 9.6 shows a constant demand of 2000 hectare metre/year. The largest intercept shows reservoir capacity as 1950 hectare metres/year. The following inferences can be drawn from the Fig. 9.6.

1. The vertical distance between successive tangents A_1B_1. A_2B_2 represent water wasted over the spillways. The spillway must have sufficient capacity to discharge this flood volume. In our case spillway capacity works out to be 600 hectare metre and reservoir capacity as 1950 hectare metre.

2. Assuming the reservoir to be full at A_1, it is depleted to 1950 – 1500 = 450 hectare-metres at C_1 and is again full at B_1.

3. The reservoir is full between B_1 and A_2 and the quantity of water spilled over spillway is equal to 600 hectare-metres.

4. From A_2 the water stars reducing in the reservoir till it becomes empty at point C_2.

5. The water again starts accumulating in the reservoir and it is full again at point B_2.

6. If somewhere tangentially drawn demand line does not intersect the mass curve it indicates that reservoir will not get filled to its full capacity.

7. If demand curve is not a straight line, then mass inflow curve and demand curves should be superimposed over each other. While superimposing it should be ensured that origin and axis of both the curves should coincide. The largest ordinate between the two gives the required storage capacity.

9.12 DETERMINE YIELD FOR A RESERVOIR OF GIVEN CAPACITY

In this case also first of all prepare a mass inflow curve. Near the same diagram draw various straight lines from a common origin representing various rates of demand. From the apices points A_1, A_2, A_3 etc., of mass inflow curve, draw tangents in such a way that their maximum vertical intercept from the mass curve does not exceed the given reservoir capacity. Thus C_1D_1, C_2D_2, C_3D_3 etc., are all equal to the reservoir capacity. (Say 2500 hectare-metre).

Now measure the slopes, of each of these tangents. The slopes indicate the demand or yield that can be achieved in each year from the reservoir of given capacity. The slope of the flattest demand line is the safe or firm yield of the reservoir.

9.13 ESTIMATION OF DEMANDS AND OPTIMAL RESERVOIR OPERATION

It has already been discussed as to how the reservoir capacity is determined for some specific demand and vice versa. The demand may remain constant throughout the year or may be varying from time to time. Demand pattern from a reservoir actually depends upon the purpose for which the reservoir has been made. Hence it is obvious that demand will vary for different types of reservoirs. Moreover, the demands are man-made, and if need be, they can be adjusted in such a way that maximum benefits are derived from the reservoir. If by slightly varying the demands, lot of savings can be made then we must go in for that. This aspect is known as optimal planning for reservoirs. It involves economic considerations.

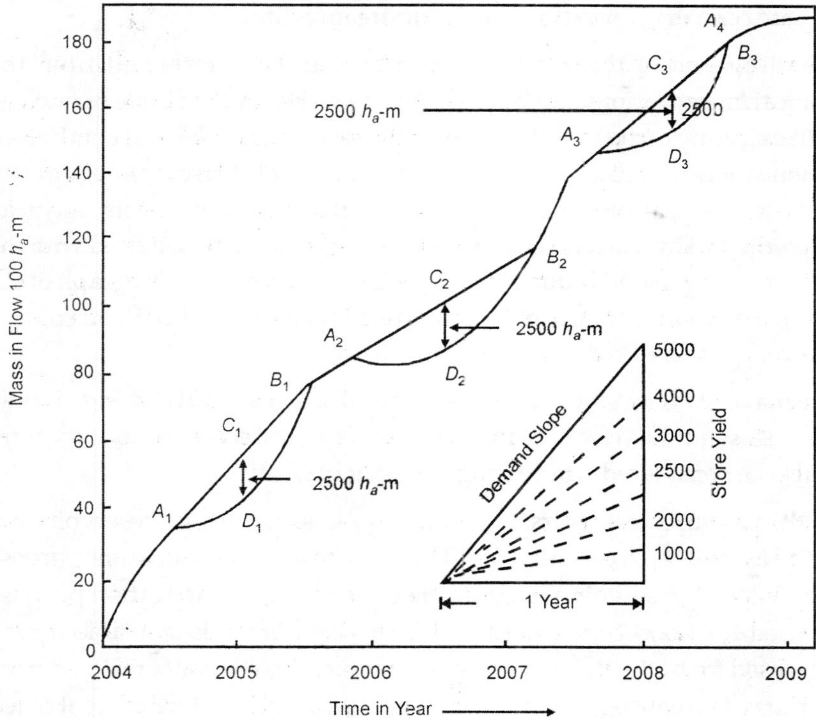

Fig. 9.7. *Determining yield from reservoir of given capacity.*

9.13.1 Demanding Patterns for Various Types of Reservoirs

1. Single purpose conservation reservoir. This reservoir stores water only to meet one particular demand. In all such cases, the demand for water can be predicted quite accurately for different parts of the year and required storage capacity can be accurately worked out as explained earlier. The worked out storage capacity should be sufficient for other uses also. The storage capacity should be justified on cost-benefit considerations also.

2. Single purpose flood control reservoirs. In such reservoirs there is no specific down stream demand. The only down stream demand, in this case, is the rate of release of water from the reservoir. The rate of release of water should be such that it does not exceed the safe carrying capacity of the down stream channel.

Since these reservoirs have to store huge quantities of flood water temporarily, their capacities should be ample so as to cause sufficient lowering of the peak of the flood discharge. The capacity of flood control reservoir is dependent upon following factors :

(i) Hydrograph of the worst flood that is likely to enter the reservoir.

(ii) Down stream permissible high flood level.

(iii) Safe carrying capacity of the down stream channel.

The hydrograph of the worst flood inflow can be determined from the hydrological investigations of the catchment area. The peak safe rate of outflow from the reservoir can be found out from the downstream channel conditions. After having found out these two factors, the capacity of the reservoir required to modify the inflow peak of a value equal to or less than the down stream safe peak can be determined by trial and error method or graphical method used with the help of *flood routing*. Flood routing is the process by which the hydrograph of the modified or moderated flood can be determined. The methods of flood routing have been explained a little later.

Hence maximum capacity of such reservoirs should be equal to design inflow less safe release in the downstream channel. However economic background should also be considered while fixing reservoir capacity.

3. Multi-purpose reservoirs. Nowadays practically all the reservoirs are constructed as multi-purpose reservoirs. They have to serve more than one purpose simultaneously. A reservoir made for conserving water for irrigation purpose may be used for flood control also and if sufficient head is available it may further be used for hydroelectric power generation also. The water released from reservoir may be running the turbine for hydro-electric power before it is led through canals for irrigation. The reservoir may also be acting as source of a water supply scheme along with its use for irrigation, flood control and hydroelectric power generation. Following may be combinations for a multipurpose reservoir :

(i) Irrigation + Power

(ii) Irrigation + Power + Flood control

(iii) Irrigation + Water supply

(iv) Flood control + Water supply

(v) Irrigation + Power + Flood control + Water supply.

Besides the above combinations, fisheries, recreation etc. also develop along the banks of reservoir.

Combination of various purposes for which water released from reservoir is used is considered as the most effectively utilized water.

But all the uses of water from a reservoir have their own characteristics. All the uses are not entirely compatible with each other. But by careful and scientific planning, all the uses may be brought into a good agreement. This planning is also known as *schedule of operations*. Hence the main feature of multi-purpose reservoirs is a scientific schedule of operations, with the help of which reservoir water can be utilized in the best possible manner. Schedule of operations is nothing but a compromise among the various uses of water. Before any schedule of operations for any particular reservoir is finally adopted, a number of tentative schedules may have to be drawn, on the basis of available data and past similar experiences. The schedule which promises maximum total benefits, from various uses, without encroaching upon the lower and upper limits of storage is accepted at the beginning. This schedule is run for sometime and if it needs any modification it is carried out and finally adopted.

The reservoir storage capacity may be used on the basis of following two assumptions :

1. Assuming that no storage is used jointly and every use of water requires separate space for storage.
2. Assuming that all the storage is jointly used in various purposes.

In the first assumption the storage spaces for all the purposes, are separately evaluated and added together to create a very large total storage in the reservoir. This storage can be economically obtained only if unit cost of storage is constant or even decreases as total storage increases. In the second assumption the same water is used for various purposes one after the other. This measure requires minimum storage which may be equal to maximum amount of water required for any one purpose out of several purposes. This second assumption is usually adopted in the operation of multi-purpose reservoirs almost everywhere in the world.

9.14 SCHEDULE OF OPERATIONS FOR A MULTI-PURPOSE RESERVOIR

In India most of the multi-purpose reservoirs are planned to perform the three functions, namely irrigation, flood control and hydro-electric power. Planning for such a reservoir requires lot of experience, past records of run-off, and other hydrological data. A typical schedule of operations for a multi-purpose reservoir constructed on a perennial snow fed river, in North India can be operated as follows :

The water level in the reservoir will normally fluctuate between normal pool level and minimum pool level. The storage between these two limits of water levels, will be maximum conservation of water which is sufficient for satisfying the irrigation needs. As water level in the reservoir shall not be allowed to go below the minimum pool level, the firm commitments of power will continue to be generated upto this level of water. Normal floods will be absorbed in the capacity lying below the normal pool level. Sever floods shall be absorbed by the capacity, lying between normal pool level and maximum pool level. Spillways start discharging water, as soon as the water level starts rising above the normal

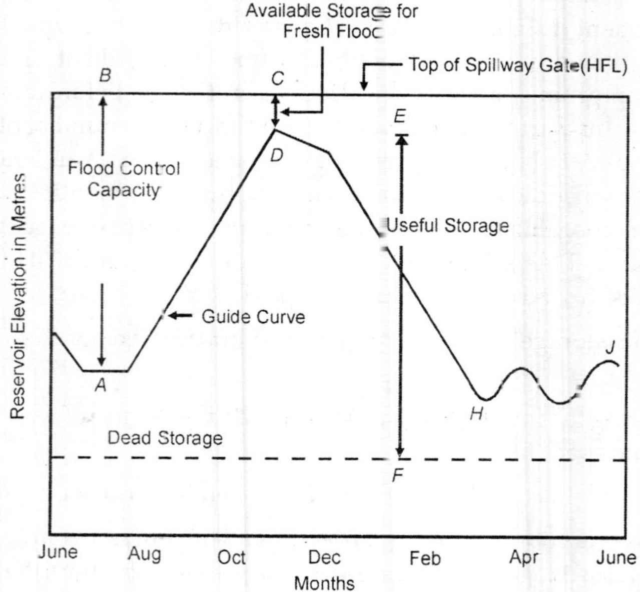

Fig. 9.8. *Guide curve for operation of a reservoir.*

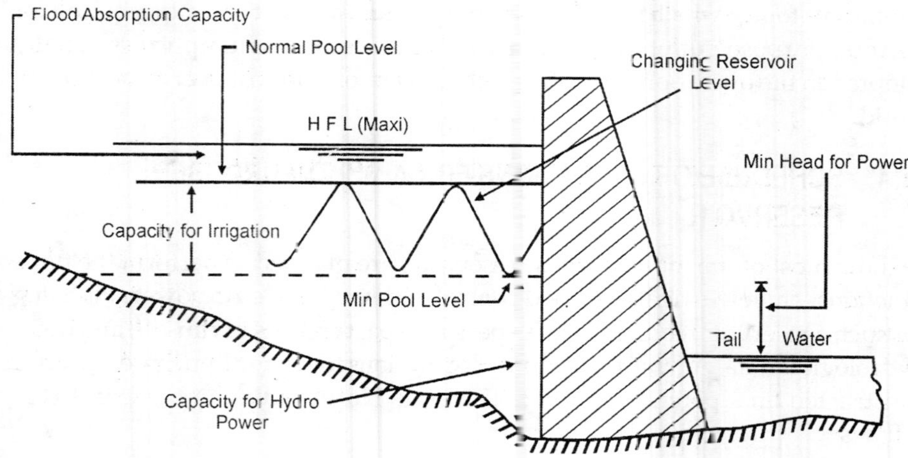

Fig. 9.9. *Various storages and fluctuation in water level in reservoir during operation.*

pool level. The discharge from the spillways depends upon the head of water above the crest of spillways. But discharge from the spill ways should not be allowed to exceed the safe carrying capacity of the down stream channel. Because of this limitation discharge from the spillway may be less than the flood inflow in the reservoir and this causes temporary accumulation of flood water between normal pool level and the maximum pool level. As the floods recede the excess temporarily stored water, between maximum pool level and normal pool level, slowly goes out within few days and reservoir becomes ready to absorb another flood.

For irrigation, we know Rabi and Kharif crops are sown in Northern India. Kharif crops are sown from April to June and harvested from September to November. Rabi crops are sown from November to January and harvested from March to April. Kharif crops require irrigation water from April to June and Rabi crops from November to February. Monsoons generally start at the beginning of July and last upto middle of October. Because of heavy outflow for Kharif crops and constant release for power generation, the reservoirs are almost in completely depleted state. Thus just before the onset of monsoon season, almost full reservoir capacity is available in empty form, not only for conservation of water but also to control the floods that may occur during rains. During the monsoons, the demand for irrigation is almost zero and water will be released only for power generation. The reservoir level will thus continue to rise steadily. The rise in water level is allowed to reach the normal pool level by the end of normal monsoon season, say middle of October. During monsoons if there happens to be unprecedented floods, the flood water will be allowed to rise upto maximum pool level and the temporarily stored flood water above normal pool level will flow out through the spillways provided for the purpose. But care should be taken that release of water from spill ways should not exceed the safe capacity of the down stream channel.

After the end of monsoon the inflow in the reservoir reduces considerably but water is constantly drawn for Rabi crops and power generation. This causes fall in the reservoir level and this process continues upto middle of February. At this stage water level in reservoir almost reaches the minimum pool level. After February snow starts melting in the head reaches of the river which augment the river discharge. Winter rains which may occur in December or January also help in augmenting the river discharge and thus reservoir storage. From March onwards, the discharge in reservoirs generally increases due to melting of snow and an equally large outflow from the reservoir is taken to satisfy the needs of Kharif crops. Thus from March to June the reservoir level may fall or rise, depending upon the amount of inflow and outflow ; but this rise or fall in the water level is very slow. At the end of June i.e. beginning of monsoons, the reservoir is almost completely depleted and ready to receive the large inflows and flood discharges of the coming rainy season. Thus operation pattern of a multi-purpose reservoir serving the needs of irrigation, power generation, and flood control is set. A curve between reservoir elevations and the time in months of a year may be prepared based on past experiences and reservoir storage may be operated according to it.

9.15 FLOOD ABSORPTION OR FLOOD ROUTING

Flood routing is a process with the help of which characteristics of hydrograph of a flood, entering the reservoir are completely changed when it emerges out of the reservoir. The change in flood hydrograph characteristics takes place because certain volume of flood is stored in the reservoir temporarily and is let off as the floods recede. The base of the modified hydrograph, which emerges from the reservoirs becomes wider and its penk gets reduced. The extent by which the inflow hydrograph gets changed due to reservoir storage can be computed by a process known as flood routing. It is an important technique with the help of which solution of a flood control problem is arrived at. In flood routing problem one has to calculate water levels in the reservoir, the storage quantities and outflow rates corresponding to a particular inflow hydrograph at various instants. It is used to determine the level upto which water may rise during floods and also the rate of the discharge in the down stream channel when a particular flood passes through it.

9.16 METHODS OF FLOOD ROUTING

There are several methods, but following two methods, are in most common use:

1. Graphical method
2. Trial and error method.

1. Graphical Method. The relation between inflow, outflow, and change in storage can be expressed mathematically as follows:

$$I = O + \Delta s \qquad (1)$$

where I = Average inflow during given time interval

O = Average outflow during same time interval

Δs = Volume of water stored or change in storage during the same time interval.

Increase in storage should be taken as positive and decrease in storage as negative.

This relation seems to be very simple but actually its evaluation is not as simple. This is because of the fact that relations between rate of inflow and time, elevation and storage of reservoir, and rate of outflow and time, cannot be expressed by simple algebraic equations.

Let I_1 and I_2 be the inflow rates and O_1 and O_2 be the outflow rates at the beginning and end of the time interval t. Let S_1 be the storage in the beginning and S_2 at the end of the time interval t. These values can be substituted in the Eq. (1) as follows :

$$\left(\frac{I_1 + I_2}{2}\right)t - \left(\frac{O_1 + O_2}{2}\right)t = S_2 - S_1 \qquad (2)$$

$$(I_1 + I_2) - (O_1 + O_2) = \frac{2(S_1 - S_2)}{t}$$

$$(I_1 + I_2) + \left(\frac{2S_1}{t} - O_1\right) = \left(\frac{2S_2}{t} + O_1\right) \quad (3)$$

Now before further analysis could be carried out, following graphs are required:

1. Inflow hydrograph. 2. Elevation-outflow curve of the reservoir and 3. Elevation-storage curve of the reservoir. Inflow rate ordinates I_1, I_2, I_3 etc. are read out from the inflow hydrograph at suitably chosen interval. Figure 9.10 shows an inflow hydrograph at 3 hours time interval and Fig. 9.12 shows an inflow hydrograph at 4 hours time interval.

Now draw the curves between $\left(\frac{2S}{t} \pm O\right)$ versus outflow O. These curves are shown in Fig. 9.14. To draw these curves values of S and O for various values of reservoir elevation are read from elevation outflow, and elevation storage curves shown in Fig. 9.13. Knowing S and O, values of $\left(\frac{2S}{t} \pm O\right)$ are found out and plotted against the corresponding outflow rates O as shown in Fig. 9 14. Here t is the any chosen time interval.

Now at the beginning both S_1 and O_1 are zero. By putting these values in Eq. (3), we get

$$I_1 + I_2 = \frac{2S_2}{t} + O_2.$$

For first time interval t, the rates of inflow at the beginning and end are respectively I_1, I_2, I_1 and I_2 are read from the inflow hydrograph curve. Hence $\frac{2S_2}{t} + O_2$ is known.

From Fig. 9.14 (a) and (b) curves enter this value of $\frac{2S_2}{t} + O_2$ and get O_2. For this value of O_2 also find $\frac{2S_2}{t} - O_2$ from the same figure curve (a). Now apply Eq. (3) for the next time interval

$$(I_2 + I_3) = \left(\frac{2S_2}{t} - O_2\right) = \frac{2S_3}{t} + O_3.$$

In this equation again I_2 and I_3 are read from mass inflow curve and $\left(\frac{2S_2}{t} - O_2\right)$ is known as explained above. Hence $\left(\frac{2S_3}{t} + O_3\right)$ is known. From Fig. 9.12 find the value of O_3 corresponding to the value of $\left(\frac{2S_3}{t} + O_3\right)$. Also find $\left(\frac{2S_3}{t} - O_3\right)$ corresponding to the value of O_3. Similarly applying Eq. (3) for the next time interval, we get

$$(I_3 + I_4) + \left(\frac{2S_3}{t} - O_3\right) = \left(\frac{2S_4}{t} + O_4\right).$$

From which $\left(\frac{2S_4}{t} + O_4\right)$ is known and the process is repeated for the complete period of flood.

Steps of Graphical Method

1. Determine total inflow for the first time interval t from the inflow hydrograph. Enter this value as AB in Fig. 9.14.

2. Through B draw a vertical line to meet the curve $\frac{2S}{t} + O$ at pint C. The point C represents the value of outflow at the end of the interval.

3. Through C draw a horizontal line $A\ CB_1$ cutting $\left(\frac{2S}{t} - O\right)$ curve in point A_1.

4. Now calculate the total inflow during second subsequent time interval t again from inflow hydrograph. Measure the total inflow from A_1 on line A_1CB_1 and mark point B_1.

5. Again draw a vertical ordinate from B_1 to meet the curve $\left(\frac{2S}{t} + O\right)$ a point C_1 and repeat the procedure as stated in steps (3) and (4) until the entire flood is routed.

6. The outflow discharge at any time interval is given by the total vertical ordinate. The largest of these ordinates will indicate the value of the peak outflow rate. The spillways are designed based of the largest discharge.

7. After determining the outflow discharge at various intervals as described above, the reservoir water level for these can be determined from (Fig. 9.13) graph II. Example 9.1 with clarify the graphical method.

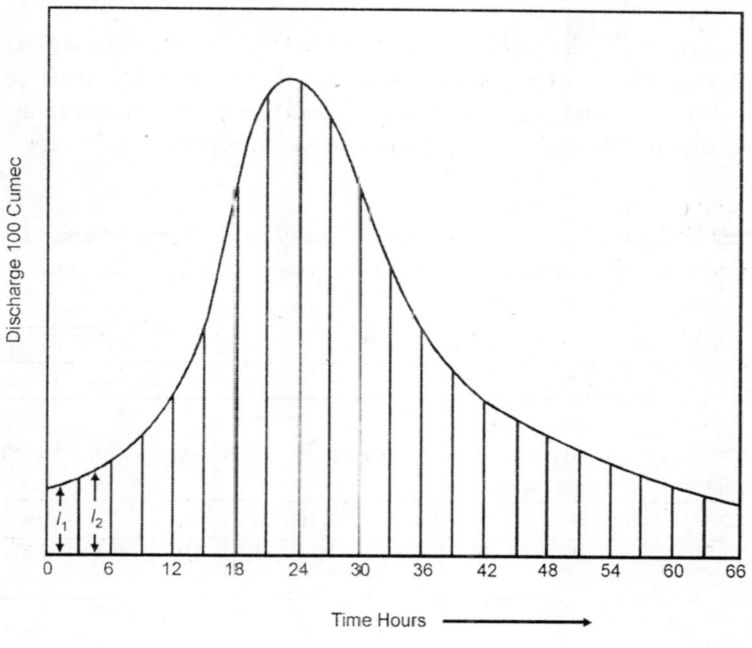

Fig. 9.10

Elevation-storage curve and elevation. Outflow curve may be drawn separately as shown in Fig. 9.11. They can be drawn in one diagram also. Figure 9.13 shows

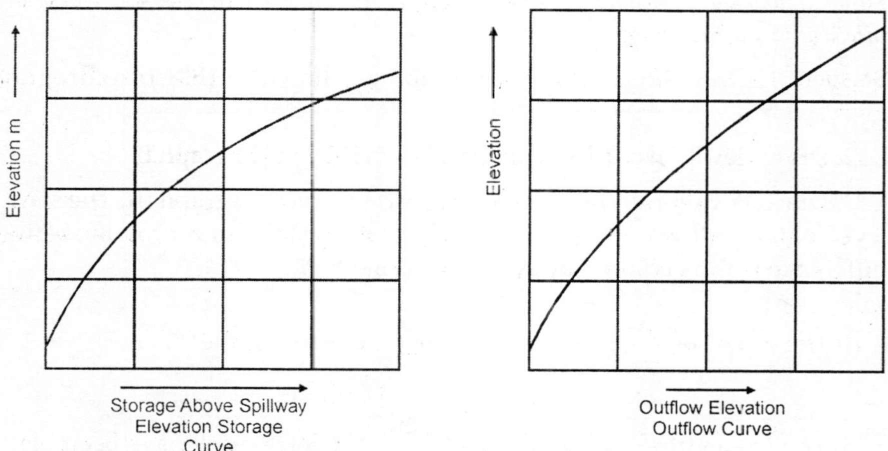

Fig. 9.11

elevation-storage curve and elevation-outflow curves on the same plot. Figure 9.13 shows cumulative storage on top n-axis time and outflow in bottom x-axis line. Example 9.1 has been solved graphically and solution has been shown on Figs 9.12, 9.13, and 9.14.

One thing should be clearly understood that to account for the worst condition, the water surface in the reservoir is normally taken to be at the maximum conservation level or the full reservoir level and hence at the beginning of the first interval of time t, both storages and outflow O, are taken equal to zero.

Example 9.1. *The following is the data of an inflow flood hydrograph whose flows in 100 cumecs have been recorded at 4 hours interval starting from zero hour on May 1, 2008 on a certain stream.*

0.55	0.65	0.75	0.85	1.25	1.95	2.50	3.35	4.25
4.20	3.95	3.50	2.75	2.25	1.75	1.25	1.15	1.05

This flood is reaching a reservoir having ungated spillways of which elevation area and Elevation outflow data are as follows :

Elevation	100.00	100.5	101.0	101.5	102.0	102.5	103.0	103.5	104.0	104.5
Area in hectares	355	365	395	405	410	430	440	455	470	490
Overflow in cumecs	0	12.5	32.5	60.0	90	130	170	225	275	350

It may be assumed that water level in the reservoir reaches the crest level (100.00) of the spillways at zero hours of May 1. 2008.

Determine the maximum reservoir level and maximum discharge over the spillway. Draw in flow and routed hydrograph indicating the reduction in peak flow and peak lag introduced due to routing.

Solution. 1. Draw the inflow hydrograph as shown in (Flood hydrograph) Fig. 9.12.

2. Draw elevation outflow curve as shown in Fig. 9.13 graph II.

3. Draw elevation storage curve as shown in Fig. 9.13 graph III. The starting level for the spillway is the crest level. For plotting this curve, volume between different contours is found by average area method.

4. The values of $\frac{2S}{t} \pm O$ have been found out in the Table 9.1.

5. The values of $\frac{2S}{t} - O$ (Col. 7) and $\frac{2S}{t} + O$ (Col. 8) each have been plotted against the outflow discharge (Col. 2) in Fig. 9.14.

Table 9.1

Elevation	*Outflow in cumecs*	*Area between contours* $\times 10^4 m^2$	*Storage volume between successive contours* $\frac{(A_1 + A_2)}{2} H \times 10^4 m^3$	*Cumulative storage above spillway crest in* $\times 10^4 m^3$ *(S)*	$\frac{2S}{t}$ *for t =4 hes* $\frac{S}{7200}$	$\frac{2S}{t} - O$	$\frac{2S}{t} + O$
(1)	(2)	(3)	(4)	(5)	(6)	(7)	(8)
100.0	0	355	—	—	—	—	—
100.5	12.5	365	180.0	180.00	250.00	237.5	262.5
101.0	32.5	395	190.0	370.00	513.83	481.39	546.39
101.5	60.0	405	200.0	570.0	791.67	731 67	851.67
102.0	90.0	410	203.75	733.75	1074.65	984.65	1164.65
102.5	130.0	430	210.00	985.75	1369.10	1239.10	1499.10
103.0	170.0	440	217.5	1201.25	1668.40	1498.40	1838.40
103.5	225.0	455	223.75	1425.00	1979.17	175417	2204.17
104.0	275	470	231.25	1656.25	2300.35	2025.35	2575.35
1045	350	490	240.0	1896.25	2633.68	2283.68	2983.68

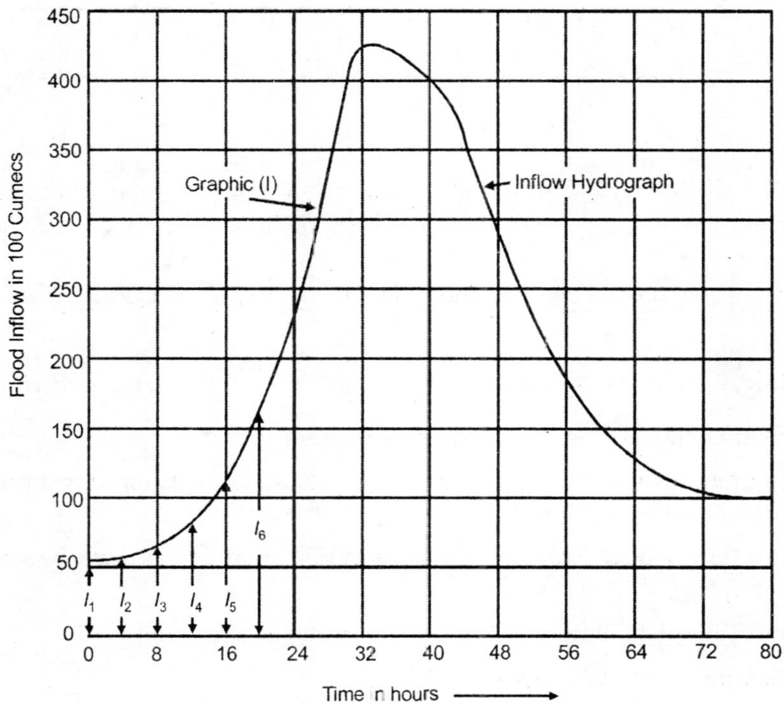

Fig. 9.12. *Inflow hydrograph.*

The routing method is same as described under the graphical method. Instead of drawing vertical ordinates and doing purely graphically, the values can be

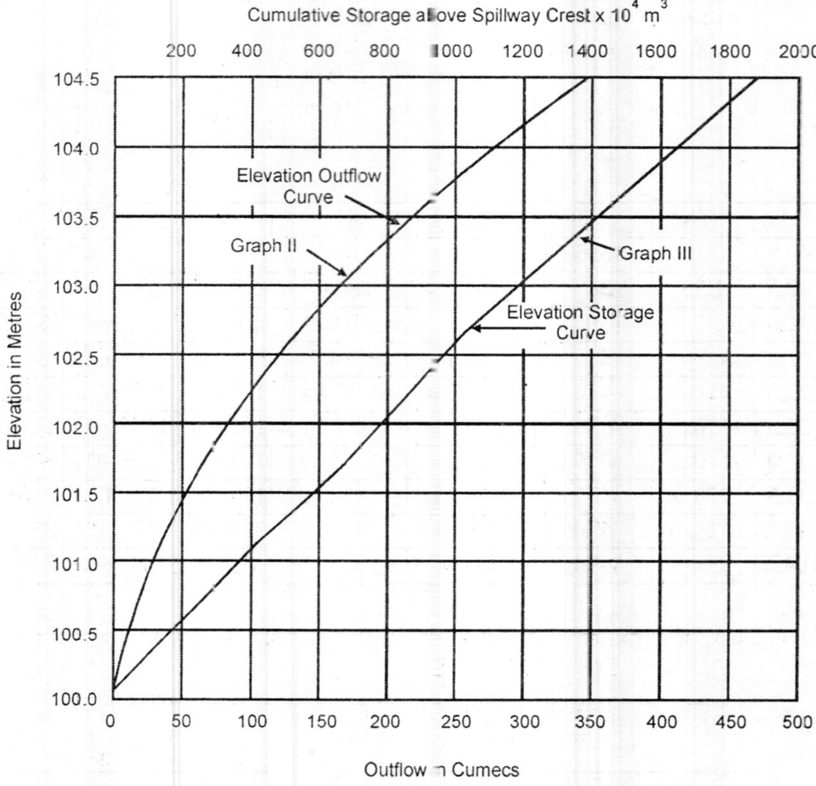

Fig. 9.13. *Storage elevation and outflow relationship.*

tabulated as given in Table 9.2. The values of $\left(\frac{2S}{t}-O\right)$ and O are read each time from Fig. 9.14 curve (A). For different values $\frac{2S}{t}+O$, the tabulated values are given here in the Table 9.2. Time interval is 4 hours.

The method of drawing vertical ordinates and corresponding horizontal ordinates are shown in Fig. 9.14.

The inflow flood hydrograph and outflow flood hydrograph have been plotted together in Fig. 9.15.

From Table 9.2 following results can be immediately noted.

1. Maximum reservoir level = 104.37 m.

2. Maximum discharge over spillway = 320 cumecs.

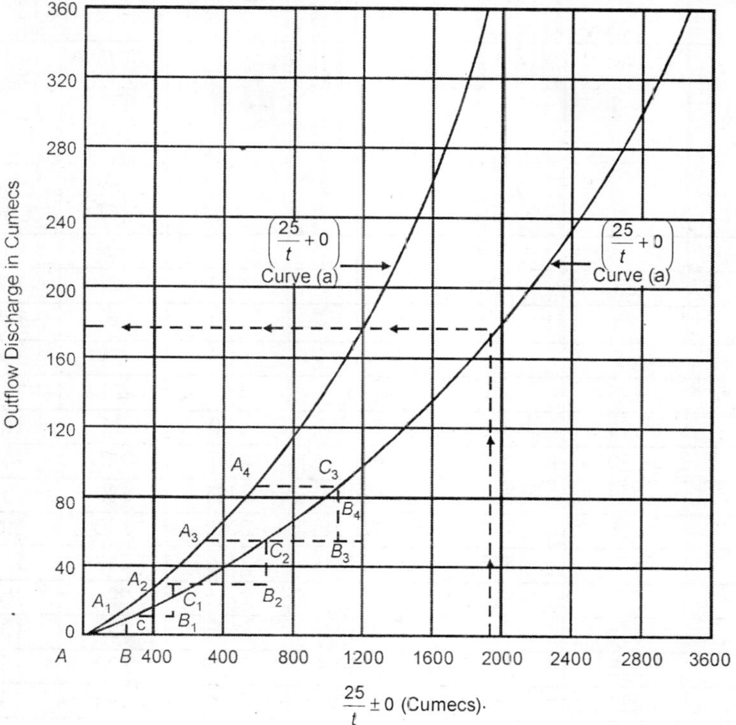

Fig. 9.14. $\left(\frac{2S}{t}\pm 0\right)$ *curves.*

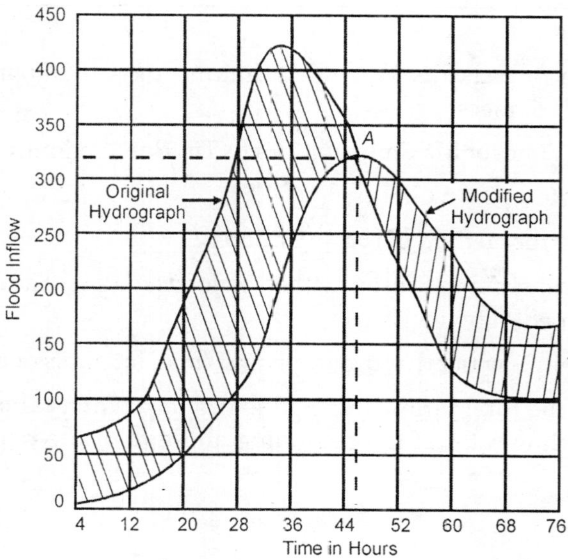

Fig. 9.15. *Modified hydrograph.*

TABLE 9.2

Time from start May 1 2008 (1)	Inflow in cumecs I (2)	$I_1 + I_2$ (3)	$\frac{2S}{t} - O$ Read from Fig. 9.14 (4)	$\frac{2S}{t} + O$ (5)	Outflow (6)	Reservoir elevation Read from Fig. 9.13 (7)
0	55				0	
4	65	120	0	120	2	100.10
8	75	140 →	60 ⇄	200	6	100.20
12	85	160 →	140 ⇄	300	14	100.50
16	125	210 →	220 →	→ 430	24	100.80
20	195	320	380 ←	700	44	101.20
24	250	445	600	1045	80	101.82
28	335	585	900	1485	130	102.50
32	425	760	1250	2010	190	103.17
36	420	845	1620	2465	248	103.75
40	395	815	1900	2715	294	104.18
44	350	745	2100	2845	320	104.37
48	275	625	2200	2825	320	104.37
52	225	500	2200	2700	294	104.18
56	175	400	2100	2500	262	103.87
60	125	300	1980	2280	232	103.55
64	115	240	1800	2040	196	103.25
72	105	220	1640	1860	174	103.05

3. Reduction in peak discharge (Maximum inflow – Maximum outflow) (425 – 320) = 105 cumecs.

4. Peak lag = Time for Maximum outflow – Time of Maximum inflow

= 46 – 32 = 14 hours.

2. Trial and error method of routing floods

This is the most commonly used method of floods routing. The method may be enumerated as per following steps :

1. Divide the inflow flood hydrograph into small intervals of time.

2. Calculate the total inflow during the first time interval by multiplying the average inflow rates at the beginning and end of the period with time interval t.

i.e. $\left(\frac{I_1 + I_2}{2}\right)t.$

3. The level of water at the start of the flood is known. It is taken at the crest level of the spillways.

4. Assume a trial level of water in the reservoir at the end of the first period t.

5. Using elevation outflow-curve, calculate the outflow at the start and end of a particular time interval corresponding to water levels. When average value of outflow is multiplied by the time interval we get the net outflow during the period.

$$\left(\frac{O_1 + O_2}{2}\right)t.$$

6. Using the elevation storage curve find out the storage at the beginning and end of the time interval for the corresponding known water levels in the reservoir. Find out the difference in storages at the start and end of the time interval i.e. $(S_2 - S_1)$. This difference represents the amount of flood stored in the reservoir during that period.

7. Add the value of outflow obtained in step (5) to the value of storage obtained in (6). The value should be equal to the total inflow during the period as calculated in step (2). If the value i.e. sum of (5) and (6) and (2) are equal the assumed trial elevation of the reservoir is correct. If not assume some other value of water elevation and trial repeated, till value of sum of (5) and (6) becomes very nearly equal to (2).

8. Repeat from steps 2 to 7 for all other little intervals till the flood is completely routed.

The Problem 9.1 solved by graphical methods has been solved by this method also.

9.16.1 Solution by Trial and Error Method

Make a table as shown in Table 9.3. Choose time interval of 4 hours and enter these values in column (1) of this table. In column (2) enter reservoir levels at start of each time interval. Calculate inflow of flood and enter its values in column 3 for each time interval. Now using discharge-elevation curve of Fig. 9.13 find out the discharge over spillway at the start of each time interval corresponding to level of water in column (2) and enter them in column 4. Assume trial elevations, and enter them into column (5). For these trial elevations, calculate spillway discharge and enter them in column (6). Determine the mean outflow rate by taking mean of column (4) and column (6). Enter mean values in column (7). Convert the values of column (7) into outflow volume in m^3 and enter in column (8). The volume of outflow will be less than inflow volume so long as the peak of flood has not been routed. Once the peak of the flood has been routed, the volume of outflow exceeds the inflow volume. At this Lillie it is considered that maximum flood conditions have occurred.

Using storage elevation curve Fig. 9.13 find the storage capacity between the reservoir elevation at the beginning and end of the period. Enter all such values in column (9) Add columns (8) and (9) and enter in column (10). Figures in column (10) show the sum of outflow volume and storage capacity. Find out the difference between corresponding figures of column (3) and column (10) and enter in column (11). This difference between figures of column (3) and (20) should be negligible. Any remark may be written in column (12).

The trial and error method in Tables 9.3 and 9.4 can be replaced by Table 9.5 also. In Table 9.5 the reservoir elevation for each time interval is first adjusted before proceeding for subsequent time interval. Method of applying trials is better understood in Table 9.5 rather than Tables 9.3 and 9.4.

9.17 RESERVOIR SEDIMENTATION

All the waters entering the reservoir carry some amount of silt. The amount of silt in the water coming into the reservoir depends upon the topography, nature of the soil, vegetation cover over the catchment, and also upon the intensity of rainfall. If soil in the catchment is soft there is always possibility of sheet erosion and amount of silt in the water coming to reservoir will be more. If catchment has large slope, it will cause rain water to flow with greater velocity and possibilities of increased amount of slit in the water will occur. Similarly higher intensities of rain fall also cause greater run off and greater erosion and consequently increased amount of silt in water approaching the reservoir. If catchment has good cover of vegetation, the amount of silt in water would decrease.

The silt load held in river water approaching the reservoir may be classified into two categories namely (a) *bed load* and (b) *suspend load.* The large sized particles comprising bed load, remain concentrated near the bed of the stream. The smaller particles remain suspended in water in upper layers. The bed load is actually dragged along the bed of the stream. Bed load is generally very small (hardly 10 to 15%) as compared to suspended load. When this silt laden water reaches a reservoir in the vicinity of a reservoir the velocity is generally considerably reduced. The most of the bed load of bigger sized particles gets deposited in the head reaches of the reservoir. Fine particles may travel some more distance and may finally deposit farther down in the reservoir. Some very fine particles may remain in suspension for much longer time and even finally escape from the dam along with the water discharged through sluice ways, spill ways etc. This deposition of silt in the reservoir, inform of bed load and suspended load is known as 'Reservoir sedimentation' or 'reservoir silting'. This deposition of the sediment, will reduce the water storing capacity of the reservoir. If the rate of sedimentation is more, the reservoir will get silted up in less time and thus the life of the reservoir would be considerably reduced. In order to prolong the life of the reservoir such measures are adopted which would slow down the rate of sedimentation. Such measures are known as sediment control measures. These measures have been enumerated a little later.

Table 9.3

Time period from start	Reservoir level at start of period in metres	Inflow in $10^6 \times m^3$	Spillway discharge at start of time interval (cumecs)	Trial reservoir level at end of time interval in metre	Spillway discharge at the end of period in (cumecs)	Mean out flow rate (cumecs)	Outflow volume $10^6 \times m^3$	Storage between start and end of the time interval in reservoir $10^6 \times m^3$	Outflow + storage in $10^6 \times m^3$	Col. (3) – col. 10	Remarks
(1)	(2)	(3)	(4)	(5)	(6)	(7)	(8)	(9)	(10)	(11)	(12)
0–4	100.00	0.864	0	100.10	2	1	0.014	0.30	0.314	0.553	
4–8	100.10	1.008	2	100.20	6	4	0.058	0.30	0 358	0 650	
8–12	100.20	1.152	6	100.50	14	10	0.150	1.10	1.25	0.10	
12–16	100.50	1.512	14	10080	24	19	0.295	1.10	1.395	0.117	
16–20	100.80	2.304	24	101.20	44	34,	0.500	1.70	2.20	0.104	
20 24	101.20	3.204	44	101.82	80	62	0.910	2.70	3.61	0.406	
24–28	101.82	4.212	80	102.50	130	105	1.475	2.60	4.075	0.137	
28–32	102.50	5.472	130	103.17	190	160	2.210	3.20	5.41	0.062	
32–36	103.17	6.084	190	103.75	248	219	3.130	2.5	5.63	0.454	
36–40	103.75	5.868	248	104.18	294	271	3.915	2.0	5.915	0.047	
40–44	104.18	5.364	294	104.37	320	307	4.43	1.0	5.43	0.066	
44–48	104.37	4.500	320	104.37	320	320	4.604	0.0	4.604	0.104	
48–52	104.37	3.600	320	104.18	.294	307	4.43	–1.00	3.43	0.170	
52–56	104.18	2.88	294	103.87	262	278	4.016	–1.50	2.516	0.364	
56–60	103.87	2.16	262	103.55	232	247	3.55	–1.50	2.05	0.11	
60–64	103.55	1.728	232	103.25	196	214	3.082	–1.25	1.832	0.1	
64–68	103.25	1.584	196	103.05	174	185	2.664	– 1.05	1.614	0.030	
68–72	103.05		174								

Column (11) shows substantial difference between inflow rate and outflow + storage and hence one more trial may be made with some increased values of assumed elevations. See Table 9.4

Table 9.4

Time period from start	Reservoir level at start of period in metres	Inflow in $10^6 \times m^3$	Spillway discharge at start of time interval (cumecs)	Trial reservoir level at end of time interval in metre	Spillway discharge at the end of period in (cumecs)	Mean out flow rate (cumecs)	Outflow volume $10^6 \times m^3$	Storage between start and end of the time interval in reservoir $10^6 \times m^3$	Outflow + storage in $10^6 \times m^3$	Col. (3) – col. 10	Remarks
(1)	(2)	(3)	(4)	(5)	(6)	(7)	(8)	(9)	(10)	(11)	(12)
0–4	100.00	0.864	0	100.25	6	3	0.043	0.80	0.843	0.021	
4–8	100.25	1.008	6	100.50	12	9	0.13	0.90	1.030	0.022	
8–12	100.50	1.152	12	100.70	28	20	0.29	0.90	1.19	0.038	
12–16	100.70	1.512	28	101.00	32	30	0.430	1.08	1.51	0.002	
16–20	101.00	2.304	32	101.45	60	46	0.66	1.70	2.46	0.06	
20–24	101.45	3.204	60	101.95	88	74	1.06	2.20	3.26	0.054	
24–28	101.95	4.212	88	102.60	137	112.5	1.62	2.70	4.320	0.108	
28–32	102.60	5.472	137	103.30	206	171.5	2.37	3.10	5.47	0.00	
32–36	103.30	6.084	206	103.90	265	235.5	3.39	2.70	6.09	0.006	
36–40	103.90	5.868	265	104.30	310	288.5	4.05	1.81	5.86	0.00	
40–44	104.30	5.364	310	104.37	312	311	4.52	0.60	5.22	0.124	
44–48	104.37	4.500	312	104.37	312	312	4.55	0	4.55	0.050	
48–52	104.37	3.60	312	104–30	310	311	4.02	–0.40	3.62	0.02	
52–56	104.30	2.88	310	104.00	275	292.5	4.23	–1.70	2.52	0.36	
56–60	104.00	2.16	275	103.75	245	260	3.74	–1.0	2.74	0.48	
60–64	103.75	1.728	245	103.30	205	225	3.24	–2.00	1.24	0.488	
64–68	103.30	1.584	205	103–10	180	192.5	2.77	–1.00	1.77	0.12	
68–72	103.10		180								

In the same way more trials can be performed.

Table 9.5

Time interval (t) from start in hrs.	Reservoir elevation at the start of interval	Inflow at the start of interval (I_1) cumecs	Inflow at the end of interval (I_2) cumecs	Volume of inflow $\left(\frac{I_1+I}{2}\right)t$ $10^6 \times m^3$	Outflow at the start of interval (O_1) cumecs	Trial reservoir elevation at the end of Interval	Outflow at the end of interval (O_2) cumecs	Mean outflow volume $O=\left(\frac{O_1+O_2}{2}\right)t$ $10^6 \times m^3$	Storage capacity at the start of interval (S_1)	Storage capacity at the end of Interval (S_2)	Change in storage (Δ_y)	Outflow volume + change in storage ($O+\Delta_y$)	Compare ($O+\Delta_y$) with col. (5)
(1)	(2)	(3)	(4)	(5)	(6)	(7)	(8)	(9)	(10)	(11)	(12)	(13)	(14)
0–4	100.00	55	65	0.864	00	100.25	5.0	0.072	0.00	0.90	0.90	0.972	diff. is large
"	"	"	"	"	"	100.20	4.0	0.0576	0.00	0.72	0.72	0.777	diff. is large
"	"	"	"	"	"	100.22	4.0	0.0596	0.00	0.792	0.792	0.852	o.k.
4.8	100.22	65	75	1.008	4.00	100.45	12.5	0.119	0.792	1.50	0.708	0.827	diff. more
"	"	"	"	"	"	100.50	13.5	0.126	0.792	1.70	0.908	1.034	large
"	"	"	"	"	"	100.48	13.0	0.122	0.792	1.680	0.888	1.010	o.k.
8–12	100.48	75	85	1.152	13.0	100.75	22.5	0.256	1.680	2.600	0.920	1.176	large
"	"	"		"	"	100.73	22.0	0.252	—	2.580	0.900	1.152	o.k.
12–16	100.73	85	125	1.512	22.0	101.10	37.5	0.428	2.580	4.100	1.520	1.948	large
"	"	"	211	"	"	101.0	32.5	0.392	—	3.700	1.120	1.512	o.k.
16–20	101.00	125	195	2.304	32.5	101.30	49.0	0.587	3.700	5.00	1.300	1.887	large
"	"	"	"	"	"	101.375	60.0	0.666	—	5.25	1.550	2.216	o.k.
"	"	"	"	"	"	101.40	61.5	0.676	—	5.30	1.610	2.286	
20–24	101.40	195	250	3.204	61.50	101.875	83.50	1.044	5.300	7.50	2.20	3.244	o.k.

Contd.

Table 9.5 *Contd.*

(1)	(2)	(3)	(4)	(5)	(6)	(7)	(8)	(9)	(10)	(11)	(12)	(13)	(14)
20–24	101.40	195	250	3.204	61.50	101.875	83.50	1.044	5.300	7.50	2.20	3.244	o.k.
24–28	101.375	250	335	4.212	83.50	102.375	115.50	1.433	7.50	9.50	2.00	3.433	large
"	"	"	"	"	"	102.50	135.50	1.577	—	9.90	2.40	3.977	large
"	"	"	"	"	"	102.57	137.00	1.581	—	10.20	2.70	4.281	o.k.
28–32	102.57	335	425	5.472	137.0	103.325	207	2.477	10.20	13.50	3.300	5.777	large
"	"	"	"	"	"	103.275	203	2.448	—	13.23	3.03	5.478	o.k.
32–36	103.275	425	420	6.084	203	104.00	275	3.442	13.23	16.50	3.27	6.712	large
"	"	"	"	"	"	103.90	265	3.370	—	16.10	2.87	6.24	large
"	"	"	"	"	"	103.875	262.5	3.355	—	16.00	2.77	6.125	o.k.
36–40	103.875	420	395	5.868	262.5	104.25	307.5	4.104	16.00	17.80	1.80	5.904	o.k.
40–44	104.25	395	350	5.364	307.5	104.375	325	4.554	17.80	18.50	0.70	5.254	large
"	"	"	"	"	"	104.40	325	4.574	—	18.60	0.80	5.374	o.k.
44–48	104.40	350	275	4.500	325	104.40	325	4.680	18.60	18.60	0.00	4.680	large
"	"	"	"	"	"	104.375	322	4.658	—	18.45	–0.15	4.508	o.k.
48–52	104.375	275	225	3.600	322	104.20	300	4.478	18.45	17.50	–0.95	3.528	o.k.
52–56	104.20	225	175	2.88	300	103.875	262	4.046	17.50	16.00	–1.50	2.546	large
"	"	"	"	"	"	104.00	275	4.140	—	16.50	–1.00	3.140	large
"	"	"	"	"	"	103.90	270	4.104	—	16.30	–1.20	2.904	o.k.
56–60	103.90	175	125	2.160	270	103.60	232	3.614	16.30	14.50	–1.80	1.814	large
"	"	"	"	"	"	103.65	236	3.643	—	14.80	–1.50	2.143	o.k.
60–64	103.65	125	115	1.730	236	103.25	200	3.140	14.80	13.40	–1.40	1.74	o.k.
64–68	103.25	115	105	1.584	200	103.05	175	2.700	13.40	12.25	–1.15	1.55	o.k.

9.18 LIFE OF RESERVOIR

Any reservoir cannot last for ever. Ultimately all the reservoirs get silted up. Silting of the reservoir starts from the day it is created. When reservoirs are created some of its capacity is left unused. This is the capacity of the reservoir lying below the crest level of the bottom most under sluices. This storage capacity which remains unused is known as dead storage. This dead storage capacity is used to accommodate deposited silt so that effective storage of the reservoir is not affected. So long as dead storage capacity of the reservoir is not silted completely, effective storage or useful storage capacity is not affected. The process of silting continues even after complete silting of dead storage. The further silting affects the effective storage of the reservoir and the reservoir does not have enough water to fully fulfil its obligations. Generally useful life of the reservoir is considered terminated when its effective storage is reduced by 20% of the designed capacity of the reservoir.

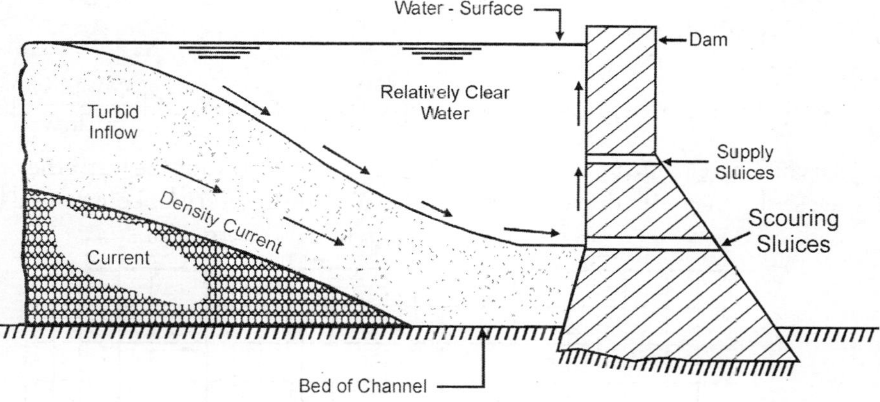

Fig. 9.16. *Density currents.*

9.19 DENSITY CURRENTS

The water stored in the reservoir is generally free from silt. The flood inflow is generally muddy and heavily charged with silt. Thus during floods there are two waters having different densities present is the reservoir. Heavily silt charged water flows along the bottom of the channel towards the dam under the influence of gravity. This current of heavily silt charged water flowing along the bottom of the reservoir towards the dam is known as *Density current.* If the density currents, are directly discharged out of the reservoir by suitably positioning and operating the sluice gates, the rate of reservoir silting can be considerably reduced.

9.20 TRAP EFFICIENCY (E)

It is defined as the percentage of the sediment deposited in the reservoir even inspire of taking precautions and measures to control its deposition.

$$E = \frac{\text{Total sediment deposited in the reservoir}}{\text{Total sediment flowing in the river}}.$$

Most of the reservoirs tap 95% to 100% of the sediment load flowing into them. It has not been possible to reduce the trap efficiency below about 90% inspire of several sediment control measures now known to us. Trap efficiency is a measure of reservoir sedimentation.

9.21 CAPACITY-INFLOW RATIO

Detailed observations have clearly shown that the trap efficiency is a function of the ratio of the reservoir capacity to the total inflow

$$E = f \times \frac{\text{(Capacity of reservoir)}}{\text{(Inflow in the reservoir)}}$$

where f is a constant.

Figure 9.17 shows a graph between trap efficiency and log of $\left(\frac{\text{capacity}}{\text{inflow}}\right)$ for a reservoir. It is clear from this curve that for a given inflow rate, the trap efficiency reduces with the reduction in reservoir capacity, due to reservoir being silted up.

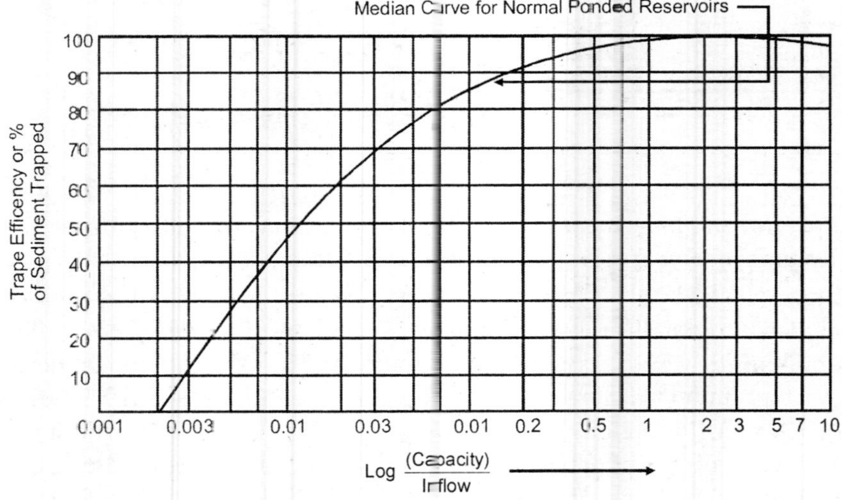

Fig. 9.17. *Trap efficiency of reservoir.*

Since capacity of the reservoir is maximum at the beginning, the rate of silting or in other words the trap efficiency, is maximum. As the reservoir goes on silting, trap efficiency also goes on decreasing and consequently the rate of silting also goes on decreasing. Thus the complete filling of a reservoir, may take a very long time.

From the same capacity-inflow ratio it can also be inferred that in inflow is increased in relation to reservoir capacity the trap efficiency is bound to decrease. Thus small capacity reservoirs constructed on large rivers have less rate of silting. On the other hand, large capacity reservoirs on smaller rivers shall have larger rate of silting or in other words greater trap efficiency. The trap efficiency for large reservoir is more because inflow water has to stay in the reservoir for considerable time before being discharged down stream as it will almost completely shed its silt load in the reservoir. In the case of smaller reservoirs the inflow is almost immediately discharged downstream from the reservoir and hence does not get time to shed its silt load.

9.22 SILTING POWER OF RESERVOIRS

For reservoirs constructed only for power generation, silting is not as important as for the reservoirs constructed for irrigation, and water supply. This is due to the fact that for running turbines a minimum head of water is required. So long as this head is available the working of power reservoir is not affected. Obviously, if the silting has affected the minimum capacity of the reservoir required for generation of power, the generation of power will not be maintained, as it should be as revervoir will deplete very fast and there will be no water to run the turbines.

Hence while silting of the reservoir is not affecting the head of water over the turbines, it is certainly affecting the reservoir capacity.

9.23 SILT CONTROL OF RESERVOIRS

Entrance of silt into the reservoir may be controlled by adopting following preventive measures.

1. Reservoir site should be properly selected. If nature of catchment soil contributing water to the reservoir is very soft, the rate of silting will he more, as soft sand will be easily carried by run-off. If slope of the catchment is steep, it will also cause more of silting, as steepness would cause increased velocity of flow of run-off and thus more erosion of the catchment soil.

2. The tributaries carrying more of silt load to the reservoir should be provided with check bunds some where upstream of the tributaries. Check bunds would cause sedimentation of silt load upstream and comparatively clear water will reach the main reservoir. The small reservoirs created at the back of the check bunds also supply stored water to the main reservoir when supplies in the tributaries have receded considerably.

3. By increasing the vegetation growth on the catchment, silt entry into the reservoir can be reduced.

4. Silt deposition in the reservoir can be considerably reduced and deposited silt can be scoured by operating sluice gates properly. During floods the inflow of silt is more and scour sluices should be kept open.

5. The deposited silt should be dredged from time to time.

6. If dam is constructed in stages, the rate of silting in the reservoir is kept low. In this measure the dam should be constructed of smaller than the required height and operated for five years. After this, dam is further increased in height and again operated for few years. By this method $\left(\frac{\text{capacity}}{\text{inflow}}\right)$ rate is kept small and trap efficiency is kept under check.

9.24 ECONOMIC HEIGHT OF THE DAM

It is that height of the dam, corresponding to which the cost of the dam per unit of storage is minimum. This requires estimates of construction cost for several heights of dam and also the corresponding storages in the reservoir. For each dam height, the reservoir storage is known. Draw the curve between dam heights and their corresponding costs of construction as shown in Fig. 9.18 (a). Now work out the construction cost per unit storage for all the dam heights and plot a curve as shown in Fig. 9. 18 (b). The lowest point on this curve gives the height of dam for which the cost per unit of storage is minimum and hence adopted as the most economical.

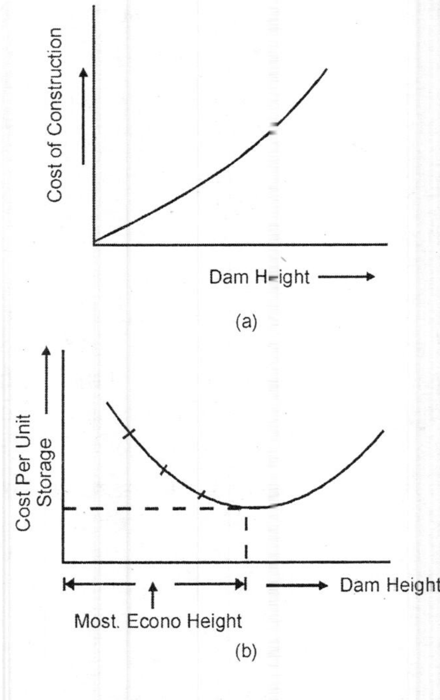

Fig. 9.18

9.25 RESERVOIR LOSSES

Reservoir losses may be classified under following three heads.

1. Evaporation losses

2. Absorption losses
3. Reservoir leakage or percolation losses.

1. Evaporation Losses. These are the major losses from a reservoir. This loss is affected by exposed surface area of the reservoir, wind velocity, temperature, relative humidity etc. This loss is expressed in cm of water depth and varies from place to place depending upon the local conditions such as temperatures, relative humidity, wind etc. Average values of losses in cm for North and South India for various methods of the year are given as follows.

Month	*Jan.*	*Feb.*	*Mar*	*Apr.*	*May*	*June*	*July*	*Aug.*	*Sept.*	*Oct.*	*Nov.*	*Dec.*
Loss of North India in cm	7	9	13	16	27	24	18	14	14	13	9	8
Loss for South India in cm	10	10	18	23	25	18	15	15	15	13	10	10

2. Absorption Losses. This loss of water is considerable in the beginning, but falls to very small values after some time when pores get saturated. These losses depend upon the soil forming the reservoir. These losses are not considered while planning a reservoir.

3. Reservoir Leakage or Percolation Losses. Reservoirs being very large, their banks are permeable. But permeability of the soil is generally very low and hence these losses do not carry any importance. But in certain cases the banks of the reservoir may be made of badly fractured rocks or having continuous seams of porous strata. Such conditions may cause serious leakage. To stop such leaks grouting with cement may have to be resorted. Hence while selecting the site of the reservoir, this aspect of the site is investigated in great length and if any such fault is noticed, it is rectified before reservoir is constructed.

9.26 FLOOD CONTROL

Flood is a relatively high flow. The floods cause overtopping of natural as well as artificial banks of the streams. The flood is caused by run off from rainfall and/ or melting snow, too large to be confined in the banks of streams. When banks are overtopped the flood water spreads over the adjoining area and causes heavy damage to crops and property. It is not possible to prevent occurrence of floods but it is possible to prevent or reduce the damage due to floods by controlling the floods. This flood control or flood management may be defined as the prevention or reduction of the flood damage.

In India about 40 million hectares of area has been identified as flood prone. Out of this about 12 million hectares have been provided with flood protections measures by March 1982. Govt. of India set up National Flood Commission in July 1976. The main function of this Commission is to study in depth the effectiveness of the works so far undertaken and recommend measures to be adopted.

9.26.1 Classification of Methods for Flood Content

The loom flood control is also sometimes referred as flood management. According to National Flood Commission 1976 the broad classification of method of flood control are as follows.

1. Methods by which attempt is made to modify the flood.
2. Methods by which attempt is made to modify the susceptibility of flood damage.
3. Methods by attempt is made to modify the loss burden.
4. Bearing the loss.

The National Flood Commission has given a very broad Table which gives very elaborate classification of the method of flood control but of this table, the methods of flood control commonly adopted has been listed below.

1. Reducing the peak flood flow by constructing reservoirs.
2. Diverting the flood waters through flood ways or bypasses.
3. Confining the flood flow within predetermined channel by levees, flood walls.
4. Reducing of peak stage by a closed conduit improving outflow channels.
5. Temporary or permanent evacuation of the flood plains.
6. Reducing flood run-off by land management.
7. Flood proofing of specific properties.

9.26.2 Design Flood and its Estimation

The *design flood* is the flood adopted for the design of a flood control project. This flood may either be the maximum probable flood or the standard project flood or a flood corresponding to some desired frequency of occurrence depending upon the degree of protection that should be provided by the flood control project.

The *maximum probable flood* is the flood which may be expected from the most severe combination of critical meteorologic and hydrologic conditions which are most likely in the region. It is estimated from the maximum probable storm applying the principle of unit hydrograph.

The *standard project flood* is the flood which is expected from the most severe combination of meteorologic and hydrologic conditions which are reasonably possible in the region. It is determined from the standard project storm rainfall applying the principle of unit hydrograph. The standard project flood is about 50% of the maximum probable flood for the area.

If it was feasible within acceptable limits of cost it would be ideal to provide protection against the maximum probable flood. But it is seldom practicable. Hence some risk has to be accepted in the selection of the design flood. Following are the methods which can be used for estimation of design flood.

1. Empirical flood formulae.

2. Applying suitable factor of safety to maximum historical flood ever occurred.

3. Envelope curves.

4. Frequency analysis.

5. Rational method of derivation of design flood from storm studies and application of unit hydrograph principle.

Dicken's formula, Ryve's formula and Inglis formula used as already explained earlier are used to estimate peak flow. These formulae are the functions of catchment area and a coefficient. Safety factor method depends on the judgement of the designer and hence its use is limited. In the envelope curve method maximum flood is obtained from the envelope curve of all the observed maximum floods for a number of catchments in a homogeneous meteorological region plotted against drainage area. This method is unreliable. Frequency method involves the statistical analysis of observed data of a fairly long (at least 25 years) period. As this method is purely statistical it should be used with caution. Rational method involves several studies and use of unit hydrograph and hence is the most reliable. However the principle of unit hydrograph cannot be used for catchments less than 25 km^2 and over 5000 km^2. For the results to be realistic large number of raingauge stations have to be established in the entire catchment so that true weighted rainfall of the catchment is reflected.

9.26.3 Flood Control Reservoirs

The main purpose of flood control reservoir is to temporarily or permanently store a portion of the flood flow so as to lower the flood peak at the point to be protected. This is achieved by discharging all reservoir inflow until the outflow reaches the safe capacity of the d/s channel. All flow above this rate is stored until inflow drops below the safe d/c channel capacity. The stored water thereafter is released to recover storage capacity for the next flood. Since the reservoir is located immediately *U/S* from the point to be protected, the hydrograph at that point is same as that released at the dam, and the peak has been reduced. The most effective location of reservoir for flood control of an area is immediately *U/S* from the area. Such a reservoir would often be located in a broad flood plain and very large area of valuable land would be submerged in the reservoir. Besides it will also need a very long dam. On the other hand the sites farther *U/S* require smaller dams and less valuable land but are less effective in dropping flood peaks. The loss is the effectiveness results from the influence of channel storage and also from the lack of control over the local inflow between the reservoir and the area to be protected. If the local area is between the reservoir and the point to be protected in sufficiently large, it may produce a flood over which the reservoir would have little or no control. It should also be noted that a single reservoir cannot give equal protection to a number of points located at different distances d/s. Hence criterion for evaluating a flood control reservoir or a system of reservoirs is the percent of the low drainage area controlled by the reservoirs.

Normally, at least one-third of the total discharge area should be under reservoir control for effective flood reduction. Economic analysis generally favours the *U/S* site despite its lesser effectiveness. Moreover, it is often preferred to establish several small reservoirs rather than having a single large reservoir.

The possible drop is peak flow by reservoir operation increases as reservoir capacity increases, since a greater portion of the flood water can be stored. Hence, a second criterion for evaluation of a flood-control reservoir is its storage capacity.

The flood-control reservoirs are of two types:

1. Detention reservoirs, and
2. Retarding reservoirs.

The detention reservoir is provided with outlets and spillways controlled by gates and valves which are operated on the judgement of the project engineer. The detention reservoirs for flood control differ from conservation reservoirs only in the sense that former used a layer sluice way capacity to permit rapid drawdown in advance of or after a flood. The retarding reservoir consists of ungated outlets which automatically regulate the outflow according to the volume of water in the storage. The outlet usually consists of a large ungated spillway or one or more ungated sluice ways. The discharge capacity of a retarding basin with full reservoir capacity should equal the maximum flow which the channel d/s can pass without causing serious flood damage.

9.26.4 Diverting the Flood Waters through Flood Ways

Flood ways are the large depressions into which a part of flood water is diverted through natural or artificial channels. The diverted flood water remains temporarily stored during the rising flood. Once the flood recedes in the river, the water stored in this depressions is allowed to flow back to the river. The flood ways thus serve two purposes firstly they create large, shallow reservoirs which store a part of the flood water and hence decrease the flow in the main channel below the diversion. Secondly, they provide an additional outlet for water from *U/S*, thus increasing velocity and decreasing stage for some distance above the point of diversion.

9.26.5 Confining Flood Flow between Levees and Flood Walls

A levee is an earth dyke, while a flood wall is usually made of masonry or concrete. Levees and flood walls are the oldest and most widely used methods of protecting land from flood water. These are nothing but longitudinal dams aligned roughly parallel to the river. They serve as artificial high banks of the river and thus prevent the river water from spilling over during floods. The flood water remains confined between the levees or flood walls and is made to flow down the river without causing any damage to the adjoining area.The levees should be so located that a sufficient channel is available to pass out the design flow with a reasonable free board against wave action. The channel width between levees and the height of the levees are closely related. When a city or any area is to be protected a ring levee which completely encircles the area may be provided.

Reducing the peak stage by improving the outflow channel. Outflow channels if improved for its hydraulic properties carry higher flood waters. This improvements may be

(a) Increasing the velocity of flow in the channel

(b) Realigning the channel along the shorter route

(c) Increasing the cross-section of the channel

(d) Enlarging the water ways to drainage crossings.

All these measures tend to reduce the prevailing water level for the corresponding flood discharge. It may however, be stated that although the channel improvement is a well recognised method for flood control, it should be used with caution. This is so because the measures adopted in this method are essentially local protection measures which may increase flood magnitude at d/s points on account of accelerated flow of run-off from the upper improved reaches of the channel.

Evacuation from flood plain. Emergency evacuation from flood threatened area is the most effective method of reducing damage due to floods. With reliable flood forecasts this method may be adopted for sparsely populated areas where property values do not justify the use of other methods of flood control and the loss of life can be avoided by prompt evacuation. However the effectiveness of this method depends on the hydrologic characteristics of the stream as well as the availability of good flood forecasting service so that sufficiently in advance warning of floods could be given to the area to complete evacuation. The term *Flood plain zoning* is defined as the method of dividing the entire flood plain area into different zones and to restrict the occupancy of the different zones of the flood plain to uses which will suffer little or no damage during floods. The objectives of zoning of the flood plains are discussed as follows:

1. Avoiding encroachment of waterway of the river. Such encroachments aggravate the flood problem and result in heavy flood damage.

2. Directing activities in other parts of the flood plain in such a manner that they are less susceptible to damages in case of floods.

Like any other measure of flood control, the flood plain zoning must also be adopted only if it is economically justified.

Reducing flood run-off Land Management. It is known that vegetal cover removes moisture from the soil by transpiration. It also promotes loose organic soil which is favourable for the infiltration of rainfall. Further a heavy vegetal cover also results in high interception loss during floods.

Water conservation measures help to conserve or store water in the soil and thus reduce surface run-off and the flood flows. Following conservation measures can be used.

1. use contour farming
2. create farm ponds to retain the flow of small creeks for irrigation
3. use of cover crops is field to avoid bare followed field during non-growing season.

However all these measure of water conservation are quite useful for reduction of soil erosion and preservation of soil-moisture, they are invariably used and any flood reduction taking place from these is an incidental benefit.

Flood proofing of specific properties. Isolated units of high value may sometimes have to be individually flood proofed. The property unit may be encircled by a high levee of as flood wall.

Flood plain management. The flood plain of a river is formed by deposition of sediment. Since the flood plains are nearly level they prove good sites for cities. This results is the occupancy of the flood plains. This however requires large annual expenditure for flood protection and such expenditure continues to increase unless a systematic approach to flood plain management is adopted. The flood plain management should reduce the costs of flood plain occupance. The costs involved are

1. The initial cost of development
2. The cost of flood protection
3. The residual flood damage and
4. The costs of relief and rehabilitation

The tools used for flood plain management are as follows.

1. Flood Hazard Surveys. The detailed survey of the flood plain is conducted and flood marks of previous floods are noted. Surveys help to assess the probability levels of floods at various points and provide the basic information for defining flood risks.

2. Flood Plain Zoning. It is a device by which occupancy of the flood plain is prohibited. The land is left either in their natural state or used for parks and other recreational purposes.

3. Flood Proofing. It is method which can be used to project individual properties from damage due to floods.

4. Flood Insurance. In order to restrict or discourage the unwise use of flood plains a proramme of flood insurance may be adopted. The premium of insurance should reflect the risk involved in undertaking the construction in the flood prone area.

5. Economics of Flood Control. Adoption of only one type of flood control method for the entire river is rare. Economic analysis also speak use of combination of different flood control methods.

The project is which several flood control methods are jointly used is known as *combined project.* Often reservoir construction is combined with construction of levees and improvement in outflow channel at key points along the stream. The cost analysis for such a project is shown in Fig. 9.19.

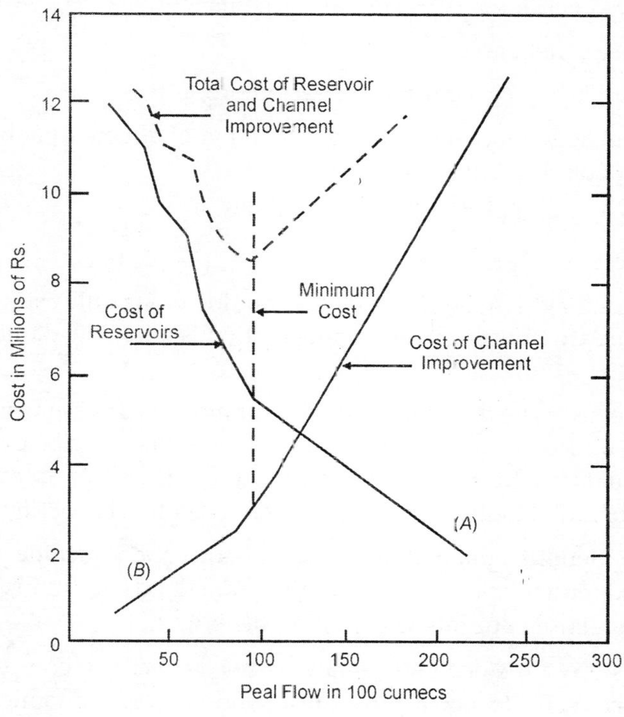

Fig. 9.19

If several reservoir sites are available they should all be exploited. It is fact that with more reservoirs being constructed, smaller regulated peak flow is obtained as shown in curve *A* of Fig. 9.19.

Since only reservoirs would not provide complete protection, levees and other channel improvements would be necessary. Curve *B* reflects the costs of these works for protection against various flows. The sum of curves *A* and *B* gives the total cost of protection against the design flood.

The construction of curves of this Fig. 9.19 require the design of channel works for protection against several different flows.

Benefits of Flood Control. The benefits of flood control measures may be classified under the following two heads:

1. Tangible benefits
2. Intangible benefits.

1. Tangible benefits. Tangible benefit of flood control measures may be considered under following two heads:

(a) Benefits arising from prevention of flood damage

(b) Those occurring from more intensive use of protected land.

The tangible benefit may also be classed as

(i) Primary benefits and (ii) secondary benefits.

The primary benefit are

1. Cost of replacing or repairing of affected property

2. Cost of evacuation, relief and rehabilitation of affected people and other emergency flood projection measures.

3. Losses as a result of disruption of business etc.

4. Loss of crops.

The flood damages in the affected area should be carefully estimated. This is generally done after the occurrence of major floods when the damage is clearly evident.

Land protected from floods may be used for more useful use which may be in form of agriculture or industry. Increase in land value may be converted into average annual benefit. Secondary benefits are difficult to assess and hence usually not included in the cost estimates of flood control benefits.

Intangible Benefits. Intangible benefits of flood control include prevention of loss of life and reduction in disease resulting from flood conditions. It is difficult to place a monetary value on the intangible benefit.

The final step in the economic analysis of a project is a comparison of the benefits and costs. Protection against the rare floods is uneconomical because of the large investment and infrequent flood occurrence. Actually for many flood control works the design criteria cannot be based on a direct economic analysis. Thus only those portions of a project which are in themselves beneficial should be constructed. Any porject must show economic feasibility before it is undertaken for construction.

9.27 RECOMMENDATIONS OF NATIONAL POLICY ON FLOOD CONTROL

Board recommendations are as follows :

1. In order to deal the problem of floods on a comprehensive basis very sustained and systematic efforts are needed, both at administration level and general public level. These tasks are very huge is magnitude and hence measures of protection from floods should be progressively undertaken so that increased protection from floods is achieved every year.

2. Even the best flood control measures do not provide complete protection and are aimed at providing a reasonable degree of protection. Reason for this being economically unjustified.

3. State governments is are primarily responsible for flood control measure projects, Central Govt. should also take initiative in the matter.

4. Flood control measures should form an integrated part of the overall development of the region in respect of water resources, economic perspective, ecological and humanitarian considerations.

5. The flood management in a particular areas should form a part of package of measures of comprehensive plan for that area. Different alternatives with various degree of min of measures should be worked out and analysed for optimisation.

6. Reservoirs must he considered as an important component in any package of measures provided they are technically and economically viable.

7. Use of natural detention basins for flood moderation should be used if condition permit.

8. Channel improvement measures should be considered with due caution because they are costly in initial as well as maintenance cost.

9. While using embankments, their side effects and associated problems should be carefully studied.

10. Measures like flood plain regulation, flood forecasting and warning, flood proofing, disaster preparedness should be adopted to reduce the susceptibility to life and property. This is essential as engineering measures by themselves cannot provide protection from all floods.

11. Disaster relief, remission of Tanes, provision of taccavi loan etc. should be continued on humanitarian considerations. Crop insurance and insurance schemes for damage to properties should be liberally continued.

12. Anti-erosion works should be taken up for protection of railway lines, roads, industrial areas, towns, group of villages where the benefit cost ratio justifies such works.

13. Afforestation and soil conservation measures are useful complements to other measures and should be considered specially in the watersheds of streams charged with heavy silt load.

QUESTIONS

9.1 Explain the terms, reservoir, single purpose reservoir, multi-purpose reservoir. Enumerate the purposes of reservoir.

9.2 How the reservoirs are classified ? Describe the characteristics of conservation reservoir, flood control reservoirs, distribution reservoirs, and multi-purpose reservoirs.

9.3 What are the different investigations required while carrying out reservoir planning ?

9.4 Enumerate the points which should be considered while choosing the site for the reservoir.

9.5 Explain the following terms.

(i) Normal pool level (ii) Minimum pool level
(iii) Maximum pool level (iv) Dead storage
(v) Useful storage (vi) Surcharged storage
(vii) Bank storage (viii) Valley storage.

9.6 Explain the various yields of the reservoir.

9.7 Explain mass inflow curve and its utility in respect of determining the reservoir capacity. Also explain the use of mass inflow curve to determine the safe yield of the reservoir.

9.8 Describe in details the demand patterns of various types of reservoirs.

9.9 What do you understand by the term schedule of operations for a multi-purpose reservoir ? Explain its adoption for reservoir located in Northern India.

9.10 What do you understand by the term flood routing? Explain various methods which are used for flood routing.

9.11 Explain reservoir sedimentation. How reservoir sedimentation affects life of the reservoir ?

9.12 Explain the following terms:

(i) Trap efficiency (ii) Capacity inflow ratio
(iii) Density currents (iv) Economic height of the dam.

9.13 Enumerate various silt control measures that if adopted would reduce the rate of silting in the reservoir.

9.14 (a) Explain various reservoir losses. (b) Silting characteristics of power generation reservoir.

❑❑❑

10

Measurement of Discharge

10.1 INTRODUCTION

The word 'stream flow' is used to represent the discharge flowing in the river. It is also sometimes represented by term stream gauging. The discharge in the stream represents the run-off in the stream at a given section and includes surface run-off as well as ground water flow.

Discharge has to be measured for canals also. Methods of measuring discharge for canals are more or less the same as those adopted for streams. Hence when we talk of measurement of discharge, it means measurement of discharge for both streams and canals. There may be little bit of difference in the application of the methods. Canals and streams differ in following respects.

Canal	*Natural stream*
1. It is artificial channel.	1. It is a natural channel.
2. It has definite longitudinal slope.	2. Its slope depends upon the topography of the area it traverses.
3. It has regular cross-section.	3. It does not have regular section.
4. Controlled supplies are left in the canal.	4. Supplies depend upon the rainfall, snow, etc. in the catchment.
5. Alignment is mostly maintained at ridge.	5. Alignment follows the valley drains in the area.

10.2 PURPOSE OF MEASUREMENT OF DISCHARGE OF STREAMS AND CANALS

The discharge of streams and canals has to be measured for following reasons.

1. The irrigation potential of a particular river at a particular point can be estimated only if discharge characteristics of the river at that particular point of the stream are known. If flow in the stream falls too low during particular season, it may become essential to construct dam or weir. Even height of the dam or weir depends upon the discharge of the river. Storage of water is further dependent upon the available discharge which again necessitates measurement of discharge.

2. Because water in the canal is left proportional to the area it has to irrigate, measurement of discharge again becomes necessary at all works. Regulation of canal system is almost fully based on the discharge measurement, as water in each canal has to be left equal to the capacity of the canal.

3. To conduct research work as regards to the behaviour of streams and canals, discharge in them has to be measured.

4. River discharge is also required to be known in the design of the river training works.

5. Discharge may have to be measured, when water of a particular river has to be distributed amongst different countries. A canal supplying water to more than one state has to be measured and distributed.

10.3 SELECTION OF SITE FOR DISCHARGE MEASUREMENT

Following points should be considered while choosing site for measurement of discharge.

A. For Canal

1. The canal should be straight for at least 10 times the width of the canal.
2. The canal should be in its true section. If possible the bed and the bank slopes should the pitched so as to maintain a true design section.
3. The site should be away from the bridge, fall, or regular, as flow of water at such places is not uniform.
4. The canal should neither be scouring nor silting at the point i.e. the canal should be in regime condition.

B. For Rivers

1. The river should be straight for substantial length both at downstream and upstream of the site.
2. The river section should be neither too narrow nor too wide.
3. The river section should neither be silting nor scouring.
4. The velocity of flow in the river should be laminar and not turbulent.
5. The river should flow in one stream only.
6. The site should be away from dam, weir or barrage.
7. Other small streams and tributaries should not be meeting the main river in the vicinity of the site.

10.4 MEASUREMENT OF DISCHARGE

While measuring discharge of a river or canal, one has to measure:

(i) The area of cross-section and

(ii) The average velocity of flow.

If both these elements are known, the discharge flowing through any particular point of the canal or river is found out by multiplying both of them. Before different methods of discharge measurement are discussed, let us discuss the methods of measuring area of cross-section and those for measuring velocity of flow.

10.5 MEASUREMENT OF AREA OF CROSS-SECTION

In the case of canals there is no difficulty in measuring the cross-sectional area. Canals are always of some regular shape whose area can be found out by any geometrical formula. But in case of natural streams and rivers, it is not very easy to determine the area of cross-section as their sections are not regular. In this case area of cross-section of the river can be determined as follows.

See Fig. 10.1. A river is flowing from left to right. Select *AB, CD* and *EF* sections of the river and pull cables at all these sections across the river. The distance between sections *AB, CD,* and *CD, EF* may be anything, say 100 m. Wooden battens are attached to all the cables and sections of river at the cable points are

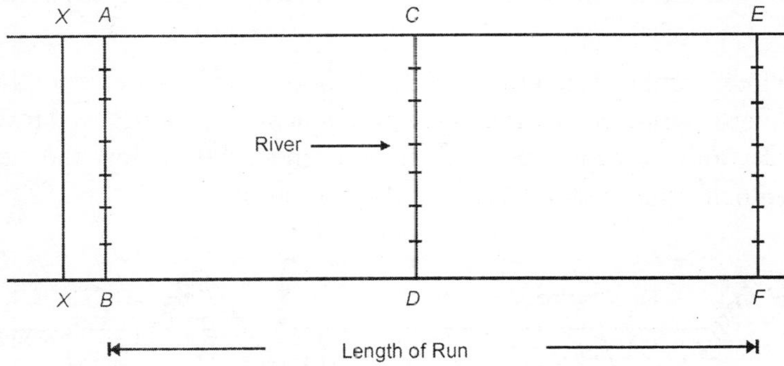

Fig. 10.1. *River length.*

divided into equal parts. In our case we have divided each cable into seven equal parts as shown in Fig. 10.2. Thus the river width gets divided into seven equal parts longitudinally.

Now take a staff or any other graduated rod and measure the depth of the river at the centre points of the each of the seven parts. Compute the areas of all seven compartments by multiplying the width of the water of each compartment by the depth of the water at the centre of each corresponding compartment. The sum of all these areas is the cross-sectional area (A) of the flowing water of the river.

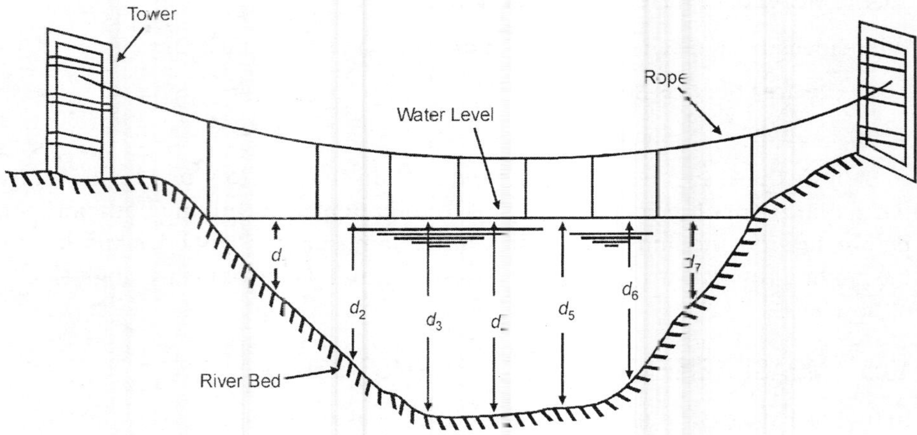

Fig. 10.2. *Cross-section of river.*

If it is not possible to measure the depth of water with staff or graduated rod, it can be measured by using measures like Kelvin's tube, Echo sounder etc.

10.6 MEASUREMENT OF VELOCITY OF FLOW

Before we actually discuss various methods to measure the velocity of flow, it is worthwhile to explain the variation in the velocity along the cross-section of the river.

The velocity of flow is not uniform throughout the cross-section of the river. It is almost zero at the bed and increases parabolically as we proceed towards the surface of water. But the velocity is maximum slightly below the surface. At surface velocity is somewhat less than the maximum.

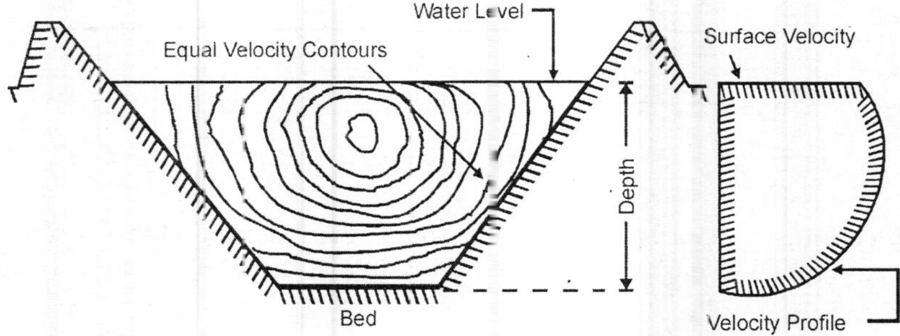

Fig. 10.3. *Velocity contours and variation of velocity along the depths.*

Now if we proceed along the width of the river, the velocity is almost zero along the banks. The velocity goes on increasing as we proceed towards the centre of the river and it reaches maximum at the centre.

Now if we consider, both along depth and along the width of the river, we can easily pin point that velocity of flow is maximum at centre width of the river but slightly below the surface of water. The velocity is zero along banks and bed of the river. If we draw contours of equal velocity they will be as shown in Fig. 10.3 (*a*). Seeing variation in velocity along depth as well as along width of the river, it can be easily understood that finding the average velocity of flow is not very easy. One has to take average of the velocities along depth as well as width for all the seven compartments of Fig. 10.2.

10.7 METHODS OF MEASURING THE AVERAGE VELOCITY OF FLOW

Following six methods can be used for determining the average velocity of flow :

1. Surface floats or simply floats,
2. Double floats,
3. Velocity rods,
4. Current meter,
5. Pitot tube, and
6. Travelling screens.

1. Surface Floats. A float or surface float is nothing but a piece of cork or anything which can float on water. Such a float gives surface velocity of flow only. The distance travelled by such a float divided by the time taken to travel a

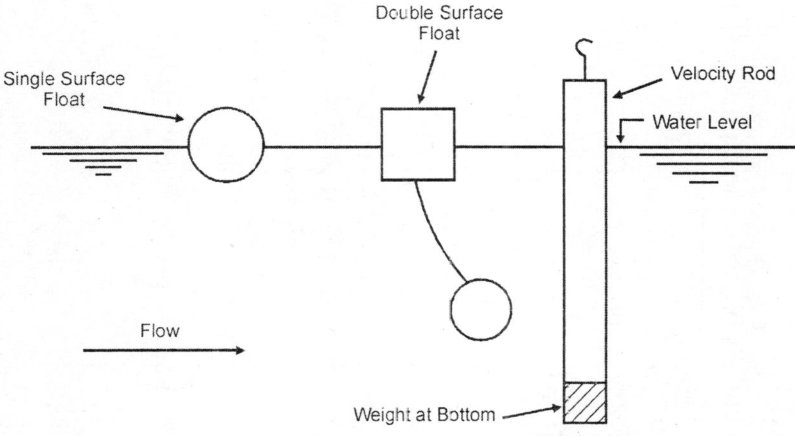

Fig. 10.4. *Single, double floats and velocity rod.*

specified distance, gives the surface velocity of flow. Surface floats are simultaneously released in each of the seven compartments of the river and time taken by the float of each compartment in reaching section *EF* from *AB* is noted. Surface velocity for each compartment can then be computed by dividing the distance between *AB* and *EF*, by the time taken by the float of the corresponding compartment. After having found out the surface velocities, the mean velocities can be calculated for each compartment by multiplying the surface velocities by a suitable coefficient 0.85. Average of all the mean velocities can then be found out, which is overall average velocity (*V*) of the flow.

2. Double Floats. It is just like a surface float with the difference that it is attached with a steel ball. The steel ball remains suspended in water. The ball is maintained at such depth that the resulting velocity of the float is not the surface velocity but average velocity. Time of travel of the float is noted for all the compartments dividing width of the river and average of the velocities for all the zones is found out which is considered as average velocity of flow.

3. Velocity Rods. These are wooden rods painted with some water proofing paint. A weight is attached to the bottom of the rod. The weight is so adjusted that rod length equal to 0.94 × depth remains inside the water. The top about 4 cm length of the rod remains out of water and remains visible. These rods when floated in flowing water give the mean velocity of flow directly.

The use of floats and velocity rods is generally restricted to straight rivers having almost uniform cross-section throughout. They are not of much use for natural stream, especially when depth of water exceeds 1.5 m or so.

4. Current Meter. It is one of the best methods for measuring the velocity of natural stream. Various types of current meters are in use these days, but the most common and widely used in India is the Price's current meter.

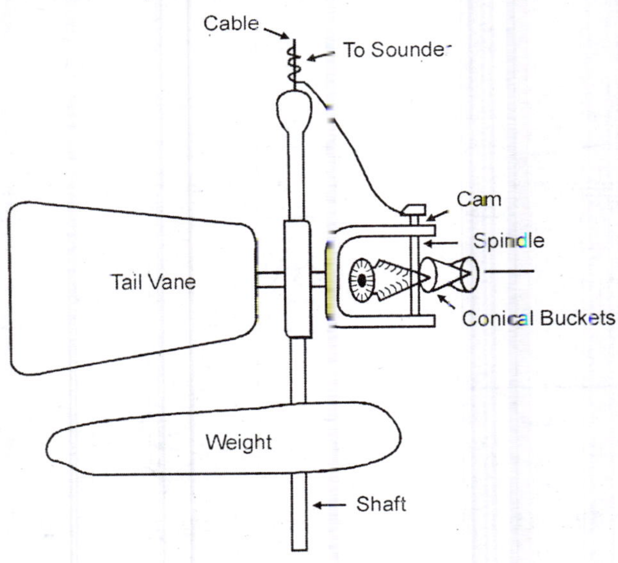

Fig. 10.5. *Current meter.*

It is a differential type of meter and consists of a horizontal wheel carrying a series of cups that rotate on a vertical axis. Half of the cups behave convex and half as concave to the current. There is a tail-vane and a counter weight at bottom to balance the meter and to keep it steady. When current meter is suspended in water, the velocity of flow causes the wheel to rotate. The current meter is fitted with a device so as to record the number of revolutions of the horizontal wheel due to velocity of flow. The rate of revolutions is proportional to the velocity of flow.

Since the velocity for different compartments is different, hence the meter is lowered in the centre of each compartment separately and individually. The wheel of the meter should face the current and is generally kept at a depth of 0.6 depth from the free surface of water. The velocity measured by keeping the meter at a depth of 0.6 depth, gives the mean velocity. Sometimes the velocities at a depth of 0.2 depth and 0.8 depth are measured and their average value is taken to be the mean velocity. The current meter is used by suspending it with the help of a boat. For narrower streams the meter can be suspended by a cable stretched across the river and at the mid-points of each compartment.

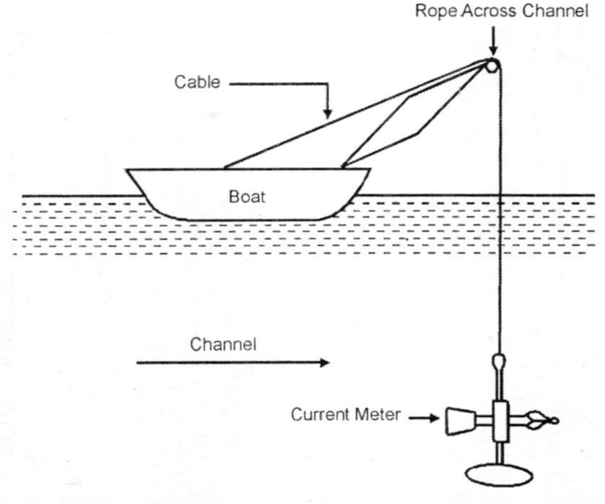

Fig. 10.6. *Current meter being used by boat.*

The boat can be conveniently used for small velocities of the order of 1 m/sec but for higher velocities and rough waters, it becomes difficult to keep the boat steady. In such a case the boat is generally anchored. See Fig. 10.6.

5. Pitot Tube. It is a device with the help of which velocity head of flowing water is converted into pressure head. The pressure head can be measured and the velocity of flow calculated by equating their pressure head to $\frac{V^2}{2g}$. Pitot tube is an ordinary tube having both ends open. The bottom end is bent at 90°. It may be having two rods one rod facing the current of water and other against the current. Velocity of flow can be determined by using Bernoullis equation

$$H + h = H + \frac{V^2}{2g}$$

or

$$V = \sqrt{2gh}$$

6. Travelling Screen Method. This method cannot be used for rivers or streams. This method can be used for canals and that too for lined canals. The method consists of a canvass screen mounted on a frame work of iron angles. The size of the screen is so adjusted that some open space remains between bottom and banks, and the canvass bottom and edges respectively. The whole frame work is mounted on a travelling trolley which can move on the rails laid at

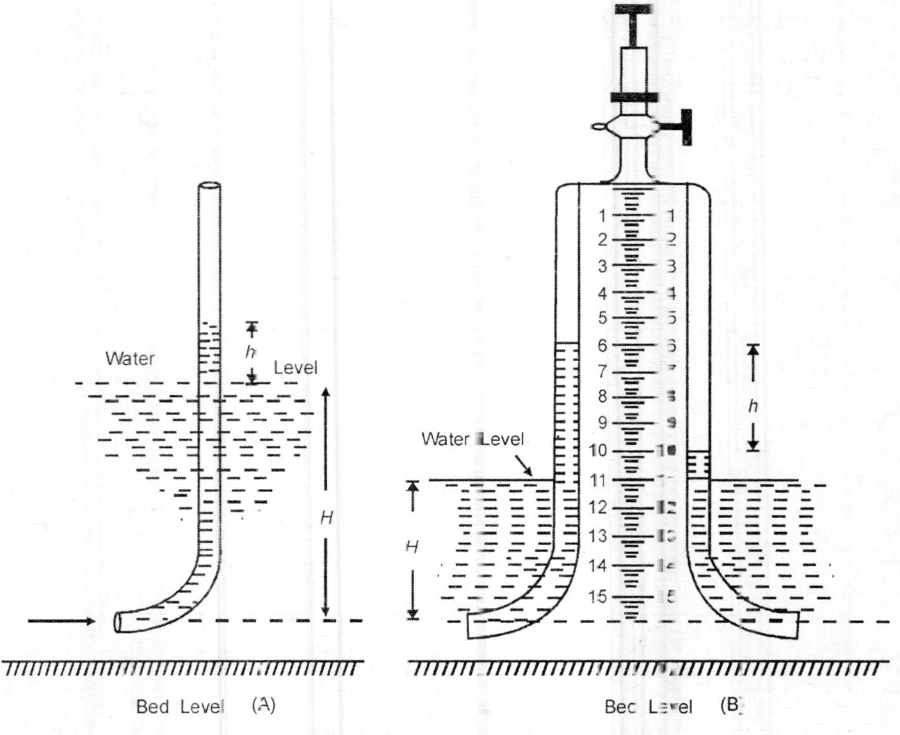

Fig. 10.7. *Pilot tubes.*

banks and parallel to the canal. When canvass screen is put against the current of flowing water it starts moving because of thrust of water. The velocity of moving screen is taken as the average velocity of flow. The method is not very common.

10.8 METHODS OF MEASURING THE DISCHARGE

Discharges of rivers, streams and canals can be measured by following methods:

1. Area velocity method,
2. Weir method,
3. Metric flumes method,
4. Chemicals method,
5. Stage discharge curve method,
6. From power plant records.

1. Area Velocity Method. This is the direct method of computing the discharge in a stream by measuring its velocity and area of flow. The methods of computing average velocity of flow and also those of area of cross- section of flow, have been discussed earlier in this chapter. The discharge is nothing but multiplication of the average velocity of flow and that of area of cross-section of water.

2. Weir Method. Various kinds of weirs are installed across the river streams to measure the discharge. The head of water over the weir crest is measured and the discharge is calculated by using the appropriate formula. Weirs are generally used for measuring discharge of small rivers and canals. For large streams the discharge is generally measured using spillway of dams as weirs. Following formulae may be used for different types of weirs.

(i) Cippoletti weir

$$Q = 1.83\, LH^{3/2}$$

(ii) 90° V - Notch.

$$Q = 1.45\, H^{5/2}.$$

3. Flumed Meters. In this method the canal or river is flumed i.e. reduced in section, either by raising the bottom of the channel or by decreasing the width of the channel or by both ways. Let B be the normal width and b the width at the flumed part of the channel. Let V and v be the velocities of flow at normal section and throat respectively. If there is no head loss between normal section and throat, then total heads at normal section and throat, will be as follows. Let H and h be the depths of water respectively at normal and throat section. See Fig. 10.8.

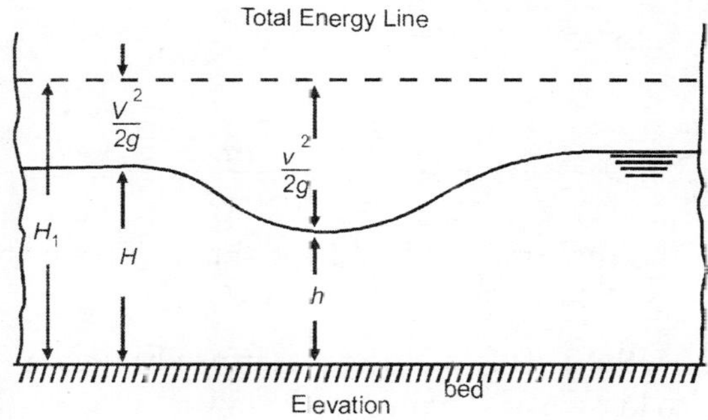

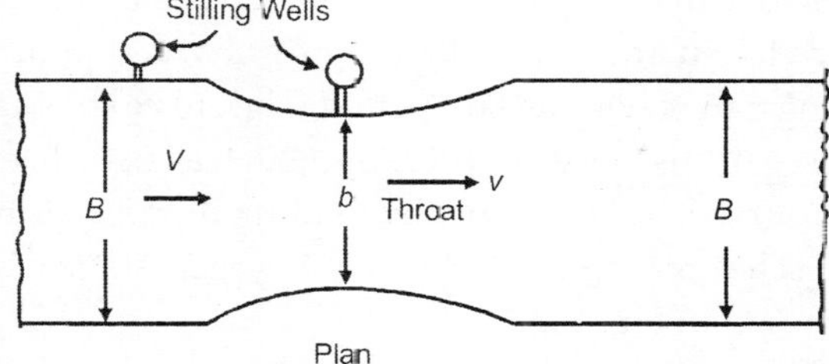

Fig. 10.8. *Flumed metres.*

$$\text{Total head at normal section} = H + \frac{V^2}{2g}.$$

$$\text{Total head at throat} = h + \frac{v^2}{2g}.$$

When there is no head loss from normal section to throat both these total heads must be equal

$$H + \frac{V^2}{2g} = h + \frac{v^2}{2g} \qquad (1)$$

If A and a are cross-sectional areas at normal and throat sections respectively then

$$AV = av \quad \text{or} \quad v = \frac{AV}{a} \qquad (2)$$

Putting value of v from (2) in (1) we get

$$H + \frac{V^2}{2g} = h + \frac{A^2V^2}{a^2 2g}$$

$$V = \frac{a\sqrt{2gH - h}}{\sqrt{A^2 - a^2}}$$

$$Q = AV = \frac{Aa\sqrt{2g(H - h)}}{A^2 - a^2} \qquad (3)$$

The length of the throat should be kept small to keep the energy losses minimum.

4. Chemicals Method. This method is not used much and as such is not important. It is also known as salt titration or salt solution method.

5. Stage Discharge Curve. It is a curve which tells us the relation between the discharge and the depth of water. In this case several observations at different gauge heights are taken and corresponding discharges computed. Discharges and corresponding gauge heights are plotted. The curve so obtained is known as station rating curve or discharge curve. After such a curve has been drawn, it is very easy to find out discharge simply by reading the gauge and finding out the corresponding discharge from such a curve A typical stage discharge or station rating curve is shown in Fig. 10.9.

This curve gives us a relationship between the stage of the river at a given time (gauge height) and the corresponding discharge. Hence it is also known as stage discharge curve. The relation expressed by this curve is known as stage discharge relationship.

If river bed has changed due to silting or scouring the same stage discharge curve will not hold good. The stage of a river may be constant, left rising or falling. During the rising stage, the measured discharge is more than that for a constant stage. Similarly during a falling stage, the measured discharge will be less than that for a constant stage.

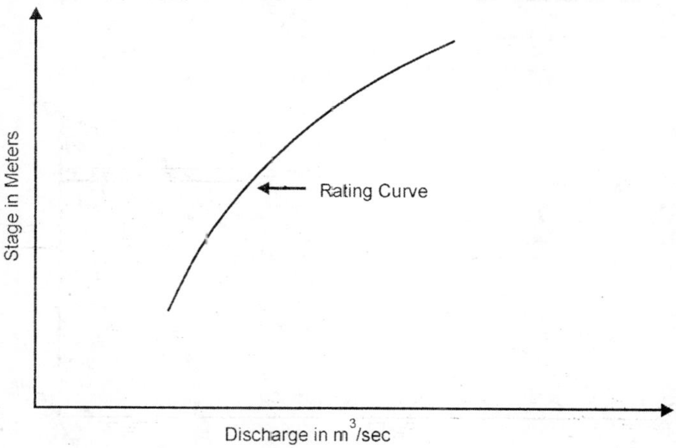

Fig. 10.9. *Stage discharge curve.*

6. From Power Plant Records. This method is an indirect method of computing discharge of the river. It is most accurate method and is generally adopted at places where no direct method such as weir station, or other measuring station is available.

Power plants are generally installed on the rivers and their operating records help us in determining the mean daily discharge passed through those plants. The total flow passing through a power plant is the summation of

(i) Flow through turbines

(ii) Flow over spillways

(iii) Flow through various sluices and leakages.

The quantities of all these flows can be computed separately and totalled to obtain the total discharge of the river. This method is especially used during flood peaks as it gives the most accurate result.

10.9 MEASUREMENT OF DISCHARGE BY AREA VELOCITY METHOD FOR BIG RIVERS

We had said earlier that cables may be stretched across the river and river width divided into small compartments by fixing tags to the cable. But in case of very large and wide rivers it is not possible to stretch cables across the river. In that case pivot method is used. It consists of a long high pole installed at one bank of the river. A long wooden plank is fixed to this pole at some height. The plank consists of a number of equidistant holes. P is the point of observation. The last hole B_7 on the plank is made in such a way that the observer from point P when views through hole B_7 strikes point B_7 at the opposite bank of the river. B_7 is the last point where river water at H.F.L. is touching the opposite bank. See Fig. 10.10.

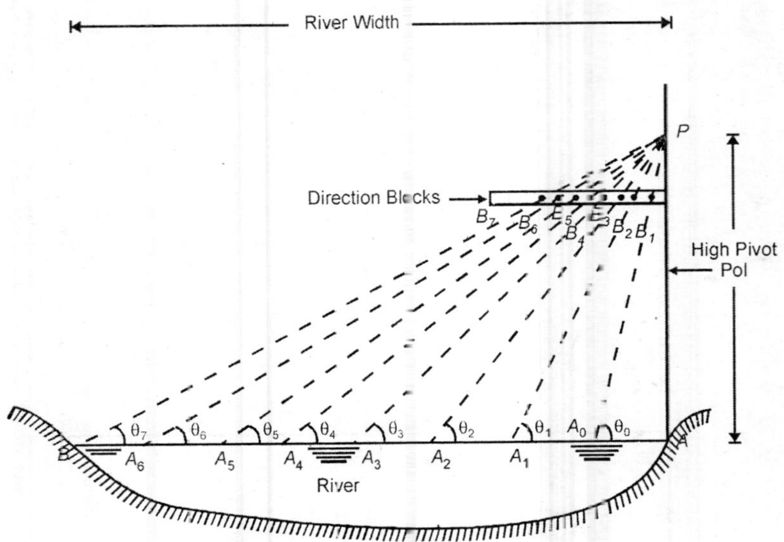

Fig. 10.10. *Dividing big river into small parts.*

Now by observing from P through points B_1, B_2, B_3......B_7, corresponding points A_1, A_2, A_3, in the river can be fixed. The velocity of flow for each compartment is then found out with the help of a current meter. Depths of water at the centre of each compartment are found out by going in boat. The discharge of the river is finally found out by adding together the discharges flowing in all the compartments separately.

QUESTIONS

10.1 Explain the term stream gauging. What are the differences between the canal and the river ?

10.2 (a) What are the purposes of measuring the discharges of rivers and canals ?

(b) Enumerate the points that should be considered while selecting the site for river gauging.

10.3 Explain the methods of measuring the areas of cross section for

(a) Smaller river

(b) Big river

(c) Trapezoidal canal.

10.4 How the velocity of flow changes according to the depth of water and according to width of the channel ?

10.5 Enumerate the different methods of measuring the velocity of a river. Give brief description of each method.

10.6 What are the different methods of measuring the discharge of a river ? Give brief description of each method.

❑❑❑

11

Dams I (General)

11.1 INTRODUCTION

The dam is a barrier constructed across the river to store water on its upstream side. Due to construction of a dam the water level of the river on its upstream is very much raised. Due to rise in water level large areas lying upstream of the dam get submerged. Dams are constructed to store the river water in form of an artificial lake or reservoir. The stored water can be utilized for generation of hydro-electric power, water supply, irrigation or for any other purpose.

11.2 CLASSIFICATION OF DAMS

Dams may be classified in several ways as follows :

1. Classification based on materials of construction.
2. Classification based on flow over its top.
3. Classification based on use of the dam.
4. Classification based on the mode of resistance offered by the dam to external forces.
5. Classification based on rigidity.

1. Classification Based on Materials of Construction. According to this classification the following may be the types of dams :

(i) Earth fill dams

(ii) Rock fill dams

(iii) Concrete dams.

(iv) Masonry dams.

(v) Steel dams.

(vi) Timber dams.

(*i*) *Earth Fill Dams.* Earth fill dams are the oldest types of dams. This dam structure consists of local soil mainly. This dam is not 100% impervious. Water percolates through the body of the dam but rate of percolation is kept under check and percolation line of hydraulic gradient line of percolating water is not allowed to get exposed anywhere. This dam does not require much of skill. They also prove economical as mostly local soil is used in their construction. The biggest drawback of these dams is that they cannot be constructed in large heights. These dams have been dealt in details in Chapter 12.

(*ii*) *Rock Fill Dam.* These dams mostly comprise rock boulders. Only boulders cannot prevent the flow of water and as such some impervious layer is laid on the upstream face of the Dam, which may be in form of cement concrete slab or earth fill covered by rip rap or any other arrangement. Rock fill provides the stability whereas impervious layer provides imperviousness to the dam to prevent flow of water through them. These dams are not much in use.

(*iii*) *Concrete Dams.* They are made either from plain cement concrete or reinforced concrete. They may be gravity dams or may be in form of arch, buttress or any other form of dams.

(*iv*) *Masonry Dams.* They are made from stone masonry in cement mortar. They are mostly gravity dams. Masonry dams and concrete dams have been discussed in details in Chapter 13 under common heading 'gravity dams'.

(*v*) *Steel Dams.* They are not used much. They are used for secondary purpose small dams.

(*vi*) *Timber Dams.* They are also of secondary nature and used for storing small amounts of water temporarily.

2. Classification Based on the Flow Over its Top. Under this classification the dams may be

(i) Over flow dams and (ii) Non overflow dams.

(*i*) *Overflow Dams.* In the case of over flow dams, the water may flow over the top. Spillways and weirs, are the examples of over flow dams. The dam after storing its designed capacity allows surplus water to flow down stream of the dam by passing over the top of the dam. This dam is mostly made of plain concrete or masonry and not of earthfill or rockfill.

(*ii*) *Non Overflow Dams.* In this case top of the dam is kept higher than the maximum expected high flood level. Water is never allowed to over top the dam. As water is not allowed to over top the dam, they may be constructed from any material.

In practically all the river valley projects overflow and non-overflow dam are combined. The main dam is constructed as non-overflow dam but a part of the dam is kept as overflow dam to act as spillway.

3. Classification Based on the Use of the Dam. Depending upon the use of the reservoir developed by the dam, the dams may be classified as follows :

(i) Storage dam

(ii) Diversion dam

(iii) Detention dam.

(*i*) *Storage Dam.* This dam is constructed mainly to impound the excess river water on its upstream side during floods. The stored water is used when river runs with deficient supply. Stored water may be used for irrigation, water power generation, water supply schemes, or for multi-purpose projects. The dam may be made of concrete, earth, masonry etc.

(*ii*) *Diversion Dam.* The purpose of this dam is not to store excess water behind it, but to raise the level of water so that water may be diverted into a system of canals. The height of this dam is, therefore, relatively small as no reservoir is to be formed to store water. During floods water passes over or through the dam to down stream side, while during normal flow the river water is wholly or partly diverted to the irrigation canals. The common examples of diversion dam are weirs and barrages.

(*iii*) *Detention Dam.* This dam is constructed to control the flood water. The flood water is stored behind the dam during floods. The stored water is released gradually at a safe rate when floods recede. Figure 11.1 (a), (b), (c), (d) shows flood hydrograph before the construction of detention reservoir. This flood hydrograph gets modified to *AB′CD* after the construction of the dam. In such dams water is temporarily stored and released through suitable outlet structures. The temporarily held water seeps into the banks and foundation and thus causes rise in ground water table in adjoining area, which may be exploited in from of lift irrigation. The submerged area during floods holds usually sufficient amount of water which may be enough for growing a crop without any irrigation. Such a detention dam is known as *water spreading dam* or *dike.* Detention dams are also sometimes constructed across tributaries carrying large silt and sediment. Such a detention dam is then called as *debris dam* whose main function is to trap the sediment before water of that tributary meets the main reservoir.

Today in all the river valley projects, the dam may serve the purposes of storage, flood protection, irrigation, power generation and other purposes.

4. Classification Based on the Mod or Resistance Offered by the Dam against External Forces. Under this classification, dams may be of following types :

(i) Gravity dams

(ii) Buttress dams

(iii) Arch dams.

Gravity dams resist all the external forces acting on the dam by virtue of its weight. Every effort is made to make dam more heavy so as to increase its stability.

Arch dams resist effect of external forces by arch action. The dams are curved in plan and are subjected to compressive stresses only.

Buttress dam is also subjected to same forces as gravity and arch dams. The total uplift pressure is reduced to a very small magnitude because of the gaps between the buttresses.

5. Classification Based on Rigidity of the Dam. According to this classification dams may be

(i) Rigid dams.

(ii) Non-rigid dams.

Rigid dams are those which are constructed of rigid materials like concrete, masonry, steel or timber.

The dams made from materials like earth, rockfill etc. are known as non-rigid dams.

11.3 ADVANTAGES AND DISADVANTAGES OF DIFFERENT TYPES OF DAMS

11.3.1 (A) Gravity Dams

Advantages

1. Maintenance cost is negligible.
2. They are specially suitable for deep steep valley conditions where no other dam is possible.
3. If suitable foundation is available, such dams can be constructed for very large heights.
4. Because they can be constructed in very large heights, they can store more amount of water.
5. If suitable separate place is not available for installation of spillways, they can be installed in the dam section itself.
6. This dam gives prior indication of instability. If remedial measures are taken in time, unsafe dams may even be rendered safe. Even if they cannot be made sale they give sufficient time for the people to move out the area likely to be submerged due to failure of the dam.
7. Silting rate of the reservoir can be reduced considerably by installing undersluices in the dam near the bed of the reservoir. Sluices can be operated from time to time and silt may be scoured out of the reservoir.
8. They are not affected by very heavy rainfall. Earth dams cannot sustain very heavy rainfall because of heavy erosions.

Disadvantages

1. They are very costly in initial construction.
2. They take lot of time of construct.
3. They require skilled labour for construction.

4. Such dams can be constructed only on good foundation.
5. If height of the dam is to be raised, it cannot be done unless provision for it had been made in the construction of the lower part of the dam.

11.3.2 (B) Earthfill and Rockfill Dam.

Advantages

1. They can be constructed on any type of foundation
2. They can be constructed in comparatively less time.
3. They do not require skilled labour.
4. Initial cost of construction is low as locally available soils, and rock boulders are normally used.
5. Their height can be increased without any difficulty.
6. They are specially suitable for conditions where slopes of river banks are very flat. Gravity dams under such conditions are not found suitable.

Disadvantages

1. They fail all of the sudden without giving any pre-warning.
2. Flood waters affect the dam safety.
3. Spillway have to be located independent of the dam.
4. They cannot be constructed as overflow dams.
5. They require continuous maintenance.
6. They cannot be constructed in narrow steep valleys.
7. They cannot withstand heavy rains unless properly protected.
8. They cannot be constructed in large heights. The usual height is 30 m for which most of the earthen dams are constructed.

11.3.3 (C) Arch Dams

Arch dam is curved dam in plan. It transmits major part of the water pressure to the abutments by arch action. The remaining part of the horizontal pressure is transferred to the foundation by cantilever action. Since most of the horizontal thrust is transmitted to abutments through arch action, it is very essential for the abutments to be very strong. The weight of the arch itself is not considered as resisting any horizontal water pressure. For this reason, the uplift on the base is not an important design factor. Early arch dams were used to be built of rubble ashlar masonry. But now almost all die arch dams are constructed in cement concrete. The arch dams may be of the following three types.

1. Constant Radius Arch Dam. This dam is shown in Fig. 11.1. The face of the dam coming in contact with water or upstream face is maintained vertical. In plan dam is circular. Thickness of arch is minimum at the top of the dam but goes on increasing as we proceed downwards. Taper is always given towards the inside of the arch curvature. In this arch dam, the radius of the arch with respect to outer face is kept constant. The centre point for all the curves remains same as shown in Fig. 11.1. This type of arch dam is adopted to U-shaped valleys. This arch dam is less economical than constant angle arch dam, which has been discussed ahead. However form work for constant radius arch dams is much simplex.

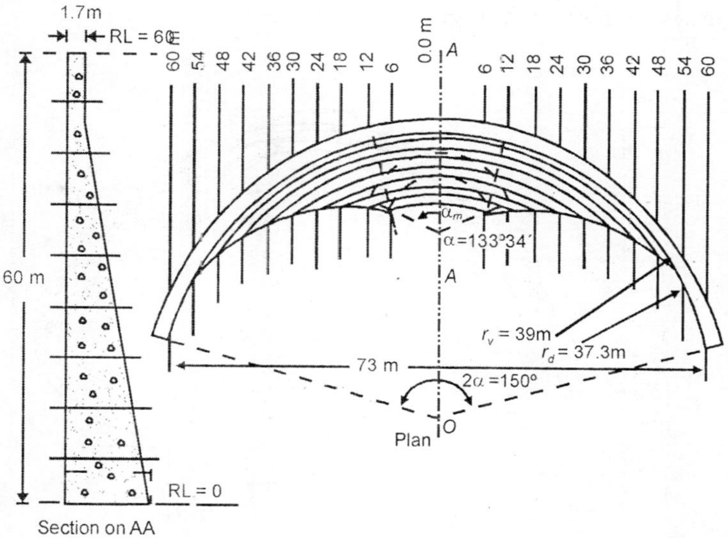

Fig. 11.1. *Constant radius arch dam.*

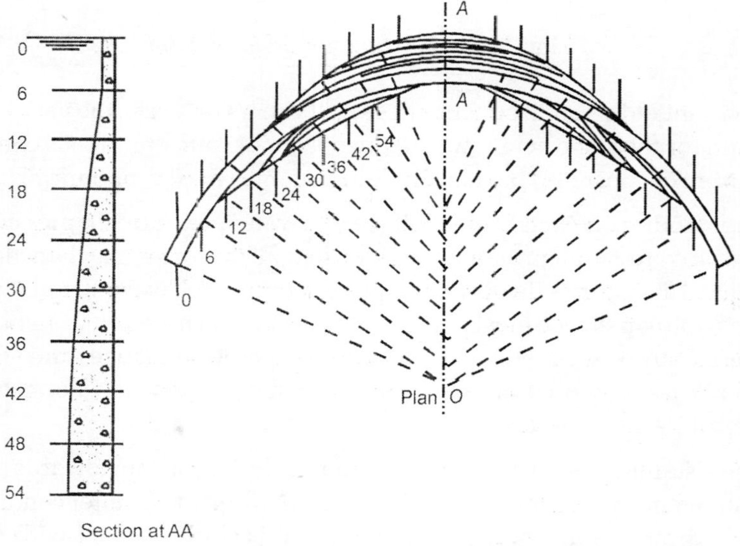

Fig. 11.2. *Variable radius arch dam.*

2. Variable Radius Arch Dam. Figure 11.2 shows this type of arch dam. This dam is adopted for narrow V-shaped valleys. In this dam radii of the intrudes curves and extrudes curves vary at various elevations which is maximum at the top and minimum at the bottom.

3. Constant Angle Arch Dam. It is a special variable radius arch dam in which central angle of the horizontal arch rings is of the same value at all the elevations. This dam is shown in Fig. 11.3. It has been found that the volume of concrete is minimum when the central angle is 133° 34'.

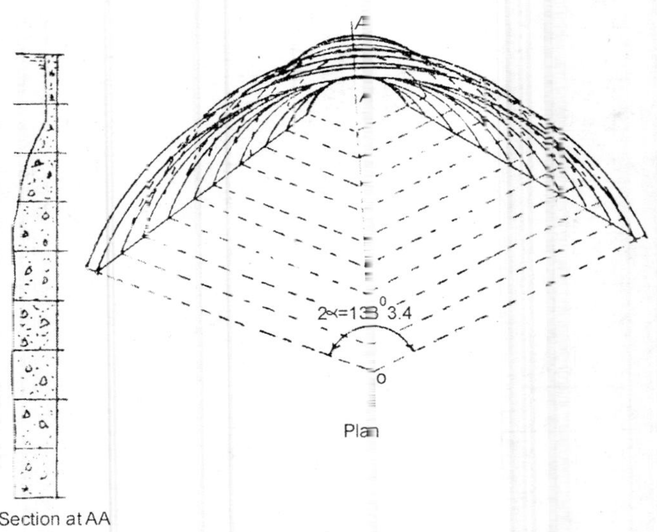

Fig. 11.3. *Constant angle arch dam.*

If constant radius arch dam requires say 100 m^3 of concrete, the variable radius arch dam for the same height will require 58 m^3 of concrete. Hence considering the amount of concrete, constant angle arch dam is most economical.

Forces Acting on Arch Dams. Self weight, water pressure, silt pressure, uplift pressure, ice pressure, earthquake pressure etc. all the forces may be acting on the dam. But all these forces do not carry equal important. Self weight, up-lift pressure, are negligibly small. The most important forces to be considered in arch dams are the internal stresses caused by ice pressure, temperature changes and yielding of abutments. Ice pressure causes a continuous concentrated load along the arch at the level of ice i.e. water level.

During summer, temperature change causes shifting of dam towards upstream direction while in winter towards down stream direction. Winter conditions are considered more appropriate for stress analysis since they act with reservoir loads. The slight yielding of abutment may also develop high internal stresses in the arch. Stress analysis has not been given here.

Advantages

1. They are particularly suitable for narrow deep gorges. In such conditions arch dams prove even more suitable than gravity dams.

2. They require comparatively very small amount of construction material. The section or thickness of the dam is very small in comparison to the gravity dam.
3. Uplift pressure is not very large as base width of the dam is quite small.
4. They can be constructed on moderate foundations.

Disadvantages

1. They require very complicated form work which is very costly.
2. They require very skilled labour.
3. Design of the dam is also difficult.
4. Speed of construction is very slow.
5. They require very strong solid rocky abutments to resist the thrust of the arch.

11.3.4 (D) Buttress Dams

This dam consists of a number of piers that divide the total length of the dam into a number of spans. All the spans are then covered either with inclined concrete slab or arches on the upstream side of the piers. If spans are covered with flat inclined slab it is known as deck type buttress dam. If spans are covered by arches the resulting dam is known as multiple arches buttress dam. The elements of this dam are the sloping slab, buttress, mat foundation, and lateral braces. Braces are provided to reinforce the buttress. Cut-off walls may also be provided to prevent seepage.

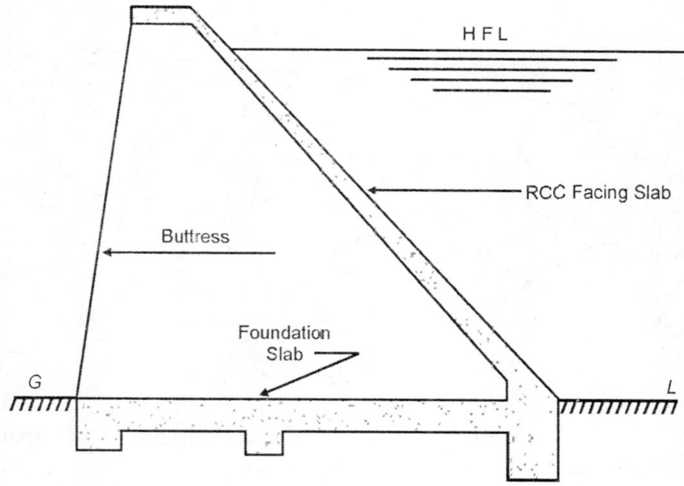

Fig. 11.4. *Deck type buttress dam.*

Advantages

1. They can be constructed on weak foundation because they are less massive than gravity dams.

2. The load of water lying on the inclined deck slab adds to the vertical component. of the dam and as such helps in increasing the stability of the dam. The factor of safety in case of buttress dams is far greater than for gravity dams.

3. Since ice tends to slide over the inclined U/S surface, ice pressure does not carry any significance.

4. Height of buttress dams can be increased by extending both buttresses and deck slab as shown in Fig. 11.5.

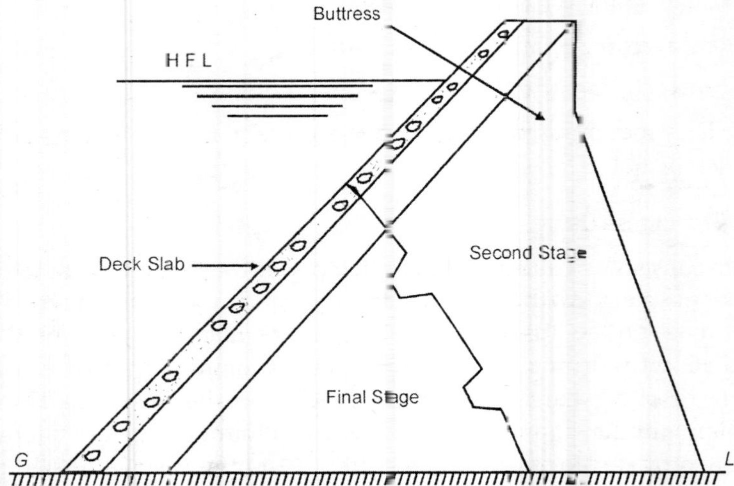

Fig. 11.5

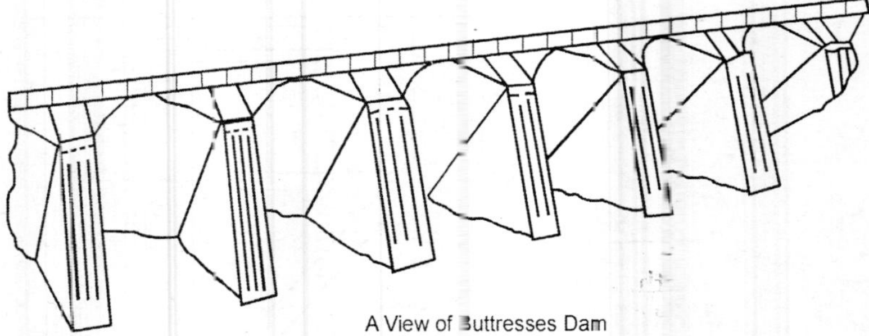

Fig. 11.6. *Buttress arch dam.*

5. Power houses or other plants can be located in the empty space between buttresses.

6. They require only half to one-third amount of cement concrete in relation to the concrete required for gravity dam for the same height. The cost is not reduced by the same proportions as increased cost is required for form work and reinforcement.

7. Because of space between buttresses, it is possible to reach the back space of deck slab. By reaching here periodic inspection of foundations etc. can be carried out.

8. Since exposed surface of concrete is more but volume of concrete is less than the gravity dam, heat dissipation during construction is achieved better. Also speed of construction can be increased as thinner sections do not cause any problem for cooling.

Disadvantages

1. More skilled labour is required.
2. Shuttering cost is more.
3. The slab being very thin, its face in contact with water is likely to deteriorate and cause damage to the dam.
4. It is more susceptible to delibrated damage. It is so because thickness of slab is very small and also there is approach to reach back of the slab through the space between buttresses.

Steel dams and timber dams have been discussed in Chapter 14 where arch dams have also been discussed in brief.

Types of Buttress Dams. Buttress dams may broadly be categorised under three heads.

(i) Rigid type.
(ii) Articulated type.
(iii) Semi-rigid or intermediate type.

In the case of rigid type buttress dam sloping deck slab is cast monolithically with the buttresses. No allowance for any settlement of foundation is made. Multiple arch dam and multiple dome dams are the examples of this type of dam.

Articulated buttress dams are quite flexible. The slabs is not cast monopolithically with the buttresses. Flat slab buttress dam is the example of this type.

Round head buttress dam and the diamond head buttress dam are the examples of semi-rigid type dams. No undesirable rigidity is considered in the design of such dams.

Following are the types of buttress dams which fall under the above three categories :

1. Flat slab buttress dam. It is also known as Ambursen type of buttress dam named after its inventor. See Fig. 11.7.

2. Multiple arch type buttress dam. It consists of series of arches transmitting water pressure to buttresses.

3. Multiple dome type dam. It consists of dome shaped deck instead of arch or flat slab.

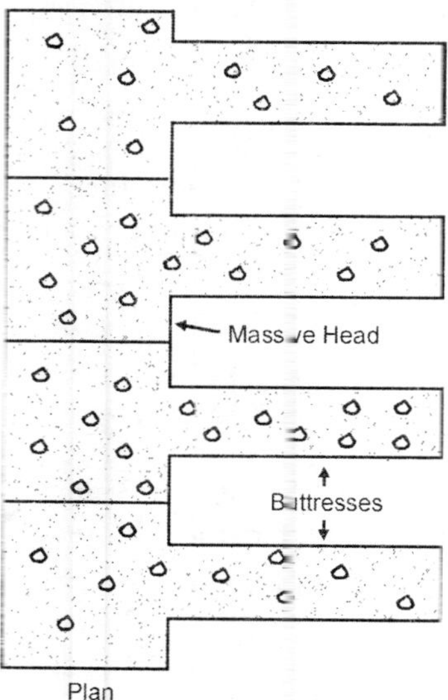

Fig. 11.7

4. Cylinder or massive head type dam. In this water supporting member is formed by enlarging the upstream end of the buttresses until they meet adjacent members.

5. Columnar buttress dam. In this case deck slab is supported on inclined Columns rather than buttresses.

6. Truss buttress dam. In this, deck slab is supported on reinforced concrete trussed buttresses.

11.4 FACTORS THAT GOVERN SELECTION OF TYPE OF DAM

It is very difficult to make decision as to what type of dam will suit best to the particular site condition. Topographical, geological and foundation conditions, availability of materials, suitable site for location of spillways, etc. influence the selection of type of dam. Availability of labour, equipment, accessibility of site etc. also affect selection of the dam to some extent. The most common physical factors have been discussed as follows.

1. Materials of Construction. Dams require very large quantities of construction materials. Only that dam may prove economical whose materials are available locally or in the vicinity of the site. If gravel, sand an crushed stone are available locally, concrete gravity dam may be most suitable. If only fine and coarse grained soils are available earth dam may prove suitable.

2. Geological Conditions. If available foundation is solid sound rock, having no faults or fissures, any type of dam can be constructed on it. If rocks are available but they have faults or fissures, they will have to be grouted first before founding gravity dam over them. Poor rock or gravel foundations are suitable for earth or rockfill dams but not for gravity dams. Deep cut-offs may be constructed under such conditions of foundation to check the seepage. Fine sand or silt, foundations are suitable only for either earth dam or for low gravity dams but not for rockfill dams. Gravity dams and rockfill dams are not suitable on clay foundations. Clay foundations are suitable for earth dams, but only after giving proper treatment.

3. Topography. This is one of the most important considerations. A narrow deep valley suggests adoption of gravity dam. If however the width of the valley at top is less than about one fourth the height and separate site for spillway is available, arch dam may suit best. In case of plain country, having very flat slopes of banks, earth dam is the best choice.

4. Spillways. For the safety of the dam, safe and efficient disposal of flood discharge is very essential. This requires a suitable site for the spillways. If they large spillway discharge is required, an overflow concrete gravity dam may be adopted. If separate site for spillway is available and spillway capacity required is also small, earth dam should be preferred. If no separate site for spillway is available and it has to be accommodated in the dam proper, then concrete gravity dam may be most preferred choice. Sometimes even earth dam may be constructed but some part of the dam which is to be used as spillway may be constructed of concrete.

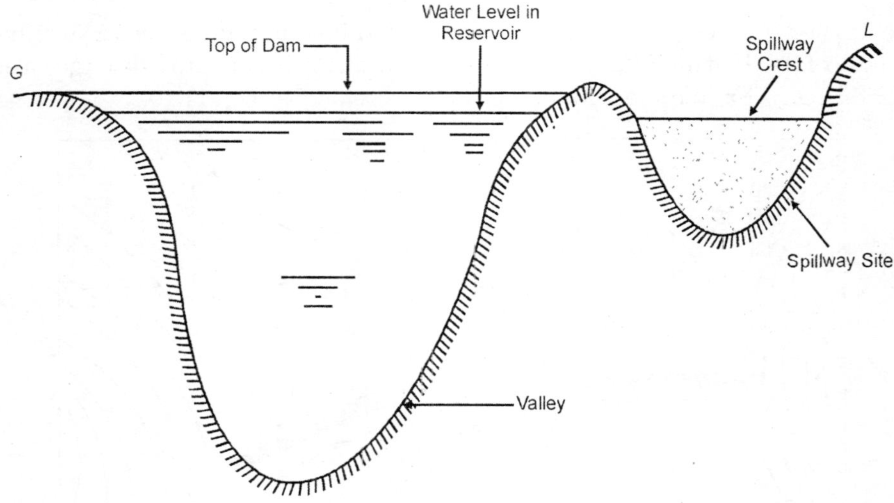

Fig. 11.8

5. Roadway Over the Dam. If a roadway is to be passed over the top of the dam, it can be none else than earth or gravity dam.

6. Life and Height of the Dam. If length of the dam is quite large, but height not much, an earth dam would be the best choice. On the other hand if length is small but height is more gravity dam should be the only choice.

7. Life of Dam. If life of the reservoir has to be very large, concrete or gravity dam should be adopted. Earth dams and rock fill dams require constant supervision and in long life they prove uneconomical.

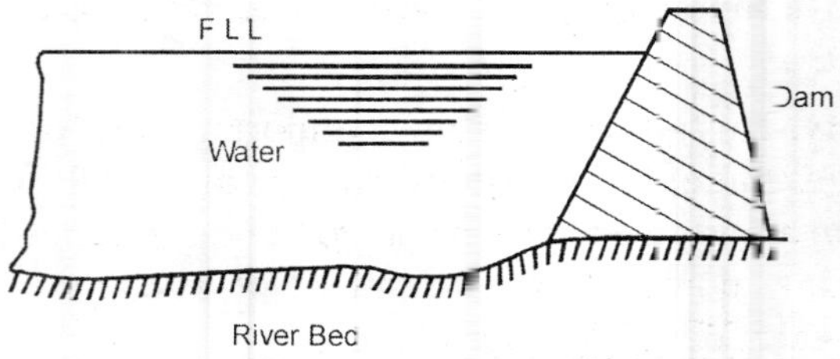

Fig. 11.9

8. Generation of Hydroelectric Power. In this case dams are exclusively made of concrete or masonry gravity dams. Earth dams, being not very high, do not develop sufficient head for running the turbines.

11.5 SELECTION OF SITE FOR DAM

Following points should be considered while selecting the site for the dam.

1. Suitable foundation for the dam should be available.

2. The width of the river at dam site should be minimum so as to keep the length of the dam minimum and thus economical. In order to provide large basin for the reservoir, the site should open out upstream. See Fig. 11.10.

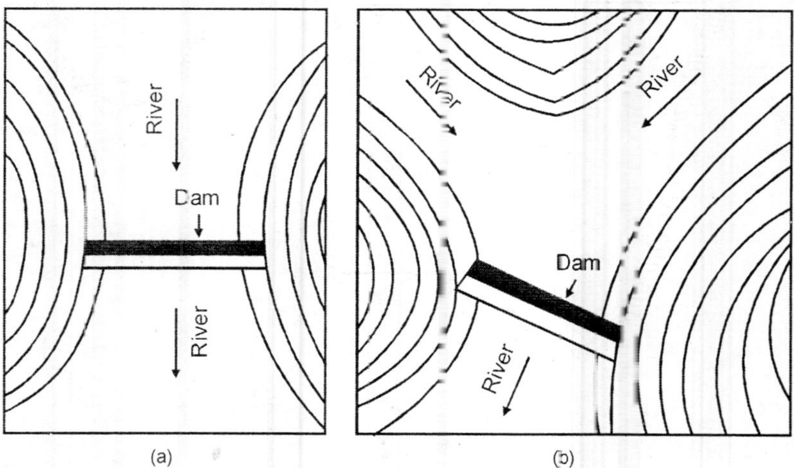

Fig. 11.10

3. If river is deep, but there is raised bed level existing, the dam should preferably be located at the raised bed. This will help in keeping height of the dam less. Such a situation has been shown in Fig. 11.9.

4. Proper separate site for location of spillways should be available. If separate spillway site is not available, the width of the river should be sufficient to locate spillway in the dam proper.

5. The materials of construction for the dam should be locally available.

6. Very costly land should not submerge in water because of the construction of the dam.

7. The site should be well connected either with railway or highway or both. This will facilitate transport of cement, machinery and other materials required in the construction.

8. The site should be such that leakage through its sides and bed is minimum. Highly permeable rocks should be absent from the site.

9. Site should such that the reservoir developed is deep and having minimum spread of water.

10. Site should be such that the tributaries carrying very high content of silt do not lead to the reservoir developed by the dam.

11.6 SOME MINOR DAMS

Arch dam, and buttress dams come under the category of medium dams, although their use is not very common. They have been discussed quite in details. Earth dams and gravity dams are the most commonly used dams. They will be discussed in their full perspective in subsequent chapters separately. Steel dams and timber dams are also sometimes adopted but only as a temporary measure of storing water. A brief description of steel and timber dams which are considered as minor dams have been given here.

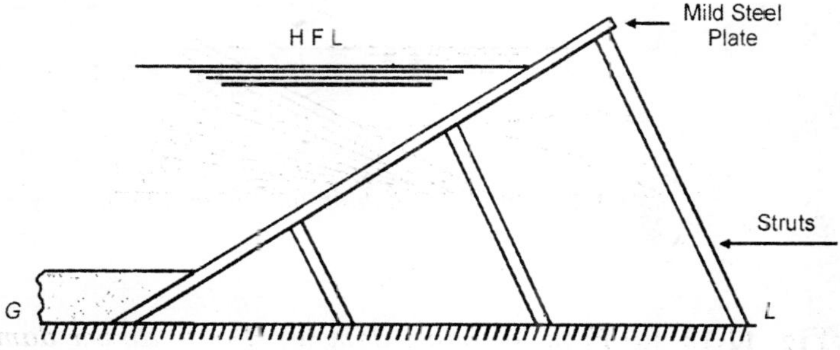

Fig. 11.11. *Steel dam.*

1. Steel dams. These dams are not in common use. No dam in Indian has yet been constructed of steel. In USA some dams had been constructed in steel in early part of this century. These dams consist of a deck slab made of steel plates. These plates are supported by inclined struts. These dams may also be of cantilever type. In this, a frame work of steel sections is installed in the rigid foundation and deck plate is fixed on them. In this case tensile forces one induced in deck plate and girders. See Fig. 11.11.

Advantages

1. They are cheaper than other rigid dams.
2. They can be built in very short time with modern methods of fabrication.
3. They are more adaptable in case of unequal settlement of foundation.
4. They are more leak proof.
5. Frost action does not have any adverse effect on steel.
6. They can be repaired easily by welding.

Disadvantages

1. Life is shorter than concrete or masonry dam.
2. They require constant maintenance.
3. They are not as adaptable to vibrations due to spilling water. This is because they are too light.
4. They have to be anchored with foundation.
5. Bearing stresses are excessive as bearing are in this case is very small.

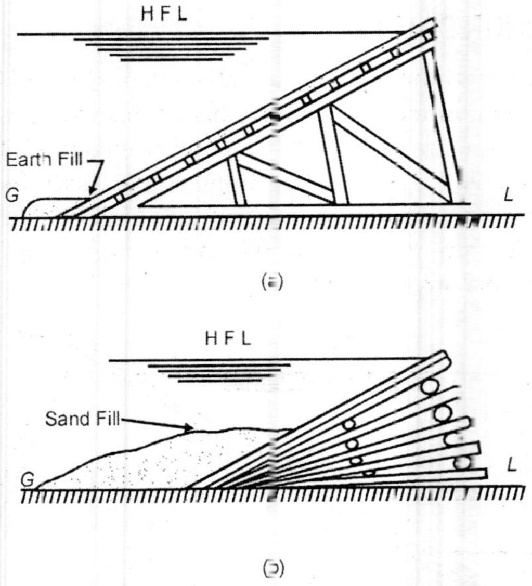

Fig. 11.12. *(a) Framed timber dam (b) Beaver timber dam.*

11.6.1 Timber Dams

They are also a sort of temporary dams. They consist of a frame work of wooden bullies and planks. Timber dams are usually of three types as follows.

1. Framed Timber Dam. In this type of dam a triangular frame work of timber bullies is formed and face which has to come in contact with water is covered by fixing studs and planks. Such dams are usually of very short heights, seldom exceeding 10 m.

2. Rock Filled Crib Timber. In such dams, crib piers of wooden members are erected at centre to centre distances of 2 to 2.5 m. In order to impart stability to the piers, rock boulders are filled in the open spaces left in the wooden members. After having erected all the crib piers, along the centre line of the dam, timber planks are fixed to prepare the face for retaining water. If rock foundation is available at the site, the bottom of the cribs is anchored with the rock. In case of earth foundation, sheet piles are driven both at the *U/S* as well as *D/S* side of the cribs. System of sheet piles or that of anchoring with rock foundation is essential to maintain the cribs in position. See Fig. 11.13.

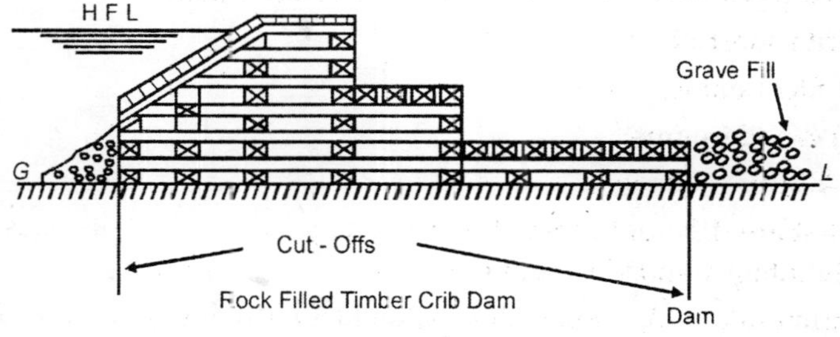

Fig. 11.13

3. Beaver Type Timber Dam. This dam does not involve even driving of sheet piles or anchoring with foundation rocks. This dam is suitable for small heights and also where plentry of timber is easily available. An inclined stack of timber logs is developed as shown in the Fig. 11.12 (b). The stock is finally covered on water face by wooden planks. To avoid scouring and also to add to the stability of the dam, the lower *U/S* end should be covered with boulders and sand.

Advantages

1. Initial cost is low.
2. Can be erected in very short time.
3. Can be constructed on any foundation.
4. Very much used for temporary works.

Disadvantages

1. Have very small life.
2. Suitable for very small heights.
3. Maintenance cost is high.

QUESTIONS

11.1 Give different classifications of dams. Explain the classification based on material of construction in details.

11.2 Explain the terms:
(i) Gravity dams.
(ii) Overflow dams.
(iii) Rigid dams.
(iv) Diversion dams.

11.3 Give advantages and disadvantages of gravity dams, earthfill dams, and arch dams.

11.4 What are different types of arch dams ? Explain all of them in details. Give advantages, and disadvantages of arch dams.

11.5 Write short notes on:
(i) Steel dams.
(ii) Wooden dams.
(iii) Buttress dams.

11.6 What are different types of buttress dams ? Give their sketches. Also give advantages and disadvantages.

11.7 Enumerate and explain the various factors that govern the selection of type of dam.

11.8 What points should be considered while selecting site for the dam?

❑❑❑

12

Dams II (Earth Dams)

12.1 INTRODUCTION

As the name of this dam indicates, it is constructed by using locally available soil predominantly. These dam have been constructed since early days of civilisation. Even today, earthen dams are the most common types of dams. The following are the main reasons due to which these dams have been and still continue to be the most commonly used types.

1. The basic material used in the construction, is the local soil. This aspect makes these dams economical.

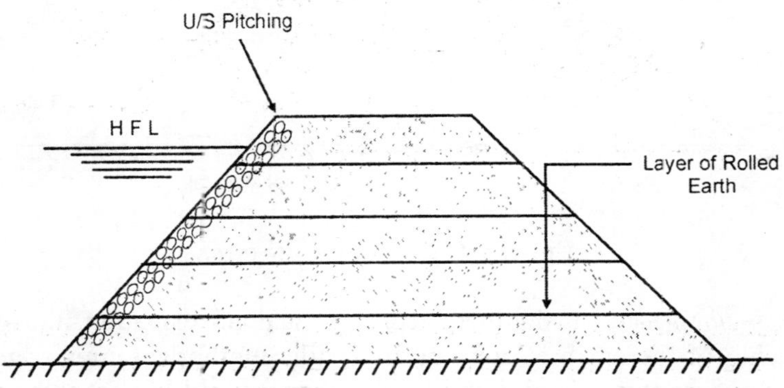

Fig. 12.1. *Rolled fill earth dam.*

2. The science of soil mechanics has recently developed to such a stage that dams constructed from earth by adopting the modern theories of soil mechanics, can be as equally relied upon as those of gravity dams.

3. No specific type of foundation is required. They can be constructed on any type of foundation by making little bit changes for different foundation conditions.
4. No skill of very high standard is required.
5. No materials has to be imported from outside as locally available soil is the main material of their construction.

12.2 CLASSIFICATION OF EARTHEN DAMS

Earthen dams may be classified in two ways :

1. Classification based upon the method of construction.
2. Classification based upon the section of the dam.

1. Classification Based Upon the Method of Construction. Under this classification earthen dams can be

(i) rolled fill earth dams and

(ii) hydraulic fill earth dams.

(i) Rolled-fill earth dams. In the construction of such dams successive layers of moistened earth are laid one after the other. Each layer is laid after the layer of earth previously laid is adequately compacted. Sheeps foot roller is mostly used for carrying out the process of compaction. Bulldozers, scrapers, draglines, like heavy earth moving machinery is used for the purpose of earth work. Moisture content during compaction is maintained at the level of optimum moisture content. Proper slope of the dam is maintained during construction. See Fig. 12.1.

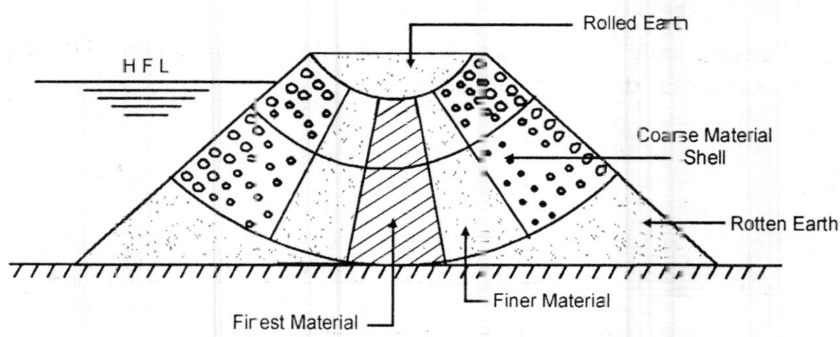

Fig. 12.2. *Hydraulic fill dam.*

(ii) Hydraulic fill dams. In this method of construction, excavation, transportation, and placing of the earth is done by hydraulic methods. Outer edges of the embankment are maintained slightly higher than the middle part of the dam. The mixture of excavated materials with water, is pumped and discharged at the outer edges of the dam. The mixture of excavated material andwater consists of coarse material and fine material. When this mixture is discharged near the outer edges, the coarser material of the mixture would settle at the edges while finer material settles at the centre. No compaction is required in this construction. The

process is continued till dam of required height is constructed. In this dam coarser material lies near outer edges which is not very impervious. The finer material deposited in the central part of the dam consists of clay mostly and thus provides imperviousness to the dam.

2. Classification Based Upon the Section of the Dam. According to this classification the dams can be of the following three types:

(i) Homogeneous earth dams,

(ii) Zoned earth dams, and

(iii) Diaphragm type earth dams.

(*i*) *Homogeneous earth dams.* This dam is made from a single material. It is suitable for low heights only. Purely homogeneous section can be modified a little by constructing rock toe at the down stream lower end of the dam and also by putting horizontal filter drains. Both these measures control the seepage and thus enable to construct much steeper slopes of the dam. These measures also keep the phreatic line of seeping water, well within the body of the dam. The homogeneous dams are made from impervious or semi-impervious soils. This is essential to put a barrier against the flow of seeping water. Upstream slope of the dam is generally kept flat so as to reduce the path of seeping water and to counter act the effect of sudden draw downs (Fig. 12.1).

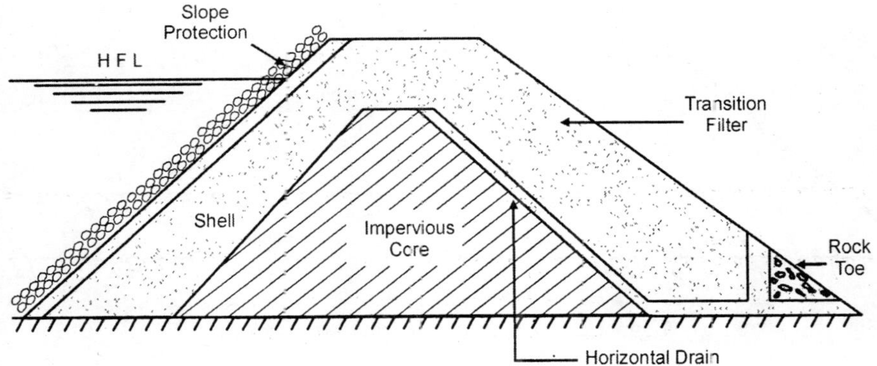

Fig. 12.3. *Zoned earth dam.*

(*ii*) *Zoned earth dam.* This dam is made by using more than one material. In this case, the central part of the dam, which is known as core is made from impervious material. Considerably more pervious material is used on both the sides of core. The dam also consists of a rock toe, a system of horizontal drains, and sometimes even system of inclined filters to carry out proper drainage of seeping water from the dam. If, at certain place, a variety of soils is available the dam should always be of zoned type. Impervious material should be used for core, while pervious soils at flanks of core. *U/S* pervious soil provides stability against rapid drawdown, while that on *D/S* side acts as a drain to control the seepage line. Central core checks the seepage. See Fig. 12.3.

(*iii*) *Diaphragm type earth dam.* This is a sort of combination of both the previously, mentioned earth dams. The bulk of the embankment is made from pervious soil, but a thin diaphragm of impermeable material is provided at the central part of the dam to check the seepage. The impervious diaphragm may be made of impervious clayey soil, cement concrete, masonry or of any other impervious material. It can be located somewhere in the central part of the dam or it may be put on the upstream face of the dam also. The difference in zoned and diaphragm type earth dams is due to thickness of impervious core or diaphragm only. If thickness of impervious core is less than 10 m or less than the height of the dam above that level the dam is known as diaphragm type, otherwise it would be zoned earth dam.

12.3 SEEPAGE LINE OR PHREATIC LINE

Seepage line and phreatic line means same thing. This is such a line in the body of the dam below which there are positive hydrostatic pressures. On the line itself the hydrostatic pressure is zero. Above the line, there is a zone of capillary saturation. The effect of capillary is however neglected in dams. Analysis of seepage has been given ahead.

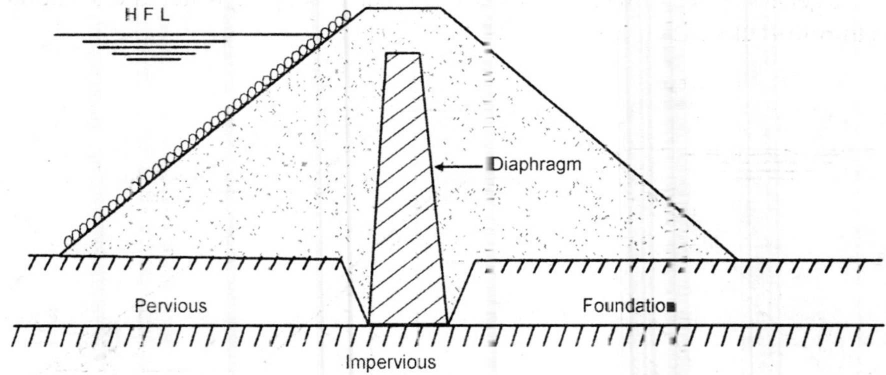

Fig. 12.4. *Diaphragm type earth dam.*

12.4 CAUSES OF FAILURE OF EARTH DAMS

One of the investigation reports on past failures of earthen dams states that 40% failures were due to hydraulic failures and 30% each due to seepage and structural failures. The causes of failure of earthen dams may be divided into three categories as follows:

1. Structural failures.
2. Hydraulic failures.
3. Seepage failures.

1. Structural Failures. Causes of structural failure of the earthen dams are the following.

(*i*) *Failure due to pore pressure.* The drainage capacity of impervious compressible soils is very slow. When earth dams are made from such soils, excessive pore pressures are developed in the soil, during and immediately after construction of the dam. The amount of pore pressure is dependent upon the permeability of the soil. Lesser the permeability greater is the pore pressure. If the permeability of the soil being used in the construction of the dam is very low, it is possible that there may be no substantial drop in pore pressure in the central part of the dam, by the time the construction of the dam finishes. The amount of pore pressure, equal to weight of the soil has been noticed above the point where pressure is being measured. The pore pressure thus may cause failure of the slope and may even impair the stability of the dam. Hence construction stage is more critical when soil being used for construction is almost impervious. It has been further noticed that pore pressures cause two types of construction slides which differ in speed and magnitude of the movement. The first type of slide is uniform and slow and continues for a period of about two weaks. Failure of North Ridge Dam in Canada had been attributed to this cause.

The second type of slide is sudden and rapid. The most of this sliding is completed within few minutes. The failure of Marshal Creek Dam in USA is the example of such a failure. It is stated that this movement took 15 to 20 minutes only.

(*ii*) *Sudden drawdown on the upstream face.* When reservoir is emptied suddenly, the pressure due to water suddenly vanishes from the water face. But this action being sudden, the saturated soil up to which water was filled before emptying, does not get time to release water so as to develop equilibrium conditions. This phenomenon is known as sudden drawdown. Due to drawdown the hydrostatic pressure due to reservoir water is removed, without leaving any counteraction against the pressure due to water held in the soil. This unbalanced outward pressure due to water held in the soil of upstream face causes upstream face to slide. Such a slide of upstream face normally does not cause failure of the dam, but soil thus slided may block the outlets of conduits and cause difficulties. When such slides of upstream face take place, the pore pressure along the surface of slide gets dissipated to a large extent. This aspect does not give birth to the process of further sloughing and sliding and hence chances of failure due to sliding are very small.

(*iii*) *Down stream slope slide.* The chances of down stream slope sliding are maximum when reservoir is full and rate of percolation is at its maximum rate. The earthen dam can easily fail by this reason. The slides that may occur on *D/S* side of the dam may be deep slides or shallow slides. Shallow slides occur in sandy soils only and they never exceed 1 to 2 in in depth in direction normal to the slope. Deep slides take place in clayey dam and clayey foundations. When a deep slide has taken place there is no relief to pore pressure and the unstable vertical slide scrap, left standing often sloughs or slides. This process of slides keeps on repeating until dam reaches the point of breach. One strong wave of water will now cause breach and failure of the dam.

(*iv*) *Foundation slide.* The dam can slide as a whole if it is founded on fine silt or soft soil. Presence of soft weak clayey seem or that of silt and sand both can cause failure of foundation. Expansion of soil on saturation may cause lifting of the slopes and thus may cause failure of foundation.

(*v*) *Failure by spreading.* When dam is constructed on stratified deposits containing layers of soft clay, the failure by spreading of the fill may take place.

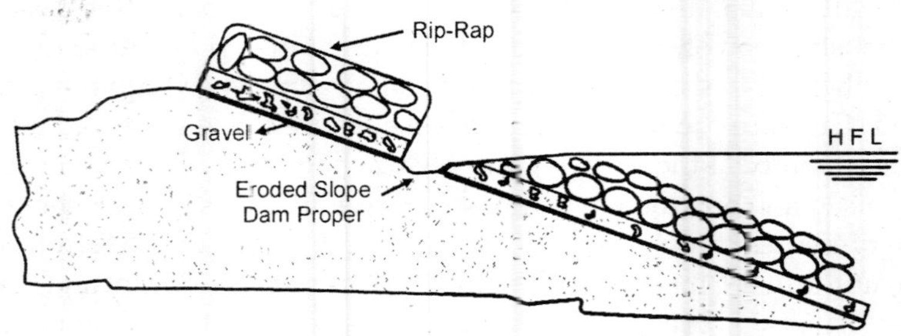

Fig. 12.5. *Failure of rip-rap.*

(*vi*) *Slope protections failure.* The slopes of the dam are generally protected by pitching or rip rap. Pitching or rip rap is a protective layer of large boulders. This layer is generally laid over a layer of gravel or filter blanket. Due to repeated striking of water waves to the riprap at level of the reservoir the rip rap may be dislocated. This exposes the embankment to wage erosion. The failure of the rip rap may ultimately lead to failure of the dam.

(*vii*) *Failure due to holes caused by burrowing animals.* The burrowing animals may dig holes through the dam or through foundation and may cause failure of the dam by piping. Normally such animals do not dig in moist soils i.e. below the seepage line. However, if water level in the dam is very low for a number of years, the burrowing animals may honey comb the dry upper part of the dam. As soon as water level rises, the water may escape through the honey combed holes and may cause failure of the dam.

(*viii*) *Failure due to earthquake.* Following may be causes of failure of earthen dam due to the effect of earthquakes.

(a) The core may develop cracks and may lead to leakage and then to ultimate failure.

(b) Dam itself or its foundation may settle down due to shaking action. This may lead to over topping of the dam.

(c) Sliding of hill tops into the reservoir and causing rise in reservoir level. This phenomenon may again cause over topping of the dam.

(d) Horizontal component of the acceleration force due to earthquake may cause shear slide of the appreciable portion of the slope of the dam and may cause dam failure.

(ix) *Failure due to soluble material in dam.* If some material which is soluble in water, is present in the dam or in its foundation, it may cause failure of the dam.

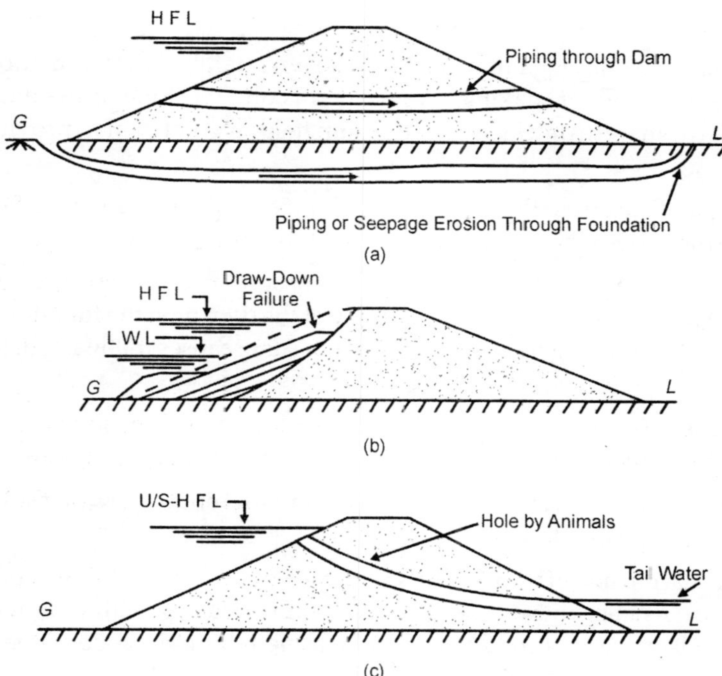

Fig. 12.6. *(a) Piping failure of dam; (b) Draw down failure; (c) Holes by burrowing animals.*

Such materials are washed away with time and may cause settlement of the dam and thus failure. Such materials if get deposited in the filters, designed for drainage purpose, may cause clogging of the filters and this may ultimately lead to dam failure, as drainage of the dam will not be proper in that case.

2. Hydraulic Failures. Hydraulic failure of the dam may take place due to following reasons.

(i) *Over topping.* If amount of flood has been under estimated, the capacity of the spillways will not be adequate. This may lead to the rise in water level in the reservoir above the maximum estimated level. This may ultimately lead to the over topping of the dam.

(ii) *Wave erosion.* If *U/S* face is not properly protected by rip-rap, the wave erosion of the upstream force will occur. Roller motion in the soil is set and thus easily gets scooped out when wave returns. If this action of wave erosion is not checked in time, it may lead to scooping out whole of the free board soil and may ultimately cause failure of the dam. Waves can also cause *u/s* slips and consequent failures.

(*iii*) *Toe erosion.* Toe erosion of the dam may occur if water on the *D/S* side of the dam is some how coming contact with the Toe of the dam. The failure due to this reason can be avoided by providing a thick riprap on *D/S* Toe of the damp up to a height slightly above the tail water level

(*iv*) *Gullying.* Gullying failure is due to heavy rainfall. Gully formation can be avoided by turfing, or making counter berms on *D/S* slope of the dam. Good system of drainage from *d/s* side of the dam helps a great deal in providing this failure.

3. Seepage Failures. Piping and sloughing area the two seepage failures of the earth dam. Both these phenomena of failures are as follows.

(*i*) *Piping.* Piping is nothing but process of progressive erosion of concentrated leaks. The piping trouble can be in the body of the dam or in the foundation of the dam. The water seeping though the body of the dam causes following detrimental effects.

(a) Seeping water develops erosive force which dislodges soil particles from soil structure and cause rearrangement of void between larger grains.

(b) The seeping flow, along with associated differential pore pressure can lift soil particles, causing boiling.

(c) Internal erosion of the soil starts from the point of exit and slowly progresses backwards. As concentration of leak goes on increasing with more and more movement backwards a conduit or pipe is formed through the soil. Formation of this conduit is known as piping.

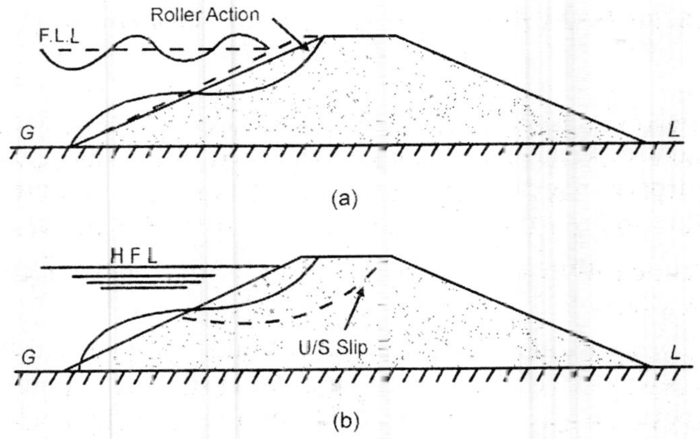

Fig. 12.7. *(a) Roller action of waves (b) Wave action on U/S face.*

(d) The internal pressures in the soil water reduce internal friction and thus weaken the soil mass.

The piping trouble can easily occur if soil around the concrete pipe is pervious and is not properly compacted. Piping may also take place if there is poor bond or compaction between the dam fill and the foundation or abutment.

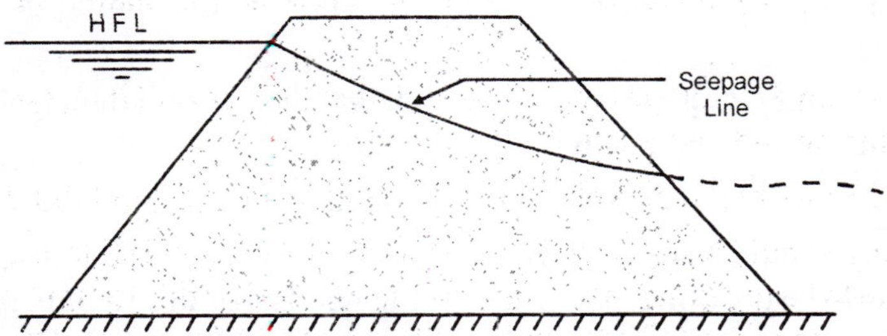

Fig. 12.8. *Seepage line in the dam.*

(*ii*) *Sloughing*. The phenomenon of sloughing is very closely related to the process of piping. When reservoir is full, the Toe of the dam is generally saturated. The Toe may erode producing a small slide. This slide makes down stream face more steep which again becomes saturated by seepage and slides again. This again causes still more steepness of *D/S* face. This process of seepage and sliding continues till the dam section is rendered too thin to withstand any water pressure and thus cause failure of the dam. This process of repeated saturation of the Toe and then its sliding is known as *sloughing* which may also be named as *ravelling*.

12.5 ESSENTAIL REQUIREMENTS FOR THE SAFE DESIGN OF THE EARTH DAM

If an earth dam has to be safe and stable during construction and in its subsequent operation, it must be designed considering following points.

1. The earth dam should never be allowed to be over topped during heavy flood. This object can be achieved by providing spillways of adequate capacity.

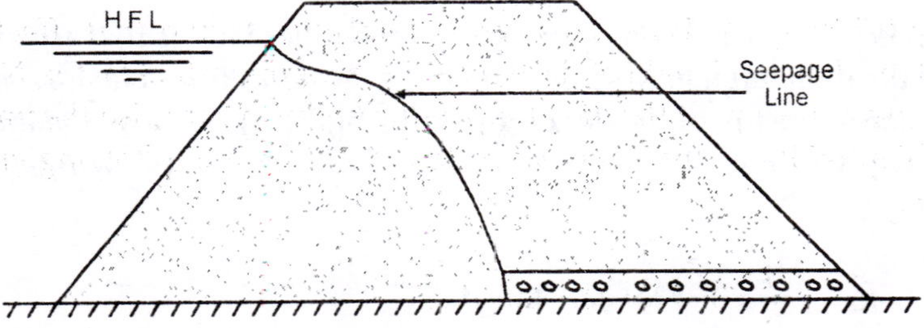

Fig. 12.9. *Effect or filter on seepage line.*

2. Sufficient free board should be provided.

3. Seepage line should remains buried sufficiently on the *D/S* face when reservoir is full. This will prevent occurrence of sloughing of the *D/S* face.

4. The dam portion laying *D/S* of the impervious core should be properly drained.

5. The *U/S* and *D/S* slopes should be such that they are not damaged by pore pressure during and immediately after construction.

6. The *U/S* face of the dam should remain stable during rapid draw downs.

7. *U/S* face should be protected against wave erosion and *D/S* face against the erosion due to heavy down pours and strong winds. For this rip rap, revetment or pitching is down on *U/S* face and counter booms and turfs are developed on *D/S* face.

8. *U/S* and *D/S* slopes of the dam should be flat. This will ensure safety of the dam against shear failure of the foundation.

9. In order to prevent internal erosion of the soil the seepage flow through the dam and foundation should be suitably controlled.

10. Free passage of water, wither through holes made by burrowing animals or by piping should be allowed.

11. In areas, frequently subjected to earth quakes, the dam should be designed considering earthquake resistant measures.

12.6 DESIGN OF THE SECTION OF AN EARTH DAM

Design of the section of an earth dam envolves fixing the sizes of the following elements.

1. Top width of the darn
2. Fixation of free board
3. *U/S* and *D/S* slopes
4. Central force of the dam
5. Drainage from *D/S* side of the dam.

1. Top Width of the Dam. There are several other factors that affect the top width of the dam but height of the dam is the most important factor. Nature of soil and use of the top of the dam for passing highway, are also the important factors. Top width of the dam can be found out by the following empirical formulae.

$$B = \frac{H}{5} + 3 \rightarrow \text{for very low dams}$$

$$B = 0.55H^{0.5} + \frac{H}{5} \rightarrow \text{(Applicable for dams less than 30 m height)}$$

$$B = 1.65\,(H + 1.5)^{0.33}. \text{ Applicable for dams higher than 30 m.}$$

$$B = 1.67\sqrt{H} \text{ is also a formula most commonly used in India.}$$

Width of the top of the dam, should not be less than 3 m for any height of the dam.

2. Free Board. The vertical height lying between the top of the dam and maximum reservoir level is known as the free board. Free board may further be stated as *normal free board* and *minimum free board.* Difference between top of the dam and normal reservoir level is known as normal free board. The difference between top of the dam and the H.F.L. when all the spillways and other outlets are working as planned, is known as minimum free board. Amount of free board depends upon the wage height of the reservoir. The free board should be at least 1.5 times the wave height. The wave height (h_w) can be found out by formula given by Molitor Stevension and reproduced as follows.

$$h_w = 0.032\sqrt{F.V} + 0.763 - 0.271\sqrt[4]{F} \quad \text{when } F < 32 \text{ km}$$

and
$$h_w = 0.032\sqrt{F.V} \quad \text{when } F > 32 \text{ km}$$

where F = Fetch in kilometres

V = Velocity of wind in km/hr

h_w = Wave height in metres measured between trough and crest of the wave.

Settlement allowance of 2% should be considered for the dams of heights upto 30 m and 3% for larger heights and earthquake prone areas.

U.S.B.R. has recommended minimum 2 m and maximum 3 m as the free board, for any dam when spillways are not controlled by gates. If spillways are provided with gates and height of the dam is say up to 60 m, the value of free board should be 2.5 m above the top of gates. For larger height dams and gated spillways free board should be 3 in above the top of gates.

3. *U/S* and *D/S* slopes. The *U/S* and *D/S* slopes of the earth dam depend upon following factors.

(i) Type of soil used in the construction of the dam,

(ii) Foundation conditions and

(iii) Height of the.dam.

Flatter slope should be provided on *U/S* side so as to remain stable under the action of sudden drawdown conditions. If soils are quite permeable steep slopes can withstand the action of sudden drawdown as pore pressure gets dissipated almost immediately. Terzaghi gave following slopes for fixing a tentative section of the earth dam.

Table 12.1. *Slopes as recommended by terzaghi*

Type of material	*U/S slope*	*D/S slope*
1. Well graded homogeneous soil.	$2\frac{1}{2}:1$	$2:1$
2. Homogeneous course silt.	$3:1$	$2\frac{1}{2}:1$
3. Homogeneous silty clay, or clay.		
(a) Height less than 15 m	$2\frac{1}{2}:1$	$2:1$
(b) Height more than 15 m	$3:1$	$2\frac{1}{2}:1$
4. Sand or sand and gravel with clay core.	$3:1$	$2\frac{1}{2}:1$
5. Sand or sand and gravel with R.C.C. core wall.	$2\frac{1}{2}:1$	$2:1$

The dimensions fixed by Strange for earth dam are as follows :

Height of dam	*Minimum Free board in metres*	*Minimum Top width in metres*	*U/S slope*	*D/S slope*
Upto 4.5 m	1.5 m	1.8	$2:1$	$1\frac{1}{2}:1$
4.5 to 7.5 m	1.5 to 1.8 m	1.85	$2\frac{1}{2}:1$	$1\frac{3}{4}:1$
7.5 to 1.5 m	1.85	2.5	$3:1$	$2:1$
1.5 m to 22.5 m	2.1	3.0	$3:1$	$2:1$

4. Central Core of the Dam. The thickness of the impervious core depends upon the following factors.

(i) Rate of seepage through the core.

(ii) Type of material available for the construction of the core.

(iii) Minimum width required, that will permit proper construction.

(iv) Design of proposed filter that is to be provided on the *D/S* side of the core.

It should be understood clearly that shear strength of core material is always smaller than the rest of the dam section. It is due to the fact that core is made of clayey material. Hence from stability point of view thinner core is preferrable. But thinner core walls are more likely to suffer from piping trouble. The thickness of the core wall should not be less than the height of the dam above the point under consideration. Width of the core at the top should be minimum 3 m so that construction of core can be carried out by heavy earth machines. Minimum width of the core at base is when foundation is either impervious or pervious but having positive cut-off. The slope of the core is decided automatically.

In case foundation is pervious and there is no positive cut off trench, the core size is fixed by giving 1 : 1 slope on both of its sides and adopting 3 m as the minimum thickness of the core at the top.

Maximum size of the core may be fixed by giving side slope of $(x - \frac{1}{2}) : 1$ on both the sides where x is the side slope of the dam itself. If side slope of the dam is 3 : 1, the value of x is 3 and maximum slope of the core will be $(x - \frac{1}{2})$: i.e. $(3 - \frac{1}{2}) : 1$ or $2\frac{1}{2} : 1$.

Sloping Core. The core in the earth dam may have its axis inclined rather than vertical.

Advantages

(a) The pressure at the contact surface of foundation and the dam is maximum. This reduces the possibilities of leakage along the contact.

(b) Height of the vertical core being less than the inclined length of the inclined core, thicker vertical core is constructed with the same quantity of core soil. This reduces the possibilities of piping failure of the core.

Advantages

(a) *D/S* part of the dam can be constructed first and inclined core may be laid later.

(b) Grouting of foundation and construction of embankment can be done simultaneously.

Slope of the sloping core is made 1 : 1 on the upstream face and $\frac{1}{2}$: 1 on the *D/S* face. The slope of the down stream face is not opposite to the slope of *U/S* face, but in same direction.

5. Drainage System from Downstream Side of the Dam. Proper drainage of the earth dam is essential for its successful working. Drainage of the dam can be provided by following methods.

(i) By providing a Rock Toe, enclosed in a filter at the *D/S* end of the dam. This measure is adopted to prevent softening of *D/S* Toe. By rock toe sloughing of the *D/S* slope is prevented.

(ii) *Horizontal blanket filters.* This measure is wisely used for the drainage of earth dams of moderate heights. It has been found that horizontal filter materials stratify with time and thus flow of seeping water through the filter is prevented. Due to this difficulty, the seeping water may start flowing horizontally at some higher level above the filter level. This causes seeping water to come out, at some higher level on *D/S* side than at rock toe level. This may ultimately lead to sloughing.

(iii) *Provision of chimney drains or filters.* It is a measure to eliminate the trouble of stratification in the dam. In this case high inclined or vertical filter drains are provided in continuation of horizontal filter drains, to intercept the seeping water along any horizontal surface before embarking at the *D/S* slope of the dam. These inclined filters are known as chimney drains. Chimney drains are used in high earth dams, to cause effective drainage for full height of the dam.

For very high dams the spread of the dam on D/S side becomes quite large and system of inclined and horizontal filter drains may not perform their intended function effectively. Effectiveness of the drainage system can be increased by constructing one or more longitudinal drains and connecting these drains with horizontal system of filter drains.

12.7 FILTER DESIGN

Filter drains should be designed in such a way that all the seeping water through the dam is effectively drained off. The filter consists of more than one layer. The filter layer which comes in contact with the seeping water first is of fine material. Subsequent layers of filter are made of sand of increased coarseness. The last layer from which water comes out of the filter is made of gravel. The filters of filter drains are also known as reverse or inverted filter. The soil of earth dam and foundation surrounding the filter is known as the base material. According to earth manual of US Bureau of Reclamation Washington 1960 following four main requirements should be satisfied.

(a) Filter material should be fine and poorly graded so that the voids in the filter are small and thus prevent base material from entering the filter.

(b) The filter material should be coarse and pervious in relation to base material. This aspect facilitates rapid removal of seeping water without building up any seepage force within the filter.

(c) The filter material should be coarser than the perforations or openings in the drain pipes, so that filter material is not lost in the drains. The openings or perforation in the pipe drain should be adequate to admit all the seeping water safely.

(d) The thickness of filter material should be sufficient to provide a good distribution of all particle sizes, throughout the filter. The thickness should also be adequate to provide safety against piping.

According to Mr. Terzaghi, the filter material should fulfil the following two requirements.

1. The D_{15} size of filter material must not be more than 4 to 5 times D_{85} size of the base material. This prevents the foundation material from carrying through the pores of the filter material.

2. The D_{15} size of the filter material should be at least 4 to 5 times the D_{15} size of base material. This keeps seepage forces within the filter to permissible limit.

The two criteria can be expressed as follows also :

$$\frac{D_{15} \text{ of filter}}{D_{85} \text{ of base material}} < 4 \text{ to } 5 < \frac{D_{15} \text{ of filter}}{D_{15} \text{ of base material}}$$

The requirements cited above should be satisfied between any two adjacent layers of the filter. Me Terzaghi's criteria has been modified as per U.S. Bureau of Reclamation Design of Small Dams'. The modifications are as follows :

(i) $\dfrac{D_{15} \text{ of filter}}{D_{15} \text{ of base material}}$ = 5 to 40, provided that the filter does not contain more than 5% of material finer than 0.075 mm.

(ii) $\dfrac{D_{15} \text{ of filter}}{D_{85} \text{ of base material}}$ = 5 or less.

(iii) $\dfrac{D_{85} \text{ of filter}}{\text{Maximum opening of perforation of the pipe}}$ = 2 or more.

(iv) The grain size curve of the filter material should be about parallel to the curve of the base material.

12.8 DESIGN OF EARTH DAMS TO SUIT AVAILABLE MATERIALS

In order to achieve economy, the design of an earth dam should be adopted to make full utilization of the materials available at or near the site. This will cause minimum transportation of other materials. The available material, at or near the site may be any one of the following types:

1. Coarse sand and gravel
2. Coarse sand, gravel and clayey silt
3. Only silt clay is available.

Foundation conditions at the site may be any of the following types :

1. Impervious of large depths.
2. Pervious for moderate depth and thereafter impervious.
3. Pervious for indefinite depths.

In all the three conditions of foundations earth dams of different materials can be constructed as shown in Figs. 12.10 to 12.14.

Case 1. *Gravel, coarse sand and clayey silt available.*

1. Impervious Foundation for Large Depth. In this case no treatment of any type is required for foundation. The dam proper is made of coarse sand andgravel but central core is made from the available clayey silt. The central core may be vertical or inclined. See Fig 12.10.

2. Pervious Foundation for Moderate Depth and then Impervious. Construction of this dam is similar as in case (1) except that central core is carried down to the level of impervious layer. See Fig. 12.11.

3. Foundation Pervious for Large Depth. The main problem in this case is seepage through foundation. The construction procedure above ground level is same as explained in (1). Following measures may be taken to prevent seepage of water through the foundation.

(i) 1 to 3 m thick impervious blanket may be introduced at the upstream of the core at junction of earth dam and foundation. This blanket may be extended beyond the upstream end of the dam. See Fig. 12.12 (upper).

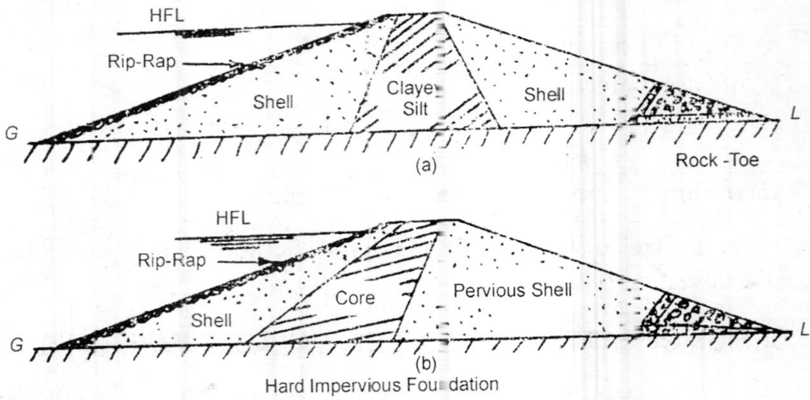

Fig. 12.10

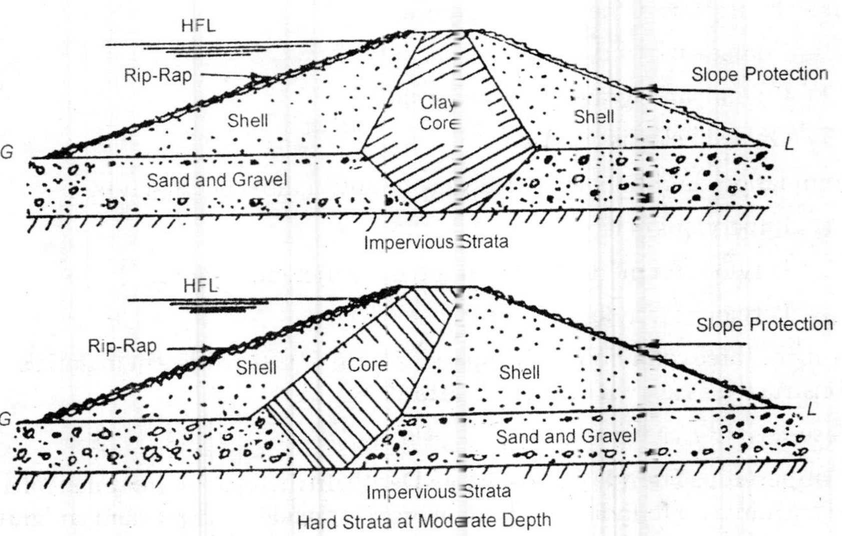

Fig. 12.11

(ii) Vertical Cut-Off in form of sheet pile is provided below the dam. See Fig. 12.12 (Lower).

Case 2. *Only coarse sand and gravel is available.*

1. Impervious Foundation for Large Depth. Dam is made of coarse sand and fine gravel but an impervious core of cement concrete or masonry is installed at the centre of the dam. See Fig. 12.13 (upper).

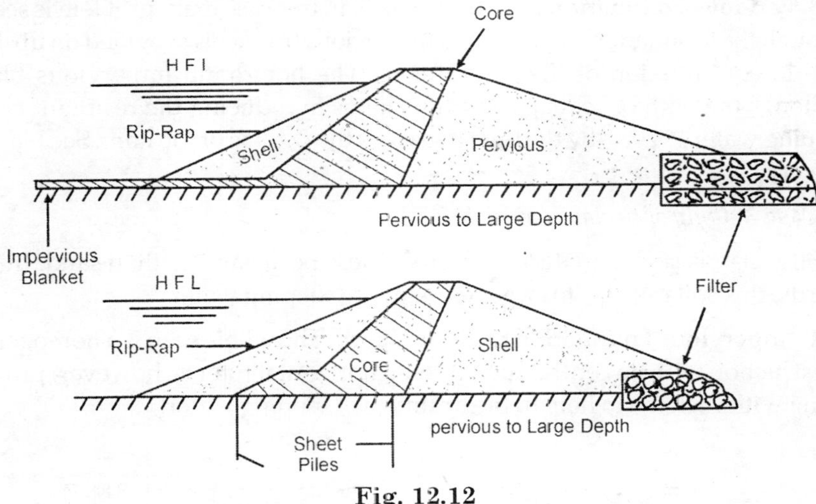

Fig. 12.12

2. Foundation Pervious for Some Depth and then Impervious. Construction is similar to (1) except that core is started from the level of the impervious foundation, Fig. 12.13 (middle).

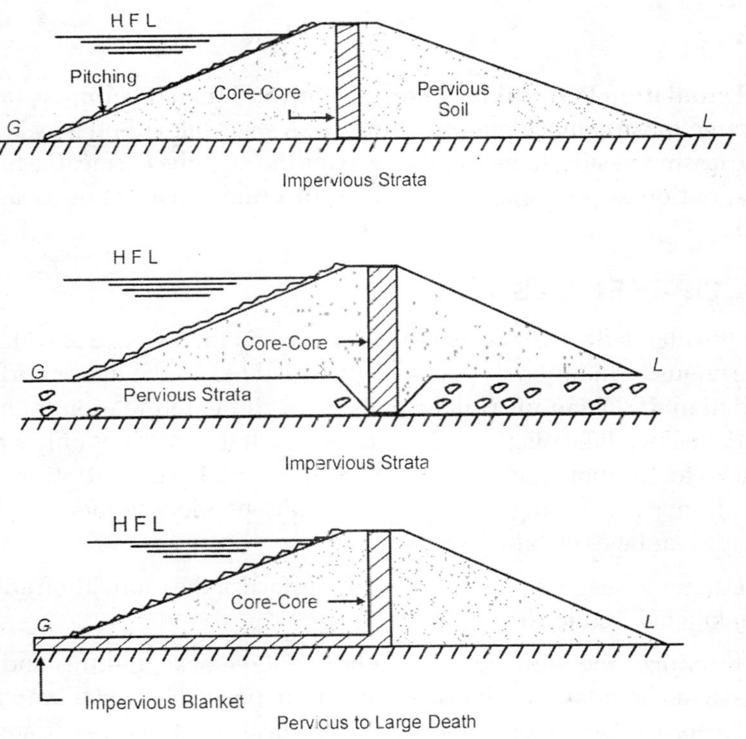

Fig. 12.13

3. Pervious Foundation for Large Depth. In this case main problem is seepage through the foundation. A horizontal impervious blanket is provided on upstream side in continuation of the central core. The horizontal impervious blanket prolongs the path of seepage and thus helps in reducing the resultant head of seeping water to zero, by the time it reaches the *D/S* toe of the dam. See Fig. 12.13 (lower).

Case 3. *Only silty clay is available.*

Silty clay is such a material which has less permeability than sand. In other words, this soil puts quite large resistance to seeping water.

1. Impervious Foundation to Large Depth. The whole dam is a homogeneous construction with no central core. A deep horizontal filter is, however, provided along with rock toe to help in the drainage of seeping water.

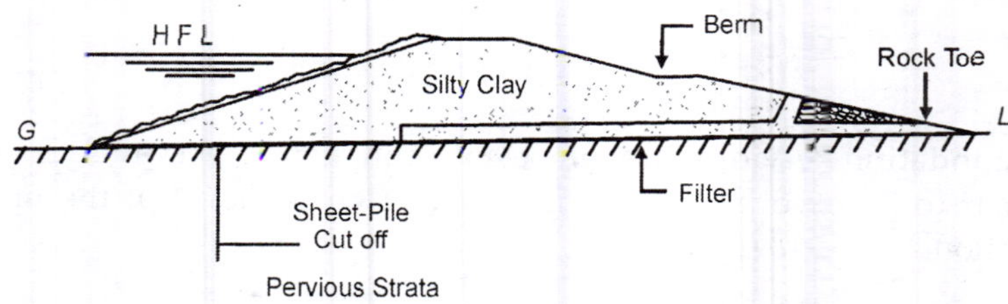

Fig. 12.14

2. Foundation Permeable to Some Depth and then it is Impervious. In this case same construction is adopted as for (1). A sheet pile cut-off may be provided on the upstream side, to reduce the seepage through the permeable foundation. If foundation is pervious for large depth same construction is adopted as for (2).

12.9 CUT-OFF WALLS

Cut-off wall is a device used to reduce or prevent the seepage of water through the pervious foundation. It is always provided below the upper surface of the foundation. If the cut-off wall extends through the foundation to impervious material below, it is called *complete cut-off wall.* If it penetrates only a part of the pervious foundation, it is known as partial cut-off. Partial cut-off walls are not very effective in reducing the seepage through a pervious foundation. Following are the advantages of cut-off walls.

1. Danger of seepage along the contact surface of the foundation and the dam embankment is reduced.

2. It reduces the seepage through horizontal cracks, fissures and pervious seams in the foundation. Almost all the soils deposited by water are stratified or have layers of materials of various sizes. A vertical cut-off will penetrate a number of these layers or strata and thus force the seepage to pass through different strata. The cut-off walls may be of any form from the following:

(a) Cut-off trenches back filled with impervious material.

(b) Masonry or concrete walls.

(c) Sheet piling – mostly made of concrete or steel.

(d) Grout curtain – A barrier is developed by grouting cement deep below the foundation.

Before choosing the type of cut-off to be used it would be appropriate if foundation conditions are fully investigated. If seepage is moderate, sheet pile treatment may be adequate. If seepage is very heavy and foundation is pervious for large depth, masonry or concrete cut-off is found more appropriate. Grout curtain is a sort of preventive measure. It is adopted mostly when after construction it is observed that seepage is excessive and it has to be controlled.

12.10 SLOPE PROTECTION OF THE DAM

For the success of the earth dam, its *U/S* and *D/S* slopes have to be properly maintained. *U/S* slopes have to be protected against wave action, and sudden draw down conditions whereas *D/S* slopes against heavy down pours, and action of burrowing animals. Following are some of the measures used for slope protection.

1. Turfing. Down stream slopes of the dams can be protected against erosion by developing turf. This measure is quite effective for slope protection.

2. Counter Berms. Down stream protection of the dam can be achieved to some extent by providing counter-beams. This measure is usually adopted for very high dams. Counter beams break the rhythm of the velocity of rain water, flowing down the berm.

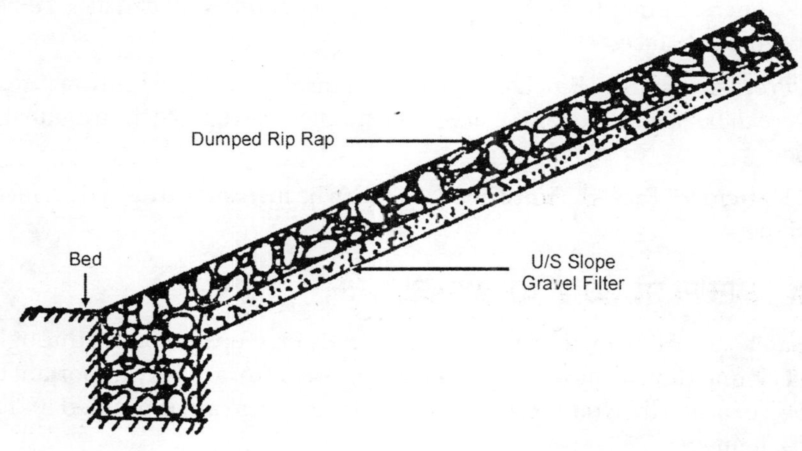

Fig. 12.15. *Details of rip-rap on U/S slope of the dam.*

3. Rip rap. This is the only measure of protecting the *U/S* slope against the action of wave erosion and sudden draw down. This measure can equally be used for protection of *D/S* slopes also. Rip rap is a layer of systematically dumped blocks on the slope. A gravel filter layer 40 to 50 cm thick is applied before blocks are laid over it.

A survey conducted by US Corps of Engineers revealed that dumped rip-rap failed in 5% cases, hand packed rip-rap in 30% cases, and concrete pavement in 36% cases. It can very easily be inferred from this that dumped rip-rap is the most successful measure of slope protection. Thickness of rip-rap excluding the thickness of gravel filter is about 1 m. The performance of rip-rap depends upon the weight and size of blocks, quality of rick of which blocks are made, thickness of rip-rap, and slope of the embankment on which rip-rap is laid. The rip-rap thickness is adopted from following criteria. Rip-rap, revetment and pitching all mean the same thing.

Height of the earth dam	*Rip-rap thickness*
Upto 5 m height	No pitching is required.
5 m to 10 m height	25 cm thick pitching over a filter layer of gravel 15 cm thick.
10 m to 15 m	30 cm pitching over 15 cm gravel filter.
15 m to 25 m	50 cm pitching over 25 cm gravel filter.
25 m to 50 m	50 cm pitching over 30 cm gravel filter.
50 m to 75 m	75 cm stone pitching over 50 cm graded gravel filter.
Above 75 m	1 m stone pitching over 75 cm graded gravel filter.

The entire *U/S* face should be pitched right upto top of the dam.

4. Cement Concrete Slab. A plain cement concrete slab can also be used to protect the *U/S* face.

5. Precast Concrete Blocks. In place of natural stone blocks, precast concrete blocks can be used to cover the *U/S* face and provide it protection against wave erosion.

6. Flatter *U/S* Face. Stability of *U/S* face can be increased by give this face very flat slope.

12.11 SEEPAGE CONTROL MEASURES

The safety of earth dam depends almost entirely on seepage control through dam and its foundation. Hence measures of seepage control are very important for the success of an earthen dam. Seepage control measures can be divided under two heads, namely :

1. Seepage control through the dam.

2. Seepage control through the foundation.

Various measures of seepage control under each categoray have been discussed in brief.

12.11.1 Seepage Control Through the Dam

1. Rock Toe. Rock toe keeps seepage line well within the dam section. It also helps a great deal for the drainage purposes. The height of the rock toe is kept about one fourth of the height of the dam. Rock toe should be designed like filter.

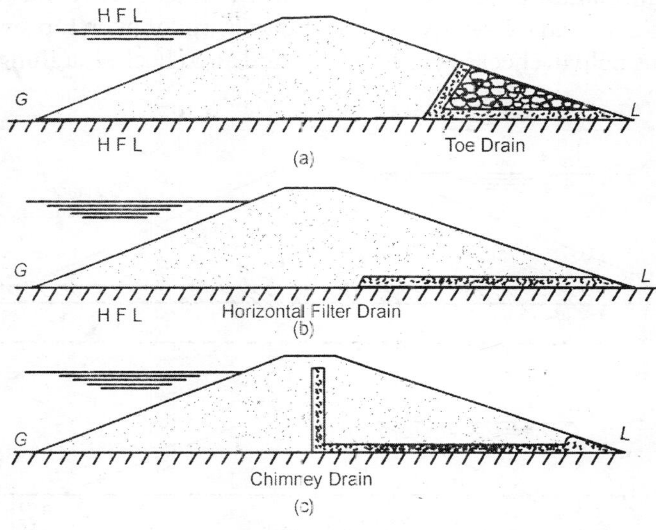

Fig. 12.16

2. Horizontal Drainage Filter. It is provided at the base of the dam, starting from down stream end of the dam and extending backwards into the dam. Backward extension of the filter depends upon so many factors. But this extension may at the most be up to centre line of the dam. This filter controls seepage line and does not allow it to get exposed on *D/S* face of the dam. It also accelerates the process of consolidation. It also causes drainage of foundation.

If seepage pressure at the *D/S* end of the dam is still excessive, the horizontal filter drain may be continued even beyond the *D/S* toe of the dam. Sometimes rock toe and horizontal filter drains are dispensed with and entire *D/S* portion of the dam may be made from coarse grained soil.

3. Chimney Drains. Under conditions of large stratification, the permeability in horizontal direction is more than in vertical direction. This causes greater speed of horizontal seepage than vertical seepage. Chimney drain or filter, if correctly built, intercepts all the seepage from the dam regardless of the stratification in the dam. Chimney drains also render earth dam earthquake resistant. See Fig. 12.16 (c).

12.11.2 Seepage Control Through the Foundation

1. Impervious Cut-off. Cut-off is a wall of relatively impervious material. It is used to prevent seepage through the foundation. It is always led down into the foundation from the ground surface. If possible, cut-off should extend upto the impervious strata lying below the ground level. Partial cut-offs do not prove much effective in preventing seepage. A 90% depth of cut-off reduces about 25% seepage.

2. *D/S* Seepage Berms. Additional berms may be built on the *D/S* side of the dam in continuation of the *D/S* end. Such a bean is useful in controlling seepage where *D/S* top strata is relatively thin and uniform or even top strata is absent. Such berms help in checking uplift of soil and also check sloughing of *D/S* slope of the dam.

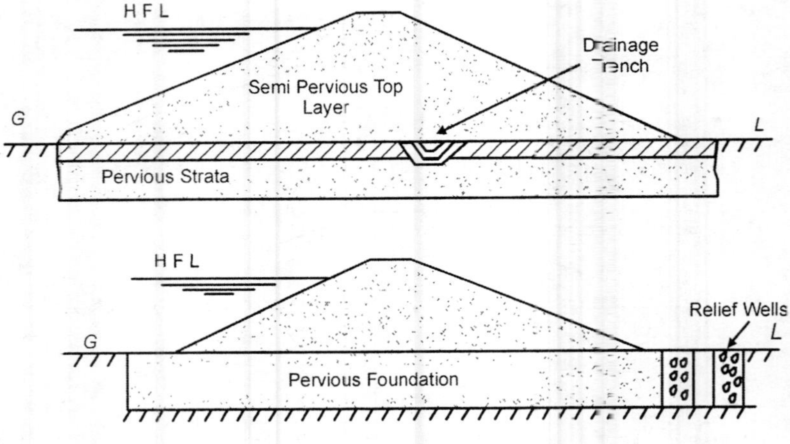

Fig. 12.17

3. Drainage Trenches. This measure is adopted when top stratum is pervious and thin. They are provided along the axis of the dam. They may be more than one depending upon the spread of the dam. They have a porous drain having longitudinal slope. Porous drains remain enclosed in gravel filters. See Fig. 12.17.

4. Relief Well. It is such a well which if not constructed would cause formation of sand boils and possibly sub-surface piping. They reduce the sub surface uplift pressure *D/S* of the dam. They intercept the seepage through the foundation and control the outlet for seepage. Relief wells become necessary when impervious layer overlies a pervious layer and the thickness of overlying impervious layer is less than the depth of water impounded. Relief wells consist of 10 cm to 15 cm diameter holes filled with filter material. See Fig. 12.17.

5. Upstream Impervious Blanket. Such a blanket when constructed over a pervious foundation reduces the quantity of seepage on *D/S* side. It also causes reduction in uplift pressure throughout the *D/S* side.

The provision of *U/S* blanket is found economical and more effective, when the depth of pervious over burden is large and provision of cut-off wall is uneconomical. Blankets are particularly effective when there are cracks and fissures in the foundation beneath the dam structure. In such cases they seal such openings and reduce the seepage considerably. The blanket should be composed of such material which is at least 100 times less pervious than the foundation material. Thickness of the blanket varies usually between $1\frac{1}{2}$ m and 3 m.

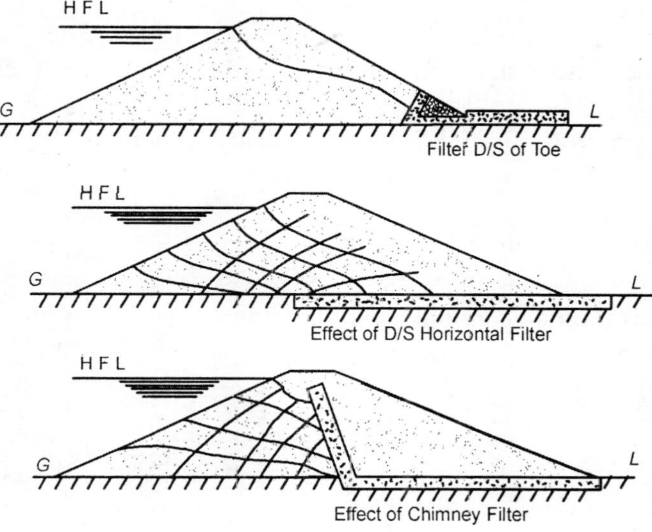

Fig. 12.18

12.12 SEEPAGE ANALYSIS

Seepage takes place through and under all dam, both earth and concrete. The problem is to minimize and control seepage so that it will have no harmful effects. Control of the total quantity of seepage and seepage pressure is an essential part of earth dam design. When water seeps through the dam it fails depending upon the resistance offered by the material to fow. The seepage line in the body of the dam is such a line below which there is positive hydrostatic pressure. On the line itself, the hydrostatic pressure is zero. Above the line, there is a zone of capillary saturation. The effect of capillary is neglected in dams.

We have seen in the discussion in this chapter that earth dam fails in most of the cases due to seepage. Hence it is utmost important of know the characteristics of seepage flow as well as distribution of water pressures. While analysing the seepage flow, the following assumptions must be kept in mind.

1. Water is incompressible.
2. Earth dam and natural Inundation soils are incompressible porous media.
3. Pore spaces do not change with time regardless of water pressure.
4. Flow of water through porous medium follows Darcy's law. In other words the seeping water flows under the hydraulic gradient due to gravity head loss only.

5. The hydraulic boundary conditions at entry and exit are known.
6. The quantity of water entering and leaving the soil element is same.

12.13 LAPLACE EQUATION FOR TWO DIMENSIONAL FLOW

The flow in earth dam is two dimensional. See Fig. 12.19 (a). It is a soil element of unit thickness and of size Δx and Δy. Let V_x and V_y be the velocity components in x and y direction at entrance. Velocity of exit in y direction will be $\left(V_y + \frac{dV_y}{dy}\Delta y\right)$ and that in x-direction $\left(V_x + \frac{dV_x}{dx}\Delta x\right)$.

Since in saturated condition quantity of water entering and leaving the soil element is same,

hence $V_x(\Delta_y \times 1) + V_y(\Delta_x \times 1)$

$$= \left(V_x + \frac{dV_x}{dx}\Delta x\right)(\Delta y \times 1) + \left(V_y + \frac{dV_y}{dy}\Delta y\right)(\Delta x \times 1).$$

Solving this, we get

$$\frac{dV_x}{dx} + \frac{dV_y}{dy} = 0. \tag{1}$$

This is the equation of continuity.

From assumption 4

$$V_x = k_x i_x = k_x \frac{dh}{dx} \tag{2}$$

and

$$V_y = k_y i_y = k_y \frac{dh}{dy} \tag{3}$$

where h = hydraulic head under which water flows

K_x and K_y = coefficient of permeability in x and y directions.

Substituting in (2) and (3)

$$\frac{d^2(K_x h)}{dx^2} + \frac{d^2(K_y h)}{dy^2} = 0 \tag{4}$$

For an isotropic soil $K_y = K_x = K$, Eq. (4) reduces to following form

$$\frac{d^2h}{dx^2} + \frac{d^2h}{dy^2} = 0 \quad (5)$$

By substituting ϕ for Kh, we get

$$\frac{d^2\phi}{dx^2} + \frac{d^2\phi}{dy^2} = C \quad (6)$$

$\phi = Kh$, is known as velocity potential. Equation (6) is called Laplacian equation for two dimensional flow.

ϕ the velocity potential may be defined as a scalar function of space and time such that its derivative with respect to any direction gives the fluid velocity in that direction.

$$\phi = Kh$$

$$\therefore \quad \frac{d\phi}{dx} = K\frac{dh}{dx} = Ki_x = V_x$$

$$\text{Similarly} \quad \frac{d\phi}{dy} = K\frac{dh}{dy} = Ki_y = V_y$$

Equation (6) can be obtained by analytical, graphical and experimental methods also.

The solution gives two sets of curves. Curves in one direction are known as equipotential lines and those in other direction as stream lines. These curves are mutually orthogonal. System of these lines is shown in Fig. 12.19 (b). Equi-potential lines represent contours of equal head. The direction of seepage is always perpendicular to the equi-potential lines. The path along which the individual particles of water seep through the soil are called stream lines or flow lines.

12.14 COMPUTATION OF RATE OF SEEPAGE FROM FLOW NET

A net work of flow lines and equipotential lines is known as a *flow net*. Figure 12.19 (b) shows a portion of such a flow net. The space between any two successive flow lines is known as flow channel. The area enclosed between two successive flow lines and successive equipotential lines is known as 'field'.

Let B and L be the width and length of a field:

Δh = head drop through the field,

Δq = Discharge passing through the flow channel

h = Total head causing flow.

According to Darch's law of flow through soils

$$\Delta q = K \frac{\Delta h}{L} B \times 1 \text{ (considering unit thickness)}$$

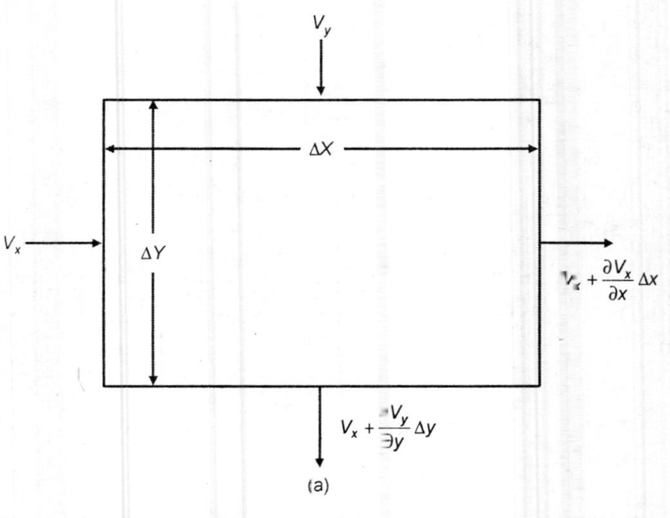

(a)

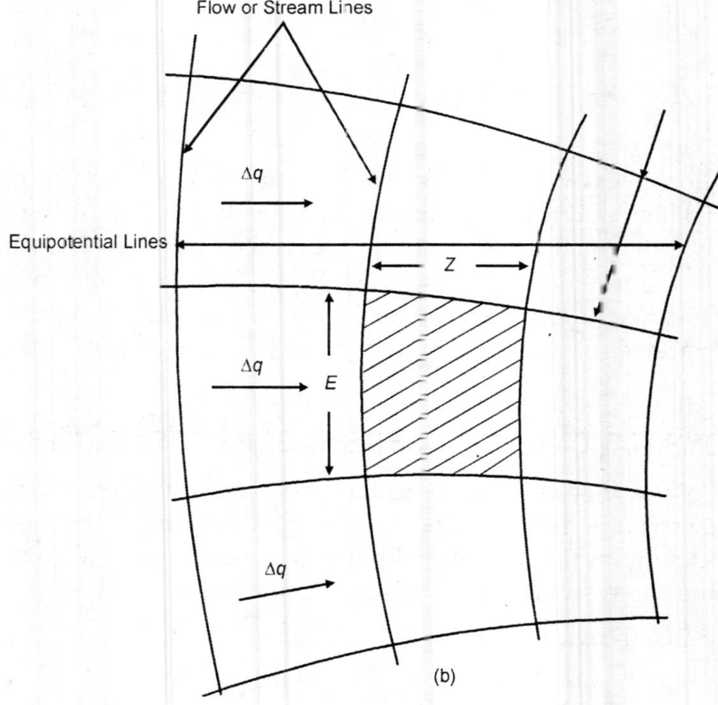

(b)

Fig. 12.19

If Nd = total number of potential dress in the complete flow net.

$$\Delta h = \frac{h}{Nd}$$

$$\Delta q = K\frac{h}{Nd}\left(\frac{B}{L}\right)$$

Total discharge through the complete flow net is given by

$$q = \Sigma \Delta q = K\frac{h}{Nd}\left(\frac{B}{L}\right)N_f$$

$$= Kh\frac{N_f}{Nd}\frac{B}{L}$$

N_f = number of flow channels in the net.

When field is square then $B = L$

$$q = Kh\left(\frac{N_f}{Nd}\right) \quad (7)$$

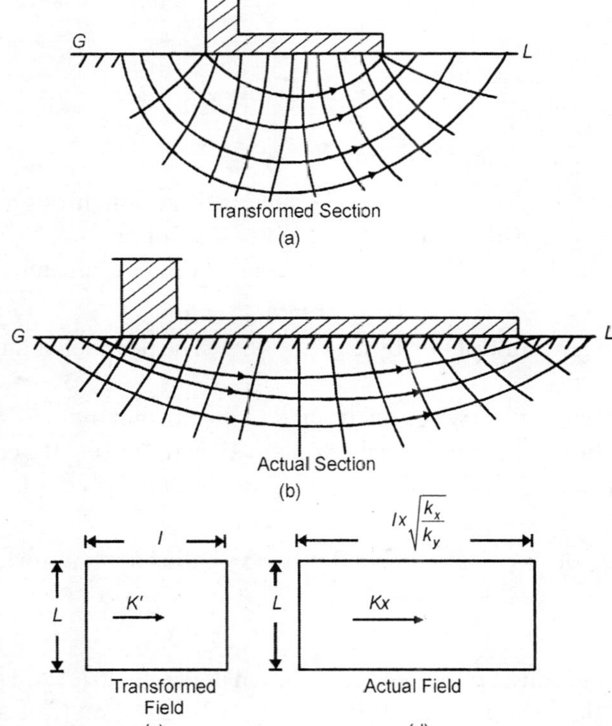

Fig. 12.20

Equation (7) is the required expression for discharge passing through a flow net. This equation is valid for isotropic soils in which $K_x = K_y = K$.

12.15 SEEPAGE DISCHARGE FOR ANISOTROPIC SOIL

In such soils K is not equal to K_y. For such soils Eq. (4) becomes

$$K_x \frac{d^2h}{dx^2} + K_y \frac{d^2h}{dy^2} = 0.$$

This equation is not a Laplacian equation and as such flow net cannot be directly drawn from it. Rewrite this equation as follows

$$\frac{Kx}{Ky}\frac{d^2h}{dx^2} + \frac{d^2h}{dy^2} = 0 \qquad (8)$$

Let us put $\quad x_n = x\sqrt{\frac{K_y}{K_x}}$

where x_n is the new co-ordinate variable in the x-direction.

The Eq. (8) becomes as follows

$$\frac{d^2h}{dx_n^{\ 2}} + \frac{d^2h}{dy^2} = 0 \qquad (9)$$

This is again a Laplacian equation.

In order to plot the flow net in such a case the cross-section through anisotropic soils is plotted in y-axis direction on natural scale. Bill, in x-axis direction the cross-section is plotted on a transformed scale. All the dimensions parallel to x-axis are reduced by multiplying by a factor $\sqrt{\frac{K_y}{K_x}}$. The flow net obtained for this transformed section will now be constructed in the normal manner as if the soil were isotropic. The actual flow net is thus obtained by re-writing the cross-section including the flow net, back to the natural scale by multiplying the x-axis co-ordinates by a factor $\sqrt{\frac{K_x}{K_y}}$. The actual flow net will thus not have orthogonal set of curves.

Actual field and transformed fields are shown in Fig. 12.20 (c) and (d). Field of transformed section is square, while field of the actual section (re-transformed) will be rectangular having its length in x-axis direction equal to $\sqrt{\frac{K_x}{K_y}}$ times the idth in y-axis direction.

Let K_x = permeability coefficient in x-direction of the actual anisotropic soil field.

K' = equivalent permeability of the transformed field.

Then for actual field

$$\Delta q = K_x \frac{\Delta h}{l\sqrt{\frac{K_x}{K_y}}}(L \times 1) \tag{10}$$

For transformed field

$$\Delta q = K' \frac{\Delta h}{L}(L \times 1) \tag{11}$$

Since quantity of flow net is the same

$$K' \frac{\Delta h}{L} \times L = K_x \frac{\Delta h}{L\sqrt{\frac{K_x}{K_y}}} \times L$$

Hence $K' = \sqrt{K_x K_y}$

Hence the distance is given by Eq. (12) as follows

$$q = K' h \frac{N_f}{N_d} = \sqrt{K_x K_y} \times h \times \frac{N_f}{N_d} \tag{12}$$

12.16 SEEPAGE OF PHREATIC LINE IN AN EARTH DAM HAVING A HORIZONTAL FILTER

Before a flow net is drawn, it is essential to find the location and shape of the phreatic line separating saturated and unsaturated zones. This can be done by graphical, analytic or experimental methods.

(*i*) *Graphical method.* This method was suggested by Casagrande. He assumed Phreatic line as the base parabola with its local point at the starting point (A) of the filter.

See Fig. 12.21 *AB* is the horizontal filter with *A* as the focal point of base parabola. *EF* is the *U/S* face of the dam. Let horizontal projection of *EF* is say *L*. From *F* measure a length *FC* equal to 0.3 *L* at the level of water as shown in figure. Point C thus obtained is considered as the starting point of the base parabola.

With C as the centre and CA as radius draw an arc cutting the horizontal line through CF in point D. Draw a vertical line DH tangential to the curve AD at point

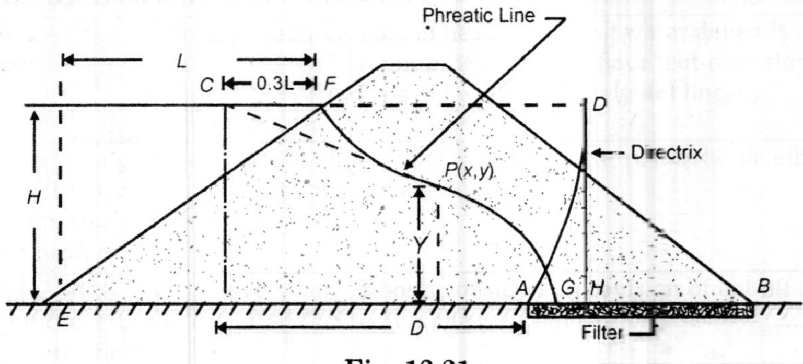

Fig. 12.21

D. Evidently CD will be equal to CA and thus vertical line DH would be directrix of the parabola. Bisect AH by point G. G is the last point of the parabola.

In order to locate intermediate points on the parabola, the fact that distance of any point on parabola are equal both from focus A and directrix point H, is used.

For example, to locate nay point P on parabola draw vertical line at any distance say x from A. Measure the distance QH. Taking A as centre and radius QH draw an arc cutting the vertical line through Q in point P. Several point like P can be located and base parabola drawn by joining all these points.

However some correction has to be made at the point of entry. The phreatic line must start from point F and not from C. Also phreatic line must start in direction perpendicular to the U/S face, EF which is also a 100% equipotential line. Hence part of parabola from F is drawn free hand, keeping in view that it must start from F at right angle to face EF and meet the rest parabola tangentially. Hence starting point of parabola is F and end point is at G. The parabola meets horizontal filter vertically at point G.

In order to develop an expression for base parabola consider any point P having xy as its co-ordinates, with respect to the focus point A as the origin.

For parabola $AP = QH$ (distance of point from H)

But $AP = \sqrt{x^2 + y^2} = QH$

If focal distance AH is taken as f then

$$QH = QA + AH = x + f.$$

$$\therefore \quad \sqrt{x^2 + y^2} = x + f \qquad \text{(i)}$$

Squaring $$x^2 + y^2 = x^2 + 2xf + f^2 \qquad \text{(ii)}$$

$$x = \frac{y^2 - f^2}{2f} \quad (13)$$

or

$$y^2 = 2xf + f^2 \quad (14)$$

This is the equation of base parabola.

For discharge (q) seeping through the body of the dam when horizontal filter is installed we consider the section PQ.

$$q = KiA = K\frac{dy}{dx}(y \times 1) \quad \text{(iii)}$$

But from expression (14) $y = (2xf + f^2)^{1/2}$

$$\frac{dy}{dx} = \frac{f}{(2xf + f^2)^{1/2}} \quad \text{(iv)}$$

Substituting in Eq. (iii), we get

$$q = K\frac{f}{(2xf + f^2)^{1/2}} \times (2xf + f^2)^{1/2}$$

$$= Kf \quad (15)$$

Equation (15) is very simple expression for discharge (q) in terms of focal distance f.

The value of f can be determined either graphically or analytically by considering co-ordinates of point C as follows

$$f = \sqrt{x^2 + y^2} - x.$$

At point C $\quad x = D$ and $y = H$ (the depth of water).

$$\therefore \quad f = \left(\sqrt{D^2 + H^2} - D\right)$$

$$\therefore \quad q = Kf = K\left(\sqrt{D^2 + H^2} - D\right) \quad (16)$$

12.17 PHREATIC LINE FOR A DAM WITH NO HORIZONTAL FILTER

12.17.1 General Solution by Casagrande

In this case there is no horizontal filter and as such D/S end point (A) of the embankment is considered as the focal point for base parabola BJG. The base parabola would evidently cut the D/S slope at point J and extend beyond the limits of the dam as shown in Fig. 12.22 (dotted lines). However, as per exit conditions shown in Fig. 12.24, the phreatic line must come out of the dam at some point M meeting the D/S face tangentially. The D/S MA part of the slope is known as discharge face and always remains wet. The correction Δa by which the parabola should be shifted downwards is found by the values of $\frac{\Delta a}{a + \Delta a}$

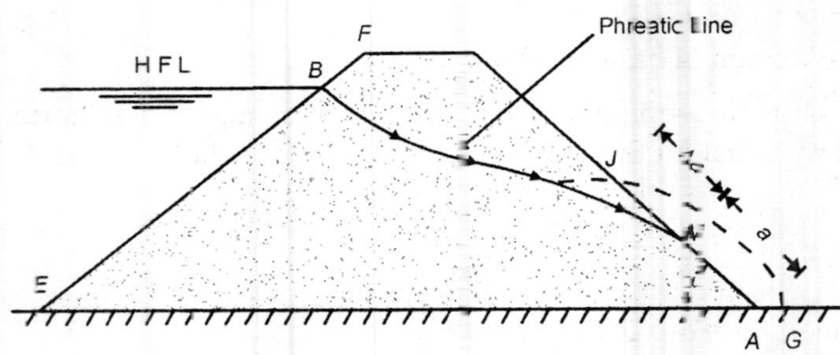

Fig. 12.22

given by Casagrande for various values of slope x of the discharge face or D/S face of the dam. The slope x can be even more than the value of 90° and it happens where dam, is provided with a rock fill toe as shown in Fig 12.24 (c).

Here $a + \Delta a = AM + MJ = AJ$ (I)

Value AJ is obtained from Fig. 12.22.

Value of $\frac{\Delta a}{a + \Delta a}$ (b) is obtained from Fig. 12.23.

Solving (I) and (II) the values of a and Δa can be easily found out.

12.7.2 For Slope Angles of D/S Slope Less Than 30°

Mr. Schaffernak and Van Iterson gave the following equation to get value a analytically

$$\alpha = \frac{d}{\cos \alpha} - \sqrt{\frac{d^2}{\cos \alpha} - \frac{H^2}{\sin^2 \alpha}} \quad (16)$$

where α is angle in degrees that D/S slope makes with the horizontal.

Casagrande gave following equation for finding value a for D/S slopes lying between 30° and 60°

$$\alpha = S - \sqrt{S^2 - \frac{H^2}{\sin^2 \alpha}} \quad (17)$$

If S is taken approximately equal to $\sqrt{H^2 - d^2}$, we get this equation in following form

$$a = \sqrt{H^2 + d^2} - \sqrt{d^2 - H^2 \cot \alpha^2} \qquad (18)$$

where a = as shown in Fig. 12.24 (a)

α = angle in degrees made by *D/S* slope of the dam with horizontal

H = depth of water at entry point

d = Total horizontal length from *D/S* end point of the dam to starting point C of parabola.

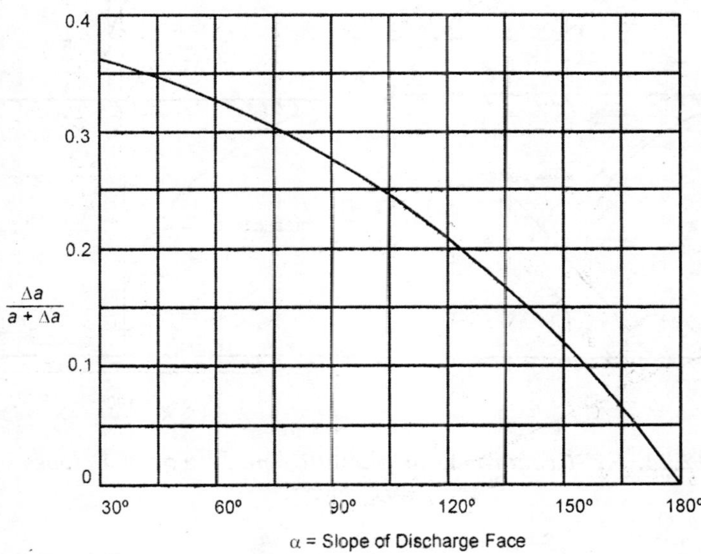

Fig. 12.23. *Relation between* α *and* $\left(\dfrac{\Delta a}{a + \Delta a}\right)$.

12.18 CHARACTERISTICS OF PHREATIC LINE OR SEEPAGE LINE

Based upon the discussions, the following inferences about phreatic line can be easily drawn.

1. Phreatic line must be normal to the slope at the point of entry. This is essential because *U/S* face represents 100% equipotential line. For vertical as well as inclination to opposite side, the phreatic line must start tangentially to the water surface.

2. Pressure along phreatic line is atmospheric. Change in head along this line is due to drop in elevation at various points. Due to this, the successive equipotential lines will meet it at equal intervals. See Fig. 12.27.

3. The phreatic lines would tend to emerge vertically if Rock Toe or horizontal filter is provided.

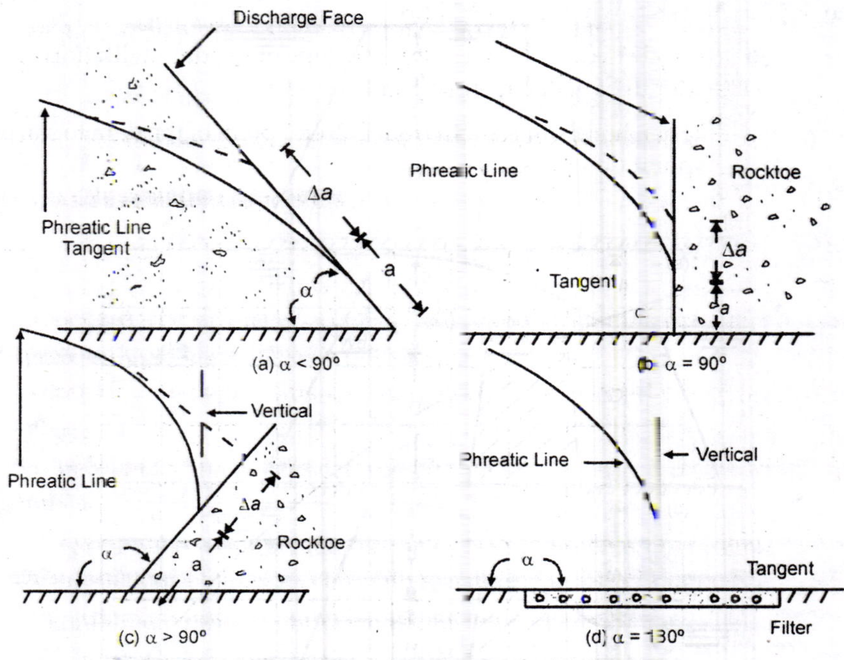

Fig. 12.24. *Exist conditions or phreatic line for various slopes of the D/S face.*

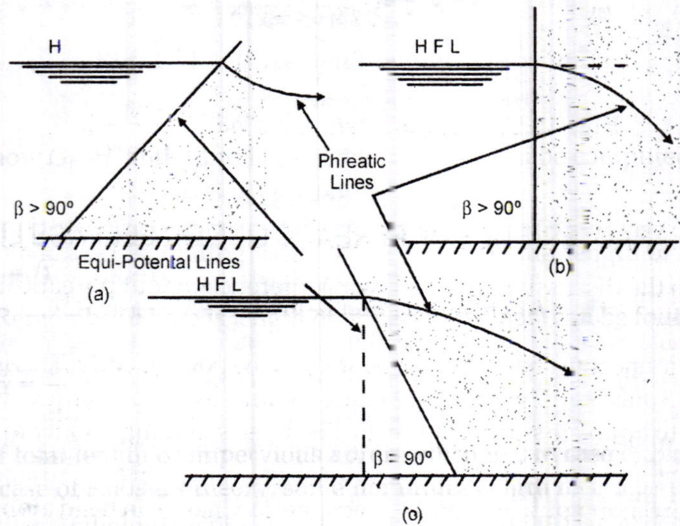

Fig. 12.25. *Entry conditions of phreatic lines.*

4. Focal point of base parabola is taken slightly with the length of the horizontal filter.

5. The presence of Pervious Foundation below the dam does not affect the location of phreatic line.

6. If filter or Rock Toe is not provided the phreatic line would cut the *D/S* slope at some point above the base. The position of this point and the phreatic line itself are independent from all other parameters like permeability or any other property so long as dam is homogeneous. They are decided by geometry of the dam only.

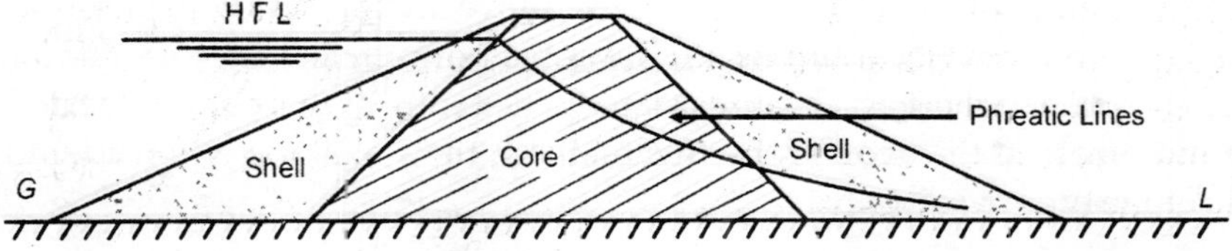

Fig. 12.26. *Effect of central core on phreatic line.*

7. In the case of zoned dams having impervious core at the centre the outer pervious shell may be neglected altogether and focus of the base parabola in this case is located at the *D/S* toe of the central core Fig. 12.26.

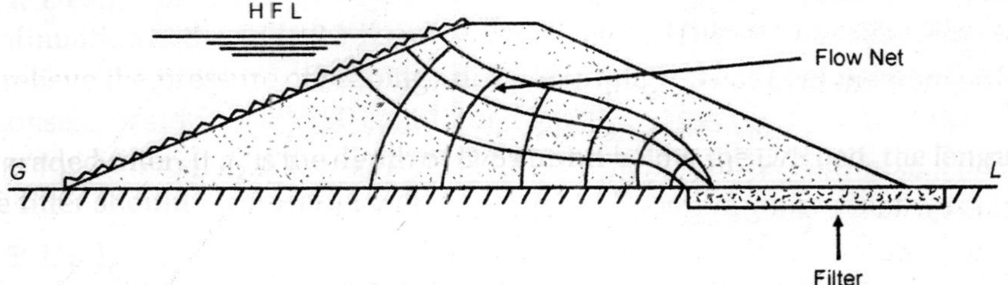

Fig. 12.27. *Flow net by graphical method.*

12.19 GRAPHICAL DETERMINATION OF FLOW NET

After having drawn the position of phreatic line, the flow net for an earth dam can be drawn keeping in view the following points suggested by Casagrande.

1. Equipotential lines and flow lines intersect each other at right angles.

2. The field enclosed between successive equipotential lines and successive flow lines is square so that circle drawn would touch all the four sides of the square.

3. Some potential drop takes place between two successive equipotential lines.

4. The quantity of water flowing through each flow channel is the same.

5. Smaller the fields, greater will be the hydraulic gradient and velocity of flow through it.

6. Transition through a homogeneous soil is smooth and curved. The curve may either be parabolic or elliptical in shape.

12.20 STABILITY ANALYSIS OF EARTH DAMS

There are numerous methods of analysing the slope stability of the earth embankments. The most commonly used method is Swedish slip circle method devised in 1922.

In this method, the curved slip surface is taken to be an arc of a circle. There will be a number of such likely slip circles but we are interested to pick out the most dangerous or critical slip circle. The critical slip circle is such a circle along which the soil had the least shear resistance. The centre of this circle is located by trial and error. In the analysis by this method, the soil slope is considered as made ofone type of soil only.

(*i*) *Cohesive soils.* Let *ABC* be any one slip circle with *O* as its centre. Let centroid of area *ABCD* of the soil lies at *G* and let x be the horizontal distance of point *G* from point *O*. The position of *G* can be obtained by any method used for irregular planes.

Let *W* be the weight of the soil mass *ABCD* and length one metre. For cohesive soils $\phi = 0$ and hence its shearing resistance is entirely dependent on cohesion only. The resistance is same along the entire surface *ABC*.

The moment (M_c) causing slip along surface *ABC* is

$$M_c = W \times x \text{ kg.m}$$

The maximum resisting moment (M_r) is the moment developed by cohesive shear resistance along surface *ABC* multiplied by radius *R*.

$$M_r = L \times 1 \times C \times R \text{ kg.m}$$

$$= LCR \text{ kg.m}$$

where L = Length of arc *ABC*

C = unit cohesion in kg/m²

R = Radius of slip circle in metres

$$L = \frac{2\pi R\delta}{360}$$

where δ = the angle of arc *ABC*, in radians and centre point *O*

$$\therefore \quad M_r = \frac{2\pi R\delta}{360} \times CR$$

$$= \frac{2\pi}{360} \times C\delta R^2$$

$$\text{Factor of safety } (FS) \text{ against sliding} = \frac{M_r}{M_c}$$

$$= \frac{2\pi C\delta R^2}{360 \times W \times x}$$

Various other slip circle can similarly be drawn and factor of safety is found for all of them. The slip circle giving the minimum factor of safety is considered as the critical slip circle. The critical slip circle is located as follows.

The values of factor of safety as worked out for different slip circles, say four, are plotted as shown in Fig. 12.28 and a curve is drawn passing through the toes of the ordinates respresenting the four values of the factors of safety corresponding to four slip circles. The lowest point on this curve is noted and a vertical line is drawn through it, meeting the top of soil mass at a point say Z. A and Z mark the two points through which the critical slip circle will pass.

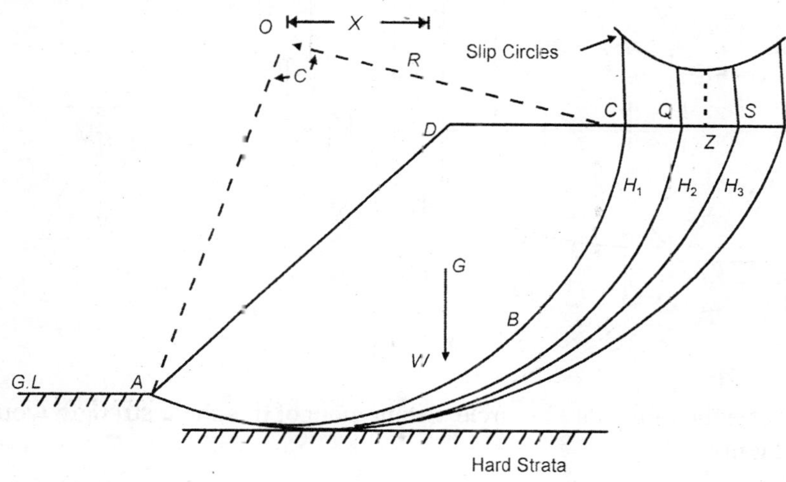

Swedish Method for Cohesive Soil

Fig. 12.28

(*ii*) *C-*ϕ *soil.* Such soils do not have constant shear strength. The shear strength of such soils which is due to cohesion and internal friction, changes from point to point along the periphery of slip circle. The Coulomb's equation $S = C + p \tan \phi$ is used to find out the shear strength of such soils, where $C = a$ constant along the slip circle but p_n the pressure normal to the slip circle due to weight of the soil which varies from point to point along the slip circle. The method is as follows.

The whole soil mass enclosed in $ABCD$ is divided into vertical strips or slices as shown in Fig. 12.29. All the slices are of equal width and thickness but different in height. Let us study say 6th strip shown in Fig. 12.29. Take one running metre of the strip, its stability is practically due to two forces only, viz. its weight W_6 and its sheering strength S_6 along the curved surface of length l_6 metres at the bottom of the strip. The weight W_6 has two components one normal and other tangential to the curved bottom of the strip. Let normal component be denoted by W_{n_6} and tangential compcnent by W_{t_6}.

The Coulomb's equation for the strip will take the form of

$$S_6 = (c.l_6 + W_{n_6} \tan \phi) \text{ kg.}$$

where $c.l_6$ = cohesive strength of strip per metre length

$W_{n_6} \tan \phi$ = frictional strength per metre length of the strip

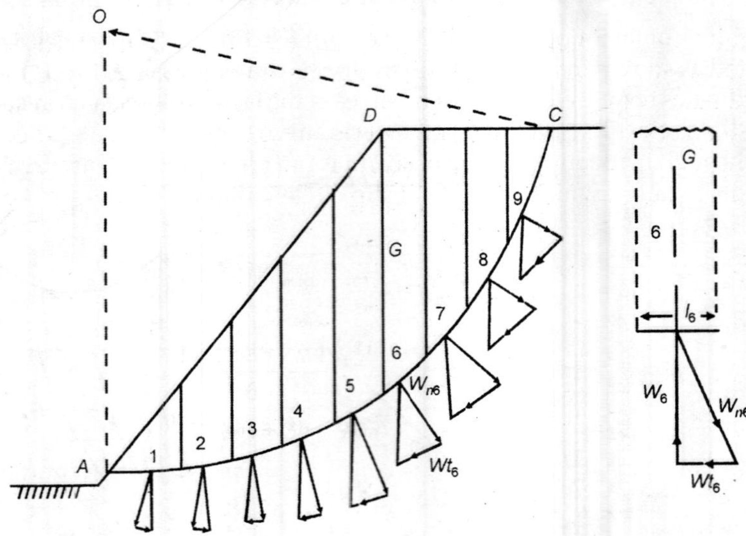

Fig. 12.29

Let R be the radius of this circle, the moment of this shear strength would be $S_6 \times R$ kg.m

Moment of shear strength (M_r) due to entire length of slip circle ABC

$$M_r = R \times \Sigma(cl_6 + W_{n_6} \tan \phi)$$

$$= R(c.L + \tan \phi \, \Sigma W_{n_6}) \text{ kg.m}$$

L = length of slip circle i.e. ABC.

ΣW_{n_6} = Sum of normal components of all the strips enclosed in a slip circle.

The moment causing slip by a single strip

$$= R \times W_{t_6} \text{ kg.m}$$

Total moment (M_c) caused by all the strip

$$= R\Sigma W_{t_6} \text{ kg.m}$$

Factor of safety against shear failure along slip circle ABC

$$= \frac{M_r}{M_c}$$

$$= \frac{R(cL + \tan\phi\Sigma W_{n_6})}{R\Sigma W_{t_6}}$$

$$= \frac{cL + \tan\phi\Sigma W_{n_6}}{\Sigma W_{t_6}}$$

General form of equation will be

$$F.S. = \frac{cL + \tan\phi\Sigma W_n}{\Sigma W_t}.$$

The weight of each strip one metre long is proportional to its area which can be found approximately by trapezoidal formula and accurately by a planimeter. Normal and tangential components for each strip can be found graphically by drawing the triangle for forces for each strip.

The factor of safety for several slip circles can be found out and the circle giving minimum *F.S.* is taken as the critical slip circle. It should be noted that tangential components of weights of some strips near the toe of the slope would be actually resisting the slipping tendency and as such they should be adopted with opposite sign.

12.21 STABILITY OF *U/S* SLOPE

The *U/S* slope of the dam is subjected to most adverse condition during sudden draw down. During steady seepage, the seepage pressure acts inwards the dam from the slope and thus tends to increase the stability of the *U/S* slope. Thus steady seepage conditions do not represent the critical state. When water is suddenly drawn away from the *U/S* slope by emptying the reservoir, the hydrostatic force acting along the *U/S* slopes is completely removed. Since the drainage from the soil is not as rapid as the draw down, the weight of drainable water held in the soil tends to help a sliding failure, as hydrostatic pressure is absent to counter act this tendency.

The stability of the *U/S* face can be analysed by developing a flow net. The factor of safety is calculated by following formula

$$F.S. = \frac{cL + \tan\phi\Sigma(W_n - U)}{\Sigma W_t}$$

where c, L, ϕ, W_n and W_t are the same as given in $c\phi$ soil. ΣU is the total uplift pore pressure on the slip surface.

12.22 POINTS TO BE CONSIDERED IN EARTHQUAKE REGIONS

The dams are likely to crack and settle due to earthquakes. Hence in earthquake prone regions following additional measures should be adopted.

1. Graded filter must be provided just to the *D/S* of the impervious core. This filter is not horizontal but along the slope of core.

2. A graded filter having fine sand should be laid along the *U/S* slope of impervious core. This filter will help seal cracks in the core developed due to earthquake shocks.

3. Both sand filters provided at *U/S* and *D/S* slopes of the impervious core should be covered with pervious coarse sand or crushed gravel thick layers. These layers offer stability to the dam.

4. Flatter slopes near the top of the dam should be adopted.

5. The core walls of manonry or concrete should be avoided. The core should be made of more impervious material and should be relatively thick.

6. The dam should be founded on more solid foundation

7. Sufficiently extra free board should be provided.

8. While analysing the stability of the slopes appropriate horizontal acceleration force should be considered along with other usual forces.

12.23 ROCK FILL DAM

This dam is made predominantly from rock boulders. Only boulders cannot prevent the flow of water through them and as such some impervious layer is

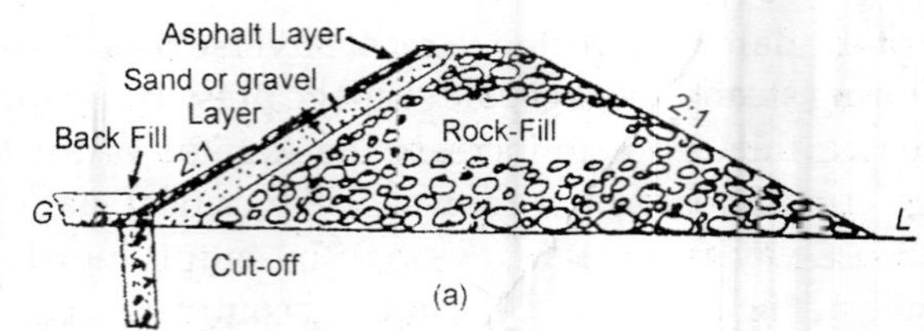

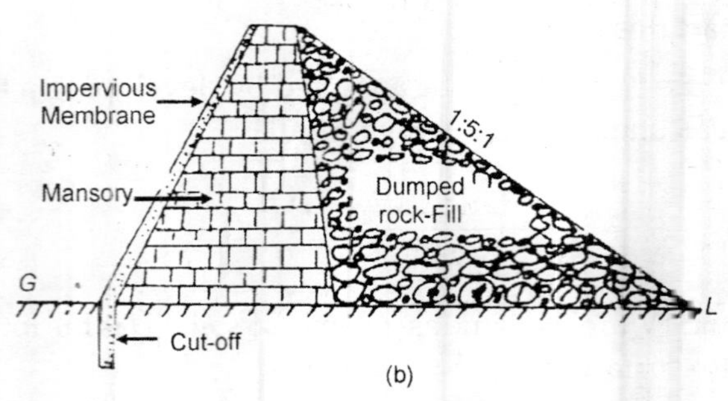

Fig. 12.30. *Rock fill dams.*

laid on the *U/S* face of the dam. The impervious layer may be cement concrete slab, or masonry work. A deep cut-off of concrete should also be provided in continuation with the inclined impervious cement concrete slab. Sometimes Asphalt mixed with sand membrane is provided on inclined *U/S* face in place of cement concrete slab. But Asphalt membrane should be laid on a thick gravel membrane. The gravel membrane acts as a cushioning layer to cause equal stabilization of the *U/S* face. Two conventional type rock fill dams are shown in Fig. 12.30.

Example 12.1 *An earth dam of homogeneous section having horizontal filter is shown in Fig. 12.31 the coefficient of permeability of the soil used in the dolt is 5 × 10^{-3} cm/sec, establish the seepage line through the dam.*

Solution. Considering starting point filter as the origin, the equation of parabola is given by

$$f = \sqrt{x^2 + y^2} - x$$

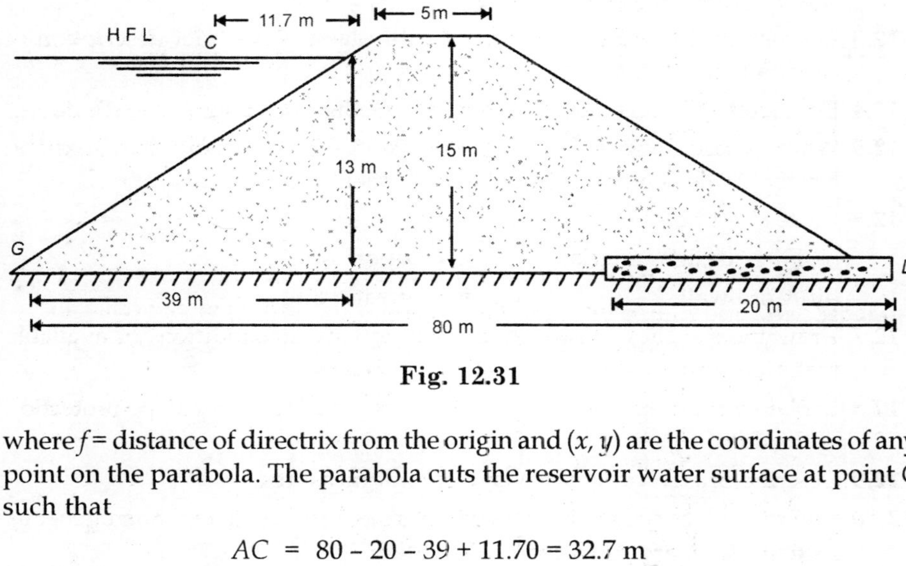

HFL, C, G, L, 11.7 m, 5 m, 13 m, 15 m, 39 m, 20 m, 80 m

Fig. 12.31

where f = distance of directrix from the origin and (x, y) are the coordinates of any point on the parabola. The parabola cuts the reservoir water surface at point C such that

$$AC = 80 - 20 - 39 + 11.70 = 32.7 \text{ m}$$

$$\text{Depth at entry point} = 13 \text{ m}$$

$$f = \sqrt{(32.7)^3 + (13)^2} - 32.7 = 2.49 \text{ m}$$

$$y = \sqrt{2fx + f^2}$$

$$= \sqrt{4.98x + 6.20}$$

x	0	10	20	30	32.7
y	2.49	7.48	10.286	12.47	13 m

Seepage flow per unit length of the dam

$$q = Kf$$
$$= 5 \times 10^{-3} \times 2.49 \times 100 \text{ cm}^3/\text{sec/cm}$$
$$= 1.245 \text{ cm}^3/\text{sec/cm}$$
$$= 1.245 \times 10^{-4} \text{ cumecs/m length of the dam}$$

QUESTIONS

12.1 State the reasons that earth dams have been and still continue to be in very common use.

12.2 In how many ways earth dams can be classified ? Explain each classification in detail.

12.3 Enumerate the causes of failure of earth dams. Give brief description of each failure.

12.4 Enumerate the essential requirements for the safe design of earth dams.

12.5 What elements have to be designed in the design of an earth dam ? Explain briefly the design procedure of each element.

12.6 Write short notes on :

(i) Filter design criteria. (ii) Cut-off wall.

(iii) Core wall. (iv) Phreatic line.

12.7 Draw the sketches of earth dams suiting various conditions of available materials and existing foundation conditions.

12.8 Explain various measures briefly that may be adopted for slope protection of the dam.

12.9 Describe briefly various seepage control measures.

12.10 Derive Laplace equation for seepage through the earth dam homogeneous section. What are its methods of solution ?

12.11 Explain the method of plotting phreatic line for an earth embankment having horizontal filter at its *D/S* side.

12.12 Explain the method of plotting phreatic line for an earth dam having no filter or Rock Toe on *D/S* side.

12.13 Explain the method of stability analysis of *D/S* slope during steady seepage using Swedish slip circle method.

12.14 Explain the terms:

(i) Piping, (ii) Sloughing.

(iii) Flow net. (iv) Pore pressure.

(v) Sudden draw down.

❑❑❑

13

Dams III (Gravity Dams)

13.1 INTRODUCTION

Gravity dam is such a dam, which resists all the forces acting on it, with the help of its own weight. This dam is considered as one of the most permanent dams. It requires very little maintenance. This is the most common type of dam, both for low heights as well as very large heights.

Gravity dam may be made of either masonry or cement concrete. However with the advent of most improved methods of construction, quality control, and curing, cement concrete is mostly used. Gravity dams mostly have their alignment straight, but they can have their alignment a bit curved also. In both the cases, the dam resists all the forces acting on it, by the aid of its own weight. Arch action developed, in slightly curved dams, is not considered in its design. The gravity dams are mostly solid. However they can be constructed hollow also. Favourable conditions for the construction of gravity dams are the following :

(i) The foundation at the site should be solid.

(ii) The valley, across which dam is to be constructed, should be deep, narrow gorge, with very steep side slopes. Earth dams might slip on such steep slopes.

(iii) The materials required for the construction should be locally available.

(iv) Skilled labour in cement concrete should be available.

(v) The height of the banks should be large so that dam of required height and capacity could be constructed.

(vi) The valley should be narrow so that length of the dam is the shortest.

(vii) Site for installation of power house should be available.

(viii) The basin of the reservoir that will be formed *U/S* due to dam, should be free from faults like fissured rocks etc.

13.2 FORCES ACTING ON A GRAVITY DAM

A detailed sketch of a gravity dam is shown in Fig. 13.1. All the predominant forces, that act on the dam, have been shown in the figure itself. The forces that act on the dam are the following :

1. Weight of the dam
2. Horizontal hydrostatic pressure due to water
3. Uplift pressure due to water percolated under the dam
4. Earthquake pressure
5. Wind pressure
6. Ice pressure
7. Wave pressure
8. Pressure due to silt deposited on *U/S* face.

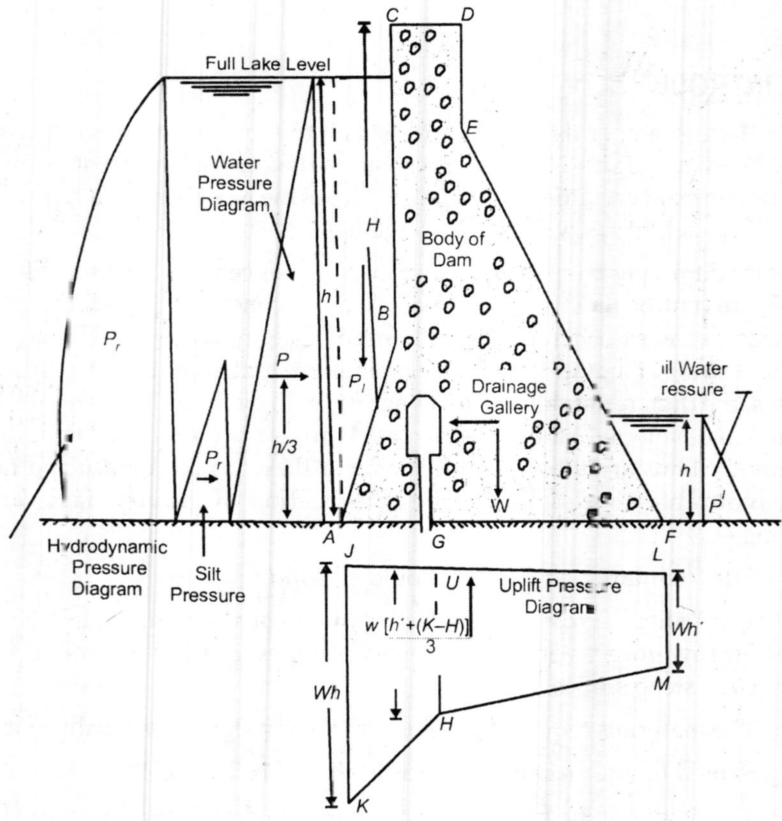

Fig. 13.1. *Water pressure, hydrodynamic pressure, silt pressure and tail water pressure distribution.*

Out of above eight forces, acting on the dam, first three forces are the major forces that are considered in the design. All other forces are not of much significance and are considered only under specific conditions.

1. Weight of the Dam. It is the most important force, particularly for gravity dams. Stability of the dam largely depends upon this force. For the design purpose only unit length of the dam is considered. Cubic content of the cement concrete is determined for unit length of the dam. This cubic content when multiplied by the density, gives the total weight (W) of the dam. The total weight (W) is considered acting at the C.G. of the dam section. The position of C.G. of the dam section can be found out by dividing the dam section into several triangles, rectangles, and trapeziums and by taking moments of these weights about any point at the base of the dam.

2. Horizontal Hydrostatic Pressure due to Water. This is the largest external force acting on the dam. It has the largest capacity for disturbing the stability of the dam. It is a horizontal force which acts at the C.G. of the pressure distribution diagram, due to water. The pressure distribution diagram is always triangular with zero value at surface of the water and increasing linearly to maximum at the base of the dam. The value of maximum horizontal pressure at base of the dam is wh where w is the density of water in kg/m^3 and h the depth of water in meters. Since pressure distribution diagram due to water is triangular, the value of the total horizontal pressure (P) due to water, will be area of the triangle.

This force P will act at C.G. of the pressure distribution triangle i.e. $\frac{h}{3}$ from the base of the dam.

Hence total horizontal pressure $P = wh \times \frac{h}{2} = \frac{wh^2}{2}$.

This force P will act at $\frac{h}{3}$ from base of the dam.

Similarly if there is tail water of height h' on the D/S side, it exerts a horizontal pressure (P') of opposite nature.

$$P' = \frac{wh'^2}{2}$$

where P' = Horizontal force due to tail water.

h' = Depth of tail water from the base of the dam.

P' would be considered as acting at $\frac{h'}{3}$ from the base of the dam.

For water, density is considered as 1000 kg/m^3.

3. Uplift Pressure due Seeping Water under the Dam. The water that seeps through the pores of the material comprising dam and foundation, causes uplift pressure and tries to tilt or topple the dam. A part of the weight of the dam would get neutralized by uplift pressure and thus net foundation reaction due to vertical forces will be reduced. Intensity of uplift pressure is maximum at *U/S* end of the dam and it goes on decreasing towards the *D/S* end. As some water can see into the concrete dam also, the uplift pressure may occur anywhere in the dam also. It is very difficult to find out the value of uplift pressure accurately. It depends upon the factors like, cut off on *U/S* side, fissures in the foundation rocks, drainability of the foundation etc.

There are two schools of thought as to on how much area uplift pressure should be considered as acting. According to one thought one-third to two-third area of foundation should be considered effective. The second thought, propagated by Terzaghi recommends consideration of full area as the effective area.

U.S.B.R. Recommendations. According to U.S.B.R. intensity of uplift pressure at *D/S* (Toe) and *U/S* (heal) is taken equal to the hydrostatic pressure of water. The variation of uplift pressure from heal to toe is linear or straight line. In order to release uplift pressure, drainage galleries are provided in the body of the dam.

The magnitude of the uplift pressure at the face of the gallery is taken as equal to hydrostatic pressure at the Toe plus one third the difference of the hydrostatic pressures at the heal and Toe.

Uplift pressure at heal $A = wh$

Uplift pressure at toe $F = wh'$

Uplift pressure at gallery point $G = w\left[h' + \frac{1}{3}(h - h')\right]$

13.2.1 Indian Standard Recommendations (IS : 7894 –1975)

According to the IS : 7894–1975 there are two constituent elements in uplift pressure.

I. The area factor or the percentage of area on which uplift acts.

II. The intensity factor or the ratio which the actual intensity of uplift pressure bears to the intensity gradient extending from head water to tail water at various points.

Effective *D/S* drainage, whether natural or artificial, will generally limit the uplift, at the toe of the dam to tail water pressure. Formed drains in the body of the dam and drainage holes drilled subsequent to grouting in the foundation, where maintained in good repair, are effective in giving a partial relief to the uplift pressure intensities under and in the body of the dam. The degree of effectiveness of the system will depend upon the character of the foundation and the dependability of the effective maintenance of the drainage system. In any case, observation of the behaviour of the dam will indicate the uplift pressures actually acting on the structure and when the uplift pressure are seen to approach or exceed design pressures prompt remedial measures should necessarily be taken to reduce the uplift pressures to values below the design pressures.

Criteria for Design. The following criteria are recommended for the calculation of uplift forces.

(a) Uplift pressure distribution in the body of the dam shall be assumed, in case of both preliminary and final design, to have an intensity which at the line at the formed drains exceeds the tail water pressure by one-third the differential between reservoir level and tail water level. The pressure gradient shall then be extending linearly to heads corresponding to reservoir level and tailwater level. The uplift shall be assumed to act over 100 per cent of the area.

(b) Uplift pressure distribution at the contact plane between the dam and its foundations and within the foundation shall be assumed for preliminary designs to have an intensity which at the line of drains exceeds the tail water pressure by one-third the differential between the reservoir and tail water heads. The pressure gradient shall then be extended linearly to heads corresponding to the reservoir level and tailwater level. The uplift shall be assumed to act over 100 per cent area. For final designs, the uplift criteria in case of dams founded on compact and unfissured rock shall be as specified above. In case of highly jointed and broken foundation, however, the pressure distribution may be required to be basesd on electrical analogy or other methods of analysis taking into consideration the foundation condition after the treatment proposed. The uplift shall be assumed to act over 100 per cent of the area.

(c) For the extreme loading conditions *F* and *G* given later the uplift shall be taken as varying linearly from the appropriate reservoir water pressure at the *U/S* face to the appropriate tail water pressure at the *D/S* face. If the reservoir pressure at the section under consideration exceeds the vertical normal stress (computed without uplift) at the *U/S* face, horizontal crack is assumed to exist and to extend from the *U/S* face towards the *D/S* face of the dam to the point where the vertical normal stress (computed on the basis of linear pressure distribution without uplift) is equal to the reservoir pressure at the elevation. The uplift is assumed to be the reservoir pressure from the *U/S* face to the end of the crack and from there to varying linearly to the tail water pressure at the *D/S* face. The uplift is assumed to act over 100 per cent of the area.

(d) No reduction in uplift is assumed at the *D/S* toe of spillways on account of the reduced water surface elevation (relative to normal tail water elevation) that may be expected immediately down stream of the structure.

(e) It is assumed that uplift pressures are not affected by earthquakes.

4. Earthquake Pressure. Primary, secondary, Releigh and Love waves, are set up in the earth's crust, due to earthquakes. These waves impact acceleration to the foundation under the dam and cause its movement. In order to prevent cracking or rupture, the dam must also move along with the movements of the foundation. The acceleration generated due to earthquake develops inertia force in the dam, due to which, stresses are first induced in lower layers and then in whole of the dam. The earthquake waves can travel in any direction. They are however resolved in vertical and horizontal directions. The effects of acceleration in horizontal and vertical directions are given as follows.

(*i*) *Effect in vertical direction.* Due to vertical component of acceleration, a vertical inertia force $F = \alpha W$ is developed on the dam. The direction of this force is in opposite direction to the acceleration. In other words inertia force acts downwards when acceleration is upwards, thus increasing the downward weight momentarily. Similarly when acceleration is vertically downwards the inertia force $F = \alpha W$ acts upwards, thus decreasing the downward weight temporarily. Hence unit weight of the dam and water get altered as follows, due to vertical acceleration.

For dam material $= w_e(1 \pm \alpha)$

For water $= w(1 \pm \alpha)$

where w_e = unit weight of dam material.

α = acceleration.

(*ii*) *Effect in horizontal direction.* Two forces are developed due to horizontal effect of acceleration. They are :

(a) Inertia force in dam

(b) Hydrodynamic pressure of water.

(a) Inertia Force. Inertia force is equal to the product of the mass of the dam and the acceleration.

$$F = \text{Mass} \times \text{acceleration}$$

where F = Inertia force

W = Weight of the dam

α = Acceleration coefficient

g = Acceleration due to gravity.

$$\text{Mass} = \frac{W}{g}$$

and acceleration $= \alpha g.$

$$\therefore \quad F = \frac{W}{g} \times \alpha g = W\alpha$$

Inertia force acts in a direction opposite to the acceleration imparted by earthquake forces. This force should be considered at the C.G. of the mass regardless of the shape of the cross-section and acts horizontally downstream.

(b) Hydrodynamic Pressure. Horizontal acceleration towards the reservoir causes a momentary increase in water pressure. The theoretical computations reveal that the distribution of hydrodynamic pressure due to earth on the upstream face of the dam is elliptical-cum-parabolic. The following formula of U.S.B.R. and given by C.N. Zargar may be used to evaluate pressure intensity at a depth y below the maximum water level

$$P_{ey} = C_y \times \alpha wh$$

where P_{ey} = Intensity of hydrodynamic pressure

a = Acceleration coefficient

w = Unit weight of water

h = Depth of water in the reservoir

C_y = A dimensionless pressure coefficient at a depth y below water level. The value of C_y is found out by formula

$$C_y = \frac{C_m}{2}\left[\frac{y}{h}\left(2-\frac{y}{h}\right)+\sqrt{\frac{y}{h}\left(2-\frac{y}{h}\right)}\right]$$

where h = Maximum depth of water in the reservoir.

y = Depth from the reservoir surface to the elevation in question.

C_m = Maximum value of the pressure coefficient for a given constant slope.

$$= 0.735\frac{\theta}{90^\circ}$$

θ = Angle in degrees that the slope of the U/S face of the dam makes with the horizontal.

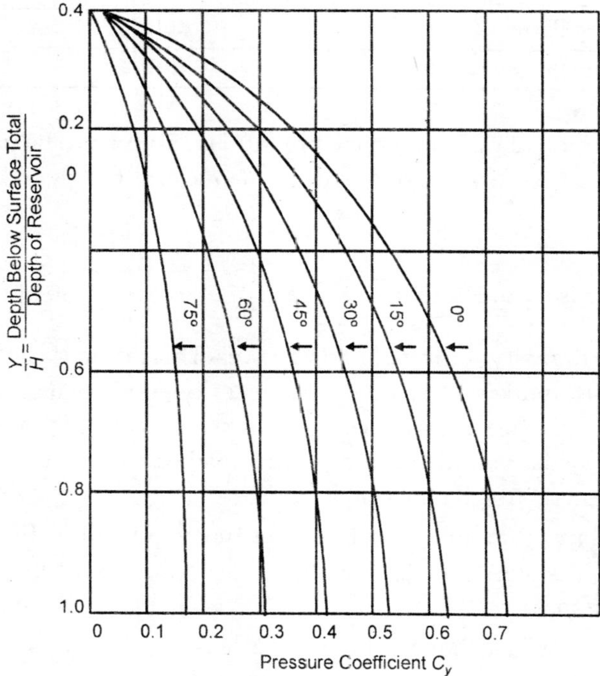

Fig. 13.2

The variation of pressure along the depth is shown in Fig. 13.1.

If the *U/S* face of the dam is inclined, but inclination does not extend to more than half the depth of the reservoir, the face may be considered as vertical for computation of C_m. If slope extends to more than half the depth, overall slope θ should be considered upto full reservoir level for finding the value of C_m as shown in Fig. 13.3.

The total pressure P_e on the portion of the dam upto depth y form top is given by

$$P_e = 0.726\, P_{ey} \times y.$$

The moment M_e about the joint upto which pressure is taken is given by

$$M_e = 0.3\, P_{ey} \times y^2.$$

The intensity of the earthquake pressure depends on the acceleration of vibratory motion and is expressed in terms of a fraction of (g) the acceleration due to gravity. Such intensity maps are available for the whole of the country. A detailed map of different zones of earthquake is given in I.S. : 1893–1970. Following accelerations have been adopted for some pojects in India.

Name of project	*Value of acceleration*
1. Bhakra Dam	0.15 g
2. Kosi Barrage	0.15 g
3. Ramganga Dam	0.20 g

If the distribution of hydrodynamic pressure is assumed to be parabolic, the increase in water pressure (P_e) due hydrodynamic effect is given by

$$P_e = 0.555\, \alpha w h^2$$

This pressure is assumed to be acting at $\frac{4h}{3\pi}$ above the base of the dam.

Rossi-Forel Intensity Rating of Earthquakes. This rating classifies different intensities of earthquakes according to the severity and extent of damage. The rating also provides the values of acceleration in terms of g for various types of earthquakes. For dams and buildings earthquakes from 7 to 10 rating are important and these are given in Table 13.1.

There are maps available which divide the region into different zones according to the severity of the earthquake. India has been divided into following five zones.

Zone I. In this zone earthquake intensity is least. Parts of western Rajasthan and central Deccan plateau are covered by this zone.

Zone II. Here intensity and frequency of earthquake is greater than zone I. This area consists of the outer periphery of Deccan plateau and consists mostly of Tamil Nadu and Kerala districts.

Table 13.1. *Rossi-Forel Rating*

Intensity class	*Effect*	*Acceleration*
7	Strong shock, fall of plaster, swing of church bells, overthrow of objects, general panic but no real damage to buildings	0.02 to 0.05 g
8	Very strong shocks, cracks in walls, chimney break shatters old buildings, or produce cracks in ground	0.05 to 0.1 g
9	Extremely strong shock, shatter or uproots buildings, cause cracks in ground or land slides	0.1 to 0.2 g
10	Shocks of extreme intensity which results in real destruction, fissures in ground, rock fills, total destruction in the region of occurrence	0.2 to 0.3 g

Zone III. Intensity of earthquake in this region is twice the intensity of zone II. Southern U.P., southern Punjab, southern Bihar, southern Bengal and parts of Tamil Nadu near Vijayawada. Region also includes western ghat except Koyana and Poona region.

Zone IV. The intensity of earthquake is quite severe. Indo-gangetic plains, Delhi, north Punjab, northern U.P. Bihar and Bengal, Runn of Kutch and western ghat near Poona or Koayana are included in this zone.

Zone V. Intensity of earthquake is maximum in this region. Himalayas on North, Arakan and Naga hills on the East, Mallar range in the West and a strip of foothills is included in this zone.

Basic horizontal seismic coefficients *recommended* for these zones are as follows

Zone	*I*	*II*	*III*	*V*	*V*
Basic horizontal seismic coefficient (α_h)	0.01	0.02	0.04	0.05	0.08

IS : 1983–1975 recommends the seismic coefficient method for dams upto 140 m height and the response spectrum method for dams greater than 100 m height.

For dams upto 100 m height the horizontal seismic coefficient is taken 1.5 times seismic coefficient at top of the dam reducing linearly to zero at the base vertical seismic coefficient is taken 0.75 times the value of α_h at top reducing linearly to zero at the base.

5. Wind Pressure. It is generally not considered in the design. If it has to be considered, it should be considered only on that portion of the superstructure which is exposed to wind. It is normally taken as 100 to 150 kg/m^2, acting on the exposed area.

6. Ice Pressure. This pressure is considered only when dams are located either at very high altitudes or in very cold regions. In very cold conditions, the reservoir gets covered with a layer of ice. When this layer is subjected to expansion and contraction due to variation in temperature, a force is developed which acts on the dam at level of water in the reservoir. This force acts linearly along the length of the dam at the reservoir level. The average magnitude of this force may be taken as 5 kg cm^2 of the contact area of the ice layer with *U/S* face of the dam. The thermal expansion of ice is about five times that of concrete.

7. Wave Pressure. Waves develop on the surface of the reservoir due to wind. The waves exert pressure on the dam. The magnitude of wave pressure depends upon the height of the waves, developed in the reservoir. The height of the wave as per D A. Molitor can be found out by formula

$$h_w = 0.032\sqrt{VF} + 0.763 - 271\sqrt[4]{F} \quad \text{for } F < 32 \text{ km}$$

and

$$h_w = 0.032\sqrt{VF} \quad \text{for } F > 32 \text{ km}$$

where h_w = Height of wave in metres from trough to crest.

V = Wind velocity in km per hour.

F = Fetch or straight length of water expanse in km.

After having found out the value of h_w the pressure intensity due to wages is found out by formula $p_w = 2.4\, w\, h_w$ (t/m^2).

The p_w is considered as acting at $\frac{h_w}{8}$ metres above the still water surface.

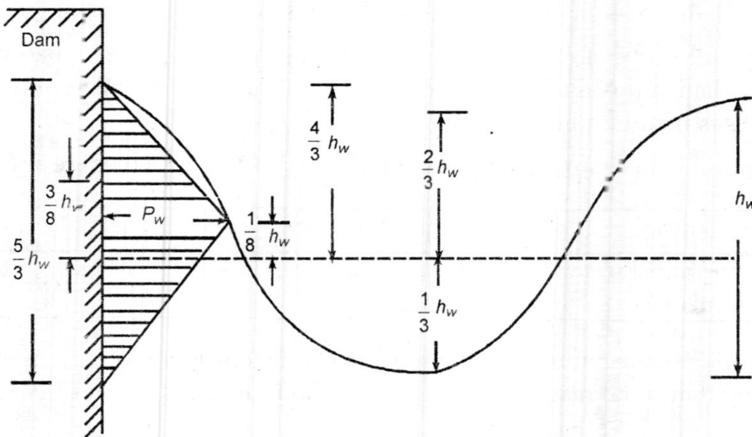

Fig. 13.3. *Wave pressure.*

The pressure distribution due to wave pressure is curved but for all design purposes it is considered as triangular as shown in Fig. 13.3.

The height of the triangle is not considered as h_w but $\frac{5}{3} h_w$.

Hence total pressure p_w due to waves is given by

$$p_w = 2.4\, wh_w \times \frac{1}{2}\left(\frac{5}{3}h_w\right)$$

$$= 2wh_w^2\ (t/m)$$

This force is considered as acting at a distance of $\frac{3}{8}h_w$ above the reservoir surface.

8. Pressure Due to Silt. Silt gets deposited at the bottom, along the *U/S* face of the dam. This force does not act when reservoir is new as no silt is there. But after some months or years silt is deposited at the bottom, which exerts additional horizontal pressure on the dam. If w_e is the submerged unit weight of the deposited silt, ϕ is its angle of internal friction and h is the height to which the silt has deposited, the silt pressure (p_s) is calculated as follows

$$p_s = \frac{1}{2}w_s h^2\left(\frac{1-\sin\phi}{1+\sin\phi}\right)$$

According to U.S.B.R. recommendation which are mostly followed in the design, pressure due to silt is considered as fluid pressure which develops a horizontal pressure with its density as 360 kg/m^3 and a vertical pressure with its density equal to 920 kg/m^3, so that additional pressure in horizontal and vertical directions are as follows.

$$\text{Horizontal pressure} = 360\frac{h^2}{2} = 180h^2.$$

$$\text{Vertical pressure} = 920 \times \frac{h^2}{2} = 460h^2.$$

Increase in silt pressure due to earthquake is generally not considered.

According to IS: 6512–1972 the silt pressure and water pressure exist together in submerged silt. The following criteria is recommended for calculating forces due to silt.

(a) Horizontal 'silt and water pressure' is assumed to be equivalent to that of a fluid weighing 1360 kg/m^3.

(b) Vertical 'silt and water pressure' is found as if silt and water together have a density of 1925 kg/m^3.

13.3 LOAD COMBINATIONS

In Section 13.2 various forces that are likely to act on the dam have been discussed. But all these forces seldom act simultaneously on the dam. There are some combinations of the loads that may act simultaneously. Design of the dam should be based on the most adverse combination of "probable load condition". USBR has recommended two load combinations.

1. Standard load combinations.

2. Extreme load combinations.

1. Standard Load Combinations. Under this heading three possible combinations are given.

(i) Horizontal water pressure, at normal reservoir level – ice pressure + silt pressure + normal uplift pressure.

But this combination is adopted where ice pressure is considerable.

(ii) Horizontal water pressure at normal reservoir level + earthquake pressure + silt pressure + normal uplift pressure.

(iii) Horizontal water pressure at H.F.L. + silt pressure + normal uplift pressure.

2. Extreme Load Combination. Maximum flood water elevation + silt pressure + extreme uplift with all the drains choked.

While designing the dam, whatever load combination might have been considered, the design must be checked for reservoir empty condition also. This condition is primarily important for testing stresses inside the dam and around the openings provided in the body of the dam.

13.2.2 Indian Standard Recommendations of Loading Combination for Design

IS has specified *A* to *G* seven types of load combinations. Gravity dam design should be based on the most adverse load conditions and using prescribed safety factors. Depending on the scope and details of the various project components, site conditions and construction programme, one or more of the loading combinations *A*, *B*, *C*, *D* and *E* may not be applicable in reality and may require suitable modifications.

I. Load combination *A* (construction condition) — Dam completed but no water in reservoir and no tail water.

II. Load combination *B* (normal operating condition) — Full reservoir elevation (or top of gates at crest), normal dry weather tailwater, normal uplift, ice and silt, (if applicable).

III. Load combination *C* (Flood discharge condition) — Reservoir of maximum flood pool elevation, all gates open, tailwater at flood elevation, normal uplift and silt (if applicable).

IV. Load combination *D* — Combination of *A* and earthquake

V. Load combination *E* — Combination *B* with earthquake but no ice.

VI. Load combination *F* — Combination *C*, but with extreme uplift (drainage inoperative).

VII. Load combination G—Combination E but with extreme uplift (drainage inoperative).

13.4 MODES OF FAILURE OF GRAVITY DAMS

A gravity dam can fail due to the following reasons:

1. Overturning of the dam
2. Sliding of the dam
3. Crushing of the dam or Foundation
4. Development of tension in the dam.

1. Overturning of the Dam. If the resultant of all the possible forces (internal as well as external) acting on the dam cuts the base of the dam downstream of the Toe, the darn would overturn unless it can resist tensile stresses. To safeguard the dam against overturning, the resultant of the forces should never be allowed to go down stream of the Toe. If resultant is maintained within the body of the dam, there will be no overturning. All the forces acting on the dam cause moments. Some of the forces help maintain stability of the dam, while others try to disturb the stability. The moments of the forces, helping dam to maintain its stability are known as resisting moments (M_r). The moments of the forces that try to disturb the stability of the dam are known as overturning moments (M_0). The moments of all the forces are generally taken about the Toe of the dam. Till $M_r \leq M_o$ there will be no overturning. As soon as M_o exceeds M_r overturn of the dam would take place. Factor of safety (F.S.) against overturning can be found out as follows.

$$\text{F.S.} = \frac{\text{Resisting moment}}{\text{Overturning moment}} = \frac{M_o}{M_r}.$$

The value of F.S. against overturning should not be less than 1.5.

2. Sliding of the Dam. In this mode of failure, the dam fails in sliding. The dam as a whole slides over its foundation or one part of the dam may slide over the part of the dam itself, lying below it. This failure occurs when the horizontal forces causing sliding are more than the resistance available to it, at that level. The resistance against sliding may be due to friction alone or it may be due to combination of friction and shear strength of the joint. Shear strength develops at the base if benched foundation is provided. At other joints the shear strength is developed by laying joints carefully so as to obtain good bond. Interlocking of stone blocks in stone masonry also helps increase shear strength.

In case of low masonry dams shear strength is not taken into account. In that case factor of safety against sliding is obtained by dividing net vertical forces by net horizontal forces and multiplying the resultant by coefficient of friction μ.

$$\text{Safety factor (F.S.)} = \frac{\mu\Sigma(V-U)}{\Sigma H}$$

where μ = Coefficient of friction.

$\Sigma(V - U)$ = Net vertical force.

ΣH = Sum of the horizontal forces causing sliding

The value of coefficient of friction varies from 0.65 to 0.75. The value of F.S. should always be greater than one.

In case of large high dams, the shear strength of the joint should also be considered along with static coefficient of friction. In this case factor of safety is known as *shear friction factor* (S.F.F.)

Hence

$$\text{S.F.F.} = \frac{\mu\Sigma(V-U)+bq}{\Sigma H}$$

where $\mu\Sigma(V - U)$ and ΣH are the same as stated earlier.

b = Width of the joint or section.

q = Shear strength of the joint which is usually taken as 14 kg/cm^2.

From safety point of view, the value of shear friction factor (S.F.F.) should lie between 4 and 5.

Factor of safety against sliding and shear friction factor as per IS: 6512–1972 are as follows.

S. No.	*Load in condition*	*Factor of safety against sliding*	*Shear friction factor*
1.	A, B, C	2.0	4.0
2.	D, E	1.5	3.0
3.	F, G	1.2	1.5

3. Crushing or Compression Failure. If the compressive stress developed anywhere in the dam exceeds the safe permissible limit, the dam may fail by crushing of the dam itself or of foundation. The maximum compressive stress can develop at toe when reservoir is full of water. If reservoir is empty the maximum compressive stress is likely to develop at the heal of the dam. The magnitude of maximum compressive and minimum compressive stresses can be found out by using following equation.

$$p_n = \frac{V}{b}\left(1 \pm \frac{6e}{b}\right)$$

where p_n = Value normal stress.

V = Total vertical force.

b = Width of the dam base at level of consideration.

e = Eccentricity of resultant force R from the centre of the base.

H = Total horizontal force.

R = Resultant force.

Plus sign is used to evaluate the amount of maximum compressive stress which will occur at Toe of the dam when reservoir is full and at heel when reservoir is empty.

Hence to safeguard the dam against failure due to compression or crushing, the value of maximum compressive stress i.e. $\frac{V}{b}\left(1+\frac{6e}{b}\right)$ should not be allowed to exceed the allowable, compressive stress (f) for the foundation material.

Hence $\frac{V}{b}\left(1+\frac{6e}{b}\right) \geqq f$

When eccentricity e becomes equal to $\frac{b}{6}$ we get maximum compressive stresses as $\frac{2V}{b}$.

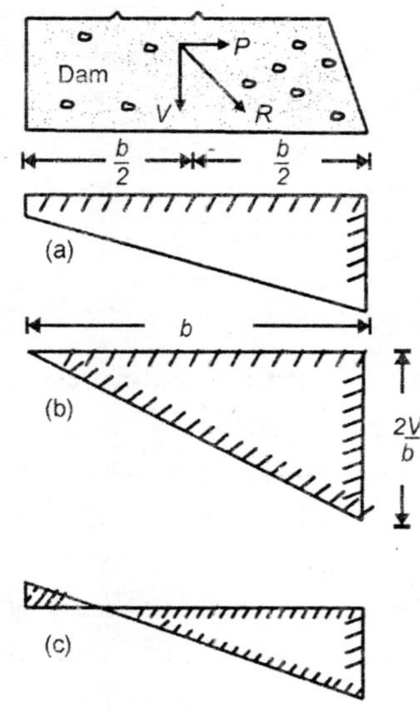

Fig. 13.4

4. Failure due to Development of Tension. Both cement concrete and masonry, are very weak in tension. Hence from safety point of view the tension is not allowed to be developed in the dam anywhere. We know that minimum compressive stress in the dam is $\frac{V}{b}\left(1-\frac{6e}{b}\right)e$. The nature of this stress remains compressive till value of e is less than $\frac{b}{6}$. When e becomes equal to $\frac{b}{6}$ the value of $\frac{V}{b}\left(1-\frac{6e}{b}\right)$ becomes zero. When e exceeds $\frac{b}{6}$ the value of $\frac{V}{b}\left(1-\frac{6e}{b}\right)$ changes sign and becomes negative. Negative sign of this stress indicates that nature of this stress is tensile rather than compressive. Hence as soon as e exceeds $\frac{b}{6}$ tension is developed in the dam and dam fails by opening of the joints, as concrete and masonry are almost nil in tension. When reservoir is full of water, tension is likely to develop at heal and when reservoir is empty tension is likely to develop at Toe of the dam. In other words it can be stated that until resultant of the forces lies within the middle third width of the base tension cannot develop anywhere in the dam.

In the case of gravity dams, having moderate height, no tension is allowed to be developed anywhere. However in case of very high dams, a small tensile stress may be permitted to be developed, but only for short durations during heaving floods or earthquakes.

Once tensile cracks develop at heel, the dam can not be rendered safe. Due to tension cracks, uplift pressure gets increased and consequently net vertical downward force is reduced. This causes further shifting of the resultant towards the Toe and this leads to further lengthening of the cracks. Due to lengthening of the cracks, effective width of base is further reduced and compressive stress at toe further increases. Ultimately compressive stress at toe increases to such an extent that dam fails by crushing at Toe.

13.5 PRINCIPAL STRESSES AND SHEAR STRESS

The vertical stress intensity is not the maximum direct stress intensity within the dam section. It is therefore necessary to determine the intensity of the maximum normal stress. The maximum normal stress occurs on the major principal plane. The *D/S* face of the dam which is inclined at an angle α to the vertical is a principal plane, as no shear stress acts on it. If any tail water is there it contributes a pressure p' entirely normal to the plane and is one of the principal stresses. The second principal plane will be inclined at right angles to the face of the dam and hence at an angle α to the horizontal. The stress acting on such plane is the major principal stress p_{max}. All the stresses can be represented by Mohr's circle. See Fig. 13.6.

The vertical stress p_n at toe of the dam in terms of principal stresses may be given as follows.

$$p_n = p' + (p_{max} - p')(1+\cos 2\theta)$$

$$p_n = p' + (p_{max} - p')\cos^2\theta$$

$$p_{\max} = \frac{p_n}{\cos^2\theta} - \frac{p'\left(1-\cos^2\theta\right)}{\cos^2\theta}$$

$$= p_n \sec^2\theta - p'\tan^2\theta$$

For maximum value of compressive stress, the tail water depth is assumed as zero.

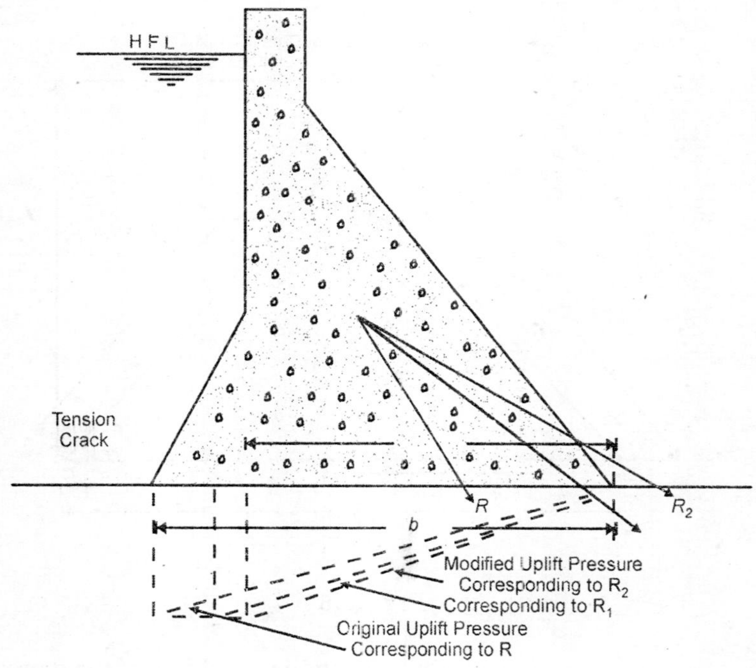

Fig. 13.5

Hence $\quad p' = 0$

$\therefore \quad P_{\max} = p_n \sec^2\theta \quad (1)$

Shear stress p_t from Mohr's circle is given by

$$p_t = \left(\frac{p_{\max} - p'}{2}\right)\sin 2\alpha$$

$$= (p_{\max} - p')\sin\theta\cos\theta.$$

Putting $\quad p = 0$ and $p_{\max} = p_n \sec^2\theta$ in this equation we get

$$p_t = p_n \sin\alpha\cos\theta\sec^2\theta$$

$$= p_n \tan\theta \quad (2)$$

Equation (1) gives the maximum intensity of direct stress which will occur near the *D/S* toe and which will act on a plane inclined at an angle θ to the horizontal. Equation (2) given intensity of shear stress on a horizontal plane near Toe.

The maximum intensity of stress at heel can also be similarly computed. For *U/S* face let

p'_{max} = Major principal stress on a plane inclined at β° with the horizontal. (Maximum intensity of direct stress).

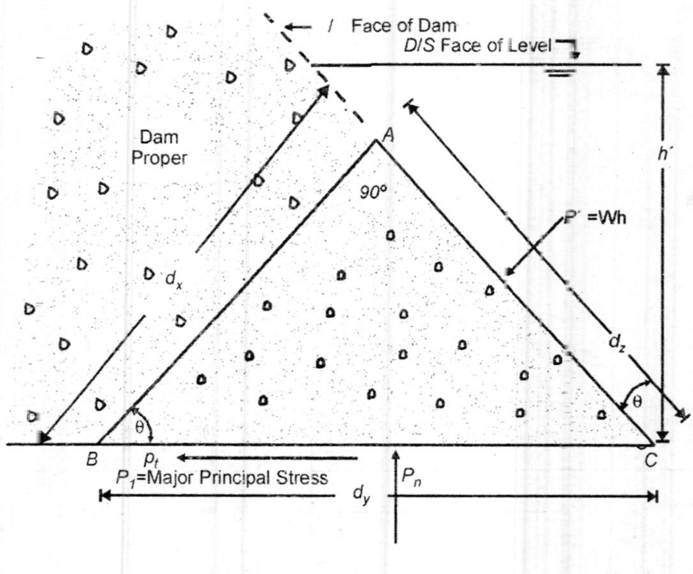

Fig. 13.6

β = Angle of inclination of *U/S* face of the dam with vertical.

p'_n = Intensity of the vertical stress on horizontal plane.

h = Depth of water impounded in *U/S*. Use following equation for calculation of maximum intensity of direct stress (p'_1)

$$p'_{max} = p'_n \sec^2 \beta - wh \tan^2 \beta \quad (3)$$

13.6 STABILITY ANALYSIS METHODS

The following are the various methods which can be adopted for carrying out the stability analysis of gravity dams.

1. Gravity method.
2. Trial load twist method.
3. Slab analogy method.

4. Lattice analogy method.

5. Experimental methods.

Brief description of each method is given here.

1. Gravity Method. This method of stability analysis, is also sometimes known as two dimensional method. In this method, the dam is considered to be composed of parallel sided vertical cantilevers. Each cantilever is considered free to act without any attachment with the adjoining cantilevers. The loads acting on the dam are resisted entirely by the weight of the individual cantilever. All the loads are ultimately transferred to the foundation by cantilever action. This method of checking the stability of the dam may be further divided into two parts.

(*a*) *Graphical method.* In this case the dam is divided into different sections according to height by drawing horizontal section lines 1–1, 2–2, 3–3 and 4–4. Section lines are generally drawn at places where the slope of the dam face changes. Starting from the top each part of the dam is analysed separately. For each section. the sum of all horizontal forces ΣH, and sum of all the vertical forces, ΣV, *acting above that section* are calculated and their line of action are graphically drawn. For each part of the dam resultant (R) of all the forces acting on any part is calculated and its line of action on section line lying immediately below that section is located. Finally a line is drawn joining the points at which the resultant cut the various section lines. This line should evidently lie within the middle third, so that tension may not develop in the dam. Such resultant lies are drawn for both the conditions, namely, reservoir full as well as reservoir empty. Both the lines of resultant pressure so drawn should lie in the middle third portion of the dam. See Fig. 13.6.

(*b*) *Analytical method.* Stability analysis by this method is done as per following steps :

(i) Consider unit length of the dam.

(ii) Find out the algebraic sum of all the vertical forces, acting on the dam. The forces that come in this category are weight of the dam, weight of water acting on inclined faces of the dam, uplift pressure etc. Let this algebraic sum be denoted by ΣV.

(iii) Determine the algebraic sum of all the horizontal forces acting on the dam. Let it be denoted by ΣH.

(iv) Determine the overturning moment (ΣM_o) and the resisting or stabilizing moment (ΣM_r) about the toe of the dam. Find out the algebraic sum of moments ΣM as follows

$$\Sigma M = \Sigma M_r - \Sigma M_o$$

(v) Determine the position of resultant force R, from the. toe ($\bar{x}$) as follows

$$\bar{x} = \frac{\Sigma M}{\Sigma V}.$$

(vi) Determine the eccentricity (e) of resultant R from the centre of the base (b)

$$e = \frac{b}{2} - \bar{x}$$

where b is the base width of the dam.

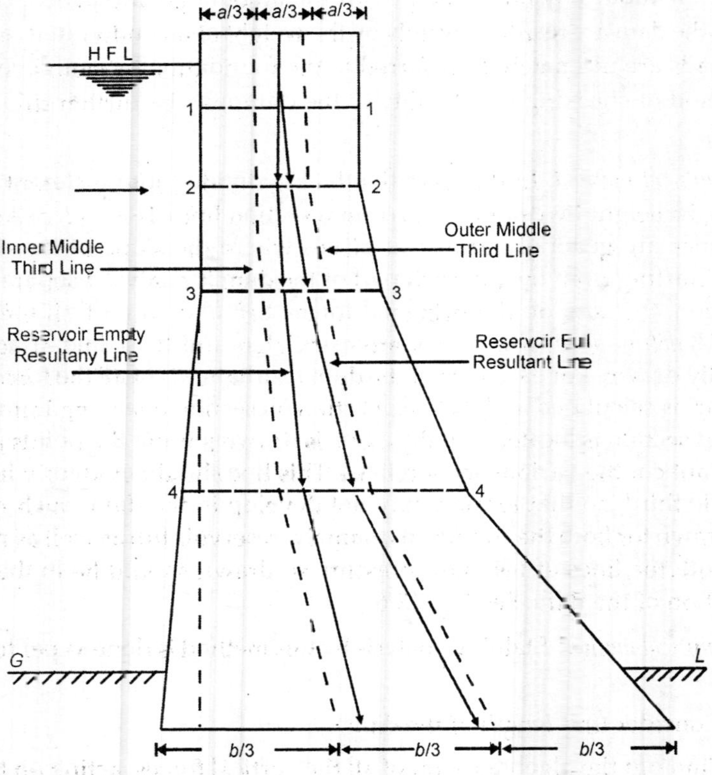

Fig. 13.7. *Graphical method.*

(vii) Find out the normal stress at toe and heel, by the following equation

$$p_n = \frac{\Sigma V}{b}\left(1 \pm \frac{6e}{b}\right)$$

(viii) Determine the principal and shear stresses at toe and heel as explained in Sec. 13.5.

(ix) Find out factor of safety against over turning as follows :

$$\text{F.S.} = \frac{\Sigma M_r}{\Sigma M_o}.$$

(x) Find out the factor of safety against sliding by following expressions

(a) Sliding factor $= \frac{\mu \Sigma V}{\Sigma H}$.

(b) Shear friction factor $= \frac{\mu \Sigma V + bq}{\Sigma H}$

It should be remembered that ΣV is the net vertical load, inclusive of the uplift.

2. Trial Load Twist Method. In this method of analysis, the entire dam is divided into a number of vertical cantilevers and horizontal beams. The water pressure is shared between vertical cantilevers and horizontal beams in such a way that the deflections at the common points on the dam face are equal. In this method, twisting effect is also developed, as all the cantilevers are of different heights. The vertical cantilever is of maximum height at the centre of the dam length. The length of vertical cantilevers goes on reducing as we proceed towards the ends of the dam, as valley banks rise quite steeply. Each cantilever element is thus dragged on its one face to greater deflections, by deeper adjacent cantilevers and is held on by the shorter cantilever on hill side. This develops a twisting effect on each cantilever.

3. Slab Analogy Method. In this method the analysis is made by dividing the analogous slab into horizontal and vertical beams. Horizontal and vertical beams are brought into slope and deflection agreement by trial loads. This is a very laborious method and hence not used much.

4. Lattice Analogy Method. This method is, though, simpler than slab analogy method, but still quite cumbersome. In it, the dam is considered as a frame work of interconnected square frames, each square being diagonally connected at the corners.

5. Experimental Methods. Experimental methods may he direct method and indirect method. Direct method is also known as three dimensional, model analysis, whereas indirect method as photoelastic model analysis.

(i) Three dimensional model analysis. This method of analysis has been found very useful in the case of high dams. In it, three dimensional models of dams are made from elastic materials. The size of the models is made proportionate. The models are located in similar manner as the prototype and also subjected to loading of the prototype. The structural actions, stress conditions, and deformations at various points of the model, are thereafter measured and a correlation is developed between the model and prototype, in regard to stress and deformation. From this correlation, the stresses and deformations for the prototype can be easily obtained.

(ii) Photoelastic models. In this method, photoelastic models of dam are prepared from elastic materials. The materials used for photoelastic models should be elastic, transparent, isotropic and free from initial stresses. Bakelite,

celluloid, gelatin, and glass are the usual photoelastic materials. The stresses in the photoelastic models are determined by a special optical instrument known as photoelastic polariscope. Polarised light is passed through it and stress and strain conditions in the model are studied. This method locates such points or regions in the model, where stresses become concentrated. This method has limited application.

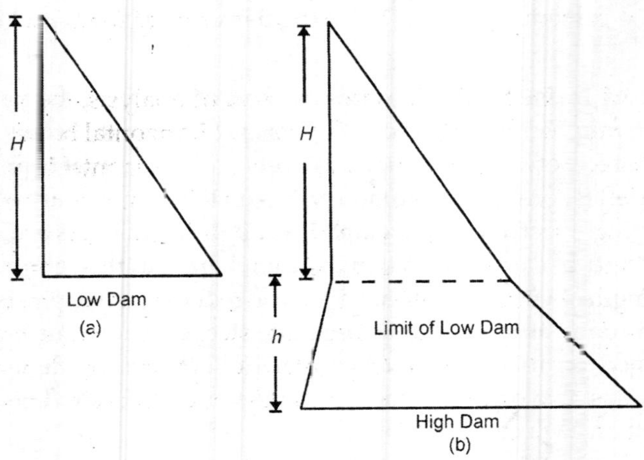

Fig. 13.8. *Low and high dam theoretical profiles.*

13.7 ELEMENTARY PROFILE OF A GRAVITY DAM

If a gravity dam is subjected to horizontal pressure due to water along with its own weight, the elementary profile of the dam would be a right angled triangle. The thickness of the dam at the level of water surface will be zero as there is no hydrostatic pressure to be resisted. The thickness at base would be maximum as there will be maximum hydrostatic pressure to be resisted by the dam. As the variation in hydrostatic pressure from water level to the bottom is linear; variation in elementary profile from top to the dam to its bottom would also be linear.

Elementary profile of the low dam is shown in Fig. 13.8 (a). The triangular elementary profile also provides the maximum possible stabilizing force against overturning without causing tension in the base when reservoir is empty. This is because in this case weight of the dam acts at $\frac{b}{3}$ from the *U/S* vertical face. Should there be any other triangular profile other than rt. angled, the stability of the dam is further increased but tension would develop at the toe when reservoir is empty.

Elementary profile of the dam is also sometimes known as theoretical profile of the dam.

Following forces are considered as acting on the dam while determining the elementary profile.

(i) Weight of the dam (W)

(W) = $\frac{1}{2}b H \rho w$. It acts vertically downwards and acts at C.G. of the dam.

(ii) Water Pressure (P)

$$P = \frac{1}{2} w H^2$$

Force P is horizontal and acts at $\frac{H}{3}$ from the bottom of the profile.

(iii)Uplift Pressure (U)

$$U = \frac{1}{2} cwb \times H$$

Uplift pressure acts vertically upwards. Its direct effect is that it reduces the gravity effect or load of the dam.

In all the three expression given above,

b = Base width of the dam.

H = Height of the elementary profile.

ρ = Specific density of the dam material.

w = Unit weight or density of water.

C = Uplift pressure intensity coefficient.

U = Uplift pressure.

P = Horizontal water pressure.

W = Weight of the dam as a whole.

Base Width of Elementary Profile. The base width of elementary profile is to be determined for following conditions.

1. Stress basis.

2. Stability of sliding basis.

1. Stress Basis. We know that for no tension to develop the resultant should pass through the inner middle third point, when reservoir is empty and through outer middle third point when reservoir in full of water. Take moments of all the forces acting on the profile about the outer middle third point, and equate it to zero.

Hence $\frac{1}{2} wH^2 \times \frac{H}{3} + \frac{1}{2} cwbH \times \frac{b}{3} - \frac{1}{2} bHrw \times \frac{b}{3} = 0.$

On solving $b = \frac{H}{\sqrt{\rho - c}}$

If uplift pressure is not considered $c = 0$ and hence $b = \frac{H}{\sqrt{\rho}}$.

2. Stability on Sliding Basis. In order to prevent sliding of the dam, the horizontal force causing sliding should be either equal to or less than the frictional resistance opposing the sliding of the dam. Critical condition occurs when horizontal force (ρ) is equal to the frictional resistance.

i.e.
$$\rho = \mu(W - U)$$

$$\frac{1}{2}wH^2 = \mu\left(\frac{1}{2}bH\rho w - \frac{1}{2}cbwH\right)$$

Solving
$$b = \frac{H}{\mu(\rho - c)}$$

If uplift is neglected

$$b = \frac{H}{\mu\rho}.$$

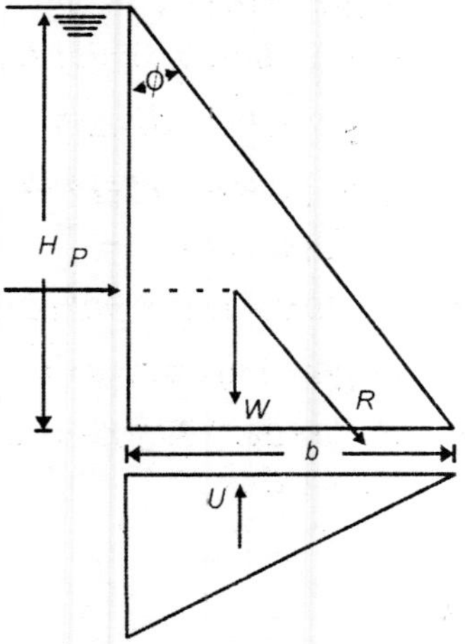

Fig. 13.9. *Stress in elementary profile of a low gravity dam.*

The value of b should be adopted greater than the values obtained on stress basis and sliding basis.

13.8 STRESSES DEVELOPED IN ELEMENTARY PROFILE

Maximum and minimum normal stresses in an elementary profile of the dam are given by

$$p_n = \frac{V}{b}\left(1 \pm \frac{6e}{b}\right)$$

where $V = (W - U)$ the algebraic sum of vertical forces

$e = \frac{b}{6}$

When reservoir is full the maximum normal stress will be at toe and minimum at heel.

Stress at Toe

$$p_n = \frac{V}{b}\left(1+\frac{6e}{b}\right) = \frac{W-U}{b}(1+1) = \frac{2(W-U)}{b}$$

$$= 2[\tfrac{1}{2}bH\rho w - \tfrac{1}{2}cbwH]$$

$$= wH(\rho - e)$$

The corresponding stress at heel

$$p_n = \frac{V}{b}\left(1-\frac{6e}{b}\right) = \frac{W-U}{b}(1-1) = 0$$

Principal stress at toe

$$p_{max} = p_n \sec^2 \phi$$

$$= wH(\rho - c)\left[\left(\frac{b}{H}\right)^2 + 1\right]$$

but $$\left(\frac{b}{H}\right)^2 = \frac{1}{\rho - 1} \qquad \left(\because b = \frac{H}{\sqrt{\rho - c}}\right)$$

$$p_{max} = wH(\rho - c)\left(\frac{1}{\rho - c} + 1\right)$$

$$wH(\rho - c + 1)$$

Shear stress at toe

$$p_t = P_n \tan \phi$$

$$= wH(\rho - c)\frac{b}{H}$$

$$= wH(\rho - c)\frac{1}{\sqrt{P - c}}$$

$$= wH\sqrt{\rho - c}$$

At heel normal stress is zero and as such principal and shear stresses will also be zero.

When Reservoir is Empty

In this case, the only force acting on the elementary profile is its weight. This weight acts at inner middle third point of the base. There will be no uplift pressure as reservoir is empty.

$$\text{Maximum normal stress at heel} = \frac{2W}{b}.$$

Minimum normal stress at toe = 0.

13.9 PRACTICAL PROFILE OF A GRAVITY DAM

The elementary profile cannot be adopted as such. Some modifications have to be incorporated, to make it adaptable in practice. Modifications in elementary profile are necessitated due to following reasons :

1. Some free board is essential whereas elementary profile does not provide any free board.

2. Road way is generally provided at the top of the dam. This necessitates quite thick top of the dam, whereas elementary profile does not provide any thickness at the top.

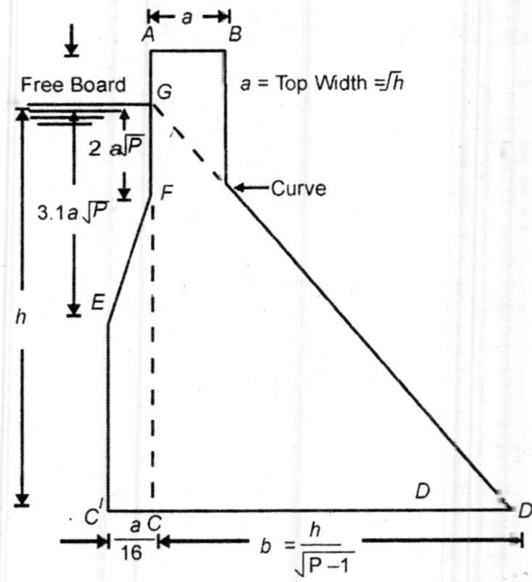

Fig. 13.10. *Practical profile of low gravity dam.*

3. Additional load due to extra height, provided for free board and also due thick top of the dam, induces some additional stress in the dam section. Some extra darn section has to be provided at the base of the dam along the *U/S* face to counter act such additional stresses. The amount of free board usually provided is 1.5 *hw* where *hw* is the height of waves in metres between through and crest. Minimum top width of the dam should be about one-seventh the

13.10 LOW DAM AND HIGH DAM

Maximum principal stress at the toe is given by

$$p_{max} = wH(\rho - c + 1)$$

In this formula variable element is only H the height of the dam. The value p_{max} should not exceed the allowable stress (p_a) for the material of which dam or its foundation is made.

Hence $wH(\rho - c + 1) = p_{max} = p_a$

or
$$H = \frac{p_a}{w(\rho - c + 1)}$$

While determining the limiting height of the dam, uplift is generally not considered.

Hence putting $c = 0$

$$H = \frac{p_a}{w(\rho + 1)}$$

If height of the dam exceeds the value given by expression $H = \frac{p_a}{w(\rho+1)}$, the value of maximum compressive stress would exceed the safe limit of the stress for the material and dam would fail if measures for lowering the value of maximum compressive stress are not taken. Now low dams and high dams can be easily differentiated. If the height of the dam is equal to or less than height as given by equation $H = \frac{p_a}{w(\rho+1)}$ the dam is called a *low gravity dam*. In case of low dams, the value of maximum compressive stress, at the most, reaches the safe allowable stress for the material. If the height of the dam is more than that, given by the equation $H = \frac{p_a}{w(\rho+1)}$, it is known as *high gravity dam*. For high dams, the *U/S* and *D/S* faces will have to be given extra slopes below the limiting height required for low dams so as to bring the excessive compressive stress within safe limits of the stress. Elementary profiles for low and high gravity dams are shown in Fig. 13.8.

13.11 PROFILE OF HIGH MASONRY GRAVITY DAM

Mr. G. Molesworth gave the following formulae for fixing the profile of a high gravity dam, made of masonry. These formulae are applicable for masonry gravity dams only.

$$x = \sqrt{\frac{1.76y^3}{p_a + 1.06y}}$$

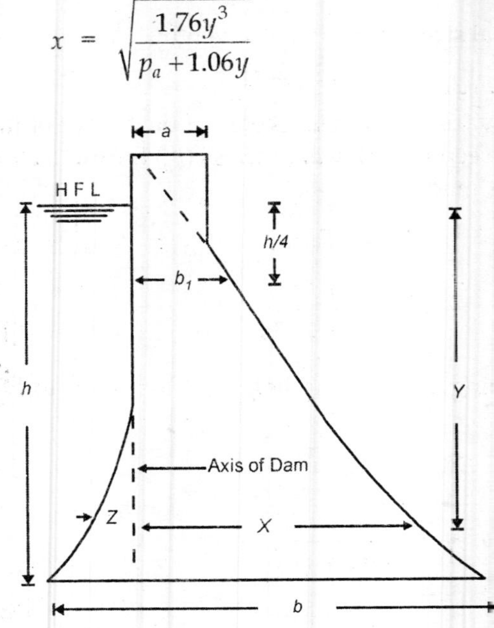

Fig. 13.11. *Molesworth profile.*

But value of z should not exceed $\frac{y}{\sqrt{\rho}}$ at any cost

$$Z = \left(\frac{y}{36{\cdot}5\,p_a}\right)^4$$

where x = D/S offset from the vertical line also known as axis of the dam at a depth y below the maximum reservoir level.

Z = U/S offset from the vertical line at a depth y metres.

b_1 = Base width of the dam at $\frac{h}{4}$ below the full reservoir level.

a = Thickness of the dam at the full reservoir level. It is taken as $0.4\,b_1$.

p_a = Allowable compressive stress for masonry in t/m^2, the value of which may vary between 77 to 110 t/m^2.

13.12 DESIGN OF GRAVITY DAMS

Elementary profile of the gravity dam has already been discussed. It is not possible to adopt elementary profile as such, because of certain practical requirements. The preliminary design of gravity dams is done by two dimensional gravity method by considering the dam as being made of a number of cantilevers of unit length and acting independently of adjacent cantilevers.

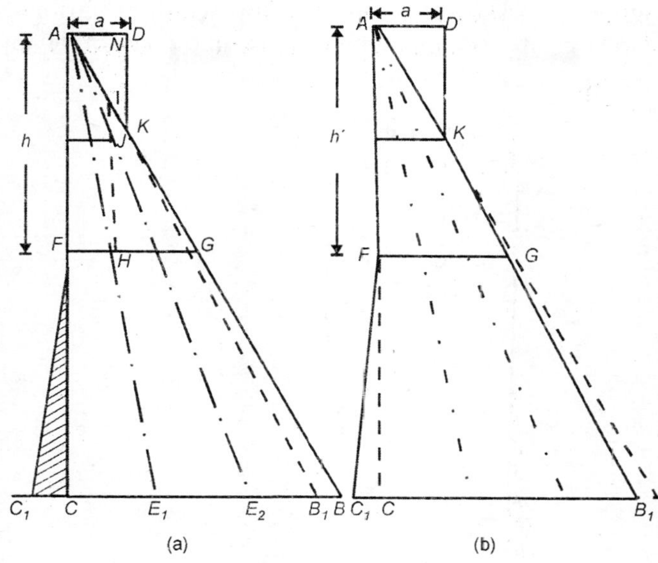

Fig. 13.12. *Effect of added top width on elementary profile of the dam.*

(*a*) *Effect of top width added at the apex of triangular profile.* ABC is an elementary profile of the dam. Let a be the top width added at the apex. The effect of this added top width is that the resultant of the dam section would shift slightly towards *U/S* side when reservoir is empty. Let AE_1 and AE_2 be the inner and outer-third point lines respectively. Let NI be the vertical line passing through the C.G. of the added triangular concrete ADK. Let NI and AE_1 lines intersect at point H. For all the dam sections lying below point H or section FHG the resultant would be shifted towards the *U/S* face of the line AE_1. This shift in resultant causes development of tension at the toe when reservoir is empty. Hence to avoid the possibilities of tension at toe when dam is empty, some batter has to be provided to *U/S* face below the plane FHG. The depth (h') of plane FHG can be found out as follows :

$$FH = AN = \tfrac{2}{3}a$$

$$FG = 3 \times FH = 3 \times AN = 3 \times \tfrac{2}{3}a = 2a$$

$$h' = AF = FG\sqrt{\rho - c} = 2a\sqrt{\rho - c}$$

c is the uplift pressure intensity coefficient. Thus upto height h' no batter in the *U/S* face is required, but for larger depths *U/S* face has to be battered.

Now, let us consider the case when reservoir is full of water. The vertical line NJ cuts the outer-third point line AE_2 at point j. Hence resultant of all the dam section lying below the plane JK are again shifted to the *U/S* side. To bring the resultant back to the outer third point line, the slope of D/S force have to be flattered. Figure 13.13 shows the effect of top width of various sizes. From diagram it can be easily seen that with increase in the size of the top width, the

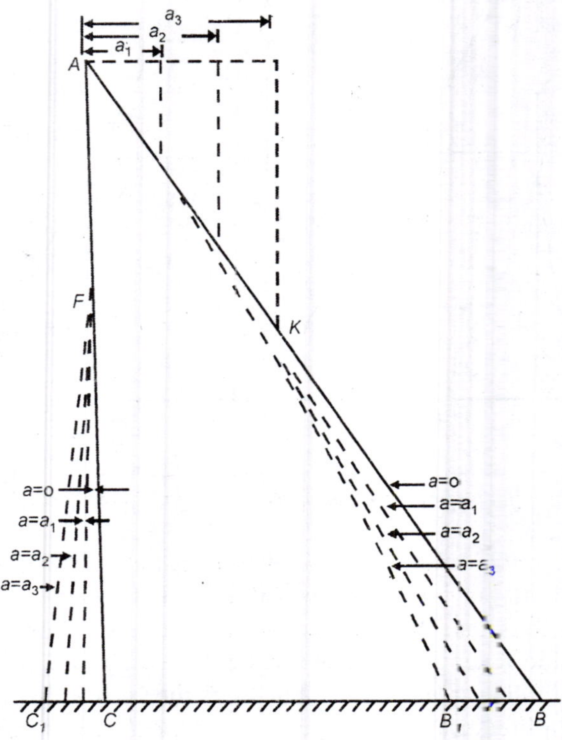

Fig. 13.13. *Effect of increase of top width of the dam.*

batter of U/S face of the dam is increased while that of D/S face decreased. The total effect of addition of masonry at the top, actually causes reduction in overall volume of the masonry of the dam rather than increasing it.

The most economical top width of the gravity dam is about 14 per cent of the height of the dam. In case of low dams increased top width may be provided from practical considerations such as provision of roadway on the top etc.

(b) *Multiple step method of design of gravity dam.* In this case, the total dam height is divided into various zones with the help of various horizontal sections. Each zone, starting from top of the dam, is deigned in such a way that all the stability requirement for it are fully satisfied. The total height may be divided into seven zones as shown in Fig. 13.14. Brief description of conditions for each zone are given one by one.

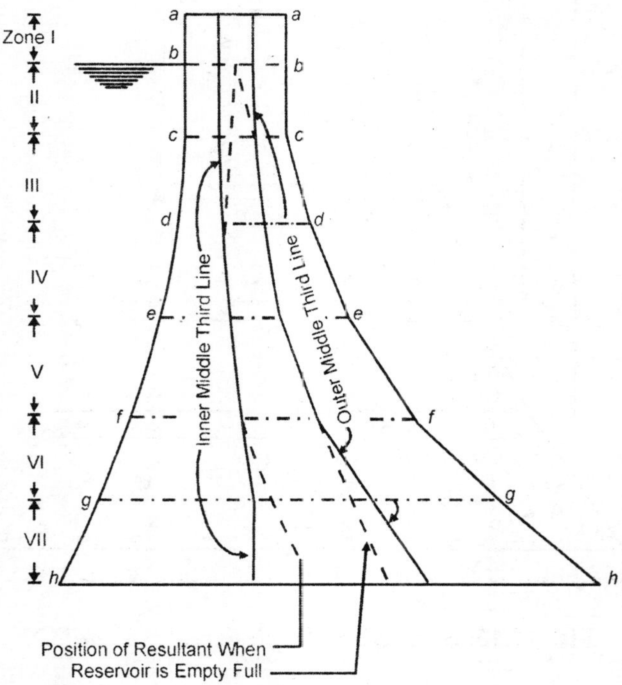

Fig. 13.14. *Multiple steps method of design of gravity dam.*

Zone I. This is the top most rectangular part of the dam lying above the full reservoir level. The height of this zone is controlled by the free board requirements and thickness is fixed on the basis of practical considerations. If ice is likely to deposit at the surface of the reservoir, the thickness and height of this zone is fixed on the basis of sliding of the zone due to ice pressure.

Zone II. This is the zone which has both of its faces vertical. The position of bottom of this zone (c.c.) is fixed, such that the resultant, passes through the outer third point of the plane c.c.

Zone III. In this zone *U/S* face is maintained vertical, whereas *D/S* face is given the slope. The position of the bottom of this zone (d.d.) is fixed in such a way that the resultant for reservoir full and empty conditions, pass through the outer middle third point and inner middle third point respectively.

Zone IV. In this zone *U/S* face of the dam is also battered like *D/S* face but batter of *U/S* face with vertical is very small. The position of the bottom of the zone (e.e.) is fixed in such a way that the maximum stress developed at toe for the condition when reservoir is full, reach the allowable limit of the stress for the material. The height of the dam above the bottom of zone IV (e.e.) is the height of the low dam.

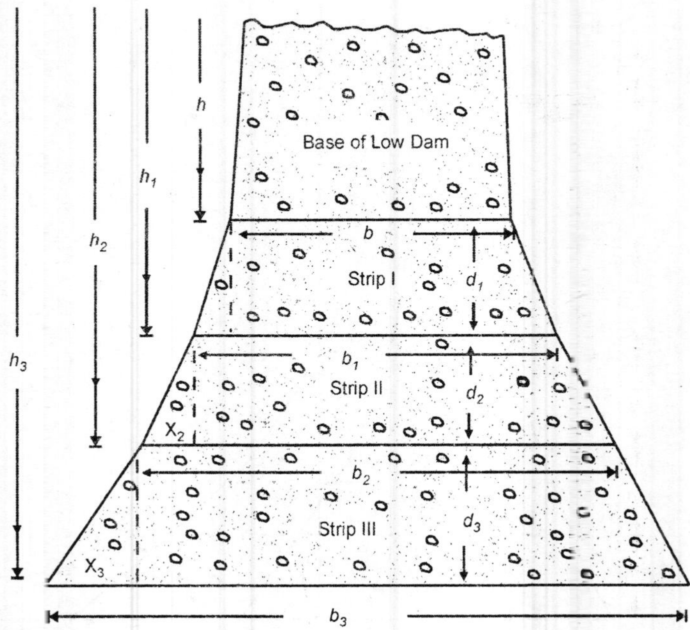

Fig. 13.15. *Strip method of design of high dam.*

Zone V. *D/S* slope is further flattened in this zone so that the maximum pressure at *D/S* toe remains within the working stress under reservoir full condition. The resultant for reservoir full, remains well within middle third point of the base of this zone. For the reservoir empty condition, the resultant cuts the *U/S* middle third point of the base of this zone in section *ff* and stress at heel reached the permissible limit.

Zone VI. In this zone, the resultant lines for both the conditions of reservoir, full and empty, well within the middle third points of the base of this zone. maximum stresses developed at toe for reservoir full condition and at heel for reservoir empty condition reach the maximum permissible value.

Zone VII. At the bottom of this zone the maximum compression at the *D/S* toe exceeds the permissible limit. This zone is generally eliminated by revising the design of the dam. If change in design is not possible, then this zone is made from superior materials so that it may sustain the increased stresses developed at toe and heel.

(c) *Single step method for design of high dams.* We have seen in multiple step method of design of the dam that upto bottom of zone IV, the dam is low gravity dam. But if we go beyond the height of low dam, *U/S* slope has very steep slope whereas *D/S* slope has convex shape outwards as shown in Fig. 13.16. Convexity outwards is not allowed. This convexity is avoided by designing the dam as a single block but conforming to the conditions laid down in zone VI. The shape of the high dam designed according to single step method is shown in Fig. 13.16.

In this method the *U/S* face of the dam is maintained vertical for some depth which may be determined as given in Fig. 13.12 (a). Below this section, both *U/S* and *D/S* faces are given such slopes that for both the conditions of reservoir full and empty, the stresses at toe and heel at all the sections reach their maximum values. This condition can be accomplished by performing several trials at all the sections. After having done this the dam is checked again forstability requirements for reservoir full and empty both, about the base of the dam.

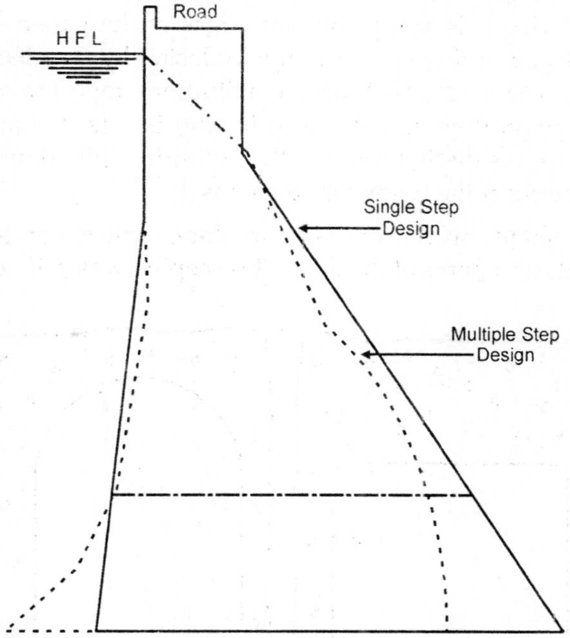

Fig. 13.16. *Super-position of Multi-step designed dam over single step design.*

Figure 13.16 shows super-position of dam section, designed according to single step method over that designed according to multi-step method. It is very clear that cubic content of the dam designed by multi-step method in economical. Dam designed by multi-step method is fully stressed at all the heights lying below the height of the low dam. But in case of dam, designed according to single step, the section is understressed everywhere except at the base. From the discussions done uptill now, following inferences can be easily drawn.

1. Low dams should be designed by multi-step method as they prove economical.

2. High dams beyond zone VI should be designed by single step method so as to avoid convex curvature on the *D/S* face.

3. It may be economical to eliminate the zones V and VI by using more superior materials in the lower part of the dam, as in that case, permissible compressive stress will be more and hence higher stresses can be sustained by the dam.

13.13 GALLERIES IN GRAVITY DAMS

Galleries have to be left in the gravity dams during their construction. The size of the galleries depends upon the purpose, they have to perform. The galleries may be aligned both along the axis and across the axis of the dam. They are provided at all the levels of the dam. All the galleries are given some longitudinal slope and small channels along both the edges of the galleries are formed. Seeping water through the dam section is collected by the channels running along the galleries. Since channels have longitudinal slope, the collected water in channels keeps on flowing automatically and is collected at some central place from where it is discharged into the *D/S* side of the dam. Galleries are constructed to perform the following functions.

1. It may be cement concrete or masonary dam, some water definitely seeps through the joints and pores of the dam. This seeping water, if not intercepted,

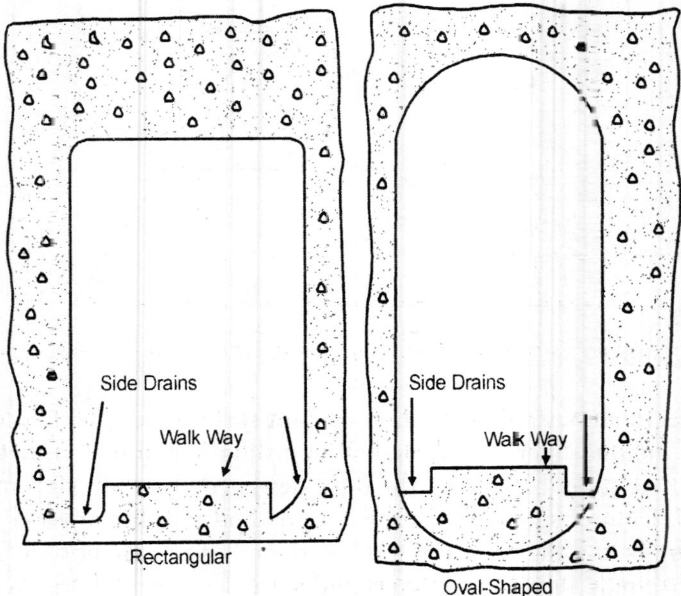

Fig. 13.17. *Galleries in gravity dams.*

would create internal stresses in the dam and may cause its failure. Galleries intercept the seeping water and relieve the dam from internal stresses.

2. They form a measure to reach each and every part of the dam.

3. They are used for dissipation of heat, developed by setting action of the cement concrete during construction of the dam.
4. They are used for plugging the longitudinal joints of the dam, after the dam has been completed.
5. They are used to carry out drilling and grouting of the dam or the foundations.
6. They are used to study the behaviour of the dam after completion.
7. They are used for fitting mechanical equipments like pumps etc. in them.
8. Galleries are also essential for operation of sluice gates, outlet gates etc.

Shape of the gallery is generally rectangular but it may be oval shaped also. Side channels on both the sides of the gallery are provided. See Fig. 13.17.

Shafts. Dams are also provided with shafts. Shaft is a vertical opening in the dam. Shafts connect galleries at various levels. Shafts are generally provided with lifts so as to conduct effective inspection of the dam and also to facilitate quick approach any where in the dam. They are sometimes used to measure the deflections of the dam also.

13.14 JOINTS IN GRAVITY DAMS

The joints that are necessarily provided in the dam may be classified under two heads.

1. Construction joints, and
2. Contraction joints.

1. Construction Joints. The concreting of the dam is not done at a stretch but in stages. Each stage of the concreting is known as lift. Lift is nothing but thickness of a horizontal layer of concreting laid once. In concrete dams the thickness

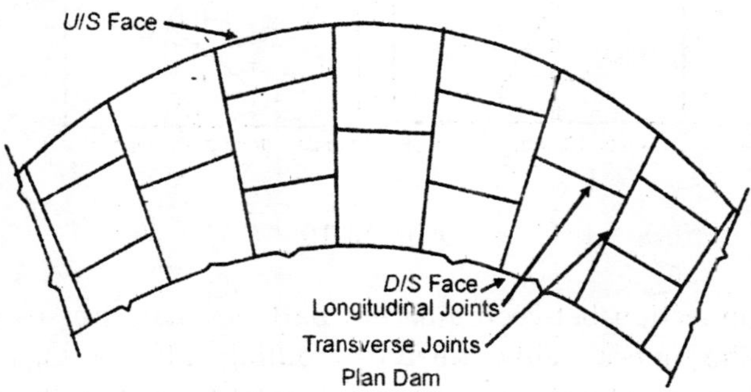

Fig. 13.18. *Joints in gravity dam.*

of each layer or in other words the lift, is kept about 1.5 m. The horizontal joint between two successive lifts is known as construction joint. The lift or thickness of each concreting layer is decided so that cooling of the concrete, while setting, may be effectively accomplished by natural as well as artificial agencies. Thickness of lower most layer of concreting is kept half i.e. about 0.75 m. Modern techniques of treatment of surface before laying the new layer of concrete, has almost eliminated the necessities of providing keys or water stops in the construction joints.

2. Contraction Joints. The main purpose of providing contraction joints is to avoid development of shrinkage cracks in the dam due to changes in the temperature. Shrinkage cracking of the concrete can be controlled to some extent by properly controlling the temperature and adopting proper methods of curing. But provision of contraction joints in mass works like dams is invariably essential. Construction joints may further be classified as transverse contraction joints and longitudinal contraction joints.

(*i*) *Transverse contraction joints.* Direction of the joints is at right angles to the axis of the dam. The spacing of these joints depends on factors like topographical features, location of the dam, type of cement, climatic conditions, but the general practice is to restrict spacing to about 15 m or the height of the dam whichever is small. These joints extend throughout the height of the dam. To insure proper

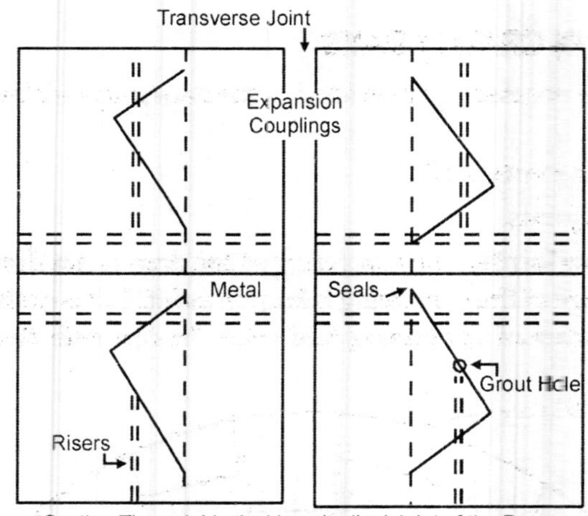

Fig. 13.19

contact or connection between adjacent parts so that stresses are properly transmitted, the joints are either filled by grouting or by leaving a slot which is filled later when shrinking ceases. The joints may not be grouted and rendered water tight by introducing water stops or key ways in the joints.

(*ii*) *Longitudinal contraction joint.* These joints are provided along the direction of the axis of the dam. These joints are considered objectionable from safety point of view as they may coincide very closely with the planes of maximum shear. Longitudinal joints do not run continuously and they are staggered in plan as shown in Fig. 13.19. However they are continuous in vertical direction. They are laid between two adjacent transverse joints. The spacing of these joints is also limited to about 15 m. These joints are provided with key ways so as to transmit the principal stresses. Spacing of key ways vertically is 1.5 m or one key way is to be provided in each lift.

Keys. It is a device, with the help of which shearing stresses from one block are transferred to the adjacent block. The provision of keys is essential for longitudinal joints but optional for transverse joints. The adjoining surfaces of each block are given such a shape that they together with transfer of stresses,

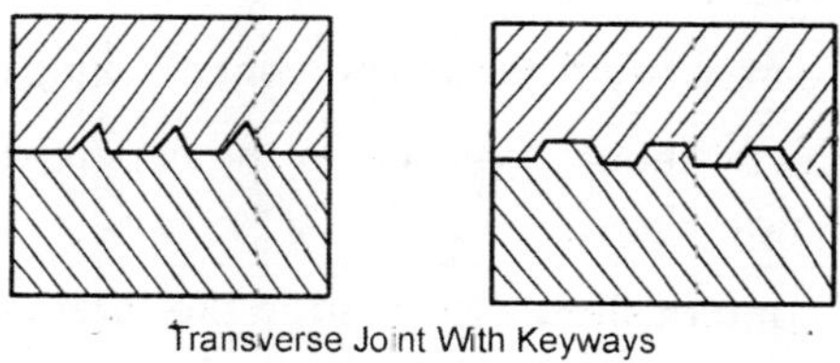

Fig. 13.20. *Key ways.*

cause effective interlocking also. Key ways may be of several shapes such as triangular, trapezoidal or trough shaped.

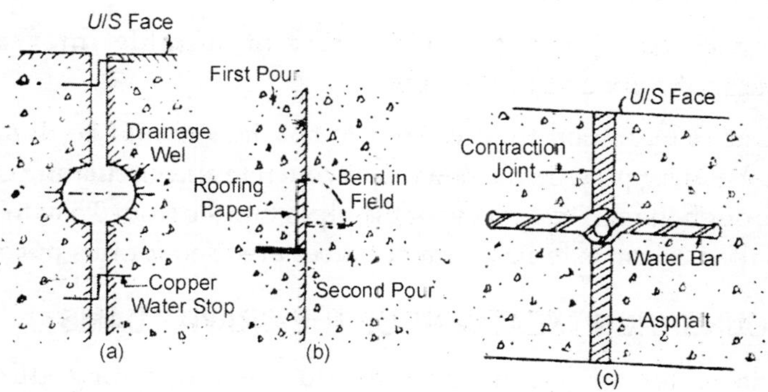

Fig. 13.21 *Various methods of sealing transverse joints*

Water Stops. The main job of water stops is to prevent leakage from the dam. They may be made of copper, steel, sheet, lead or of natural rubbers and plastics. Metal water stops are provided only in the case of non-yielding foundations. Rubber water stops are used in case the foundation is of yielding type. Drainage wells filled with asphalt are also sometimes used along with along water bars. Water bars are also a type of water stops.

13.15 CONTROL ON CONCRETE DAM CRACKING

When cement starts setting, heat of hydration is developed. Concrete dams are large masses of concrete. Heat of hydration is dissipated easily from the surface of the dam as surface remains exposed to the atmosphere. It is very difficult to dissipate the heat of hydration from interior of the dam. If some arrangement for this is not made, very high temperatures may develop inside the dam. Surface of the dam being cool and temperature in the interior being very high, cracks are likely to develop and cause difficulties. Dam may also develop surface cracks due to daily variation of the temperature at the surface. Surface cracks are not harmful for short durations but cause very bad effect in the long run. Water enters the surface cracks and solidifies to ice if temperature fall down to zero. Ice thus formed expands and causes the surface cracks to deepen.

Following measures may be adopted to prevent cracking of the concrete dams:

1. Low heat cement should be used in the concrete being used for the dam construction.
2. Low cement content should be used in the interior parts of the dam.
3. Ice cold or refrigerated water should be used for the preparation of cement concrete.
4. Thickness of lift of the concrete, should not be allowed to exceed 1.5 m.
5. Considerable (at least 5 days) time should be allowed for a lift to set before next lift is laid over it.
6. Contraction joints should be laid at suitable intervals, both longitudinally and transversely.
7. Heat of hydration from interior of the dam should be dissipated by embedding pipes in the newly laid concrete and circulating cold water through them. The spacing of the pipes may be from 0.5 m to 2 m. Thin walled 2.5 cm external diameter tubes are used for this purpose.

13.16 FOUNDATION TREATMENT OF THE GRAVITY DAMS

The foundation of the gravity dam should be hard, strong, durable and impervious. Imperviousness of foundation is very important as uplift pressure depends greatly upon the seepage. Uplift pressure is increased when seepage is more. Hence to render the foundation water proof or impervious, it has to be suitably treated. All the loose overlying soil from the site is removed and solid rocky foundation is reached. The rocky foundation should also be excavated for some depth so that the proposed dam fits in the rock. This aspect will prevent the sliding of the dam over its foundation. A 3 cm thick layer of rich cement mortar should be laid on the excavated rocky foundation before concreting is done over it. All the faults, seams, cavernous rocks, crushed zones etc., should be either made good or removed from the foundation site. In order to prevent seepage through foundation, trenches may be excavated near heal of the dam and filled with cement concrete. Holes are drilled covering whole of the foundation and filled with cement grout. If foundation is to be treated for large depths, the holes are also drilled for large depth. The holes are filled with cement grout in stages. Any other fault noticed in the foundation is rectified.

13.17 ADVANTAGES OF CONCRETE DAMS

There are some advantages of concrete dams described as following:

1. Maintenance cost of gravity dams is very small.
2. Spill ways can be installed in the dam itself and no separate site for them is required.
3. They can be constructed for very large heights.
4. Ice and other outer effects do not affect the stability of the dam.
5. Water is not lost by seepage.
6. Outlet sluices may be installed in the dam.
7. At valleys where side slopes are very steep, only this dam is found as the most suitable choice.
8. This dam gives pre-warning before failure. If timely measures are taken, the dam may even be made safe.

13.18 DISADVANTAGES OF GRAVITY DAMS

There are some disadvantages of concrete dams as below :

1. They are very costly.
2. They require very skilled labour for construction.
3. They have to be continuously cured during construction.
4. Large calculation work is envolved in their design.
5. They require solid hard foundation.
6. The height of the dam cannot be increased later.

13.19 SUPPLY SLUICES IN GRAVITY DAMS

Some openings in the dam have to be provided so as to pass the excess flow *D/S* of the dam. These openings are known as outlet sluices or supply sluices. If water from the reservoir is to be released for irrigation purpose at a controlled rate, it is done with the help of these supply sluices. All the supply sluices are fitted with gates which can be raised or lowered. The control on the gates is exercised from the top of the dam. The gates are fitted in the grooves formed at the sides of the openings in the dam. When gates are lowered they stop flow of water and when raised they again start discharging water D/S. Sluices may be provided at more than one depth. By this, water can be drawn from different

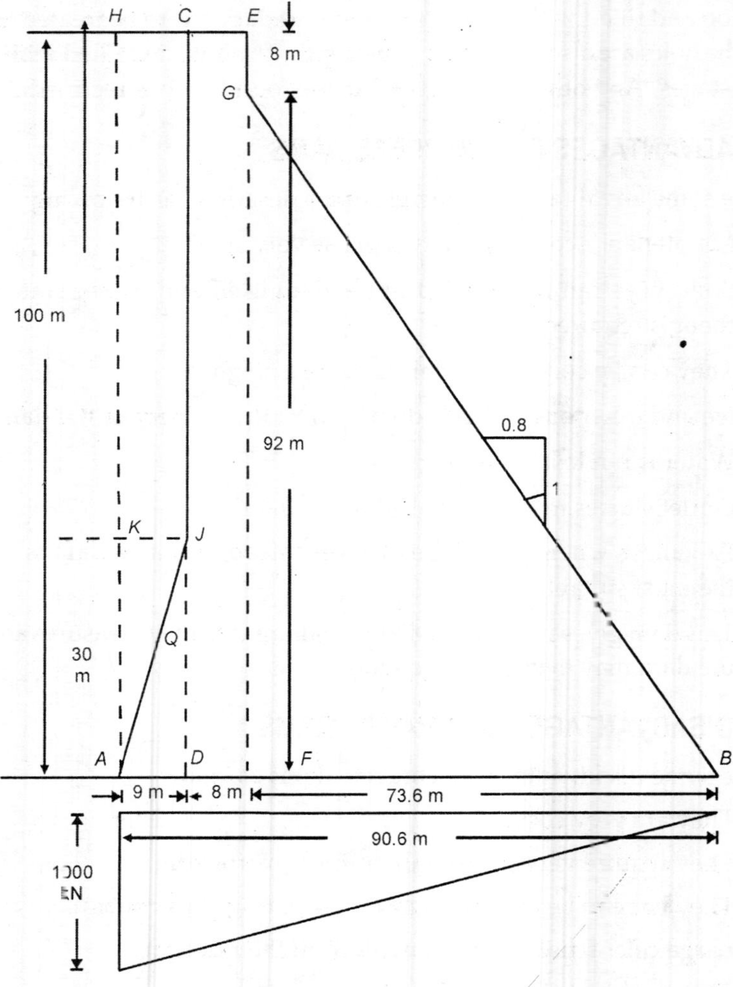

Fig. 13.22. *Supply sluices.*

elevations or depths. Supply sluices are also sometimes used to scour out the silt deposited in the vicinity of the U/S of the dam. See Fig. 13.22.

Example 13.1. *Determine the equations for base width of an elementary profile of gravity dam so that resultant passes through the outer middle third points. Consider earthquake force, hydrostatic pressure and uplift pressure for computations.*

Solution. (*a*) *Vertical forces*

I. Force due to self weight of dam $W = \frac{1}{2} b h w_p$

This force acts downwards.

II. Force due to vertical acceleration of earthquake

$$F = \alpha W = \alpha\, bh\, w_p$$

This force acts upwards.

III. Force due to uplift $u = \frac{1}{2} bwh$

This force also acts upwards.

Hence, total vertical force = $\Sigma V = W - \alpha W - U$

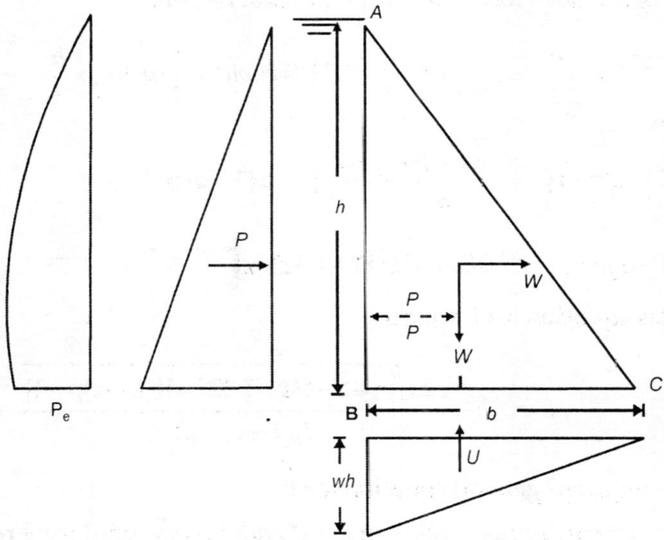

Fig. Ex. 13.1

$$= \frac{1}{2} bh\, w_p - \alpha\, bhw - \frac{1}{2} bhw$$

$$= \frac{1}{2} bhw[(1-\alpha)p - 1]$$

where w = unit weight of water

p = specific weight of concrete

a = coefficient of earthquake acceleration.

(*b*) *Horizontal forces*

I. Force due to water pressure $p = \frac{1}{2} wh^2$

II. Force due to hydrodynamic pressure of water at base

$$C_m = 0.735\frac{\theta}{90} = 0.735 \text{ since } \theta = 90^\circ$$

$$p_e = C_m\,\alpha\, wh = 0.735 \times \alpha\, wh$$

$$P_e = 0.726\, p_e\, h$$

$$M_e = 0.299\, pe\, h^2 = 0.299 \times 0.725\,\alpha\, wh^3 = 0.2205\,\alpha\, wh^3$$

III. Inertia force horizontal = $\alpha w = \frac{1}{2}\alpha\, bhwp$

If the resultant of all the forces has to pass through the outer third point M_2, the moment of all these forces at this point must be zero

$$\Sigma V \times \frac{b}{3} = \frac{1}{2}wh^2 \times \frac{h}{3} + 0.2205\,\alpha wh^3 + \frac{1}{2}\alpha bhw_1 \times \frac{h}{3}$$

$$\frac{b^2hw}{6}[(1-\alpha)p-1] = \frac{bh^2wp\alpha}{6} + \frac{wh^3}{6}[1+\alpha(1.323)]$$

$$b^2[(1-\alpha)p-1] = bhp\,\alpha + h^2[1+1.323\,\alpha]$$

Solving this equation for b we get

$$b = h\frac{p\alpha \pm \sqrt{p^2\alpha^2 + 4(1+1.323\alpha)\{(1-\alpha)p-1\}}}{2\{(1-\alpha)-1\}}$$

This is the required general equation for b.

When there is no earthquake, value of $\alpha = 0$ and the equation for b reduces to the form $b = \dfrac{h}{\sqrt{p-1}}$.

Example 13.2. *Height of gravity dam is 120 m, free board 2 m and slope of U/S face is 0.15:1. If the value of α is 0.1 find out*

1. Hydrodynamic earthquake pressure

2. Its moment at a joint located 60 m below the maximum water surface 0.

Solution. Let ϕ be the angle, the *U/S* face of the dam makes with vertical

$$\phi = \tan^{-1}\frac{0.15}{1} = 8^\circ 30'$$

Hence $$\theta = 90 - 8^\circ\,30' = 81^\circ\,30' = 81.5^\circ$$

$$C_m = 0.735 \times \frac{81.5}{90} = 0.666$$

Hence $h = 120 \text{ m}, y = 60 \text{ m}$

$$C_y = \frac{C_m}{2}\left[\frac{y}{h}\left(2-\frac{y}{h}\right)+\sqrt{\frac{y}{h}\left(2-\frac{y}{h}\right)}\right]$$

$$= \frac{0.666}{2}\left[0.5\,(2-0.5)+\sqrt{0.5(2-0.5)}\right]$$

$$= 0.538$$

Now $p_{ey} = C_y \alpha wh = 0.538 \times 0.1 \times 1000 \times 120 = 6456 \text{ kg/cm}^2$

$P_{ey} = 0.726\, p_{ey}\, Y = 0.726 \times 6456 \times 60 = 2.81 \times 10^5 \text{ kg}$

$M_{ey} = 0.299\, p_{ey}\, y^2 = 0.299 \times 6456 \times (60)^2 = 168 \times 10^7 \text{ kg} = \text{m}$

6.9×10^6 kg-m.

Example 13.3. *A concrete dam 12 m high is trapezoidal in shape. Its top width is 2 m and bottom width 10 m. The exposed face to water has a batter of 1:10. Test the stability of the dam. Also find out the principal stresses at the toe and heel of the dam. Assume unit not of the concrete 25000 N/m³, unit weight of water is 10000 N/m³ and permissible shear stress is 1.4 N/mm². It is assumed that water is likely to reach to the top of the dam.*

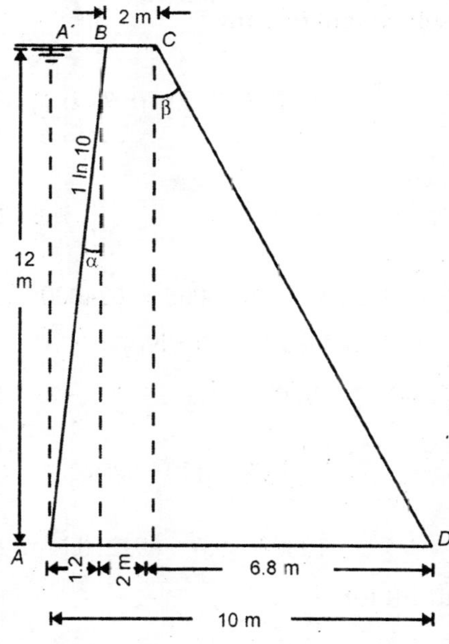

Fig. Ex. 13.3

Solution. Consider 1 m length of the dam

1. Vertical forces

(a) self weight of the dam.

$$= \frac{10+2}{2} \times 12 \times 1 \times 25000$$

$$= 1800000 \text{ N(acts downward)}$$

(b) Weight of water enclosed is triangle $AA'B$.

$$= \frac{12 \times 1.2}{2} = 10000 = 72000 \text{ N (acts downward)}$$

(c) Uplift force $\quad U = \frac{12 \times 10000}{2} \times 10 = 600000 \text{ N (acts upward)}$

Net downward $\Sigma V = 1272000$ N

2. Horizontal water pressure

$$\Sigma H = \frac{wh^2}{2} = \frac{1000 \times (12)^2}{2} = 720000 \text{ N}$$

Moments due to various force at Toe point D

Consider overturning moment at negative and restoring moment as positive

(a) Moment due to self weight of dam

$$= \frac{1}{2} \times 1.2 \times 12 \times 25000 \times (8.8 + 0.4) + 2 \times 12 \times 25000 \times 7.8$$

$$+ \frac{1}{2} \times 6.8 \times 12 \times 25000 \left(\frac{2}{3} \times 6.8 \right)$$

$$= 1656000 = 4680000 + 4624000 = 10960000$$

$$= 10960000 \text{ N-m. (+ ive)}$$

(b) Moment due to water triangle $AA'B$

$$= \frac{12 \times 1.2}{2} \times 10000 (10 - 0.4)$$

$$= 7.2 \times 10000 \times 9.6 = 691200 \text{ N-m (+ ive)}$$

(c) Moment due to uplift force

$$= 600000 \times \frac{2}{3} \times 10 = 4600000 \text{ N-m (– ive)}$$

(d) Moment due to horizontal water pressure

$$= 720000 \times \frac{12}{3} = 2880000 \text{ N-m (–ive)}$$

$$\Sigma M = 10960000 + 691200 - 4000000 - 2880000$$

$$= 4771200 \text{ N-m}$$

Calculation of factor of safety

Factor of safety against over turning

$$= \frac{M_r}{M_o} = \frac{+M}{-M} = \frac{11654200}{6880000} = 1.693 < 2 \text{ (Unsafe)}$$

Factor of safety against sliding

$$= \frac{\mu \Sigma V}{\Sigma H} \quad \text{Assume } \mu = 0.7$$

$$= \frac{0.7 \times 1272000}{720000} = 1.236 > 1 \text{ (Safe)}$$

Safe friction factor $= \dfrac{\mu \Sigma V + bq}{\Sigma H}$

$$= \frac{0.7 \times 1272000 + 10 \times 1.4 \times 10^6}{720000}$$

$$= \frac{14890400}{720000} = 20.68$$

Calculation of stresses

The resultant acts at a distance $\bar{x}$ from toe

$$\bar{x} = \frac{\Sigma M}{\Sigma V} = \frac{4771200}{1272000} = 3.75$$

Its distance from centre is

$$e = \frac{b}{2} - \bar{x} = 5 - 3.75 = 1.25 \text{ m}$$

Compressive stress at Toe is

$$p_n = \frac{\Sigma V}{b}\left(1 + \frac{6e}{b}\right)$$

$$= \frac{1272000}{10}\left(1 + \frac{6 \times 1.25}{10}\right)$$

$$= \frac{1272000}{10} \times 1.75 = 222600 \text{ N/m}^2$$

Compressive stresses at heel

$$= \frac{\Sigma V}{b}\left(1-\frac{6e}{b}\right) = \frac{1272000}{10}\left(1-\frac{6\times1.25}{10}\right)$$

$$= \frac{1272000}{10} \times 0.25 = 31800 \text{ N/m}^2$$

From Fig. Ex. 13.3 $\tan\alpha = \frac{1}{10}$, threfore, $\sec\alpha = \sqrt{1.0}$

$$\tan\beta = \frac{6.8}{12} \quad \because \sec\beta = \frac{13.70}{12} = 1.149$$

Principal stress at the toe of the dam

$$p_n \sec^2\alpha = 222600 \times (1.149)^2$$

Principal stress at heel

$$= p_n \sec^2\alpha - p\tan^2\alpha$$

$$= 31800 \times 1.01 - 10000 \times 12 \times 0.01$$

$$= 32118 - 1200 = 30918 \text{ N/m}^2$$

Shear stress at Toe $Z = p_n \tan\beta$

$$= 222600 \times \frac{6.8}{12}$$

Shear stress at heel $= (p_n - p)\tan\alpha$

$$= -(31800 - 10000 \times 12)\frac{1}{10}$$

$$= +8820 \text{ N/m}^2$$

Example 13.4. *A concrete dam is shown in Fig. Ex. 13.4. Check the stability of the dam find out the values of sliding factor and shear factor. Also determine the magnitude and direction of principal stresses normal stress and shear stress at toe and heel. Effect of only horizontal earthquake is to be considered for which $\alpha = 0.1$ and $C_m = 0.70$. Reservoir is full of water unit weight of concrete 25 kN/m^3 unit shear for concrete = 1.4 N/mm^2. Uplift pressure at the base of any horizontal section varies from full reservoir level pressure at U/S to zero or tail water pressure at d/s and is considered to act over $\frac{2}{3}$ area of section.*

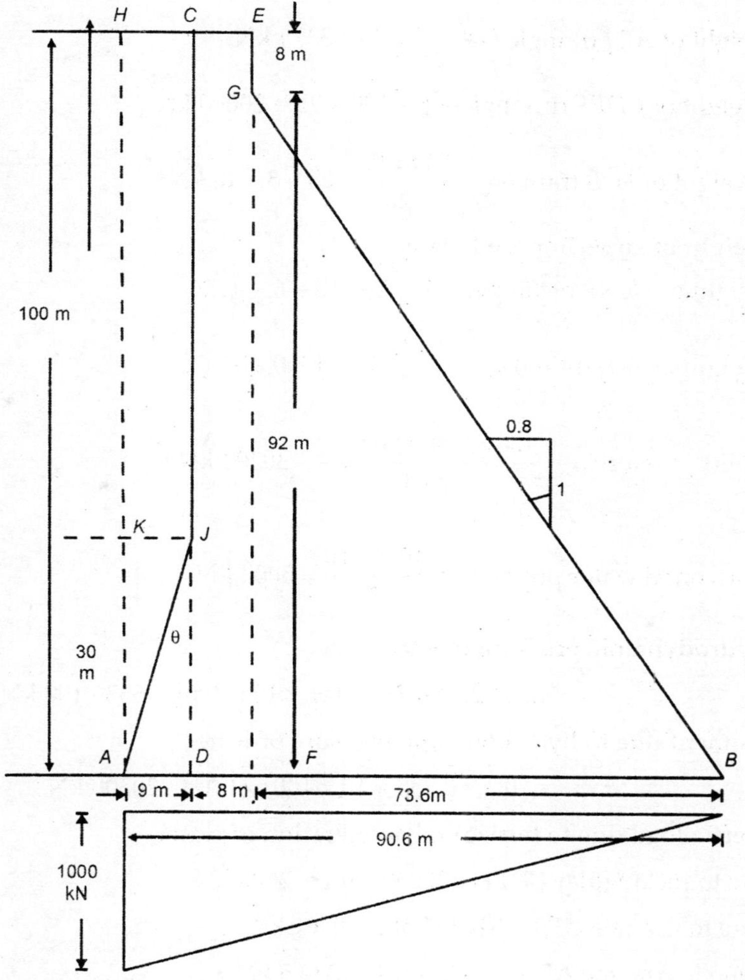

Fig. Ex. 13.4

Solution

Calculation of stresses

$$\tan \phi = 0.8$$

$$\sec \phi = \sqrt{1.64}$$

$$\tan \theta = \frac{9}{30} = \frac{3}{10} = 0.3$$

$$\sec \theta = \frac{31.32}{30}$$

1. Self weight of the dam

I. Weight of *ADJ* traingle = $\frac{9 \times 30}{2} \times 25 = 3375$ kN.

II. Weight of *CDEF* rectangle = 8 × 100 × 25 = 20000 kN

III. Weight of *FGB* traingle = $\frac{73.6 \times 92}{2} \times 25 = 84640$ kN.

2. Weight of superimposed water

I. Weight of *HCKS* rectangle = 9 × 70 × 10 = 6300 kN.

II. Weight of *AJK* traingle = $\frac{30 \times 9}{2} \times 10 = 1350$ kN.

3. Uplift pressure $u = \frac{100 \times 1000 \times 90.6 \times 2}{2 \times 3} = 30200$ kN.

4. Horizontal water pressure = $\frac{100^2 \times 10}{2} = 5000$ kN.

5. Hydrodynamic pressure due to water

$= 0.726 \times 0.735 \times 0.1 \times 10 \times 100^2 = 5336.100$ kN

6. Moment due to hydrodynamic pressure of water

$= 0.3 \times 0.735 \times 10 \times 100^3 \times 0.1 = 220500$ kN

7. Inertia load due to horizontal acceleration of water

I. Due to rectangular *CDEF* = 20000 × 0.1 = 2000 kN

II. Due to triangle *GFB* = 84640 × 0.1 = 8464 kN

III. Due to triangle *ADS* = 3375 × 0.1 = 337.5 kN

Distance where resultant acts from Toe

$$\bar{x} = \frac{\Sigma M}{\Sigma V} = \frac{2577.725 \times 10^3}{854.65} = 30.16 \text{ m}$$

Its distance e from the centre of the base is

$$e = \frac{b}{2} - \bar{x} \; \frac{90.6}{2} - 30.16 = 15.14 \text{ m}$$

Distance where resultant acts from Toe

$$\bar{x} = \frac{\varepsilon M}{\varepsilon V} = \frac{2577.725 \times 10^3}{85465} = 30.16 m.$$

Its distance e from the centre of the base is

$$e = \frac{b}{2} - \bar{x} = \frac{90.6}{2} - 30.16 = 15.14\text{m}$$

Normal stress at Toe

$$p_n = \frac{\Sigma V}{b}\left(1 + \frac{6e}{b}\right)$$

$$= \frac{85465}{90.6}\left(1 + \frac{6 \times 15.14}{90.6}\right)$$

$$= \frac{85465}{90.6} \times 2.002649 = 1889.14 \text{ kN/m}^2 \text{ (Compressive)}$$

Normal stress at heel

$$p_n = \frac{\Sigma V}{b}\left(1 - \frac{6 \times 15.14}{90.6}\right)$$

$$= \frac{85465}{90.6} \times 0.002649 = 2.5 \text{ kN/m}^2 \text{ (Tensile)}$$

Principal stress at toe (there is no Tail water)

$$\sigma_1 = p_n \sec^2(\phi) = 1889.14 \times 1.64 = 3098.2 \text{ kN/m}^2$$

Its direction is parallel to downstream force

Principal stress of heel is

$$\sigma_1 = p_1 \sec^2\theta - (p + p_c)\tan^2\theta$$

where p = water pressure 100 ´ 10 = 1000 kN/m²

p_e = hydrodynamic pressure at base

= $c\alpha wh$

$c = c_m = 0.70,\ \alpha = 0.1$

$p_e = 0.7 \times 0.1 \times 10 \times 100 = 70 \text{ kN/m}^2$

$$\sigma_1 = -2.5\left[\left(\frac{31.32}{30}\right)^2\right] - (1000 + 70)(0.3)^2$$

$$= -2.725 - 96.3 = -99.02 \text{ kN/m}^2 \text{ (Tensile)}$$

Its direction is perpendicular to U/S face.

$$\sigma_1 = 99.02 \text{ kN/m}^2 = \frac{99020}{(1000)^2} = 0.099 \text{ N/mm}^2$$

It is less than 0.4 N/mm² permissible tensile stress. Hence dam is safe since developed tensile stress 0.099 N/mm² is much less than the permissible value

$$\text{Shear stress at toe } c = p_n \tan \phi = 1889.14 \times 0.8$$

$$= 1511.3 \text{ kN/m}^2$$

Shear stress at heel

$$= p_n - (p + p_e) \tan \theta$$

$$= [2.5 - 1070] \times 0.3 = 321.75 \text{ kN/m}^2$$

Safety Factors

$$\text{(a) Against sliding} = \frac{\mu \Sigma V}{\Sigma H} = \frac{0.75 \times 85465}{66137.6} = 0.97 < 1 \text{ (Unsafe)}$$

(b) Shear friction factor

$$SSF = \frac{\mu \Sigma V - bq}{\Sigma H} = \frac{0.75 \times 85465 + 90.6 \times 1400}{66137.6}$$

$$= 2.89 < 4 \text{ or } 5 \text{ (hence unsafe)}$$

(c) Against overturning

$$= \frac{\Sigma M_r}{\Sigma M_o} = \frac{6651.16 \times 10^3}{4073.435 \times 10^3} = 1.63 < 2$$

Hence dam is unsafe for the pressure loading condition.

Example 13.5. *A diagram of a gravity dam is shown in Fig 13.4. Check the dam stability, shear friction factor and sliding. Find out the principal stresses, normal stresses and shear stresses at Toe and heel of the dam. Assume only horizontal effect of the earthquake. Following data is also given*

$$C_m = 0.73$$

$$\alpha = 0.1$$

Density of concrete 24 kN/m³

Safe shear stress for concrete 1.4 N/mm²

Uplift at U/S face is equal to pressure of water (hydrostatic pressure) and zero at the toe.

Uplift should be taken as acting on $\frac{2}{3}$ *area*

Factor of safety against sliding = 1.

Shear friction factor = 3.

Safety factor against overturning = 1.5

Safe compressive stress for concrete 2.5 N/mm².

S.N.	Classification of area	Forces (kN)		Lever Arm in mm	Moment about Toe kN-m	
		Vertical	Horizontal		+ive	–ive
	Self weight of the dam					
1.	Triabgle ADJ	3375		84.6	285.5×10^3	
2.	Rectangle CDEF	20000		77.6	1552×10^3	
3.	Triangle FGB	84640		$73.6 \times 2/3$	4153×10^3	
	Weight of superimposed water					
4.	Rectangle HCKJ	6300		86.1	542.4×10^3	
5.	Triangle AJK	1350		87.6	118.26×10^3	
6.	Uplift pressure	– 30200		60.4		1824×10^3
7.	Horizontal water pressure		50000	100/3		1666×10^3
8.	Horizontal Hydro-dynamic pressure due to water		5336.100			
9.	Moment due to hydro-dynamic pressure of water					
	Inertia load due horizontal acceleration					220.5×10^3K
10.	Due to CDEF rectangle		2000	50.0		10×10^3
11.	Due to GFB triangle		8464	92/3		259.56×10^3
12.	Due to ADJ triangle		337.500	10.0		3.375×10^3
		85465 kN	66137.6		6651.16×10^3	4073.435×10^3
					$\Sigma M = 2577.725 \times 10^3$ kN-m	

Solution Case I. Reservoir is full and uplift is also acting

1. *Weight of the dam and its moment about point A.*

(a) Rectangular part *GHKJ* = 8 × 90 × 1 × 24

= 17280 kN.

Distance of C.G. of rectangular section from *A* = 14 m.

Moment about point *A* = 17280 × 14

= 241920 kg-m.

(b) Weight of triangular part *JHF*

$$= 70 \times \frac{83}{2} \times 24 = 69720 \text{ kN}$$

Distance of C.G. from point $A = 18 + \frac{70}{3} = \frac{124}{3}$ m.

Moment about point $A = 69720 \times \frac{124}{3}$

= 2881760 kN-m.

(c) Consider triangle *ABG*.

$\text{Weight} = 27 \times \frac{10}{2} \times 2400 = 3240 \text{ kN}.$

Distance of C.G. from point $A = \frac{20}{3}$ m.

Moment about point $A = 3240 \times \frac{20}{3}$

= 21600 kN-m.

(d) Consider weight of water of rectangular part *BCDE*

Weight = 60 × 10 × 10 = 6000 kN.

Distance of C.G. from *A* = 5 m.

Moment about point *A* = 6000 × 5

= 30000 kN-m.

(e) Consider weight of water of triangular part *ABE*.

$\text{Weight} = 10 \times \frac{27}{2} \times 10 = 1350 \text{ kN}.$

Distance of C.G. from $A = \dfrac{10}{3}$ m.

Moment about point $A = 1350 \times \dfrac{10}{3}$

$= 4500$ kN-m.

2. Horizontal pressure due to water $P = \dfrac{wh^2}{2}$

$= \dfrac{10 \times (87)^2}{2}$

$= 37845$ kN.

Its moment about point $A = 37845 \times \dfrac{87}{3}$

$= 1097505$ kN-m.

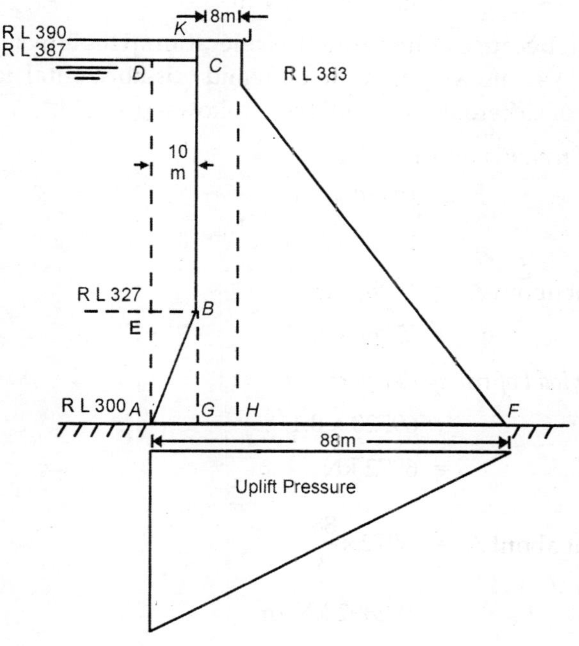

Fig. Ex. 13.5

3. Uplift pressure $= \dfrac{87 \times 10 \times 88}{2} \times \dfrac{2}{3}$

$= 25520$ kN.

Its moment about point $A = 25520 \times \frac{88}{3}$

$= 748586.67$ kN-m.

4. Earthquake effect

Hydrodynamic pressure due to water

$$P_e = 0.555\ \alpha w h^2$$

$$= 0.555 \times 0.1 \times 10 \times (87)^2$$

$$= 4200.80 \text{ kN.}$$

Distance of C.G. from the base of the dam

$$= \frac{4h}{3\pi} = \frac{4 \times 7}{22 \times 3} \times 87 = 37 \text{ m. (Approx.)}$$

Moment due to hydrodynamic pressure about point A

$$= 4200.80 \times 37$$

$$= 155429.60 \text{ kN-m.}$$

Inertia load because of horizontal acceleration. This effect is taken as equal to αW, when W is the weight of the dam and α is horizontal acceleration. The effect of load of different parts will be as follows. $\alpha = 0.1$.

(*a*) *Effect of rectangular part GHKJ,*

$= 17280 \times 0.1$

$= 1728$ kN

Its moment about $A = 1728 \times 45$

$= 77760$ kN-m

(*b*) *Effect of load of triangular part HJF*

$= 69720 \times 0.1$

$= 6972$ kN

Its moment about $A = 6972 \times \frac{88}{3}$

$= 192892$ kN-m

(*c*) *Effect of small triangular part ABG* $= 3240 \times 0.1$

$= 324$ kN

Its moment about point $A = 324 \times \frac{27}{3}$

$= 2916$ kN-m

Net vertical force $\Sigma V = 17280 + 69720 + 3240 + 6000 + 1350 - 25520$

$= 72070$ kN

Algebraic sum of all the moments ΣM

$= 241920 + 2881760 + 21600 + 30000 + 4500$

$+ 1097505 - 74858667 + 155429.60 + 77760$

$+ 192892 + 2916$

$= 395769.53$ kN-m

Distance (z) of noint of resultant from A

$$z = \frac{3957695.93}{72070} = 54.9 \text{ m}$$

Eccentricity of load $e = 54.9 - 44 = 10.9$ m

$$\text{Compressive stress at toe} = \frac{V}{b}\left(1+\frac{6e}{b}\right)$$

$$= \frac{72070}{88}\left(1+\frac{6\times 10.9}{88}\right)$$

$$= \frac{72070}{88}\times\frac{153.4}{88} = 14.28 \text{ kg/cm}^2$$

$$\text{Stress at heel} = \frac{72070}{88}\left(1-\frac{6\times 10.9}{88}\right)$$

$= 2.1$ kg/cm^2 (Compressive)

Principal stress $p_1 = 14.28 \sec^2$

$$\sec^2\phi = \frac{11789}{6889}$$

$$\therefore \quad p_1 = 14.28\times\frac{11789}{6889} = 24.42 \text{ kg/cm}^2 \text{ (Safe)}$$

It is not necessary to find out the principal stresses at heel.

Shear stress at toe $= p_n \tan\phi$

$$= 14.42\times\frac{70}{83} = 12 \text{ kg/cm}^2 \text{ (Safe)}$$

The dam will be safe in shear at heel also.

1. *Check against sliding :*

Assume value of $\mu = 0.75$

Factor of safety against sliding

$$= \frac{\mu \Sigma V}{\Sigma H} = \frac{72070 \times 0.75}{51069.80}$$

$$= 1.06 > 1 \text{ (hence safe)}$$

2. *Check against overturning*

When reservoir is full of water, the dam can overturn about toe i.e. point F. Overturning moment about

$$F, \Sigma M_o = 3023675.94 \text{ kN-m}$$

Resisting moment about

$$F, \Sigma M_r = 5408140 \text{ kN-m}$$

Factor of safety $= \dfrac{5408140}{3023675.94} = 1.79(\text{Safe})$

3. *Shear friction factor (S.F.F.)*

$$\text{S.F.F.} = \frac{\mu \Sigma V + bq}{\Sigma H}$$

$$= \frac{0.75 \times 72070 + 88 \times 1 \times 14 \times 100}{51069.80}$$

$$= 3.5 > 3 \text{ (Safe)}$$

Case II. *When reservoir is empty.*

The most critical condition takes place when reservoir has emptied but uplift is still acting at its full pressure. In this condition self weight of the dam and uplift both will be simultaneously acting.

In that case :

$$\Sigma V = 17280 + 69720 + 3240 - 25520$$

$$= 64720 \text{ kN}$$

Moments about point A i.e. ΣM

$$\Sigma M = 241920 + 2881760 + 21600 - 748586.67$$
$$+ 77760 + 192892 + 2916$$

$$= 2670261.33 \text{ kN-m.}$$

The position of that point from A where resultant ΣV will be acting.

$$Z = \frac{2670261.33}{64720} = 41.26 \text{ m.}$$

Eccentricity $e = 44 - 41.26 = 2.74$ m.

Maxi. stress at point $A = \frac{\Sigma V}{b}\left(1+\frac{6e}{b}\right)$

$$= \frac{64720}{88}\left(1+\frac{6\times 2{\cdot}74}{88}\right)$$

= 9.73 kg/cm^2 (Compressive)

Mini. stress at toe point $F = \frac{\Sigma V}{b}\left(1-\frac{6e}{b}\right)$

= 8.9 kg/cm^2 (Compressive)

In this case it is not essential to find out the principal as well as shear stresses. It is because magnitudes of the stresses are quite small in relation to the permissible stresses. Dam will not fail in sliding when reservoir is empty. Eccentricity being very small the dam is safe against overturning also.

Case III. *When reservoir is full but there is no uplift.*

Sometimes it so happens that value of stresses at toe reach maximum when reservoir is full and uplift is not there. Such a condition may happen when foundation of the dam is so impervious that seepage is just absent from it.

ΣV = 17280 + 69720 + 3240 + 6000 + 1350

= 97590 kN

ΣM = 241920 + 2881760 + 21600 + 30000 + 4500 + 1097505 + 155429.60 + 77760 + 192892 + 2916

= 470628260 kg-m.

Distance of resultant from point $A = \frac{470628260}{9759000}$

= 48.225 m.

Eccentricity = 48.225 – 44 = 4.225 m.

Normal compressive stress at toe

$$= \frac{\Sigma V}{b}\left(1+\frac{6e}{b}\right)$$

$$= \frac{9759000}{88}\left(1+\frac{6\times 4.225}{88}\right)$$

$$= \frac{9759000}{88\times 10^4}(1+0.288)$$

= 14.28 kg/cm^2.

Normal compressive stress at heel $= \frac{\Sigma V}{b}\left(1-\frac{6e}{b}\right)$

$$= \frac{9759000}{88\times 10^4}\left(1-\frac{6\times 4.225}{88}\right)$$

= 7.90 kg/cm^2

Principal stress at toe = $p_n \times \sec^2 \alpha$

$$= 14.28 \times \frac{11789}{6889}$$

$$= 24.42 \text{ kg/cm}^2 \text{ (Safe)}$$

Maxi. shear stress $= p_n \tan \alpha$

$$= 14.42 \times \frac{70}{83} = 12 \text{ kg/cm}^2 \text{ (Safe)}$$

The dam need not be checked for sliding and overturning as it would be safe in that if it is safe for condition (1) i.e, reservoir full and uplift acting on the dam.

Example 13.6. *Design the practical profile of a gravity dam of cement concrete. R.L. of base of dam 1050 m, R.L of HFL is 1092 m. S.G of cement concrete is 2.5 and safe compressive stress for cement concrete 1600 kN/m².*

Height of waves is 0.90 m.

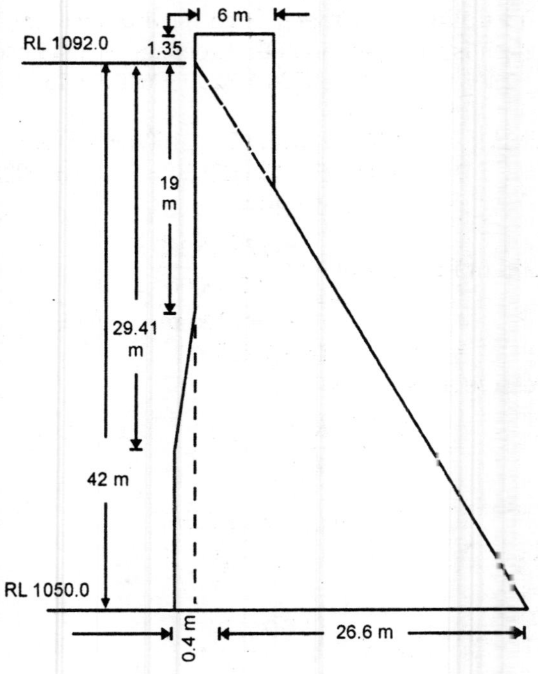

Fig. Ex. 13.6

h_w = 0.9 m; Free board = 1.5 h_w = 1.35 m

R.L. of top of dam = 1092.0 + 1.35 m = 1092.35 m

Height of dam = 1093.35 – 1050.0 = 43.35 m

Limiting height of dam

$$= \frac{f}{w(p+1)} = \frac{1600}{10(2.5+1)} = 45.71 \text{ m}$$

Since actual height of the dam is smaller than limiting height, hence dam is a low gravity dam.

Depth of water = 1092.0 – 1050.0 = 42 m

Top width can be determined on the following criterion

(i) a = 14% of h for economy

$0.14 \times 4335 = 6.07$ m = 6 m say

(ii) Width of road way if any

However keep top width a = 6 m

Base width of elementary profile

$$= \frac{h}{\sqrt{p}} = \frac{42}{\sqrt{2.5}} = 26.56 \text{ m}$$

$$= 26.6 \text{ m say}$$

U/S offset $= \frac{a}{16} = \frac{6}{16} = 0.4$ m

Total base width = 26.6 + 0.40 = 27 m.

Distance upto which *U/S* slope is vertical from the *U/S* water level

$$= 2a\sqrt{p} = 2 \times 6 \times \sqrt{2.5} = 18.97 \text{ m}$$

$$= 19 \text{ m say}$$

Designed profile is shown in Fig. 13.6

QUESTIONS

13.1. What do you understand by a gravity dam ? Enumerate various forces acting on it, with brief description of each force.

13.2. What do you understand by term hydrodynamic pressure ? What is effect of this pressure and how is it evaluated?

13.3. What are various modes of failure of a gravity dam ? Discuss each mode of failure briefly.

13.4. What are the various methods by which stability of the gravity dam is analysed? Explain each method briefly.

13.5. Explain the terms :

(i) Elementary profile of a low gravity dam.

(ii) Practical profile of a low and a high gravity dam.

(iii) Low gravity dam.

(iv) High gravity dam.

13.6. Explain the multiple step method of design of a gravity dam. How multiple step method differs from the single step method.

13.7. Why galleries are provided in the gravity dams ? What are the essential requirements of galleries ?

13.8. Explain various types of joints which are essentially provided in gravity dam.

13.9. Explain with the help of diagram various joints, water scales, and keys used in gravity dam.

13.10. How do you control the cracking of concrete gravity dam?

13.11. Find out the critical height of alow gravitydam if specific gravity of the concrete is 2.5 and allowable compressive stress as 400 t/m^2.

[**Ans.** 114.285 m]

13.12. Design the practical profile of a gravity dam made of stone masonry from following data :

R.L. of the base of dam	= 103.00 m.
R.L. of H.F.L.	= 183.00 m.
Specific gravity of masonry	= 2.4.
Safe compressive stress of masonry dam	= 150 t/m^2.

14

Spillways

14.1 INTRODUCTION

Spillway is a structure, used to pass surplus flood water from reservoir to the downstream side of the dam. It is sometimes also known as surplussing work, as its main job is to handle the surplus water. It is a sort of safety valve for a dam. During floods, the reservoir level goes on increasing, as more and more water enters the reservoir. If spillways had not been there, the level in the reservoir will go on rising even beyond the maximum reservoir level. This may lead to over topping of the dams. Additional stresses may be induced in the dam and this may cause failure of the dam. More rise in water level beyond the maximum flood level will cause submergence of vast additional areas and may cause suffering to the people living on the *U/S* side. Spillways act as safety valve for the safety of the dam and also for the adjoining areas. When water level in the reservoir reaches some specified level, the surplus water automatically starts escaping through the spillways. As more and more flood water enters the reservoir, there is more and more rise in water level and consequently more and more water starts escaping over the spillways. At maximum flood level (H.F.L.) of reservoir, the outflow rate over spillways reaches the inflow in the reservoir and thus further rise in the water level of the reservoir is stopped. The spillways should be of sufficient capacity so as to be capable to effectively handle the largest floods, without causing any damage to the dam or any appurtenant structure. Spillways are also designed according to the safe capacity of the *D/S* channel, otherwise *D/S* areas may get flooded. The design flood discharge required to be handled by the spillway can be found by flood routing as explained in Chapter 9.

14.2 TYPES OF SPILLWAYS

The spillways may be broadly classified as :

1. Emergency spillways, and
2. Main spillways.

14.3 EMERGENCY SPILLWAYS

Main spillway is the spillway which is called upon to work under normal conditions of flood. Emergency spillway is provided to deal with abnormal conditions of flood. Under normal flood conditions, the emergency spillways are not required to work. They may be called upon to work under following abnormal conditions.

(i) Rise is water level in the reservoir is continuing above the maximum permitted level even when main spillway and other outlet works are working at their full capacity. This happens when flood greater than design flood has occurred.

(ii) When, due to certain reasons, enforced shut down of the outlet works have to be applied and flood water is entering the reservoir with full vigour.

(iii) If spillway gates have got struck down and cannot be opened immediately.

In all the above three conditions the level of reservoir may rise above the maximum contemplated level and thus may cause failure of the dam and extra submergence on the *U/S* side. In all such conditions, the emergency spillways are opened. Emergency spillways are generally formed by lowering the crest of a dike section below that of the main embankment. Natural saddle, if available, is an ideal site for emergency spillway. Emergency spillways are constructed as low height earth embankments, but before earth embankments are made the surface of the saddle or lowered dike section is made pucca, so that when emergency spillway is opened it is not unnecessary eroded and widened.

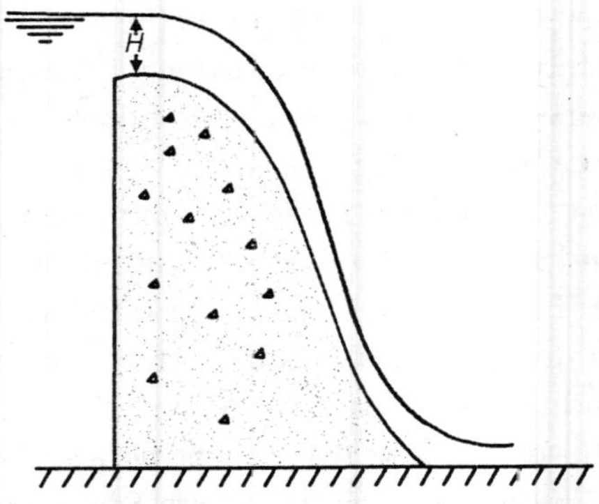

Fig. 14.1. *Overflow or Ogee spillway.*

The crest or top of the emergency spillway is kept at or little above the designed maximum reservoir water surface. When rising water level in the reservoir reaches the maximum high flood level and is still rising, the emergency spillway, which is an ordinary earth fill low dam, gets either automatically washed away or is deliberately cut. This causes very large amount of flood water to escape, and further rise in reservoir level stops. When flood waters are discharged *D/S*, and the reservoir level has dropped below the maximum flood level, the flow through the emergency spillway stops. After this main spillways are capable to handle the situation. When flow through emergency spillway has stopped, it is again filled with earth embankment and emergency spillway again restored. Since emergency spillway section either breaches itself or is deliberately cut, it is sometimes also known as *breaching section.*

14.4 MAIN SPILLWAYS

We have already stated that these spillways work under the designed flood conditions. Emergency spillways only help the main spillways to save the situation created by abnormal conditions. There are several types of spillways which may be used as main spillways. They are as follows :

1. Overflow or ogee spillway.
2. Free overfall or straight drop spillway.
3. Side channel spillway.
4. Chute spillway, which is also called trough or open channel spillway.
5. Tunnel spillway.
6. Siphon spillway.
7. Shaft, or drop inlet or morning glory spillway.

The spillways may be gated or ungated. Gated spillways are referred as controlled spillways while others as uncontrolled spillways. Shape, discharge features, and discription of each type of spillway is given one by one.

14.5 OVERFLOW OR OGEE TYPE SPILLWAY

This spillway consists of a control weir whose shape is ogee or *S*-shaped. The shape of the ogee spillway conforms to the profile of lower nappe of water, falling from a sharp crested weir. The profile of the spillway is so shaped that the overflowing water remains in touch with the spillway surface throughout the fall, when the rate of overflow corresponds to the maximum designed capacity.

When head on the spillway crest is less than the maximum designed head, the falling nappe would remain adhering to the profile of the spillway. This causes positive hydrostatic pressure and thus reduces the discharging capacity of the spillway. On the contrary, when head on the crest exceeds the maximum design head during unexpected large floods, the lower nappe of falling water may not remain in touch with the spillway surface. This phenomenon causes negative pressure and cavitation. Cavitation has a damaging effect which is caused at points where lower surface of the falling nappe does not remain in touch with the profile. At such points vacuum is developed which results in cavitation effect. In addition to cavitation, vibrations are also set up in the dam due to alternate making and breaking of the contact between falling nappe of water and profile of the spillway.

The D/S curve of the ogee follows following equation

$$x^{1.85} = 2H^{0.85} \times y \tag{14.1}$$

where x and y are the co-ordinates of the crest profile measured from the apex of the crest. H is the design head. Other elements of the crest as shown in Fig. 14.2 are as follows :

$$\left.\begin{aligned} r_1 &= \frac{H}{2} \\ r_2 &= \frac{H}{5} \\ a &= \frac{7}{40}H \\ b &= \frac{141}{500}H \end{aligned}\right\} \tag{14.2}$$

It should be taken care of, that the upper curve at the crest should be neither too sharp nor too broad.

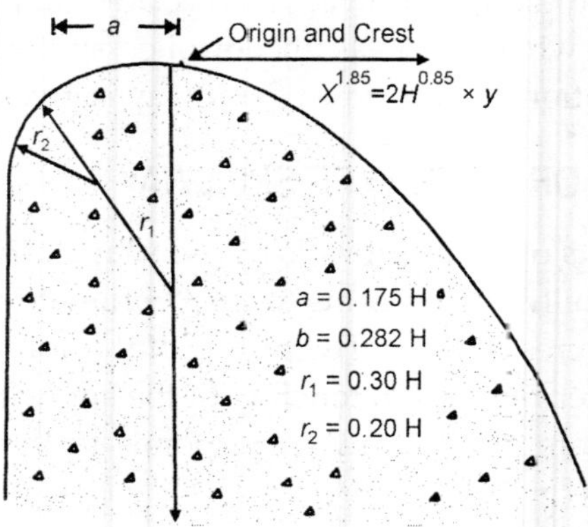

Fig. 14.2. *Crest details of ogee type spillway.*

Discharge over an ogee spillway is found out by following equation

$$Q = CL_eH_e^{1.5}$$

where Q = Discharge in cumecs

C = A factor whose value is influenced by several factors. Its value varies from 2.1 to 2.5.

L_e = Effective length of the crest.

H_e = Total head on crest including velocity head.

$$= H + \frac{V_a^2}{2g}$$

V_a = Velocity of approach in m/sec.

Effective length of the crest can be determined by following formula

$$L_e = L - 2\,(NK_p - K_a)\,H_e$$

where L = Net length of crest.

N = Number of piers.

K_p = Pier contraction coefficient whose value for rectangular or square nose is taken as 0.02 and for circular nose 0.01.

K_a = Abutment contraction coefficient whose value for square abutment is taken as 0.2 and for circular abutment as 0.1.

This spillway is the most common type of spillway used in gravity dams.

14.6 FREE OVERFALL OR STRAIGHT DROP SPILLWAY

It is a low height weir, whose D/S face is either vertical or nearly vertical. In this spillway the flow drops freely from the crest. The under side of the falling nappe is sufficiently ventilated to prevent a pulsating jet. The crest is also sometimes extended in the form of an over hanging slab on D/S side so as to keep the nappe of falling water away, from coming in contact with the D/S surface of the spillway. The falling jet of water hits the D/S floor almost vertically and thus lot of energy is generated due to impact. This energy can be dissipated by adopting any of the following measures.

(i) Provide concrete apron for some length on the *D/S* side.

(ii) Develop a small pool of water, below the falling jet of water, by constructing a small secondary dam at some distance on the *D/S* side.

(iii) Impact blocks of small heights are constructed on the *D/S* side. The energy is dissipated by means of turbulences caused by the impingement of the flow against the impact blocks.

(iv) By developing hydraulic jump.

This type of spillway is recommended for large heads.

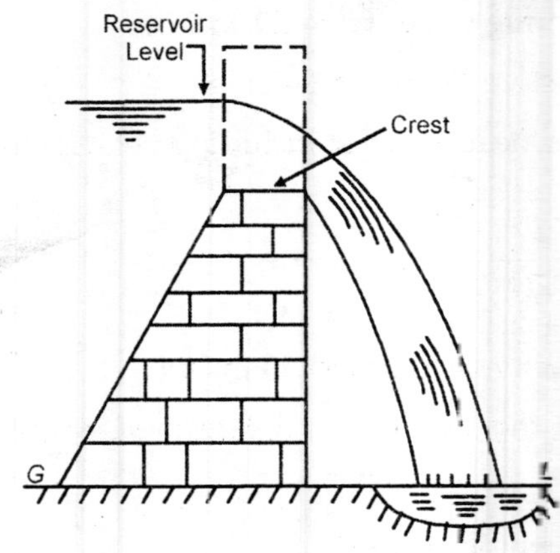

Fig 145.3. *Free overall or straght drop spillway.*

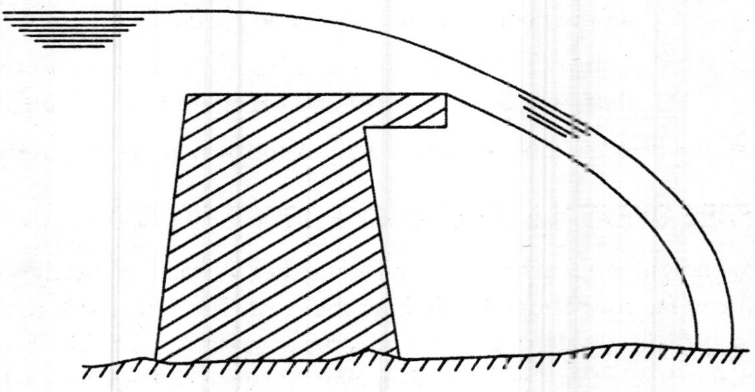

Fig. 14.4. *Straight drop spillway with over-hanging lip.*

The difference between ogee type and free overfill spillway can now be easily understood. In case of free overfall spillway, the water jet falls clearly away from the D/S face and space between D/S face and underside of the jet is ventilated.

In ogee spillway nappe of falling water remains in touch with the D/S surface of the spillway. In case of ogee spillway a smooth gradual reverse curvature is given on the D/S face of the spillway so as to cause dissipation of excess energy.

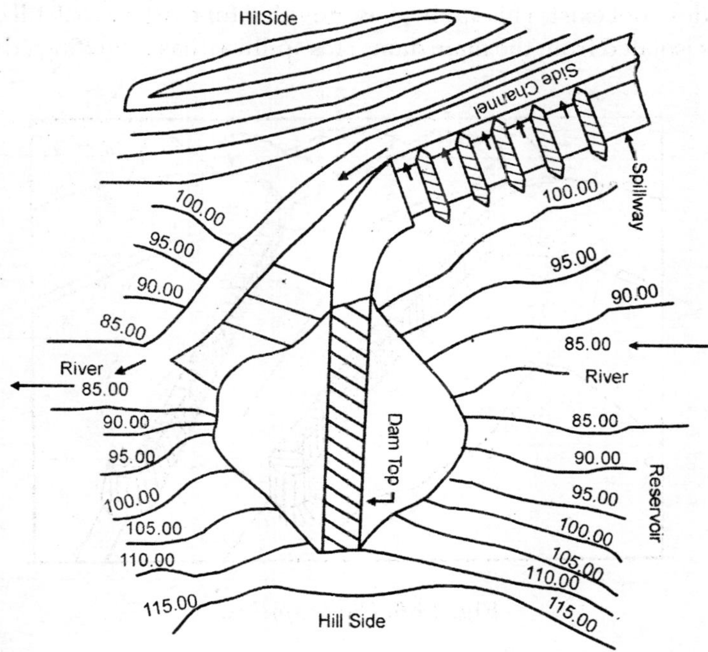

Fig. 14.5. *Side channel spillway.*

14.7 SIDE CHANNEL SPILLWAY

This spillway is suitable for earth or rock fill dams, in narrow canyons and other situations, where installation of direct overflow spillway is not possible. This spillway is also the best choice where a long overflow crest is required so as to limit the surcharge head and also under the situation where abutments are steep and precipitous. In this spillway the control weir is kept along the side and approximately parallel to the upper portion of the spillway. The flow after passing over the crest turns at approximately 90°. Flow may enter the side channel from one side only or from both the sides and one end also. Discharge characteristics are similar to ordinary overflow weir, except that at high discharge its crest may be partly submerged. Generally a trough is formed on *D/S* side of the crest and water enters it before taking a turn to flow in the side channel.

14.8 CHUTE SPILLWAY

It is also called open channel or trough spillway. It consists of a steep sloped open channel which is called trough or chute. It is placed in the dam abutment, or in a saddle. The discharge flown over the chute, is carried to the *D/S* of the dam through a natural channel which may be meeting the river on D/S side. The channel may be formed artificially or by excavating the trench, if natural channel does not exist. This spillway is provided for earth or rock fill dams and is always isolated from the main dam. This spillway has following advantages.

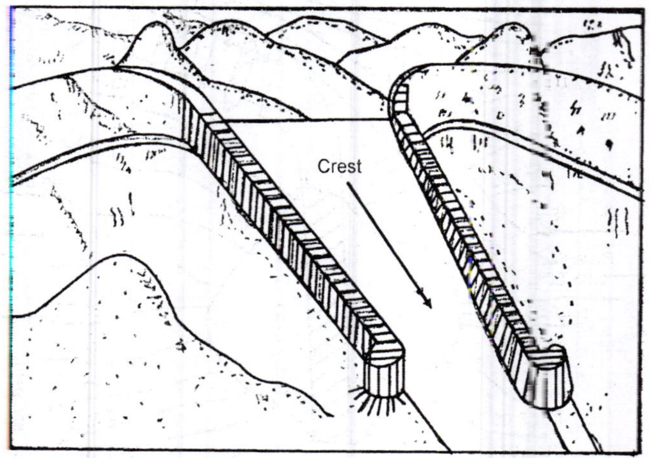

Fig. 14.6. *Chute spillway.*

(i) It is simple in design and construction.

(ii) It is adoptable to almost any foundation condition.

(iii) It is economical. See Fig. 14.6.

14.9 TUNNEL SPILLWAY

In this spillway a closed channel is used to convey surplus water from *U/S* to *D/S* of the dam. The closed channel may be vertical, inclined, or horizontal. When the closed channel is carried under the dam, it is known as conduit spillway. Conduit spillway is suited best for darn in wide valley. Tunnel spillway is considered best when dam site is narrow canyon. This spillway is designed to flow partly full. Full flow is not allowed as it will develop siphonic action and negative pressures may be created in the conduit.

14.10 SIPHON SPILLWAY

This spillway consists of one or a number of units of conduits formed by inverted U-tubes. The initial discharge through the conduits is similar to that of a weir, but when air is drawn away from above the end over the crest, the discharge starts flowing by siphonic action. The inside of the bend of inverted U-tube is at normal reservoir level. When water level in the reservoir rises, the water first flows as for the weir spillway and when air inlet is blocked by rising water, siphonic action starts and spillway starts discharging with full capacity. Spillway

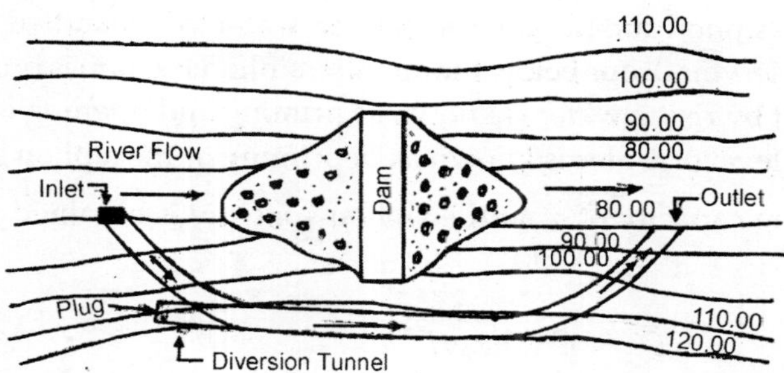

Fig. 14.7. *Tunnel spillway.*

continues to discharge with full capacity unless the reservoir level drops below the air inlet pipe, when air enters the conduit and breaks the siphonic action. Siphon spillway can be classified into two categories.

(i) Saddle siphon spillway.

(ii) Volute siphon spillway.

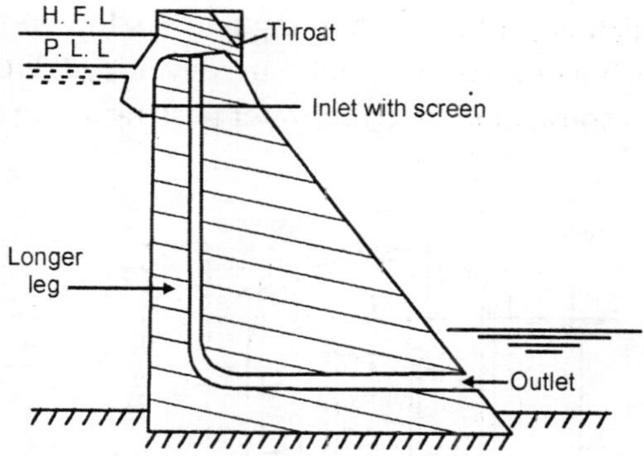

Fig. 14.8. *Saddle siphon spillway.*

1. Saddle Siphon Spillway. Saddle siphon spillway can be constructed in many designs but basic concept for all of them is the same. One types of saddle siphons is shown in Fig. 14.8. The diagram consists of a U-shaped siphon pipe whose D/S end remains submerged in tail water. An air inlet is provided near the top of the bend. Crest of the inside of U-tube is maintained at the normal reservoir level and air inlet slightly above this level. At normal reservoir level no waterflows through the siphon. During floods as the water level in the reservoir rises above the normal reservoir level, water will start flowing through the conduit of the spillway. When rising water level blocks the air inlet, the flow through the conduit immediately starts under siphonic head. A number of such siphonic units may be installed depending upon the diameter of the conduit of the siphon and the amount of flood water to be handled. When water level in the reservoir drops below the air inlet, siphonic action is broken. Closing of the air inlet by rising water is known as priming and opening of the air inlet by dropping level of water is known as depriming of the siphon spillway.

Discharging capacity of a saddle siphon spillway is obtained by following formula

$$Q = CA\sqrt{2gH}$$

where A = Area of cross-section at crown.

H = Operating head. It is taken as vertical height from reservoir level to the centre of the outlet if outlet is discharging in atmosphere and upto tail water level if outlet is submerged.

C = Coefficient of discharge whose average value is taken as 0.65.

Siphon spillways can be further classified as high head, medium head and low head siphons.

2. Volute Siphon Spillway. It is special type of siphon spillway. It comprises a vertical shaft which is bent at discharging end. Its top end consists of funnel shaped lips in which a number of volutes are developed. Due to volutes in the funnel shaped lip, a spiral motion to the water passing along them is set up. The

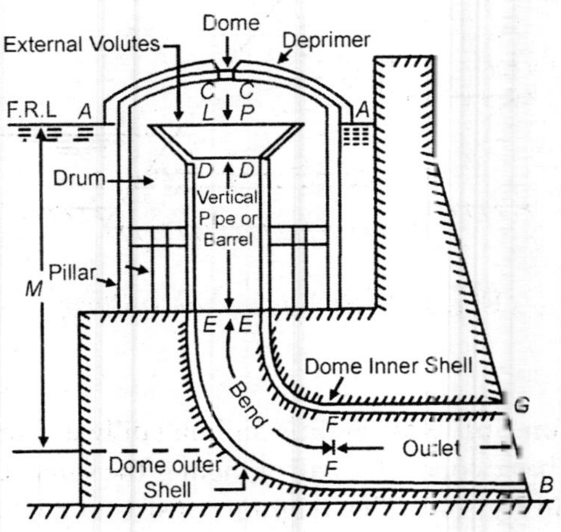

Fig. 14.9. *Volute siphon spillway*

funnel shaped lip remains covered by a concrete dome supported on number of pillars. A deprimer is also installed over the dome which connects the outer air with the funnel shaped lip. The outer end of the deprimer is kept just at the reservoir level. When water level in the reservoir rises, it closes the deprimer and entry of air is prevented. Now water flowing over the volutes sucks off the entrapped air due to spiral flow of water, and siphonic action starts and siphon start discharging at its full capacity. When reservoir level falls below the air inlet, air enters and breaks the siphonic action. Discharging capacity of this siphon is found out by following formula:

$$Q = A\sqrt{2g(H - H_L)} = CA\sqrt{2gH}$$

A = area of the cross-section of the pipe.

C = coefficient of discharge.

H = Maximum operating head.

H_L = Head loss through the siphon which is generally very small and may be ignored.

Advantages of Siphon Spillway

1. Since its discharge does not depend upon the head above the crest, it starts discharging with full capacity once siphonic action starts.
2. Since these spillways starts discharging with full capacity right from the beginning, the reservoir level will not rise much. This will cause less submergence of *U/S* areas during floods.
3. River training works of low heights will have to be constructed.

Disadvantages

1. They do not allow additional waters to be stored in the reservoir to augment the supplies at a later stage.
2. They cause lot of vibrations when in action. This may cause cracks in the dam and may lead to their ultimate failure.
3. They have to be maintained regularly.

14.11 SHAFT SPILLWAY

This spillway consists of a vertical shaft and a horizontal conduit. The top of the shaft is specially designed, through which water enters and then drops through a vertical shaft and finally gets carried to the *D/S* side by a horizontal conduit. This spillway is also sometimes known as drop inlet or morning glory spillway. The shaft spillway may be *standard crest type or flat crest type.* Both types are shown in Fig. 14.10. Small shaft spillway may be made from steel or concrete pipes, but large spillways are made from heavily reinforced pipes. Diversion tunnels used for diverting the river water during the construction of the dam may be plugged and then used as shaft spillway by attaching vertical shafts to the horizontal tunnels. Top flared inlet of the vertical shaft is known as *morning glory* and is considered as an essential feature of the shaft spillway.

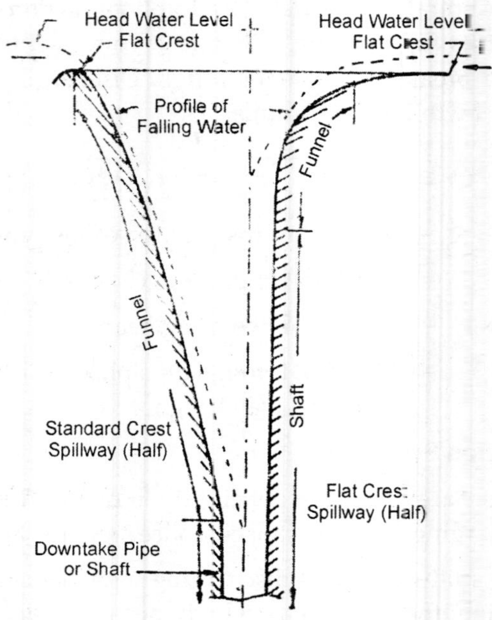

Fig. 14.10. *Shaft spillway.*

14.12 DYNAMIC FORCE ON SPILLWAYS

There is a continuous change in the velocity of water when water flows over the curved surface of ogee spillway. The continuous change in velocity phenomenon induces change in momentum from section to section and thus causes a force on the spillway structure. This force is known as dynamic force.

Consider two sections A and B of water on the curved surface. The resultant of the forces on the element of water is as follows.

$$\Sigma F = \rho Q \Delta V$$

or

$$\Sigma F = \rho Q(V_2 - V_1)$$

where $r = \dfrac{w}{g}$ = Mass density of water

Q = discharge

$V_2 - V_1$ = change in velocity from section B to A

Resolving equation $\Sigma F = \rho Q\,(V_2 - V_1)$ horizontally and vertically we get

$$\Sigma F_h = \rho Q\,(V_2 h - V_{1_h})$$

$$\Sigma F_v = \rho Q\,(V_{2v} - V_{1_v})$$

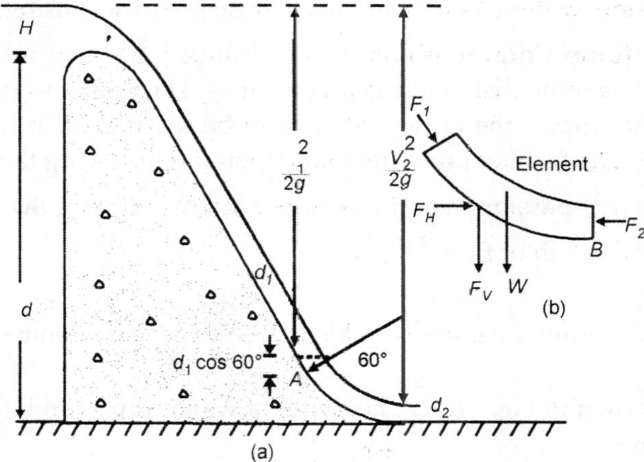

Fig. 14.11

The forces F_h and F_v are the forces acting on free body of water. These forces include gravity force, hydrostatic pressures and reactions of any object in contact with water.

14.13 ENERGY DISSIPATION BELOW SPILLWAYS

When stream of water moving with hyper-critical velocity, meets an other stream, moving with sub-critical velocity there is abrupt rise in the water level of the stream. This abrupt rise of water is known as hydraulic jump. Hydraulic jump is a very useful measure of dissipating the energy of flowing water.

Flood water at the level of the crest of the spillway has potential energy proportionate to the rise of the crest above the *D/S* floor of the spillway. When flood water passing over the crest of spillway reaches the bottom of spillway on *D/S* side, the potential energy of water is converted into kinetic energy, as velocity is greatly increased by the time water reaches from crest to *D/S* floor of the spillway. This high kinetic energy may cause very deep erosions on *D/S* if measures to disscipate it are not taken. There are several methods of dissipating the energy of the shooting flow of water. The following are the most important measures :

1. Formation of hydraulic jump.
2. Developing water cushion on *D/S* side.
3. Stilling basins with or without blocks of different sizes and shapes arranged in different manners.
4. Bucket type energy dissipators.

Where excess energy of gliding water is to be dissipated before the flow joins the *D/S* main river channel, the hydraulic jump is considered one of the most effective measures. If spillway discharge is directly being spilled into the river on *D/S* side, water cushion and other measures are considered best.

Hydraulic Jump Computations. The hydraulic jump that is formed at the stilling basin has some distinctive characteristics. The jump assumes a definite form depending upon the energy of flow to be dissipated in relation to the depth of flow. The form of jump depends upon the following factors :

(i) Discharge passing over the crest per metre length of the spillway (g),

(ii) Critical depth of flow (d_e), and

(iii) Froude number parameter which is $\frac{V}{\sqrt{gd}}$, various elements of the jump are shown in Fig. 14.12. The depth of water jump can be computed as follows.

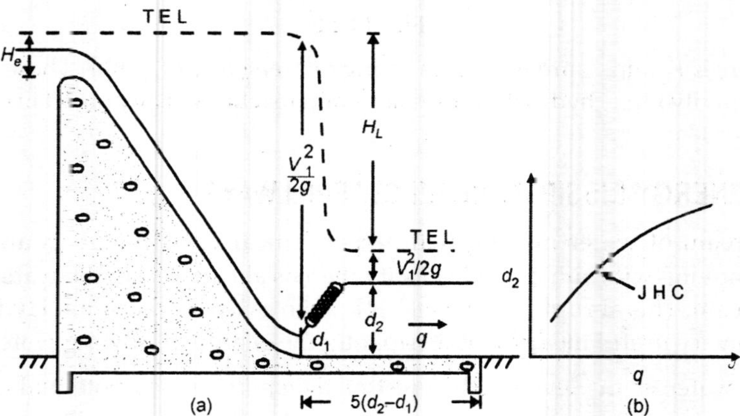

Fig. 14.12. *Hydraulic jump and J.H.C. curve.*

(a) For a given discharge q per metre length of the spillway, find out the total head of water (H_e) at the crest of the spillway including head due to velocity of approach.

$$H_e = \left(\frac{g}{c}\right)^{2/3}$$

U/S total energy line (T.E.L.) = crest level + H_e. If we consider that no loss of energy has taken place while transition from crest to toe, the specific energy (E_1) at toe of the spillway will be equal to the T.E.L. at *U/S*.

$$E_1 = U/S \text{ T.E.L.}$$

After having found out E_1 and q the depth (d_1) of water at pre-jump position can be found by trial and error method.

$$E_1 = d_1 + \frac{V_1^2}{2g} = d_1 + \frac{q^2}{2gd_1^2}$$

Determine Froude number F_1

$$F_1 = \frac{V_1}{\sqrt{gd_1}} = \frac{q}{\sqrt{gd_1^3}}$$

The depth of water at post jump (d_2) can be found out from following equations

$$d_2 = -\frac{d_1}{2} + \sqrt{\frac{2V_1^2 d_1}{g} + \frac{d_1^2}{4}} \qquad (1)$$

or

$$d_2 = \frac{d_1}{2}\left[\sqrt{1 + \frac{8q^2}{gd_1^3}} - 1\right] \qquad (2)$$

The values d_2 can be plotted for various values of q and a curve known as jump height curve (J.H.C.) is plotted as shown in Fig. 14.12 (b).

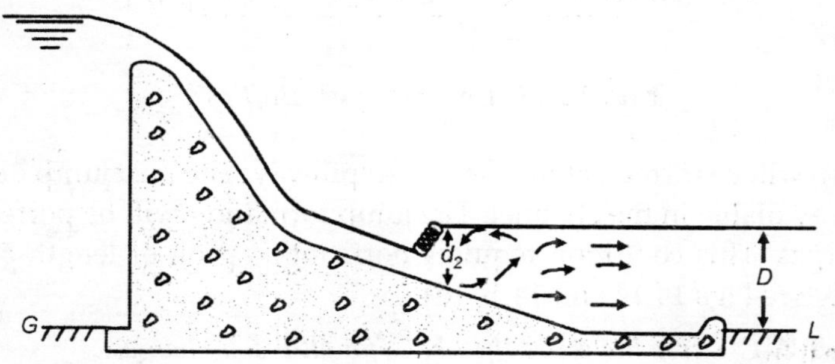

Fig. 14.13 *Slopping Glacis Above Bed*

The height of the tail water for each discharge, may or may not correspond to the height of perfect jump.

A curve relating depth of tail water (D) with discharge (q) may be drawn. Such curves are known as tail water curves (T.W.C.).

By comparing J.H.C. and T.W.C. following five conditions of jumps are possible.

1. Both the curves coincide.
2. T.W.C. curve lies above the J.H.C. for all the discharges.

3. T.W.C. lies below the J.H.C. for all the discharges.
4. T.W.C. lies above for small discharges and then below at higher discharge.
5. Reverse of case 4.

Case I. *Both the curves (J.H.C. and T.W.C.) coincide for all the discharges.*

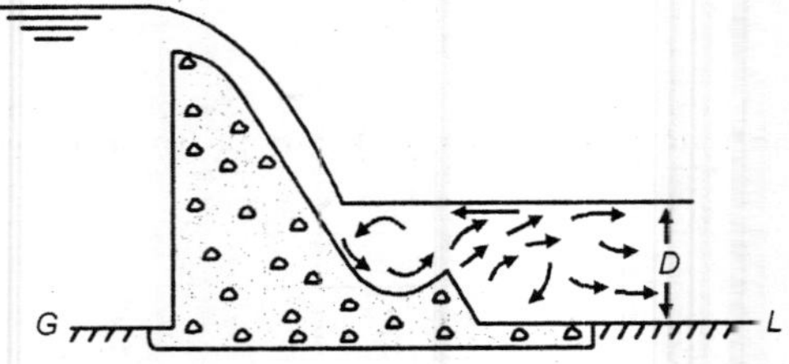

Fig. 14.14 *Upturned Bucket*

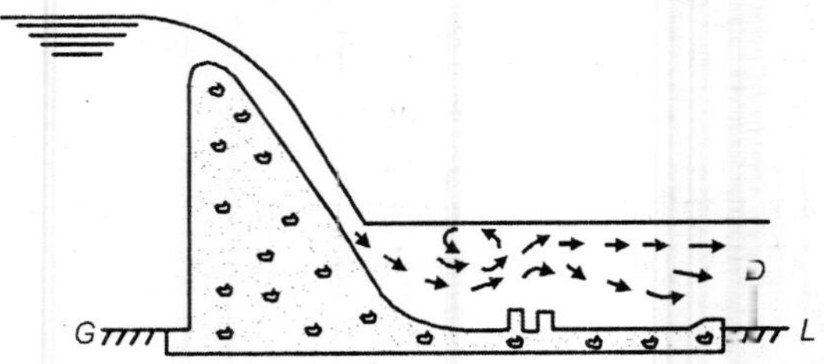

Fig. 14.15 *End Sill with Baffles*

The jump will be formed at the toe of the spillway. The post jump depth will be already available in the channel. The jump formation will be perfect for all the discharges. This condition requires horizontal apron of length $5(d_2 - d_1)$ beyond toe See Figs 14.12 and 14.16 (d).

Case II. *T.W.C. curve lies above the J.H.C. for all the discharges.*

In this case the value of jump height (d_2) is less than the tail water depth. In this case jump will be completely submerged. In this case very little energy will be dissipated. But this condition can be made effective by reducing the depth of tail water at the point of formation of the jump. See Figs. 14.13, 14.14, 14.15 and 14.16 (b). The protection works in this case may be in the form of :

(i) A sloping apron instead of horizontal apron. See Fig. 14.13.

(ii) Using a low level bucket which is sharply turned up. It acts as a deflector. See Fig. 14.14.

(iii) Providing end sills with baffles. In this case energy is dissipated by impact and friction. See Fig. 14.15.

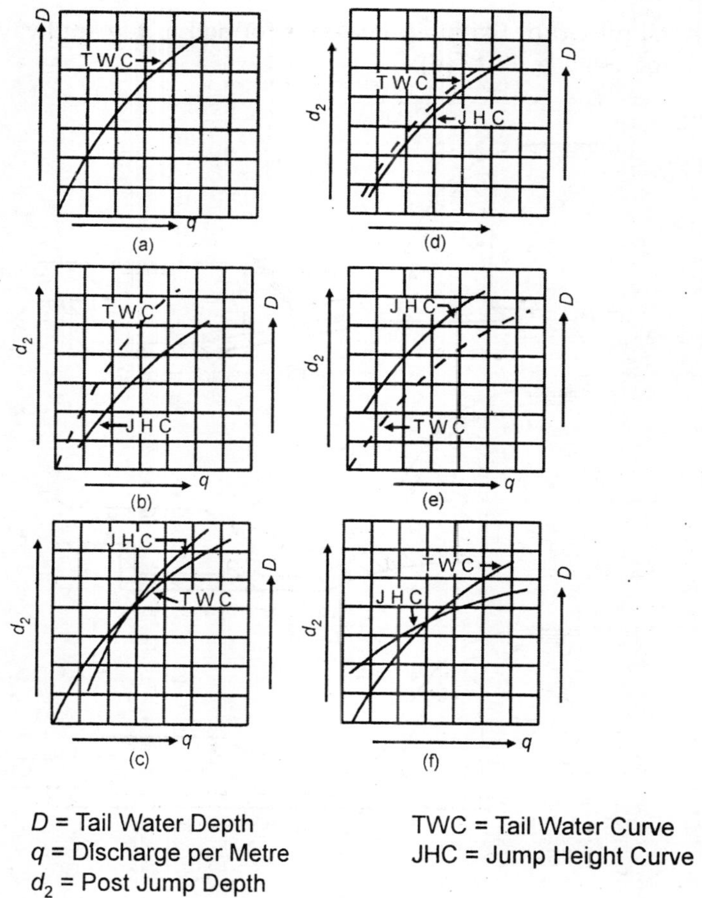

Fig. 14.16. *Positions of TWC and JHC curves depending upon the hydraulic jump.*

Case III. *T.W.C. lies below the J.H.C. for all the discharges.*

In this case depth of tail water level is smaller than the depth of the jump for all the discharges. The protection of the bottom can be accomplished by any of the following measures. See Fig. 14.16 (e).

(i) Provide a depressed cistern having its bed below the level of the bed level of the river. Toe of the spillway and bed of cistern are joined by slopping glacis. See Fig. 14.17 (a).

(ii) Providing a cistern like (i) but instead of joining toe and bed of cistern by sloping glacis, baffle walls are provided in the cistern. See Fig. 14.17 (b).

(iii) Baffles and low weir may be provided. See Fig. 14.17 (c).

(iv) Upturned or Ski-jump bucket is formed in the spillway structure at toe. See Fig. 14.17 (d).

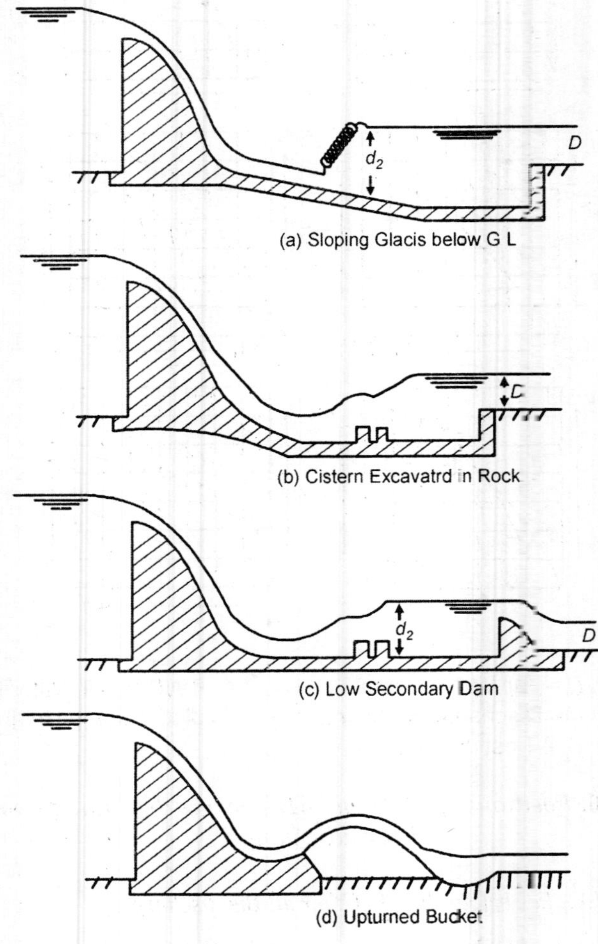

Fig. 14.17

Case IV. *T.W.C. lies above for small discharges and below for large discharges.*

In this case, jump depth d_2 is smaller than D for small discharges and larger for large discharges. Thus jump will remain submerged at low discharges whereas at high discharges tail water depth will be insufficient.

In this case a slopping apron partly below and partly above the bed level of river is constructed. Horizontal length of the apron is sufficiently provided. If need be, end sill is also provided. In this case jump will be formed at higher up point on the sloping apron for small discharges. At higher discharges the jump will be carried further on D/S side. See Figs 14.16 (c) and 14.18.

Case V. *T.W.C. lies below for small discharges and above for large discharges.*

This case is just the reverse of case IV. Protective measure is also the same as in case IV. The formation of jump will also be reverse in order i.e. at high up points for high discharges and at lower points for low discharges. See Fig. 14·16 (f).

IS 4997—1968 has given the criteria for design of hydraulic jump type stilling basins with horizontal as well as for sloping aprons.

14.14 SPILLWAY GATES

The spillways may be uncontrolled or controlled types. Uncontrolled spillways do not require any gate. Controlled spillways are provided with gates. Controlled spillways are considered superior to uncontrolled ones, as better control on the reservoir level can be exercised with gates. During low flows,

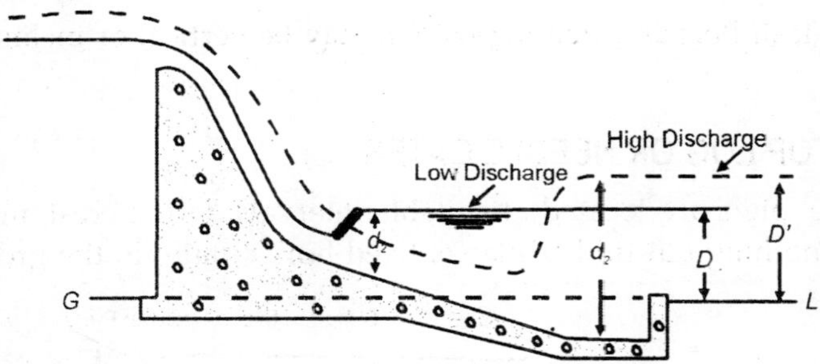

Fig. 14.18

additional storage can be done in the reservoir with the help of gates. However, during high floods gates are fully lifted or opened so that the full discharging capacity of the spillway may be used to dispose of flood water. In case of earth dams, spillways are not provided with gates. Slakness on the part of lifting the gates, may cause over-topping of the earth dam and may lead to its failure. Following are the various types of gates, which are commonly used :

1. Flash Boards
2. Stop log or needle gates
3. Tainter gates or radial gates
4. U.S.B.R. Drum gates
5. Bear tap gates
6. Vertical lift gates
7. Rolling gates
8. Trash rack
9. Flap gates
10. Fish Belly Flap gates

Brief description and sketches of all the gates are given here one by one.

14.15 FLASH BOARD

It is a temporary arrangement which is adopted to store extra water during low flows in the river. At large floods they are either removed or allowed to drop on

D/S side if hinged at the crest. Flash Boards are made by joining wooden planks which are fixed against the pins. The pins are installed over the crest at suitable

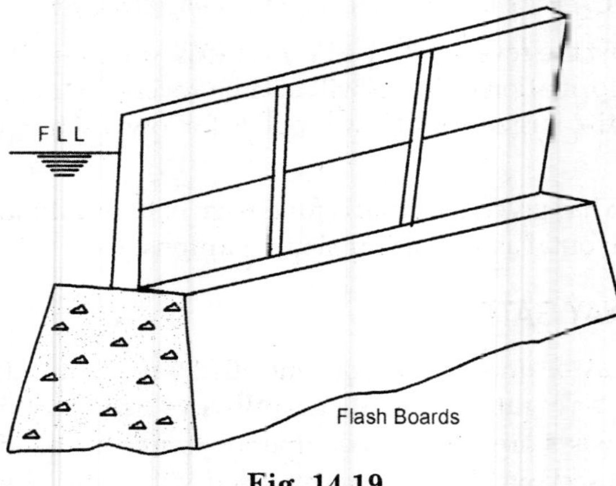

Fig. 14.19

intervals. Flash Boards when in position may be vertical or inclined. See Fig. 14.19.

14.16 STOP LOG OR NEEDLE GATES

In this case piers are located at suitable intervals on the crest and stop logs which are nothing but timber planks fixed horizontally in the grooves of the

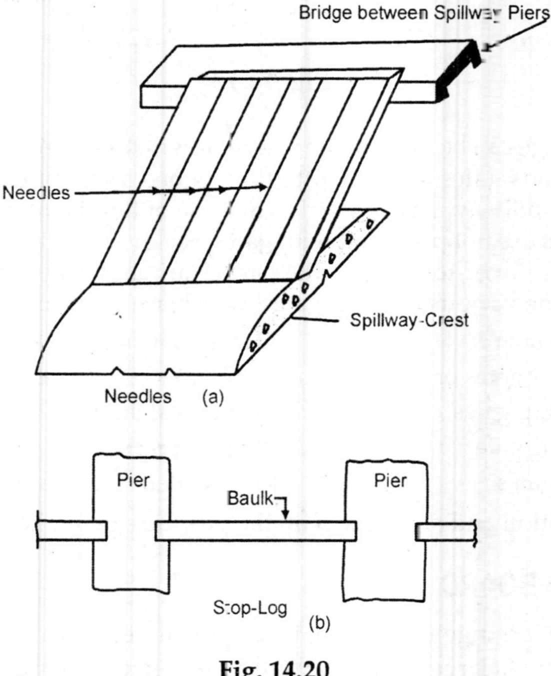

Fig. 14.20

piers, one above the other. While opening, the stop logs are removed one by one. See Fig. 14.20 (b). In the case of needle gates, wooden planks are not used horizontally but vertically. A groove is developed at the crest of the spillway. Bridging girder is also put parallel to the crest of the spillway. The wooden needles are fixed with their bottom in a groove in the crest and top against the bridging girder. See Fig. 14.20 (a).

14.17 TAINTER GATES OR RADIAL GATES

This gate is in the form of a sector of a circle. A curved plate is used to support water. The curved plate is given adequate support by a frame work of steel sections. The gate remains hinged at the piers, on both the ends. This gate can thus rotate about horizontal axis. Water pressure is ultimately transferred to the piers, through bearings fixed in piers. The tales are generally very heavy and can be lifted or lowered with the help of power driven winches only.

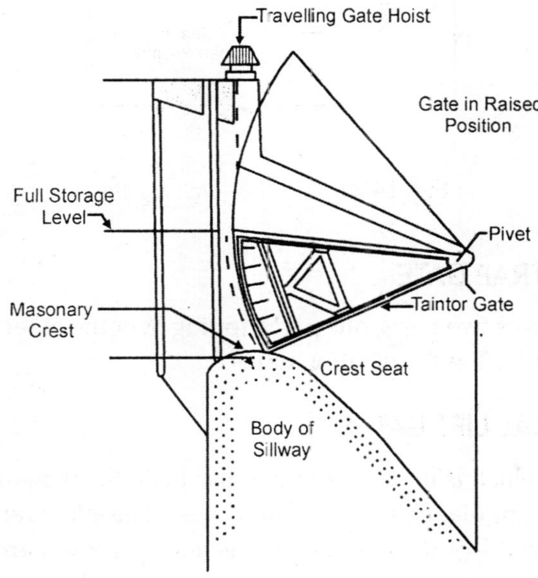

Fig. 14.21. *Radial gates.*

14.18 U.S.B.R. DRUM GATES

This gate remains hinged at the top of the crest of the spillway. This gate when open i.e. when not in position, remains swung into a ditch formed at the crest of the spillway. In open position drum of the gate itself forms the crest of the spillway while frame work supporting the skin plate remains in the ditch. See

Fig. 14.22 dotted position. The gate remains hinged at point A. The gates is put in closed position with the help of power driven winches only.

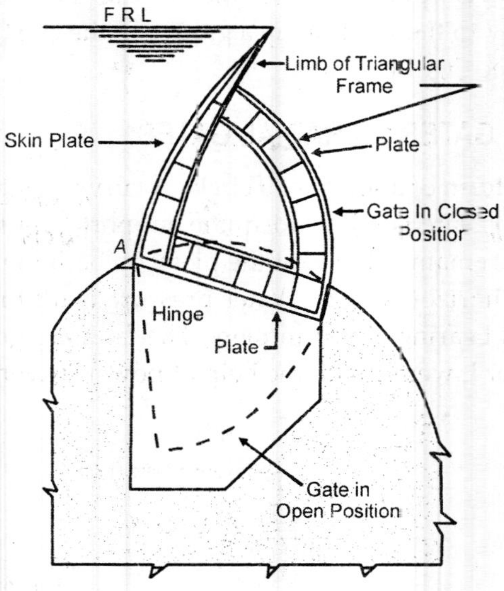

Fig. 14.22. *U.S.B.R. Drum gates.*

14.19 BEAR TRAP GATE

This gate consists of two parts, one part slipping over the other for some length. It is mostly used for low Navigation dams.

14.20 VERTICAL LIFT GATES

This is the gate which is in most common use in all the irrigation structures. It consists of a Rectangular skin plate of mild steel. This plate remains supported on a frame work of I-girders or channel sections. Grooves are constructed on the crest as well as along the inside of piers. The gates slide, or move, vertically up end down in these grooves. Water pressure acting on the gate is transferred to the crest and piers. In order to facilitate lifting and lowering of the gates, rollers are fitted in the grooves. The gates are lifted with the help of winches. Counter weights are also used to ease the lifting of the gates. Counter weights remain suspended with one end of the ropes. The other end of ropes being connected to the gates. If gates are very big in size, they may be constructed in more than one part and each part moves separately. These gates may be constructed in sizes varying form 6 m to 15 m height and 15 m width.

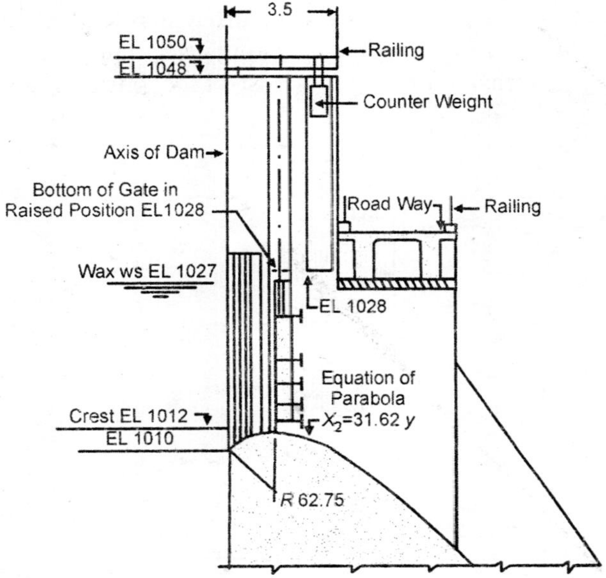

Fig. 14. 23. *Vertical lift gate.*

14.21 ROLLING GATE

This grate is shown in Fig. 14.24. It consists of a steel cylinder as large in diameter as the height of the opening and spanning between the piers. Each pier has an

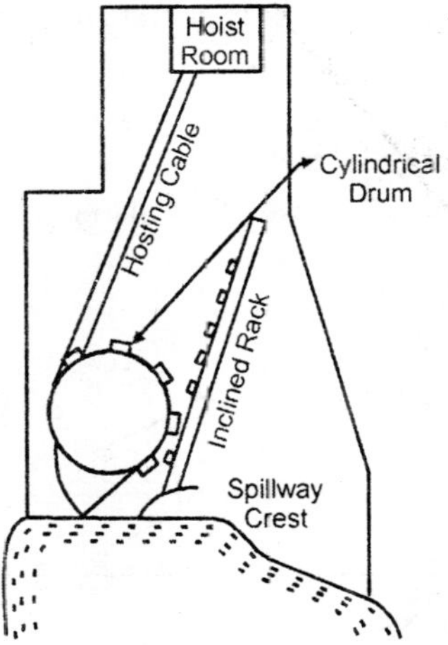

Fig. 14.24. *Rolling gate.*

inclined rack. The gate is wound round the periphery of the cylinder. The gate is also attached to the inclined rack with the help of toothed gears. The gate can be rolled while opening and unrolled which closing the gate.

14.22 TRASH RACK

It is not a gate but a sort of permanent structure which is installed at the mouth of the opening. Its main purpose is to prevent entrance of debris into the openings. A framed trash rack structure of steel sections or of R.C.C. members

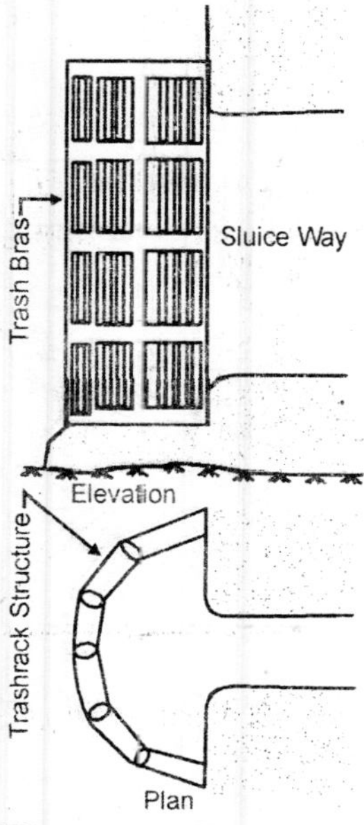

Fig. 14.25. *Trash rack*

is constructed and iron bars are put in the openings of the frame work to prevent entrance of debris. See Fig. 14.25. This structure is usually constructed at the entrances of sluice ways or penstocks etc.

14.23 FISH BELLY FLAP GATE

This gate is shown in Fig. 14.26. This gate is also known as Bascule type of gate.

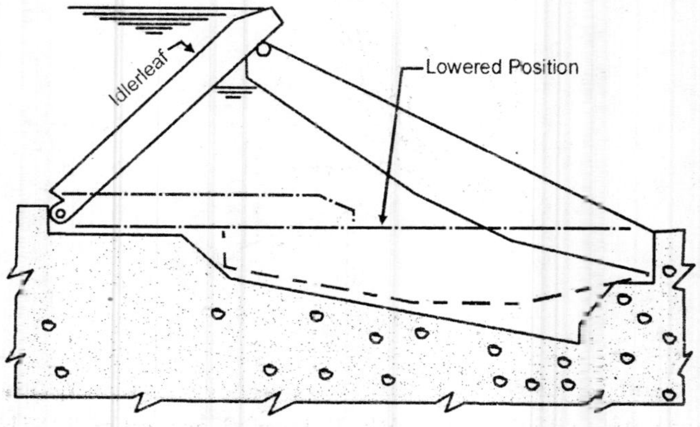

Fig. 14.26. *Fish-belly flap gate.*

This is used at the top of the weir crest to store extra storm water. The gate is fitted on the crest with the help of hinged joint. It is operated with the help of a lever rod.

Example 14.1 *Head of water over the crest of ogee spillway is 3 m and coefficient of discharge 2.5. Weir is 100 m long and height of crest above the base of the approach channel is 10 m. Width of approach channel is equal to length of the weir. Find out the discharge passing over the spillway.*

Solution. $Q = CLH^{3/2}.$

Effect of approach velocity and end contractions has been neglected.

$$Q = 2.5 \times 100 \times 3^{3/2} = 1300 \text{ cumec.}$$

Velocity of approach

$$V_a = \frac{Q}{\text{Head} \times \text{width of channel}}$$

$$= \frac{1300}{(10+3)100} = 1 \text{ m}$$

$$H_a = \frac{V_a^2}{2g} = \frac{(1)^2}{2 \times 9.81} = 0.05 \text{ m}$$

$$H = h + H_a = 3 + 0.05 = 3.05 \text{ m.}$$

Modified discharge $Q = 2.5 \times 100 \times (3.05)^{3/2}$

$= 1330$ cumecs.

Example 14.2 *Find out the discharge of a siphon spillway from following data :*

Number of siphon units = 4

Area at throat in m^2 = 3

Full reservoir level = 150 *m*

R.L. of centre of outlet = 128

Tailwater level on D/S side during rains = 130 *m*

Tailwater level during winter = 125 *m*

Discharge coefficient = 0.60 ·

Solution.

Case 1. In rainy days outlet remains submerged and hence discharge depends upon the Tail water level.

Working head = R.L. of reservoir – R.L. or T.W.L.

$= 150 - 130 = 20$ m

$$Q = CA\sqrt{2gh}$$

$$= 0.60 \times 3 \sqrt{2g \times 20}$$

$$= 35.75 \text{ cumecs.}$$

Discharge of Four units = 4 × 35.75 = 143 cumecs.

Case II. In winter T.W.L. falls down and spillways discharge free in the air.

Available head = Reservoir level – R.L. of centre of outlet

= 150 – 128 = 22 m.

$$Q = CA\sqrt{2gh}$$

$$= 0.60 \times 3 \times \sqrt{2 \times 9.81 \times 22}$$

$$= 37.5 \text{ cumecs.}$$

Discharge of four units

= 4 × 37.5 = 152 cumecs.

14.24 OUTLET WORKS

The impounded water in the reservoir has to be regulated through outlet works to use it usefully. Outlet works are required for following purposes

(i) Feeding water for power generation.

(ii) Discharging heavily laden silty water D/S.

(iii) Supplying water to irrigation channels.

(iv) Supplying water for water supply purpose.

(v) To evacuate the water from the reservoir in anticipation of flood inflow.

Outlet works consist of three component parts.

1. Water way.
2. Gates to control the flow of water.
3. Intake structures.

A pipe or tunnel passing though the dam is the water way. It is sometimes also known as sluice way. In concrete gravity dams the sluice ways may be made in the body of the dam whereas in case of earth dam they are set up outside the limits of the embankment. If sluice way has to be provided in the body of the earth dam, the sluice way should have projecting collars at regular interval. Collars prolong the seepage path and prevent the failure of the dam.

The inlet of the sluice way should further be placed at minimum reservoir level.

Gates. Rectangular vertical lift gates are mostly used for exercising control on the entry of water. Sometimes side way sliding gates may also be used.

Intake Structures. It is a structure placed at the entrance of the sluice way.

The main purpose of intake structure is to allow only that water to pass through the sluice way which is clear of grit, floating debris etc. It has several openings at various heights. All the openings are covered with screen. Trash rack is also one of the devices which prevents the entrance of objectionable material in the sluice way. It also prevents any heavy body hiring the outlet.

QUESTIONS

14.1. What do you understand by term spillway ? How the spillways are broadly classified ? On what factors the discharging capacity of the spillway is fixed.

14.2. What are the functions of a spillway ? Prepare a list of main spillways.

14.3. Make a neat sketch of an Ogee type spillway and describe its working in brief.

14.4. Why emergency spillway is installed ? Is it made of earth or cement concrete?

14.5. Explain the formation of hydraulic jump. Derive the expression for height of hydraulic jump on D/S side.

14.6. What are various methods of energy dissipation ? Compare jump height curve with tail water curve and suggest the protective works which would be most effective for dissipation of excess energy of flowing water.

14.7. Discribe various types of siphon spillways with the help of neat sketches.

14.8. Draw the sketches of

(i) Radial gates.

(ii) Vertical lift gates.

(iii) Needle gates.

❑❑❑

15

Diversion Head Works

15.1 INTRODUCTION

A structure whose main purpose is to supply water to the off taking canal is known as *head works.* When off taking canal is to supply water for irrigation purpose, its head works is called *irrigation head works.* Irrigation headworks may be divided into two categories.

1. Diversion head works or simply head works.
2. Storage headwork.

1. Diversion head works. The purpose of this head work is to divert river water into the off taking canal. It is located across the river near the point of take off of the canal. A diversion head work performs following functions:

(i) It regulates the flow in the off taking canal.

(ii) Silt entrance into the canal is controlled.

(iii) It raises the level of water in the river. This causes diversion of the river water into the off taking canal, under gravity flow.

(iv) By raising the level of water in the river with the help of diversion work, and consequently raising the off take level of the canal, more area maybe brought into the command of the proposed canal system.

(v) Level of water in the rivers remains fluctuating due to changing flow. Diversion works reduce such fluctuations in river water level.

(vi) Although the purpose of diversion head work is not to store water but still some water definitely gets stored. This stored water can be used to augment the short supplies of the river. This type of

PLATE 15.1 (To be given in between 406-407)

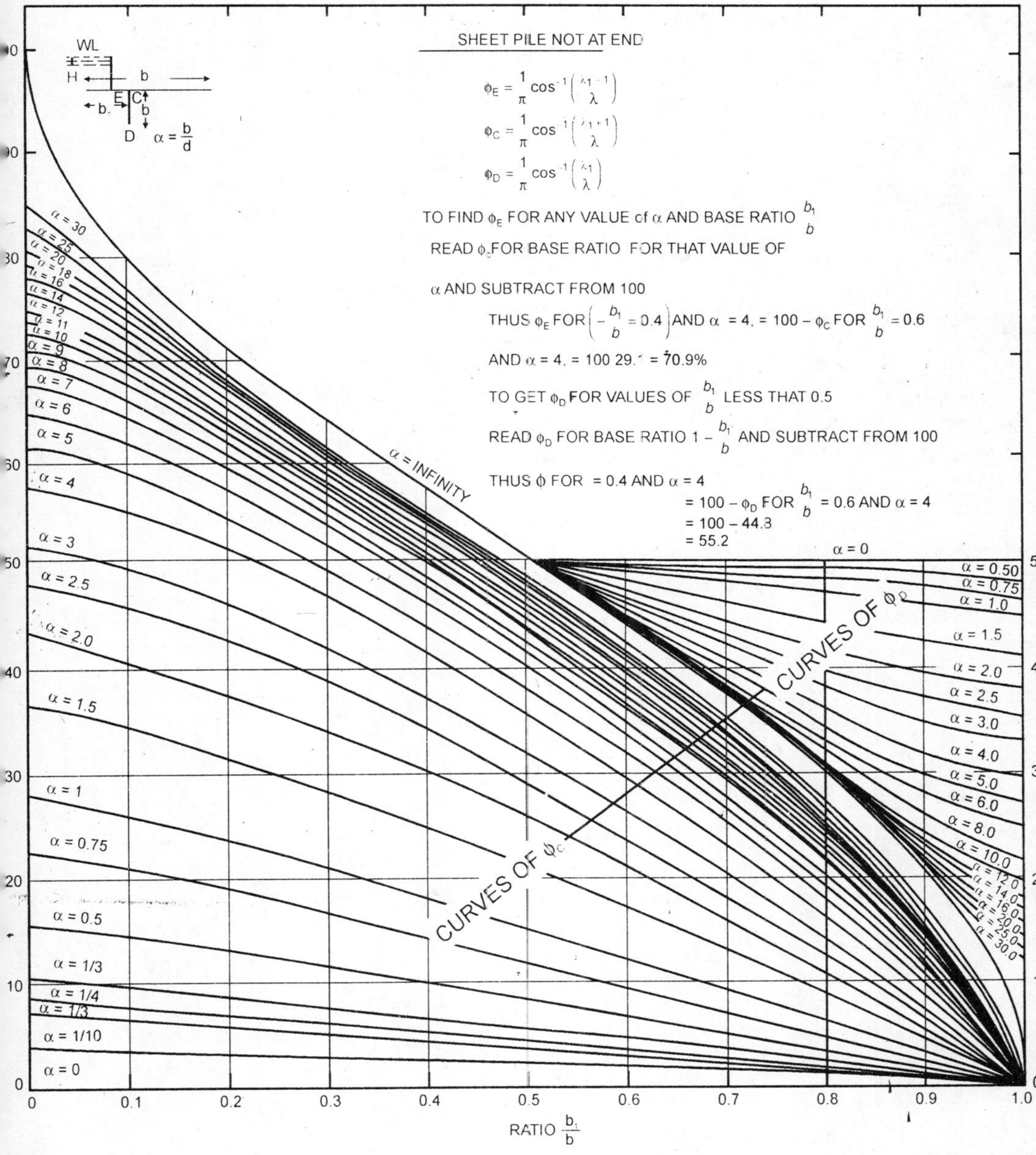

augmentation is possible for few days only, because stored water is not much in amount.

2. Storage head works. Storage works are the works which store surplus water when available in the river. This stored water is used to supplement the available flow when flow in the river falls short of the demand. Storage head work comprises the construction of dams across the river. Various aspects of storage of water have already been discussed in chapter on Reservoir Planning Construction Aspects. Various types of days also have been discussed in earlier chapters. In this chapter only diversion head work will be dealt with.

15.2 TYPES OF DIVERSION HEAD WORKS

Diversion works can be classified under following two heads:

1. Temporary bunds or spurs.
2. Permanent weirs and barrages.

Temporary bunds or spurs are temporary works, which have to be constructed every year after floods. These bunds cannot sustain the assault of floods and get washed away. Immediately after floods they are again constructed. Temporary bunds can be constructed on small streams only. Such works are generally carried out by the nearby villagers jointly, so as to make arrangements for irrigation for their fields for lean months of flow. Such temporary spurs or bunds can be economically constructed in boulder reaches of the river. However, for important diversion works permanent weirs and barrages have to be designed and constructed.

Weir. Weir is a solid obstruction, constructed across the river. It is used to raise the water level in the river and then divert it into the canal. The weirs can also be used to store surplus flood water, to tide over the shortages likely to occur during lean months. In such a case, the weir is known as storage weir. The main difference between storage weir and dam, is only is regard to the height and duration of storage. Dams store large amounts of water and for longer durations than storage weirs. One more difference is that water in the reservoir never overtops the dam, but in case of weirs the water can flow over the weir crest. In case of dams the surplus flood water is disposed off *D/S* through spillways.

Barrage. Functions of barrage are the same as those of weirs; the only difference being in mode of heading up of water. The weirs cause heading up of water by obstructing the flow by themselves, but in case of barrages water is headed up with the help of gates. In case of weirs the high flood water passes over the crest but in case of barrages, gates are lifted up and flood water passed *D/S*.

In case of barrage, crest level is maintained at low water level which is almost in level with bed level of the river. Because gates can be lifted for passing the flood water and shut down for storing water, control on the water level in the river is better exercised with the help of barrage. However barrages are much more costlier than the weirs. A road bridge is generally constructed over the barrage, with a little additional cost. Hence a barrage may be used to act as bridge also.

15.3 LAYOUT OF A DIVERSION HEAD WORKS

A typical layout of a diversion head work is shown in Fig 15.1.

Following is the list of component parts of a diversion head works. The design and description details of all these components have been given one by one in subsequent articles.

1. Weir or barrage.
2. Scouring sluices or under sluices.
3. Divide wall.
4. Fish ladder.
5. Log chutes.
6. Canal head regulator.
7. River training works.
8. Silt control devices

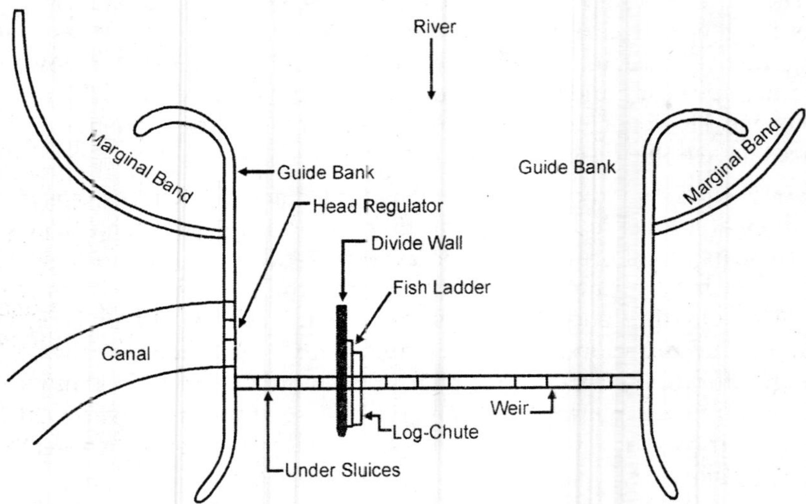

Fig. 15.1. *Layout of a diversion head works.*

Out of all these component parts river training works include lot of other elements also and hence have been discussed in a separate chapter later on. Before we take up description and design of all these component parts of a diversion headworks, let us first of all discuss *location of head works.*

15.4 LOCATION OF DIVERSION HEAD WORKS

The total length of the river may be divided into following four stages.

1. Mountainous stage. The first length of the river originating from hills is known as mountainous stage of the river. The longitudinal slope in this stage is very steep and velocity of flow of water, very fast. The width of the river in this region is generally very small. This region is found most suitable for the construction of storage works like high dams. Large storages are possible as very deep valleys are easily available here. Moreover the length of the dam is also relatively small. Materials of construction are locally available. This stage is however not suitable for diversion works.

2. Sub-mountainous stage. The bed and banks of the river in this stage are made of gravel. The velocity of flow is considerable though smaller than first stage. The river water contains lot of silt load. The river generally does not flow in a single channel, but is form of small several channels. Longitudinal slope is less than first stage but still considerable. Diversion works can be located in this region of the river.

3. Alluvial or trough stage. This length of the river is in plains. The section of the river is made of alluvial soils or silt. The longitudinal slope of the river is small and velocity of flow moderate. This length of the river is the largest of all other lengths. Diversion works are mostly located in this length. The area to be irrigated lies very near to this length and thus reduces the overall length of the canal network.

4. Delta stage. This is the last length of the river before it falls into the sea. This length of river is only a few kilometres long. Here velocity of flow is very small and silting rate very high. Because of high rate of silting, the rivers frequently change their course. No diversion work is constructed in this region. Secondly available command area near this region is also small.

Canals cannot be taken from mountainous stage, as lot of cuttings and fillings are involved and thus canal construction becomes very costly. Lot of falls will have to be constructed adding still more to the cost.

After having decided the region in which diversion head work is to be located, following additional points should also be taken into account :

(i) All the construction materials should be available in the vicinity.

(ii) In order to keep the length of the canal minimum, diversion head works should be located as near the area to be irrigated as possible.

(iii) The elevation of the water in the river should be such that water may flow to the area under gravity. If level of water is low, the position of the head works should be shifted a few kilometres upstream so that increased elevation of water is available.

(iv) The canal should run in such a way that canal is partly in cutting and partly in filling.

(v) The river at site of diversion, should be neither too wide nor too narrow. The river should neither be scouring nor silting.

(vi) The river should be straight and flowing in one channel. Its bed and banks should be stable.

(vii) The canal alignment should be such that minimum streams cross it. This will reduce the number of cross-drainage works to be constructed.

(viii) The canal should take off, either at 90° or greater than 90° with the river. Canal should not run parallel to the river. In such circumstances the canal may be eroded during floods.

15.5 WEIRS

Weir is a solid obstruction, put across the river to store water on its *U/S*. The stored water is diverted to the off-taking canal. Depending upon the criterion of design, the weirs may be gravity type or non-gravity type. The gravity weir is the weir in which uplift pressure below the weir due to seepage is fully resisted by the self weight of the weir. In the case of non-gravity weirs the thickness of the floor is kept relatively small and uplift pressure is largely resisted by the bending action of the reinforced concrete floor.

Plate 15.2

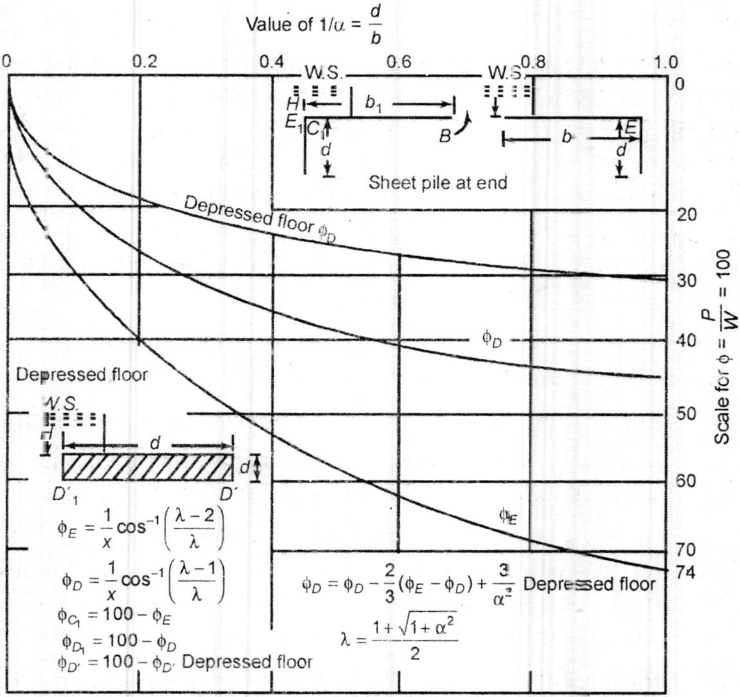

Sheet pile at end (Khosla Curves)

Depending upon the design features and available construction materials gravity weirs or simply, weirs can be further subdivided into following three categories.

1. Vertical drop weirs.
2. Rock fill weirs.
3. Concrete weirs with sloping glacis.

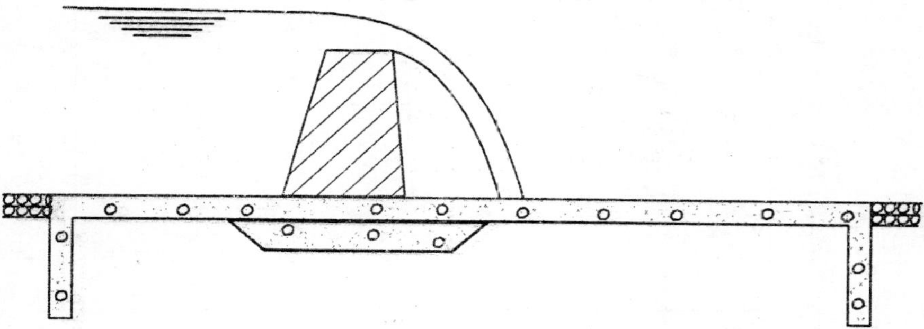

Fig. 15.2. *Vertical drop weir.*

1. Vertical drop weirs. This weir consists of a vertical drop wall or crest wall, rectangular or nearly rectangular in shape. The weir may have gates at the crest. Pucca impervious floor is provided below the crest wall and also for substantial lengths both on *U/S* and down stream side. Cut off piles are also provided both at ends of *U/S* and *D/S* pucca floors. To guard against the scouring, launching aprons are provided, both at *D/S* and *U/S* ends of the pucca floors. A graded, inverted filter is also provided on *D/S* side, in continuation of *D/S* launching apron, to relieve the uplift pressure of seeping water. This weir can be used over any type of foundation soil.

2. Rock fill weir. The body of this weir is made by the combination of masonry walls and stone boulder filling. Main weir wall is constructed in required height. A few more masonry walls, also called core walls, are constructed from masonry at suitable interval of *D/S* side of the main weir wall. The intervening space between masonry walls is filled by hand packed stone boulders. Boulder packing and masonry walls are arranged in such a way that sloping glacis both on *U/S* and *D/S* are formed. *D/S* slope is generally made very flat. It requires a very large amount of stone boulders and as such is restricted only for the places where stone is abundantly available. Such a weir exists at Okhla, near Delhi.

3. Concrete weir with sloping glacis. The design of this weir is of recent origin, and its design is done according to Khosla's theory. This weir consists of a sloping concrete slab. The parts of the slab slope downwards both on *U/S* and *D/S* sides. Deep cut-off piles are driven at the ends of *U/S* and *D/S* pucca aprons. Sometimes, even intermediate cut-off piles are also provided. Hydraulic jump is formed at the *D/S* sloping glacis to dissipate the energy of gliding water. This weir is found suitable for soft sandy foundations. This weir is mostly used when difference in weir crest and *D/S* river bed is limited to about 3 m.

4. Parabolic weirs. It is similar to the spillway section of a dam. Weir wall is made parabolic and is designed as a low dam. A cistern is developed on the *D/S* side to dissipate the energy of flowing water. *U/S* and *D/S* protection works

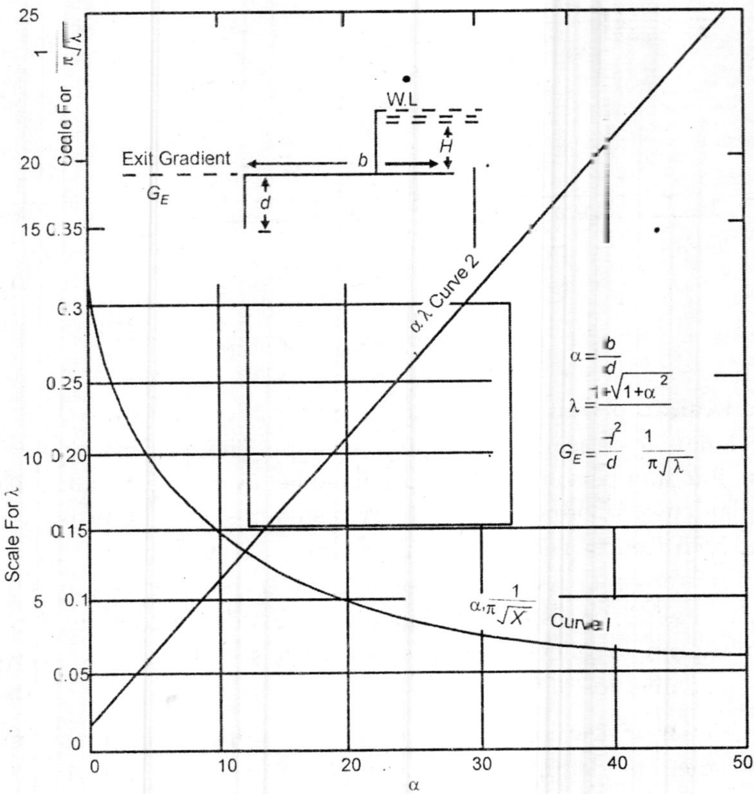

Fig. 15.3. *(Plate 15.3 Exit gradient) (Khosla Curve).*

are provided in the similar way as explained for vertical drop or sloping glacis weirs. This weir is not much in use.

15.6 CAUSES OF FAILURE OF WEIRS ON PERMEABLE SOILS AND THEIR REMEDIES

The weir failure may take place due to the following reasons

1. Piping or undermining.
2. By uplift pressure.
3. By suction due to standing wave.
4. By scour on the *U/S* and *D/S* of the weir.

1. Piping or undermining. When water seeping through the permeable foundation emerges out at the *D/S* end of the impervious floor of the weir, at hydraulic gradient or exit gradient greater than the critical value for the foundation soil, the soil starts boiling at the exit point. Boiling of the soil indicates lifting of the soil against gravity and it happens only when exist gradient of seeping water is greater than the safe limit for the foundation soil. The soil gets washed out with boiling water. With washing out of some soil from *D/S* side, the exit gradient increase and boiling of soil starts with even more vigour. This process of erosion of soil from below the foundation, progressively, works backwards towards the *U/S*. This process ultimately develops a channel or pipe, below the foundation of the weir and causes failure of the weir. Such a failure can be avoided by adopting following preventive measures.

(i) Provide deep cut-off piles at *U/S* and *D/S* ends.

(ii) Provide sufficient length of pucca floor so as to maintain the exit gradient below the critical limit, for the foundation soil.

2. By uplift pressure. Seeping water through the foundation, exerts uplift pressure, on the floor. The uplift pressure is maximum at the point, just *D/S* of the weir wall or crest wall, when water is full up to the top of the gates and there is no water on the *D/S* side. Hence critical section the floor is just at the *D/S* side of the weir's crest wall. If thickness of the floor is insufficient, its weight would be inadequate to resist the uplift pressure. This may ultimately lead to bursting of the floor and thus failure of the weir may occur. Actually, once the floor is burst due to uplift pressure the effective length of seepage is very much reduced which causes further increase in the exit gradient and consequent failure by piping. Such a failure of the weir can be avoided by adopting following measures :

(i) Provide sufficient length of impervious floor.

(ii) Provide impervious floor of sufficient thickness.

(iii) Provide a cut-off pile at *U/S* end so as to reduce the effect of uplift throughout the foundation.

3. By suction due to standing wave. Standing waves of hydraulic jump formed at the *D/S* of the weir cause suction. This suction increases the effect of uplift. If floor thickness is inadequate to sustain the combined effect of uplift and suction, it will fail. Such failures occurred at Marala weir on the Chenab and Rasul weir. Remedial measures may be the following :

(i) Provide increased thickness of floor to resist the additional effect of standing wave.

(ii) Floor should be laid in one layer of concrete, instead of several, thin layers of masonry.

4. **By scour on *U/S* and *D/S* of the weir.** The beds of alluvial rivers are scoured to considerable depths specially during floods. This scouring may take place on *U/S* side also, but it is most likely on *D/S* side of pucca floors. If this scour is allowed unchecked, it will form scour holes underneath the pucca floor. These holes may slowly progress towards the main weir and cause its failure. This

failure can be prevented by adopting the following measures :

(i) Providing deep piles at *U/S* and *D/S* ends of pucca floor. The piles should be taken much below the calculated scour depth.

(ii) Suitable lengths of launching aprons, both at *U/S* and *D/S* of impervious floors should be provided.

15.7 DESIGN OF IMPERVIOUS FLOOR OF THE WEIR FOR SUB-SURFACE FLOW

Khanki weir failed in 1895. An expert's committee under the Chairmanship of Lt. Col. Caliborn, then Principal of Thomson College, Roorkee was set up to investigate the causes of failure of this weir, which was founded on sandy soil. The committee confirmed the Darcy's law of seepage through sandy soils for low heads. Heads are generally low in case of weirs. Later Mr. W.G. Bligh put forward

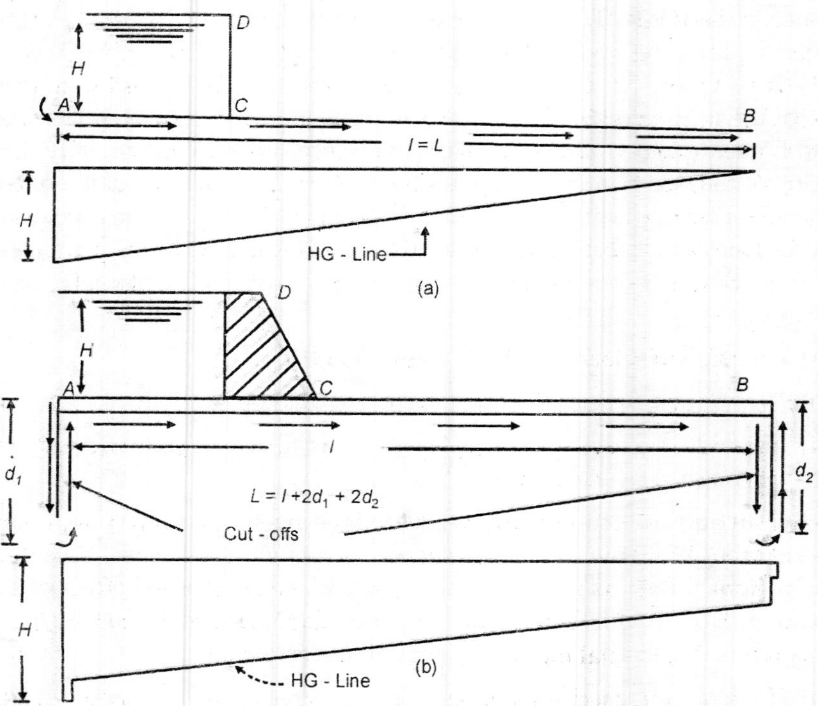

Fig. 15.4. *Creep length by Bligh's theory.*

his own theory known by his name "Bligh's creep theory" for sub-surface flows. Later on Mr. Lane put forward is own theory in 1932 after having analysed a large number of dams. This theory is known as *'Lane's weighted creep theory'*.

A most elaborate scientific study based on facts was later carried out in 1936 by Pavlovsky and A.N. Khosla. This theory is known as "Khosla theory of seepage".

Mr. Khosla carried out most of his investigations in Punjab. He published the results of his analysis in Publication No. 12 of Central Board of Irrigation and Power, New Delhi. Bligh's creep theory, Lane's weighted creep theory, and Khosla's creep theories, are being explained in details one after the other.

15.8 BLIGH'S CREEP THEORY

This theory is based on the assumption that seeping water through the soil below the weir, follows the path along the contact of the base, with the underlying sub-soil. The length of the path of seeping water from the point of entrance into the sub-soil from the *U/S* of the impervious apron, to the point at the *D/S* end of the impervious apron is known as *creep length.* Bligh also assumed that loss of head of the seeping water is proportional to the length of its travel irrespective whether length of travel is in horizontal or vertical direction. He also assumed that unless cut-off sheet piles extend to the impervious subsoil strata, no amount of sheet piling could stop the flow of percolating water.

See Fig. 15.4 (a) *AB* is the length of impervious apron l and H is the head of water filled upto the top of the weir *CD* and there is no water on *D/S* side. According to Bligh's theory $L = l$ where L is the total creep length. If vertical cut-offs are provide below the impervious apron as shown in Fig. 15.4 (b), the total creep length will be as follows

$$L = l + 2d_1 + 2d_2.$$

Length of vertical cut-off has been taken double because vertical cut-off provides the creep length equivalent to twice the length of the cut-off, as seeping water once goes down and then comes up along the cut-off.

If H is the total head causing seepage or total loss of head, and L the total creep length, the loss of head per unit length of creep (C) is given by

$$C = \frac{H}{L} = \frac{H}{l + 2d_1 + 2d_2}.$$

Loss of head per unit length of creep (C) is known as percolation coefficient. The reciprocal of percolation coefficient is known as *coefficient of creep* (C). Safe values of coefficient of creep for different soils are given below.

Coefficient of Creep

Type of soil	*Coefficient of creep* (C)	*Percolation coefficient* (C)
1. Sand mixed boulders, gravel and shingles.	5 to 9	1/5 to 1/9
2. Coarse grained sand as found in south and central Indian rivers	12	1/12
3. Sands as found in Northern Indian rivers	15	1/15
4. Light sand and mud as found in Nile river in Egypt.	18	1/18

Weir design by Bligh's theory. According to Bligh's theory, two design criteria are to be considered.

(*i*) *Safety against undermining or piping.* To safeguard the weirs against failure by piping, the creep length should be provided according to the following formula :

$$L = CH, \text{ where } C \text{ is coefficient of creep.}$$

Such a length would provide safe hydraulic gradient and seeping water emitting from *D/S* end of the impervious apron, will not be having sufficient uplift pressure to dislodge the soil particles. This will avoid boiling action of soil at the *D/S* end of the apron.

(*ii*) *Safety against uplift pressure.* To counter balance the force of uplift, sufficient floor thickness should be provided, specially on the *D/S* side of the weir.

Let at distance L_1 creep length from *U/S* end of the impervious apron h is the resultant uplift head, the net uplift pressure wh can be computed as follows

$$wh = W\left(H - \frac{H}{L} \times L_1\right)$$

Downward load of pucca impervious apron per unit area

$$= w\rho \times t$$

where w = is the specific weight of water

p = Specific gravity of floor material

t = Thickness of the floor.

Upward pressure (uplift) should be equal to downward force.

i.e. $wh = wPt$

$$h = Pt$$

$$h - t = Pt - t = t(\rho - 1)$$

$$t = \frac{h-t}{\rho-1} = \frac{h'}{\rho-1}$$

where h' = ordinate of hydraulic gradient line measured above the top of the impervious apron.

If we adopt $\frac{4}{3}$ as the factor of safety, the thickness of floor is given by

$$t = \frac{4}{3}\left(\frac{h'}{\rho-1}\right).$$

The thickness of the impervious apron on the *U/S* side of the weir wall is provided as nominal thickness. This is because uplift pressure *U/S* of the weir wall always remains counter balanced by weight of water, which always remain filled up over it. It is also advantageous if more length of impervious apron is provided on *U/S* side. This will reduce the intensity of uplift pressure on the *D/S* side of the weir wall. However some minimum length of impervious apron has to be provided on the *D/S* side also. This aspect will be explained a little later.

Example 15.1 *A weir on a sandy soil is shown in Fig. 15.5. Find out the Uplift pressures at points 4, 8, and 12 m from the U/S end of the pucca apron. Also find out the thickness of the floor at these points considering specific gravity y of floor material as 2.24. Adopt Bligh's theory for analysis and weight of water as one tonne per cubic metre of water.*

Solution Total length of creep = 2 × 4 + 16 + 2 × 6 = 36 m

Hydraulic gradient or percolation coefficient

$$= \frac{H}{L} = \frac{4}{36} = \frac{1}{9}$$

I. Uplift at Point *A*

Creep length upto point *A* = 4 × 2 + 4 = 12 m.

$$\text{Resulting uplift head} = \left(H - \frac{H}{L} \times L'\right)$$

$$= 4 - \frac{4}{36} \times 12 = 2.67 \text{ m}$$

Uplift pressure = wh = 1 × 2.67 = 2.67 t/m^2

$$\text{Floor thickness at point } A = \frac{4}{3} \times \frac{h}{\rho - 1}$$

$$= \frac{4}{3} \times \frac{2.67}{(2.24 - 1)} = 2.87 \text{ m.} \qquad \textbf{Ans.}$$

II. Uplift at point *B*

Seepage length unto point *B* = 4 × 2 + 8 = 16 m.

$$\text{Uplift head at } B = H - \frac{H}{L} \times L'$$

$$= 4 - \frac{4}{36} \times 16 = 2.22 \text{ m.}$$

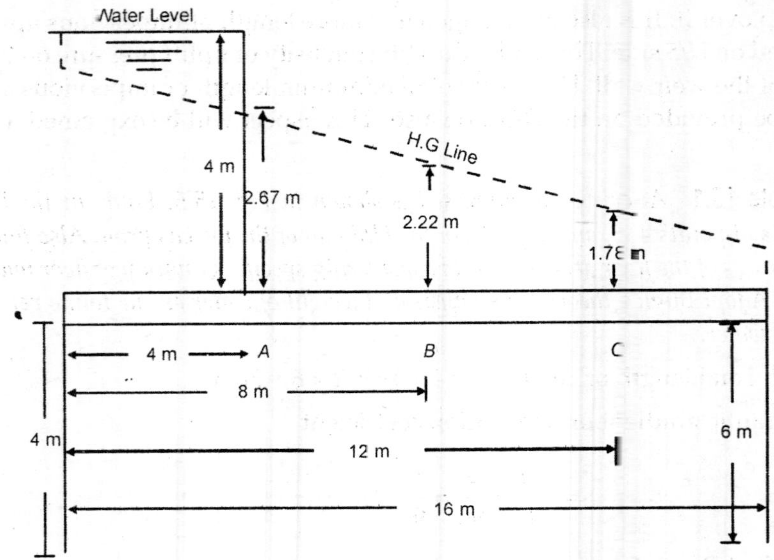

Fig. Ex.15.1

Uplift pressure = 1 × 2.22 m = 2.22 t/m^2.

Floor thickness at point $B = \frac{4}{3}\left(\frac{h}{\rho - 1}\right)$

$$= \frac{4}{3}\left(\frac{2.22}{2.24 - 1}\right) = 2.39 \text{ m. } \textbf{Ans}$$

III. Consider Point *C*

See page length upto point $C = 2 \times 4 + 12 = 20$ m.

Uplift head $= H - \frac{H}{L} \times L'$

$$= 4 - \frac{4}{36} \times 20 = 1.78 \text{ m}$$

Uplift pressure $= 1 \times 1.78 = 1.78\ t/m^2$

Floor thickness at point $C = \frac{4}{3}\left(\frac{h}{\rho - 1}\right)$

$$= \frac{4}{3} \times \frac{1.78}{(2.24 - 1)}$$

$$= \textbf{1.91 m.}$$

Limitations of Bligh's Theory. Following are some of the limitations of this theory.

1. He did not distinguish vertical cut-off from horizontal impervious floor as regards to creep length.

2. Bligh did not make any mention of exit gradient. The safety against piping or undermining cannot be guaranteed simply by providing flat gradient. The exit gradient on *D/S* end of the apron should be maintained well below the critical value for the soil laying underneath the pucca impervious apron.

3. This theory holds good so long as distance between subsequent piles is maintained substantial. The horizontal distance between subsequent piles should not be allowed to be smaller than twice the depth of the pile.

4. Bligh did not make any difference in effectiveness of outer and inner faces of *U/S* and *D/S* end piles and intermediate piles. By investigations it has been established that outer faces of both the end piles are much more effective than inner ones. The intermediate piles, if provided and their lengths are smaller than outer piles, they are ineffective and do not provide any additional creep length. They may, however, cause local redistribution of pressure.

5. Bligh did not feel necessity of providing a sheet pile at *D/S* end. It has been found absolutely essential on investigations, to have a deep cut off at *D/S* end. It is considered as most effective measure for preventing the undermining of weirs.

6. Bligh considered loss of head, as linear along the creep length whereas it is not so. Distribution of uplift pressure is also not linear but follows sine curve.

This theory was very well accepted as the basis for design of irrigation structures in earlier days. Bhim Goda weir near Haridwar and some other works constructed according to this theory, are doing very well and did not give any trouble. Some of the works designed according to this theory had however failed. When causes of such failures were investigated, it was found that only those works have failed, below the impervious aprons of which very deep cut off piles were provided, very close to each other. As deep as 12 m deep piles were found in some works. In the case of deep cut offs, the creep length is not twice the depth of the pile but much smaller than that. This fact has been proved that flow of seepage neither follows the path along the vertical cut-off piles and nor along the underside of the horizontal impervious apron. Flow of seepage follows parabolic stream lines. From this fact it follows that intermediate piles do not provide any additional seepage length if their lengths are equal to or slightly more than the length of the end piles. But Bligh considered all the piles of equal effectiveness and this fact actually led to the failure of some of the weirs designed according to his theory.

15.9 LANE'S WEIGHTED CREEP THEORY

Mr. Lane gave weights to vertical creep length and horizontal creep lengths. Horizontal length is given a weight of one and vertical length provided by cut off piles a weight of 3. Total creep length is finally found out by adding total weighted horizontal length and weighted vertically offsets. This theory is Empirical and does not provide any background, as to why vertical cut off should be given weight of three. This theory did not gain any ground and thus is not used anywhere.

15.10 KHOSLA'S THEORY

Some siphons on upper Chenab canal, which were designed according to Bligh's theory gave trouble. Khosla along with his associates, were asked by the government to investigate the causes of trouble and to suggest remedial measures. Khosla and his associates inserted some pipes on the *D/S* side of the weir through impervious aprons of some of the trouble giving works. The pipes were inserted to verify whether the pressures below the impervious apron were in accordance with Bligh's theory or not. The pressures measured in the pipes were not as they should have been according to Bligh's.

Based on his investigations A.N. Khosla drew following conclusions:

1. The outer faces of the end piles are much more effective than their inner faces and also more effective than horizontal length of the impervious apron.
2. Intermediate sheet piles, if equal to or smaller in length than the outer piles, are almost ineffective and do not provide any additional creep length. They may only cause local redistribution of pressures.
3. Piping or undermining of the impervious floors starts from the *D/S* end of the pucca impervious floor. If the exit gradient at *D/S* end was more than the critical gradient for the soil underlying the foundation, the soil particles will get lifted up and carried away with seeping water. This process if once starts progresses continuously towards the *U/S* side and ultimately a cavity is formed and failure of weir becomes imminent.
4. It is very essential to have a deep vertical cut off at the *D/S* end. This measure prevents undermining to a large extent.

In 1929 Panjnad weir was designed according to Khosla's theory. The pipes were inserted in its *D/S* floor, to verify the pressures of seeping water at various points. The pressures were found as they should have been as per Khosla's theory. This gave wide recognition to Khosla theory and since then most of the irrigation works in the world are being designed according to this theory.

Khosla proved that seeping water through permeable soils follows parabolic stream lines and not along the underside profile of the impervious floor as envisaged by Mr. Bligh. This is the fundamental difference between the two theories. Flow of water through permeable soils, as assumed by Mr. Khosla, is shown in Fig. 15.5. Since seeping water flows in parabolic stream lines their theoretical solution is possible. Flow of seeping water takes place according to Laplace equation.

$$\frac{d^2\phi}{dx^2} + \frac{d^2\phi}{dy^2} = 0$$

where $\phi = KH$ (H is head of water)

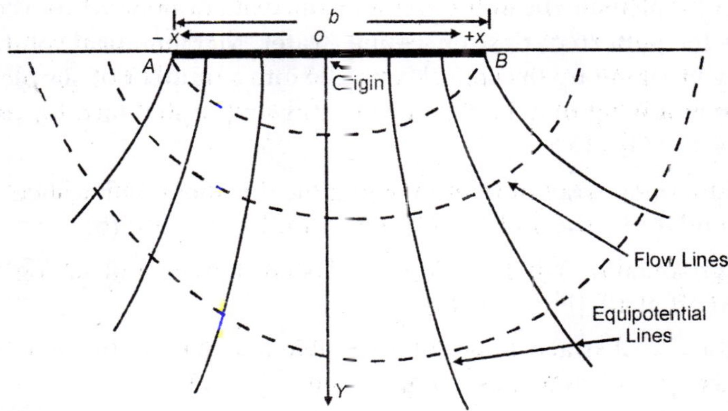

Fig. 15.5. *Flow through permeable soils as assumed be Khosla.*

This equation can be solved graphically or mathematically and a graph of equipotential lines and flow lines in form of a flownet show in Fig. 15.6 can be prepared. After having prepared the flownet, uplift pressure at any point below the weir's impervious floor, can be easily found out.

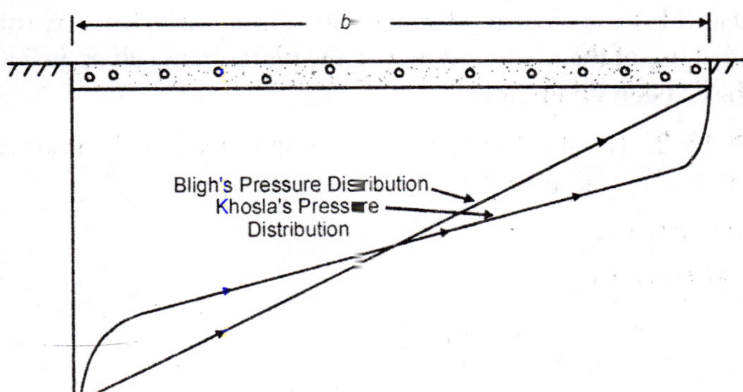

Fig. 15.6. *Uplift pressure distribution by Bligh and Khosla.*

Khosla's method of analysis pointed out following two more points of difference from Bligh's method.

1. *D/S* half of the floor of the weir is subjected to more uplift pressure than that according to Bligh's method.

2. Slope of pressure diagram as found out by Khosla is infinite at entrance and exit points. An infinite force would be acting downwards at entrance point and upwards at exit point. The infinite upward force at exit point would cause

boiling of sand. In order to prevent sand boiling a deep cut off or a depressed floor is essentially provided at the D/S end.

Distribution of uplift pressure below the impervious floor of the weir for both Khosla's and Bligh's methods are shown in Fig. 15.7.

In order to calculate the uplift pressure and exit gradient, Khosla considered flownet as the pattern of flow of seeping water. Mathematical solutions were evolved by breaking up the complex profile into a number of simple standard forms. The following may be the most useful simple standard forms of break up profiles. See Fig. 15.8.

(i) Horizontal straight floor of negligible thickness with a sheet pile at U/S end and a sheet pile at D/S end. Fig. 15.8 (a) and (b).

(ii) Horizontal straight floor depressed below the bed and having no vertical cut off at all. [Fig. 15.8 (c)].

(iii) Horizontal straight floor of negligible thickness with a cut off vertical sheet pile at some intermediate point . Fig. 15.8 (b).

All these simple profiles are shown in Fig. 15.8. All these profiles were analysed by Mr. Khosla's team with the help of Schwarz Christoffel transformation. They prepared three plates, Plate 15.1, 15.2 and 15.3. Plate 15.1 gives the values of pressures at key points *C, D* and *E*, when sheet pile is at neither of the ends but at some intermediate point. Plate 15.2 gives uplift pressures at key points for the case when sheet pile is at D/S end. Plate 15.3 gives the values of safe exit gradients. Plates 15.1 and 15.2 give pressures at key points *C, D* and *E*. Pressures at intermediate points can be linearly interpolated. The method of use of the plates is given on the plates themselves, just to illustrate, examples have been given here.

Example 15.2. *Length of horizontal floor is 15 m and 3 m deep vertical sheet pile is attached at its D/S end. Find the uplift pressures at key points.*

Head of water is 4 m.

Solution (*i*) At *D/S* end

$$\alpha = \frac{b}{d} = \frac{15}{3} = 5.$$

$$\frac{1}{\alpha} = \frac{1}{5} = 0.2.$$

From Plate 15.2 for $\alpha = 0.2$ we get values of ϕ_D and ϕ_E as follows:

$$\phi_D = 27\% \text{ or pressure at } D \;\; P_D = 0.27\, H = 0.27 \times 4$$

$$\phi_E = 40\% \text{ or pressure at } E \;\; P_E = 0.40\, H = 04 \times 4 = 1.6 \text{ m}.$$

(*ii*) *If pile is provided at U/S end.* In this case pressures at key points C_1, D_1 can be found out by first finding ϕ_D and ϕ_E for the case when pile is at *D/S* end and then by following relation.

$$\phi_{C_1} = 100 - \phi_E$$
$$\phi_{D_1} = 100 - \phi_D$$

For our case $\phi_E = 40\%$ and $\phi_D = 27\%$

$$\phi_{C_1} = 100 - \phi_E\, 100 - 40 = 60\%$$
$$\phi_{C_1} = 100 - \phi_E\, 100 - 27 = 73\%$$

Pressure at C_1 $P_{C_1} = 0.6\, H = 0.6 \times 4 = 2.4$ m

Pressure at D_1 $P_{D_1} = 0.73\, H = 0.73 \times 4 = 2.92$ m.

Example 15.3. *Thickness of the floor is 3 m and length of the floor is 15 m. Find the uplift pressures at bottom points of D/S and U/S ends of the floor. Water fill is 5 m above the floor.*

Solution
$$\alpha = \frac{b}{d} = \frac{15}{3} = 5$$

$$\frac{1}{\alpha} = \frac{1}{5} = 0.2.$$

$\phi_D' = 18\%$ from the curve of ϕ_D' from Plate 15.2. From the principal or reversibility of flow.

$$\phi_{D_1}' = 100 - 18 = 82\%$$

Pressure at $\phi_D' = 18\% = 018 \times 5 = 0.90$ m

Pressures at $P_{D1}' = 82\% = 0.82 \times 5 = 4.1$ m

Example 15.4. *Find out the exit gradients for examples 15.2 and 15.3.*

Solution For example 15.2, for D/S end the exit gradient G_E is given by

$$G_E = \frac{H}{d} \times \frac{1}{\pi\sqrt{\lambda}}$$

$$\lambda = \frac{1+\sqrt{1+\alpha^2}}{2}$$

$$\alpha = \frac{b}{d} = \frac{15}{3} = 5$$

$$\lambda = \frac{1+\sqrt{1+(5)^2}}{2} = 3.05$$

$$\frac{1}{\pi\sqrt{\lambda}} = \frac{1\times 7}{22\sqrt{3.05}} = 0.1822$$

$$G_B = \frac{H}{D} \times \frac{1}{\pi\sqrt{\lambda}}$$

$$= \frac{5}{3} \times 0.1822 = 0.3036.$$

For example 15.3 exit gradient is given by formula

$$G_E = \frac{0.84}{d} \times P_D'$$

$$= \frac{0.84}{3} \times 0.90 = 0.252.$$

15.11 EXIT GRADIENT (G_E)

At each point under the floor, there is a certain residual uplift pressure and a certain rate at which the head is being lost, indicating a pressure gradient.

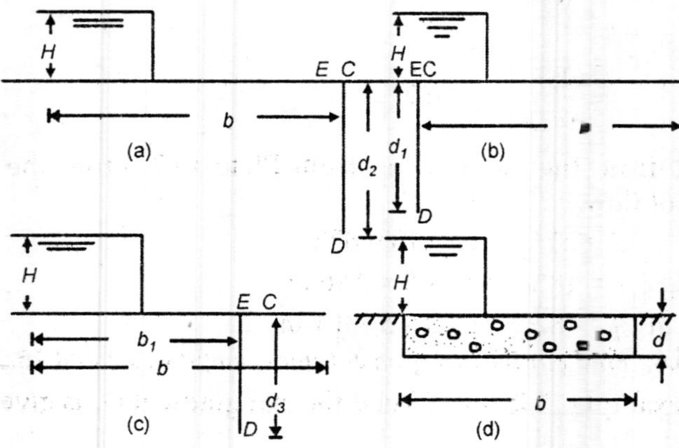

Fig. 15.7. *Break-up of complex profile to simple standard profiles*

The pressure gradient at the exit point is called exit gradient. The purpose of design of Weir's impervious aprons is to restrict the exit gradient below the critical value so that soil particles are not disturbed. For floor length b and vertical cut off of depth d at the D/S end, the exit gradient (G_E) is found out from formula

$$G_E = \frac{H}{d} \times \frac{1}{\pi\sqrt{\lambda}}$$

where H is the depth of water upto crest of the weir and λ is an element which is taken as

$$\lambda = \frac{1+\sqrt{1+\alpha^2}}{2} \text{ and } \alpha = \frac{b}{d}$$

From Plate 15.3, for any vlaue of α, or $\frac{b}{d'}$, the corresponding value of $\frac{1}{\pi\sqrt{\lambda}}$ can be read. Now applying formula

$G_E = \frac{H}{d}\frac{1}{\pi\sqrt{\lambda}}$ value of exit gradient G_E can be easily found out. For $d = 0$ value of G_E becomes infinity and hence a vertical cut-off must be provided at the *D/S* end of the floor. To guard against the undermining, the exit gradient must not be allowed to exceed the safe limit for different soils as given below :

Type of soil	*Safe exit gradient (G_E)*
Fine sand	1/6 to 1/7
Coarse sand	1/5 to 1/6
Shingle	1/4 to 1/5

The uplift pressure must be kept as low as possible consistent with the safety at the exit, so as to keep the floor thickness to the minimum.

15.12 METHOD OF INDEPENDENT VARIABLES

Break up of a complex profile of a weir to simple three elementary forms has been discussed in Section 15.10. This break up shows the theoretical profiles. Actually, the usual weir section consists of a combination of all the three elementary profiles. In addition to this, floor also has some thickness. Khosla solved the actual profile of the weir by an empirical method known as the *method of independent variables*. According to this method, actual complex profile is broken into a number of simple profiles known as elementary profiles. Each elementary profile is then treated independent of the others. Each elementary profile is independently amenable to mathematical treatment. The pressures at key points are read from the Khosla's curves. The key points are the junction points of the floor and pile and bottom points of different piles. The pressures read from the Khosla's curves are true for individual elementary profiles. But when all the profiles are combined into one complex form, which is actually the condition, corrections for pressures will have to be applied. When different elementary profiles are combined, they influence the pressures due to mutual interference of the piles. Thickness and slope of floor also affect the pressures at key points. In order to find out the pressures at various key points for the weir as a whole, following corrections for the pressures will have to be applied.

1. Correction for the thickness of the floor.
2. Correction for the mutual interference of piles.
3. Correction for the sloping floor.

1. Correction for the thickness of the floor. Figure 15.8 shows the details of key points for profiles one each at upstream end, downstream end, and intermediate point. Assume thickness of the floor negligibly small. *E* and *C* are the key point at the top of the floor. *E* point is on *U/S* side and *C* on *D/S* side of the pile.

Let E_1 and C_1 be the corresponding key points at the bottom of the floor. Pressures at points E_1 and C_1 for each profile are computed as follows:

(*i*) *Pile at the U/S end.* For point *E* no correction is required as pressure at this point is not going to interfere with pressure system of any other pile.

$$\text{Correction for point } C_1 = \frac{(\phi_D - \phi_C)t_1}{d_1} \text{(Additive)}$$

$$\therefore \text{ Pressure at } C_1 = \phi_{C_1} = \phi_C + \frac{(\phi_d - \phi_C)t_1}{d_1}$$

t_1 = thickness of floor

d_1 = depth of *U/S* pile.

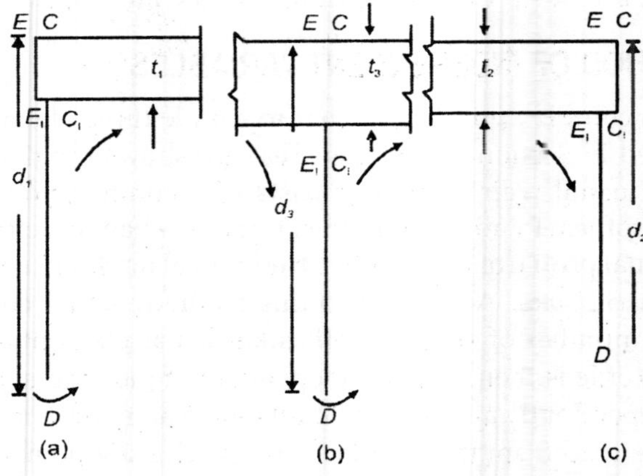

Fig. 15.8. *Correction for floor thickness.*

(*ii*) *Pile at D/S end.* Correction at E_1

$$= \frac{(\phi_E - \phi_D)t_2}{d_2} \text{ (subtractive)}$$

$$\text{Pressure at } E_1 = \phi_{E_1} = \phi_1 - (\phi_E - \phi_D)\frac{t_2}{d_2'}.$$

(*iii*) *Pile at intermediate point.* This will require corrections at both the points E_1 and C_1.

$$\text{Correction at} \quad C_1 = (\phi_D - \phi_C)\frac{t}{d} \text{ (additive)}$$

$$\text{Correction at} \quad E_1 = (\phi_E - o_D)\frac{t}{d} \text{ (subtractive)}$$

Pressure at $E_1 = \phi_{E_1} = \phi_E - (\phi_E - \phi_D)\frac{t}{d}$

Pressure at $C_1 = \phi_{C_1} = \phi_C + (\phi_D - \phi_C)\frac{t}{d}$

2. Correction for the mutual interference of piles. Correction is applied as follows :

$$C = 19\sqrt{\frac{D}{b'}}\cdot\frac{(d+D)}{b}$$

where C = correction to be applied in percentage of head.

D = the depth of the pile, the influence of which is required to be determined on the neighbouring pile of depth d.

d = the depth of the pile on which the effect of another pile of depth D is required to be found out

b' = the distance between the two piles.

b = total length of impervious pucca floor.

The sign of this correction is positive for the points in *U/S* direction, i.e. against the flow, and negative for *D/S* direction i.e. in the direction of the flow. This equation does not apply to the effect of an outer pile on an intermediate pile, if intermediate pile is equal to or smaller in length than the outer pile and also when distance between them is less than twice the length of the outer pile. Interference of any pile is only for the faces of the adjacent piles, which lie towards the interfering pile. For example pile No. 2 will interfere with the *D/S* face of pile No. 1 and *U/S* face of pile No. 3. Various dimensions marked in Fig. 15.10 are, for the point *C* on pile No. 1. *U/S* face of pile No. 2 will be influencing the pressure at point *C* of pile No. 1. Pile No. I will have no influence on *D/S* face of pile No. 2 and on pile No. 3. Mutual interference mainly extends upto a distance equal to the depth of the pile. Beyond this, it gradually falls off to negligible in a length of about twice the length of the pile.

3. Correction for slope. Correction for slope has to be applied. The correction is plus for the down and minus for the up slopes, following the direction of flow. The corrections are given in the table below for various slopes.

Slope of (V/H)	*Correction in % of pressure*
1 in 1	11.2
1 in 2	6.5
1 in 3	4.5
1 in 4	3.3
1 in 5	2.8
1 in 6	2.5
1 in 7	2.3
1 in 8	2.0

The correction is applicable to the key points of the piles fixed at the start or the end of the slope. In Fig. 15.9 slope correction is applicable only to point E of

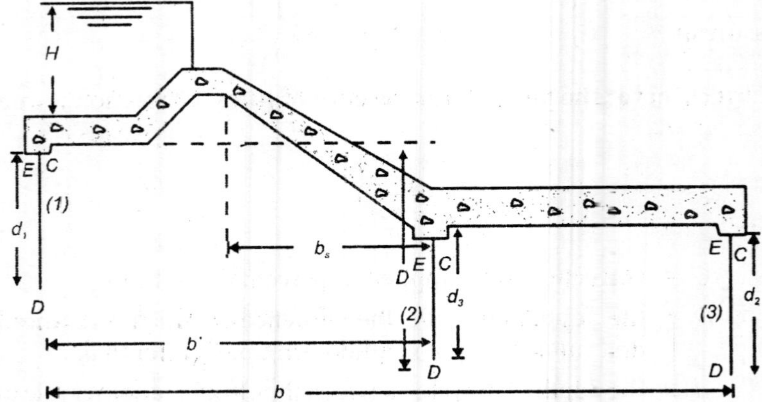

Fig. 15.10. *Correction for mutual interferences.*

the pile No. 2. The percentage correction given in the above table is to be further multiplied by the proportion of the horizontal length of the slope to the distance between the two pile lines, in between which the sloping floor is situated. The correction to be applied at point E of pile No. 2, will be obtained by multiplying the appropriate figure from the above table by a factor $\frac{b_s}{b'}$.

Example 15.5 *Figure Ex. 15.5 showns the sketch of the barrage. The various dimensions and reduced levels are shown in the figure itself. Find out the uplift pressures at important key points. Also determine the exit gradient. If barrage is founded over sandy soil, check its safety against uplift and piping. Permissible exit gradient is 1 in 8.*

Solution. Consider pile line (1) i.e. U/S end pile

$$d = 97.00 - 89.00 = 8 \text{ m}$$

$$b = 48.5 \text{ m} \quad b_1 = 0.5 \text{ m}$$

$$\alpha = \frac{b}{d} = \frac{48.5}{8} = 6.1 \text{ (say)}$$

$$\frac{b_1}{b} = \frac{0.5}{48.5} = 0.01 \text{ (almost zero)}$$

From the curves, for $\frac{b_1}{b} = 0$ and $\alpha = 6.1$

$$\phi_{C_1} = 64\%, \phi_{D_1} = 75\%, \phi_{E_1} = 100\%$$

Correction for floor thickness

$$t = 97.0 - 96.0 = 1 \text{ m}$$

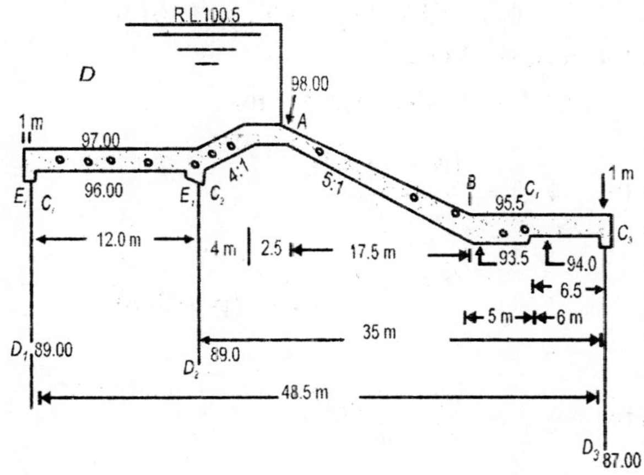

Fig. Ex. 15.5

$$\text{Correction} = \frac{(\phi_{D_1} - \phi_{C_1})t}{d} = \frac{(75.0 - 64.0)}{8} \times 1$$

$$= 1.37 \text{ (positive)}$$

Mutual interference correction for pile (1) due to pile (2)

$$c = 19\sqrt{\frac{D}{b'}}\left(\frac{d+D}{b}\right)$$

$$d = 96.0 - 89.0 = 7 \text{ m}$$

$$b' = 12.0 \text{ m}$$

$$b = 48 \text{ m}$$

$$D = 96 - 89 = 7 \text{ m}$$

$$\therefore \quad C = 19\sqrt{\frac{7}{12}}\left(\frac{7+7}{48}\right) = 4.23\% \text{ (positive)}$$

Hence corrected pressure

$$\phi_{C_1} = 64 + 1.37 + 4.23 = 69.6\%, \phi_{D_1} = 75\%$$

Consider intermediate piles

$$d = 97 - 89 = 8 \text{ m}$$

$$d = 48 \text{ m}, b_1 = 12.5 \text{ m}$$

$$\alpha = \frac{b}{d} = \frac{48}{8} = 6.0 \quad \frac{b_1}{b} = \frac{12.5}{48} = 0.26.$$

From Khosla curves

$$\phi_{C_2} = 54\%, \phi_{E_2} = 76\%, \phi_{D_1} = 64\%$$

Correction for floor thickness

$$t = 97.00 - 96.0 = 1 \text{ m}$$

Correction in $\phi_{C_2} = \dfrac{(\phi_{D_2} - \phi_{C_2})t}{d}$

$$= \frac{(64-54)1}{8} = 1.25 \text{ (positive)}$$

Correction for $\phi_{C_2} = \dfrac{(\phi_{E_2} - \phi_{D_2})t}{8}$

$$= \frac{(76-64)\times 1}{8} = 1.5 \text{ negative}$$

Correction due to interference of col. pile (1).

Because piles (1) and (2) are of same length and also fixed at same level the correction in ϕ_{E_2} due to pile (1) will be negative and its value 3.3%.

Correction for interference of pile (3) (*D/S* end pile)

$$d = 96 - 89 = 7 \text{ m}$$

$$D = 96.0 - 87.0 = 9 \text{ m}$$

Correction $\phi_{C_2} = 19\sqrt{\dfrac{9}{35}}\dfrac{(7+9)}{48} = 3.21\ \%$ (positive)

Correction due to slope

From table for slope correction, the correction works out to be 2.8% and its sign is negative since it is taken negative in the direction upward slope.

Horizontal length = 4 m

Hence actual correction for slope in ϕ_{C_2} in

$$= 2.8 \times \frac{4}{35} = 0.32$$

Hence corrected pressures are

$$\phi_{E_2} = 76 - (1.5 + 2.8) = 71.6\%$$

$$\phi_{C_2} = 54 + 1.25 + 3.21 - 0.32 = 58.14\%$$

Pile 3 at the *D/S* end

$$d = 95.5 - 87 = 8.5 \text{ m}$$

$$\frac{1}{\alpha} = \frac{d}{b} = \frac{8.5}{48} = 0.177$$

From Khosla curves

$$\phi_{E_3} = 38\% \quad \phi_{D_3} = 25\%$$

$$\text{Correction for depth} = \frac{\left(\phi_{E_3} - \phi_{D_3}\right)t}{d}$$

$$= \frac{(38-25)1.5}{8.5} = 2.3\%(\text{negative})$$

Correction for interference of pile 2 (intermediate pile)

$$d = 94.0 - 87.0 = 7 \text{ m}$$

$$D = 94.0 - 89.0 = 5 \text{ m}$$

$$\text{Correction for} \quad \phi_{E_3} = 19\sqrt{\frac{5}{35}}\left(\frac{5+7}{48}\right)$$

$$= 1.8\% \text{ (negative)}$$

Hence corrected pressure $\phi_{E_3} = 38 - 2.3 - 1.8 = 33.9\%$.

The percentage pressure and the actual pressure heads at the key points are given in tabular form as follows :

Table 15.4. Maximum percolation head = 100.5 – 95.5 = 5 m

Point	*% pressure, ϕ*	*Pressure head in metres*
C_1	69.6	3.48 m
C_2	58.14	2.904 m
E_2	71.2	3.56 m
E_3	33.9	1.695 m

The pressures at intermediate points may be computed by linear variation.

Check for thickness of floor

Let us check the thicknesses at points *A*, *B* and C.

$$\text{Pressure head at } A, P_A = P_{C_2} - \left(P_{C_2} - P_{E_3}\right)\frac{6.5}{35}$$

$$= 2.904 - \frac{(2.904 - 1.695)6.5}{35}$$

$= 2.68$ m.

Pressure head at B, P_B $= P_{C_2} - \frac{(P_{C_2} - P_{E_3})24}{35}$

$= 2.904 - \frac{(2.904 - 1.695)24}{35}$

$= 2.074$ m.

Pressure head at C, P_C $= 2.904 - \frac{(2.904 - 1.695) \times 29}{35}$

$= 1.904$ m

If specific density of concrete is 2.24

Required thickness at $A = \frac{P_A}{\rho - 1} = \frac{2.68}{1.24} = 2.2$ m

Available thickness at $A =$ 98.97 = 1 m.

Required thickness at B

$= \frac{P_B}{\rho - 1} = \frac{2.074}{1.24} = 1.7$ m

Available thickness at $B =$ 95.50 – 93.50 = 2 m

Required thickness at $C = \frac{P_C}{\rho - 1} = \frac{1.940}{1.24} = 1.54$ m

Available thickness at $C =$ 95.50 – 94.0 = 1.5 m.

Hence we see that available thickness at point A just on the D/S of the weir crest is deficient. So increase thickness of floor here from 1 m to 2.2 m. At point B thickness is adequate.

At point C available thickness is almost sufficient. It will be preferred if it is also increased to 1.6 m.

Exit gradient

Seepage head = 100.5 – 95.5 = 5 m.

Depth of cut off at D/S = 95.50 – 87.0 = 8.5 m

$$\alpha = \frac{b}{d} = \frac{48}{8.5} = 5.65$$

From exit gradient curve for $\alpha = 5.65$ we get

$$\frac{1}{\pi\sqrt{\lambda}} = 0.17.$$

$$\therefore \quad G_E = \frac{H}{d} \times \frac{1}{\pi\sqrt{\lambda}} = \frac{5}{8.5} \times 0.17 = \frac{1}{10}$$

Permissible exit gradient is $\frac{1}{8}$.

Hence $\frac{1}{10}$ exit gradient is quite safe.

15.13 CUT-OFF MATHEMATICAL SOLUTION OF KHOSLA THEORY

It is an impervious vertical membrane provided at the bottom of the impervious floor of the weir. Water cannot seep through it. Water can only cross it by taking round of its bottom end. It may be made of masonry, concrete, R.C.C., or steel. Cut-offs are provided so as to reduce the uplift pressure below the weir foundation. They are very helpful in preventing the failure of the weirs by undermining. In the case of all the irrigation structures, the impervious floors must be provided with deep cut-offs, at least one each at *D/S* and *U/S* ends. *D/S* cut-off in very effective to reduce the exit gradient and also to prevent undermining failure of the weirs. *U/S* cut-off reduces the effect of uplift pressure throughout the length of the floor of the weir.

Intermediate piles or cut-offs, can also be used but they are not found much effective. Depths of *U/S* and *D/S* cut-offs are dependent upon the scour depth. The cuts-offs should be carried deeper than the scour depth. Scour depth can be found out by following formula

$$R = 1.35 \left(\frac{q^2}{f}\right)^{1/3}$$

where R = Scour depth in metres.

q = Intensity of discharge per metre length of the weir

f = Silt factor.

15.14 COMPARISON OF BLIGH'S AND KHOSLA'S THEORIES

Both Bligh's theory and Khosla's theory have been discussed in details. Limitations and basic characteristics of each, have been explained. Comparison of these theories is as follows :

Bligh's theory	*Khosla's theory*
1. Loss of head of seeping water is linear	Loss of head of seeping water depends upon profile of weir floor, cut-offs, slope etc. Loss of head is definitely not linear.
2. Seeping water follows the path along the surface, in contact with the underside of the impervious floor profile.	Seeping water follows parabolic or elliptical stream line path.
3. Bligh did not give any significance to the cut-off at *D/S* end of the floor of the weir.	Khosla considered provision of cut-off at *D/S* end of the weir floor as a must.
4. In order to prevent undermining, reduction of hydraulic gradient was considered adequate measure.	Khosla relied upon the exit gradient for preventing undermining. He said that value of exit gradient at the *D/S* end of the floor should be less than the critical value for the soil.
5. This theory is very simple and requires very simple calculation work.	It is a very complex theory envolving lot of calculations. Khosla curves have however reduced this work, but still this theory is difficult to understand.

15.15 DESIGN OF A VERTICAL DROP WEIR

Vertical drop weir is such a weir, whose *D/S* face is given steep batter. The *U/S* face is mostly vertical, but it can also be given batter like *D/S* face. The design of such a weir consists of :

1. Fixing various elevations by suitable hydraulic computations.
2. Design of weir wall.
3. Design of impervious floor of the weir.
4. Design of inverted filter.
5. Design of *D/S* talus.

1. Fixing various elevations by hydraulic computations. Before we actually perform any hydraulic calculation, we must know the following data :

(a) Maximum likely flood discharge (Q).
(b) Level upto which water reaches during floods (HFL).
(c) Full supply level of the canal taking off from the river.
(d) Allowable rise in water level due to weir i.e. afflux.
(e) Lacey's silt factor (f).

Lacey's silt factor for the soil, through which canal and river are running, is found out by formula $f = 1.76\sqrt{mr}$, where mr is the mean particle diameter of silt in millimetres. Value off varies from 0.50 to as much as 1.50. Its value is less for finer silt and more for coarser silt or sand.

(i) Length of the water way, or in other words length of the weir (L) is found out from Lacey's following regime formula :

$$L = 4.75\sqrt{Q}$$

where L = Length of the weir in metres.
Q = Discharge in cumecs.

(ii) Find out the discharge per metre length (q) of the weir from relation $q = \frac{Q}{L}$.

(iii) Find out the scour depth by Lacey's following formula

$$R = 1.35\left(\frac{q^2}{f}\right)^{1/2}$$

(iv) Regime velocity (V) is found out as follows :

$$V = \frac{q}{R} \text{ m/sec.}$$

(v) Find out velocity head $= \frac{V^2}{2g}$.

We will be referring term T.E.L. very frequently. This term denotes *total energy line* (T.E.L.).

(vi) Level of *D/S* T.E.L. = *D/S* H.F.L. before weir construction + $\frac{V^2}{2g}$.

(vii) Level of *U/S* T.E.L. = *D/S* T.E.L. + Afflux.

(viii) Level of *U/S* H.F.L. *U/S* T.E.L. – $\frac{V^2}{2g}$

(ix) Crest level of the weir *U/S* T.E.L. – k.

where $k = \left(\frac{q}{1.70}\right)^{2/3}$ and $q = 1.70\,k^{3/2}$.

k is the height of *U/S* T.E.L. above the crest of the weir.

(x) Reservoir level at the back of weir = level of top of gates

= F.S.L. of canal + head loss through regulator.

Loss of head through regulator may be taken anything between 0.50 m to 1 m.

Height of gates = (g) = Level of top of gates – crest level

(xi) Level of bottom of *U/S* pile *U/S* H.F.L. – 1.5 *R*.

(xii) Level of bottom of *D/S* pile *D/S* H.F.L. after retrogation – 2*R*.

2. Design of weir wall

(*i*) *Top width.* Top width of the weir is usually denoted by a and bottom width by b.

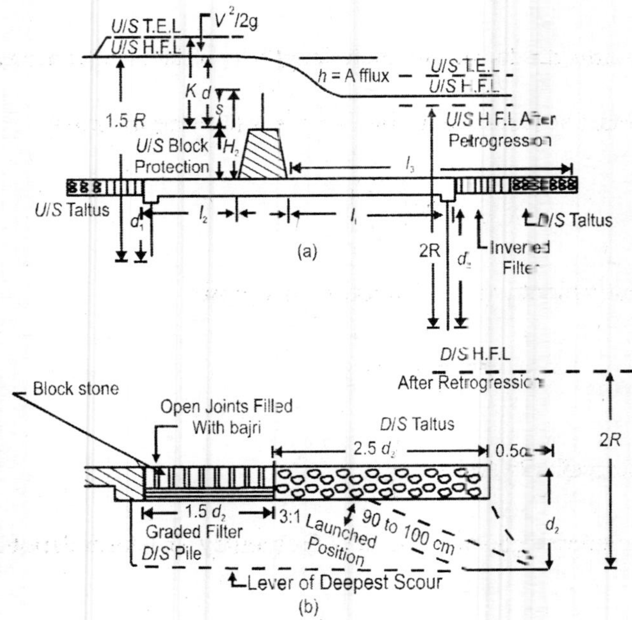

Fig. 15.10

Top width should be greater of the three values given as follows :

$$a = \frac{d}{\sqrt{\rho}}$$

$$a = \frac{3d}{2\rho}$$

$$a = g + 1 \text{ metres}$$

where g is the height of shutter fitted over the crest of the wall.

(*ii*) *Bottom width.* (b) It is calculated by equating overturning moment (M_0) to the resisting moments (M_r) taken about the outer third point of the base. There can be three states of flow.

(*a*) Water filled upto crest level or upto top of crest of gates if provided. In this case $M_o = \dfrac{w(H+g)^3}{6}$. See Fig. 15.11 (a).

$$M_r = \frac{wH\rho}{6}\left(b^2 + ab - a^2\right)$$

Equation $M_0 = M_r$, b can be found out.

(*b*) **Water passing over the crest and weir is submerged.** In this case $M_0 = \dfrac{whH^2}{2}$, and $M_r = \dfrac{wH(\rho-1)}{12}\left(b^2 + ab - a^2\right)$ equating both, value of b can be found out. See Fig. 15.11 (b).

(c) **Water is passing over the weir crest but water level on D/L side is below the crest level.** This case is a typical case, as in it, there are two variables. Depth of water above crest on *U/S* side (d) and depth of water on *D/S* side (D), both are variable. For finding a suitable relation between d and D river gauging is necessary. See Fig. 15.11 (b). According to SVK Pillai

$$d = kD, \text{ where } k \text{ is a constant.}$$

Maximum value of overturning moment (M_0) is this case is

$$M_0 = \frac{wH^3}{6}\left(1 + 2k^{3/2}\right)$$

By equating M_0 to resisting moment M_r, value of b can be found out

$$M_r = \frac{wH(\rho-1)}{12}\left(b^2 + ab - a^2\right)$$

In all the three stages of flow, *U/S* face of the weir wall has been considered as vertical.

(d) The value of b can also be found out by the following formula

$$b = \frac{H+d}{\sqrt{\rho}}.$$

where d = Depth of water above crest level.

H = Height of crest of the weir wall above the bed level.

3. Design of impervious floor of the weir wall. The critical condition for maximum percolation will exist when water is filled upto top of the crest gates and there is no tail water level. In this case maximum percolation head is H_s.

Length of impervious floor (L) according to the Bligh's theory is $L = CH_s$.

If Khosla's theory is to be used, length of horizontal impervious apron is found out by considering safe exist gradient G_E.

$$G_E = \frac{H_s}{d}\frac{1}{r\sqrt{\lambda}}.$$

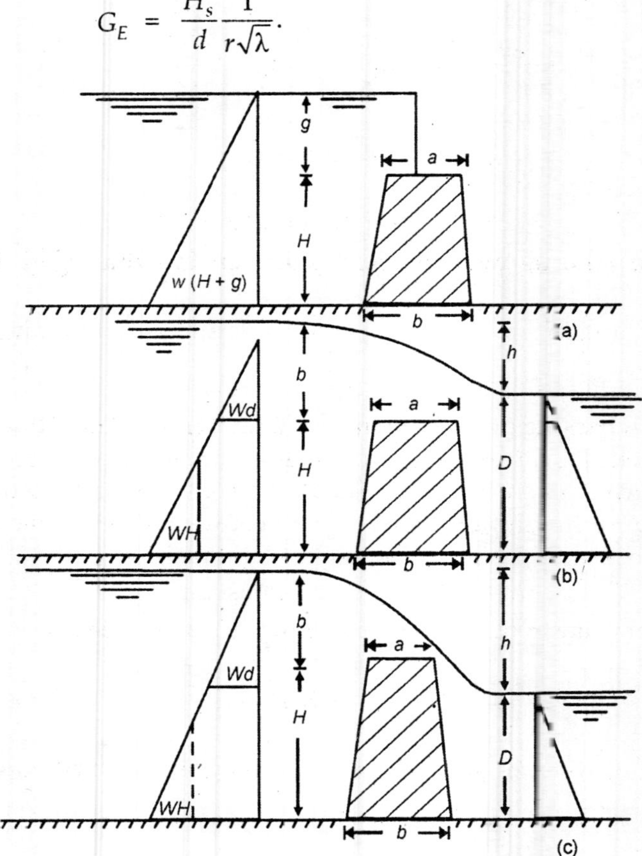

Fig. 15.13. *Three conditions of flow on weirs.*

We know G_E, H_s and d and hence value of $\frac{1}{\pi\sqrt{\lambda}}$ can be computed. Knowing value of $\frac{1}{\pi\sqrt{\lambda}}$ corresponding value of α is read out and length of impervious floor is (Remember b here is length of impervious floor) can be found out from relation $\alpha = \frac{b}{d}$.

Out of total length of impervious apron which is L in case of Bligh's theory and b in case of Khosla's theory, some minimum length has to be provided on D/S side. The remaining length of pucca apron is provided below the weir and U/S of the weir wall. Mr. Bligh gave following relations for floor lengths. He divided length of the floor in three length l_1, l_2 and l_3. See Fig. 15.12 (a).

l_1 = Length of the pucca impervious floor measured from D/S toe of the weir wall to the D/S end.

l_2 = Length of the impervious floor measured from U/S heel of the weir wall to the U/S end of the pucca floor.

l_3 = l_1 + length of D/S filter + length of D/S talus.

$$l_1 = 2.21\, C\sqrt{\frac{H_s}{13}} \text{ weir crest with no shutter.}$$

$$= 2.21\sqrt{\frac{H}{10}} \text{ weir crest with no shutter.}$$

$l_2 = L - l_1 - (b + 2d_1 + 2d_2)$, here b is bottom width of the weir wall.

$$l_3 = 18\, C\sqrt{\frac{H_s}{13} \times \frac{q}{75}} \text{ for weir wall with crest shutters.}$$

$$= 18\, C\sqrt{\frac{H}{13} \times \frac{q}{75}} \text{ for weir with no crest shutter.}$$

4. Design of inverted filter. An inverted filter is invariably provided in continuation at the end of D/S end of the impervious apron. Its main purpose is to relieve the pressure of seeping water. No design calculations are necessary. It consists of stone blocks 90 cm to 1 m thick placed over *50* to *75* cm thick layer of graded filter. If d_2 is the depth of D/S cut-off below the D/S bed, the length of the filter should be at least 1.5 d_2.

5. Design of launching apron or talus at D/S end. Immediately after D/S filter, launching apron of 2.5 d_2 length is provided. Launching apron is also sometimes called talus. When talus is launched or sloped at 3 : 1 slope its thickness should be 90 cm to 1 m. Knowing the launched length i.e. inclined length and thickness, the volume of stones can be easily calculated. Thickness in horizontal position will be slightly more than in inclined position. Launching apron prevents undermining of the weir floors and protects D/S piles.

6. Design of launching apron or talus at *U/S* end. Talus is also provided U/S of the U/S end of the impervious floor or apron. Horizontal length of this talus should be $2d_1$, where d_1 is the depth of U/S pile. In launched position (3 : 1) its thickness like D/S talus should be 90 cm to 1 m.

Example 15.6 *Design a vertical drop weir according to Bligh's theory based on following data. Check the safety of weir thus designed by Khosla's theory.*

Data. Maximum flood discharge 2770 cumecs.

Level of H.F.L. before: the construction of weir = 185.0 m.

Minimum level of D/S bed level = 178.0.

F.S.L. of canal = 184.0 m.

Permissible afflux = 1 m.

Creep coefficient = 12.

Permissible exit gradient $= \frac{1}{7}$

Value of $f = 1.0$.

If any data is missing assume its suitable value.

Solution. Length of weir $L = 4.75\sqrt{Q}$

$$= 4.75\sqrt{2770} = 250 \text{ m.}$$

Discharge passing over one metre length (q) of the weir

$$q = \frac{Q}{L} = \frac{2770}{250} = 11.2 \text{ cumecs.}$$

Scour depth $= R = 1.35\left(\frac{q^2}{f}\right)^{1/3}$

$$= 1.35\left(\frac{(11.2)^2}{1}\right)^{1/3} = 6.74 \text{ m (say)}$$

Regime velocity $= \frac{q}{R} = \frac{11.2}{6.74} = 1.66 \text{ m/sec.}$

Velocity head $= \frac{V^2}{2g} = \frac{(1.66)^2}{2 \times 9.81} = 0.14 \text{ m.}$

$\therefore$ *D/S* T.E.L. = H.F.L. before weir was constructed + $\frac{V^2}{2g}$

$$= 185.0 + 0.14 = 185.14 \text{ m.}$$

U/S T.E.L. = *D/S* T.E.L. + Afflux

$$= 185.14 + 1.0 = 186.14 \text{ m.}$$

U/S H.F.L. = *U/S* T.E.L. $- \frac{V^2}{2g}$

$$= 186.14 - 0.14 = 186.0 \text{ m.}$$

If 50 cm depth is left for retrogation then actual

D/S H.F.L. = 185.0 – 0.5 = 184.50 m

$$K = \left(\frac{q}{1.70}\right)^{2/3} = \left(\frac{11.2}{1.70}\right)^{2/3} = 3.56\,\text{m}$$

Crest level of weir = U/S T.E.L. – k

= 186.14 – 3.56 = 182.58 m.

Reservoir level = Top level of the gate.

= F.S.L. of canal + head loss at head regulator

= 184.0 + 0.5 (say)

= 184.5 m.

Height of gate g = Level of top of the gate – weir crest level

= 184.50 – 182.58 = 1.92 m.

R.L. of bottom end of U/S pile

= U/S H.F.L. – 1.5 R

= 186 – (6.75 × 1.5) = 175.88 m

Adopt R.L. of bottom end of U/S pile as 176.0 m.

R.L. of bottom end of D/S pile

= D/S H.F.L. after retrogation – 2R

= 184.50 – (6.75 × 2)

= 171.0 m.

Head of water H_s = R.L. of top of gate – R.L. of bed.

= 184.5 – 178.0 = 6.50 m.

Height of crest wall = Crest level – Bed level

= 182.58 – 178.0 = 4.58 m.

(For check $H_s = H + s =$ 4.58 + 1.92 = 6.50 m)

Design of Weir Wall

Width of top of the weir wall

Water depth above crest level

d = U/S F.S.L. – crest level of weir wall

= 186.0 – 182.53 = 3.42 m.

Thickness of crest wall top $a = \dfrac{d}{\sqrt{\rho}} = \dfrac{3.42}{\sqrt{2.24}} = 2.3$ m.

Based on sliding $a = \dfrac{3d}{2\rho} = \dfrac{3\times3.42}{2\times2.24} = 2.3$ m

Based on practical ground, $a = g + 1 = 1.79 + 1 = 2.79$ m.

$= $ say 3 m.

Hence adopt 3 m as the thickness of the top of the crest.

Thickness of Crest Wall at Bottom

First Case. *Water is head up upto top of crest gate and there is no tail water.*

$$M_0 = \frac{w(H+g)^3}{6} = \frac{1\times(6.5)^3}{6} = 45.8 \text{ t-m.}$$

$$M_r = \frac{wH\rho}{6}(b^2 + ab - a^2)$$

$$= \frac{1\times4.58\times2.24}{6}\left(b^2 + 3b - 9\right)$$

$$= 1.71\ (b^2 + 3b - 9)$$

Equation $M_0 = M_r$ we get

$1.71\ (b^2 + 3b - 9) = 45.8$

$$b^2 + 3b - 9 = \frac{45.8}{1.71} = 26.78$$

$b^2 + 3b - 35.78 = 0$

Solving $b = 4.67$ m (considering *U/S* face vertical)

Second case. *Water is flowing over the crest and weir is submerged.*

$$M_0 = \frac{whH^2}{2}$$

(For maxi. value of M_0, h is just equal to d)

For this condition $h = d\left(\dfrac{q^2}{\left(\frac{2}{3}C\right)^2\times2g}\right)^{1/3}$

Adopt $C = 0.58$

$$h = d = \left\{\frac{(11.2)^2}{\left(\frac{2}{3}\times0.58\right)^2\times2\times9.80}\right\}^{1/3}$$

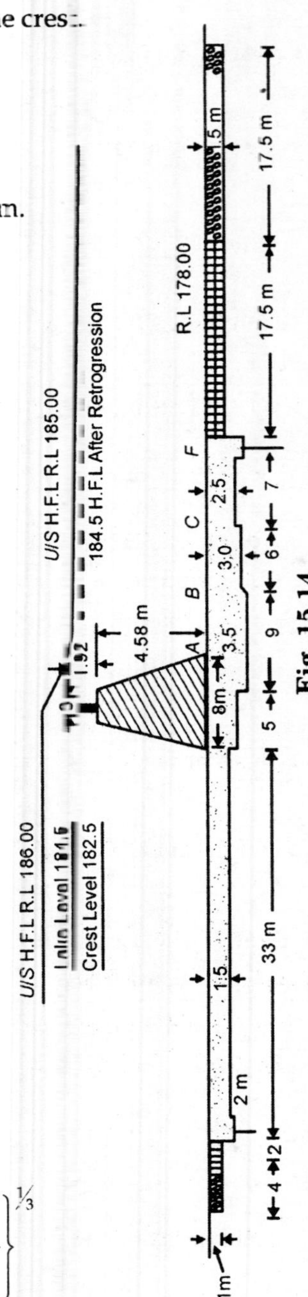

Fig. 15.14

$= 3.50$ m

$$M_0 = \frac{1 \times 3.5 \times (4.71)^2}{2} = 39 \text{ t-m}$$

$$M_r = \frac{wH(\rho - 1)}{12}(b^2 + ab - a^2)$$

$$= \frac{1 \times 4.71 \times 1.24}{12}(b^2 + 3b - 9)$$

$$= 0.4867\ (b^2 + 3b - 9)$$

Equating $M_r = M_0$

$$0.4567\ (b^2 + 3b - 9) = 39$$

$$b^2 + 3b - 9 = \frac{39}{0.4867} = 80.13$$

$$b^2 + 31b - 89.13 = 0$$

Solving $b = 8$ m.

Adopt top width 3 m and bottom width 8 m, *U/S* face should be vertical. If both the faces are given equal slope even then 3 m top width and 8 m bottom width will do.

Design of impervious apron

Seepage length according to Mr. Bligh

$$L = CH_s = 12 \times 6.5 = 78 \text{ m.}$$

D/S impervious apron $l_1 = 2.21 \times 12\sqrt{\frac{6.5}{13}} = 18.8$ m

$= 19$ m. (say).

For *U/S* impervious apron

Depth of *U/S* pile $d_1 = 178 - 176 = 2$ m.

Depth of *D/S* pile $d_2 = 178 - 171 = 7$ m

$$l_2 = L - l_1 - (b + 2d_1 + 2d_2)$$

$$= 78 - 19 - (8 + 4 + 14) = 33 \text{ m.}$$

Length of *D/S* impervious + pervious apron i.e. length (l_3)

$$l_3 = 18C\sqrt{\frac{H_s}{13} \times \frac{q}{75}}$$

$$= 18 \times 12\sqrt{\frac{6.5 \times 11.2}{13 \times 75}} = 54\,\text{m}$$

Length of filter + launching apron = $l_3 - l_1$.

= 54 – 19 = 35 m

Minimum length of filter = 1.5 d^2 = 15 × 7 =10.5 m.

Minimum length of launching apron = 2.5 d_2 = 2.5 × 7 = 17.5 m.

Minimum length of both filter and launching apron

10.5 + 17.5 = 28 m.

Since available length is 35 m. Hence adopt 17.5 m as length of filter instead of 10.5 m.

Provide 1 m thick stone blocks over 60 m graded filter

Length of launching apron 3 : 1 slope.

$$= \sqrt{10}\,d_2 = \sqrt{10} \times 7 = 22\,\text{m}$$

Volume of stones per meter length of weir = 22 m^3.

Thickness of horizontal launching apron

$$= \frac{22}{17.5} = 1.3\ \text{m}.$$

Adopt thickness of launching apron = 1.5 m.

U/S launching apron, d_1 = 2 m.

Length of launching apron = 2 m.

Provide 1 m thick stone blocks over 50 cm thick graded filter

Length of *U/S* talus = $2d_1$ = 4 m.

Volume of stone blocks 1 m thick launching apron in 3 : 1 position

$$= \sqrt{10}\,d^1 \times 1 = 6.28\ \text{m}^3$$

Thickness of launching apron in horizontal position

$$= \frac{6.28}{4} = 1.75\ \text{m} = 1.60\ \text{m}.$$

Thickness of D/S impervious floor. Provide nominal thickness of *U/S* floor 1 m thick and 1.5 m below the weir wall.

Uplift pressure at point *A* just on *D/S* side of weir wall.

$$H_A = \left(H_s = \frac{H_s}{l} \times x\right)$$

$$= 6.5 - \frac{6.5}{78}(2 \times 2 + 33 + 8) = 2.6 \text{ m.}$$

Required thickness $t = \dfrac{4}{3} \dfrac{H_A}{(\rho - 1)}$

$$= \frac{4}{3} \frac{2.6}{1.24} = 2.8 \text{ m.}$$

Hence provide floor thickness of 2.8 m for 6 m from point A.

Uplift pressure after 6 m distance on D/S side of weir wall

$$H_B = 6.5 - \frac{6.5}{78}(45 + 6) = 2.08 \text{ m}$$

Required thickness $= \dfrac{4}{3} \times \dfrac{2.08}{1.24} = 2.25 \text{ m} = 2.30 \text{ m say.}$

Provide 2.30 m thickness for 6 m from point B upto point C.

Uplift pressure at point C 12 m D/S of A

$$H_C = 6.5 - \frac{6.5}{78}(45 + 12) = 1.56 \text{ m.}$$

$$t = \frac{4}{3} \frac{1.56}{1.24} = 1.68 \text{ m.}$$

From C to D provide floor thickness of 1.70 m.

Check by Khosla's Theory

(*i*) **Check for exit gradient.** Length of impervious floor

$$b = 33 + 8 + 19 = 60 \text{ m.}$$

Depth of pile at D/S end

$$= 7 \text{ m.}$$

$$\alpha = \frac{b}{d} = \frac{60}{7} = 8.6 \text{ (approx.)}$$

From Khosla's exit curves

For $\alpha = 8.6, \dfrac{1}{\pi\sqrt{\lambda}} = 0.14$

$$\therefore \quad G_E = \frac{H_s}{d} \frac{1}{\pi\sqrt{\lambda}} = \frac{6.5}{7} \times 0.14 = \frac{1}{7.8}$$

Pressible exit gradient is 1 in 7, hence 1 in 7.8 gradient is quite safe.

Check for floor thickness. See Fig. 15.14.

Consider C_1 and D_1 points for the pile at *U/S* end.

$$\frac{d_1}{b} = \frac{1}{\alpha} = \frac{2}{60} = \frac{1}{30} = 0.033$$

From Khosla curves $\phi_{d_1} = 88\%$ and $\phi_{c_1} = 83\%$

Correction for floor thickness

$$\phi_{c_1} = \frac{88-83}{2} \times 1 = 2.5\% (+)$$

Interference correction due to pile at *D/S* end

$$C = 19\sqrt{\frac{D}{b'}}\,\frac{d+D}{b}$$

Here $b = b' = 60$ m

and $D = 7$ m, $d = 2$m

$$C = 19\sqrt{\frac{7}{60}} \times \frac{9}{60} = 0.97\% (+)$$

Corrected % pressure at

$$C_1 = \phi_{c_1} = 83 + 2.5 + 0.93$$
$$= 86.5\% \text{ (Approx.)}$$

Uplift Head at $C_1 = H_s \times \phi_{c_1}$

$= 65 \times 86.5\% = 5.63$ m (Approx.)

Uplift pressure at points E_2 and D_2 for the pile at *D/S* end.

$$\frac{d_2}{b} = \frac{1}{\alpha} = \frac{7}{60} = 0.117$$

$$\phi_{E_2} = 31\% \text{ and } \phi_{d_2} = 21\%$$

Correction in ϕ_{E_2} due to floor thickness 1.7 m.

Correction $= \left(\frac{31-21}{7}\right) 1.7 = 2.43\%$ (– ive).

Interference effect of *U/S* pile on ϕ_{E_2}.

$$= 19\sqrt{\frac{D}{b'}} \times \frac{d+D}{b}$$

$$D = 2 \text{ m}, d = 7 \text{ m}, b = b' = 60 \text{ m}$$

Correction $= 19\sqrt{\frac{2}{60}} \times \frac{9}{60} = 0.53\%$ (– ive)

Corrected value of $\phi_{E2} = 31 - 0.53 - 2.43 = 28.4\%$

Pressure head at $E_2 = 6.5 \times \frac{28.04}{100} = 1.83$ m.

If variation in pressure head is considered linear the pressure head and floor thickness at points *A*, *B* and *C* are found as follows

Point A. Pressure head at point *A*.

$$P_A = 5.63 - \frac{(5.63 - 1.83)}{60}(33 + 8)$$

$$= 3.03 \text{ m.}$$

Floor thickness at point *A*

$$= \frac{3.03}{1.24} = 2.45 \text{ m.}$$

Hence provided thickness of 2.80 m as per Bligh's method is quite safe.

Point B

$$P_B = 5.63 - \frac{(5.63 - 1.83)}{60}(33 + 8 + 6)$$

$$= 2.66 \text{ m.}$$

Thickness at $B = \frac{2.66}{1.24} = 2.14$ m.

Provided thickness of 2.30 m by Bligh's theory is quite adequate.

Point C

$$P_C = 5.63 - \frac{(5.63 - 1.83)}{60}(33 + 8 + 12)$$

$$= 2.27 \text{ m.}$$

Required thickness $= \frac{2.27}{1.24} = 1.83$ m.

We have provided a thickness of 1.7 m at point *C* and onwards which is not adequate.

Increase thickness of floor here to 2 m. For detailed design see Fig. 15.14.

15.16 SCOURING OR UNDER SLUICES

These are the openings in the weir. They are formed by depressing the weir. They have gates fitted in them. The scouring sluices are located on the side of the weir on which head regulation of the canal, taking off from here, is located. If two canals are taking off on either side of the river two sets of undersluices should be located one on either side of the weir.

Undersluices perform the following functions.

1. A clear well defined approach channel is preserved in the river just *U/S* of the canal head regulator.

2. Silt entry into the off taking canals is controlled.

3. When sluices remain closed for few days, silt gets accumulated in the get just *U/S* of the sluices. This can be easily scoured to *D/S* side by opening the undersluices time to time.

4. Low floods can be passed through them without weir shutters being dropped.

5. During high floods, they provide greater water way and thus rise in the reservoir level can be restricted.

6. They help store fair amount of receding flood water.

Capacity of undersluices depends upon many factors. However it should be fixed based upon the following considerations.

1. Twice the rated discharge of the taking off canal.

2. 10 to 20% of the high flood discharge.

3. Winter floods should pass through them without the weir crest shutters being dropped.

Design of undersluices. In this case, intensity of discharge per metre length, and crest level of the undersluices are different from the weir proper. Hence it is necessary to design them separately. The design of undersluices is done based on the principles of weir design. Following considerations, however, should be made in the design.

1. The thickness and length of impervious floor should be designed in the same way as explained for weirs.

2. Crest level of undersluices should be about 1 m below the crest level of the head regulator of the canal, if no silt excluding device is provided in the river. If silt excluding device, in form of silt excluder, has been provided in the river, the crest of head regulator should be about 2 m above the crest level of the under-sluices.

3. In the case of undersluices, the discharge is more intensive and hence more protective length on *D/S* side is required. Total length of apron and Talus on *D/S* side (l_3) should be as follows :

$$l_3 = 27C\sqrt{\frac{H_s}{13} \times \frac{q}{75}} \text{ metres,}$$

where q is the intensity of discharge in cumecs/metre.

4. Length of impervious apron on *U/S* side of the sluice gate, should be as follows

$$l_2 = 3.9C\sqrt{\frac{H_s}{13}} \text{ metres}$$

5. A bridge should be constructed on pillars of undersluices. Winches or cranes can move on this bridge, to lift or drop the gates.

6. *D/S* floor should be depressed below the *D/S* bed level. This depression is dependent upon the likely retrogression of the bed level.

15.17 DIVIDE WALL

It is also called groyne, or groyne wall. It is an embankment constructed in the river, *U/S* of the weir. Its axis is kept at right angles to the axis of the weir. The embankment is protected from all the sides with the help of stone or concrete blocks. The divide wall separates weir from undersluices. It extends a little *U/S* of canal regulator and on *D/S* end upto loose protection of the under sluices. It may be made of concrete or masonry, with top width of 1.5 m to 3 m. This wall should be designed for following conditions.

(i) Silt pressure upto full tank level on the face opposite to the face lying towards the head regulator and minimum possible or no water on the face lying towards the head regulator.

(ii) During high floods the water level behind the weir should be assumed about 1 m to 2 m above the level of water behind the undersluices.

(iii) Top width is taken 1.5 m to 2 m, while bottom width is found out on the basis that resultant of forces acting on divide wall life within the middle third of the base.

The main purposes of divide wall are as follows.

(i) To separate the floor of channel formed behind under-sluices from the floor level behind the weir proper. Level of channel is at lower level than that of weir proper.

(ii) The area enclosed between undersluices, divide wall and head regulator, is known as pocket. The divide wall provides a relatively quite pocket in front of canal head regulator. This phenomenon concentrates more silt in lower layers of water entering the pocket. The lower silt laden layers are directly passed through the undersluices on *D/S* side and thus water entering the canal is relatively clear, having very little of silt.

(iii) Divide wall serves as one of the side walls of fish ladder.

(iv) Divide wall prevents formation of cross currents and thus avoids their erosive effect. For this purpose sometimes more than one divide walls may have to be provided.

(v) When undersluices are not worked, as in case during low water in winter, silt may get deposited behind the under sluices. This silt is easily washed to *D/S* side by opening undersluice gates and thus approach channel is kept clear of silt.

15.18 FISH LADDER

Fish ladder is a fish pass provided along the divide wall to enable migrating fishes to move from *U/S* to *D/S* and *D/S* to *U/S* direction, in different seasons. Fish ladders are provided on all such works which hinder their movements. Fish ladder is always located along the divide wall as some water always remains here. Fish ladder consists of a rectangular trough having sloping floor joining water levels on *D/S* and *U/S* of the weir. The difference in water level on *U/S* and *D/S* sides of weir is divided into several water steps with the help of baffle

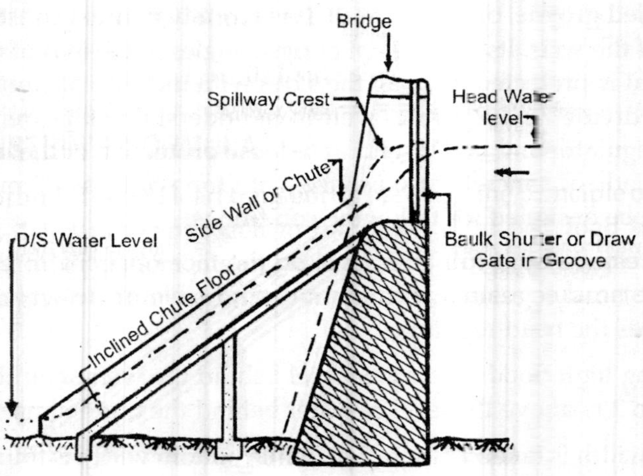

Fig. 15.12. *Log chute.*

walls, constructed across the inclined floor. The slope of the fish ladder should not be steeper than 1 : 10 so that velocity of flow in it does not exceed 2 m/sec. To exercise effective control on the flow through the fish ladder, grooved gates should be provided on *U/S* and *D/S* ends. All the baffle walls are provided with adequate sized holes so that water and fishes can smoothly move through them.

15.19 LOG CHUTE

It is an inclined platform joining *U/S* and *D/S* waters. It is provided to pass the wooden logs from *U/S* to *D/S* of the weir. Log chutes are located on diversion works of those rivers only which are being used to transport the timber from jungles. Cut timber is floated into the river, which carries them along the flow. Diversion works constructed across the river obstruct the flow of logs and as such to utilize the transportation capacity of the river all the diversion works constructed across them, are provided with log chutes. All the floating logs are assembled at the mouth of log chute. When sufficient timber logs have accumulated, the gate of the log chute is opened and all the logs are floated to *D/S* side one by one. In the case of log chute, some minimum depth of water on *D/S* side is very essential, as otherwise logs may hit the *D/S* floor and damage it. See Fig. 15.15. Log chutes may also have side walls so that logs slipping over the inclined floor may not fall at sides.

15.20 MAIN CANAL HEAD REGULATOR

It is a masonry or concrete structure, constructed at the head of the canal taking-off from the river. It is constructed *U/S* of the under sluices and located in one bank. Its alignment is kept at angle varying from 90° to 120° with the axis of the weir. The head regulator consists of a number of spans separated by piers and each span is fitted with a steel gate which can be moved up or down in the grooves made in the piers, with the help of either manual labour or winches. In old regulators, the spans used to be quite small, but the modern trend is to use larger spans of 8 to 18 m. Following are the functions of a main canal head regulator.

(i) To open or close the discharge in the canal as and when required.

(ii) To check the silt entrance into the canal.

(iii) To prevent river floods entering the canal.

Design of head Regulator

1. The water way of head regulator should be adequate. Discharge passing through a head regulator having broad crest and sloping glacis at the *D/S* is found by following formula.

$$Q = 1.7\,(L - knH)\,H^{3/2}$$

where Q = Discharge in cumecs

L = Water way length of regulator in metres

H = Head causing flow

n = Number of end contractions.

k = A constant, whose value depends upon the shape of the nose of the pier. Its value varies from 0.01 to 0.03.

2. The angle of alignment of head regulator with the axis of weir may be 90° to 120°. Greater angle is preferred from the point of view of smooth entrance of water.

3. Crest level of head regulator should be about 1 m above the crest level of undersluices. If some silt excluder device is to be located *U/S* of head regulator

in the river, this difference may be increased to about 2 m

4. Design of sloping glacis and impervious floors, is done in the similar way as for weirs D/S cut off must be provided to keep exit gradient well within the limits.

5. Piers should extend beyond the sloping glacis so as to provide support to the cistern floor against bending.

6. To prevent high flood water, spilling into the canal, R.C.C. breast wall should be provided on the *U/S* side of the regulator, from reservoir level to well above the H.F.L. This wall is supported on piers and is designed by considering its own weight and the water pressure from *U/S*.

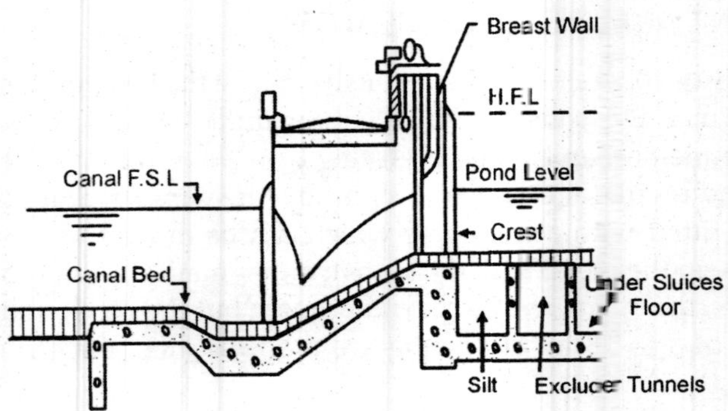

Fig. 15.13. *Main canal head regulator.*

7. A high bridge is normally provided over the piers and winches and cranes are operated from the bridge for lifting and droping the gates.

8. Length of *D/S* Talus should be about 4 to 5 times the depth of water in the canal and its thickness about 1 m. It is made of concrete or stone blocks.

15.21 REGULATION OF HEAD REGULATOR

In order to control silt entry into the canal, following two methods of regulation of a head regular may be adopted.

1. Open flow regulation.
2. Still pond regulation.

Open flow regulation. In this method undersluices are kept open to pass the surplus river water *D/S*. Bottom layers of silt laden water are lead to under sluices and passed to the *D/S*. The top layers of water having comparatively less silt, are diverted to the canal. In this operation the pocket behind the undersluices remains clear of silt.

Stilling pond Regulation. In this case a separate pocket is formed in the river and head regulator is fixed in this pocket. Only that much water is admitted in to the pocket as is required by the canal. The excess water remains passing

D/S over the weir proper. In this case water, in the pocket is almost still. This pocket is also known as still pocket or still pond. Since velocity of flow here is very small, the silt gets deposited in it. When silt reaches the level of say about 1 m below the crest of the head regulator, the gates of regulator are dropped, and gates of undersluices opened to scour off the deposited silt *D/L*.

15.22 SILT CONTROL AT HEAD REGULATOR

Excess silt entering into the main canal has to be prevented by all the possible means. If excess silt gets entered the main canal, it will cause silting of the whole of canal system and ultimately the capacity of the canal will be reduced. The silt entry into the main canal from river can be controlled by following measures.

1. Divide wall in the river creates a quite pocket behind the head regulator. The quite pocket renders top layers of water silt free and only top layers of water are admitted into the canal.

2. Provide raised crest for the head regulator. This measure automatically eliminates lower silt laden layers of water.

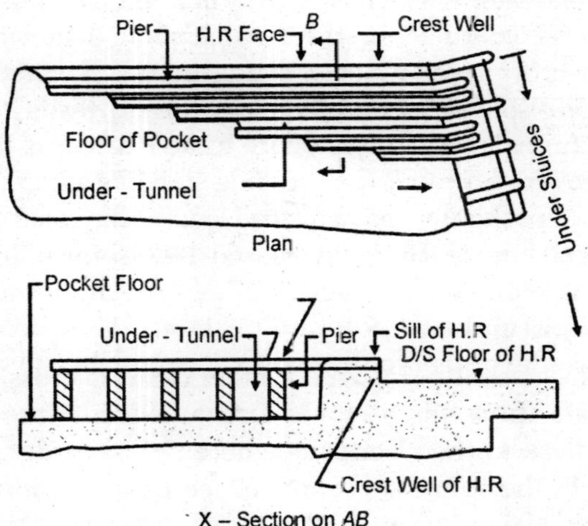

Fig. 15.14. *Silt excluder.*

3. Providing wide head regulator, also causes less entry of silt into the canal.

4. Entry of water into the canal should be smooth. Smooth entry does not cause any disturbance and as such less silt enters the canal.

5. By adopting still pond system of operation of the head regulator.

6. Installing silt excluders in the river, *U/S* of head regulator.

7. Installing silt ejectors in the head reaches of the main canal to withdraw silt laden water from the canal.

The two measures 'still excluders' and 'still ejectors' have been discussed here in details.

1. Silt excluder. It is a device by which silt laden bottom layers of river water are separated from the top comparatively clear layers of water. This device is located in the river bed just *U/S* of the head regulator. It consists of a number of rectangular tunnels, running parallel to the axis of head regulator and terminating close to undersluices. The top level of die roof slab of the excluder tunnels is kept at the same level as the head regulator crest. The silt excluder consists of a number of tunnels each of different length. The tunnel near the head regulator is almost of the same length of the width of the head regulator, but each successive tunnel decreases in length. The water which enters the tunnels is ultimately discharged to the *D/S* side by the undersluices which are kept partially open upto the level of the roofs slab on the tunnels. Usually two to three days of the undersluices are covered by the excluder. The capacity of the excluder tunnels should be about 20% of the canal discharge and minimum velocity of flow through them tunnels. 2 to 3 m/sec. Total numbers of tunnels can be worked out, once the discharge and velocity of flow are decided for the silt excluder.

Silt ejector. It is sometimes also known as silt extractor. It is a device by which the silt laden water which has already entered the main canal somehow, is extracted This is constructed in the canal at some distance *D/S* of the head regulator. It consists of curved tunnels, located across the canal. Curved tunnels start along the axis of the canal, and then take turn towards a bank. The bed of the canal, where tunnels for silt ejector are to be located is lightly depressed. The tunnels are covered by an R.C.C. roofing slab like silt excluder. The top of the roofing slab is kept slightly above the bed level of the canal. Height of tunnels is kept about the 60 cm for sandy rivers and 1.20 m for boulder stage rivers. Velocity of flow is maintained about 3 m/sec. All the tunnels are provided with gates at the exist end.

Silt ejector is located at point where main canal is crossing some natural drainage. The water from silt ejector is discharged in to the drainage, which leads this water back to the river somewhere *D/S*. In order to accelerate the velocity of flow in the tunnels, sectional area of the tunnels is reduced by streamlined vanes. The radius of bend of the tunnels varies from 10 to 15 times the tunnel width.

Bottom layers of heavily silt laden water of canal, enter the tunnels and get separated from top layers of water, having comparatively less amount of silt. The silt water entering the tunnels is led out of canal and discharged into the natural drainage by opening the gates. Silt ejectors are usually designed for 20% of the canal discharge.

15.23 EFFECTS OF CONSTRUCTION OF DIVERSION HEADWORKS ON THE BEHAVIOUR OF RIVERS

Following are the changes that take place in the behaviour of the river after the construction of a diversion head works.

1. Silt carrying capacity of river is decreased because heading up of water

causes flattening of the surface slope on the *U/S* side of the weir. Because of reduction in silt carrying capacity, the pond formed on *U/S* side starts silting.

2. Because of silting on *U/S* side, the water passing *D/S* of the weir contains less amount of silt. To make up for the deficiency of the silt, the *D/S* flowing water starts scouring the bed and banks. The scouring may lead to undermining the stability of the weir.

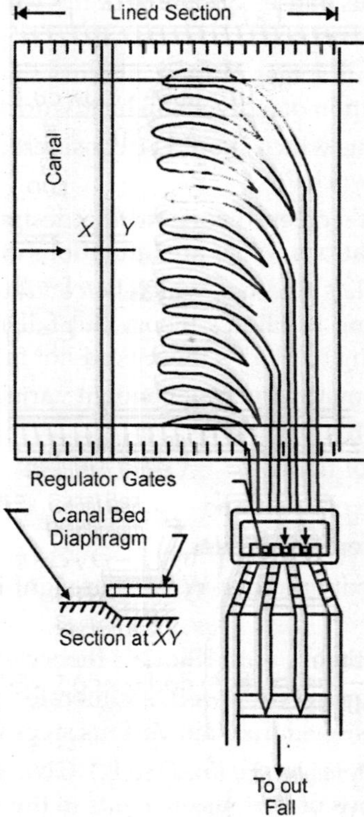

Fig. 15.15. *Silt extractor.*

3. Because of silting of the pond, the afflux goes on increasing and thus more and more areas on *U/S* are submerged. At last a stage is reached when no more silt is deposited and the normal silt charge is passed on to the *D/S* side. Devices like silt excluder also discharge more silt *D/S*.

4. The silt discharge *D/S* is more than the silt carrying capacity at low discharges. This will cause silting *D/S* side. In other words where scouring was taking place in the initial stages, silting will start. This process is known as recovery process of *D/S* bed levels. The recovery of levels on *D/S,* may lead to loss of control on silt excluding devices. Because of this, sufficient margin should be left between F.S.L. of canal and the pond level so that if need be, crest level of the head regulator could be raised later.

QUESTIONS

15.1. Draw a neat sketch of a diversion head work and name its various parts. Give brief explanation of each part.

15.2. Explain the terms weir, barrage and dam. Point out the points of difference among them.

15.3. Enumerate various causes of failure of the weirs. Discuss each cause and suggest the remedies.

15.4. Explain the various stages of river. Discuss the suitability of each stage in regard to the location of a diversion headworks.

15.5. What are the points which should be considered while selecting the site for a diversion headworks?

15.6. Explain the Bligh's creep theory for the design of impervious aprons on permeable foundations. What are limitations of this theory?

15.7. Expalin the Khosla's findings which he conducted on the weirs that were designed according to Bligh's theory but failed. Why some of the works designed according to Bligh's theory did not fail while others failed?

15.8. Explain Khosla's method of independent variables. Explain the method of applying following corrections :

(i) For thickness of floor.

(ii) For inclination of floor, and

(iii) Mutual interference of piles.

15.9. Define the term exit gradient. What is its significance in regard to the weir design ?

15.10. Compare Bligh's theory with Khosla's theory of seepage flow.

15.11. In the design of vertical drop weir, enumerate the various elements that are required to be designed. Explain various steps in the design of weir wall.

15.12. Explain the term *total energy line* (T.E.L.). Give various formulae and steps, by which elevations of various elements in the weir design are fixed.

15.13. Explain the terms :

(i) Launching apron.

(ii) Inverted filter.

(iii) Divide wall.

(iv) Log chute and

(v) Fish ladder.

15.14. What are various measures that control silt entry into the main canal at the head works ? Explain with a sketch silt excluder.

15.15. Explain the terms silt ejector, undersluices, with the help of neat sketches.

❑❑❑

16

Classification and Alignment of Canal

16.1 CANAL

Canal is an artificial channel, usually trapezoidal in section. It is constructed on the surface of the ground. It is used to convey water from river, lake, reservoir, etc., to fields for irrigation, for water supply schemes, for power generating units etc. They always flow under gravity. The canal may be Kucha or Pucca. Pucca canals are known as lined canals.

16.2 CLASSIFICATION OF CANALS

The canals can be classified in several ways. All the possible classifications are given as follows :

1. Classification based on financial output. Under this classification canals may be divided into two types:

(i) Protective canal.
(ii) Productive canal.

(*i*) *Protective canal.* The purpose of protective canal is to protect the areas most prone to famines. The canals are constructed having all the permanent works required for their regulation. No discharge of water is left in them under normal conditions. But whenever famine conditions are anticipated due to shortage or no rains, these canals are cleared by employing labour at a short notice and water is run in them to provide water for drinking as well as irrigation purposes. These canals do not give any revenue to the state.

(*ii*) *Productive canals.* These are such canals, which after deducting repair, maintenance, and supervision charges, yield revenue to the state. The revenue they yield should be instalment of initial investment plus $6\frac{1}{4}$ % interest on the total investment. Most of the irrigation canals pertain to this category of the canals.

2. Classification based upon the nature of source of supply. Under this classification, the canal can be divided into two categories:

(i) Permanent canals.

(ii) Inundation canals.

(i) *Permanent canals.* When canals are fed regularly or continuously, from a permanent source, such canals are known as permanent canals. Permanent canals have a regular, well defined section. They have permanent concrete masonry regulation works. Such canals run practically throughout the year. Such canals are also sometimes known as *Perennial canals.* These canals are closed only when either some construction is to be carried out over them or silt clearance is to be done. These canals always take off from ice fed perennial rivers.

(ii) *Inundation canals.* These are such canals which run only for the duration, during which water level in the river remains above some specified level. These canals do not have a very regular section and structures like falls etc. They are not provided with any diversion works in the river, in form of weir or barrage. They however have a head regulator. Inundation canals will be discussed in details in this chapter a little later.

3. Classification based upon the purpose of the canal. Following types of canals come under this category:

(i) Irrigation canals.

(ii) Water supply channels.

(iii) Power generating canals.

(iv) Navigation canals.

(v) Carrier canals.

(vi) Feeder canals.

All these canals are made for some specific purpose. Irrigation canals supply irrigation water to fields and water supply channels supply water to cities for drinking purpose. Power generating canals carry water to run generating unit and Navigation canals are used for the purpose of agumenting the inland transportation. Carrier canals do irrigation and side by side carry water for other canals. Feeder canals are constructed to feed two or more smaller canals.

4. Classification based upon the relative position in a given network of canals. An elaborate network of irrigation canals consists of following categories of canals:

(i) Main canal.

(ii) Branch canal.

(iii) Distributory.

(iv) Minor.
(v) Water course.

(i) Main canal. This canal takes off directly from a river or reservoir. It is generally very big. Being too big, direct irrigation is generally not done from it except in exceptional circumstances. It acts as a carrier to feed branch canals or major distributaries.

(ii) Branch canals. Irrigation area for big canals is generally very large. It may not be possible to supply irrigation water from one canal. In such circumstances the main canal is bifurcated into two or more parts, which are known as *branch canals.* Each branch canal is assigned to the task of irrigating specified area. Discharge of each branch canal is decided depending upon the area to be irrigated by each. Branches also carry quite large discharges and as such direct outlets should be given to lonely higher spots only lying along the alignment which cannot be irrigated from the distributaries. Branches act as feeder canals for distributaries.

(iii) Distributaries. Distributaries are channels carrying small discharges of say $\frac{1}{2}$ to 7 cumecs. They usually take off from branch but they can also be taken from main canal, but their discharge has to be smaller than branch canal, otherwise they will become branches. The most of the irrigation is carried out by distributaries. Outlets are located at regular intervals and water is supplied to the fields.

(iv) Minors. They are also sometimes called minor distributaries. They take off either from branch or distributaries. Mostly they take off from distributaries. Mostly area lying along the branches is quite high and cannot be irrigated by distributaties. In that case, a small minor is also taken off from the head works of some distributary and this minor is run along the branch canal. Outlets to the area lying in the vicinity of the branch are given from the minors. There may be some areas lying very low or area may be located quite far off from the distributary. In that case such areas may be irrigated by providing minors from distributary. Minors carry hardly discharge for 10–15 outlets. Hence its discharge may be from 0.25 to 0.50 cumec.

(v) Water course. They are small channels that ultimately carry water to the fields from outlets. Water courses are also sometimes known as *gools.* They may be Pucca or lined. Nowadays stress is being given for lining of canals and water courses, as lot of precious irrigation water is otherwise lost in percolation. Outlets are usually taken from distributaries and minors, but they can be taken from branches also but only in special circumstances.

5. Classification based on the alignment. Depending upon the alignment they follow, the canals can be classified into following three categories.

(i) Contour canal.
(ii) Ridge or water shed canal.
(iii) Side slope canal.

(*i*) *Contour canals.* These canals run nearly parallel to the contours of the country. Main canal taking off from a river is hostly contour canal for some length near the diversion head works. Even branch and distributaries can be contour canals. The contour choosen for the alignment should include all the contours of the area it has to irrigate. Contour canals, provide irrigation on one side only as contours of other side are higher and irrigation water cannot flow under gravity. However irrigation facilities can be provided to the area lying on higher side of the contour canal by lift canals.

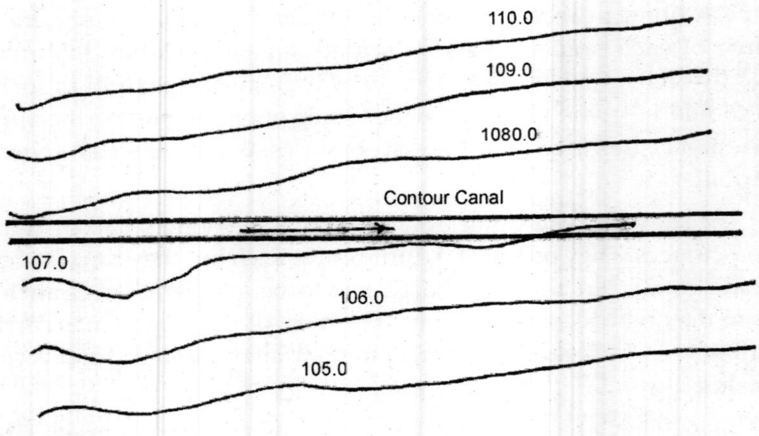

Fig. 16.1. *Contour canal*

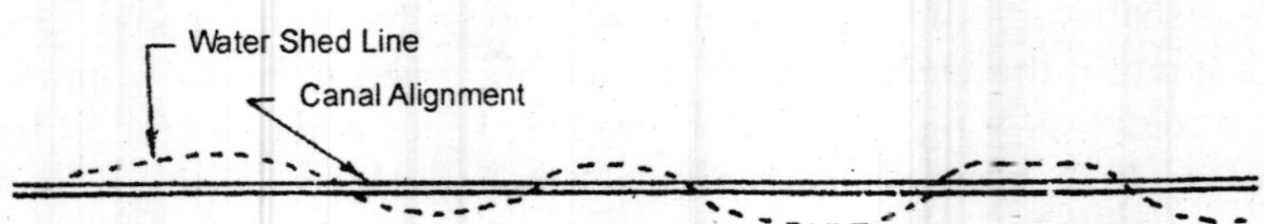

Fig. 16.2. *Water shed canal*

Contour canals have only one bank. Other side being higher does not require the second bank. These canals are also sometimes known as *single bank canals.* They may however be having two banks also. They do not follow the same contour all along. Some longitudinal slope has to be given to cause flow in the canal. Because of longitudinal slope, contour canal slowly leaves the higher contour and adopts the next lower contour.

(*ii*) *Ridge canal.* The canal which follows the ridge of the country is known as ridge canal. It generally takes off from a contour canal. It irrigates on both sides. Since this canal can irrigate areas along both the banks it commands the largest area with minimum length of canal. They do not cross any drainage and hence construction of cross-drainage works are obviated. If ridge takes a very sharp turn the canal should be aligned straight. This reduces the length of the canal but envolves construction of cross drainage work to pass run-off from enclosed area, to the other side of the canal. Also irrigation in this enclosed area cannot be done. Canals may also have to leave the watershed to bye-pass the towns and villages located on the water shed. Most of the irrigation canals are ridge canals.

(*iii*) *Side slope canals.* The side slope channels are aligned roughly at right angles to the contour canals, along the slope between the ridges and the valleys. They are roughly parallel to the natural drainage of the country. They do not intercept any cross-drainage and hence no cross-drainage works have to be constructed. Side slope canals have to be lined, as they have very steep bed slope and Kucha canal may not withstand the erosive effect of increased velocities.

6. Classification based upon the material of construction. Under this category the canals may be

(i) Kucha or unlined canals.

(ii) Lined canals.

(*i*) *Kucha or unlined canals.* The canal which runs through the natural soil of the region, is known as Kucha canal or unlined canal. The section of such a canal is trapezoidal. The side slope of the banks depends upon the nature of soil. Slopes vary from 1 : 1 to 2 : 1 in cutting and 2 : 1 to 3 : 1 in filling for general soils like soft clay, alluvial soil, sandy loan etc. These canals have to be run with restricted velocity so that erosion or sour may not take place. Large amount of water is lost by percolation. Most of the canals in India are Kucha canals. But government is aware of the shortcomings of such canals and laying more and more emphasis on lining the existing as well as new canals.

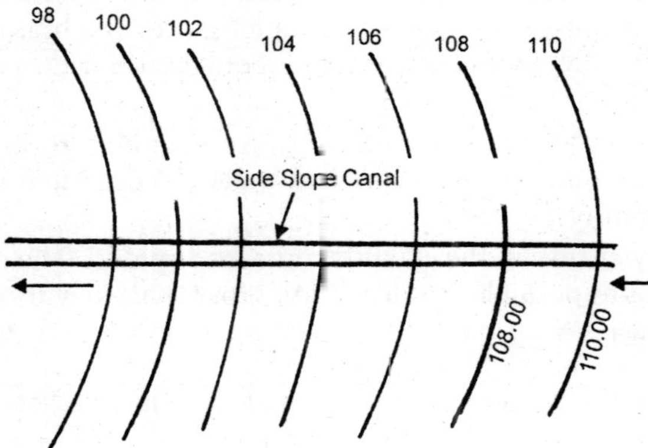

Fig. 16.3. *Side slope channel*

(*ii*) *Lined canals.* The section of such a canal in pucca section, made of some strong and impervious material. Lined canals can be run with large velocities and as such section of the canal can be considerably reduced, thus causing economy in the earth work. Lined canals do not allow any percolation loss, and wore ever smaller areas being exposed, evaporation losses are also considerably reduced. The irrigation water saved by lining of the canals can be used to provide irrigation facilities to additional areas. Lined canals have more bed slope and thus lot of command, is lost as the level of water is depressed faster. Sources of water are limited and to provide irrigation facilities to larger areas, judicious use of available water is very essential. Hence lined canals is the need of the time. If we analyse the benefits of lining of canals on long term basis, we can easily conclude that benefits would out weight the expenditure incurred for lining.

16.3 POINTS TO BE CONSIDERED WHILE FIXING CANAL ALIGNMENT

Following points should be considered while fixing the alignment of a canal :

1. The canal should be straight. Such a canal would be minimum in length. Length being minimum losses due to percolation and evaporation are bound to be minimum.

2. The canal should follow the ridge as far as possible. This will cause irrigation of the areas lying on both the sides of the canal. If somewhere canal leaves the ridge, our effort should be to catch the ridge again as soon as possible.

3. Cross-drainage works should be minimum, as such works are very costly.

4. The canal should not pass through a village or town ; but by the side of it.

5. Canal taking off from a river is a contour canal for some length. Every effort should be made to mount if the main water shed, as soon as possible.

6. Alignment should avoid deep cuttings and high embankments.

7. Idle length of the canal i.e. length of the canal not providing any irrigation, should be minimum.

8. The alignment should avoid rock, fissured, and brakish formations. Fissured formations cause loss of water through fissures and brakish formations render water useless. Rocky tracks envolve lot of labour for the construction of canals.

9. Suitable foundations for works like falls, cross regulators, head regulators etc. should be available. Besides this, materials of constructions should be available in the vicinity.

10. Unnecessary curves in the canals should be avoided. The curves should be of as large radius as possible. Radii of curves for canals, of various discharges should be as follows :

Discharge of canal in cumec	upto 0.3	0.3 to 3	3 to 15	15 to 30	30 to 80	80 and above
Radius of curve in metres	100	150	300	600	900	1500

11. The canal section should be partly in cutting and partly in filling. Cutting should equal filling if work has to be most economical.

12. Diversion works should be so located that idle length of the canal is minimum. Moreover the elevation of water being diverted to the canal, should be such that maximum possible available area is brought under command of the proposed canal.

13. To avoid excessive percolation losses the canal should not be passed through sandy tracks.

16.4 ALIGNMENT OF WATER COURSES

Though the responsibility of maintenance of the water courses is that of farmers, still in new irrigation works, the alignment, of the water courses is fixed by the government. This work is being done these days by canal area development (C.A.D.) department or collonisation department. Following points should however be considered while fixing the alignment of water courses.

1. Separate water courses should be provided for high lands and low lands of the same village.

2. They should not cross through field but should be laid along the boundries of the fields.

3. Sufficient water should be available to fields lying at the farthest point from the outlet.

4. The water courses should be aligned along the ridges of the area of the village.

5. If low lying area is near the outlet and high land is at the further end of the village, the water course which has to carry water for high lands should be made in heavy embankments and such water courses should preferably be lined.

6. In order to ensure equitable distribution of water to all the fields, the water course levels for each square of land, should be fixed and made permanent. If this is not done, the farmer whose land is low dig the water courses deep and when a farmer whose land is high takes water from the same deep water course, has to fill the water curse by the water from his own time of watering, because otherwise water will not rise to higher fields. This causes less availability of water to high fields. Again if the turn of watering of a former whose land is low, he gets water course full for water and this causes advantage to him. To curb this injustice, it is very very important to fix the water course levels for each square and these levels should be made pucca. This will make irrigation water available to all the fields at desired levels. Deep excavation of water courses will not give any advantage to the farmers whose land is low.

16.5 INUNDATION CANALS

Inundation canals are made under the following circumstances.

1. If river water level during floods remains high for considerable length of time.

2. If sufficient flood water reaches the river in March or April, the flood water may be used to submerge the land, so that Kharif crop could be sown.

3. If flood water remains available say up to late September, the Kharif crops can be irrigated upto this time and, even, areas may be submerged to sow early Rabi crops in October.

4. Areas to be irrigated when in the vicinity of the river banks the inundation canals may serve the purpose of irrigation.

5. When soil has good stabilizing power so that bed and banks of the canal remain stable.

Inundation canals are more or less similar to the permanent canals. The major difference is that, in the case of permanent canals, permanent masonry or concrete works like weirs, barrage, head regulator, fall. cross regulator, are constructed to regulate the supplies, but in case of inundation canals all these works are not there. In order to take water into the canal, river bank is cut. The rising flood water in the river enters the inundation canal through the cut, made in the bank. Sometimes in worst floods, the cut made in the river bank is eroded and develops into a large deep cut. This may cause flooding of the adjoining areas as capacity of the canal may not be that much to deal with the situation. To avoid such possibilities of flooding, a crude type of head regulator or weir may be made at the off take point.

16.6 CHARACTERISTICS OF INUNDATION CANALS

Inundation canals are mostly used in deltic and alluvial regions of the river, as river course here is generally at a higher level and course is maintained between dykes or embankments.

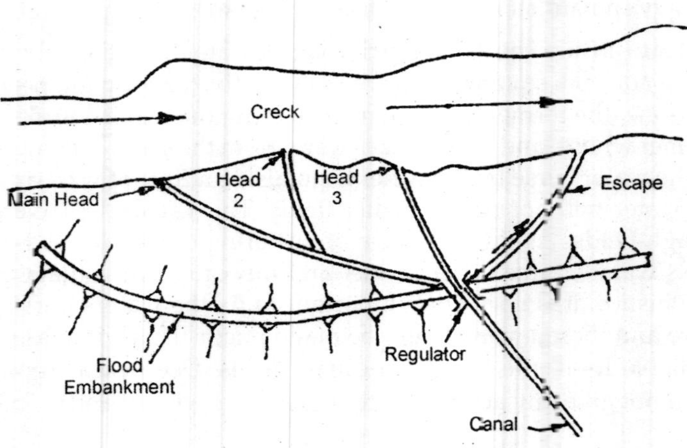

Fig. 16.4. *Inundation canal.*

The section of the canal is not regular. The banks are not very strong and may breach if not properly looked after. These canals have longitudinal, slope varying from 25 cm to as much as 1 m per kilometre length. Depth of the canal may be 1.5 m to 3.5 m. Like regular canal system inundation canal system may also have main canal and then smaller canals.

The bed level of these canals should be kept at the most at the level of minimum water level in the river. Lower the bed level of the canal, more will be the time for which water will remain available. Inundation canal may be provided with two or three heads. It is done that if one head is washed away water may be taken from the other head.

The head reach of the canal may extend to several kilometres. At the end of the head reach, a flood regulator is constructed. An escape should be provided near the regulator, so that if more water gets entered into the canal, it may be taken out and redischarged into the river *D/S*. Lot of silt is deposited in the canals and as such silt clearance should be done after every flood.

16.7 CHOOSING OFF TAKING POINTS FOR INUNDATION CANALS

Following point should be considered while choosing the off-taking point for an inundation canal.

1. The width of the river should be normal.

2. The river should be straight and stabilized.

3. If off-take point is to be on curve it should be on the outer bank of the river.

4. Off-take point should be as near the area to be irrigated as possible.

5. If one off-take point has been rendered unservicable, there should be no difficulty in locating the second off-take point.

6. The fluctuation in the level of river water should be minimum.

7. If a river bye-pass is available, off-take point should be located on the bye-pass. This measure will cause less silting of the canal.

16.8 DESIGN CONSIDERATION OF INUNDATION CANALS

1. The head regulator should be installed a few kilometres *D/S* of the off-take point. This measure would eliminate the risk of head regulator being washed away.

2. As escape should be located *D/S* of the head regulator and escape channel should be joined to the river *D/S*. If more water enters the canal at off-take point it can be taken out through the escape.

3. In order to reduce the chances of silting of the canal, the flood regulator should be provided with vertical lift gates which could be lifted in stages.

4. The full supply level of the canal should be fixed at a level at which the river water is more or less steady for a period of 40 – 50 days during which the canal can be run full to irrigate the lands. This steady level of water in the river is called *fair irrigation level*.

5. The bed of the canal should be as low as possible. Low bed level of the canal can draw some river water even when the river is in lower stage.

6. The section of the canal should be kept liberal to carry the entire required discharge as quickly as possible within the limited time factor.

7. The longitudinal slope should be such that scouring velocities are not developed longitudinal. Slope of the canal may be 0.25 m per kilometre length. The slope of the canal depends upon the general slope of the area also.

8. The flood water is heavily charged with silt. So these canals are more likely to be silted. Silt clearance must be done after every season.

9. The canals may be aligned with a number of bends so as to trap the silt at intervals.

Advantages of Inundation Canals

Following are the advantages of foundation canals.

1. It is cheap, as no head work and other works have to be constructed.

2. Water being rich in silt, has good manurial qualities.

3. The area is less liable to be water logged.

Disadvantages of Inundation Canals

Disadvantages of foundation canals are as following.

1. As there is no head works structure the head of the canal is liable to be washed away during floods.

2. Duty of water is very low.

3. Since availability of water is dependent upon the floods, the scarcity of water is always felt. Irrigation water may be badly required and the flood water in the river may not be available.

4. Due to lack of assured supply, the farmers take little interest in their work.

5. Canals will have to be frequently cleared from silt.

6. The bigger size of the canal is generally adopted. This is done to get as much of flood water as possible within limited time

7. The alignment of the canal is not very precise. It may be silting at some section, and scouring at other section.

Maintenance of inundation canals. Inundation canals always suffer from the difficulty of silting, scouring and scarcity. In order to maintain these canals in fine shape, following provisions should be necessarily made.

1. Besides the main off-take point, subsidiary off-take points should be constructed. In case, main off-take point is closed some how, water could be admitted in the canal from subsidiary off-take points.

2. A feeder canal should be constructed linking several foundation canals, taking-off from the same river. This measure avoids the necessity of constructing

several subsidiary heads. By this method good head of water can be availed for considerable length of time.

3. Provision of a pucca head regulator, attached with an escape, a few km *D/S* of the off-take point, also enables exercise good control on the inundation canal.

16.9 BANDHARA IRRIGATION

It is special type of irrigation practised in some parts of Maharashtra. It is essentially a minor irrigation scheme consisting of diversion weir walls across small streams, and small canals taking-off from the *U/S* of the weirs and commanding small tracks of lands. The bandhara irrigation scheme is very economical and by constructing a number of such structures in series across minor streams, the irrigation facilities can be economically extended

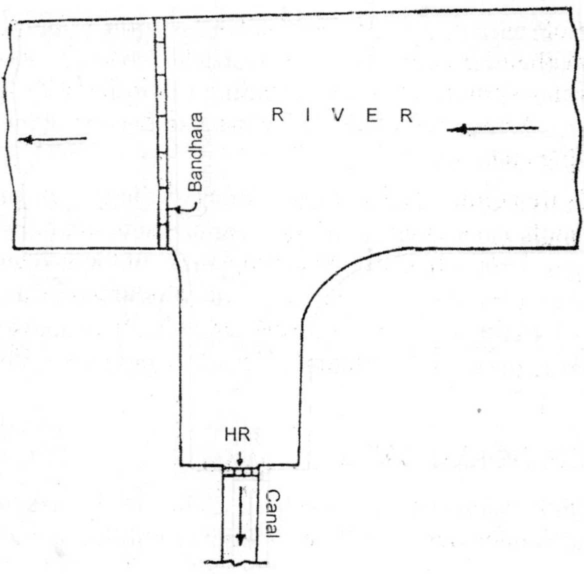

Fig. 16.5. *Bandhara irrigation*

to large areas. The irrigating capacity of each Bandhara may vary from few hectares to hundreds of hectares depending upon the availability of water. Bandhara is a name given to small weir. Bandhara irrigation schemes are applicable for small streams, and also where large irrigable area is not available at a stretch. The small amount of water of such streams is used for irrigation of the small isolated tracks of irrigable land lying in the vicinity of the streams.

16.10 LOCATION OF BANDHARA

1. The stream should preferably be running throughout the year (perennial).

2. Good foundation for the weir should be available.

3. The section of the stream should be straight, narrow and well defined.

4. H.F.L. at the site should be as low as possible.

5. The natural banks should be high so that less land is submerged on *U/S* side.

6. The canal taking off from Bandhara should not cross any important natural drainage in that area.

7. The area under the command of the canal should be adequate.

8. The land available for irrigation should be adequate and furtile.

16.11 WORKING OF BANDHARAS

A weir and an off taking channel, both form one unit of Bandhara irrigation scheme. Several Bhandara units may be constructed in series one after the other. The surplus water spilling over one bandhara is utilized by the subsequent unit of Bandhara. A head regulator is constructed to control the entry of water into the off taking channel.

The irrigable area under Bandhara is known as (Thal). A thal may be divided into phads or small blocks. One phad may contain several fields of a number of cultivators. A particular phad will be having only one type of crop in a season. For example if a particular thal has, say, three phads there can be at the most three crops, each phad having one type of crop only. In order to give rotation to the crops, the crop grown in the phad may be changed in the subsequent season or year.

16.12 DESIGN OF BANDHARA

It is designed in the same way as drop weir. H.F.L. of the stream and the F.S.L. of canal should be considered to fix the height of Bandhara and the sill level of the canal.

The crest width is kept $\sqrt{H}$ subjected to a minimum of 1.2 m. The base width is obtained by designing the section as a retaining wall. The *U/S* face is usually vertical and *D/S* face is given a slope of 1 in 2 to 1 in 5.

For calculating discharge over bandhara, the section is considered as broad-crested weir. The formula $Q = 1.7\,LH^{3/2}$ is used to find the length of the weir wall. Scouring sluices of 1 m × 1.20 m should be provided at suitable intervals to clear the silt. The canal may sometimes be provided with a scouring escape.

Advantages of Bandhara Irrigation

Advantages of bandhara irrigation are as following.

1. Initial cost of construction is very low.

2. Maintenance is generally nil as it is done by farmers jointly.

3. Water of small streams is economically used, otherwise this water would have gone waste.

4. Percolation and evaporation losses are very low as length of canals is very small.

5. Area to be irrigated being very near, duty of water is very high.

6. Small isolated irrigable areas, for which no big scheme is possible are put to most economical use.

Disadvantages of Bandhara Irrigartion

There are some disadvantages of bandhara irrigation.

1. The area to be irrigated is fixed and hence if more water is available it cannot be properly utilised.

2. The supply of water is not reliable in summer as the discharge in small streams is appreciably reduced.

3. The population living *D/S* and using stream water for its domestic purposes goes without water during dry periods.

QUESTIONS

16.1. Give various classifications of canals, with brief description of each classification.

16.2. What points should be considered while choosing alignment of a canal ?

16.3. What do you understand by inundation canal ? Under what circumstances they are constructed ?

16.4. (a) Enumerate the design considerations of an inundation canal.

(b) What points should be considered while choosing off taking point for an inundation canal ?

16.5. What is meant by Bandhara ? Describe its working. Also state the circumstances under which Bandhara is considered most suitable mode of irrigation. Enumerate advantages and disadvantages of Bandhara irrigation scheme.

16.6. Write short notes on:

(i) Ridge canal
(ii) Side slope canal
(iii) Distributory
(iv) Lined canal
(v) Main canal

❑❑❑

17

Kennedy's and Lacey's Theories

17.1 FLOW IN OPEN CHANNELS

The flow in open channels, is steady, uniform, and under gravity. In order to develop suitable velocity of flow suitable longitudinal slope has to be given to the open channels. A steady and uniform flow is said to prevail when discharge at all the sections in a particular length of the channel remain same as long as depth of flow does not change. A French Engineer Mr. Chezy gave following equation for flow of water in the channels

$$V = C\sqrt{RS} \tag{17.1}$$

where V = Mean velocity of flow in m/sec

R = Hydraulic mean depth in metres

S = Longitudinal slope of the channel

C = Chezy's constant whose value depends upon the shape and surface of the channel.

Various formulae have been given by different authors, to determine the value of Chezy's constant C. But formulae given by Kutter and Manning, have found the wide acceptance and they are mostly being used. According to Kutter, value of C is given by the following formula :

$$C = \frac{23 + \frac{1}{N} + \frac{0.00155}{S}}{1 + \left(23 + \frac{0.00155}{S}\right)\frac{N}{\sqrt{R}}} \tag{17.2}$$

Putting this value in Chezy's formula, we get

$$V = \frac{23 + \frac{1}{N} + \frac{0.00155}{S}}{1 + \left(23 + \frac{0.00155}{S}\right)\frac{N}{\sqrt{R}}}\sqrt{RS} \qquad (17.3)$$

This formula, though quite cumbersome, gives quite satisfatory results.

According to Manning, $C = \frac{R^{1/6}}{N}$

Substituting in place of C in Chezy's equation, we get

$$V = \frac{1}{N} R^{2/3} S^{1/2} \qquad (17.4)$$

where N = a roughness coefficient ; whose value for both Kutter as well as Manning's formula is same within practical ranges.

It is also known by name *Rugosity coefficient.* The value of N varies with the physical roughness of bottom and sides of the channel. The values of N recommended by I.S.I. for excavated channels are given in Table 17.1. The value of N is also influenced by channel conditions such as extent of weeds, silting, scouring, etc.

Table 17.1. *Value of N for Excavated open Channels*

Type of Soil	*Value of N*
Earth channel clean straight and uniform.	0.016 to 0.020
Earth channel clean after weathering.	0.018 to 0.025
Earth channel having short grass and few weeds	0.022 to 0.033
Rock cuts smooth and uniform.	0.025 to 0.040
Rock cuts irregular.	0.035 to 0.050

Mr. Buckley had recommended the value of N for alluvial rivers as follows.

Table 17.2. *Buckley's Values of N*

Condition of the channel	*Value of N*
Condition very good	0.0225
Condition very good	0.0250
Condition very Indifferent	0.0275
Condition very Bad	0.0300

The minimum value of N as observed in unlined channels running in alluvial soils is 0.02. General value of N may be taken lying between 0.025 and 0.03.

Central Board of irrigation has recommended value of N between 0.025 to 0.03 for alluvial soil channels. The lower values should be adopted for main and branch canals and higher values for smaller channels such as distributaries and minors.

17.2 DESIGN OF CHANNEL BASED UPON THE MAXIMUM PERMISSIBLE VELOCITY

The mean velocity of flow in an open channel can be found out, by using Manning's or Kutter's formula ; for known values of N, longitudinal slope and sectional dimensions. If maximum permissible velocity of flow is given, the channel dimensions for a particular longitudinal slope (S) can be determined.

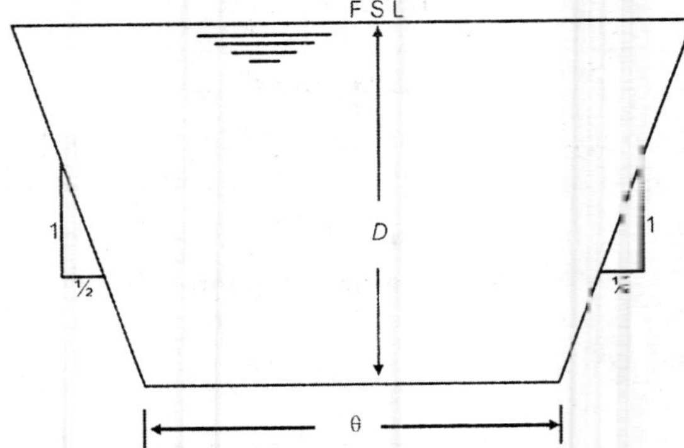

Fig. 17.1. *Trapezoidal section of canal*

The channels running in erodable soils, are designed, on the basis of maximum permissible velocity, also known as non-erodible velocity. Non-erodible velocity is the maximum mean velocity that will not cause erosion of the channel. The non-erodible velocities, for different types of soils, as recommended by CWCRS Poona are given in Table 17.3.

Table 17.3. *Recommended Maximum Permissible Velocities of Flow*

Type of soil	*Maximum permissible velocity in m/sec*
1. Disintegrated rocks, and gravel	1.5
2. Murum and hard soil	0.9 to 1.1
3. Sandy loam, black cotton soils and similar soils	0.6 to 0.9
4. Very light loose and average sandy oil	0.3 to 0.6
5. Ordinary soils	0.5 to 0.9

The channels running in alluvial soils may be designed based on the principle of minimum permissible velocity, also known as non-silting velocity. Such design criteria is adopted where silting of the channels is anticipated. Non-silting velocity is that minimum velocity which would not start sedimentation. This velocity is very uncertain and cannot be found out easily.

Plate 17.3 (To be given in between Pages 472-473)

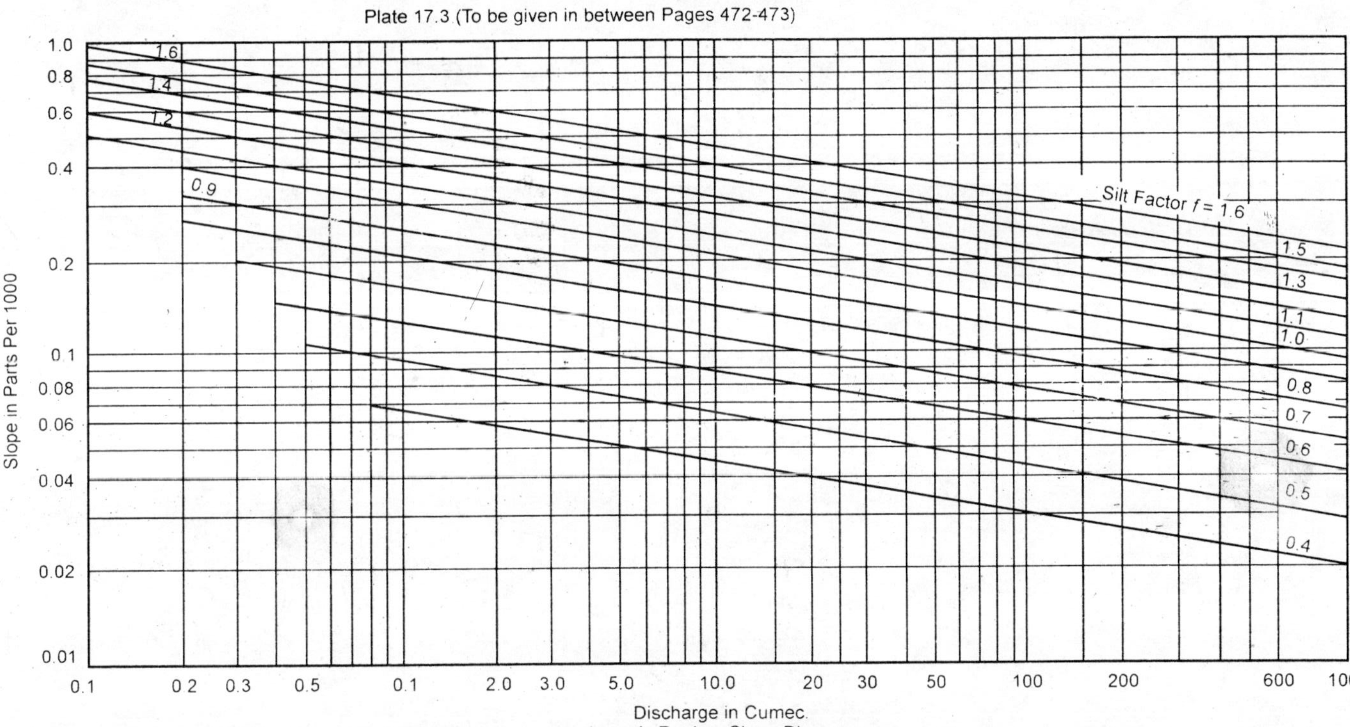

Discharge in Cumec.
Lacey's Regime Slope Diagram
Discharges 0.1 to 600 Cumec Silt Factor 0.4 to 0.6

Most of the canals in India run through alluvial soils and they are also unlined.The rivers supplying water to these channels always carry sediment rolling on their bed or held in suspension. Very high velocity will erode the bed, whereas very low velocity will cause silting. The designed canal should carry such a velocity of flow that neither the erosion nor silting takes place.

Example 17.1 *Find out normal water depth and velocity of flow in the channel carrying a discharge of 5 cumecs and having a bed width of 4 m. Side slope of the channel is 1 : 1.*

N = 0.0225 and bed slope S = 0.0015.

Solution. Area of channel $(A) = (B + D)\,D$

Wetted perimeter $(P) = \left(B + 2\sqrt{2D}\right)$

Apply Manning formula

$$V = \frac{1}{N} R^{2/3} S^{1/2}$$

$$R = \frac{A}{P} = \left[\frac{(B+D)D}{B+2\sqrt{2D}}\right]$$

$$V = \frac{1}{0.0225} \times \left(\frac{(B+D)D}{B+2\sqrt{2D}}\right)^{2/3} \times (0.0015)^{1/2}$$

$$= 1.7213\left(\frac{(B+D)D}{B+2\sqrt{D}}\right)^{2/3} \qquad \text{(From Eq. 17.1)}$$

Also $\qquad Q = A \times V$

$$= (B + D)\,D \times 1.7213\left[\frac{(B+D)D}{B+2\sqrt{2D}}\right]^{2/3}$$

$$= 1.7213\frac{\left[(B+D)D\right]^{5/3}}{\left(B+2\sqrt{2D}\right)^{2/3}} \qquad \text{(From Eq. 17.2)}$$

Value of $\qquad Q = 5$ cumecs.

Solve Eq. (2) by trial and error method

$$\frac{1.7213\left[(B+D)D\right]^{5/3}}{\left(B+2\sqrt{2D}\right)^{2/3}} = 5$$

$$\frac{1.7213[(4+D)D]^{5/3}}{(4+2\sqrt{2D})^{2/3}} = 5$$

Use first trial value of D as say 0.80 m

$$\frac{1.7213[(4+0.8)0.8]^{5/3}}{(4+2\sqrt{2\times 0.8})^{2/3}} = 4.77.$$

Let us make a trial with 0.9 m depth.

$$\frac{1.7213\times[(4+0.9)0.9]^{5/3}}{(4+2\sqrt{2\times 0.9})^{2/3}} = 5.83 \text{ cumecs}$$

Adopt value of depth as 0.85 m.

Value of discharge works out to be 5.29 cumecs.

Hence the correct value of depth will most probably be 0.82 m.

Putting values of B = 4 m and D = 0.82 m in Eq. (1), we get

$$V = 1.7213\left(\frac{(B+D)D}{B+2\sqrt{2D}}\right)^{2/3}$$

$$= 1.7213\left(\frac{(4+0.82)0.82}{4+2\sqrt{2\times 0.82}}\right)^{2/3}$$

$$= 1.7213\left(\frac{3.9524}{6.3198}\right)^{2/3}$$

$$= 1.2589 \text{ m/sec. } \textbf{Ans}$$

Example 17.2. *Design a channel with side slope 1:1 and carrying 10 cumecs at a slope of 0.0016. Maximum permissible velocity in channel is 1.5 m/sec.*

Solution. Value of N =0.025

Area $A = (B + D)\, D$.

Wetted perimeter = $B+2\sqrt{2D}$

$$R = \frac{A}{P} = \frac{(B+D)D}{B+2\sqrt{2D}}$$

$$V = \frac{1}{N} R^{2/3} S^{1/2}$$

$$= \frac{1}{0.025} \times \left(\frac{(B+D)D}{B+2\sqrt{2D}}\right) \times (0.0016)^{1/2}$$

$$1.5 = \frac{1}{0.025}\left[\frac{(B+D)D}{B+2\sqrt{2D}}\right]^{2/3} \times (0.0016)^{1/2}$$

$$\left[\frac{(B+D)D}{B+2\sqrt{2D}}\right]^{2/3} = 0.9375 \qquad (1)$$

$$A = \frac{Q}{V}$$

$$(B+D)D = \frac{10}{1.5} = 6.667 \qquad (2)$$

Substituting the value of $(B + D) D$ from (2) into (1)

$$\left(\frac{6.667}{B+D2\sqrt{2}}\right)^{2/3} = 0.9375$$

$$B + D2\sqrt{2} = 8.439$$

$$B = 8.439 - 2.828\,D \qquad (3)$$

Putting value of B from (3) in (2), we get $(B + D) D = 6.667$

$$(8.439 - 2.828D + D)\,D = 6.667$$

$$8.439D - 1.828D^2 - 6.667 = 0$$

$$1.828D^2 - 8.439D + 6.667 = 0$$

Solving $D = 1.01$ m, $B = 5.583$ m **Ans.**

17.3 SILT THEORIES

The canals taking off from rivers draw silt along with water. This silt is either carried in suspension or carried rubbing the bed of the canal. Silt poses a difficult problem in channel design on alluvial soils. The silt carrying capacity of flowing water is dependent upon the velocity of flow. If the velocity of flow is very small, silt load is dropped on the bed and consequently the capacity of the canal is affected i.e. reduced. On the contrary, if velocity is increased beyond certain limit, the flowing water would scour the canal bed to make up its silt load. This results in the scouring of the canals. Hence the canals should be designed, with such a velocity of flow that would neither cause silting nor scouring of the canal, so that capacity of the canal is maintained.

Many Engineers have worked on various existing channels so as to evolve some relationship among the parameters that affect the stability of the section of the canal.

The pioneer work in this directions was done by Mr. R.G. Kennedy in Punjab and later by Gerald Lacey in Uttar Pradesh. Work of both of them particularly that of Geraled Lacey has been universally accepted for the design of irrigation canals.

In the design of an irrigation canal, following information is known.

1. Discharge for which canal is to be designed.

2. Rugosity coefficient (N) which is dependent upon the soil properties in which canal is to be constructed.

3. Silt factor (f) which is dependent upon the coarseness of the silt, carried by the canal water.

For the complete design of an irrigation channel, following unknowns have to be determined.

1. Area of cross-section of the canal (A).

2. Hydraulic mean depth R.

3. Velocity of flow (V) in the canal.

4. Bed slope or longitudinal slope of the canal.

For determining these four elements we have only two following hydraulic equations.

1. $Q = A \times V$ (1)

2. $V = f(N \times R \times S)$ (2)

Equation (1) is known as continuity equation and equation (2) as flow equation which may be Manning's or Kutter's equation.

We have enlisted four unknowns and there are only two equations available to us. Four unknowns cannot be evaluated from two equations only. Hence either two additional equations will have to be formulated or out of four unknowns, two will have to be fixed. Following four criterions are possible by which either two additional equations are generated or values of two unknowns are fixed.

(i) Fix some ratio of B/D based on experience.

(ii) Fix available ground slope as the slope governing the slope of the canal.

(iii) Fixing the Limiting value of velocity so that no scouring or silt takes place.

(iv) Fixing a relation between A and R for a best discharging section of the canal.

Mr. Wood's prepared Table giving relation between discharge and B/D ratio. Criterion (ii) and (iii) have been utilized by Mr. Kennedy in the design of the canals. Criterion (iv) is not considered for alluvial soils for which non silting and non scouring velocity is a must. Lacey, in his work, gave four equations for the complete evaluation of the four unknowns. He did not depend upon equations by Manning or Kutter.

17.4 KENNEDY'S SILT THEORY

Mr. R.G. Kennedy was Executive Engineer of Punjab PWD. He carried out most of his investigations on some of the canal reaches in the Upper Bari Doab canal system. He selected some straight reaches of the canal section which has not posed any silting or scouring problem during the past 30 years or so. Based upon his observations he concluded that.

(i) The silt supporting power of a channel, is mainly dependent upon the generation of the eddies, rising to the surface. These eddies are generated due to the friction of the flowing water with the channel surface. The vertical component of these eddies tries to move the sediment up, while the weight of the sediment tries to bring it down, thus keeping the sediment in suspension. Since only vertical component of eddies is effective for keeping the silt in suspension the silt supporting power is therefore proportional to the bed width of the stream and not wetted perimeter. If the velocity is sufficient to generate these eddies, so as to keep the sediment just in suspension, silting of the canal will be avoided.

(ii) Kennedy also defined the critical velocity (V_0) in the channel as the mean velocity which would just keep the channel free from silting or scouring. He related critical velocity (V_0) to the depth of flow by following equation.

$$V_0 = 0.55\, D^{0.64} \quad \text{...(17.5)}$$

This formula is true for Upper Bari Doab canal system and not for all the conditions. He later realized this shortcoming and introduced a factor m known as *critical velocity ratio* (C.V.R.). The value of m depends upon the type of soil, the canal traverses, through. The modified form of Kennedy's equation is

$$V_0 = 0.55\, m\, D^{0.64} \quad \text{(17.6)}$$

where $m = \dfrac{V}{V_0}$ = C.V.R = critical velocity ratio

V_0 = Velocity of flow as obtained by Kennedy's equation.

and V = Actual velocity of flow.

Table 17.4. *Values of C.V.R. for different types of soils*

Type of silt	*Value of C.V.R. i.e. m*
1. Silt of river Indus of Pakistan.	0.70
2. Light sandy silt in the rivers of Northern India.	1.0
3. Some what coarser silt, sandy silt.	1.10
4. Sandy loamy silt.	1.20
5. Rather coarser silt or debris of hard soils.	1.30

As a general rule, for sands coarser than the standard, the values of m were given from 1.0 to 1.2 and for sands finer than the standard, m was valued between 0.9 to 0.8.

For channels carrying lot of bed and suspended loads, the value of m may be taken, as 1.1 for the head reaches and 0.85 for the tail reaches.

The Kennedy's Eq. (17.5) may be written in general form as follows.

$$V_0 = CD^n \qquad (17.7)$$

It was observed by different investigators that not only the value of C varies according to the silt grade, the value of n also varies. Based upon this concept the values of C and n for different regions are given as follows.

Table17.5. *Values of C and n for different regions*

Region	*Value of C*	*Value of n*
1. Egyptian Canals.	0.283	0.75
2. Godavari Delta.	0.391	0.55
3. Krishna Western Delta.	0.530	0.52
4. Lower Chenab Canal.	0.567	0.57
5. Upper Bari Doab Canal System.	0.550	0.64

The biggest shortcoming of the Kennedy's theory was that it did not give any equation for the slope of the canal. The slope is decided as per the general slope of the area. By giving different values to the slope, different sections for the canal can be worked out for the same discharge and Kennedy's theory does not give any indication as which of these sections would suit best for a particular discharge. This deficiency of Kennedy's theory was removed to a very large extent by Mr. Wood. He prepared Tables relating B/D ratio with the discharge of the canal. These Tables are known as *Wood's normal Tables.* These Tables were prepared for Punjab and they are still very much in use. Woods Table is given here as a specimen.

Table 17.6. *Woods Normal Design Table N = 0.0225*

Discharge		$\frac{B}{D}$ *ratio*	B	D	*Slope 1 in*	$\frac{V}{V_0}$	*Mean V in m/sec*
Q in cumecs	*Q cumecs*						
10	0.28	2.9	1.45	0.49	3333	0.92	0.344
25	0.71	3.4	2.21	0.66	3636	0.01	0.424
50	1.42	3.7	3.13	0.84	4000	1.0	0.476
100	2.84	4.2	4.42	1.04	4444	1.0	0.555
250	7.10	4.5	6.70	1.43	4444	1.01	0.700
500	14.20	5.7	9.75	1.72	5000	1.0	0.775
1000	28.40	7.6	15.28	1.98	5000	1.03	0.880
2000	57.0	11.3	25.41	2.26	5714	1.03	0.945
5000	14.0	22.5	56.50	2.50	6666	0.98	0.975
10000	28.0	41.0	105.0	2.59	666	1.02	1.03
			110.0	2.68	8000	1.93	0.955
20000	560	78.0	212.0	2.72	8000	0.94	0.975

Kennedy's theory does not consider any specific width, shape and slope of the channel. Hence before proceeding with the design of a channel by this theory, trial values of these parameters are assumed. The velocity worked out for the assumed parameters, should satisfy the Kennedy's equation and also the velocity of the flow should not be less than the critical velocity.

17.5 DESIGN OF A CHANNEL BY KENNEDY'S THEORY

Before we start design work by Kenndey's theory we know the discharge (Q) of the channel, Slope (S) of the channel, Rugosity coefficient (N), and C.V.R. i.e. m. For design purpose we use following equations.

1. $$Q = AV \tag{1}$$

2. $$V = \frac{23 + \frac{1}{N} + \frac{0.00155}{S}}{1 + \left(\frac{23 + 0.00155}{S}\right)\frac{N}{\sqrt{R}}} \times \sqrt{RS} \tag{2}$$

3. $$V_0 = 0.55 \text{ m } D^{0.64} \tag{3}$$

Design steps

1. Assume trial value of D. Put this value in equation (3) above and compute V_0. This is the critical velocity required for this trial depth.

2. Find A by equation $A = \dfrac{Q}{V_o}$.

3. Write A in terms of B and D, assuming side slope of ½:1. If some different side slope is specified, it should be adopted in place of $\frac{1}{2}:1$ From this, value of B is worked out. For assumed value of D and worked out value of B, find out the hydraulic mean depth (R).

4. Using the worked out value of R, find out the actual velocity of flow for the assumed value of D. For this use Eq. (2) i.e. Kutter's equation.

5. If the value of velocity worked by Kutter's equation equals the velocity obtained by step (1) the assumed value of D is correct. Otherwise repeat the calculations with changed value of D. This process is continued till velocities obtained by Kutter's equation and Kennedy's equation are almost equal.

Example 17.3 *Design a channel for 40 cumecs discharge by Kennedy's Theory. Other given data is as follows,*

$$N = 0.0225$$

$$m = 1$$

$$S = 16 \text{ cm/km.}$$

Solution. Assume 1.8 in depth of water as a first trial.

$$V_0 = 0.55 \text{ m } D^{0.64}$$

$$= 0.55 \times 1 \times 1.8^{0.64} = 0.8 \text{ m/sec.}$$

$$A = \frac{Q}{V_o} = \frac{40}{0.8} = 50 \text{ m}^2$$

Side slope has been taken as $\frac{1}{2}:1$

$$A = B \times D + \frac{D^2}{2}$$

$$50 = B \times 1.8 + \frac{(1.8)^2}{2}$$

$$B = 26.9 \text{ m.}$$

Perimeter $P = B + \sqrt{5}D$

$$= 26.9 + 1.8\sqrt{5}$$

$$= 30.9248 \text{ m}$$

$$R = \frac{A}{P} = \frac{50}{30.9248} = 1.62 \text{ m}$$

$$V = C\sqrt{RS}$$

$$S = \frac{16 \text{ cm}}{1000 \text{ m}} = \frac{0.16}{1000}$$

$$C = \frac{23 + \frac{1}{0.0225} + \frac{0.00155}{0.16} \times 1000}{1 + \left(23 + \frac{0.00155}{0.16} \times 1000\right)\frac{0.0225}{\sqrt{1.62}}} = 49.$$

$$V = C\sqrt{RS}$$

$$= 49\sqrt{\frac{1.62 \times 0.16}{1000}} = 0.79$$

$$m = \frac{0.79}{0.80} = 1$$

Hence assumed value of Depth i.e., $D = 1.80$ m is adequate.

Hence adopt width 26.9 m i.e. 27 m say, and depth 1.8 m.

17.6 GARRET DIAGRAMS

As we saw in the solution of Example 17.3, lot of calculation work is involved in the design of irrigation channels by Kennedy's and Kutters equations. To safe mathematical calculations, graphical solution of Kennedy's and Kutter's equations was evolved by Mr. Garret. The Garrets diagrams establish relationship among bed slope, discharge, depth of flow and critical velocity of flow. The diagrams are shown in two plates i.e. 17.1 and 17.2. Bed slope of the canal is indicated on the vertical axis on left hand side of the graph, whereas depth of flow and critical velocity of flow are shown on right hand side along the vertical axis. The discharge (Q) in cumecs is plotted along the horizontal axis.

Design procedure by Garret diagrams. The discharge (Q), bed slope (S), rugosity coefficient (N), value of C.V.R. are given. The design of the canal by Garret diagrams can be done in following steps.

1. Locate the point (A) of intersection of the given slope and the curve of given discharge.

2. From point A draw a vertical line intersecting the various bed width curves.

3. For different bed widths (B), the corresponding values of water depth (D) and critical velocity (V_0) are read on the right hand side vertical axis of the graph. Each bed width (B) and corresponding depth (D) would satisfy Kutter's equation and is capable of carrying the required discharge at the given slope and value of N. Select one such value of B and D and determine the actual velocity of flow i.e., V.

4. Find out the C.V.R. i.e. ratio $\frac{V}{V_o}$ where V is the calculated velocity of flow by Kutter's formula and V_0 is the velocity as read from the graph.

5. Value of C.V.R. should be very nearly equal to one or as specified.

If difference is substantial repeat the procedure with changed values of B and D.

Garret diagrams have been drawn for a trapezoidal channel with side slopes as $\frac{1}{2}:1$ ($\frac{1}{2}$ horizontal : one vertical). This slope is adopted on the assumption that irrigation channels adopt approximately this shape even though they were constructed on different side slopes.

A monogram has been provided on the top of these diagrams. (Plate 17.1 and 17.2). This facilitates the use of the same curves for different values of N. An arrow on the Monograph represents the value of N for which the curves have been drawn. When the same curves are to be used for some other value of N, the point of intersection of discharge and slope curves has to be shifted to the same direction and extent, by which value of N for which curves are being used are shifted from the value of N for which the diagrams have actually been drawn.

Example 17.4 *Design an irrigation channel be Kennedy's theory for the following Data.*

$$Q = 2.83 \text{ cumecs}, \ S = \frac{1}{5000}, \ N = 0.0225, \ \frac{V}{V_o} = 1.$$

Solution. The results read from the Garret diagrams can be represented in Tabular form as follows.

Q	B	D	$A = BD + \frac{D^2}{2}$	$V = \frac{Q}{A}$	V_0	$\frac{V}{V_o}$	Remarks
2.83	9	0.695	6.502	0.435	0.432	1.01	Ist trial
	6.0	0.875	5.632	0.504	0.502	1.00	IInd trial

Hence width of canal 6 m and depth 0.875 m are the most fitting values.

Example 17.5 *Design an irrigation channel by Garret diagrams for the following Data.*

$$Q = 2.50 \text{ m}^3, \ S = \frac{1}{5000}, \ N = 0.0225 \ \frac{V}{V_0} = 1.05.$$

Solution. Using Garret diagram, various possible sections are as follows :

Q	B	D	A	$V=\frac{Q}{A}$	V_0	$\frac{V}{V_0}$	*Remarks*
2.50	10	0.60	6.18	0.405	0.396	1.02	Small
	11	0.563	6.34	0.392	0.378	1.035	Small
	12	0.536	6.48	0.380	0.364	1 045	Satisfactory

In this example, based upon velocity of flow, 12 m wide and 0.536 m deep section in suitable. But this section has very small depth and too large the width. This section would cause excessive evaporation and require wide strip of land. From practical considerations ratio of B/D should remain within specified limits. The slope of the channel is adjusted thereafter.

In Uttar Pradesh (U.P.) B and D for channels having discharge upto 15 cumecs are related as follows.

$$D = 0.5\sqrt{B}$$

For channels carrying discharge more than 15 cumecs depth is taken as follows:

Q	15	30	75	150	300
D	1.70	1.85	2.30	2.60	3.00

According to recommendations of central water and power commission (CWPC) B/D ratio for discharges varying from 0.3 to 300 cumecs should be read

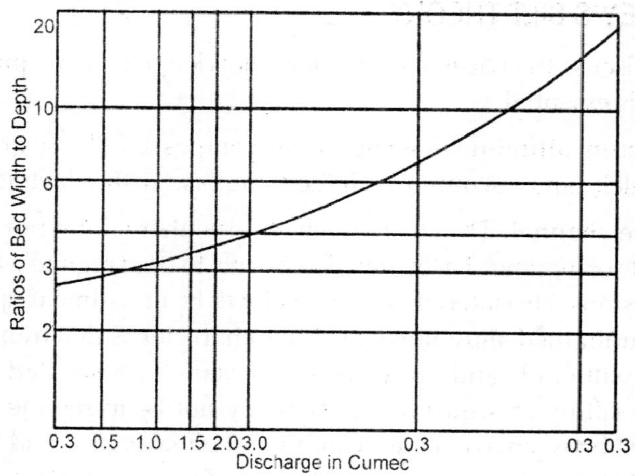

Fig. 17.2. *B/D ratio given by CWPC.*

from Fig. 17.2. This ratio of B/D is for unlined canals.

17.7 DRAWBACK'S OF KENNEDY'S THEORY

1. Kennedy did not give any significance to B/D ratio.

2. He did not make any mention of silt grade and silt charge any where.

3. He did not give any relation for longitudinal slope.

4. He used Kutter's equation to find out the actual velocity of flow. This introduced the drawbacks of Kutters equation also in this theory.

5. The regime acquired by Kennedy was only average regime and not the final.

6. The effect of concentration of silt in water on flow was represented by m. But actually this effect is not as simple as he considered.

17.8 LINDLEY'S FORMULAE

According to Kennedy's Theory, for a specific discharge, there can be a number of canal sections, for different longitudinal slopes. But all these sections cannot be equally effective. It is not possible to select the best section by this theory. Lindley of Punjab Irrigation Department gave his two formulae by which he related D and B separately with critical velocity of flow. The formulae are

$$V_0 = 0.567\, D^{0.57} \text{ and} \quad (17.8)$$

$$V_0 = 0.274\, B^{0.35} \quad (17.9)$$

If V_0 is eliminated from these two formulae, we get following relation between B and D.

$$B = 7.80\, D^{1.61} \quad (17.10)$$

17.9 LACEY'S SILT THEORY

Before we discuss the Lacey's silt theory, knowledge of some important terms given here, is essential.

1. In coherent alluvium. It is such a soil, composed of loose granular graded material which can be scoured with the same case with which it is deposited.

2. Regime channel. The channel which is neither silting nor scouring may not be in state of 'regime'. Lacey stated a channel to be in regime when it acquired ideal dimensions. He stated that a channel will be in regime if it flows through in coherent unlimited alluvium of the same character as that transported by the channel, the silt grade and charge being constant. Lacey stated that a channel showing no silting or scouring may actually not be in regime. He therefore, differentiated between two regime conditions I Initial regime and II Final regime. The channel in initial regime is not a channel in regime as it appears from non-silting and non-scouring behaviour and hence regime theory is not applicable to them. Lacey's theory is applicable to channels which are either in true regime or final regime.

3. True regime. The channel is said to be in true regime when following five conditions are satisfied :

(i) Discharge of the channel remains constant.

(ii) Silt grade and silt charge remains constant.

(iii) Channel is flowing through a soil which can be scoured as easily as it can be deposited. In other words the channel has freedom to form its own section.

(iv) The channel is flowing in the soil of the same character as that being transported.

(v) The soil forming the section of the channel is unlimited incoherent alluvium.

But in practice all these conditions can never be satisfied and hence artificial channels can never be in true regime. They can only be in initial or final regime. They can be in true regime only if they are neglected like rivers in plains, which is not possible as channels have to flow in assigned dimensions.

4. Initial regime. Initial regime is established in the channels because banks and bed of channel has some rigidity and are not as readily scourable as stipulated in regime conditions. The regidity of the banks may be either due to grass on then or they may be made of clayey soil rather than incoherent alluvium.

A channel excavated with some defective slope and some what narrow section when run, get silted up in the upper reaches thereby increasing the longitudinal slope of the channel. Consequently the velocity in the channel is increased and non-silting equilibrium is established, called 'initial regime'. The channel has achieved only working stability due to the rigidity of the banks. Actually, the slope and velocity of flow are higher and cross-section narrower than what would have been if the banks were not rigid. Such channels are termed as channels in initial regime, as they do not run through incoherent alluvium. Regime theory is not applicable to such channels.

5. Final regime. A channel can achieve final regime only if there is no lateral restraint. For this, all the variables such as perimeter, depth, slope etc. are equally free to vary and finally get adjusted to permanent stability called final regime. But this final regime is for a particular silt content, and discharge. If silt content is altered, another series of changes in slope, depth and perimeter occur till a new 'final' regime is achieved. In order to attain final regime for a particular silt content and discharge, the channel first forms its section, and slope is fixed thereafter. Regime theory is applicable to such channels.

6. Permanent regime. The channels which have its banks and bed made of some protective material is said to be in *permanent regime.* In such channels, banks and bed, cannot be scoured and as such changes that have to take place while achieving final regime cannot occur. Regime theory is not applicable to such channels. In the case of Permanent regime, no change can take place either in section of slope, but in case of final regime section may be permanent for only specific silt content and discharge. Changes start taking place in section and slope as soon as silt content and discharges are varied.

A channel assumes semi-elliptical section if all the variables are equally free to vary. The coarser the silt, the flatter will be the semi-ellipse. The finer the silt, the more nearly the section attains a semi-circular shape. See Fig. 17.3.

17.10 LACEY'S SILT THEORY

According to Lacey, for a given discharge and silt factor (silt charge) all the dimensions such as width, depth, and longitudinal slope are all fixed by nature. There is only one section of a channel and only one slope at which the channel carrying a given discharge will carry a particular grade of silt. Natural silt transporting channels have a tendency to assume a semi-elliptical section. The coarser the silt the greater will be the width of water surface. In case of finer silt the section assumes almost semicircular shape. See Fig. 17.3.

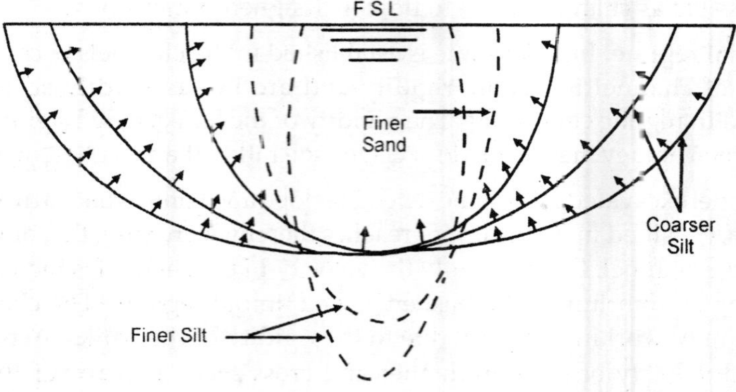

Fig. 17.3

If channel section is too small and slope too steep, scour will occur till regime is attained. On the contrary if section is too large and slope too flat, silting will take place till final regime is established.

Silt supporting eddies are generated from the perimeter of the channel and not only from the bed width. Hence Lacey considered the hydraulic mean depth as the variable for silt supporting.

17.11 LACEY'S REGIME EQUATIONS

Orginally Lacey gave two equations known as Lacey's regime equations. These equations are :

1. $$V = \sqrt{\frac{2}{5} fR} \qquad (17.11)$$

2. $$Af^2 = 140.0\,V^5 \qquad (17.12)$$

Later Lacey gave one more equation known as Lacey's flow equation. This equation is as follows.

3. $$V = 10.8\, R^{2/3} S^{1/3} \qquad (17.13)$$

This equation establishes relationship among velocity, H.M.D. and longitudinal slope.

The above three equations are the fundamental equations given by Mr. Lacey. Lacey related f(silt factor) with the mean particle diameter of silt in mm. For f, he used following equation

$$f = 1.76\sqrt{mr} \qquad (17.14)$$

where mr is the mean particle diameter of silt in mm.

Values of Lacey's silt factor f for various soils are given below:

Soil	*Silt factor (f)*	*Grains size in mm*
Coarse sand	1.50	0.725
Medium sand	1.25	0.505
Standard silt	1.00	0.323
Medium silt	0.70	0.158
Fine silt	0.4 to 0.6	0.052 to 0.120
Bajri	1.75 to 2.75	0.88 to 2.42
Gravel	4.75 to 9.00	7.28 to 26.0
Boulders	12.0 to 24	50 to 190 mm

The silt factor is determined by equation $f = 1.76\sqrt{m}$. Now by solving the Lacey's fundamental Eqs. 17.11 to 17.13 following additional equation can be found out.

Fundamental Equation of Lacey

$$V = \sqrt{\frac{2}{5} fR}$$

$$Af^2 = 140.00\, V^5$$

$$V = 10.8\, R^{2/3} S^{1/3}.$$

The Additional Equations Derieved from Fundamental Equations

1. $$P = 4.75\sqrt{Q}$$

2. $$V = \left(\frac{Qf^2}{140}\right)^{1/6}$$

3. $$S = \frac{f^{5/3}}{3340\, Q^{1/6}}$$

4. $$S = \frac{f^{5/2}}{4980\, R^{1/2}}$$

5. $$R = 0.47\left(\frac{Q}{f}\right)^{1/3}$$

6. $$R = 1.35\left(\frac{q^2}{f}\right)^{1/3}$$

7. $$R = \frac{5}{2}\frac{V^2}{f}.$$

17.12 DESIGN OF CHANNEL BY LACEY'S THEORY

Longitudinal slope and section of the canal are nowadays designed by Lacey's Theory. If discharge (Q) and silt factor (f) are known the canal section is completely worked out by this theory. Design steps are as follows:

1. Fix silt factor $f = 1.76\sqrt{m_r}$.

2. Find out velocity of flow $V = \left(\frac{Qf^2}{140}\right)^{1/6}$.

3. Find out area of cross-section, $A = \frac{Q}{V}$.

4. Find out wetted perimeter $P = 4.75\sqrt{Q}$.

5. If B and D are bed width and depth of the canal, values of A and P can be worked out as follows for $\frac{1}{2}:1$ side slope.

$$A = BD + \frac{D^2}{2} \text{ and } P = B + \sqrt{5}D$$

$$D = \frac{P - \sqrt{P^2 - 6.944\,A}}{3.472}$$

$$B = P - 2.236\,D$$

6. Find out H.M.D. $R = \frac{5}{2}\frac{V^2}{f}$

$$R = \frac{BD + \frac{D^2}{2}}{B + 2.23D}.$$

Both the values of R should be same.

7. Find out longitudinal slope (S)

$$S = \frac{f^{5/3}}{3340\,Q^{1/6}}.$$

17.13 LACEY'S DIAGRAMS

Seeing to the number of equations one has to solve the task of design of a channel, is quite cumbersome. In order to eliminate most of mathematical work, Lacey prepared two diagrams known after him as Lacey's diagrams. Both these charts or diagrams are shown in Plates 17.3, 17.4 and 17.5. From Chart 17.3 and 17.4, B and D are read corresponding to the point of intersection of discharge curve and silt factor curve. From Chart 17.5 longitudinal slope (S) is obtained corresponding to same discharge and silt factor. Method of reading Lacey's charts is not difficult and can be easily understood. These diagrams have been prepared for side slope of ½ : 1.

Example 17.6 *Design a canal carrying a discharge of 15 cumecs and silt factor 1.0 by Lacey's Theory. The side slope of the canal is* $\frac{1}{2}:1$.

Solution.

$$V = \left(\frac{Qf^2}{140}\right)^{1/6}$$

$$= \left(\frac{15 \times 1}{140}\right)^{1/6} = 0.69 \text{ m/sec.}$$

Area of cross-section

$$A = \frac{Q}{V} = \frac{15}{0.69} = 21.75 \text{ m}^2$$

H.M.D.

$$R = \frac{5}{2}\frac{V^2}{f}$$

$$= \frac{5(0.69)^2}{2 \times 1} = 1.20 \text{ m}$$

$$R = \frac{A}{P} = \frac{21.75}{B+\sqrt{5}D} = 1.20$$

$$B = \sqrt{5}D = 18.25 \quad (1)$$

Also
$$A = BD + \frac{D^2}{2} = 21.75 \quad (2)$$

Solving (1) and (2), we get

$$B = 15.1 \text{ m}, D = 1.38 \text{ m}$$

$$P = 4.75\sqrt{Q} = 4.75\sqrt{15} = 18.3 \text{ m}.$$

But
$$P = B+\sqrt{5}D = 18.25$$

Since both the values of P are almost same hence design is satisfactory.

Longitudinal slope

$$S = \frac{f^{5/2}}{3340\,Q^{1/6}} = \frac{1^{5/3}}{3340\,(15)^{1/6}} = \frac{1}{5260}.$$

Hence designed channel should be 15.1 m wide 1.38 m deep and should have a longitudinal slope of 1 in 5260.

Example 17.7 *Design a regime channel by Lacey's Theory for 40 cumec discharge and silt factor 0.9.*

Solution.
$$V = \left(\frac{Qf^2}{140}\right)^{1/6} = \left\{\frac{40\times(0.9)^2}{140}\right\}^{1/6} = 0.783 \text{ m/sec}$$

$$R = \frac{5}{2}\frac{V^2}{f} = 2.50\,\frac{(0.783)^2}{0.9} = 1.70 \text{ m}$$

$$P = 4.75\sqrt{Q} = 4.75\sqrt{40} = 30 \text{ m}$$

$$P = B+\sqrt{5}D = 30 \quad (1)$$

$$A = BD+\frac{D^2}{2} = \frac{Q}{V} = \frac{40}{0.783} = 51.08$$

or
$$BD+\frac{D^2}{2} = 51.08 \quad (2)$$

Solving (1) and (2), we get

$$B = 26.7 \text{ m}, D = 1.925 \text{ m}$$

$$S = \frac{f^{5/3}}{3340 Q^{1/6}} = \frac{(0.9)^{5/3}}{3340 \times (40)^{1/6}} = \frac{1}{7363}$$

$$= 0.136 \text{ m}/1000 \text{ m}.$$

Almost same values are obtained from Lacey's diagrams.

17.14 DEFECTS IN LACEY THEORY

1. In true sense a trapezoidal channel, which is the usual shape of an artificial channel, cannot be a regime channel and as such regime theory is not applicable to it. Shape of Regime channel is semi-elliptical for course soils which slowly approaches to semi-circular shape with more of fineness of the soil.

2. Lacey did not define the silt grade and silt charge properly.

3. Concentration of silt varies from place to place. Lacey did not consider concentration of silt as variable factor.

4. This theory does not give any clue, as what is that thing which controls the characteristics of channel in alluvial soils.

5. The silt particles go on decreasing in size by attrition effect of particles with bed of the channel. The ultimate effect of this change is not represented any where in this theory.

6. Lot of water is lost from channels by way of absorption and evaporation. Lacey did not take into account the silt left in the channel by water, lost in this manner.

7. All the Lacey's equations are dependent upon one factor i.e. silt factor (f). Dependence on a single factor is not good.

8. Lacey assumed the section of a regime channel as made of free incoherent alluvium which is readily scoured when velocity of flow increases even slightly than the regime velocity. But actually canal banks are not so freely erodible. After a run of few months the banks of canal become so rigid that they cannot be scoured even with the twice of the regime velocity.

9. This theory does not give any clear mention of physical aspects of the problem.

17.15 COMPARISON OF LACEY'S THEORY WITH KENNEDY'S THEORY

1. The basic concept of silt transportation is the same in both the theories. Both the theories agree that the silt is carried by the vertical eddies developed by the friction of the flowing water against surface of the channel. The difference is that Kennedy considered channel of trapezoidal section and he neglected the effect of eddies generated from the sides. For this reason Kennedy's critical

velocity formula was derived in terms of Depth of flow. On the other hand, Lacey considered that an irrigation channel achieves semi-elliptical shape and that entire wetted perimeter of the channel contributes to the development of silt supporting eddies. He therefore used H.M.D. as a variable in his regime velocity formulae.

2. According to Kennedy all those channels which neither silt nor scour are in regime condition. But Lacey differentiated initial regime conditions and final regime conditions.

3. Kennedy did not given any importance of B/D ratio. But Lacey connected, wetted perimeter (P) and area of cross-section (A). Thus in case of Lacey's theory there is a definite relationship between Bed width and depth of the channel.

4. Kennedy did not give any slope formula. Lacy gave a definite slope formula in terms of discharge and silt factor.

5.According to Lacey the grain size of the soil forming the channel is an important factor. He connected the grain size (m_r) with his silt factor (f) by equation $f = 1.76\sqrt{m_r}$. This factor (f) occurs in all the Lacey's equations which are used to determine the channel dimensions. Kennedy did not give much importance to it and simply introduced a term (m) the critical velocity ratio.

6. Kennedy used Kutter's formula for determining the actual velocity of flow where in value of N is again a guess work. Lacey on the other hand gave a general regime equation stating $V = 10.8\, R^{2/3}\, S^{1/3}$. The equation was given by him on the basis of analysis of very large data obtained from regime channels.

7. Lacey differentiated between two types of resistances in alluvial channels. One resistance determined by grain size and the other due to irregularities of the channel. Kennedy did not make any such distinction.

8. Kennedy's theory envolves trial procedure for the design of canals. Lacey's theory does not involve any trial and error procedure.

Example 17.7 *Design a regime canal by Lacey's theory for a discharge of 40 cumecs. Silt factor 0.9. Verify the results so obtained by Lacey diagrams.*

Solution.

$$V = \left(\frac{Qf^2}{140}\right)^{1/6} = \left\{\frac{40 \times (0.9)}{140}\right\}^{1/6} = 0.783 \text{ m/sec}$$

$$R = \frac{5}{2}\frac{V^2}{f} = \frac{2.50 \times (0.783)^{\circ}}{0.9} = 1.70 \text{ m}$$

$$P = 4.75\sqrt{Q} = 4.75\sqrt{40} = 30 \text{ m}$$

For trapezoidal channel with $\frac{1}{2}$: 1 side slope

$$P = B + \sqrt{5}D \tag{1}$$

$$A = BD + \frac{D^2}{2} = \frac{Q}{V} = \frac{40}{0.783} \qquad (2)$$

By solving Eqs. (1) and (2), we get

$$B = 26.7 \text{ m}, D = 1.925 \text{ m}$$

$$S = \frac{f^{5/3}}{3340\,Q^{1/6}} = \frac{(0.9)^{5/3}}{3340 \times (4)^{1/6}} = \frac{1}{6670}$$

or 15 cm in 1000 m.

From Lacey's diagram for a slope of 15 cm in 1000 m we get B = 26.5 m and D = 1.92 m for discharge of 40 cumbers and silt factor 0.9.

Hence results found by analytical calculation tally with the result obtained by Lacey's diagrams.

17.16 DISTRIBUTION OF SILT OVER CANAL CROSS-SECTION

We had ready stated somewhere earlier in this book that the velocity of flow along the depth is minimum at the bed and maximum slightly below the surface of water. The velocity at surface is not the maximum but slightly less than that.

If we consider along the width of the channel the velocity is maximum at centre of the width but decreases to almost zero at banks.

Distribution of silt along the depth of the canal is shown in Fig. 17.4. If we closely examine the silt change we will find that 60 to 70% of the average silt

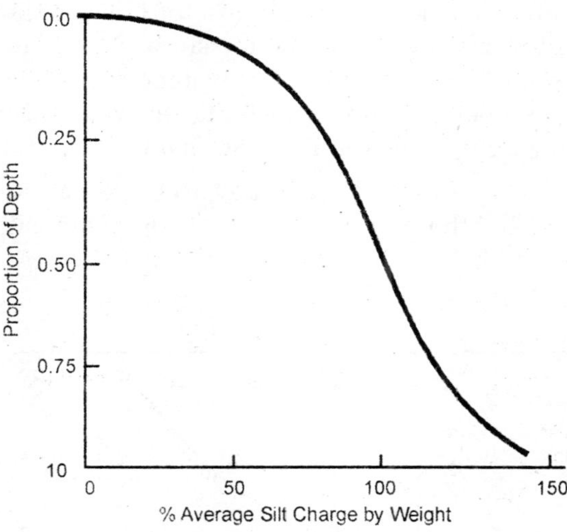

Fig. 17.4. *Silt distribution along depth*

charge, remains concentrated near the surface of the flow. At about 0.6 depth the silt charge is average and at bottom of the channel it is about 130% of the average silt charge.

The transverse distribution of silt along the section of the canal was studied by Mr. A.S. Gibbs in Punjab. He stated that when water flows in straight channel a rolling motion is developed. The tendency of top layers of water, flowing with relatively larger velocity is to topple over the bottom layers of water which

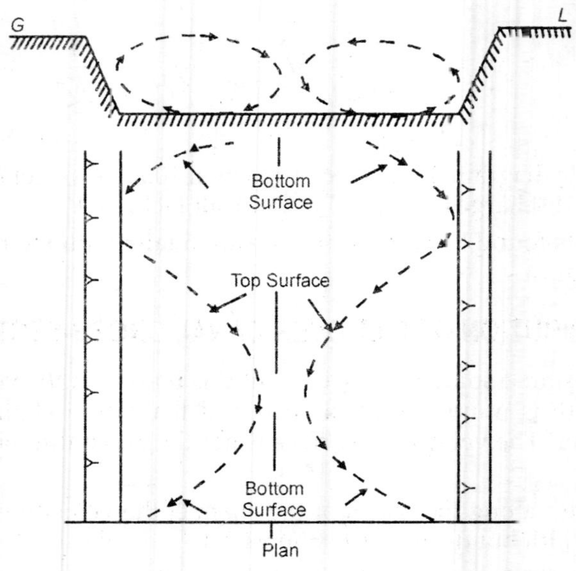

Fig. 17.5. *Silt distribution in plan*

flow with relatively smaller velocity. Because of this toppling action bottom layers are pushed towards the banks and top surface layers towards the centre of the channel. Figure 17.5 shows motion of water currents in section as well as in plan. Due to this phenomenon coarser silt remains concentrated near the banks while the finer one at the centre of the channel.

Distribution of silt in a canal section at a horizontal curve is shown in Fig. 17.6. Inside bank of the curved channel is subjected to silting and outer

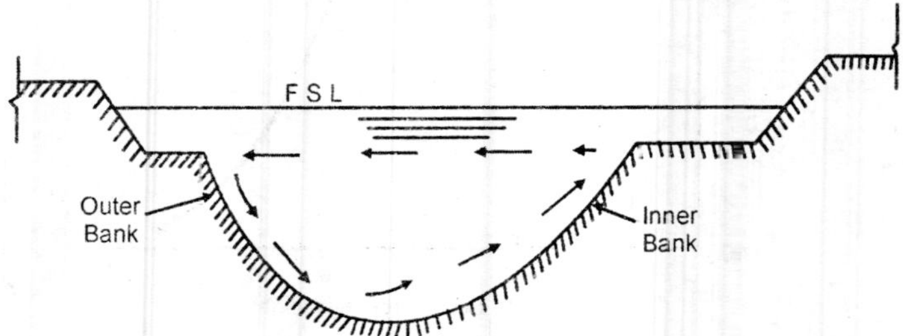

Fig 17.6. *Section of canal at curve*

bank to scouring. If some channel has to be taken off from a curved parent channel, it should be taken from outer bank and not from the internal bank.

QUESTIONS

17.1. Write short notes on :

(i) Regime section of the canal

(ii) Silt factor

(iii) Rugosity coefficient

(iv) Critical velocity ratio.

17.2. Write the followine flow formulae :

(i) Chezy's formula (ii) Kutter's formula and

(iii) Manning's formula.

17.3. Explain the salient features of Kennedy's silt theory. How he considered silt charge in his theory? Explain the defects of this theory.

17.4. Explain the terms :

(i) Final regime (ii) Initial regime

(iii) Permanent regime and (iv) True regime.

Draw regime sections of a channel flowing through a coarse grained soil and fine grained soil.

17.5. Explain Lacey's silt theory. In what way it differs from the Kennedy's theory Give the basic or fundamental equations given by Lacey.

17.6. Enumerate the defects of Lacey's silt theory..

17.7. Compare Lacey's theory with Kennedy's theory.

17.8. Explain the following

(i) Garret diagrams (ii) Lacey's diagrams

(iii) Lindley's formula (iv) Incoherent alluvium

(v) Silt distribution along the section of the channel.

17.9. Design an irrigation channel to carry 50 cumecs of discharge. The channel has bed slope of 1 in 4000 and C.V.R. for the soil is 1.1. Kutter's rugosity coefficient is 0.023.

[**Ans.** B = 14.14 m, D = 2.7 m, side slope $\frac{1}{2}$: 1]

17.10. Design a regime channel to carry a discharge of 50 cumecs; silt factor is 1.1. Use Lacey's theory.

[**Ans.** B = 29.77 m, D = 1.65 m, S = 1 in 5420]

17.11. Design an irrigation channel to carry a discharge of 40 cumecs. Adopt C.V.R. as 1.0 and B/D = 2.5 Kutter's rugosity coefficient is 0.023. Use Kennedy's theory.

[**Ans.** B = 8.4 m, D = 3.34 m, S = 1 in 4800, side slope $\frac{1}{2}$: 1]

❑❑❑

18

Tractive Force and Sediment Transport

18.1 SEDIMENT TRANSPORT

A flowing water in a channel, always tries to scour its surface. Silt, gravel or even larger boulders are first detached from its bed or banks, and then swept *D/S* by moving water. The phenomenon of detaching and then sweeping the silt or gravel from bed or banks is known is *sediment transport.* The phenomenon is of great economic importance. Following are some of the important works where sediment transport has its effects.

1. The design and execution of flood control schemes is chiefly governed by the peak flood level, which in turn depends upon the scour and deposition of sediment.

2. Silting of channels and reservoirs also depends upon the sediment transport.

3. Sediment deposited in rivers and harbours requires costly dredging.

18.2 SEDIMENT LOAD

The quantity of solids (silt) entering the channel, is known as *sediment load.* It is a single important factor which controls the shape and cross-section of the true regime channel. The sediment moving in a fluid can be broadly divided into two parts:

1. Bed load and

2. Suspended load.

1. Bed load. It is the load of bed material, in the bottom most layer of the flow. Suspension of sediment is not possible in this layer, because, of fluid dynamic reasons. The grains of bed load are not supported vertically by the flow but rest on the bed while rolling, sliding and jumping. The weight of the bed load particles is borne by the stationary grain particles of non-moving bed. The particles of bed load move regularly and exchange places with the similar particles of non-moving bed.

2. Suspended load. With increase in the velocity, smaller particles are thrown in suspension by the upward component of the turbulent velocity of flow. The particles always settle due to gravity effect. The flowing water provides an upward motion due to turbulent exchange. Due to turbulent exchange the water is regularly exchanged between various horizontal layers of water over a definite distance. The ascending water originates from lower layers of higher concentration and descending water originates from higher layer of lower sediment concentration. In this process of exchange of turbulences, there is a net upward surplus force which provides an upward motion to sediment particles and counterbalances the settlement effect of the particles due to gravity. Suspended load also causes additional hydrostatic pressure on the bed of the channel.

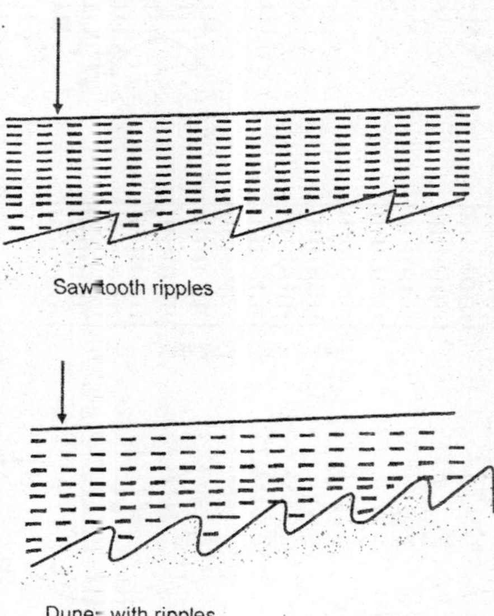

Fig. 18.1. *Development of ripples at bed.*

18.3 PRACTICAL ASPECTS OF BED FORMATION

The bed of the channel in which water is flowing, may adopt various shapes depending upon the velocity of flow.

At very low velocity of flow, bed of the channel does not move at all. When velocity is slowly increased a stage is reached when the sediment load is just at the point of motion. This stage is known as "threshold of motion". If bed is made of fine sand, having particles of less than 2 mm diameter, saw tooth type ripples develop in the bed on slightly further increasing the velocity. This phenomenon can be easily seen in sand at beaches. At still larger velocity, dunes with ripples appear at the bed, which on further increase on the velocity take the shape of rounded *dunes*. When velocity is still further increased, the dunes are eliminated and a flat surface becomes available. If velocity is increased still further, sand waves are formed in association with surface waves. When velocity is still further increased. Froude number $\left(\frac{\dot{V}}{\sqrt{gD}}\right)$ approaches almost unity and surface waves become so steep that they break and the whole wave system gradually moves upwards. The sand waves at this point of velocity are known as *anti-nodes*. Antinodes can form only in open channel flows and not in wind blown sand, because they require an interaction between the bed of the channel and the water surface for their formation. See Figs 18.1 and 18.2.

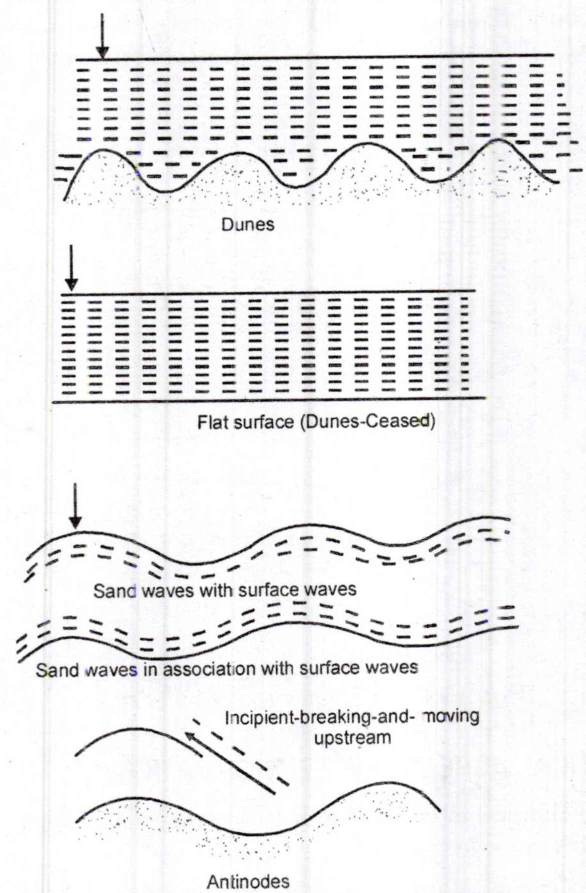

Fig. 18.2. *Development in the bed of the canal at various velocities of flow.*

18.4 TRACTIVE FORCE THEORY

In the study of mechanics of sediment transport, the soil particles are always considered as incoherent. Most of the river beds are made up of sand and gravel and as such assumption in regard to soil being incoherent is correct. The basic mechanism that controls the sediment transport, is the drag force exerted by water in the direction of flow on the channel bed. This drag force is also known as *tractive force* or *shear force*. This force is nothing but a pull of water on the wetted area of the channel.

Consider a channel of length L and cross-sectional area A. The volume of water in this length of channel would be $A \times L$. If w is the unit weight of water, then weight of water stored in this length (L) of the channel will be wAL.

Weight of water acting in vertical direction = wAL. Horizontal component of this weight = $wAL \sin\theta$. But θ is slope of the channel which is represented by S.

$\therefore$ Horizontal component of weight = $wALS$.

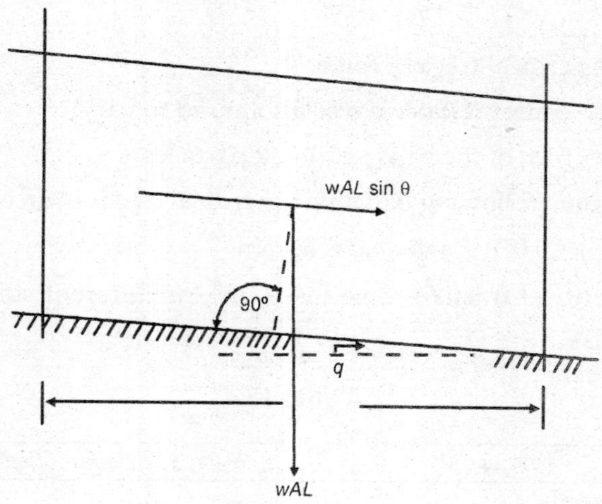

Fig. 18.3. *Flow of water in channel and forces acting on silt at bed.*

This horizontal force exerted by water is nothing but tractive force.

If average tractive force per unit of wetted area is τ, then

$$\tau = \frac{wALS}{\text{wetted area}} = \frac{wALS}{P \times L} = \frac{wSA}{P}$$

$$= wSR \qquad (1)$$

$$\left(\because \frac{A}{P} = R = \text{H.M.D.}\right)$$

Hence average unit tractive force also known as shear stress is given by $\tau = wSR$.

where w = Unit of weight of water

S = Longitudinal slope of the channel

R = Hydraulic mean depth H.M.D. $\left(\frac{A}{P}\right)$

In the case of wide open channels, the value of hydraulic mean depth (H.M.D.) is almost equal to depth of the channel. Hence tractive force can be written as follows also

$$\tau = wSD \qquad (2)$$

If tractive force becomes greater than the frictional resistance between particles, the particles are set in motion. The resistance of sediment to motion in proportional to the diameter d of the particle and the submerged weight of sediment in water. The value of critical tractive force τ_c becomes as follows :

$$\tau_c = C(G-1)\,d$$

where τ_c = critical tractive force

C = constant (later it was also found to vary)

G = S.G. of the solid sediment particles

E.W. Lane gave following equation for critical tractive force τ_c.

τ_c (in kg/m^2) = 0.078 d, where d is in mm.

Values of critical tractive force in kg/m^2 for different soils are given in Table 18.1.

Table 18.1

Type of soil	*Value τ_c in kg/m^2 (Tractive force)*
Medium sand	0.17
Sandy loam	0.20
Alluvial silt loam, coarse sand etc.	0.25
Fine gravel	0.37
Coarse gravel	1.47
Shades and hard pan	3.18

18.5 EFFECT OF SIDE SLOPES

In the last article we had discussed the stability of horizontal bed of the canal and developed equation $\tau = wSD$ or wSR where τ is the tractive or disturbing force. But on the slope of the banks there is one more disturbing force i.e. the component of the weight of the particle.

Consider a particle of weight w resting on the slope of the bank. This particle is subjected to following forces :

(i) $\tau = wSR$.

(ii) Component of w_s along the slope of the bank $w \sin \theta$.

(iii) Component of w_s normal to the slope of the bank $R \tan \phi$.

where θ = angle of slope of the bank with the horizontal.

ϕ = angle of repose.

The various forces acting on the particle and their free body diagram are shown in Fig. 18.4.

$$\tau_c = W \times \mu = W \tan \phi$$

τ_c is the critical shear stress that is required to move a similar grain on a horizontal bed.

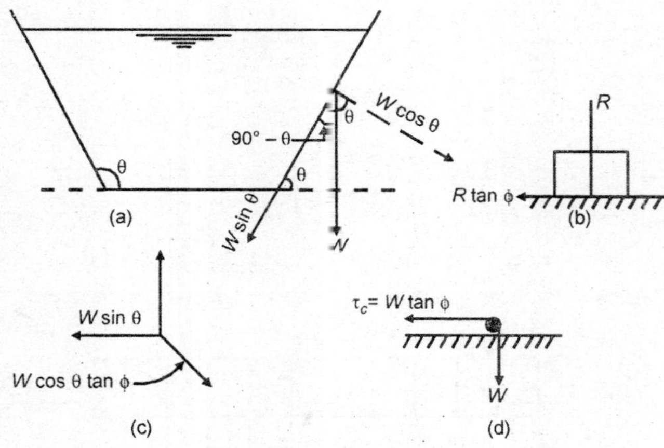

Fig. 18.4. *Forces acting on sloping banks of the canal.*

Considering the free body diagram

$$(\tau_0)^2 + (w \sin \theta)^2 = (\tau_c \cos \theta)^2$$

$$= (W \tan \phi \cos \theta)^2$$

$$(\tau_0)^2 + \left(\frac{\tau_c}{\tan \phi} \sin \theta\right)^2 = \left(\frac{\tau_c}{\tan \phi} \cos \theta \tan \phi\right)^2$$

$$\tau_0^2 + \frac{\tau_c^2}{\tan^2 \phi} \sin^2 \theta = \tau_c^2 \cos^2 \theta$$

$$\tau_0^2 = \tau_c^2 \left[\cos^2 \theta - \frac{\sin^2 \theta}{\tan^2 \phi}\right] \quad (1)$$

$$\frac{(\tau_0)^2}{(\tau_c)^2} = \cos^2\theta - \frac{\sin^2\theta}{\tan^2\phi} = \cos^2\theta\left[1 - \frac{\tan^2\theta}{\tan^2\phi}\right] \tag{2}$$

Again consider Eq. (1)

$$\left(\frac{\tau_0}{\tau_c}\right) = \left[\cos^2\theta + \sin^2\theta - \sin^2\theta - \frac{\sin^2\theta}{\tan^2\phi}\right]$$

$$= \left[1 - \sin^2\theta - \frac{\sin^2\theta}{\tan^2\phi}\right]$$

$$= \left[1 - \sin^2\theta\left(1 + \frac{1}{\tan^2\phi}\right)\right]$$

$$= \left[1 - \sin^2\theta\left(\frac{1 + \tan^2\phi}{\tan^2\phi}\right)\right]$$

$$= \left[1 - \sin^2\theta\frac{\sec^2\phi}{\tan^2\phi}\right]$$

$$= 1 - \frac{\sin^2\theta}{\sin^2\phi}$$

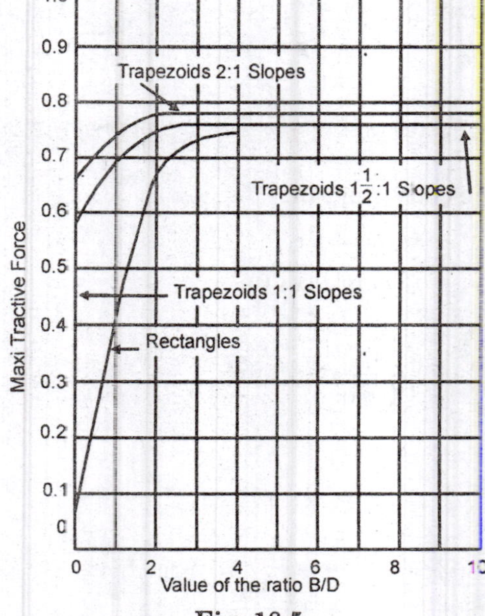

Fig. 18.5

$$\frac{\tau_0}{\tau_c} = \sqrt{1-\frac{\sin^2\theta}{\sin^2\phi}} \qquad (3)$$

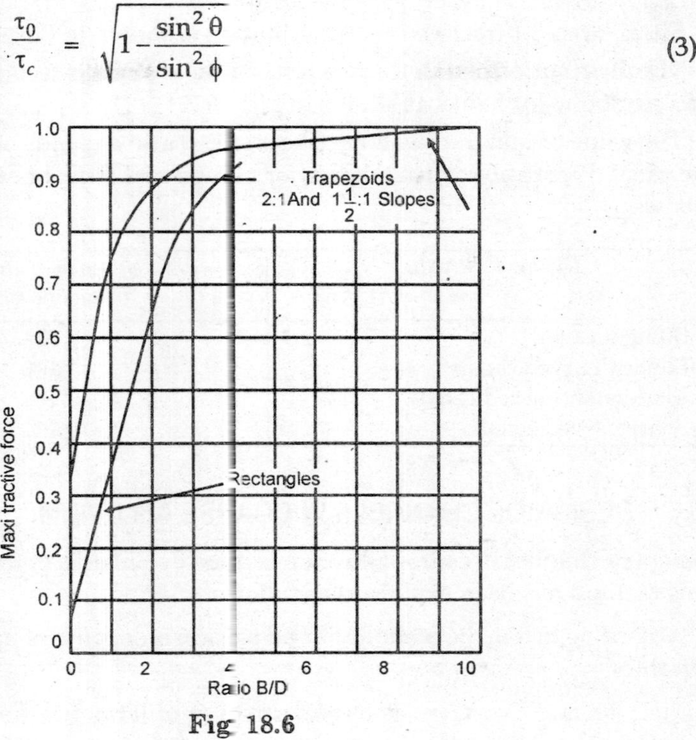

Fig. 18.6

This equation shows that τ_c is greater than τ_0 which means that tractive force or shear stress required to move a grain on the slope is less than that required to move the grain on horizontal bed.

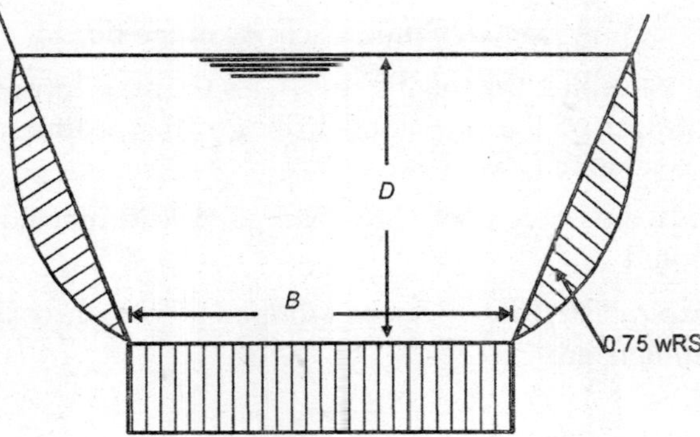

Fig. 18.7. *Distribution of shear force.*

On the horizontal bed, average value of shear stress or tractive force, that will be generated by the flowing water in a channel is given by $\tau_0 = wRS$, while on banks it has been established to be

$\tau_c = 0.75 \ \tau_0 = 0.75 \ wRS.$

Shear stress or tractive force distribution is shown in Fig. 18.7.

The distribution of unit tractive force or shear stress is rectangular for bottom and parabolic for banks as shown in Fig. 18.7.

The value of unit tractive force (shear stress) also depends on the curvature of the canal. Permissible tractive force for curved and straight canals are given as follows :

Degree of curvature	*Percentage of relative critical tractive force*
Straight canal	100%
Slightly curved canals	90%
Moderately curved canals	75%
Sharp curved canals	60%

18.6 DESIGN OF CHANNELS IN COARSE ALLUVIUM

Design of channel in coarse alluvium is done by Shield's entrainment method. This method has been explained as follows:

According to him the motion of a bed particle is dependent upon the following variables

τ_0 = The shear stress or unit tractive force of the base.

P_s = Density of the grain or particle

P_s = Density of fluid.

P_f = Diameter of particles.

d = Acceleration due to gravity.

μ = The dynamic viscocity of the fluid.

Out of all these variables, the shear stress (τ_0) is the most important factor as it measures the power of the flow that dislodges the sediment. This factor depends upon the velocity (V) of flow.

In order to study the movement of the bed particle, dimensionless method of analysis is adopted.

All the six variables listed above can be grouped to form three dimensionless numbers. The numbers are

$$\frac{\tau_0}{P_f g d}, \ \frac{P_s}{P_f}, \ \frac{d\sqrt{P_f \tau_0}}{\mu}.$$

Although velocity term is not required, yet it is convenient to have a parameter with the dimension of velocity. We introduce a new term known as shear friction velocity and denote it by V'.

$$\therefore \quad V' = \sqrt{\frac{\tau_0}{P_f}} \quad \text{or} \quad \tau_0 = (V')^2 P_f$$

The first dimensionless number i.e. $\frac{\tau_0}{P_f gd}$ becomes

$$\frac{V'^2 P_f}{P_f gd} = \frac{V'^2}{gd}$$

The three dimensionless numbers are

(i) $\frac{V'^2}{gd}$

(ii) $\frac{P_s}{P_f}$ and

(iii) $$\frac{d\sqrt{P_f \tau_0}}{\mu} = \frac{d\sqrt{P_f (V')^2 P_f}}{\mu} = \frac{dV'T_f}{\mu} = \frac{dV'}{\frac{\mu}{P_f}} = \frac{d \times V'}{v}$$

where v is the kinematic viscosity of the fluid.

Combining (i) and (ii) dimensionless numbers, we get another dimensionless number as $\frac{V'^2}{gd(S_s - 1)}$.

It can be further simplified as follows :

$$\frac{V'^2}{gd(S_s - 1)} = \frac{\frac{\tau_0}{P_f}}{gd(S_s - 1)} = \frac{\tau_0}{P_f gd(S_s - 1)} = \frac{\tau_0}{wd(S_s - 1)}$$

$P_f g$ = w = Unit weight of water.

S_s = Specific gravity of the grain.

This number is known as Shield's Entrainment function and is denoted by F_s.

$$F_s = \text{Shield's entrainment function} = \frac{\tau_0}{wd(S_s - 1)}.$$

The third dimensionless number which $\frac{dV'}{V}$ is known as Particle Reynold's number R_e where V is the kinematic viscosity.

Now we can easily say that motion of a particle is dependent upon these two numbers i.e. F_s and R_e.

Graphs may be plotted between F_s and R_e as shown in Fig. 18.8. The curve obtained forms the basis for the design of channels, where it is desired to prevent bed movement or to keep it to a minimum. At Reynold number 400 and above the value of F_s, becomes constant and thus application of this aspect becomes more simple and equal to 0.056. It has been seen that when particle size exceeds 6 mm diameter. Particle Reynold number has been found to be above 400.

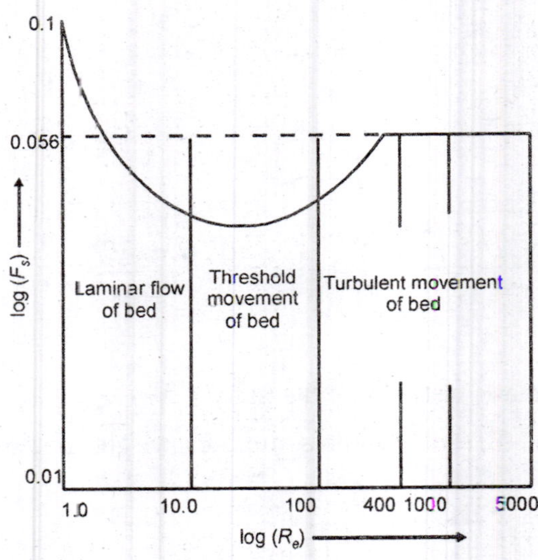

Fig. 18.8. *Relation between F_s and R_e.*

Hence for channel in coarse alluvium (above 6 mm)

$$\frac{\tau_0}{wd(S_s - 1)} = 0.056$$

For water $\quad w = 1$

$$\tau_0 = wRS$$

$$\because \quad \frac{wRS}{wd(S_s - 1)} = 0.056$$

$$\therefore \quad \frac{RS}{d(2.65-1)} = 0.056$$

On solving, $d = 11\ RS.$

This equation gives the maximum size of the particle that will remain at rest in a channel of given R and S. If d is less than this the particle will set in motion.

Since with passage of time the smaller particles are flushed out of the surface, lined with coarse stones, actual size to be adopted should be somewhat larger than what is calculated from the equation $d = 11\ RS$.

Example 18.1 *An irrigation canal is to be constructed in coarse alluvium gravel 5 cm size. Channel has to carry 5 cumecs of discharge and longitudinal slope is 0.01. The banks of the canal are protected by grass. Find out the minimum width of the canal.*

Solution. $d = 5$ cm $= 0.05$ m.

Calculate value of Rugosity coefficient N by Strickler's formula. This formula is used for rivers with beds of coarse materials practically free from ripples. The formula is

$$N = \frac{d^{1/6}}{24} \text{ where } d \text{ is diameter of particle in metres.}$$

$$\therefore \quad N = \frac{d^{1/6}}{24} = \frac{1}{24}\left(\frac{5}{100}\right)^{1/6} = 0.0258$$

Critical value of R is found from formula $d = 11\ RS$

$$R = \frac{d}{11S} = \frac{0.05}{11 \times 0.01} = 0.4551$$

This is the maximum value of R. Value of R less than 0.455 is also applicable but value more than 0.455 is not applicable.

$$V = \frac{R^{2/3} S^{1/2}}{N}$$

$$V_{max} = \frac{1}{0.0258} \times (0.455)^{2/3} (0.01)^{1/2}$$

$$= \frac{1}{0.0258} \times 0.592 \times 0.1 = 2.29 \text{ m/sec}$$

$$= 2.29 \text{ m/sec.}$$

Assume $R = D,\ Q = AV = B \times D \times V = BRV$

If R and V is taken maximum, bed width B will be minimum.

$$Q = BRV$$

$$5 = B_{min} \times 0.455 \times 2.29$$

$$B_{min} = \frac{5}{0.455 \times 2.29} = 4.80\,\text{m}$$

Use base width of 4.80 m or say 5 m.

18.7 ESTIMATION OF SUSPENDED LOAD

Suspended load of the channel can be estimated either by stream sampling or by the analytical method. Suspended load samplers may be used to measures the suspended load in the flowing water. Following equation is used to evaluate suspended load

$$\frac{C}{C_a} = \left[\frac{D-y}{y} \times \frac{a}{D-a}\right]^{\frac{v_f}{KV'}}$$

where C = Sediment concentration at a height y above the bottom which is to be determined by the above equation.

C_a = Known concentration at a known height a above the bottom of the channel.

v_f = Fall velocity of the grain in still water.

D = Depth of water.

K = Von Karman's constant whose value is 0.4.

V' = Shear friction velocity = $\sqrt{\frac{\tau_0}{P}} = \sqrt{gRS}$

where τ_0 = Intensity of shear stress or tractive force at the bottom of the channel.

The above equation is known as suspended load equation. The limitation of this equation is that, it cannot be used directly to predict the sedimentation concentration at any point, unless, the sediment concentration at some known distance is pre-known.

18.8 ESTIMATION OF BED LOAD

The most important factor responsible for the movement of the bed grains is the tractive force or drag force, already explained earlier

$$\tau_0 = wRS$$

where w = Unit weight of water = 100 kg/m^3.

We also know that certain minimum value of shear stress or unit tractive force is required to move the grain, depending upon the internal friction of the soil particles. This force is called critical shear stress and is represented by τ_c. For normal turbulent flow and for quartz grains, the value of τ_c, has been estimated as $\tau_c = 0.07\ d$ where τ_c is in kg/m^2 and d is in mm. When the unit tractive force caused by the flowing water (τ_0) exceeds the critical unit tractive force (τ_c), the sediment starts moving. The rate of bed load transported must therefore be a function of ($\tau_0 - \tau_c$). But the problem becomes complex when we take into account the fact that as soon as the grain starts moving, the channel bed develops ripples and a large part of shearing force is absorbed in the *form resistance* caused by these ripples. A part of the tractive force is therefore lost in overcoming the ripples and does not play any role in transporting bed material. The amount of shear stress lost in this process is not known and no perfect mathematical solution has been put forward to work out this quantity.

It has been widely accepted that the tractive force is reduced, due to ripples in following ratio.

$$\tau_c' = \tau_0 \left(\frac{N'}{N}\right)^{3/2}$$

where τ_0' = Unit tractive force left after the ripple resistance is overcome. Itmay also be stated as τ_0 minus tractive fore lost in overcoming ripples.

τ_0 = wRS, or original tractive force exerted by flowing water.

N' = Rugosity coefficient that should come into play theoretically ina channel of given R and S. Its value may be obtained by Strickler's formula

$N' = \dfrac{d^{1/6}}{24}$ where d is the effective grain size in metres.

N = Rugosity coefficient actually observed by experiments on the rippled bed of the channel and its value is generally taken as 0.02 for discharges over 11 cumecs and 0.0225 for smaller discharges.

Let us expand the equation $\tau'_0 = \tau_0 = \left(\frac{N'}{N}\right)^{3/2}$

$$\tau_0' = \tau_0 \left(\frac{N'}{N}\right)^{3/2} = wRS\left(\frac{N'}{N}\right)^{3/2}$$

$$= w\left[R\left(\frac{N'}{N}\right)^{3/2}\right]S$$

$$= wR'S$$

where $R' = R\left(\frac{N'}{N}\right)^{3/2}$

R' is the corresponding H.M.D. that would exist in the channel if the bed were unrippled. In other words if we use the value of R' in all our calculations, instead of R, we can forget about bed ripples i.e. the effect of ripples is only to reduce the hydraulic mean depth to a value R' from R.

18.9 CERTAIN FORMULAE FOR BED LOAD TRANSPORT

Some formulae suggested by various investigators in regard to bed load transport are given here.

1. Du-Boy's formula

$$q_s = C_s \tau_0 (\tau_0 - \tau_c)$$

where the q_s = Volume of bed load transported per second per unit width of channel.

τ_0 = Average shear stress on the channel boundary.

τ_c = Minimum shear stress required to move the particle, called the critical shear stress.

C_s = A constant which depends upon the grain size and is related to it by following relationship

$$C_s = \frac{0.173}{(d)^{3/4}}$$

Here d is in mm but all other elements q_s, τ_c, τ_0 are in F.P.S. units.

2. Shield's Formula

$$\frac{q_s S_s}{qS} = 10\left[\frac{\tau_0 - \tau_c}{w(S_s - 1)d}\right]$$

where S = Bed slope.

S_s = Specific gravity of the stone or grain.

q = Discharge per unit width.

w = Unit weight of water.

d = Diameter of stone or grain.

3. Meyer-Peters formula. On the lines already discussed Meyer and Peter had suggested that the unit tractive force causing bed load to move, is reduced by ripples in the ratio of

$$\tau_0 = \tau_0'\left(\frac{N'}{N}\right)^{3/2}$$

The effective unit tractive force going to cause bed load transportation, is then given by

$$\tau_{eff} = \left[\tau_0 \left(\frac{N'}{N}\right)^{3/2} - \tau_c\right]$$

On these concepts, Meyer and Peter had suggested the following formulae for calculating quantity of bed load transport

$$g_s = 4700 \left[\tau_0 \left(\frac{N'}{N}\right)^{3/2} - \tau_e\right]^{3/2}$$

where g_s is the rate of bed load transport (by wt) in kg/m per hour i.e. $g_s = g_s\, w\, S_s$.

where g_s = Volume of sediment transported per meter width of channel perhour.

w = Unit weight of water.

S_s = S.G. of the grain.

N' = Manning's coefficient pertaining to grain size on an unrippled bed and given by Strickler's, formula, $N' = \frac{d^{1/6}}{24}$ when d is in metres

N = Rugosity coefficient actually observed on rippled channels. Its value is taken as 0.02 for discharges of more than 11 cumecs and 0.0225 for lower discharges.

τ_v = Critical shear stress required to move the grain and given by $\tau_c = 0.07\, d$ where τ_c is in kg/m² and d in mm.

τ_0 = Unit tractive force produced by the flowing water i.e. wRS. Truelyspeaking, its value should be taken as the unit tractive force produced by flowing water on bed i.e. 0.97 wRS.

Example 18.2 *Design a channel which has to carry 20 cumec discharge with a bed load concentration of 50 ppm by weight. The average grain diameter of the bed material is 0.4 min. Use Lacey's regime perimeter and Meyer-Peter's formula.*

Solution. 50 P.P.M. $= \frac{50}{10^6}$ wt. of water

Quantity of Bed load transported per second

$$= \frac{50}{10^6} \text{ (20 cumec} \times 1000 \text{ kg/m}^3)$$

Quantity of bed load transported per hour

$$= \frac{50}{10^6}(20 \times 1000) \times 3600 \text{ kg/hr.}$$

$$= 3600 \text{ kg/hr.}$$

Lacey's Regime Perimeter

$$= 4.75\sqrt{Q} = 4.75\sqrt{20} = 21.24 \text{ m}$$

Assume bed width (B) say 18 m.

Rate of bed load transport per unit width

$$= \frac{3600}{18} = 200 \text{ kg/m/hr.}$$

$$g_s = 200 \text{ kg/m/hr.}$$

Meyer-Peter's equation is

$$g_s = 4700\left[\tau_0\left(\frac{N'}{D}\right)^{3/2} - \tau_c\right]^{3/2}$$

$$N' = \frac{1}{24}d^{1/6} \text{ when } d \text{ is in metres}$$

$$= \frac{1}{24}\left(\frac{0.4}{1000}\right)^{1/6} = 0.0113.$$

$$N = 0.020 \text{ as discharge is more than 11 cumecs.}$$

$$\frac{N'}{N} = \frac{0.0113}{0.02} = 0.565$$

$$\tau_c = 0.07\,d = 0.07 \times 0.3 = 0.021.$$

Putting these values in Meyer-Peter's formula we get

$$g_s = 4700\left[\tau_0\left(\frac{N'}{N}\right)^{3/2} - \tau_c\right]^{3/2}.$$

$$\tau_0 = wRS.$$

$$200 = 4700\,[1000 \times R \times S\,(0.565)^{3/2} - 0.021]^{3/2}$$

$$\left(\frac{200}{4700}\right)^{2/3} = [11000\,RS\,(0.565)^{3/2} - 0.021]$$

$$0.1219 = [424.69\,RS - 0.021]$$

$$424.69\,RS = 0.1429$$

$$RS = \frac{0.1429}{42469} = 0.00033648 \tag{1}$$

Use Manning's equation

$$V = \frac{1}{N} R^{2/3} S^{1/2}$$

$$Q = AV = \frac{AR^{2/3} S^{1/2}}{N}$$

But $P = 21.24$ m.

$$A = PR = 21.24R$$

$$Q = \frac{(21.24R)R^{2/3}S^{1/2}}{N'}$$

$$20 = \frac{1}{0.02} 21.24R\,R^{2/3}S^{1/2}$$

$$\frac{20 \times 0.02}{21.24} = R^{5/3}S^{1/2}$$

$$R^{5/3}S^{1/2} = 0.0188 \tag{2}$$

From (1) $S = \dfrac{0.00033658}{R}$

$$\therefore \quad R^{5/3} \times \left(\frac{0.00033658}{R}\right)^{1/2} = 0.0188$$

$$R^{7/6} = \frac{0.0188}{\sqrt{0.00033658}} = \frac{0.0188}{0.0183} = 1.023$$

$$R = (1.023)^{6/7} = 1.020 \text{ m}$$

$$S = \frac{0.0003365}{1.02} = \frac{1}{3031}$$

Let D be the depth of water in trapezoidal channel of slope $\frac{1}{2}:1$.

$$P = 18 + \sqrt{5}D$$

$$A = 18D + \frac{D^2}{2}$$

$$R = \frac{A}{P} = \frac{18D + 0.5D^2}{18 + 2.24D}$$

But, value of worked out $R = 1.02$

$$\therefore \quad \frac{18D + 0.5D^2}{18 + 2.24D} = 1.02$$

$$18\,D + 0.5\,D^2 = 18.36 + 2.285\,D$$

$$D^2 + 31.43\,D - 36.72 = 0$$

$$D = 1.13 \text{ m.}$$

Hence use the following channel dimensions

$$B = 18 \text{ m}, D = 1.13 \text{ m.}$$

$$S = 1 \text{ in } 3031 \text{ side slope } \tfrac{1}{2}:1.$$

18.10 EINSTEIN'S FORMULA

Einstein's formula for bed load transport is based on the principle of probability. It is a semi-theoretical approach to the problem of bed load transport. He assumed a number of elements and also a number of experimental coefficients. The formula given by him correlates two dimensionless parameters ϕ and Ψ as follows.

$$\phi = F(\Psi)$$

$$\phi = \frac{q_s}{v_f d} \quad (1)$$

$$\phi = F_s = \frac{\tau_0}{wd(S_s - 1)} \quad (2)$$

where ϕ = Einstein's bed load function

Ψ = F_s = Shield's entrainment function which has already been explained

q_s = Volume of sediment transported per second per unit width

v_f = Fall velocity

d = Diameter of grain

ϕ and Ψ relationship is represented by a curve shown in Fig. 18.9. This curve can be used to design stable channels for a given bed load transport.

ϕ and Ψ relationship is represented by a curve shown in Fig. 18.9. This curve can be used to design stable channels for a given bed load transport.

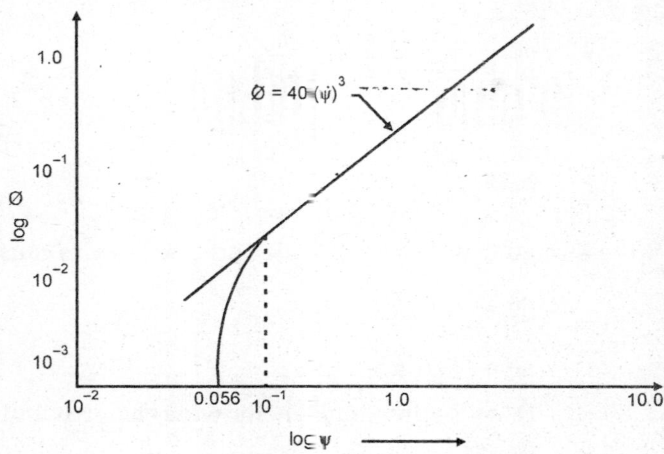

Fig 18.9

The straight line portion of the curve follows the relation

$$\phi = 40\,(\Psi)^3.$$

Putting value of ϕ and Ψ we get

$$\frac{d_s}{v_f d} = 40\left[\frac{\tau_0}{w \times d(S_s - 1)}\right]^3$$

But according to Einstein $v_f = G\sqrt{gd(S_s - 1)}$

where $$G = \sqrt{\frac{2}{3} + \frac{36\nu^2}{gd^3\,(S_s - 1)}}$$

when $$d = 1 \text{ mm } G = \frac{2}{3}$$

$$Gd\sqrt{gd(S_s - 1)} = 40\left[\frac{\tau_0}{w_d\,(S_s - 1)}\right]^3$$

But $$\tau_0 = w\,RS.$$

$$\frac{q_s}{Gd\sqrt{gd(S_s-1)}} = 40\left[\frac{R^3S^3}{d^3(S_s-1)^3}\right]$$

or

$$q_s \propto \frac{R^3S^3d\sqrt{d}}{d^3}$$

$$q_s \propto \frac{R^3S^3}{d^{3/2}}$$

Now if it is assumed that channel is wide and Chezy's C is constant then

$$Q = CA\sqrt{RS}$$

$$q = CD\sqrt{RS}$$

But $D = R$ approximately for wide channels and river.

$\therefore$ $q = CR\sqrt{RS} = CR^{3/2}\,S^{1/2}$

$$q^2 = \alpha\, R^2S$$

$$q_s \infty \frac{R^3SS^2}{d^{3/2}} = \frac{q^2S^2}{d^{3/2}}$$

$$\frac{q_s}{q} \propto \frac{qS^2}{d^{3/2}}$$

This equation is an important form of Einstein equation. It shows that sediment carrying capacity of the channel depend upon the discharge per unit width. Following important conditions can be drawn from this equation.

(i) With increase in width, the discharge per unit width (q) decreases (Q remaining constant) and thus $\frac{q_s}{q}$ i.e. sediment carrying capacity will reduce and deposition of sediment will start which will increase the bed slope. The silting will continue till the slope S is increased to such a value that qS^2 becomes constant.

(ii) Since the sediment carrying capacity depends upon the discharge, the floods will carry more sediment. Floods are in fact mainly responsible for most of the annual sediment load.

(iii) When branch channel is taken out from the main channel sediment carrying capacity of the main channel immediately D/S of take off point is reduced, because discharge per unit width (q) is reduced. This process would divert more and more of water to off taking channel as with more and more silting of D/S of main channel, bed level will be continuously rising.

The most general equation given by Einstein for bed load rate is as follows

$$\phi_* = \frac{i_B}{i_b}\phi = \frac{i_B}{i_b}\frac{q_s}{G(G-1)^{1/2}\left(gd^3\right)^{1/2}}$$

where f* = Einstein's bed load function.

q_s = Rate or volume of sediment transported per second per unit width.

i_B = Fraction of q_s which is of diameter d.

i_b = Proportion of grains of diameter d in the bed

d = Diameter of sediment particle

G = Specific gravity of sediment particle

g = 9.81.

For uniform soils

$$i_B = i_b \text{ and } \phi_* = \phi$$

Putting these values in general equation above we get following form of Einsten's equation

$$\phi = \frac{q_s}{G\sqrt{G-1}\left(gd^3\right)^{1/2}}$$

$$\phi_* = \psi = G-1\frac{d}{R'S}$$

R' is the H.M.D. which would exist in the channel with bed unrippled. When N represents granular roughness only. Manning's equation can be used for determining R'.

$\phi^* - \Psi_*$ relationship has been plotted by Einstein in form of a graph shown in Fig. 18.9. If the value of Ψ_* in known ϕ_* can be found out and so q_s can be calculated from general equation given above.

For uniform bed material ϕ and Ψ relation can be expressed as follows

$$\phi = \frac{1}{0.465}E^{-0.391\Psi}$$

Example 18.3. *A wide channel is 2.5 m depth and has a slope of* $\frac{1}{4000}$. *The bed material consists of uniform grain size of 0.3 m and having a fall velocity of 0.03 m/sec,* *in still water. Determine the quantify of bed load moved by the channel according to Einstein equation and Peter-Meyer's equation. Also determine the concentration of suspended load 1 m above the channel bed if concentration at 0.3 m above the bed is 300 pm. S. G. = 2.65.*

Solution. Quantity of bed load moved in the channel according to Meyer-Peter's equation,

$$g_t = 4700\left[\tau_0\left(\frac{N'}{N}\right)^{3/2} - \tau_c\right]^{3/2}$$

$$N' = \frac{d^{1/6}}{24} = \frac{\left(\frac{0.3}{1000}\right)^{1/6}}{24} = 0.0108$$

$$N = 0.02$$

$$\left(\frac{N'}{N}\right)^{3/2} = \left(\frac{0.0108}{0.020}\right)^{3/2} = 0.397$$

$$\tau_0 = wRS = wDS = 1000 \times 2.5 \times \frac{1}{4000}$$

$$= 0.625 \text{ kg/m}^2$$

$$\tau_c = 0.074 \quad d = 0.021$$

τ_c can also be found out by following formula

$$\tau_c = 0.47\,(S_s - 1)\,d$$

$$= 0.47(2.65 - 1)\,0.3$$

$$= \frac{0.47 \times 1.65 \times 0.3 \times 10^3}{104}$$

$$= 0.023$$

$$g_s = 4700\,[0.625 \times 0.397 - 0.023]^{3/2}$$

$$= 4700\,[0.248 - 0.023]^{3/2}$$

$$= 4700\,(0.225)^{2/2} = 501.62 \text{ kg/m/hr.}$$

Quantity of bed load moved according to Einstein equation,

$$\Psi_* = \Psi = \left(\frac{G-1}{1}\right)\frac{d}{R'S}$$

$$R' = R\left(\frac{N'}{N}\right)^{3/2}$$

Assuming $R = D = 2.5$ m

$$R' = 2.5 \times 0.397 = 0.993 \text{ m.}$$

$$\Psi = \left(\frac{2.65-1}{1}\right)\frac{0.3\times 4000\times 10^{-3}}{0.993\times 1}$$

$$= \frac{1.65\times 0.3\times 4000\times 10^{-3}}{0.993}$$

$$= 1.994.$$

From Einstein's curves,

For $\Psi_* = \Psi = 1$ we get

$$\phi_* = \phi = 3$$

$$\phi = \frac{q_s}{G\sqrt{G-1}\left(gd^3\right)^{1/2}}$$

$$q_s = \phi\times G\sqrt{G-1}\left(gd^3\right)^{1/2}$$

$$= 3\times 2.65\times\sqrt{1.65}\left\{9.81\times\left(\frac{0.3}{1000}\right)^3\right\}^{1/2}$$

$$= 3\times 2.65\times 1.2845\times\left[9.81\times\left(\frac{0.3}{1000}\right)^3\right]^{1/2}$$

$$= 3 \times 2.65 \times 1.2845 \times 0.000016275$$

$$= 0.0001662 \ t/\text{sec/m.}$$

$$= 0.0001662 \times 1000 \times 3600 \text{ kg/m/hr.}$$

$$= 598.32 \text{ kg/m/hr.}$$

Hence rate of bed load transport.

By Einstein equation = 598.32 kg/m/hr.

By Meyer-Peter's equation = 501.62 kg/m/hr.

Concentration of suspended load

$$C_a = 300 \text{ p.p.m. at } 0.3 \text{ m above bed.}$$

$$= 300 \times 10^{-6} \times 10^{3}$$

$$= 0.3 \text{ kg/m}^2.$$

$$V_f = \text{Fall velocity} = 0.03 \text{ m/sec.}$$

$$V' = \sqrt{\frac{\tau_0}{S_s}} = \sqrt{gR'S}$$

$$= \sqrt{\frac{9.81 \times 0.993}{4000}}$$

$$= 0.0493$$

$$k = 0.4$$

$$\frac{v_f}{V'k} = \frac{0.03}{0.0493 \times 0.4} = 1.52$$

$$\frac{C}{C_a} = \left[\frac{(D-y)a}{y(D-a)}\right]^{1.52}$$

$$= \left[\frac{(2.5-y)0.3}{y(2.2)}\right]^{1.52}$$

$$C = 0.3\left(\frac{(2.5-y)0.3}{2.2y}\right)^{1.52}$$

at 1 m above the bed $y = 1$.

$$\therefore \quad C = 0.3\left\{\frac{(2.5-1)0.3}{2.2 \times 1}\right\}^{1.52}$$

$$= 0.3\left(\frac{0.45}{2.2}\right)^{1.52}$$

$$= 0.0269$$

$$= 26.9 \text{ P.P.M.}$$ **Ans.**

Example 18.4 *A channel is 30 m wide and 3 m deep having a bed slope of 1 in 5000. The representation bed material size is 0.55, while mean diameter of the bed material is 0.35 mm. The value of Manning's coefficient N = 0.02. If friction from sides is neglected. Find the bed load transported by*

1. *Meyer-Peter's equation.*

2. *Einstein's equation.*

Solution. *1. Meyer-Peter's method*

$$N' = \frac{d^{1/6}}{24} = \frac{0.55 \times 10^{-3}}{24} = 0.011$$

$$\frac{N'}{N} = \frac{0.011}{0.02} = 0.55$$

$$\tau_0 = wRS \qquad \text{put } R = D$$

$$= 1000 \times 3 \times \frac{1}{5000} = 0.6$$

$$\tau_c = 0.07 \times 0.35 = 0.0245$$

$$q_s = 4700\,[0.6 \times (0.55)^{2/2} - 0.0245]^{3/2}$$

$$= 4700\,[0.2447 - 0.0245]^{3/2}$$

$$= 4700 \times (0.202)^{3/2}$$

$$= 485.65 \text{ kg/m/hr}$$

$$= \frac{485.65 \times 24 \times 30}{1000}$$

$$= 699.34\ t/\text{day}. \qquad \textbf{Ans.}$$

2. Einstein's formula

$$R' = R\left(\frac{N'}{N}\right)^{3/2}$$

$$= R\left(\frac{0.011}{0.02}\right)^{3/2} = 0.408\,R$$

Side friction has been neglected

$$R = 3 \text{ m}$$

$$R' = 0.03 \times 0.408 = 1.224 \text{ m}$$

$$\Psi = \frac{G-1}{1}\,\frac{0.35 \times 5000 \times 1}{1.224 \times 1 \times 1000}$$

$$= \frac{(2.65-1)}{1}\,\frac{0.35 \times 5000}{1.224 \times 1000}$$

$$= \frac{1.65 \times 0.35 \times 5000}{1.2254 \times 1000} = 2.359$$

corresponding to $\Psi = 2.359 \quad \phi = 2.8$

Also
$$\text{¢} = \frac{q_s}{G\sqrt{(G-1)}\left(gd^3\right)^{1/2}}$$

$$q_s = \phi G\sqrt{G-1}\left(gd^3\right)^{12}$$

$$= 2.8 \times 2.65 \times \sqrt{1.65}\left\{9.81 \times \left(\frac{0.35}{1000}\right)^3\right\}^{1/2}$$

$$= 2.8 \times 2.65 \times 1.284 \times 0.0000205$$

$$= 0.000195 \; t/\text{sec}/\text{m}$$

$$= 0.00195 \times 3600 \times 24 \times 30$$

$$= 505 \; t/\text{day}. \qquad \textbf{Ans.}$$

QUESTIONS

18.1. Explain the terms sediment transport, sediment load, bed load, suspended load.

18.2. Show the various forms of bed formation as they develop when velocity of flow is increased from very small value to turbulent value.

18.3. Explain the theory of tractive force.

18.4. Explain the effect of side slope on the theory of tractive force at banks.

18.5. Explain the method of channel design, by Shield's entrainment method.

18.6. Give the formulae and the significance of various terms in formula used for the estimation of suspended load.

18.7. Give various formulae used in the estimation of Bed load.

19

Design of Canals

19.1 LONGITUDINAL SECTION OF A CANAL

The points which should be considered in fixing alignment of the canal, have been given in Chapter 16. Actually the whole of the area where irrigation is proposed, is surveyed, and contour plans prepared. The other features of the area are also marked over the plans. The contour plan on which other features of the area are also marked in known as *shajra sheet*. Alignment of the main canal is fixed on the main ridge of the area proposed to be irrigated, so that irrigation is possible on both the sides of the canal. Branch and distributory channels are aligned along the main ridges of the area allotted for their command. In this way, the whole of the area to be irrigated is divided into several parts and each part is commanded by a branch, distributory or minor depending upon the extent of the area.

The area under the command of each distributory or minor, is further sub-divided into small areas surrounded by small drainages and each area is known as *chak*. The chak is that area which is generally surrounded by minor drainages. One outlet for each chak is provided from the distributory. The outlet is located in such a way that water may flow to all the areas of a chak under gravity. Size of the outlet depends upon the command areas available in a chak for irrigation.

After having fixed the canal alignment, detailed levelling is done along the alignment, and longitudinal section is plotted on a drawing sheet. For preparing maps of the area, Horizontal scale of 1 cm 160 m and vertical scale of 1 cm = $\frac{1}{2}$ m is used. If the area is very much undulated, the vertical scale may be different.

After drawing the longitudinal section, along the alignment, bed level of the proposed canal is marked. While marking bed level it should be ensured, that it will involve either too much cutting nor filling and secondly the full supply level of the canal will remain above the ground level so that irrigation is possible on both the sides, along the alignment.

Bed level and F.S.L. of the off-taking channels should be decided in relation to the Bed level and F.S.L. of the parent channel. When water is diverted to the off-taking channel, some head loss is bound to occur. To overcome this head loss, F.S.L. off-taking channel should be about 30 cm below the F.S.L. of the parent channel.

All the irrigation channels are given some longitudinal slope as these are gravity channels and water can flow only if some longitudinal slope is given to them. The slope of the canal is also decided in relation to the general slope of the area, in which canal is to run. If slope of the channel is almost same as general slope of the area, no fall will have to be constructed. But such ideal conditions are seldom available. Most of the canals are designed by Lacey's theory and this theory states that for a particular discharge and silt factor (f), there is a fixed slope for the canal. If this slope is changed to suit the general slope of the area, the canal will not remain regime canal. This fixed slope given by Lacey may not be same as general slope of the area. Generally designed slope of the channel is much less than the general slope of the area, and, as such a number of falls have to be constructed.

If somewhere, general slope of the ground is smaller than designed slope, the slope of the channel is changed to general slope of ground and section of the canal is accordingly modified.

After deciding the longitudinal slope and also the bed level at the head regulator, the bed levels of the channel at all the points are known. F.S.L. is marked parallel to the bed level. F.S.L. should always remain above the ground level so that water may flow to fields under gravity.

After having marked Bed levels on the longitudinal section of the proposed channel, mark the lengths where cuttings or filling are to be done. Extent of cutting or filling at a particular point is determined from the difference of ground level and Bed level. In order to economize the work, cutting should be just equal to filling at a particular point.

Following points should be taken care of, while drawing longitudinal section of canal.

1. Cutting and filling at all the points should be equal. In other words cutting and filling should balance each other.

2. F.S.L. of off-taking channel should be below the F.S.L. of parent channel. The difference in F.S.L. of off-taking channel and parent canal should be minimum 30 cm for distributory, 70 cm for branch and 1 m for main canal.

3. F.S.L. of the channel should remain above the ground level for most of the length. At isolated high spots, it may remain below the ground level.

4. F.S.L. should be above ground level only by 15 to 30 cm. This is considered sufficient because canals being aligned on water shed, will develop sufficient cross-slope and water will be flowing to fields under sufficient head.

5. Canals should not be too much in filling. The canals are always subjected to breaches at such reaches.

6. Bed slope as obtained by Lacey's theory if equals general slope of the ground it will be an ideal situation. If design slope of the canal is less than general slope of ground, canal falls will have to be provided, at suitable intervals. The fall or drop structure should be such that F.S.L. of the canal *D/S* of fall remains below G.L. for about $\frac{1}{2}$ km distance and then emerges out of G.L. The high ground on both the sides, *D/S* of fall is irrigated by taking outlets from *U/S* of the fall.

7. If designed slope of the channel is greater than general slope of the ground, the channel would go deep in cutting after running for a short distance running. In such a case general ground slope should be adopted and section of the channel should be accordingly amended.

19.2 BALANCING DEPTH OF CANAL

The canal section is considered to be most economical when cutting at a particular section equals the filling. For such a section payment has to be made only for one operation. More so the formation of borrow pits, or spoil banks is completely eliminated. The balancing depth is worked out as follows. See Fig. 19.1.

Let H = depth of the canal from top of the bank to bed of the canal.

D = full supply depth of the canal.

d = depth of cutting.

B = bed width of the canal.

P = top width of the canal bank.

$n:1$ = side slope of bank filling.

$m:1$ = side slopes of canal cutting.

Area of cutting $= B \times d + md^2$

Area of filling $= [(H-d)\,P + n\,(H-d)^2]2$ equate the area of cut and fill.

$$B \times d + md^2 = 2[H-d)P + n(H-d)^2]$$

Usual slope of canal in cutting is 1:1 and that of filling 1.5:1.

Therefore $m = 1$ and $n = 1.5$

$$B \times d + d^2 = [H-d)P + 1.5(H-d)^2]2$$

After solving this equation reduces to

$$d^2 - \left(\frac{B}{2} + P + 3H\right)d + \left(P + \frac{3}{2}H\right)H = 0$$

The values of d and B are found out from this equation.

19.3 DESIGN OF THE CANAL

In the design of a canal one has to find out bed width (B) depth (D), longitudinal slope (S), and velocity of flow (V). If discharge (Q) and silt factor (f) are given, the method of finding out all these elements either by Kennedy's theory or by Lacey's theory has been explained in Chapter 17. This aspect limits only finding the sectional dimensions of the canal. But how to compute the discharge at a particular reach is an important aspect of canal design.

Discharge in the off-taking canal does not remain constant throughout the length. Outlets fixed on the canal at regular intervals draw discharge from the canal and supply it to the fields for irrigation. Evaporation and percolation losses also go on increasing with length of the canal. Hence because of discharge withdrawn by outlets, and also continuous evaporation and seepage losses, the remaining discharge in the canal goes on decreasing as canal flows towards the tail. As discharge at various points on the canal is not constant, the section of the canal will also be changing. While fixing the discharge for any canal to be withdrawn from a head regulator, one must know the following data.

(i) Gross command area of which the proposed canal is going to be incharge.

(ii) Percentage by which G.C.A. is to be multiplied, to determine the cultural command area.

(iii) Various crops that will grow after the commissioning of the proposed canal.

(iv) Duties of water for various crops.

(v) Intensities of irrigation during Rabi and Kharif crops.

(vi) Losses due to seepage and evaporation.

The discharge required at a particular point on the canal depends upon the area to be irrigated lying *D/S* of that point and also upon the seepage and evaporation losses occurring in the canal itself, lying *D/S* of that point. Since area under irrigation, and losses due to evaporation and seepage, go on increasing as we proceed towards the *U/S* side, we have to design the canal sections at different points. Design of the canal is always started from the tail end of the canal and proceeded step by step towards *U/S* side till head regulator of the canal is reached. An example as to how discharge is determined at various points on the canal is given here.

Let a 5 km long minor distributory is to be designed. Let discharge required at the tail, on irrigated area basis, is 2.02 cumec.

Let canal length of 5 km be divided into 5 parts each 1 km long. All the 5 parts we denote by lengths measured from head regulator in kilometre. Thus first kilometre length canal is denoted by 0 – 1 kilometre and second kilometre length by 1 – 2 kilometre. Similarly other lengths are 2 – 3, 3 – 4 and 4 – 5 kilometres.

Discharge at the end of each kilometre is worked out as follows.

At 5 kilometre point. Discharge for which canal is to be designed = 2.02 cumec.

At 4 km point. Let from 4 km to 5 km point there be five outlets, each of 0.06 cumec. Also evaporation and seepage losses are say 0.08 cumec.

The discharge required at 4 km point should be as follows so that 2.02 cumec water remains available at 5 km point.

Required discharge at 4 km point

$$= 2.02 + 5\,(0.06) = 0.08 = 2.40 \text{ cumec}$$

At 3 km point. Let there by five outlets between 3 km and 4 km points, each of 0.03 cumec discharge. Let losses in the same length are 0.05 cumecs.

Total discharge required at 3 km point

$$= 2.40 + 5\,(0.03) = 0.05$$

$$= 2.60 \text{ cumecs}$$

At 2 km point. Let there be 0.02 cumec sized six outlets between 2 km and 3 km points and losses are 0.08 cumec.

Total discharge required at 2 km point

$$= 2.\,60 + (0.\,02)\,6 + 0.08$$

$$= 2.80 \text{ cumecs}$$

At 1 km point. Let there be 5 outlets each of 0.06 cumec, between 1 km and 2 km points and losses are 0.07 cumec.

Total discharge required at 1 km point

$$= 2.80 + 5\,(0.06) + 0.07 = 3.17 \text{ cumecs}$$

At zero km point i.e. at head regulator. Let there be 6 outlets, each of 0.05 cumec discharge between zero km and 1 km points and losses be 0.03 cumec.

Total discharge required at Head regulator

$$3.17 + (0.05)\,6 + 0.03 = 3.50 \text{ cumecs}$$

If say a minor of 1 cumec is taking off from this distributory between 2 km and 3 km points. The discharge in the distributory *D/S* of 3 km point will continue to be same as calculated above but discharges at 2 km, 1 km, and head regulator will increase by I cumec. In this case discharge at Head regulator, 1 km point, and 2 km points, will be 4.50, 4.17 and 3.80 cumecs respectively. All the information regarding discharge, canal section, slope, area under irrigation, losses at each point etc. is filled in a standard table known as *schedule of area statistics and channel dimensions.* This table is shown at the end of Solved Example 2 of this chapter.

Now discharge in the canal at various points is known. Fixing appropriate value of silt factor (f), channel dimensions at various points can be easily found

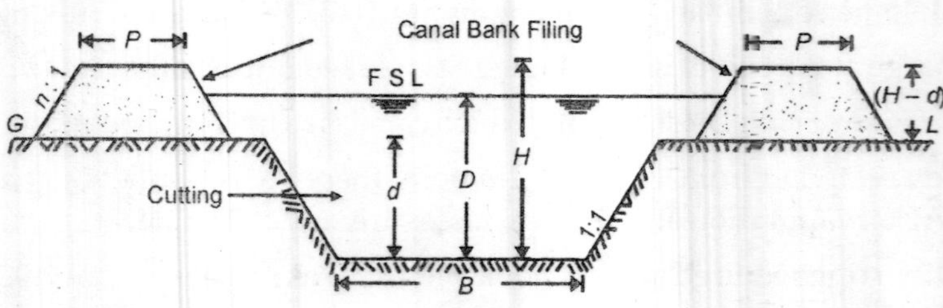

Fig. 19.1 *Balancing depth of canal*

out, using garret diagrams or Lacey's charts. Garret charts and Lacey's charts have been given in Chapter 17. Examples 19.1 and 19.2 have been solved assuming suitable data.

19.4 BED WIDTH AND DEPTH RELATIONSHIP

Kennedy's theory does not give any importance to B/D ratio Very large number of sections with varying B/D ratio and all satisfying the C.V.R. are possible by this theory. But all these sections cannot be equally satisfactory. This drawback of Kennedy's theory was made good to some extent by Mr. Woods, who gave B/D ratio table for various discharges. In Uttar Pradesh (U.P.) bed width, and depth are related by following equation.

(i) For discharge of the channel up to 15 cumec

$$D = \frac{1}{2}\sqrt{B}$$

(ii) For discharge of 15 cumecs and above depths for various discharges should be as follows.

Discharge Q	15	30	75	150	300
Depth (D) in metres	1.70	1.85	2.30	2.60	3.00

19.5 CHANNEL CROSS SECTION

The channel sections for an irrigation canal may be of following four types.

1. Canal in cutting.
2. Canal in filling.
3. Canal in heavy filling.
4. Canal partly in cutting and filling.

1. Canal in Cutting. This canal does not require any bank as F.S.L. lies below the G.L. If F.S.L. is just at the G.L., small banks may have to be provided. In this section F.S.L. of canal lies just at G.L. or slightly below it. See Fig. 19.2 (a).

2. Canal in Filling. In such a section the bed level of the canal lies at the G.L. Section whose bed level is slightly above the G.L. also comes under this category. See Fig. 19.3 (a).

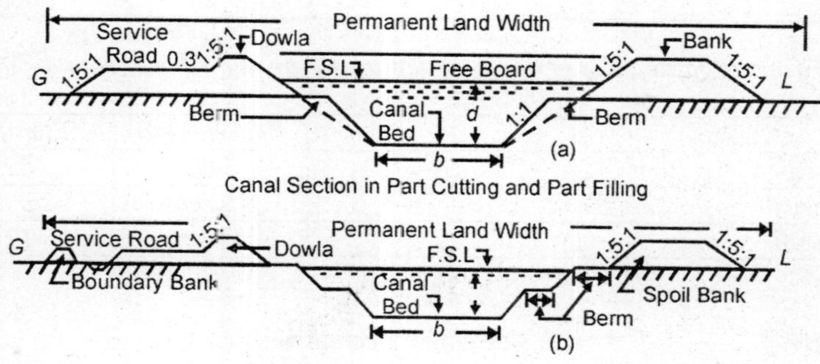

Fig. 19.2

3. Canal in Heavy Filling. In this section, the bed level of the canal lies substantially above the G.L. Such a section should as far as possible be avoided.

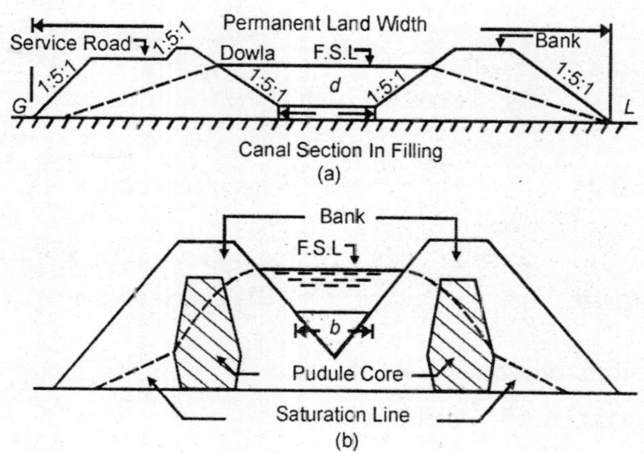

Fig. 19.3

Such sections are easily liable to breach and cause lot of damage to canal itself and surrounding areas. See Fig. 19.3 (b).

4. Canal Partly in Cutting and Filling. In such a section the ground level lies in between the F.S.L and bed level of the canal. See Fig. 19.2 (a).

Service road is usually provided on the left bank of the canal. One bank of the canal always has service road and the other bank is made banked section. Service road and canal berms are separated by a small bund called Dowla.

Example 19.1 *G.C.A. and C.C.A. for two kilometre length of a distributary is given in following table. Evaporation and percolation losses in this 2 km. length is 0.5 cumecs.*

Distance from Head regulator in km	*G.C.A. in Hectare*	*C.C.A. Hectare*
0	14000	10000
1	12000	8000
2	10000	6000

Intensity of irritgation, Kor period, Kor depth for Rabi and Kharif crops are as follows:

	Rabi (wheat)	*Kharif (rice)*
Intensity of irrigation	30%	12 %
Kor period	4 weeks	2.5 weeks
Kor depth	14 cm	19 cm

Silt factor $f = 0.8$, *side slope of canal* $\frac{1}{2}:1$

Design a suitable canal section

Solution Outlet discharge factor = O.D.F.

$$= \frac{8.64 \times B}{\Delta}$$

$$B = 4 \times 7 = 28 \text{ days}$$

$$\Delta = 14 \text{ cm} = 0.14 \text{ m}$$

$$\therefore \text{O.D.F. for Rabi} = \frac{8.64 \times 28}{0.14} = 1728 \text{ hectares/cumec}$$

$$\text{O.D.F. for Kharif} = \frac{8.64 \times 2.5 \times 7}{0.19} = 760 \text{ hectares/cumec}$$

Maximum discharge controls the C.C.A.

Total C.C.A. = 10000 hectare

30% intensity of irriagation for Rabi

$$= 10000 \times \frac{30}{100}$$

$$= 3000 \text{ hectare}$$

Area under Kharif at $12\frac{1}{2}$ % irrigation intensity

$$= \frac{10000 \times 12.5}{100} = 1250 \text{ hectare}$$

$$\text{Discharge for Rabi crop} = \frac{3000}{1728} = 1.74 \text{ cumec}$$

$$\text{Discharge for Kharif crop} = \frac{1250}{760} = 1.645 \text{ cumec}$$

Hence O.D.F. 1728 hectare/cumec should be adopted.

Design of canal section at diffecrent sections

Design of canal is always started from tail end of the canal. It is so because losses at the tail end are know.

1. At 2 km D/S of Head regulator

C.C.A. = 6000 hectare

Intensity of irrigation in Rabi = 30%

Area under irrigation during Rabi season

$= 0.30 \times 6000 = 1800$ hectares

$$\text{O.D.F.} = \frac{1800}{1728} = 1.04 \text{ cumec}$$

Total losses = 0.5 cumec

Hence total discharge at 2 km from head

$= 1.04 + 0.5 = 1.54$ cumec

Adopt design discharge 10% in excess

Hence design discharge at this point

$= 1.1 \times 1.54 = 1.694$ cumec

$f = 0.8$

From Lacey's diagrams depth of canal = 0.8 m

Width of canal = 4.5 m.

Longitudinal slope 20 cm in 1000 m.

2. At 1 km D/S of Head regulator

C.C.A. = 8000 hectares

Area under irrigation in Rabi = 8000×0.30

= 2400 hectares

$$\text{Outlet discharge factor} = \frac{2400}{1728} = 1.39 \text{ cumec}$$

Total losses D/S of 2 km = 0.5 cumec

Let losses of water from 1 km to 2 km is 0.01 cumec.

Hence total amount of water which should be available at 1 km D/S = 1.39 + 0.50 + 0.1 = 1.90 cumec.

Adopt 10% extra discharge.

Total discharge at 1 km = 1.1 × 1.90 = 2.09 cumecs

$$Q = 2.09 \text{ cumec}, f = 0.8$$

From Lacey's diagrams

Width of canal	5 m.
Depth of water	0.84 m.
Slope	0.2 m in 1000 m.

3. At Head regulator

C.C.A. = 10000

Area under Rabi crop = 10000 × 0.30 = 3000 hectares

$$\text{O.D.F.} = \frac{3000}{1728} = 1.74 \text{ cumec}$$

Water lost upto 1 km point D/S = 0.5 + 0.01

= 0.51 cumec

Water lost from head regulator to 1 km point = 0.01 cumec

Required discharge at head regulator

= 1.74 + 0.51 + 0.01

= 2.26 cumecs

Design regulator for 10% extra discharge.

Hence total discharge from a regulator

= 1.1 × 2.26 = 2.486 cumec

Section of canal just D/S of head regulator as read from Lacey's diagrams for 2.486 cumec discharge and 0.8 silt factor will be as follows.

Bed width = 5.5 m.

Depth = 0.88 M.

Slope = 18 cm per 1000 m.

All the dimensions of the section and other informations are filled in a Table known as schedule of area statistics and channel dimensions. A typical such Table is given on page 542.

Since discharge of the canal everywhere is less than 5 cumecs adopt free board throughout the length as 45 cm and berm width 1 m.

Example 19.2 *A distributary has been taken off from a branch canal. Design three kilometre length of the distributary from head regulator. Assume outlet discharge factor as 1800 hectare cumec. Gross command area below each kilometre is as given in the following Table.*

D/S of kilometres	*G.C.A. in hectares*
From 0 km	*30000*
D/S of 1 km	*25000*
D/S of 2 km	*20000*
D/S of 3 km	*18000*

Assume C.C.A. 70% of G.C.A. Design distributary considering area under Rabi 30% of C.C.A. Assume channel losses 2 cumec for 1000000 m^2 wetted area and Total losses D/S of 3 km point as 0.4 cumec. Adopt C.V.R. 0.95 and N = 0.0225. The Reduced levels of the centre line of the distributary at 200 m interval are as follows.

Distance from head	*R. L.*
0.0	*223.35*
200	*223.25*
400	*223. 20*
600	*223.05*
800	*222.90*
1000 (1 km)	*222.80*
1200	*222.65*
1400	*222.50*
1600	*222.55*
1800	*222.40*
2000 (2 km)	*222.30*
2200	*222.25*
2400	*222.20*
2600	*222.15*
2800	*222.05*
3000 (3 km)	*222.00*

Solution

At kilometer three

G.C.A. = 18000 hectares

C.C.A. = 70% × 180000 = 12600 hectares

Intensity of irrigation of Rabi = 0.30 × 12600

= 3780 hectares

O.D.F. = 1800 hectare/cumec.

$$\therefore \text{ Discharge of outlet} = \frac{3780}{1800} = 2.1 \text{ cumec}$$

Loss D/S of 3 km = 0.4 cumec.

Total discharge at 3 km = 2.10 + 0.40 = 2.50 cumec

Let slope is 1 in 5000.

For $Q = 2.50$ cumec, $\frac{V}{V_0} = 0.95, N = 0.0225$

following section is found as the most suitable section, using Garret diagrams:

Q	S	B	D	A	$V = \frac{Q}{A}$	V_0	$\frac{V}{V_0}$
2.50	1 in 5000	4.30 m	1 m	480	0.521	0.550	0.946

For 1 in 5000 slope following relation should also be satisfied.

$$D = \frac{1}{2}\sqrt{B}$$

$$D = \frac{1}{2}\sqrt{4.30} = 1.03 \text{ m}$$

1.03 m is almost equal to the depth of 1 m given by Garret diagram. Hence this section is the most suitable section.

2. At 2 km from head regulator

G.C.A. = 20000 hectares.

C.C.A. = 70% = 14000 hectares

Intensity of irrigation in Rabi = 30%

Area under irrigation during Rabi = 0.30 × 14000

= 4200 hectares

$$\text{Discharge of outlet} = \frac{4200}{1800} = 2.33 \text{ cumec}$$

Total loss D/S of 3 km = 0.4 cumec.

To find out loss from 2 km to 3 km, we must know the section of the canal. But it is not yet known. To find out wetted perimeter adopt same section as laying D/S of 3 km.

$$\text{Wetted perimeter} = B + \sqrt{5D}$$

$$= 4.30 + \sqrt{5} + 1.00 = 6.54 \text{ m}$$

Wetted area in 1 km length = 6.54 × 1000

= 6540 m^2

$$\text{Losses} = \frac{6540 \times 2}{1000000} = 0.01308 \text{ cumec}$$

$$= 0.015 \text{ cumec (say)}$$

Hence total discharge required at 2 km point

$$= 2.33 + 0.4 + 0.015 = 2.745 \text{ say } 2.75 \text{ cumecs}$$

Adopt slope 1 in 5000. For $\frac{V}{V_0} = 0.95$ and $N = 0.225$ $B = 4.75$ m and $D = 1$ m.

(From Garret diagrams)

Q	S	B	D	A	$V = \theta/A$	V_0	$\frac{V}{V_0}$
2.75	1/5000	4.75	1.00	5.25	0.524	0.550	0.952

D sholud be $\frac{1}{2}\sqrt{13}$.

$D = \frac{1}{2}\sqrt{4.75} = 1.08$ m which is very nearly equal to adapted depth of 1 m.

At 1 km point from the head regulator

G.C.A. = 25000 hectares

C.C.A. = 0.70 × 25000 = 17500 hectares

Area to be irrigated in Rabi = 0.30 × 17500

= 5250 hectares

$$\text{Discharge required} = \frac{5250}{1800} = 2.92 \text{ cumec}$$

Total losses D/S of 2 km point 0.4 + 0.015

= 0.415 cumec

For losses between 1 km and 2 km point adopt same section as found out at 2 km.

$$\text{Wetted perimeter} = 4.75 + \sqrt{5} \times 1 = 6.99 \text{ m}$$

$$\text{Losses} = \frac{6.99 \times 1000 \times 2}{10^6} = 0.01398 \text{ cumec}$$

$$= 0.015 \text{ cumec (say)}$$

Required discharge at 1 km point

$= 2.92 + 0.415 + 0.015 = 3.35$ cumec

$= 3.40$ cumec (say)

For $Q = 3.40$ cumec, $S = \frac{1}{5000}$, $\frac{V}{V_0} = 0.95$, $N = 0.0225$

read Garret diagram

$$B = 6.0 \text{ m}, D = 1.0 \text{ m}$$

$$A = 6.50 \text{ m}^2, V = \frac{Q}{A} = \frac{3.45}{6.50} = 0.523 \text{ m/sec}$$

$$V_0 = 0.55 \text{ m/sec}$$

$$\frac{V}{V_0} = \frac{0.523}{0.55} = 0.95$$

$$D = \frac{1}{2}\sqrt{13} = \frac{1}{2}\sqrt{6.0} = 1.22 \text{ m}$$

There is substantial difference between 1.22 m depth and 1 m depth. In order to make $\frac{B}{D}$ ratio adoptable use slope $\frac{1}{4550}$ instead of $\frac{1}{5000}$. From Garret diagram again read section for changed slope of $\frac{1}{4550}$.

Q	S	B	D	A	$V = Q/A$	V_0	$\frac{V}{V_0}$
3.40	1/4550	5.25	1.07	6.17	0.550	0.575	0.95

$$D = \frac{1}{2}\sqrt{13} = \frac{1}{2}\sqrt{5.25} = 1.15 \text{ m}$$

This depth of 1.15 m is quite near to 1.07 m and hence this revised section is adequate.

Hence adopt $B = 5.25$ m and $D = 1.07$ m at 1 km point.

At head regulator i.e. zero km

G.C.A. = 30000 hectares

C.C.A. = $0.70 \times 30000 = 21000$ hectares

Area to be irrigated in Rabi = 0.30×21000

= 6300 hectares

Required discharge $= \frac{6300}{1800} = 3.50$ cumec

Losses D/S 1 km point = 0.43 cumec

Losses between zero and 1 km point ; adopt section as at 1 km point.

$$\text{Wetted perimeter} = 5.25 + \sqrt{5} \times 1.07 = 7.64 \text{ m}$$

$$\text{Losses} = \frac{7.64 \times 1000 \times 2}{10^6} = 0.01528 \text{ cumec}$$

$$= 0.016 \text{ cumec (say)}$$

$$\text{Total losses} = 0.43 + 0.016 = 0.446 \text{ cumec}$$

Total discharge required at head

$$Q = 3.50 + 0.446 = 3.945 \text{ cumec} = 4 \text{ cumec (say)}$$

For $\quad Q = 4 \text{ cumec}, S = \frac{1}{4550}, N = 0.0225, \frac{V}{V_0} = 0.95$

See Garret diagram and note the section as follows:

Q	S	B	D	A	$V = \theta/A$	V_0	$\frac{V}{V_0}$
4.0	1/4550	5.50	1.156	7.02	0.570	0.632	0.947

$$D = \frac{1}{2}\sqrt{13} = \frac{1}{2}\sqrt{5.50} = 1.17 \text{ m}$$

$$B = 5.50 \text{ m}, D = 1.156 \text{ m is a suitable section}$$

All the channel dimensions and area statistics are filled in a table known as schedule of area statistics and channel dimensions. Such a table and various dimensions filled in it are shown on page 542..

19.5 SIDE SLOPE

The side slope of the canal depends upon the type of the soil. The canals in alluvial soils, are designed assuming $\frac{1}{2}$:1 side slope, irrespective of the actual initial side slope. It is assumed that after due course of run, the canal section would ultimately acquire $\frac{1}{2}$: 1 slope. This happens because silt gets deposited on the berms. Had section been designed with $\frac{1}{2}$: 1 slope initially, the section

Table 19.1 *Schedule of Area Statistics and Channel Dimension*

Below kilometer	G.C.A.	C.C.A	Area to be irrigated			Outlet discharge factor	Losses D/S of the point	Losses in reach itself	Total losses	Total discharges	Bed slope	Bed width	Depth of water	Free board	Height of banks	Width of banks	Velocity	$\frac{V}{V_0}$ = 0.95
			Rabi	Sugar cane	Rice													
(1)	(2)	(3)	(4)	(5)	(6)	(7)	(8)	(9)	(10)	(10)	(12)	(13)	(14)	(15)	(16)	(17)	(18)	(19)
0.00	30000	21(100	6300	–	–	3.50	0.430	0.016	0.446	3.946	1/4550	5.50	1.156	0.50	0.826	2 m	0.57	0.947
1.00	25000	17500	5250	–	–	2.92	0.415	0.015	0.430	3.35	1/4550	5.25	1.070	0.50	1.070	2 m	0.55	0950
2.00	2000	14000	4200	–	–	2.33	0.400	0.015	0.415	2.745	1/5000	4.75	1.00	0.50	0.724	2 m	0.524	0.952
3.00	18000	12600	3780	–	–	2.10	0.40	–	0.40	2.500	1/5000	4·30	1.00	0.50	0.824	2 m	0.521	0.946

would be reduced in due course of time due to silting and the section remaining would be inadequate.

As per recommendation of *Central Water and Power Commission* (C.W.P.C.) the side slopes for various soils should be give Table 19.2.

Table 19.2. *Recommended side slope for Kucha Canals*

Type of soil	*Side slope (Hor : Vert)*	
	Cutting	*Filing*
1. Rock	1/4 : 1 to 1/2 : 1	2:1
2. Hard soil murum	3/4 : 1 to 1.5 : 1	
3. Alluvial soils, soft clay, black cotton soil	1 : 1 to 1.5 : 1	
4. Sandy loam, average sandy soil, very light loose sand	1.5 : 1 to 2 : 1	2 : 1 to 3: 1

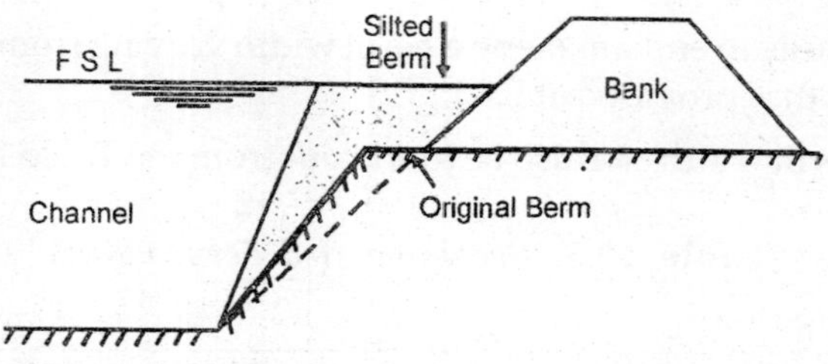

Fig. 19.4. *Berms*

19.6 BERMS

It is a narrow strip of land, left on either side of a channel at G.L., between upper edge of the cut and the inside toe of the bank. The width of the berm depends upon the size of the channel. If canal is in partial cutting, the original width of the berm will be small but it becomes wider after silting over the berms and side slopes. The canals while constructing area excavated with 1 : 1 slope, but after a run for few months the section automatically acquires a side slope of $\frac{1}{2}$:1. Berm performs following functions.

1. The width of the canal can be easily increased if required.

2. Slipping soils and boulders are held up at berms and do not allow them to be dropped into the channel.

3. It acts as a storage space for materials if some repair or construction work is to be clone in the canal.

4. Barrow pits may be made on banks for taking soil. These berms get silted up very soon. Such a need for barrow pits arises during canal breaches.

5. They strengthen the channel banks.

6. Rise in water level is marginal with substantial increase in discharge above full capacity of the canal. If by mistake excess discharge enters the canal, they do not allow water to rise much and thus possible breach of the canal is averted.

7. Because of silting of inside edge and top, the terms become impervious, and as such, loss of water by seepage is reduced.

8. Waves developed in the canal do not come in direct contact of the banks and hence possibilities of bank erosion are reduced.

9. They also provide easy path for inspection.

10. They increase the width of the bank, and thus, seepage line is not likely to be exposed.

For channels in full cutting, a berm of width equal to depth of water is provided at 50 cm above the F.S.L. In channels, partly in cutting, the berm provided is such that after silting its width at F.S.L. will not exceed twice the depth of water.

In channels, fully in embankment, a berm width varying from twice the depth to thrice the depth is provided at F.S.L.

Minimum berm width can also be found out from the Table 19.3.

Table 19.3. *Recommended berm width*

Discharge in cumecs	*Berm width in terms of depth of water, D*
4.25	$0.6 + 0.5\ D$
4.25 to 28	$1.25 + 0.5\ D$
28 and above	$1.25 + \dfrac{D}{28} + 0.5D$

19.7 FREE BOARD

The vertical distance between F.S.L. and top of the lowest bank of the channel is known as *free board.* It is provided to prevent waves or fluctuations in water surface from overtopping the banks. Free board depends upon the canal size, wind action, soil characteristics and location. According to *USER,* free board may be worked out from following formula, under ordinary conditions.

$$F = \sqrt{CD}$$

where F = free board

C = a constant whose value varies from 0.46 to 0.76

D = the depth of water in metres

Lacey gave following formula for the free board

$F = 0.20 + 0.15\, Q^{1/3}$

According to CWPC the free board should be as follows:

Discharge in (Q) cumecs	0.7	0.7 to 1·4	14 to 8.5	above 8.5
Free Board in metres	0.46	0.61	0.76	0.92

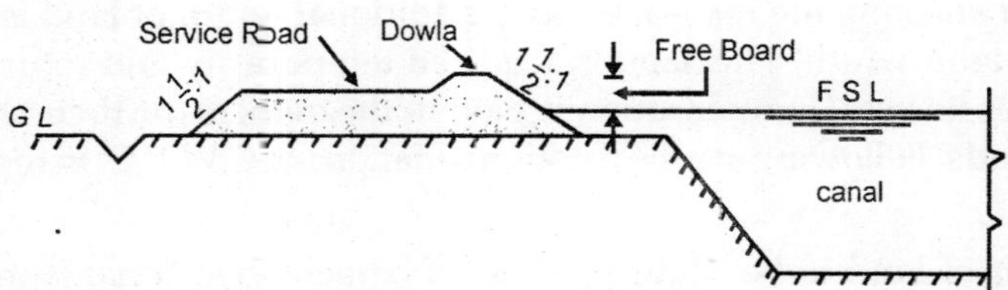

Fig. 19.5. *Free board, Dowla, and service road.*

19.8 CANAL BANKS

The purpose of banks is to prevent spread of water beyond the specified limit. The width of the banks should be enough so that a minimum cover of 0.5 in soil is available everywhere above the saturation line. Hydraulic gradient in ordinary soils is kept 1 in 4 and for light soils 1 in 6. If banks become very high counter-berms may be provided on the outer slopes of the banks. Banks should be properly compacted while making. CWPC has given following bank widths depending upon the discharge.

Discharge in (Q) cumecs	Upto 0.28	0.28 to 1.4	1.4 to 4.2	4.2 to 10	10 to 14	14 to 28	28 to 140
Top width of bank in	0.92	1.22	1.53	1.83	2.44	3.61	4.38

19.9 SERVICE ROAD AND DOWLA

A service road is provided usually on the left bank of the channel. In case of very small channel i.e. minor, it may or may not be provided. The top width of the bank carrying service road should not be less than 5 m. According to CWPC minimum road width should be 61 m. On large canals service roads may be provided on both the banks.

Service road level should be 50 cm to 1 in above the F.S.L. depending upon the size of the canal.

A Dowla is provided along the service road, separating service road from berms. This is provided as a measure of safety. Dowla is an earthen bond 50 cm high and 50 cm wise at the top. Side slope of Dowla is 1.5 : 1. It also prevents erosion of the slope due to rain.

19.10 LAND WIDTH

The width of land, required to accommodate the canal cross-section and its connected elements, is known as land width for the canal. The distance between outer toes of canal banks plus a few metres on both the sides for the construction of side rain water drains or for growing tree rows, is known as the *permanent land width* of the canal. This land has to be acquired before canal construction is started.

During construction of the canal, some additional land is required for borrow pits and for stacking the materials. Such additional width of land is known as temporary land width. This land is acquired temporarily and returned to the owners after its use. Compensation is paid to the owners for their temporarily acquired lands. Following are the recommendations of C.W.P.C. in regard width of the land.

(i) Width of land to be acquired clear of banks when canal is in less than balancing depth of cutting.

(*a*) *For major canals.* Width due to full height of bank + 5 m.

(*b*) *For minors and distributaries.* Width due to full height of banks above ground + 1.5 m.

In this case (i) extra land is required for borrow pits.

(ii) Width of land to be acquired clear of banks when canal cutting is more than balancing depth.

(*a*) *For major canals.* As per actual drawing + 5 m.

(*b*) *For minors and distributaries.* As per actual requirements + 1.5 m.

19.11 COUNTER BERM

It is also known as back berm. It is provided on the outer slope of the banks. It is required only in case of high banks and very permeable soils. Its main purpose is not to allow the seepage line expose on the outer slope of the bank.

19.12 SPOIL BANKS

It is a method of disposal of surplus excavated soil from very deep reaches of the canal. When quantity of surplus excavated soils is not much it is used either to widen the banks or to raise the height of banks. If amount is large it is disposed of, by constructing spoil banks parallel to canal banks, but slightly away from the banks. The area enclosed between canal banks and spoil banks is properly drained.

19.13 BORROW PITS

When amount of soil obtained from cutting is not enough to complete the banks of the canal extra earth is required. This extra earth is obtained from the borrow pits. Borrow pits may be constructed out of canal section or within the bed of the canal. Outside borrow pits are not preferred as they may become mosquito breeding centres during rains. The outside borrow pits should not be deeper than 30 cm so that they may be easily reclaimed by the owners when land having borrow pits is returned to them. Outside borrow pits should be located at least 5 m away from the toe of the bank, in case of small canals, and 10 m in case of large canals.

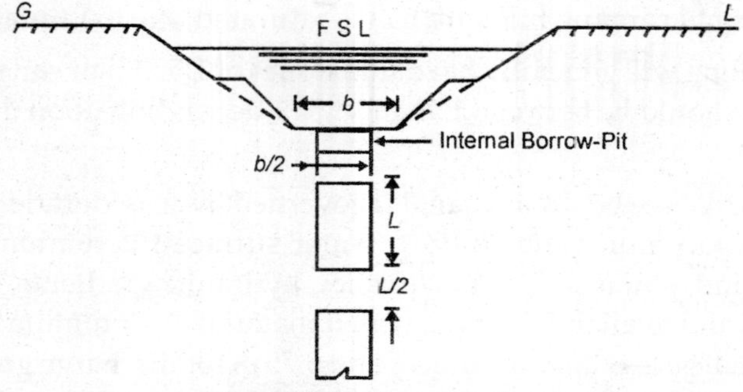

Fig. 19.6. *Borrow pits.*

Inside borrow pits are preferred to outside ones. The reason being that inside borrow pits get silted up during course of time, automatically. Inside borrow pits should not cover area more than half the bed width and suitable unexcavated bed should be left after each section so that water remains held up in them during running of the canal for silting purposes. Depth of inside borrow pits should not exceed 1 m.

19.14 STANDARDS OF CANAL CROSS-SECTION

Usual dimensions of canal cross-section elements have been given here and there in appropriate articles. The standards as suggested by CWPC are given as follows.

Element	*Discharge of canal in cumecs*					
	Below 0.3	*0.3 to 1*	*1.0–5*	*5–10*	*10–30*	*30–150*
1. Minimum width of bank	0.6 in	1.0 m	1.5 m	2.0 m	3.5 m	5 m
2. Free board	0.3 in	0.4 in	0.5 in	0.6 in	0.75 m	0.90 m
3. Width of road way	nil	3.5 m	3.5 m	5 m	5 m	6 m
4. Depth of cover over saturation gradient	0.5 m	0.5 m	0.5 m	0.5 m	0.75 m	1 m

Indian standard criteria for design of unlined canals in alluvial soil (IS : 7112–1973)

IS : 7112–1973 has given following criteria for design of unlined canals in alluvial soils.

1. General. A trapezoidal section is recommended for the canal. The longitudinal slope of the canal is determined depending upon the average slope of the natural ground along the proposed alignment. This is the maximum average slope which can be provided on the canal.

2. Side Slopes. These are dependent on the local soil characteristics and are designed to withstand the following conditions during the operation of the canal.

(a) The sudden drawdown condition for inner slopes and

(b) The canal running full with banks saturated due to rainfall.

Canal in filling will generally have side slopes of 1.5 : 1. For canals in cutting, the side slope should be between 1 : 1 and 1.5 : 1 depending upon the type of the soil.

3. Free Board. Free board in a canal is governed by consideration of the canal size and location, rain water inflow, water surface fluctuations caused by regulators, wind action, soil characteristics, hydraulic gradients, service road requirements and availability of excavated material. A minimum freeboard of 0.5 m for discharge less than 10 cumecs and 0.75 m for discharge greater than 10 cumecs is recommended. The free board shall be measured from the FSL to the level of the top of bank; the height of dowel portion should not be used for free board purposes.

4. Bank Top Width. The minimum values recommended for top width of the bank are as follows.

Discharge in m³/sec	*Minimum bank top width*	
	Inspection bank (m)	*Non-inspection bank (m)*
0.15 to 7.5	.5	1.5
7.5 to 10.0	5.0	2.5
10.0 to 15.0	6.0	2.5
15.0 to 30.0	7.0	3.5
30.0 and above	8.0	5.0

The banks should invariably cover the hydraulic gradient. The width of the non-inspection bank should be checked to see that cover for hydraulic gradient is provided.

5. Berms. Berms along earthen canal are usually provided to reduce bank loads which may cause sloughing of earth into the canal section and to lower the elevation of the service road for easier maintenance. Berms are to b provided in all cuttings when the depth of cutting is more than 3 m where a canal is constructed in a deep through cut requiring waste banks, berms should be provided between the canal section cut and the waste bank. Various other factors may be involved in determining whether berms should be used and case should be taken that their use is justified by results obtained. However, the following practice is recommended.

(a) When the full supply level is above ground level but the bed is below G.L. i.e. the canal is partly in cutting and partly in filling, berms may be kept at natural surface level equal to 2*D* in width where *D is* the full supply depth.

(b) When FSL and the bed level are both above the G.L. i.e. the canal is in filling the berm may be kept at the F.S.L. equal to 3 D in width.

(c) When the F.S.L. is below the G.L, i.e. the canal is completely in cutting the berm may be kept at the F.S.L equal to 2D in width.

In embankments, adequate berms may be provided so as to retain the minimum cover over the hydraulic gradient line.

6. Dowels. Dowels having top width of 0.5 m, height above road level of 0.5 m and side slopes 1.5 : 1 shall be provided on the service road side between the road and the canal.

7. Bed Width, Depth and Slope. There shall be designed for the various reaches to carry the required discharges according to the best prevalent practice. For the design of alluvial channels, Lacey's regime equations have been in use for nearly four decades. However, the divergence from the dimensions given by Lacey's equations in existing stable canals has been found significant in many cases. Hence, regime type-fitted equations have been recommended. Another method of design is by *tractive force approach*.

8. Hydraulic Grade Line. When water runs against fill banks the line of saturation slant downwards from the water surface through the embankment material. The gradient depends mainly on characteristics and relative placement of the different types of material in the embankment. For embankments more than 5 m high, the true position of the saturation line shall be worked out by laboratory tests and the stability of the slope checked. However, the following empirical values for the hydraulic gradient (horizontal to vertical) may be used for banks less than 5 m high

For silty soils	4 : 1
For silty sands	5 : 1
For sandy soils	6 : 1

The H.G line shall have a cover of 0.3 m. When counter berms are required for this purpose, top level of the same shall be 0.3 below F.S.L. and the top width of the same shall be 2 m for branch canals and 1 m for distributories. In case of canals is very high filling a second counter berm may be provided so as to cover the H.G. line.

Lacey's Method

According to Lacey, a canal is said to have attained regime condition when a balance between silting and scouring and dynamic equilibrium in the forces generating and maintaining the canal cross-section and gradient are obtained. If a canal runs indefinitely with constant discharge and sediment charge rates, it will attain a definite stable section having a definite slope. If a canal is designed with a section too small for a given discharge and its slope is kept steeper than required, scour will occur till final regime is obtained. On the other hand, if the section is too large for the discharge and the slope is flatter than required, silting will occur till true regime is obtained. In practice true regime conditions do not develop because of variations in discharge and sediment rates.

Lacey postulated that the required slope and channel dimensions are dependent on the characteristics of the boundary material which he quantified in terms of the silt factor (f) defined as

$$f = \frac{2.29\,V^2}{R}$$

or
$$f = 1.76\sqrt{D_{50}}$$

where V = The mean velocity of flow in m/sec.

R = The hydraulic mean depth of an existing stable canal, and

D_{50} = The average particle size of the boundary material in mm.

The following three relationships may be used for determining required slope and canal dimensions

$$S = \frac{0.003\,f^{5/3}}{Q^{1/6}}$$

$$P = 4.75\sqrt{Q}$$

$$R = 0.47\left(\frac{Q}{f}\right)^{1/3}$$

where S = longitudinal slope of the canal

Q = discharge in cumecs

P = wetted perimeter of the section in metres

R = hydraulic mean depth in metres.

Knowing the desirable values of P, R the curves given in Fig 19.7 given on next page may be used for determining the corresponding canal bed width (B) and depth (D) for a canal having internal side slopes of 0.5 : 1 (it is assumed that the canal attains a slope of 0.5 : 1 after running in regime.

Regime type fitted equations for Design of unlined canals in alluvial soils

The regime type fitted equations evolved on the basis of data collected from various states in India are given in Table 19.4. In these equations, average boundary conditions is taken care of by fitting different equations to data obtained from different states and assuming similar average boundary conditions in a state.

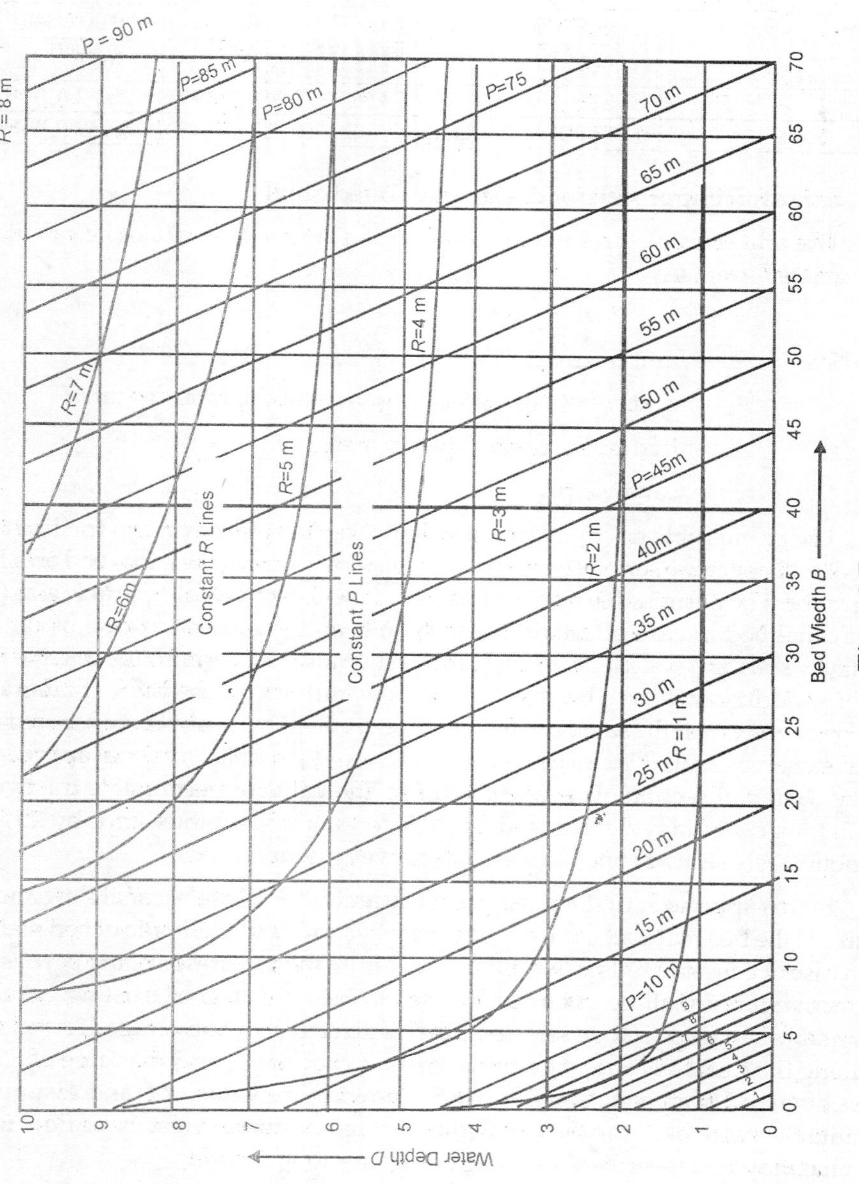
Water Depth D
Bed Wiedth B
Constant R Lines
Constant P Lines
R = 8 m
R=7 m
R=6m
R=5 m
R=4 m
R=3 m
R=2 m
R = 1 m
P = 90 m
P=85 m
P=80 m
P=75
70 m
65 m
60 m
55 m
50 m
P=45m
40m
35 m
30 m
25 m
20 m
15 m
P=10 m

Fig. 19.7

Table 19.4 *Regime type fitted equations.*

S. No.	*Hydraulic parameter*	*All India canals*	*Punjab canals*	*U.P. canals*	*Bengal canals*
1	*S*	$\frac{0.00315}{(Q)^{0.065}}$	$\frac{0.000251}{(Q)^{0.051}}$	$\frac{0.00035}{(Q)^{0.045[illegible]}}$	$\frac{0.0001346}{(Q)^{0.0375}}$
2	*P*	$4.30\ (Q)^{0.53}$	$7.0 (Q)^{0.6019}$	$3.98\ (Q)^{0.5026}$	$5.52\ (Q)^{0.4190}$
3	*R*	$0.515\ (Q)^{0.3106}$	$0.466\ (Q)^{0.2389}$	$0.448\ (Q)^{0.3642}$	$0.438\ (Q)^{0.4463}$

Tractive force approach for design of unlined canals

The unit tractive force exerted on bed of a running canal can be calculated from the formulae.

$$Z = Y_w RS$$

when Z = unit tractive force in kg/m^3

Y_w = unit weight of water in kg/m^3 usually 1000 kg/m^3.

R = hydraulic mean radius in metre and

S = canal slope.

The permissible tractive force may be defined as the maximum tractive force that will not cause serious erosion of the material forming the canal bed on a level surface. The permissible tractive force is a function of average particle size (D_{50}) of canal bed in case of canals in sandy soils and void ratio in case of canal in clayey soils and sediment concentration. The values of permissible tractive force for straight canal have been given by some authors on the basis of laboratory experiments, but the same can better be determined by analysis of observed data on existing canals. Once this is done, this would provide a rational approach to the design of section of regime channels. The value of permissible tractive for sinuous canal may be reduced by 10% for slightly sinuous once by 25% for moderately sinuous ones and by 40% for very sinuous ones.

In this approach, first the sediment concentration X of the canal flow and D_{50} size of the bed material in case of non-cohesive soils and void ratio of bed material in case of cohesive soils is determined and from these corresponding permissible tractive force shall be obtained by use of observed data of existing canals. A suitable bed slope is then selected either with reference to average ground slope along the canal alignment or on the basis of experience and the value of R shall be obtained from equation $Z = \gamma_w RS$. Knowing the value of R and assuming a suitable value of N for the canal, the average desirable velocity of flow in the canal may be determined by using the Manning's formula

$$V = \frac{R^{2/3}\ S^{1/2}}{N}$$

Thus the area of cross-section may be determined and knowing R and A, the desirable canal bed width (B) or depth (D) may be calculated.

Values of Rugosity Coefficient (N) for unlined canals as per IS : 7112–1973 is as follows.

Table 19.5. *Value of N for unlined canals as per IS: 7112—1973*

Nature earth and surface	Value of N		
	Minimum	Normal	Maximum
1. Earth, straight and uniform			
(a) Clean, recently completed	0.016	0.018	0.020
(b) Clean, after weathering	0.018	0.022	0.025
(c) Gravel, uniform section, clean	0.022	0.025	0.030
(d) With short grass, few weeds	0.022	0.027	0.033
2. Earth winding and sluggish			
(a) No vegetation	0.023	0.025	0.030
(b) Grass, some weeds	0.025	0.030	0.033
(c) Dense weeds or aquatic plant in	0.030	0035	0.035
deep channels	0.030	0.035	0.035
(d) Earth bottom and rubble sides	0.030	0.035	0.040
(e) Stone bottom and weedy banks	0.025	0.035	0.040
(f) Cabble bottom and clean sides	0.030	0.040	0.050
3. Dragline excavated or Dredged			
(a) No vegetation	0.025	0.028	0.033
(b) Light brush on banks	0.035	0.050	0.060
4. Channels not maintained			
(Weed and brush uncut)			
(a) Dense weed, high as flow depth	0.050	0.080	0120
(b) Clean bottom, brush on sides	0.040	0.050	0.080
(c) Same, highest stage of flow	0.045	0.070	0.100
(d) Dense brush, high stage	0.080	0.100	0140

Note 1. For normal alluvial soil, it is usual in India to assume a value of N = 0.02 for bigger canals (Q > 15 cumecs) and N = 0.0225 for smaller canals Q < 15 cumecs.

2. A suitable value of N should be adopted keeping in view the local conditions and the above values as a guide.

QUESTIONS

19.1. What information should be given in L-section and C-section of a canal ? What points should be specially taken care of ?

19.2. What do you understand by *schedule of area statistics and channel dimensions* ? Give different columns of this table and write down its importance.

19.3. How many types of cross-sections of a channel are possible ? Draw a neat sketch of a cross-section partly in cutting.

19.4. Explain the term berm. Why berms are provided in canals ? On what basis the width of the berm is decided ?

19.5. Write short notes on :

(a) Free-board

(b) Dowla

(c) Land width

(d) Borrow pits

(e) Counter berms.

❑❑❑

20

Lined Canals and Their Design

20.1 INTRODUCTION

Canal lining is a treatment given to the canal bed and banks, so as to render the canal section impervious Since imperviousness of the canal section is achieved mostly by making canal section pucca, either by cement concrete, or bricks, the lined canals are also sometimes known as pucca canals. Lined canals are mostly referred as pucca canals. It has been estimated that seepage losses in irrigation canals may amount from 30% to 50% of the water admitted in the canals at diversion head works. This much loss of water in unlined channels cannot be afforded, as resources of water in India are limited in relation to irrigable area available. Hence in order to reduce or rather eliminate the seepage losses, lining of canals is the need of the hour.

Advantages of Lining

Following are the benefits of lined canals:

1. Seepage losses are practically eliminated or reduced to minimum.
2. Maintenance cost of the canals is reduced.
3. Canals can be run with increased velocity.
4. Section of the canal is considerably reduced as due to increased velocity, same discharge can be passed through smaller sectional area.
5. Cost of earth work is reduced, as smaller section will have to be constructed.
6. Silting does not take place, as canals are run at considerably larger velocities than silting velocity.
7. No scouring occurs any where in canal section.

8. Less width of land is required for the canal.
9. Weed growth in the canal is reduced.
10. Harmful salts from adjoining soils do not get dissolved in the canal water.
11. More head for power generation becomes available as lined canals can be laid at flatter gradient. This measure also helps in bringing more areas under command.
12. Possibilities of canal breaches are eliminated.
13. Additional areas may be brought under irrigation from the water saved by lining.
14. Possibilities of water logging of adjoining areas are reduced.
15. Losses due to evaporation are reduced, as due to increased velocity, water takes less time to reach the outlet.

Disadvatnages of Canal Lining

Following are the disadvantages of the canal lining:

1. It requires a very heavy initial expenditure.
2. When lining gets old it generally develops cracks. Leakage through cracks is very difficult to check.
3. Joins in lining always create problems.
4. Position of outlets cannot be shifted easily.
5. Lined canals do not have berms. If some person or animal happens to fall in the canal, it is very difficult to pull him out. To overcome this difficulty there should be stepped sections at regular intervals along the length of the canal.
6. Lined canals are generally seep. Only that man can enjoy bath in the canal who knows swimming.
7. As seepage of water is almost completely stopped, rows of trees cannot be grown along the canal as in case of Kucha canals.
8. In order to run with greater velocity, lined canals may have to be given increased longitudinal slopes. This may result in loss of command in the area.
9. Lining materials like clay, bitumen, etc. get easily destroyed under the feet of animals crossing the canal.

Financial viability. In order to justify the lining of canal it is essential to analyse on one hand the extra capital cost of providing lining and on the other hand an evaluation of the benefits of lining. For lining to be economically feasible, the capitalized value of benefits should be equal to or greater than extra cost involved in providing lining. In other words the capitalized value of benefits divided by extra capital cost of providing lining should always be more than one.

20.2 ESSENTIAL PROPERTIES OF A GOOD LINING

Following are the essential properties of a good lining:

1. The lining should be completely water tight.
2. The rugosity coefficient of the lining material should be low, so as to make the section more efficient, hydraulically.
3. The lining should be strong and durable.
4. Initial cost of lining and its subsequent maintenance cost should be low.
5. The lining should not get damaged by tramping of cattles.
6. It should resist growth of weeds and attack of burrowing animals.
7. Lining should not get damaged when flow in the canal is stopped.

20.3 TYPES OF LINING

The linings may be classified under the following four heads:

1. Hard surface lining
2. Buried and protected type membrane lining
3. Earth lining
4. Porous type lining,

Each class of lining is discussed in details one by one.

20.3.1 Hard Surface Lining

Following types of linings come under this category:

1. Cement concrete lining
2. Shot-crete lining
3. Precast concrete lining
4. Cement mortar lining
5. Brick or tile lining
6. Stone masonry lining
7. Asphaltic concrete lining.

1. Cement concrete lining. This lining has excellent hydraulic properties but being costly its use in India has been limited. The success of this lining depends upon the stability of foundation layer to a large extent. Bed sleepers and precast sleeper on sides are provided at regular intervals. Bed and side sleepers act as gauges for the dressing and preparation of the sub-grade. The joints of concrete lining are always located at sleepers so as to reduce seepage to minimum.

The thickness of lining depends upon the desired imperviousness, and structural strength. In case of ordinary M_{150} concrete thickness of lining varies from 5 cm to 10 cm depending upon the discharge the canal is to carry. If lean concrete say M_{100} has been used the thickness may vary from 7.5 cm to 15 cm.

Concrete lining is liable to crack on account of shrinkage and temperature changes. The cracking of lining may be prevented either by providing reinforcement or by providing contraction joints. In case of continuous lining operations only transverse or transverse and longitudinal grooves at 3 m to 5 m interval should be formed.

When lining is cast in panels, slabs are laid in alternate compartments at an interval of at least one day. The grooves in the joints should be filled with suitable sealing compound. Various types of joints provided in concrete lining are shown in Fig. 20.1.

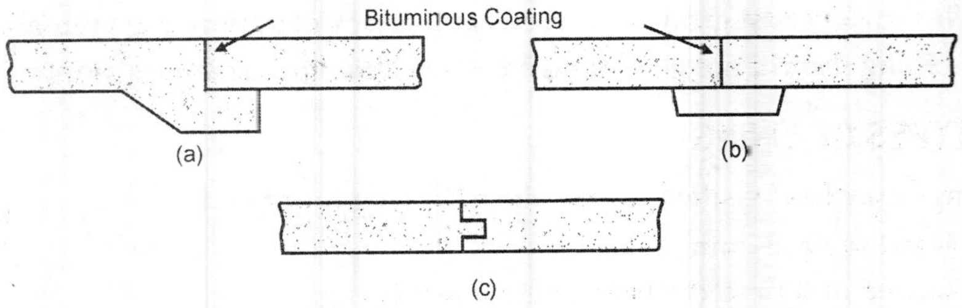

Fig. 20.1. *Concrete lining joints.*

2. Shot-crete Lining. In this lining a cement sand (1 : 4) slurry is prepared and is shot on the prepared sub-grade with the help of pneumatic pressure. This lining is costlier than cement concrete lining of equal thickness.Its utility lies in plugging the joints of old concrete lining and fissures in rock formations. This lining is not common.

3. Precast concrete lining. In this lining, the precast cement concrete slabs are prepared at factory site and then used for lining the canals. Concrete blocks most commonly used are 50 cm long, 25 cm wide and 5 cm thick. The blocks have interlocking arrangement with the help of which different blocks can be inter-locked and sealed by sealing compound. The slabs are laid on well compacted sub-grade. For precast slabs, the concrete should not be of leaner grade then M_{150}.

4. Cement mortar lining. It has never been used all alone as lining material. It is mostly used as a sandwich material between brick layers. Thickness of this layer may vary from 1 cm to as much as 4 cm. It is also not in very common use.

5. Brick of tile lining. This lining has been extensively used for lining of canals in India. Rajasthan canal project is also adopting this lining. The lining maybe singled layered or double layered. The subgrade is prepared well and first layer of tiles or bricks is laid on 15 mm thick layer of 1 6 cement mortar. A 15 mm layer of 1 : 3 cement mortar is placed on the first layer of tiles and then second layer of tiles is laid. The tiles of second layer should be laid so as to break the joints of the bricks of first layer. The usual size of tiles used for lining is 30 × 15 × 5 cm. Following are inherent advantages of brick lining.

(i) Ordinary mason can lay tiles.
(ii) Rounded work at corners can be easily done without any form work.
(iii) Strict quality control is not required.
(iv) Expansion joints are not required to be provided.

Single tile lining consists of a single layer of tiles, laid on 10 cm thick 1 : 5 cement mortar, on well compacted sub-grade. A 20 mm cement plaster 1 : 3 is laid over tiles and given a smooth finish.

If this lining fails at few spots it can be easily repaired. Sub-sided lining is excavated and new lining is placed in its place.

6. Stone masonry lining. This lining consists of undressed stone blocks. The subgrade is prepared as usual and stone blocks are laid over it in cement mortar. Lining may be made from dressed blocks but it proves very uneconomical. This lining offers a very large resistance to flow as its rugosity coefficient is very high.

7. Asphaltic concrete lining. A mixture of asphalt with graded aggregate is prepared and placed at a high temperature of about 200°C. The layer so laid is covered with earth layer of at least 30 cm thickness. Earth layer provides protection to the asphaltic concrete layer.

20.3.2 Buried and Protected type Membrane Lining

These linings are such lining in which water proof thin membrane is place on the prepared subgrade and thereafter it is covered by a protective layer of earth or gravel. Protective layer provides protection to the lining against damage due to outside effects and membrane itself provides imperviousness. The commonly used buried membranes may be :

1. Sprayed asphaltic lining. In this hot asphalt is sprayed on the subgrade which acts as water proofing barrier.

2. Prefabricated asphaltic membrane lining. In this, asphalt lined papers, clothes, mats etc. are used to put a barrier against seepage. All these fabrications are available in marked in form of rolls. The membrane is laid on smooth well prepared subgrade and covered with a fine soil or earth.

3. Plastic or rubber membrane lining. In this case, plastic or synthetic rubber membrane is used as water proofing membrane. Out of numerous such membranes polyethylene film has shown encouraging results. This lining is liable to be easily ruptured by sharp stones or weed growth and as such the subgrade should be prepared smooth and treated with herbicide so as to prevent weed growth.

4. Bentonite and clay membrane lining. It has not been used on irrigation canals as yet. The main characteristic of Bentonite is swelling due to absorption of water. On swelling the membrane becomes perfectly water-tight and controls seepage from canals. Bentonite layer is formed by laying 2.5 to 3 cm thick layer of bentonite over a prepared subgrade. The layer is lastly covered with 15.30 cm protective blanket of suitable earth or gravel.

20.3.3 Earth Lining

Under this category of lining following linings come.

(*i*) *Soil cement lining.* In this lining, 2 to 5% cement is mixed dry with fine soil of which 35% passes through 200 sieve. This mixture is prepared in dry state of the soil. The water is sprinkled on this mixture and compacted on the prepared subgrade. The lining is kept wet for about a week before water is allowed to flow in the canal.

(*ii*) *Clay puddle lining.* The clay puddle is prepared by first excavating the clay and then exposing it to weathering. After a week or 10 days of exposure, water is added and clay is pugged throughly. The pugged clay is known as clay puddle. The pugged clay is put along the perimeter of the canal to act as lining. The lining of clay puddle may be about 30 cm. It is generally protected by earth layer so that it is not exposed and cracked when canal is closed for few days.

(*iii*) *Sodium carbonate lining.* A mixture is prepared with local soil by mixing about 10% clay and 6% sodium carbonate. This mixture is added with water and laid in a layer of 10 cm thickness. It may be used on water courses or other small channels. It is not suitable for larger canals as it is not durable.

20.3.4 Porous Type Lining

In head reaches, the ground water table is generally much higher than the bed level of the main canal. Porous lining is advisable in such circumstances. Porous lining allows water pressure to be released and thus occurrence of back pressure is eliminated.

15 cm inverted filter is spread evenly on the prepared sub-grade and stone pitching is done by hand packing. If stones are not available near the site, brick pitching may be done. This lining does not provide any imperviousness but is used for the drainage of the banks.

20.4 DESIGN OF LINED CANAL

Design of lined canals is always done by Kennedy's Theory. Following equations given by Kennedy are used in the design.

$$V_0 = 0.54\, m\, D^{0.64}$$

$$V = \frac{R^{2/3} S^{1/2}}{N}.$$

The values of N for different types of linings are given in Table 20.1. These values of N are as per IS : 4745—1968.

Table 20.1. *Values of Rugosity Coefficients N*

Surface of lining	*Value of N*
1. Surface finishing of concrete surface	0.013 – 0.017
(i) By Form work	0.012 – 0.014
(ii) Trowel finished	0.013 – 0.015
(iii) Float finish	0.016 – 0.019
(iv) Gunned section	0.015 – 0.017
2. Bottom concrete float finished and sides as follows	0.017 – 0.020
(i) Dressed stone in mortar	0.020 – 0.025
(ii) Random stone in mortar	0.016 – 0.020
(iii) Cement rubble masonry	0.020 – 0.030
(vi) Cement rubble masonry plastered	
(v) Dry rubble (rip rap)	
3. Brick lining with gravel bottom	0.014 – 0.017
4. Asphalt smooth and rough	0.013 – 0.016
5. Concrete lined excavated rock good section	0.017 – 0.022

The value of N for protected type of linings is taken same as for natural soil. The value of N of natural soil varies from 0.02 to 0.025.

In order to obtain the most economical section, it is necessary to adopt the best discharging section. The flow will be greater when the friction is least. This happens when wetted perimeter is least for any particular given area of the

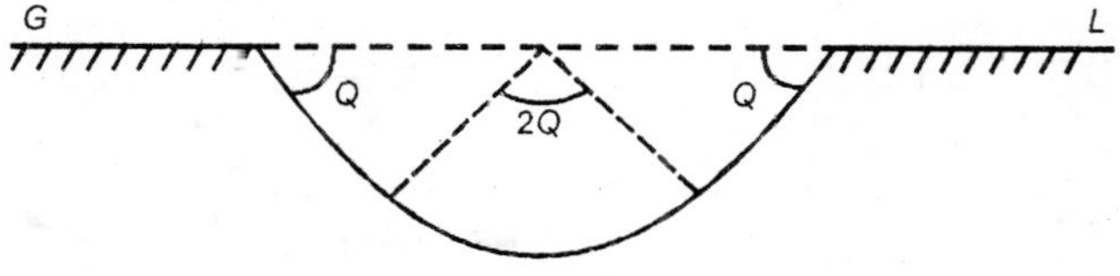

Fig. 20.2. *Mehboob section lined canal.*

channel. In other words, the discharge will be maximum when Hydraulic mean depth (H.M.D.) is maximum. A semi-circular section of channel is considered as the best theoretical section. But Trapezoidal channel section is mostly adopted on practical grounds. In order to increase the H.M.D., the angle subtended at the corners at bottom should be same as side slopes of channel make with the horizontal. See Fig. 20.3.

Design steps. Following data should be known before design of the canal can be carried out :

(i) Discharge of the channel (Q).

(ii) Rugosity coefficient (N).

(iii) Longitudinal slope (S).

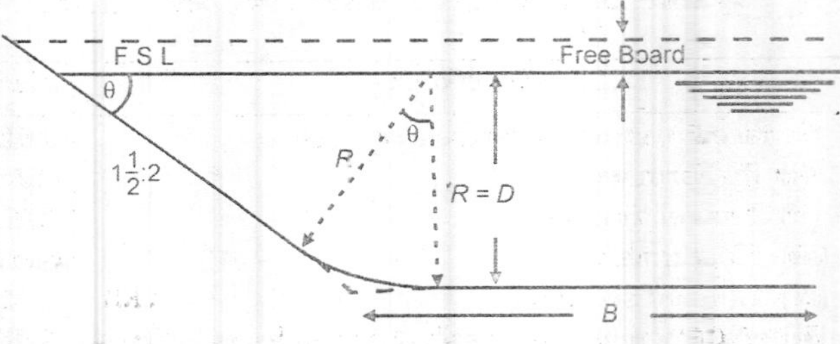

Fig. 20.3. *Trapezoidal section of lined canal.*

(iv) Slope of banks.

(v) Permissible velocity of flow (V).

Use following equations.

$$Q = AV$$

$$V = \frac{R^{2/3} S^{1/2}}{N}$$

Steps. (i) Knowing the limiting values of V, N, and S find out the H.M.D. (R) from equation $V = \dfrac{R^{2/3} S^{1/2}}{N}$.

(ii) From an expression in terms of Bed width (B) and depth (D) by pudding values in $R = \dfrac{A}{P}$.

(iii) From another equation in terms of B and D from equation $Q = AV$.

(iv) By solving equations formed in step (ii) and (iii), find out the unknown values of B and D.

For different values of side slopes, the sectional area, wetted perimeter, in terms of B and D are given as follows. The radius of arc for monitoring bottom corners is taken equal to depth of water.

Side slope	*Sectional area in*	*Wetted perimeter in m*
1:1	$D^2\left(\frac{B}{D}+1.785\right)$	$D\left(\frac{B}{D}-3.571\right)$
$1\frac{1}{4}:1$	$D^2\left(\frac{B}{D}+1.925\right)$	$D\left(\frac{B}{D}-3.850\right)$
$1\frac{1}{2}:1$	$D^2\left(\frac{B}{D}+2.09\right)$	$D\left(\frac{B}{D}-4.18\right)$

Note. If channel alignment is not straight loss of head by resistance would increase.

Example 20.1 *Design a concrete lined channel to carry a discharge of 200 cumecs. Slope of canal is 22.5 cm for every 1000 m. Side slope of banks is $1\frac{1}{2}$: 1 and value of N is 0.016. Permissible velocity of flow is 2 m/sec.*

Solution

$$S = \frac{22.5\,\text{cm}}{1000\,\text{m}} = \frac{22.5}{100000} = \frac{1}{4450}$$

$$V = \frac{R^{2/3}S^{1/2}}{N}.$$

Putting values, we get

$$2 = \frac{R^{2/3}}{0.016}\left(\frac{1}{4450}\right)^{\frac{1}{2}}$$

$$R = 3.10\text{ m.}$$

Side slope $1\frac{1}{2}$:1

$$\theta = \cot^{-1}\frac{3}{2} = 34.1 = 0.59\,\text{radians}$$

$$A = BD + \pi D^2\frac{\theta}{\pi} + 2\frac{D^2\cot\theta}{2}$$

$$A = BD + D^2\theta + D^2\cot^2\theta$$

$$P = B + 2D\theta + 2D\cot\theta$$

$$R = \frac{A}{P} = \frac{BD + D^2\theta + D^2\cot\theta}{B + 2D\theta + 2D\cot\theta} = 3.10$$

By putting value of θ and cot θ, we get

$$3.10 = \frac{BD + 0.59D^2 + 1.5D^2}{B + 2D \times 0.59 + 3D}$$

$$BD + 2.09\,D^2 = 3.10\,B + 12.95\,D \qquad (1)$$

$$A = \frac{Q}{V}$$

$$BD + 2.09D^2 = \frac{200}{2} = 100 \qquad (2)$$

$$B = \frac{100 - 12.95D}{3.10}$$

Eq. By putting value of B in Eq. (2)

$$D\left(\frac{100-12.95D}{3.10}\right)+2.09D^2 = 100$$

$$D = 4.23 \text{ m}$$

$$B = 100 - \frac{12.95 \times 4.23}{3.10} = 14.65 \text{ m.} \quad \textbf{Ans.}$$

Hence adopt 14.65 m × 4.23 in trapezoidal section.

20.5 USE OF DESIGN CHART

Design of lined canals can be done by using Plate 20.1 on page 567. This chart is made by adopting $1\frac{1}{2}:1$ side slope, $NI = 0.018$, and slope of 1 in 1000. This chart can be made use of for other values of N and S. The method of use of this chart for other values of N and S is explained as follows.

1. For other values of N chart is used as follows :

(i) To determine R_1, read bed width (B) and depth (D) for the given discharge and multiply both B and D by factor $\frac{N_1}{N}$. It should be remembered that value of N is 0.018 and value of N_1 is for which chart is to be used or $R = \frac{A}{P} \times \frac{N_1}{N}$.

(ii) To determine value of silt factor f_1, read value of f from the chart corresponding the given discharge and multiply the read value of f by $\left(\frac{N}{N_1}\right)^2$ factor or $f_1 = f\left(\frac{N}{N_1}\right)^2$.

(iii) To determine the velocity of flow V_1, read the velocity V from the chart for given value of discharge and multiply the read value by $\frac{N}{N_1}$ factor.

$$V_1 = V \times \frac{N}{N_1}.$$

2. For other values of slope (S), the chart can be made useful as follows :

(i) To determine R_1, read bed width (B) and Depth (D) for the discharge Q_1 where $Q_1 = \frac{Q}{(S_1/S)^{\frac{1}{2}}}$, S is the slope of $\frac{1}{10000}$ and S_1 is the value of slope for which chart is to be used.

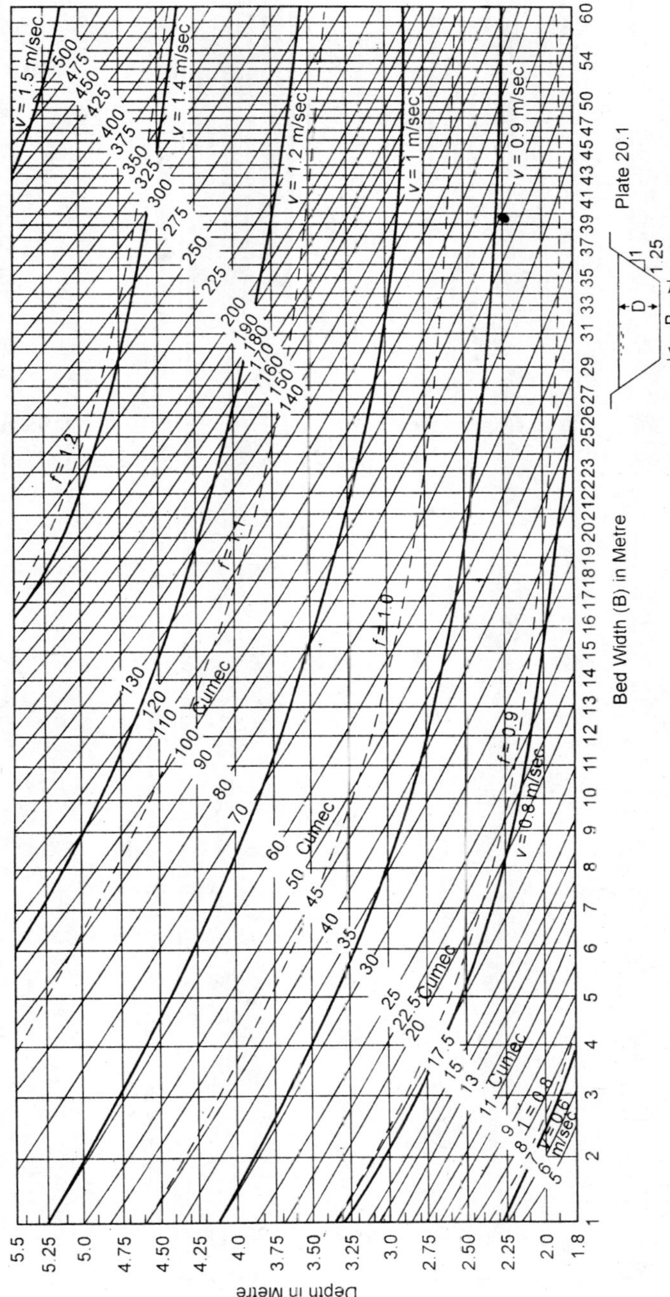

Plate 20.1 (To be given in between Pages 560-561)

(ii) Calculate the changed discharge as follows:

$$Q_1 = \frac{Q}{(S_1/S)^{\frac{1}{2}}}$$

where Q_1 = changed or calculated discharge

Q = actual discharge

S_1 = changed slope

$S_1 = \frac{1}{10000}$.

In order to determine value of f_1, read value of f for the discharge Q_1 and multiply it by $\left(\frac{S_1}{S}\right)$

$$f_1 = f \times \left(\frac{S_1}{S}\right).$$

(iii) To determine the value V_1. Read value of V for the discharge Q_1 and multiply the read value by $\left(\frac{S_1}{S}\right)^{\frac{1}{2}}$

$$V_1 = V \times \left(\frac{S_1}{S}\right)^{\frac{1}{2}}.$$

Example 20.2. *Design a lined channel for a design discharge of 250 cumecs. Slope of the canal is* $\frac{1}{10000}$. *Side slope* $1\frac{1}{4}:1$. *The value of* $N = 0.015$ *and* $V = 1.6$ *m/sec.*

Solution The given discharge C = 250 cumec.

The discharge taken for finding width and depth is

$$Q \times \frac{N_1}{N} = 250 \times \frac{0.015}{0.018} = 208.33 \text{ cumec.}$$

The velocity to be taken for finding B and D

$$= 1.6 \times \frac{N_1}{N} = 1.60 \times \frac{0.015}{0.018} = 1.33 \text{ m/sec.}$$

From the graph, moving along 208.33 cumec line and stopping when velocity is 1.33 m/sec, we find

B = 27 m

D = 4.50 m

Sectional area

$$= D^2\left(\frac{B}{D}+1.925\right)$$

$$= (4.5)^2\left(\frac{27}{4.50}+1.925\right)$$

$$= 20.25\ (6 + 1.925)$$

$$= 20.25 \times 7.925 = 160.48\ \text{m}^2.$$

Wetted perimeter $= D\left(\frac{B}{D}+3.85\right)$

$$= 4.5\left(\frac{27}{4.50}+3.85\right)$$

$$= 4.5 \times 9.85 = 44.325\ \text{m}$$

$$R = \frac{A}{P} = \frac{160.48}{44.325} = 3.62$$

Velocity $= \frac{1}{N}(3.62)^{2/3}\left(\frac{1}{10000}\right)^{\frac{1}{2}}$

$$= \frac{1}{0.015}(3.62)^{2/3}\frac{1}{100}$$

$$= \frac{(3.62)^{2/3}}{1.5} = \frac{2.36}{1.5} = 1.57\ \text{m/sec.}$$

Discharge $= 160.48 \times 1.57 = 251.95$ cumec.

It is almost 250 cumecs. Hence designed section is OK.

Velocity as read = 1.33 m/sec. Multiply it by factor $\frac{0.018}{0.015}$ i.e., 1.2, we have actual velocity = 1.33 × 1.2 = 1.596 m/sec which is same as evaluated and adopted.

Hence bed width of 27 in and depth 4.5 m will pass 250 cumec discharge safely at 1.6 m/sec velocity, side slope of channel being $1\frac{1}{4}:1$.

QUESTIONS

20.1. Enumerate the advantages and disadvantages of lining of canals.

20.2. (a) What do you understand by financial viability of the canal lining

(b) What are the characteristics of good lining ?

20.3. What are the usual types of linings ? State briefly their method of construction.

20.4. On what theory the lined canals are designed ? Enumerate the steps.

❑❑❑

21

Canal Regulation Works

21.1 INTRODUCTION

To exercise control on discharge, full supply level, velocity of flow, silting etc. various masonry or concrete structures have to be constructed over the canals. All these structures are known as *Canal Regulation Works*. Main Regulation works may be listed as follows:

1. Canal falls
2. Head regulator
3. Cross regulator
4. Metres and Flumes
5. Canal escapes.

Canal outlets, also come under the category of regulation works, but they have been discussed separately.

21.2 CANAL FALLS

21.2.1 Why Falls are Provided

When natural slope of the country is steeper than the designed longitudinal slope of the irrigation canal, the falls have to be provided. If falls are not provided, the canals run so much in filling that it will be almost impossible to even construct the canals. If canal is constructed very much in filling, it will be a constant danger to the adjoining people, and also it will be impossible to mantain it. For example, let slope of a particular canal be say 1 in 5000 and general slope of the ground say 1 in 2500. If canal bed at the start i.e. zero distance, is say 1 m in cutting the bed of canal will just coincide with ground level at a distance of 5000 m and thereafter the bed of the canal will come in filling. The fill of the canal will go on increasing at the rate of 1 m per 5000 m. If total length of the canal from head regulator to tail is say 30 km, the canal bed which was 1 m below the G.L. at head, will come 5 m in filling at the end of the canal. The construction and maintenance of a canal in this much filling will be very costly. Besides this percolation and seepage losses will also be excessive. To overcome the difficulty of heavy embankments and also to reduce the cost of construction and maintenance, falls are provided. The falls lower down the bed level of the canal. Since bed is lowered, the water surface, at fall is also equally lowered. Due to fall in the water surface excessive energy is developed on the *D/S* side which is dissipated by adopting suitable measures discussed in this chapter later.

21.2.2 Location of Falls

The location of the fall on a channel depends upon the general slope of the country through which channel passes. Following points should however by considered while deciding the location of the fall.

1. In the case of main canals which do not do irrigation directly, the position of the fall should be located by considering economy in cost of excavation of the canal with regard to balancing depth and the cost of the fall itself. Main canal should not be allowed to run into much filling. The canal being very large, any breach in it may cause lot of damage to life and property.

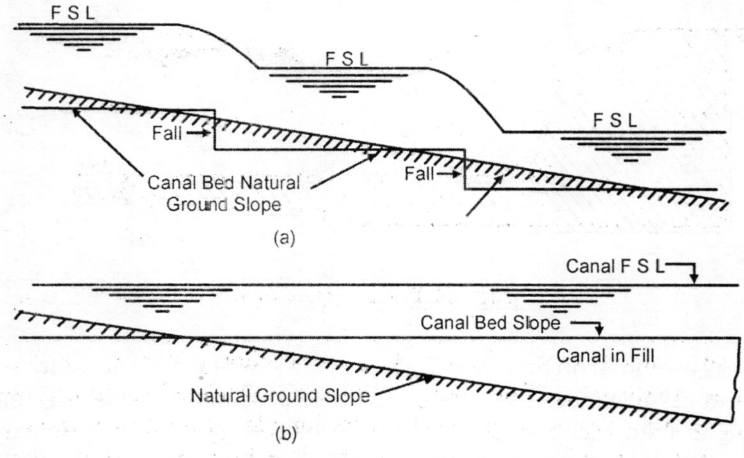

Fig. 21.1. (*c*) *Canal with falls* (*b*) *Canal without falls.*

2. In case of branches and distributaries, which do direct irrigation, the falls are located considering command area. The F.S.L. of the canals should remain above the G.L. for most of the length. However the F.S.L. of channel *D/S* of fall may remain below G.L. for some length. The area *D/S* of the fall for which

F.S.L. of channel remain below G.L., is commanded by taking outlets from *U/S* of the fall.

3. Location of the fall may be slightly varied so that its construction may be combined with a regulator or bridge.

4. The falls may be provided large in number but of smaller fall or they may be smaller in number but of larger fall. Both the alternatives should be weighed and adopted.

21.2.3 History of Falls

In old days, modern scientifically designed falls were not in existence. The excess fall of the general country was used to be adjusted or dispensed with by increasing the length of the canal by giving circuitous alignment to the canals. Eastern Yamuna canal in India, constructed during Mughal period, does not have any fall and canal alignment is very much circuitous. Development of falls started late in the 19th century when large irrigation projects like Ganga, Cauvery, Eastern and Western Yamuna canals were constructed. A brief description of each fall which came into being one after the other is being given here.

1. Ogee Type Fall. This fall was first constructed on Ganga canal. The crest of the fall was kept at level with the *U/S* bed of the canal. A smooth concave

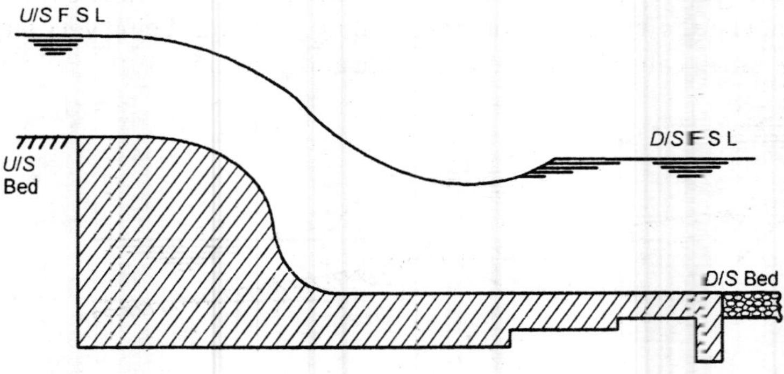

Fig. 21.2. *Ogee type fall.*

surface was provided on the *D/S* side, to provide smooth transition and to reduce impact and disturbance. The energy of water passing over the fall, remained preserved and thus deep scourings for very long lengths on *D/S* side were used to develop. Crest of the fall being at the *U/S* bed level, there used to be considerable draw down effect on the *U/S* side. See Fig. 21.2.

2. Rapids. This fall consists of a sloping glacis. The slope is kept from 1 in 10 to 1 in 20. Because of long sloping glacis, the formation of hydraulic jump is assured. It is hydraulic jump formed on *D/S* side, which carries out the dissipation of excess energy. This fall worked very well but the major drawback of this fall was its high cost (Fig. 21.3).

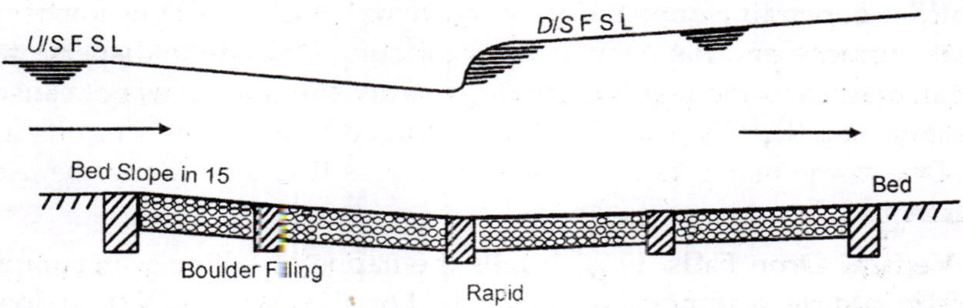

Fig. 21.3. *Rapid.*

3. Stepped Fall. As its name indicates, it consists of a number of steps. The total fall is broken into a number of steps of smaller falls. This fall also has the disadvantage of being very costly. As far as its performance is concerned it is very satisfactory.

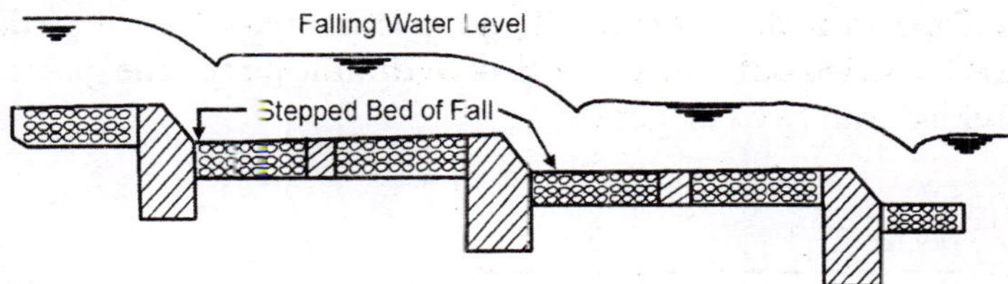

Fig. 21.4. *Stepped fall.*

4. Notch Type Fall. This fall consists of notches situated in a high crested wall, built across the channel. The fall may be having only one notch or a number

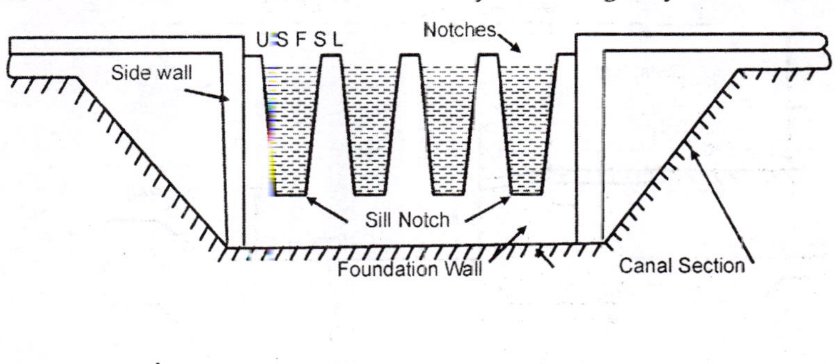

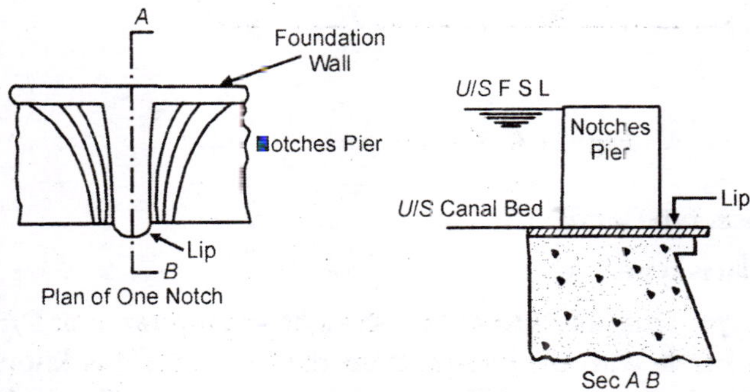

Fig. 21.5. *Notch fall.*

of notches depending upon the discharge to be handled. All the notches have smooth entrance and flat circular lips projecting D/S side to disperse water. This fall maintains the depth discharge relationship and does not cause any draw down on the U/S side. This fall continued to be in use for quite a long time. One of the major defect of these falls was that they cannot be used as regulators.

5. Vertical Drop Falls. In such falls the nappe of falling water impinges vertically into the water cushion developed on the D/S side. Nappe does not remain in contact of the crest wall on D/S side. In such falls, excess energy of falling water is dissipated by turbulent diffusion in the pool of water. Sarda type fall and C.D.O. type fall, are the usual examples of vertical drop falls. Design of this fall has been given in details later on in this chapter.

6. Glacis Type Fall. In such falls a sloping ramp is provided D/S of the crest. Because of slope, formation of hydraulic jump is assured. The energy dissipation in such falls is carried out with the help of hydraulic jump. The glacis type falls may be divided into two categories:

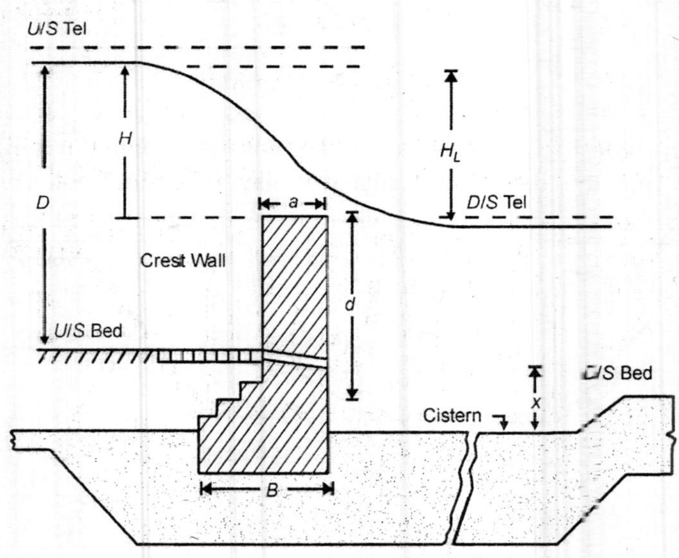

Fig. 21.6. *Vertical fall.*

(i) Straight glacis type, and

(ii) Parabolic glacis type.

(i) Straight glacis type. This fall consists of a straight sloping ramp at D/S. If a baffle platform and baffle wall are provided on the D/S side, it is known as *Inglis fall.* In such falls the formation of jump is a must on the baffle platform.

(*ii*) *Parabolic glacis type.* The fall having parabolic *D/S* glacis is known as Montague type fall.

21.3 METER AND NON-METER FALLS

The falls may be divided into following two types :

1. Meter falls
2. Non-meter falls.

1. Meter Falls. The falls which can be used to measure the discharge flowing over them, are known as meter falls. Such falls must have broad crest so that the discharge coefficient remains constant under variable head. Glacis type falls are most suitable as meter falls. If section of the channel at the site of the fall has to be flumed, from economic considerations, smooth *U/S* transition should be provided to avoid turbulences and also to maintain accurate stage discharge relationship. See Fig. 21.7.

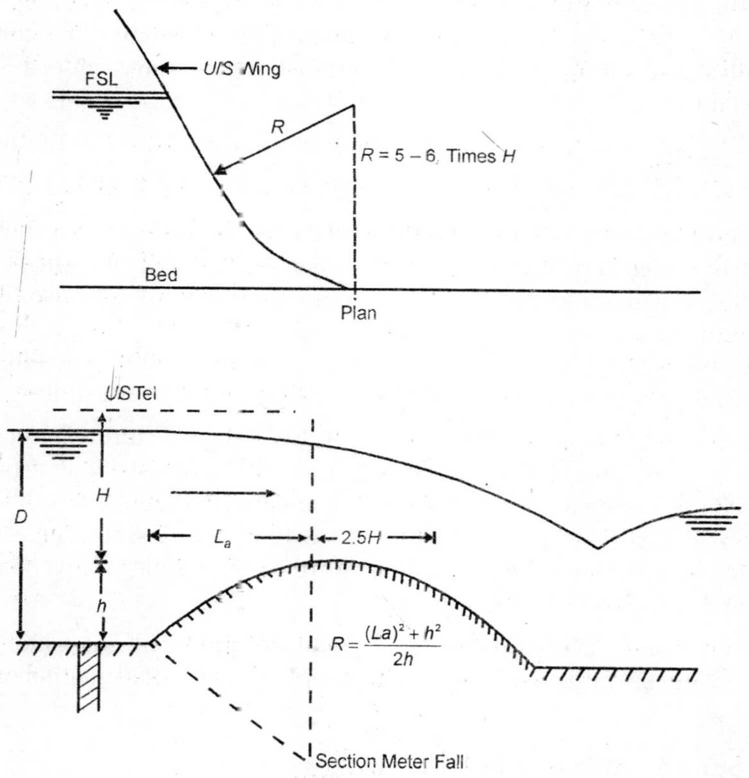

Fig. 21.7. *Sloping meter fall.*

2. Non-meter Falls. The falls which cannot be used to measure the discharge passing through them, are known as non-meter falls. Vertical drop falls cannot be used as meter falls, as some vacuum is developed under the nappe of falling water, which causes non-uniform flow conditions.

21.4 CLASSIFICATION OF FALLS

The falls can be classified into following four categories according to the approach conditions.

1. Falls which maintain relation between depth and discharge.
2. Falls which maintain level of *U/S* water constant.
3. Combination of fall and regulator.
4. Miscellaneous falls.

1. Falls in which D-Q relationship is maintained. In such falls there is neither drawdown nor heading up of water on *U/S* side. The Depth-discharge relationship for the channel is maintained. Trapezoidal notch falls and low crested rectangular notch falls, fall under this category.

2. Falls which maintain fixed F.S.L. on the *U/S* side of the fall. Siphon fall or Siphon spillway high crested weir fall are the falls which fulfil this condition. Such falls cause silting on *U/S* side. Such falls are necessary under following circumstances.

(i) When a sub-channel has to take off at some distance *U/S* of the fall.

(ii) When Hydro-electricity is being generated using drop in the fall.

3. Combination of fall and regulator. In such falls the level of water on *U/S* side, can be varied at will. The regulation is done with the help of vertical needles, horizontal stop logs and sluice gates. They are used as regulators also. They are installed at locations where some sub-channel is to take off from the parent channel and also fall is to be provided on the parent channel. The falls of this category are exclusively the rectangular notch falls fitted with sluice gates.

4. Miscellaneous type falls. The falls of this category suit to some specific conditions. Well or cylindered fall is one such fall. It consists of a vertical well connected to *D/S* side with the help of a horizontal pipe. Water enters the vertical well from *U/S* side and escapes into Tail water on *D/S* side. The energy is dissipated by turbulences in the well. They are quite suitable for low discharges but large drops. See Fig. 21.8.

Fall may be inform of a steep sloped channel or pipe. The *U/S* water surface and *D/S* water surface are connected by steeply sloped open channel or closed pipe.

21.5 SELECTION OF THE TYPE OF FALL

The type of fall should be chosen considering following points.

(i) Vertical drop fall is found most suitable for discharges upto 15 cumecs and drops upto 1.5 m. It should not be flumed.

(ii) Glacis fall with baffle wall has been found suitable for all discharges and drops of more than 1.5 m. It can be flumed.

(iii) Straight glacis fall, without baffle wall, is found suitable for drops upto 1.5 m and discharges upto 60 cumec. They may be flumed on economic considerations.

Hence knowing the drop and discharge, the most appropriate fall may be selected for use.

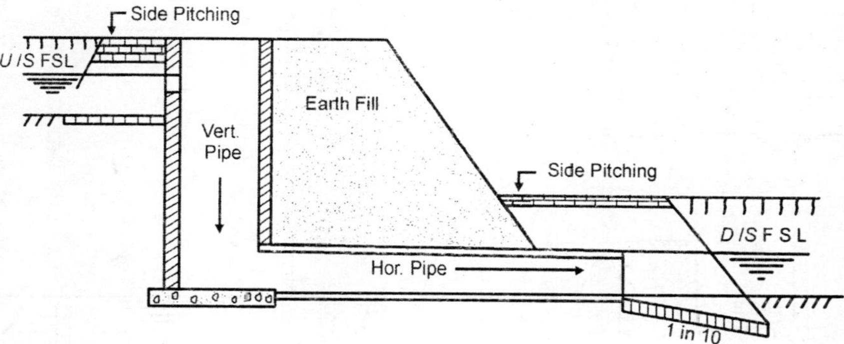

Fig. 21.8. *Vertical well fall.*

Out of all the falls discussed earlier only vertical drop and straight glacis falls are found in most common use. Others have become obsolete and out of date.

21.6 CISTERN DESIGN

The depressed portion of the floor lying just *D/S* of the crest wall, of the fall is known as *cistern*. The main purpose of cistern is to dissipate the surplus energy of water falling or gliding from the crest of the fall.

Slope of the glacis (if any), the cistern, and roughening devices, are the elements which are used is one way or the other, to dissipate the surplus energy of water.

Cistern for Vertical Drop Fall. In this case there is no sloping glacis and water falls vertically after passing over the crest. The falling water nappe impinges against the water filled in the cistern and its direction gets changed from vertical to horizontal. The energy is dissipated by means of impact and deflection of flow suddenly from vertical to horizontal direction. Cistern performs following two functions. (Fig. 21.9)

(i) It helps reduce intensity of impact of falling nappe of water and thus protects the *D/S* floor from being damaged.

(ii) The water in the cistern acts as a cushion against falling nappe of water and destroys its energy.

Following area some of the formulae which are used to fix the length and depth of the cistern.

(a) U.P. Irrigation Research Institute Formulae (Fig. 21.9)

$$x = \frac{1}{4}(EH_L)^{2/3}$$

$$L_C = 5\sqrt{EH_L}$$

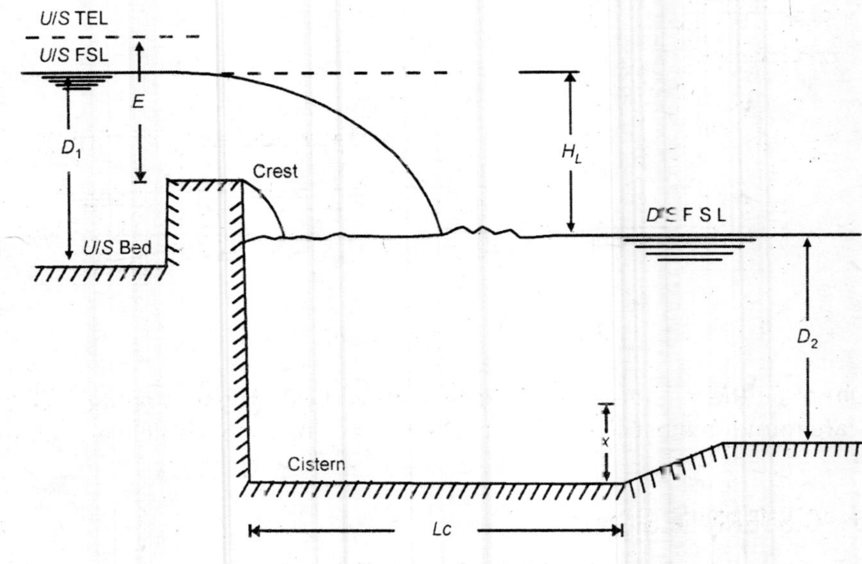

Fig. 21.9. *Cistern.*

(b) Etchevery Formulae. These formulae were developed in U.S.A.

$$L_C = 3\sqrt{H_L E}$$

$$x = \frac{1}{6}L_C$$

(c) Montagu's Formula. They were developed in Punjab.

$$L_C = 4E_{f2}, x = \frac{1}{2}E_{f2}$$

(d) E.L. Glass Formulae. They were developed for Bihar and Orissa.

$$x = 1.85\, E^{1/2}\, H^{1/3} - D_2$$

$$L_C = x + D_2$$

In all the above formulae

E = U/S Total energy above the crest

L_C = Length of the cistern

X = Depth of cistern below general floor level on D/S side

E_{f2} = Energy of flow D/S for the discharge intensity q and the fall H_L

H_L = Fall in the U/S and D/S water levels. (Fig. 21.9)

Cistern for glacis type falls. In case of glacis type falls the energy is dissipated by hydraulic jump and not by impact as in case of vertical drop falls. It has been explained earlier that for a given drop the energy line H_L, and discharge intensity q, there will be a definite value of D/S specific energy or D/S depth required for the formation of jump.

In straight glacis falls, the glacis is sloped usually at 2 : 1 slope. Theoretically the cistern should be provided at the lowest level of the glacis where the jump may be formed. The position of the jump in a horizontal bed is not stable and there may be instances when jump is not formed on the bed but quite distant on D/S side. This may cause bed and bank scours. To safeguard against such eventualities, the depth of the cistern is increased to 1.25 E_{f2}. Thus level of the cistern is fixed by deducting 1.25 E_{f2} from the total energy line on D/S side. If, however, the bed level on D/S is lower than the bed level of the cistern so obtained, the cistern should be provided at D/S bed level. Length of the cistern is kept 5 E_{f2} for normal soils, to 6 E_{f2} for sandy soils. If bed level of cistern is lower than the D/S bed level of the channel, bed level of the cistern and that of channel should be connected by giving a slope of 5 : 1.

Cistern for glacis fall with baffle platform and baffle wall. In this case dimensions of cistern, baffle platform and baffle wall are fixed as follows. In such falls baffle platform is formed just D/S of sloping glacis, then baffle wall, and

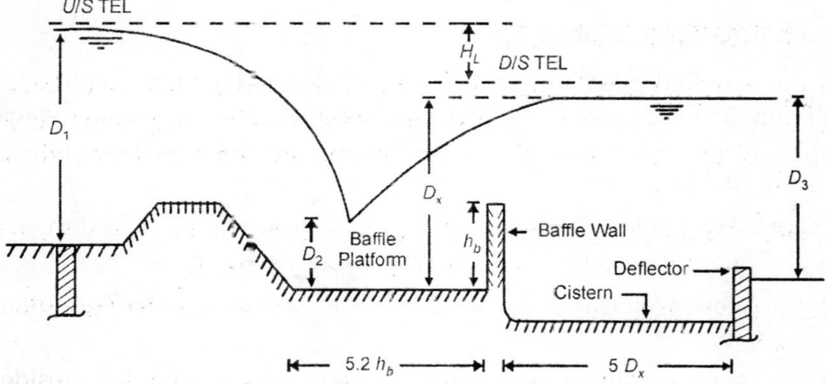

Fig. 21.10. *Baffle platform cistern in glacis fall.*

lastly the cistern as shown in Fig. 21 .10. Discharge intensity (q), fall (H_L) and depth of subcritical flow for unflumed fall are connected by

$$D_n = 0.98\, q^{0.52} \times H_L^{0.21}$$

For flumed falls, subcritical depth (D) is given by

$$D = H_x + D_x - H_L$$

where H_x = the calculated drop in metres which is calculated by $H_x = \frac{H_L}{k^{0.125}}$ where k is the fluming ratio.

R. L. of Baffle platform is given as follows:

For unflumed fall = D/S F.S.L. – D_x

For Flumed fall = D/S F.S.L. – D

Baffle wall

Determine D_C, critical depth = $\left(\frac{q^2}{g}\right)^{1/3}$

D_2 the pre-jump hyper critical depth = $0.183\, q^{0.89}\, H^{-0.35}$

Height of Baffle wall = $D_C - D_2$

Thickness of Baffle wall = $\frac{2}{3}$ × Height of Baffle wall

Length of Baffle platform = 5.25 × height of Baffle wall.

Depth of the cistern below the D/S bed should be 10% of the D/S water depth D_3.

R.L. of cistern = D/S bed level – 0.1 D_3

Length of cistern $L_C = 5 \times D_x$.

21.7 ROUGHENING DEVICES

In case where hydraulic jump is not formed, and vertical fall conditions are also not there, the energy dissipation can be achieved by roughening devices, provided on the D/S side of the fall. Following are some of the roughening devices mostly used.

1. Baffle Wall. It is a weir type wall which is constructed across the channel at the D/S end of the cistern. Its main purposes are two-fold.

(i) It keeps water head up on U/S side so as to ensure the formation of jump.

(ii) To withstand the thrust of high velocity which has not yet subsided.

2. Friction Blocks. They are rectangular concrete blocks, adequately anchored with the floor. They are provided just D/S of the crest. They form a most simple and useful device of energy dissipation. They act as follows:

(i) To divide the bottom high velocity of water laterally.

(ii) To reduce the velocity of water leaving the pucca floor of the fall.

(iii) To help in the formation of jump in case of glacis falls.

Recommendations about friction blocks for various falls are:

(*a*) *Vertical fall.* Two rows of friction blocks in staggered fashion are found adequate. They should be provided at 1.5 D_c from *D/S* toe of the crest where D_c is critical depth.

Length, breadth and height of the blocks should be $2D_c$, D_c, D_c respectively.

Clear spacing between rows should be D_c and between blocks of the same row $2D_c$.

In addition to friction blocks, cube blocks may be provided at the end of the pucca floor. Size of cube blocks is kept about $\frac{1}{8}$ th of the depth of water on *D/S* side.

Arrows may be used in place of friction blocks. Arrows are triangular in plan with tapper towards the top. Top is also sloped a little.

(*b*) *Glacis fall.* In case of flumed glacis fall, the friction blocks are generally provided in four rows, the first row being at a distance of 5 times the height of the blocks from toe of glacis. Height of the block is kept $\frac{1}{8}$ th depth of water on *D/S* side length $3h$ where h is the distance between rows.

Glacis Blocks. Glacis blocks are just like friction blocks. They are provided just before the *D/S* toe of the glacis in one row. The effect of these blocks is to reduce the turbulences. The size of these blocks is $D_c \times 2D_c$ in plan.

Montagu gave following formula for finding the distance (L) upto which roughening of bed by friction or arrow blocks is required.

$$L = C \frac{D_2^{3/2} H_L^{1/2}}{D_1}$$

D_1 = Depth of cistern

D_2 = Depth on *D/S* side

H_L = Drop or fall.

C = A constant whose value is 1 for vertical fall, 3 for horizontal impact, 4 to 6 for inclined impact and 8 for no impact falls.

3. Dentated Sill. If high velocity still persists after cistern, a dentated sill as shown in Fig. 21.11 should be provided at the end of the cistern. The object of such a sill is to deflect up the high velocity from near the bed and to break it.

4. Deflector. It is a wall of height $D_3/10$ provided at the *D/S* end of glacis falls. Its object is to deflect up the high velocity jet near the bed causing a reverse roller action. See Fig. 20.11.

5. Biff Wall. It is also provided at the end of cistern. Its object is to deflect back the water from the cistern.

6. Cellular or Ribbed Pitching. It is an arrangement of projecting bricks on the sides of the channel. This device roughens the perimeter of the channel to destroy surplus energy, D/S of the fall. Fig. 21.11.

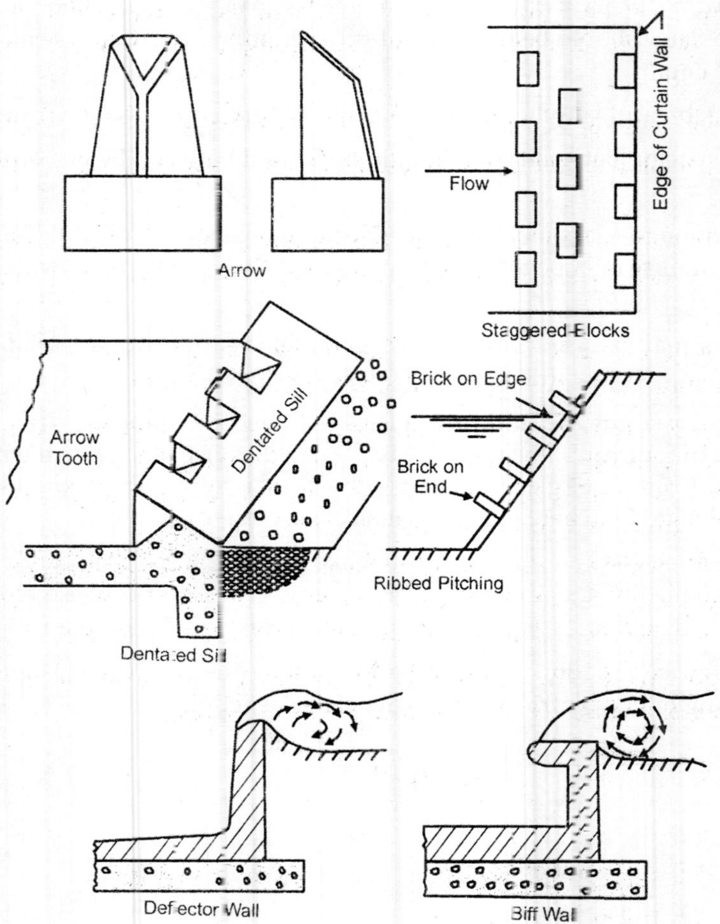

Fig. 21.11. *Roughening devices of dissipation of surplus energy.*

21.8 DESIGN OF SARDA TYPE FALL

This fall was used for the first time on Sarda canal in U.P. and hence named Sarda type fall. In the area where Sarda canal runs, a thin layer of sand-clay layer lies above the stratums of pure sand. In such conditions falls of large drop are not feasible to be constructed. Large number of falls with small drops were constructed so as to avoid deep cuttings. A complete design of Sarda type fall is shown in Fig. 20.12.

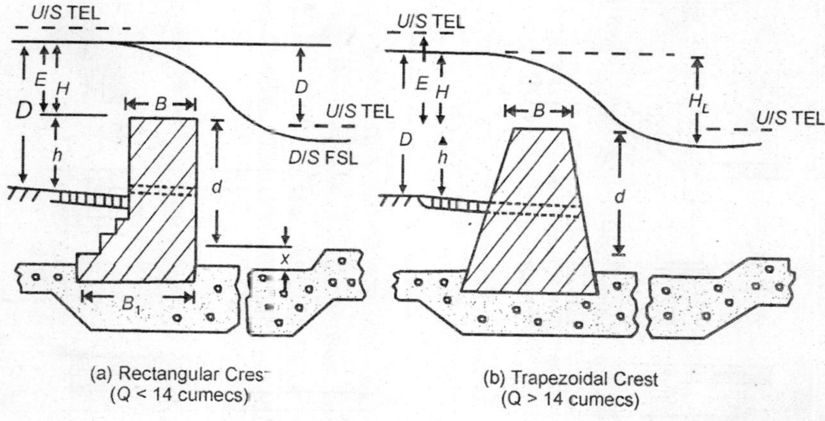

Fig. 21.12. *Crest of Sarda fall.*

In the design of Sarda type fall, following elements are required to be designed:

1. Crest wall
2. Cistern
3. Impervious floor
4. *D/S* protection
5. *U/S* approach.

1. Design of Crest Wall. Length of the crest wall is not flumed. Crest length may be equal to bed width of the canal. To allow for possible future increase in discharge, the length of crest wall may be kept equal to bed width plus depth. For discharges up to 14 cumecs the shape of the crest wall is kept rectangular with *D/S* side absolutely vertical. For more than this discharge shape of the crest wall is kept trapezoidal with *D/S* face batter of 1 : 8 and *U/S* batter of 1 : 3.

Rectangular Crest

Top width of crest $B = 0.55\sqrt{d}$

Base width $B_1 = \dfrac{H+d}{\rho}$

Discharge formula for rectangular crest is

$$Q = 1.835\, LH^{3/2}\left(\frac{H}{B}\right)^{1/6} \qquad (1)$$

where Q = Discharge in cumecs

L = Length of the crest in metres

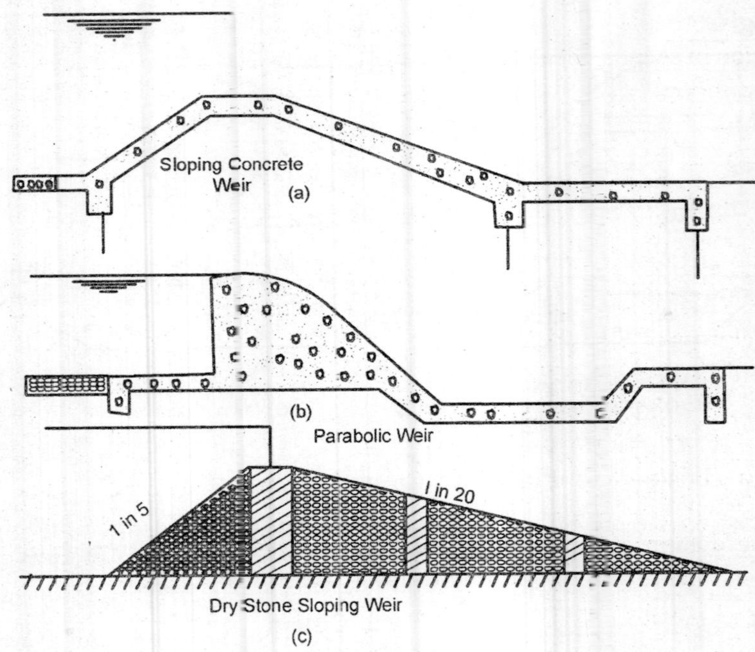

Fig. 21.13. *Various types of weirs.*

Trapezoidal Crest

Top width $B = 0.55\sqrt{H-d}$

U/S batter 1 : 3 and *D/S* batter 1 : 8.

Discharge formula for trapezoidal crest wall is

$$Q = 1.99\, LH^{3/2}\left(\frac{H}{B}\right)^{1/6} \tag{2}$$

Determining Crest Level. Calculate the value of *H* from formula (1) or (2) as the case may be.

$$\text{R.L. of crest} = U/S \text{ F.S.L.} - H$$

Height of crest above bed $h = D - H$.

For falls over 1.5 m the stability of the crest wall should be tested by actual analysis.

2. Design of Cistern. Find out length (L_c) and depth (x) of the cistern by following formulae.

$$L_c = 5\,(EH_L)^{1/2}$$

$$x = \frac{1}{4}(EH_L)^{2/3}$$

3. Design of Impervious Floor. Total length of impervious floor is determined by Bligh's theory or by Khosla.s theory as already discussed. The critical condition of seepage head occurs when water is filled upto crest level and there is no water on *D/S* side. Out of total impervious length, a minimum length (l_d) determined as follows must be provided on *D/S* side of the crest.

$$l_d = 2(D + 1.3) + H_L \text{ in metres.}$$

$$H_L = \text{Drop in metres and } D \text{ is water depth on } D/S \text{ side.}$$

Some of the remaining impervious floor automatically comes under crest wall, and rest on *U/S* side of the crest.

Thickness of Floor. The thickness of floor is found out on the basis of uplift pressure as discussed in chapter on *weir desig*. *U/S* impervious floor does not theoretically require any thickness as uplift is counterbalanced by weight of water. However a minimum thickness of 30 cm to 40 cm is provided from practical considerations.

Thickness of the *D/S* floor is worked out on the basis of actual uplift pressure. If thickness worked out on the basis of uplift pressure is very small, minimum thickness of 40 cm for small falls and 60 cm for large falls must be provided.

Full uplift pressure should be allowed for working out the thickness of impervious floors in case of falls resting on permeable soils, while 75% may be taken in case of ordinary or black cotton soils.

The top of cistern and *D/S* impervious floor should in addition he lined with bricks on edge either in lime or cement mortar. A vertical cut off, at least 1.5 m deep, must be provided at the *D/S* end of the impervious floor.

4. Up Stream Protection. Brick pitching on the U/S side of the crest is laid on a slope of 1 in 10 upwards on U/S side. The length of this protection may be 2 to 4 m. The minimum length should be equal to U/S water depth. A drain hole should be provided in the crest wall at level with the *U/S* protection. This hole drains off water from *U/S* side to *D/S* side during canal closures.

5. *U/S* Approaches. For falls with less than 15 cumec discharge, the *U/S* approach wings may be splayed straight at an angle of 45°. For larger discharges, the wings are kept segmental. The radius of the segment is kept 5 to 6 times *H* and its angle at the centre should be 60°. After this, the wing walls are carried into the berm tangentially and embedded in it for a length of about 1 m. The foundation of the wings is kept on the impervious floor itself.

6. *D/S* Protection. The *D/S* protection consists of following elements.

(*i*) *Bed protection*. It is a protection to the bed, provided by dry brick pitching 20 cm thick laid on 10 cm thick ballast. The lengths of bed protection for different heads along with curtain wall details are given in Table 21.1.

Table 21.1. *Details of Bed Pitching and Curtain Walls*

S.No.	*Head of water on crest of wall in metres*	*Total length of bed pitching on D/S side in metres*	*Curtain walls*	
			Number	*Depth in metres*
1	Upto 0.30 m	3m	1	0.3
2	0.30 m to 0.45 m	$3 + 3H_L$	1	0.3
3	0.45 m to 0.60 m	$4.5 + 2H_L$	1	0.45
4	0.60 m to 0.75 m	$6 + 2H_L$	1	0.60
5	0.75 m to 0.9 m	$9 + 2H_L$	1	0.75
6	0.9 m to 1.05 m	$13.5 + 2H_L$	2	0.90
7	1.05 m to 1.20 m	$18.0 + 2H_L$	2	1 05
8	1.20 m to 1.50 m	$22.5 + 2H_L$	3	1.35
9	1.50 in to 1.75 m	$27.0 + 2H_L$	3	1.50
10	1.75 m to 2.80 m	$33.0 + 2H_L$	4	1.50

(*ii*) *Side protection*. After the wing walls, the side slopes of the channel are pitched with one brick on edge in the length equal to three times the *D/S* water depth. The *D/S* masonry wing walls are warped from vertical to a slope of 1 : 1. If banks are sloped at $1\frac{1}{2}$: 1 the side pitching has to be warped from a slope of 1 : 1 to $1\frac{1}{2}$: 1. The pitching is supported on toe wall $1\frac{1}{2}$ brick thick. The depth of the toe wall is kept half of the depth of water on *D/S* side.

(*iii*) *D/S wing walls*. They are kept vertical for a length of 5 to 8 times the $\sqrt{EH_L}$ from the crest. Thereafter they are gradually warped or flared to a slope of 1 : 1 or $1\frac{1}{2}$: 1. Average splay of 1 in 2.5 to 1 in 4 for achieving the required slope is given to the top of the wings. The wings generally follow the circular path starting tangentially from the starting point of warp in plan.

Wing walls should be designed as retaining walls. For heavy structure they should actually be designed, but for normal conditions the thickness at any depth should not be less than $\frac{1}{3}$ the height of the wing wall above that depth.

Example 21.1 *Based upon following data, design a Sarda type fall.*

Full supply discharge	= *80 cumecs*
R.L. of F.S.L. on U/S side	= *103.30 m*
R.L. of F.S.L. on D/S side	= *101.80 m*
Depth of canal both U/S and D/S	= *2.5 m*
Width of the canal both on U/S and D/S side	= *36 m*
Drop	= *1.5 m*

Design impervious floor of the fall by assuming Bligh's creep coefficient of 8 and check the design by Khosla's theory. Safe exit gradient is 1 in 5.

Solution

1. Find out value of *H* and *d*. Discharge is more than 14 cumecs and hence assume crest wall of trapezoidal shape

$$Q = 1.99\, LH^{3/2} \left(\frac{H}{B}\right)^{1/6}$$

$$L = 36 \text{ m}, Q = 80 \text{ cumec}$$

$$H + d = \text{Depth of canal + fall in bed level}$$

$$= 2.5 + 1.5 = 4 \text{ m}$$

$$B = 0.55\sqrt{4} = 1.1 \text{ m}$$

$$\therefore \quad 80 = \frac{1.99 \times 36 H^{3/2+1/6}}{(1.1)^{1/6}}$$

on solving $\quad H = 1.078 \text{ m} = 1.1$ say

$$H + d = 4 \text{ m}$$

$$d = 4 - H = 4 - 1.10 = 2.90 \text{ m}$$

Height of crest wall above *U/S* bed level

$$= D - H = 2.5 - 1.1 = 1.4 \text{ m}$$

Design of crest

Thickness at top =1.1 m

Assume *D/S* slope 1 in 3 and *D/S* slope 1 in 8.

Velocity of approach

$$V_a = \frac{Q}{(\text{Bed width} + D)D}$$

(slope of banks is 1 : 1)

$$= \frac{80}{(36 + 2.5)2.5} = 0.83 \text{ m/sec}$$

Velocity head $\quad = \dfrac{V_a^2}{2g} = \dfrac{(0.83)^2}{2 \times 9.8} = 0.033 \text{ m}$

Total energy line (T.E.L.) on *D/S* side

$$= U/S \text{ F.S.L.} + \text{velocity head}$$

$$= 103.30 + 0.033 = 103.333 \text{ m}$$

R.L. of top of crest = D/S F.S.L. – H

= 133.30 – 1.1 = 102.2 m

Value of E = U/S T.E.L. – R.L. of top of crest

= 103.333 –102.20

= 1.132 m

Design of Cistern

Cistern depth $x = \frac{1}{4}(E \times H_L)^{2/3}$

H_L is 1.5 m, which is always difference in U/S and D/S F.S.L.

$x = \frac{1}{4}(1.13.3 \times 1.5)^{2/3} = 0.375$ m

Length of cistern $L_c = 5\,(E \times H_L)_{1/2}$

$= 5\,(1.133 \times 1.5)^{1/2} = 2.52$ m

R.L. of bed of cistern = R.L. of D/S side of canal – x

= 101.80 – 2.50 – 0.375

= 98.925 m

Adopt R.L. of cistern as 98.92 m

Actual value of x = 101.80 – 2.50 – 98.92 = 0.38 m

Design of Impervious Floor

Max. seepage head

$H_s = d = 102.20 - 99.30 = 2.90$ m

Creep coefficient $C = 8$

Length of impervious floor = 2.9 × 8 = 23.2 m

Provide 1.5 in deep cut off at U/S end and 2 m deep at D/S end of the impervious floor.

Vertical length of creep = (1.5 + 2.0) 2 = 7 m

Horizontal length of impervious floor = 23.2 – 7.0 = 16.2 m

Adopt floor length of 17 m horizontally.

Minimum length of impervious floor on D/S side

$= 2\,(D + 1.3) + H_L$

= 2 (2.5 + 1.3) + 1.5

= 91 m

Adopt this length as 10 m.

The remaining length of pucca floor to be provided under and *U/S* of the crest wall = 17 – 10 = 7 m.

Uplift Pressure and Floor Thickness

Total creep length = 17 + 2 (1.5 + 2.0) = 24 m

(i) *U/S* of crest, does not require any pucca floor from uplift point of view because uplift pressure is counter balanced by the depth of water filled over it. However provide 50 cm thickness. There is no need of performing any calculations in this respect.

(ii) On *D/S* of crest, determine uplift pressures at various points by Bligh's theory and provide the thickness of floor according to it.

The maximum adverse effect of uplift pressure is at point *A*, just *D/S* of the crest wall toe.

$$\text{Uplift pressure at point } A = 2.90\left(1-7\times\frac{2\times1.5}{24}\right)$$

$$= 1.70 \text{ m}$$

$$\text{Required thickness at point } A = \frac{1.70}{(2.24-1)} = 1.38 \text{ m.}$$

Provide 1.6 m thick pucca floor which should be covered by 20 cm thick pitching of bricks.

(**Note.** While determining the thickness of the floor multiplication by factor $\frac{4}{3}$ has not been done because 20 cm brick pitch-gives added safety to the floor).

Provide 70 cm thickness of impervious floor at *D/S* end. Provide thickness at intermediate point according to uplift pressures.

Check for Floor Thickness by Khosla's Theory

$$\text{Exit gradient } GE = \frac{1}{\pi\sqrt{\lambda}}\frac{H_s}{d_2}$$

Value of *GE* is given as 1 in 5 or $\frac{1}{5}$.

$$\therefore \quad \frac{1}{5} = \frac{1}{\pi\sqrt{\lambda}}\times\frac{2.90}{2.0}$$

$$\text{or} \quad \frac{1}{\pi\sqrt{\lambda}} = 0.138$$

From Khosla curves, for $\frac{1}{\pi\sqrt{\lambda}} = 0.138, \alpha = 10$

$\therefore$ Base width $= \alpha d_2 = 10 \times 2 = 20$ m.

We have provided 17 m length of horizontal floor, whereas according to Khosla it should be 20 m. To provide for this added length either *D/S* cut-off will have to be increased in length or 3 m additional length of horizontal floor will have to be provided. We are adopting the first alternative. We have increased length of *D/S* cut off to 2.5 m while horizontal length has been maintained 17 m.

$$b = 17 \text{ m}, \; d_2 = 2.5 \text{ m}$$

$$\alpha = \frac{b}{d_2} = \frac{17}{2.5} = 6.8$$

For $\alpha = 6.8 \quad \frac{1}{\pi\sqrt{\lambda}} = 0.153$

$$GE = \frac{1}{\pi\sqrt{\lambda}}\frac{H_3}{d_2} = 0.153 \times \frac{2.9}{2.5} = \frac{1}{5.6} < \frac{1}{5} \text{ hence safe.}$$

Uplift Pressure

U/S cut off $\quad d_1 = 1.5 \text{ m}, b = 17 \text{ m}$

$$\frac{1}{\alpha} = \frac{d_1}{b} = \frac{1.5}{17} = \frac{1}{11} = 0.0882$$

$$\phi_{C1} = 100 - 28 = 72\%, \phi_{D1} = 100 \times 18 = 82\%$$

Thickness of *U/S* floor provided = 50 cm

Corrected value of ϕ_{C1} for floor thickness

$$= 72 + \frac{(82-72)}{1.5}\frac{1}{2} = 75.3\dot{3}\%$$

(ii) *D/S* cut off wall

$$d_2 = 2.5 \text{ m}, b = 17 \text{ m}$$

$$\frac{1}{\alpha} = \frac{2.5}{17} = 0.147$$

$$\phi_{E2} = 36\% \text{ and } \phi_{D2} = 24\%$$

Floor thickness provided at *D/S* end 70 cm.

$$\text{Corrected value of } \phi_{E1} = 36 - \frac{(36-24)}{2.5} 0.7 = 32.64\%$$

Variation of uplift pressure in from 75.33% at *U/S* pile to 32.64% at *D/S* pile.

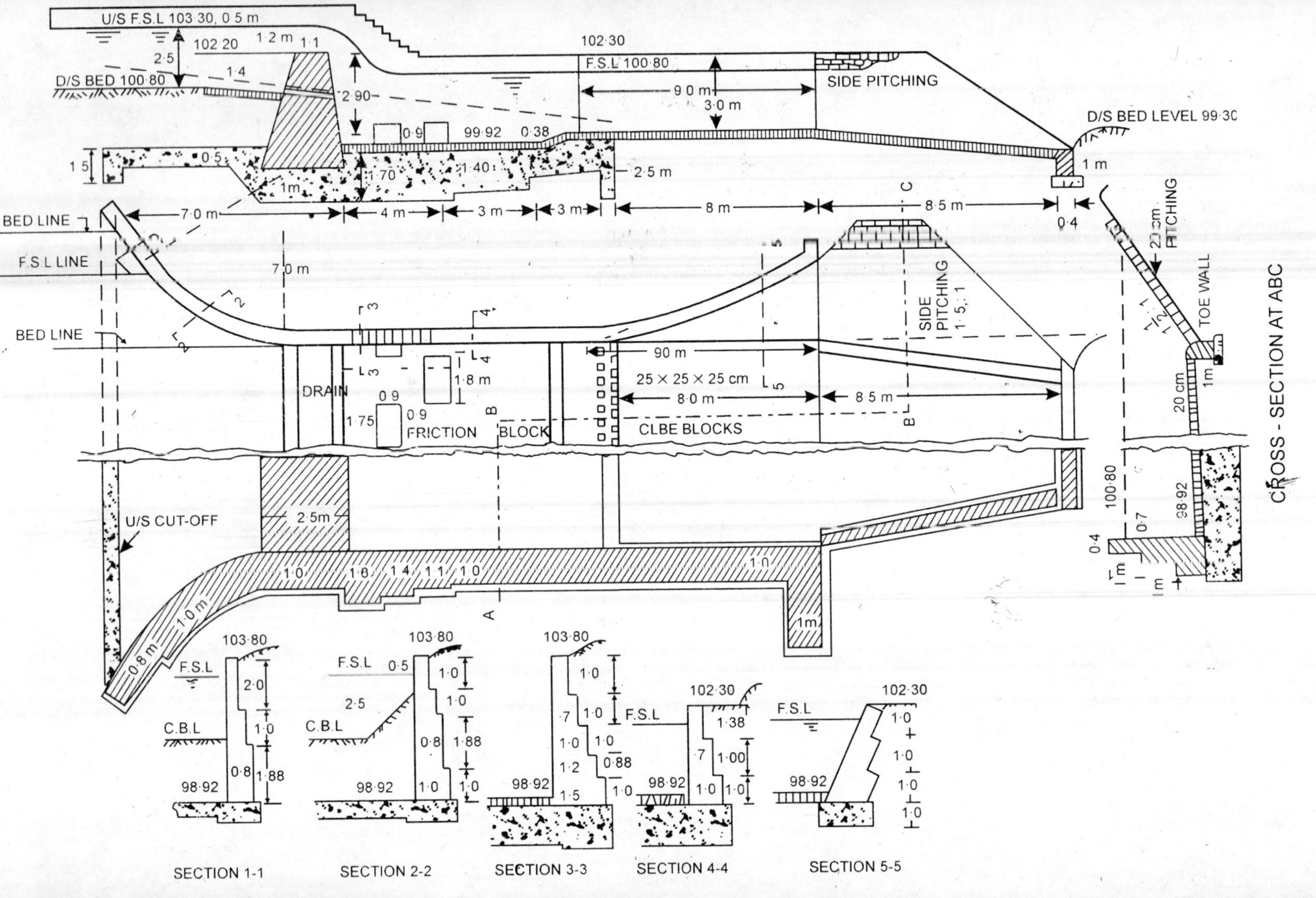

CROSS - SECTION AT ABC

Thickness of Floor by Khosla's Theory

Provide 50 cm thickness on *U/S* side

Pressure at point *A* (i.e. *D/S* of toe)

$$= 32.64 + \frac{(75.33 - 32.64)10}{17}$$

$$= 57.74\%$$

Maximum static head = 0.5774 × 2.9 + 0.38 = 2.06 m

Required thickness of floor $= \frac{2.06}{1.24} = 1.7$ m

We have already provided thickness of 1.7 m by Bligh's theory. Pitching of 20 cm will give some added safety and hence this thickness may be taken as adequate.

At point *B*, 4 m *D/S* of point *A*

$$\text{Pressure in \%} = 32.64 + \frac{(75.33 - 32.64) \times 6}{17}$$

$$= 32.64 + 15.1 = 47.74\%$$

Maximum static head $= 0.4774 \times H_s + x$

$$= 0.4774 \times 2.9 + 0.38 = 1.76 \text{ m}$$

Required thickness at point *B* $= \frac{1.76}{1.24} = 1.34$ m

Provide 1.40 m thickness of floor, covered by 20 cm brick pitching.

At point *C*, 3 m *D/S* of point *B*

$$\text{Pressure in \%} = 33.64 + \frac{(75.33 - 32.64) \times 3}{17}$$

$$= 32.64 + 7.53 = 40.17\%$$

Here water pressure of 0.38 m will not be available as cistern does not extend to this point.

$\therefore$ Available head = 0.4017 × 2.90

= 1.165 m

Floor thickness $= \frac{1.165}{1.24} = 0.95$ m

Hence beyond point C, D/S, provide thickness of pucca floor 1 m instead of 70 m provided according to Bligh. Provide 20 cm pitching also at the top.

Floor thickness at various points will be as follows:

1. U/S floor = 50 cm
2. D/S floor A to B = 1.70 m.
3. B to C = 1.40 m
4. C to D/S end of floor = 1.0 m.

Design of D/S Wing Walls

D/S wing walls may be kept vertical for a length of

$$= 6\sqrt{EH_L} = 6\sqrt{1.133 \times 1.5} = 7.8 \text{ m.}$$

Thereafter flare the walls slowly to a slope of 1 : 1. In plan slope of the wings is kept 1 in 3.

Height of D/S wing wall = 2.5 + free board

= 2.5 + 0.5 = 3.0 m

Horizontal projection will be 3 in at 1 : 1 flare.

Length along the centre line of the canal at 1 in 3 back slope.

= 3.0 × 3 = 9 m

***D/S* Pitching**

(a) *Bed pitching* $H = 1.1$ m

Length of bed pitching = 13.5 + 2 × 1.5 = 16.5 m.

This pitching should be horizontal, upto end of the wing walls. Thereafter it should be given 1 in 10 slope upwards.

Length of horizontal pitching 8 m and length of sloping pitching 8.5 m.

(b) *Toe wall.* At D/S end of pitching provide 40 cm thick and 1 m deep toe wall.

(c) *Side pitching.* This pitching is done in bricks on edge (i.e. 20 cm). Starting from end of the wing wall warp the itching slowly $1\frac{1}{2}$: 1 Curtail side pitching at 45° at the end of bed pitching. Provide a toe of 40 cm thick and 1 m deep and support the side pitching over it.

***U/S* Approach**

Radius of U/S wing walls = $6H$ = 6 (1. 1) = 6.6 m say 7 m

U/S wing walls are provided in form of a segment of 7 m radius. The segment should form an angle of 60° at the centre point. Thereafter the wings are carried into the banks and embedded in them fora length of 1 m.

Design of Energy Dissipators

They are not required for small discharges and small drops.

$$\text{Critical depth} = \left(\frac{q^2}{g}\right)^{1/3}$$

$$= \left\{\left(\frac{80}{36}\right)^2 \times \frac{1}{9.8}\right\}^{1/3} = 0.9 \text{ m}$$

Size of friction blocks= 1.80 m × 0.9 m × 0.9 m

Distance from toe of the crest = 1.5 × d_c = 1.5 × 0.9 = 1.35 m

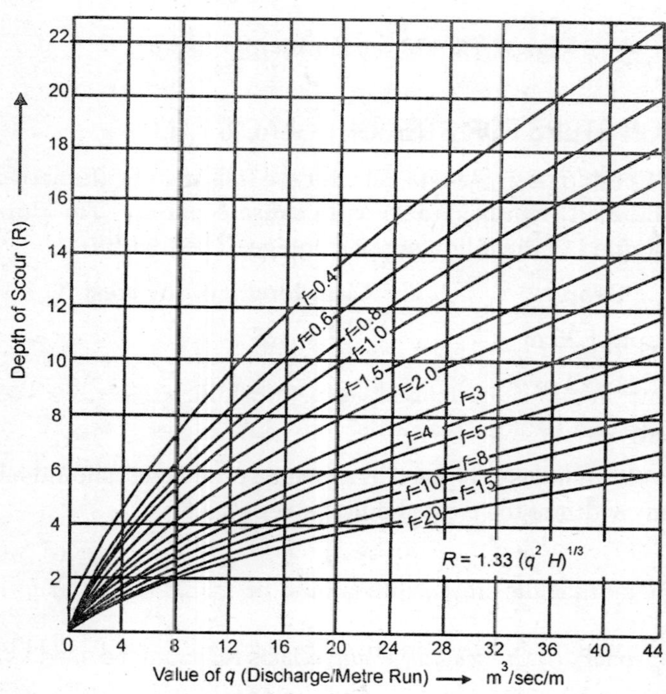

Fig. 21.14

Hence provide two rows of frictions blocks in staggered fashion. The first row should lie at 1.35 in from toe of the crest.

Size of Cube Blocks

Length × Breadth × height 25 cm × 25 cm × 25 cm

Side of a cube $\frac{D}{10} = \frac{2.5}{10} = 25$ cm.

Provide two rows of cube blocks in staggered fashion at the end of the pucca floor.

For detail see Fig. 21.15.

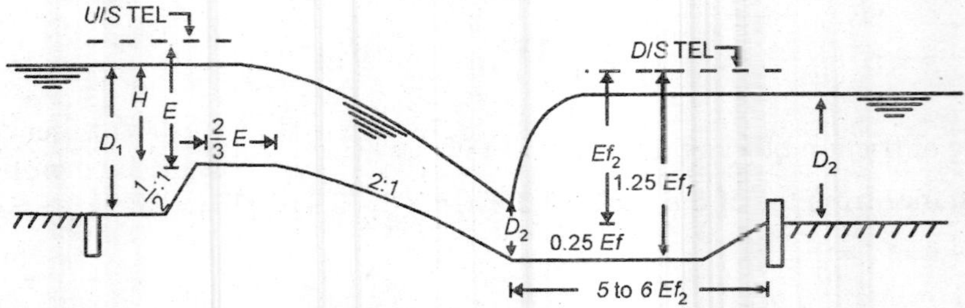

Fig. 21.15. *Non-meter fall (glacis type).*

21.9 DESIGN STEPS FOR STRAIGHT GLACIS FALL

1. Design of Crest. *(i) Crest Length.* Glacis type falls may be flumed when they are to be combined with bridge so as to cause economy. The fluming may however be limited to the following percentage of bed width.

Drop	Clear length of the crest
upto 1.25 m	65%
1.25 to 3 m	75%
above 3 m	85%

(ii) Crest Width. If fall is to be used as a meter a broad crest should be adopted. The minimum width of the crest should be kept 2.5 *E*.

For non-meter falls the crest width should be $\frac{2}{3}$ *E*.

2. U/S Approach. *(i) Curved Side Walls.* Glacis falls may be used as meters or non-meters.

For meter glacis fall, wings are curved and curvature of the wings is kept 5 to 6 time *E*. The segment of the curve forms an angle of 60° at the centre and thereafter carried straight into the embankment.

For non-metre fall, the *U/S* side walls may be splayed at an angle of 45° from *U/S* edge of the crest and carried into the embankment for about 1 m.

(ii) Hump. The bed curve of the approach may be curved or straight sloped.

In case of meter fall, bed curve should start from the same cross-section as the side curves start. The radius of hump or bed curve should be equal to

$$\frac{L_a^2 + h^2}{2h}.$$

For non-meter falls the bed approach in sloped at $\frac{1}{2}$: 1 slope and joined tangentially to the *U/S* edge of the crest with radius $\frac{E}{2}$.

3. *U/S* Protection. If glacis falls are not flumed, no protection is generally required on the *U/S* side. If fall is flumed, both bed and sides should be protected for a length equal to *U/S* water depth. It may be in form of a dry brick on edge laid at a slope of 1 in 10 in the bed.

4. Discharge Formula. The discharge passing over the crest is worked out from the following formula.

$$Q = C L_t E^{3/2}$$

where Q is the discharge in cumecs, L_t is the effective length of the crest. If there are piers on the crest the effective length of the crest is worked out by deducting 0.2 n from the total length of the crest.

$L_t = L - 0.2\,n$, where L is the total length of the crest and n is the number of piers.

Value of C for broad crested fall is Liken as 1.70 and with narrow crest i.e. equal to $\frac{2}{3}E$, its value is taken as 1.84.

Using equation $Q = 1.84\, L_t E^{3/2}$ value of E can be found out.

5. Crest Level

$$\text{Crest level} = \text{Total } U/S \text{ T.E.L.} - E$$

$$\text{Total } U/S \text{ T.E.L.} = U/S \text{ F.S.L.} + \text{velocity head}$$

If crest level works to be too high, the fall may be flumed or if it is already flumed, the fluming ratio should be further increased so that crest is not higher than 0.4 times the *U/S* water depth.

6. *D/S* Glacis. *D/S* glacis may be curved or straight sloping. The main purpose of glacis is to ensure formation of hydraulic jump at its toe. Theoretically the glacis slope should be such that it imparts maximum horizontal acceleration and thus ensures maximum dissipation of energy. Montague gave following equation for curved profile

$$x = U\sqrt{\frac{4y}{g}} + y$$

where x = horizontal ordinate measured from the *D/S* edge of the crest

y = vertical ordinate measured below crest level

$$U = \left(\frac{q^2}{g}\right)^{1/3}$$

where q = intensity of discharge per metre length of the fall.

It is very difficult and costly to construct a curved profile and as such is not used much. A straight glacis with 2 : 1 slope may be used in place of curved profile.

If baffle platform with baffle wall is to be used, the slope of glacis may be increased to $\frac{2}{3}$: 1 to ensure the formation of jump in the baffle platform.

For meter falls glacis with 2 : 1 slope should be used even if baffle wall is introduced. The sloping glacis should be joined to *D/S* edge of the crest with curve of radius *E*. The same curve should be used for joining *D/S* end of glacis with the horizontal floor.

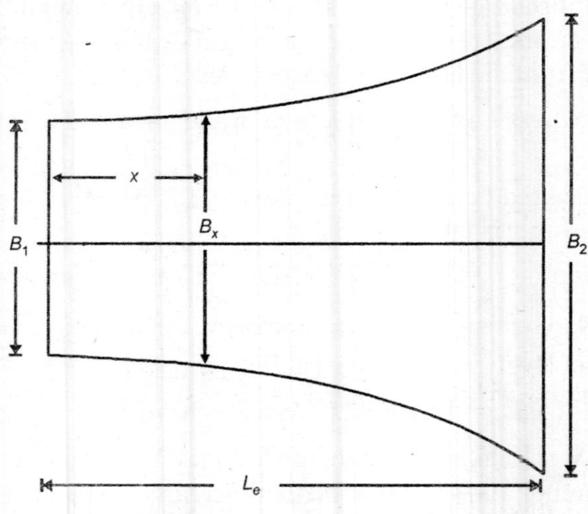

Fig. 21.16. *D/S expansion.*

7. Cistern Design

Length of cistern = 5 to 6 E_{f2}

R.L. of cistern = *D/S* T.E.L. – 125 E_{f2}

E_{f2} is the energy of flow.

8. Length and Thickness of Impervious Floor. It is designed according to safe exit gradient and uplift pressure on the basis of Khosla's theory.

9. *D/S* Protection. *(i) Bed protection.* No bed protection is required on *D/S* side as deflector wall is generally provided at the end of the floor. The height of deflector wall above D/S bed should be $\frac{D_3}{10}$ where D_3 is the depth of water on D/S side channel.

(ii) *Side protection*. Dry brick pitching should be done on sides for a length of $3D_3$. The pitching should be supported on a toe wall 40 cm wide and $\frac{D_3}{2}$ deep subjected to a minimum depth of 50 cm.

The end should be protected with curtain or profile wall 40 cm thick and $\frac{D_3}{2}$ depth.

10. *D/S* Expansion. Vertical sides are carried straight to the toe of glacis or end of baffle platform if provided. Thereafter the expansion of the walls starts. The most preferred expansion is rectangular hyperbolic. The bed width Bx at any distance x from the D/S toe of the glacis or from the beginning of the expansion,

$$B_x = \frac{B_1 \times B_2 \times L_c}{L_e \times B_2 - (B_2 - B_1)x}$$

where L_e = total length in which expansion is to be done. However in case of small falls, straight 1 in 3 expansion may be provided.

The side walls in expansion may be splayed out from vertical to 1 : 1 and extended into the berm.

11. Energy Dissipators. Friction blocks, glacis blocks and deflector walls should be used to dissipate the surplus energy. Details about these are given in Section 21.9.

Example 21.2 *Design an unflumed straight glacis non-meter fall for the data given below:*

1. *Full supply discharge* = *60 cumecs*
2. *Full supply level in U/S side* = *108.00 m*
3. *Full supply level on D/S side* = *105.50 m*
4. *Full supply depth U/S side* = *2.00 m*
5. *Full supply depth D/S side* = *2.00 m*
6. *Bed width U/S side* = *32 m*
7. *Bed width D/S side* = *32 m*
8. *U/S Bed level* = *106.00*
9. *D/S Bed level* = *103.50*
10. *Drop* = *2.5 m*
11. *Safe exit gradient* = $\frac{1}{5}$

Solution

Design of Crest

Length of crest = width of channel = 32 m

$$Q = 1.84\, L_t\, E^{3/2}$$

$$60 = 1.84 \times 32 \times E^{3/2}$$

$$E = \left(\frac{60}{1.84 \times 32}\right)^{2/3} = 1.01 \text{ m}$$

Assuming side slope of canal 1 : 1

Velocity of approach

$$V_a = \frac{Q}{(B+D)D}$$

$$= \frac{60}{(32+2)2} = 0.88 \text{ m/sec}$$

$$\text{Velocity head} = \frac{(0.88)^2}{2 \times 9.81} \simeq 0.04$$

U/S T.E.L. = U/S F.S.L + velocity head

= 108.00 + 0.04 = 108.04 m

Crest level = U/S T.E.L. – E

= 108.04 = 1.01

= 107.03 m

Width of crest = $\frac{2}{3}E = \frac{2}{3} \times 1.01 = 0.67\text{ m} = 0.70\text{ m}$

Provide D/S glacis at slope $\frac{1}{2}:1$ joined tangentially to the U/S bed with radius $\frac{E}{2} = \frac{1.01}{2}$ = 50 cm say.

Keep D/S glacis at 2 : 1 slope joined tangentially to the cistern bed and to the crest, each with radius E i.e. 1.01 m say 1 m.

Design of Cistern

$$q = \frac{Q}{L_t} = \frac{60}{32} = 1.875 \text{ cumec/m}$$

$$H_L = \text{Drop} = 2.5 \text{ m}$$

From Blench curves, corresponding to the value of $q = 1.875$ cumec/m and $H_L = 2.5$ we get $E_{f2} = 1.65$ m

$$R_L \text{ of cistern} = D/S \text{ T.F.L.} - 1.25\, E_{f2}$$
$$= 105.50 + 0.04 - 1.25 \times 1.65$$
$$= 105.54 - 2.0625$$
$$= 103.4775$$

This is slightly lower than Bed level 103.50.

Keep R.L. of cistern 103.00 m.

$$\text{Length of cistern} = 6 \times E_{f2} = 6 \times 1.65 = 9.9 \text{ m}$$
$$\text{Depth of cistern} = 103.50 - 103.0 = 50 \text{ cm.}$$

Keep cistern 10 m long and 50 cm deep.

Join cistern to *D/S* bed at a slope of 5 : 1.

Design of Impervious Floor

Minimum depth of *U/S* curtain wall

$$= \frac{\text{Depth}}{3} = \frac{2.0}{3} = 0.67 \text{ m}$$

Provide 40 cm thick and 1 m deep curtain wall.

Depth *D/S* curtain wall

$$= \frac{\text{Depth}}{2} = \frac{2.00}{2} = 1 \text{ m}$$

Provide 40 cm thick and 1 m deep curtain at *D/S* end.

This curtain wall must project by $\frac{D_2}{10}$ above *D/S* bed so as to act as deflector wall

$$\frac{D_2}{10} = \frac{2.0}{10} = 20 \text{ cm}$$

Exit gradient is $\frac{1}{5}$.

$$G_E = \frac{1}{\pi\sqrt{\lambda}} \frac{H_s}{d_2}$$

$$H_s = \text{Maximum seepage head.}$$
$$= \text{Crest level} - D/S \text{ bed level}$$
$$= 107.03 - 103.50$$
$$= 3.53 \text{ m}$$
$$d_2 = \text{depth of } D/S \text{ curtain} = 1 \text{ m}$$

Putting values

$$\frac{1}{5} = \frac{1}{\pi\sqrt{\lambda}} \frac{3.53}{1}$$

$$\frac{1}{\pi\sqrt{\lambda}} = \frac{1}{5 \times 3.53} = 0.05666$$

From Khosla's curves,

For $$\frac{1}{\pi\sqrt{\lambda}} = 0.05666$$

Value of α becomes so large that it does not come under the limits of the Khosla curves. Hence revise the calculation by adopting *D/S* curtain wall 2..5 m stead of 1 m initially adopted.

$$\frac{1}{5} = \frac{1}{\pi\sqrt{\lambda}} \times \frac{3.53}{2.50}$$

$$\frac{1}{\pi\sqrt{\lambda}} = \frac{1}{5} \times \frac{2.50}{3.53} = 0.142$$

From Khosla curve for

$$\frac{1}{\pi\sqrt{\lambda}} = 0.142, \alpha = 9$$

$$b = \alpha d_2 = 9 \times 2.5 = 22.5 \text{ m}$$

This length of 22.5 is arranged as follows:

1. Length of cistern = 10 m.
2. Length of *D/S* glacis = 2 (107.03 – 103.00) 8.06 m.
3. Width of crest = 0.70 m
4. Length of *U/S* glacis = (107.03 – 106.00) = 0.50 m

Total length of above (1) to 4 = 19.26 m.

Balance to be provided to the *U/S* of *U/S* glacis.

$$= 22.5 - 19.26 = 3.24 \text{ m}$$

Pressure Calculations

(i) *U/S* Curtain Wall = 1

$$b = 22.5 \text{ m}$$

$$\frac{1}{\alpha} = \frac{d_1}{b} = \frac{1}{22.5} = 0.0444$$

$$\phi_{C_1} = 100 - 20\% = 80\%$$

$$\phi_{E_1} = 100 - 13\% = 87\%$$

Assume minimum thickness of 0.40 m at *U/S*.

Correction for thickness = $\left(\dfrac{87-80}{1}\right)0.40 = 2.8\%\ (+)$

Corrected $\phi_{C_1} = 80 + 2.8 = 82.8\%.$

(ii) *D/S* Curtain Wall

$$d_2 = 2.5 \text{ m}, b = 22.5 \text{ m}.$$

$$\frac{1}{\alpha} = \frac{d_2}{b} = \frac{2.5}{22.5} = 0.111$$

$$\therefore \quad \phi_E = 31\%$$

$$\phi_D = 21\%$$

Assuming a minimum thickness of 0.5 m at the *D/S*

Corrected $\phi_E = \left(\dfrac{31-21}{2.5}\right) 0.5 = 2\%\ (-)$

Correct $\phi_E = 31 - 2 = 29\%$

(iii) Toe of *D/S* Glacis. Assume linear variation of pressure, the % pressure at the *D/S* end of glacis.

$$= 29 + \left(\frac{82.8-29.0}{22.5}\right) 10$$

$$= 29 + \frac{53.8}{22.8} \times 10$$

$$= 52.91\%$$

Floor Thickness

(i) Provide minimum thickness of 40 cm at the *U/S* floor.

(ii) Floor thickness at the Toe of glaci.

Maximum static head

$$= 3.53 \times \frac{52.91}{100} + (103.50 - 103.0)$$

$$= 1.87 + 0.50$$

$$= 2.37 \text{ m}$$

$$\text{Floor thickness} = \frac{2.37}{(2.24-1)} = 1.91 \text{ m}$$

Provide 2 m floor thickness at Toe of the *D/S* glacis for a length of 4 m.

(iii) Floor thickness at 4 m from Toe of *D/S* glacis

$$\% \text{ pressure} = 29 + \frac{(82.8-29.0)}{22.5} 6$$

$$= 29 + \frac{53.80}{22.5} \times 6$$

$$= 29 + 14.35$$

$$= 43.35\%$$

Maximum static head

$$= 04335 \times 3.53 +$$

$$(103.50 - 103.0)$$

$$= 1.53 + 0.50$$

$$= 2.03 \text{ m}$$

$$\therefore \text{Floor thickness} = \frac{2.30}{(2.24-1)} = 1.64 \text{ m},$$

Provide a thickness of 1.70 from 4 m to 7 m from the Toe of the glacis.

(iv) Floor thickness at 7 m and *D/S*, from Toe,

$$\% \text{pressure} = 29\%$$

Maximum static head

$$= 0.29 \times 3.53 = 1.02 \text{ m}$$

$$\text{Floor Thickness} = \frac{1.02}{(2.24-1)}$$

$$= 0.83 \text{ m} = \text{say } 0.9 \text{ m}$$

Provide thickness at *D/S* end of floor 0.9 m mminstead of 0.60 originally assumed.

(i) Design of D/S protection. No bed protection is required as a deflector wall has been provided.

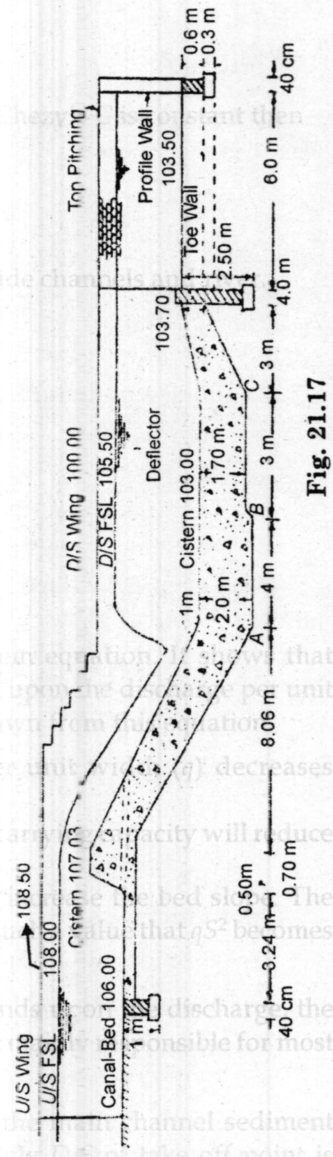

Fig. 21.17

(ii) *Side protections.* Length of side protection

$= 3D_2 = 3 \times 2 = 6$ m

Hence provide 20 cm thick brick pitching 6 m length beyond the impervious floor at 1 : 1 slope. Pitching will rest on 40 cm thick and $\frac{2}{2} = 1$ m deep Toe wall. Provide 0.4 wide profile wall at the end of the pitching to protect the side pitching.

(iii) Since no fluming has been done, no friction blocks are required.

Design of *U/S* Approach. The *U/S* wing walls should be splayed at 45° from the *U/S* end of the impervious floor and extended into the earthen banks for at least 1 m.

21.10 OFF-TAKE ALIGNMENT

When any distributary or branch takes-off from a parent canal, the alignment of the off take of both the canals should be carefully decided. Four off-take alignments have been shown Fig. 21.18. The requirements of a best off take alignment are as follows :

1. There should be minimum of disturbance water.
2. Both parent as well as off taking canals, should share the silt load according to their discharges.
3. There should be no scouring or silting on *U/S* of the off take.

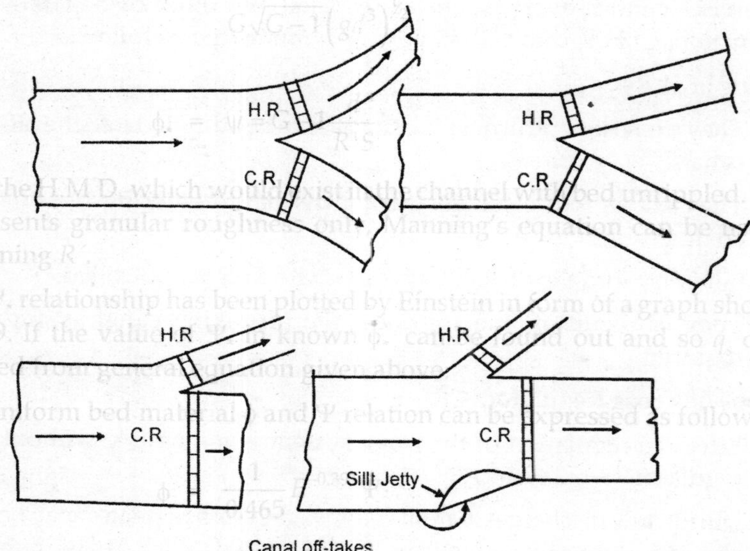

Fig. 21.18

The off-take alignment is considered best when it makes zero angle with the parent channel and then separates out form of transition curve. Transition curves, should be given to both off taking and parent canals.

If somehow it is not possible to made transition curves, then both the canals should make an angle with the *U/S* alignment of the parent channel.

If parent channel cannot be given any angle and has to be kept straight, the angle of off-take should be decided in relation to the edges of the parent channel rather than centre line. In this case section of the parent channel should be narrowed down from one edge only and not from both the edges equally. When edge of the parent channel in which off-take head regulator is not located, is narrowed down, a silt jetty is formed just *U/S* of the cross-regulator, on that edge. This alignment should be avoided as far as possible.

21.11 HEAD REGULATOR AND CROSS-REGULATOR

Head Regulator (H.R.). It is a masonry or concrete structure, constructed at the head of off-taking channel. Its main function is to admit, the regulated supplies of water in the off-taking channel.

Cross Regulator (CR). It is a masonry or concrete structure, constructed across the parent channel just Down stream of the head-regulator of the off-taking canal. Its main function is to raise the water level the parent channel, to such a level that requisite supplies the off-taking canal may be diverted through the head regulator.

Both H.R. and C.R. cannot be used alone. They have to be used together. One exercises control over the off-taking canal and other over parent canal. Main functions of H.R. and C.R. have been enumerated as follows.

Functions of H.Rs

1. They exercise control over the supplies to be admitted into the off-taking canals.
2. They act as metre for measuring discharge entering the off-taking channel.
3. Silt entry into the off-taking canal is checked or controlled.
4. They can shut-off the supplies the off-taking canal as and when required.

Functions of Cross Regulator (CR)

1. Effective regulation of the canal system is impossible without cross-regulator.
2. During low discharges parent channel, the cross regulator may be partly closed and thus water level on *U/S* side raised which can be diverted to run off taking canal full. Off-taking canals can be run rotation.
3. Flow of water *D/S* of parent channel may be completely closed for the purpose of repairs etc.

4. They absorb fluctuations at various sections of the canal system and thus help in preventing the possibilities of breaches in the Tail reaches.
5. Water can be stopped, to close the breaches the *D/S* side of C.R.
6. Cross-regulator is generally located where some road has to cross the parent canal. In that case it acts as bridge also.

Plate 21.2

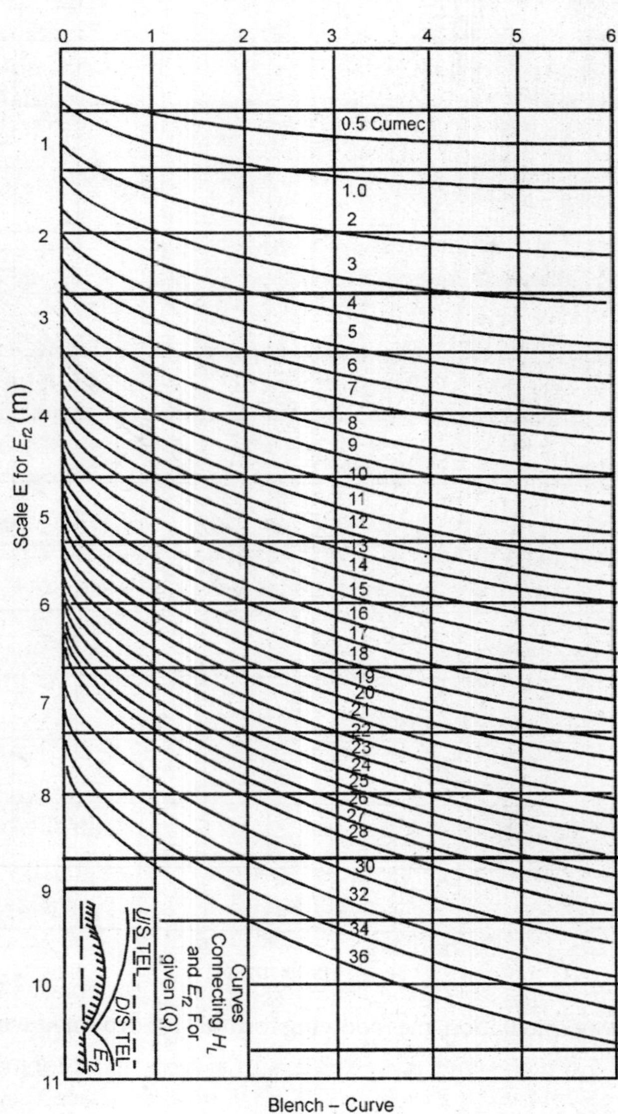

Blench – Curve

21.12 DESIGN OF CROSS REGULATOR AND HEAD REGULATOR

1. Fixing Crest Levels and Water Way. Crest level of C.R. is kept is *U/S* bed level of the parent channel and crest level of Head Regulator (H.R.) is fixed 30 cm to 1 m higher than the crest level of C.R.

Plate 21.3

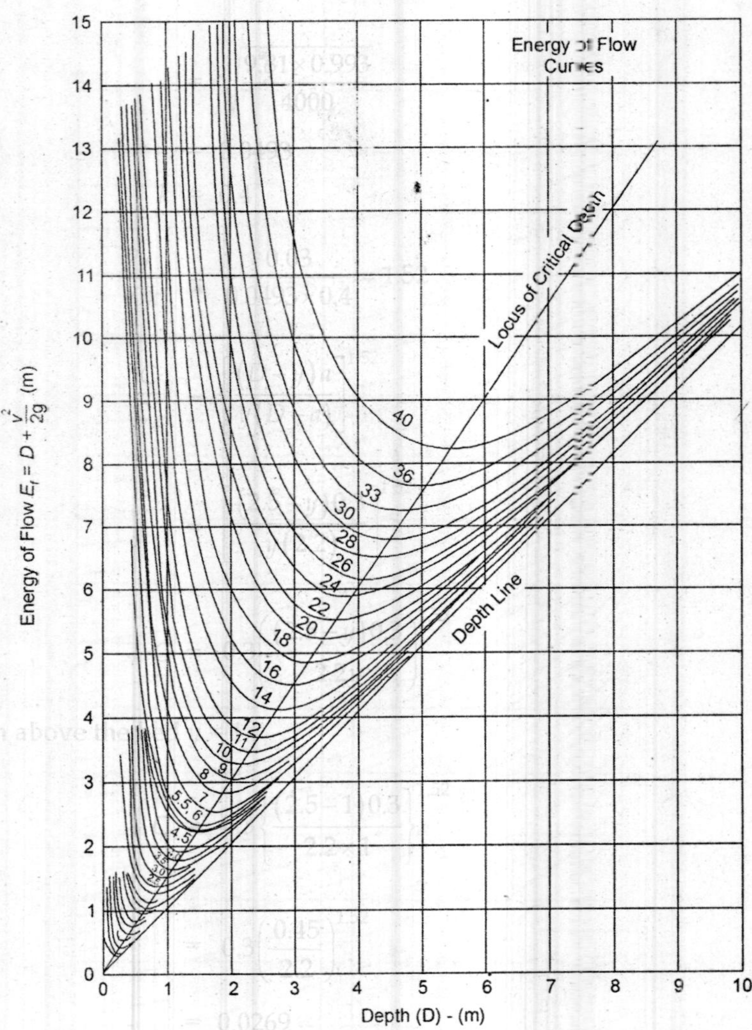

For water way calculation, the following formula for drowned weir conditions is used.

$$Q = \frac{2}{3}C_1L\sqrt{2g}\left\{(H+H_a)^{3/2}-(H_a)^{3/2}\right\}+C_2L\times D\sqrt{2g(H+H_a)}$$

C_1 = A constant whose value is taken as 0.577

C_2 = Another constant whose value is 0.80

L = Effective water way length in metres

H = Difference in U/S and D/S water levels in meters

H_a = Velocity of approach head

D = Depth of D/S water level in the channel above the crest in meters

Q = Discharge in cumecs.

Velocity of approach head is generally neglected, being very small.

After having fixed crest level, the crest is joined to D/S floor with 2 : 1 sloping glaces. U/S glacis is given a slope of $\frac{1}{2}$: 1.

2. Design of D/S Floor. The design of D/S floor is done for the following conditions of flow:

(i) Full supply discharge is passing through both C.R. and H.R.

(ii) Off-taking canal running full and C.R. partially opened.

For the conditions of flow, the intensity of discharge (q) and the head loss H_L are calculated and corresponding value of E_{f2} read from Blench curves.

$\therefore$ D/S Floor level = D/S TEL − E_{f2}.

The velocity of recession being small is neglected and the D/S TEL may be taken equal to D/S water level.

The design is usually governed by the first condition of flow. But in some cases second flow condition becomes more critical. If D/S floor level for the most critical condition works out to be higher than the bed level of the channel, the floor level is provided at the bed level and in no case higher than this.

For length of D/S floor, find out E_{f1}, $E_{f1} = E_{f2} + H_L$. From energy of flow curves read D_1 and D_2 corresponding to the values of E_{f1} and E_{f2} for different flow conditions.

Length of D/S Floor = $5(D_2 - D_1)$.

This length is sometimes very small in comparison to the total length of the floor. In such cases, the D/S length of floor is kept about $\frac{2}{3}$ of the total floor length.

3. Cut-offs. Total length of pucca floor should be found out based upon the safe exit gradient.

$$\text{Depth of U/S cut off} = \frac{U/S\ \text{water depth}}{3} + 0.6 \text{ m}$$

$$\text{Depth of } D/S \text{ cut off} = \frac{D/S \text{ depth of water}}{2} + 0.6 \text{ m}$$

4. Thickness and Length of Pucca Floor. Total floor length (b) is worked out on the basis of safe exit gradient.

Calculate Maximum static head = *D/S* F.S.L. - *D/S* Floor level. Knowing the maximum static head (H_S) and depth of *D/S* cut-off, work out $\frac{1}{\pi\sqrt{\lambda}} = G_E \frac{d}{H_S}$.

Read the value of α from Khosla curves. Total floor length

$$b = \alpha d$$

Thickness of floor is worked out, based upon the uplift pressure. A minimum thickness of about 50 cm should be provided on practical considerations.

5. Design of *U/S* and *D/S* Protection. The scour depth below floor in the *D/S* side is taken as

$$\left(\frac{\text{Depth of water on U/S side}}{3} + 0.6 \text{ m}\right)$$

and that on *D/S* side

$$\left(\frac{\text{Depth of water on D/S side}}{2} + 0.6 \text{ m}\right)$$

These scour depths are measured below the bed levels. Protection works have to be designed for them.

The block protection in the *U/S* and filter in the *D/S* are provided in a length equal to 1.5 times the scour depth. The cubic contents of material in the launching apron should be equal to 2.25 times the scour depth cubic metres/metre length.

Example 21.3. *Design a cross-regulator and head-regulator for a distributary channel taking off from a parent channel. Following data are given:*

1. *Discharge of the branch canal* = 120 *cumec*
2. *Discharge of the distributary channel* = 10 *cumec*
3. *Silt factor* = 0.9
4. *F.S.L. of Distributary* = 320.00
5. *F.S.L. of Branch on D/S side* = 321.00
6. *F.S.L. of Branch on D/S side* = 320.85
7. *Width of branch on U/S side* = 50 *m*
8. *Width of branch on D/S side* = 55 *m*
9. *Depth of water in branch both D/S and U/S* = 2 *m*

10. *Safe exit gradient* = *1 in 5*
11. *Bed width of Distributary* = *12.80 m*
12. *Depth of water in distribution* = *1.20 m*

Solution The design will have two parts. Design of cross regulator and design of Head regulator. See plate 21.4.

Part I. Design of cross-regulator

(i) *Fixation of crest level and water way.* Crest level of the C.R. is fixed at *U/S* bed level of the parent channel.

$$\therefore \quad \text{Crest level} = U/S \text{ F.S.L. of channel} - \text{Depth of water}$$

$$= 321.00 - 2.00 = 319.00 \text{ m}$$

$$Q = \frac{2}{3}C_1 L\sqrt{2g}\left[(H+H_a)^{3/2} - (H_a)^{3/2}\right] + C_2 LD\sqrt{2g(H+H_a)}$$

$$C_1 = 0.577, C_2 = 0.80$$

Since H_a is generally very small, it has been neglected

$$H = U/S \text{ F.S.L.} - D/S \text{ F.S.L.} = 321.00 - 320.85 = 0.15 \text{ m}$$

$$D = U/S \text{ F.S.L.} - \text{crest level} = 320.85 - 319.00 = 1.85 \text{ m}$$

$$120 = \tfrac{2}{3}\times 0.577\times L\sqrt{2\times 9.81}\times(0.15)^{3/2} + 0.80\times L\times 1.85\sqrt{2\times 9.8\times 0.15}$$

On solving $L = 45.7$ m

Adopt 8 m wide 6 spans, providing length of crest as 48 m. Provide 1.6 m thick five piers.

Total length of cross-regulator (C.R.) = 48 + 8 = 56 m

(ii) *Level of D/S Floor and length.*

$$Q = 120 \text{ cumecs}$$

$$q = \frac{120}{48} = 2.5 \text{ cumecs}$$

Head loss $H_L = U/S$ F.S.L. $- D/S$ F.S.L.

$$= 321.00 - 320.85 = 0.15 \text{ m}$$

$$H_L = 015 \text{ m}, q = 2.5 \text{ cumec}$$

From Blench curves

$$E_{f2} = 1.435 \text{ m}$$

$$D/S \text{ Floor level} = D/S \text{ F.S.L.} - E_{f2}$$

$$320.85 - 1.435 = 319.315 \text{ m}$$

This is higher than D/S bed level of 318.85 m. Hence adopt cistern level same as D/S bed level, i.e. 318.85 m.

$$E_{f1} = E_{f2} + H_L = 1.435 + 0.15 = 1.585 \text{ m}$$

$$D_1 = 0.534\ D_2 = 1.30 \text{ m (from energy flow curves)}$$

Required length of cistern

$$= 5\,(D_2 - D_1) = 5\,(1.30 - 0.534) = 3.83 \text{ m}$$

$$U/S \text{ cut-off Depth} = \frac{D/S \text{ depth}}{3} + 0.6 \text{ m}$$

$$= \frac{2}{3} + 0.6 \text{ m} = 1.27 \text{ m say } 1.3 \text{ m}$$

Provide 1.3 m deep cut off below the floor on U/S end.

$$D/S \text{ cut-off} = \frac{D/S \text{ depth of water}}{2} + 0.6 \text{ m}$$

$$= \frac{2}{2} + 0.6 = 1.6 \text{ m}$$

Provide 1.6 m deep cut off at D/S end of the floor.

3. Floor Length and Exit Gradient. Maximum uplift pressure occurs when water level on U/S side is filled upto F.S.L. and there is no flow on D/S side.

$$U/S \text{ F.S.L.} = 321.00 \text{ m}, \quad D/S \text{ Floor level} = 318.85 \text{ m}$$

$$\text{Maximum head} = 321.00 - 318.85 = 2.15 \text{ m}$$

$$\text{Depth of } D/S \text{ cut off} = 1.60 \text{ m}$$

$$G_E = \frac{H}{d}\,\frac{1}{\pi\sqrt{\lambda}}$$

$$\frac{1}{5} = \frac{2.15}{1.60} \times \frac{1}{\pi\sqrt{\lambda}} \quad \text{or} \quad \frac{1}{\pi\sqrt{\lambda}} = \frac{1.6}{5 \times 2.15} = 0.149$$

For $\frac{1}{\pi\sqrt{\lambda}} = 0.149$, $\alpha = 8$, from Khosla curves.

Total length of Floor = $\alpha d = 8 \times 1.60 = 12.80$ m. Provide 12.8 m length as follows:

Length of D/S floor = 6.5 m

Length of D/S glacis = 2 (319.00 – 318.85) = 0.30 m

U/S floor length including crest = 6.0 m

Total length = 6.50 + 0.30 + 6.0 = 12.8 m

4. Uplift Pressure Calculations

(a) *U/S* cut-off, d = 1.30 m, b = 12.8 m

$$\frac{1}{\alpha} = \frac{d}{b} = \frac{1.30}{12.80} = 0.101$$

$$\phi_D = 80\% \text{ and } \phi_C = 72\%$$

$$\phi_D - \phi_C = (80 - 72) = 8\%$$

Assume thickness of U/S floor 60 cm. Corrections in ϕ_C for floor thickness

$$= \frac{0.6}{1.30} \times 8 = 3.7\% \ (+)$$

Corrected ϕ_C = 72 + 3.7 = 75.7%,

(we have assumed effects of D/S and U/S cut-offs as nil).

(b) *D/S* cut-off, d = 1.60 m, b = 12.80 m

$$\frac{1}{\alpha} = \frac{1.60}{12.80} = 0.125$$

$$\phi_E = 31\ \%, \phi_D = 22\%, (\phi_F - \phi_D) = 9\%$$

Assume thickness of D/S floor 0.60 m.

Correction for floor thickness

$$= \frac{0.6 \times 9}{1.60} = 3.38\% \ (-100)$$

Corrected ϕ_E = 31 – 3.38 = 27.67%.

Effect of U/S cut-off on D/S cut-off has been left.

5. Floor Thickness. *(i) At 2 m from D/S end.*

$$\% \text{ pressure} = 27.62 + \frac{1.50}{11.80}(75.2 - 27.62) = 33.72\ \%$$

$$\text{Static head} = \frac{33.72}{100} \times 2.15 = 0.725 \text{ m}$$

$$\text{Thickness of floor} = \frac{0.725}{2.24 - 1} = 0.584 \text{ m}$$

Adopt 0.60 m thickness.

(ii) At 4.5 m from D/S end.

$$\% \text{ pressure} = 27.62 + \frac{4}{11.80}(75.7 - 27.62) = 43.93\%$$

$$\text{Static head} = 2.15 \times \frac{43.97}{100} = 0.944 \text{ m}$$

Required Floor thickness

$$= \frac{0.944}{(2.24-1)} = 0.756 \text{ m}$$

For (4.50 – 2) = 2.50 m length provide 0.8 m floor thickness.

(iii) At 6.50 m from D/S end.

$$\% \text{ pressure} = 27.62 + \frac{6}{11.80}(75.70 - 27.62) = 52.15\ \%$$

$$\text{Static head} = 2.15 \times \frac{52.15}{100} = 1.14 \text{ m}$$

$$\text{Floor thickness} = \frac{1.14}{(2.24-1)} = 0.905 \text{ m}$$

Adopt floor thickness in this 2 m length as say 1 m.

For *U/S* Floor thickness there is no need of any calculation. Provide 60 cm nominal thickness.

6. *U/S* Protection

$$U/S \text{ scour depth} = \frac{\text{Depth of water}}{3} + 0.6 \text{ m}$$

$$= \frac{2}{3} + 0.6 = 1.27 \text{ m}$$

Protection work in form of blocks = 1.27 m^3/m width.

If thickness of block protection is 1 m, then length of block protection work = 1.27 m.

Provide 80 cm × 80 cm × 60 cm concrete blocks in two rows on a 40 cm thick and 1.6 in long apron.

***U/S* Launching apron.** Volume of stone blocks

$$= (2.25 \times \text{scour depth})\ m^3/m$$

$$= 2.25 \times 1.27 = 2.86\ m^3/m.$$

If thickness of Launching apron is kept 1 m, then its length would be 2.86 m; provide say 3 m.

***D/S* Protection.** *D/S* scour depth

$$= \frac{\text{Depth of water}}{2} + 0.6 \text{ m}$$

$$= \frac{2}{2} + 0.6 = 1.6 \text{ m}$$

(i) Inverted filter. Volume required

$$= D/S \text{ scour depth m}^3/\text{m}$$

$$= 1.6 \text{ m}^3/\text{m}$$

Provide 60 cm filter thickness and above it, provide 60 cm Blocks. Total thickness 0.60 + 0.60 = 1.20 m.

$$\text{Length of filter} = \frac{1.60}{1.20} = 1.33 \text{ m}$$

Provide two rows of 80 × 80 × 60 cm concrete blocks over 60 cm thick graded filter.

(ii) Launching apron. Volume of stone blocks

$$= (2.25 \times \text{scour depth}) \text{ m}^3/\text{m}$$

$$= 2.25 \times 16 = 3.60 \text{ m}^3/\text{m}$$

If thickness of apron is kept 1 m, its length would be 3.60 m. Provide 4 m long launching apron.

Provide 40 cm thick and 1.2 m deep Toe wall between Toe filter and launching apron.

Part 2. Design of Head Regulator. Bed level of off-taking canal = 320.20 – 1.20 = 319.0 m.

(i) Fixation of crest and water way. Crest level of H.R. is kept 0.50 m above the *U/S* bed level of the parent channel.

U/S floor level = 319.0 m

Crest level = 319.0 + 0.5 = 319.50 m

Let L be the crest length of the H.R.

$$Q = \tfrac{2}{3}C_1L\sqrt{2g}\,H^{3/2} + C_2Ld\sqrt{2gH}\,.$$

∴ Velocity of approach head has been neglected

$$H = 321.00 - 320.20 = 0.80 \text{ m}, \; C_1 = 0.577$$

$$d = 320.20 - 319.50 = 0.70 \text{ m}, \; C_2 = 0.80$$

Substituting the values

$$10 = \tfrac{2}{3} \times 0.577 \times L\sqrt{2g}\,(0.8)^{3/2} + 0.8 \times L \times 0.7\sqrt{2g \times 0.8}$$

$$L = 2.925 \text{ m}$$

Provided two spans of m each with a 1 m pier in between.

Total water way $= 2 \times 3 + 1 = 7$ m.

The wing walls are sloped backwards so that full width of the channel is achieved.

2. Length of *D/S* Floor and its Level

$$Q = 10 \text{ cumecs}, L = 6.0 \text{ m}$$

Intensity of discharge

$$q = \frac{10}{6} = 1.67 \text{ cumecs m}$$

Head loss H_L = *U/S* F.S.L. – *D/S* F.S.L.

= 321.00 – 320.20 = 0.80 m

For q = 1.67 cumec and H_L = 0.8m, E_{f2} from curves 5 found out as 1.37 m.

Required floor level = *D/S* F.S.L. – E_{f2}

= 320.20 – 1.37

= 318.83 m

Keep *D/S* floor level as 318.80 m.

$$E_{f1} = E_{f2} + H_L = 1.37 + 0.80 = 2.17 \text{ m}$$

From energy charts,

$$D_1 = 0.32 \text{ and } D_2 = 1.32 \text{ m}$$

Length of cistern = 5 (1.32 – 0.32) = 5 m

Adopt cistern length 6 m.

3. Cut-off Walls

$$D/S \text{ cut-off} = \frac{\text{Depth of water on } D/S \text{ side}}{2} + 0.60 \text{ m} = 1.20 \text{ m}$$

Adopt 1.60 m as the depth of *D/S* cut-off

$$U/S \text{ cut-off} = \frac{\text{Depth of water on } U/S \text{ side}}{3} - 0.6 \text{ m}$$

$$= \frac{2}{3} + 0.6 = 1.27 \text{ m}$$

Provide depth of *U/S* cut-off as 1.30 m.

4. Total Length of Impervious Floor and Exit Gradient. The maximum static head conditions are created when water is filled upto F.S.L. of the parent channel and there is no water on *D/S* side of the head regulator.

U/S F.S.L. = 321.00 m

D/S floor level = 318.80 m

Maximum static head = 321.00 – 318.80 = 2.20 m

Depth of *D/S* cut-off = 1.60 m

Safe exit gradient is $\frac{1}{5}$

$$G_E = \frac{4}{d}\frac{1}{\pi\sqrt{\lambda}} \text{ or } \frac{1}{\pi\sqrt{\lambda}} = \frac{1}{5}\times\frac{d}{H} = \frac{1}{5}\times\frac{1.60}{2.20}$$

$$= 0.145$$

From Khosla curves, for $\frac{1}{\pi\sqrt{\lambda}} = 0.145, \alpha = 8$

Total floor length $\alpha = \frac{b}{d}$

or $8 = \frac{b}{1.60}$

or $b = 12.8$ m

adopt 13 m as the length of impervious floor.

Total length of impervious floor will be provided as follows:

L/S floor length = 6.00 m

D/S glacis length at 2 : 1 slope = 2 (319.50 – 318.80) = 1.40 m

Crest width = 1.00 m

U/S glacis at 1 : 1 slope 1 (319.50 – 319.00) = 0.50 m

U/S floor = 4.10 m

13.0 m

Total = 6.00 + 1.40 + 1.00 + 0.50 + 4.10 = 13 m

5. Pressure Calculations. *(a) For U/S cut off*

$$d = 1.30 \text{ m}, b = 13 \text{ m}, \frac{1}{\alpha} = \frac{b}{d} = \frac{1.30}{13.0} = 0.10$$

$$\phi_{D1} = 80\%, \phi_{C1} = 72\%, \phi_{D1} - \phi_{C1} = 8\%$$

Assume floor thickness of 60 cm on *U/S* side.

$$\text{Correction in depth} = \frac{0.60}{1.30} \times 8 = 3.7\% \ (+)$$

$$\text{Corrected} \quad \phi_{C1} = 72 + 3.7 = 75.7\%$$

(ii) D/S cut-off

$$d = 1.60 \text{ m}, b = 13 \text{ m}, \frac{1}{\alpha} = \frac{d}{b} = \frac{1.5}{13.0} = 0.123$$

$$\phi_E = 32\%, \phi_D\ 22\%, \phi_F - \phi_D = 32 - 22 = 10\%$$

Assume thickness of floor at *D/S* end as 0.60 m.

$$\text{Correction for depth} = \frac{0.6}{1.30} \times 10 = 4.62\% \ (-\text{ive})$$

$$\text{Corrected} \quad \phi_E = 32 - 4.62 = 27.38\%$$

Effect of *U/S* and *D/S* cut-offs on each other has been left.

6. Thickness of *D/S* Floor. *(i) Consider a point at 2 m from D/S end*

$$\% \text{ pressure} = 27.38 + \frac{1.5}{12.00}(75.70 - 27.38) = 33.40\%$$

$$\text{Static head} = \frac{33.40}{100} \times 2.20 = 0.736 \text{ m}$$

$$\text{Required thickness of floor} = \frac{0.736}{(2.24 - 1)} = 0.592 \text{ m}$$

Hence 0.6 in adopted thickness is quite sufficient.

(ii) Consider point at 4 m from D/S end

$$\% \text{ pressure} = 27.38 + \frac{3.50}{12.00}(75.70 - 27.38) = 41.50\%$$

$$\text{Static head} = 0.4150 \times 2.20 = 0.913 \text{ m}$$

$$\text{Floor thickness} = \frac{0.913}{(2.24 - 1)} = 0.74 \text{ m}$$

Adopt floor thickness of 0.80 m from 2 m to 4 m from the *D/S* end.

(iii) At 6 m from D/S end

$$\% \text{ pressure} = 27.38 + \frac{5.5}{12.0}(75.70 - 27.38) = 49.40\%$$

$$\text{Static head} = \frac{49.40}{100} \times 2.20 = 1.0850$$

$$\text{Floor thickness} = \frac{1.0850}{(2.24-1)} = 0.875 \text{ m}$$

Adopt 1 m thickness from 4 m to 6 m from *D/S* end. *U/S* floor does not require any calculation. Provide a nominal thickness of 0.6 m directly below the crest.

7. *U/S* Protection Work. Done in the similar way as has been explained for cross-regulator.

8. *D/S* Protection Work. *D/S* protection works are done for following scouring depth (*D*)

$$D = \frac{\text{Depth of water on } D/S \text{ side}}{2} + 0.6 \text{ m}$$

$$= \frac{1.20}{2} + 0.6 = 1.20 \text{ m say } 1.30 \text{ m}$$

(i) Inverted filter. Required volume for 1 m length = D m^3 = 1.30 m^3. Provide 50 cm filter thickness and above it 50 cm concrete block. Thus provide total thickness of 1 m.

$$\text{Length of the filter} = \frac{1.30}{1} = 1.30 \text{ m}$$

Provide 80 × 80 × 50 cm cement concrete blocks in two rows over 50 cm graded filter.

(ii) Launching apron

Volume of stone blocks = 2.25 *D/M*

$$= 2.25 \times 1.13 = 2.92 \text{ m}^3/\text{m}.$$

Adopt 0.80 m as the thickness of apron.

$$\text{Length of apron} = \frac{2.92}{0.80} = 3.65 \text{ m}$$

Adopt launching apron 4 m long.

Provide a toe wall between filter and launching aprom 40 cm thick and 1 m depth.

For details see Plate 21.4.

21.13 METER FLUMES

It is a very good device of measuring discharge in the canals. The normal section of the canal is reduced by providing converging slope of say 1 : 1 to the *U/S* side wall. After crest the side walls are giving diverging slope from 1 in 3 to 1 in 10. Discharge passing over a flume is measured by following formula,

$$Q = 1.70\, LH^{3/2}$$

where L is the effective length of the crest and H is the head causing flow. Discharge can be actually worked by using following formula also

$$Q = \frac{CaA}{\sqrt{A^2 - a^2}} \sqrt{2g(H-h)}$$

where C = A constant whose value varies from 0.65 to 1.0.

A = Area of normal canal section.

a = Area at throat of the flume.

H = Depth of water in normal canal section.

h = Depth of water at throat.

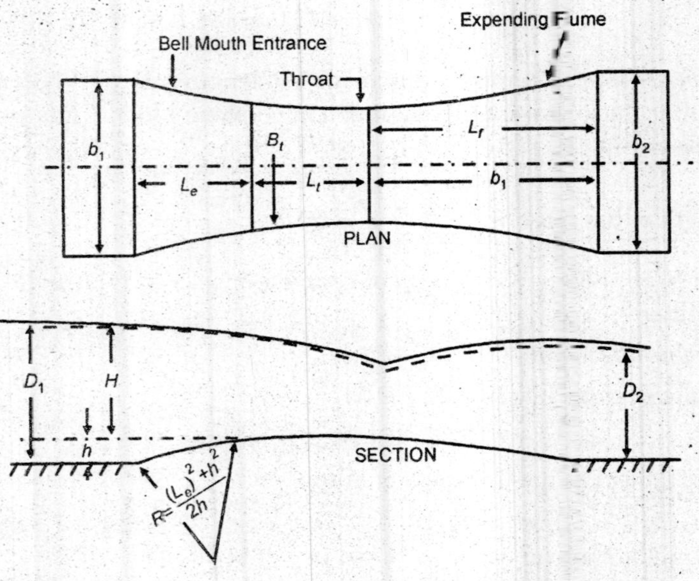

Fig. 21.19. *Standing wave flume.*

21.14 STANDING WAVE FLUME

It is also a meter flume which is designed for maximum discharge and thus implies critical conditions at the throat. All the elements of a standing wave flume are shown in Fig. 21.19. Various dimensions should be as follows:

Height of hump (h) $= \dfrac{D_1}{5}$

Radius of hump $(R) = \dfrac{L_e^2 + h^2}{2h}$

Width at throat $(B_t) = \dfrac{b_1}{3}$

Length of bell, mouth entrance $L_e = 1.87\,H$

Head over hump crest $= H$

Length of throat $L_t = 2H$

Radius of bell mouthR_1 $= 2H^{3/2}$

Length of expanding flume L_f with 4 : 1 splay $= 2\,(b^2 - B_t)$

Bed width of channel *D/S* $= b_2$

Bed width of channel *U/S* $= b_1$

Slope *D/S* of throat $\dfrac{h}{L_f}$ = 1 in 20

Radius of curvature of expanding flume $= 10\,H$

21.15 CANAL ESCAPES

Canal escape is a head regulator type structure used to extract either only excess water which has got entered into the canal by mistake or for whole of canal water in case of breach. Escape is installed at such a location where canal happens to

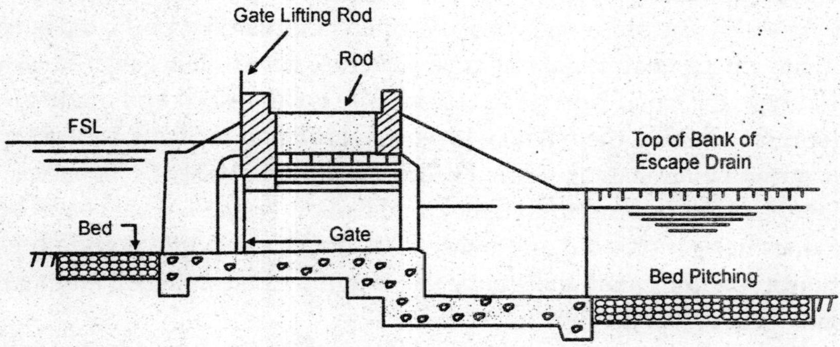

Fig. 21.20. *Canal escape.*

cross a natural drainage. The discharge from the escape is carried either to low lying areas or back to the river on *D/S* side with the help of a natural drain. Following are the circumstances when escapes have to be used.

(i) When cultivators suddenly close their outlets, the excess water from the canal is extracted through escape.

(ii) During breach water from the canal has to be immediately drained off so as to decrease water pressure at the breach. It is done by escape.

(iii) By mistake or otherwise, when excess discharge than the design discharge has entered the canal.

(iv) When due to rains in upper regions, excess rain water has entered the canal.

In all the above circumstances, the main purpose of escape is to extract excess discharge from the canal.

Theoretically the discharging capacity of escape should be equal to full discharge of the canal, but on practical grounds, the capacity is limited to half the discharge of the channel at that point.

Since escape conveys water from the water shed (canal runs on water shed) to the drain, its alignment should be chosen to take advantage of low contours. The velocity of flow in the drain should however be not allowed to reach the scouring velocity.

Escapes can be classified into following three categories:

(i) Canal scouring escape

(ii) Surplus escape

(iii) Tail escape

The escape provided for the purpose of scouring the silt from the canal is known as *canal scouring escape*. Such escape is a gated escape. If purpose of the escape is to extract only surplus water from the canal then that escape is known as *surplus escape*. Surplus escape may be gated or ungated type. Gated surplus escape has pucca head regulator type structure fitted with gates. In case of ungated type the canal bank at the escape site is depressed and made pucca. The depressed bank is then filled with soil and full fledged bank is developed. Under normal circumstances canal keeps on running between its hanks. But whenever canal is to be emptied, the canal bank is opened. Canal bank being pucca does not get scoured and hence does not develop into a wide breach. After emptying the canal, the escape site is again filled with soil and again a safe bank is developed.

Tail escape is such an escape which is installed at the tail of the canal. Its job is to discharge off excess water from tail of the canal so as to maintain F.S.L. at the tail. The structure is weir type with its crest level at the F.S.L. of the canal at the tail.

The channel which leads the surplus water from the escape to the natural drain is known as *escape channel*.

21.16 SILT CONTROL AT THE HEAD

In order to prevent excess entry of silt into the main canal from rivers, measures like silt excluder and silt ejectors are adopted. These measures have already been discussed earlier. To prevent excess entry of silt from one canal to off taking canal, following measures at head regulator should be adopted.

1. Raised Sill. The level of head regulator of off-taking canal should be kept 30 cm to 1 m above the bed level of the parent channel. This will exclude the bottom Silt laden layers of parent channel from entering the off-taking channel and thus silt entry will be reduced. But this measure will prove helpful only if disturbance in water is very small. But normally when water takes turn to off-taking channel, disturbance is bound to occur and silt of bottom layers, comes into upper layers.

2. King's Vanes. These are curved vertical concrete walls, constructed in parent channel just *U/S* of the head regulator at small intervals. If depth of water in parent channel is say d, the *c/c* distance between successive walls should

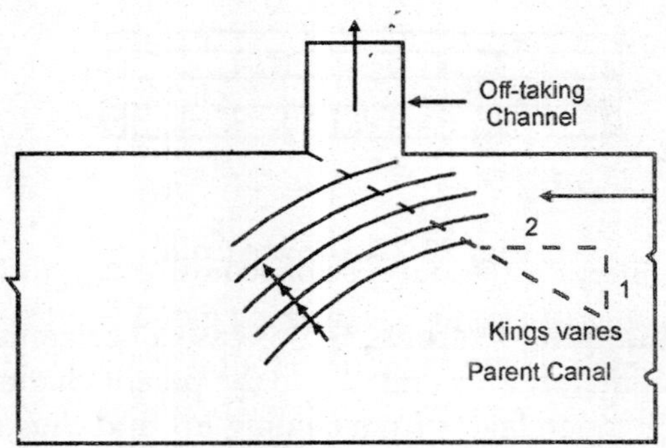

Fig. 21.21. *Kings vane.*

be $d/2$ and height of these walls should be from $d/4$ to $d/3$. The curvature of the walls should be away from the head regulator. Silt laden bottom layers of parent channel enter the spaces between these walls and are carried away from the head regulator. This measure helps in preventing silt entrance into the off-taking canal. See Fig. 21.21.

3. Gibb's Groyne Walls. *U/S* wing wall of a head regulator, lying on *D/S* side of the parent channel is extended into the parent channel in such a way that only that much water is trapped which is required to run the off-taking canal. This extended wall is always curved and is known as Gibb's groyne wall. Since disturbance in the water is reduced, the silt entry into the canal is reduced. See Fig. 21.22.

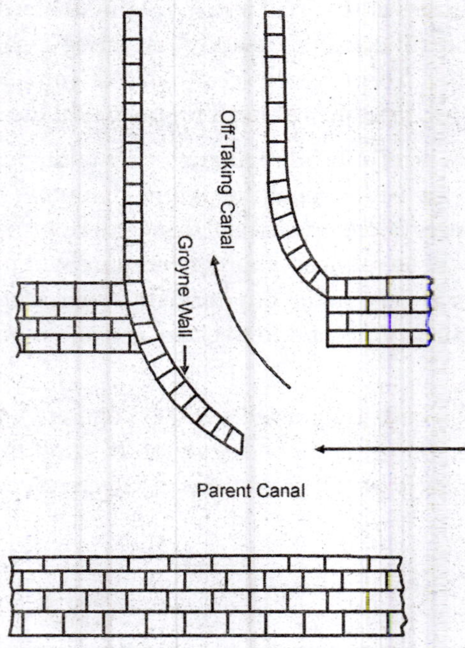

Fig. 21.22. *Groyne wall.*

4. Cantilever Skimming Platform. In this case a horizontal slab at level of crest of the head regulator, is extended into the parent channel. This measure prevents bottom silt laden layers from coining up, and thus silt entrance into off-taking canal is reduced (Fig. 21.23).

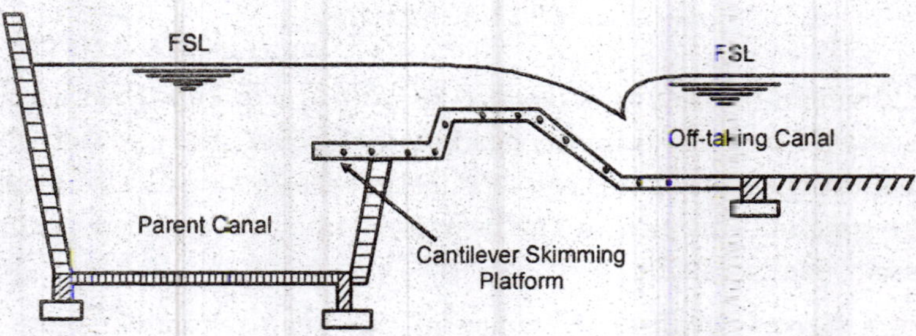

Fig. 21.23. *Skimming platform.*

QUESTIONS

21.1. Enumerate various canal regulation works and discuss in brief the purpose of each.

21.2. Why falls are provided ? Name various falls and write down their characteristics in brief.

21.3. Classify the falls according to mode of dissipation of surplus energy. What are cistern elements ? Explain how cistern is designed.

21.4. Write the procedure of designing Sarda type canal.

21.5. Explain the procedure of designing straight glacis fall.

21.6. What do you understand by head regulator and cross-regulator ? State their functions separately.

21.7. Explain the procedure for designing the head regulator of a distributary.

21.8. Write short notes on the following

(i) Canal escape

(ii) Metre flume

(iii) Cistern

(iv) Montagu type fall

(v) Inglish fall.

21.9. State various silt control measures at head regulator of a distributary.

❑❑❑

22

Irrigation Outlets

22.1. INTRODUCTION

Outlet is a small structure or device, built at the head of the water course to facilitate flow of water from distributing channel to the water course. The water course receiving water from outlet leads it to the fields for irrigation purposes. We may call outlet, as a head regulator for the water course. Government is responsible for maintenance and running of main canal, branch canals, distributaries and minors. Outlets are also controlled by the government. But responsibility of maintenance of water courses is of farmers. Outlet is a device which connects the government controlled channel with farmer controlled water course.

22.2 REQUIREMENTS OF A GOOD OUTLET

An outlet should fulfil the following requirements:

1. The outlet should be simple in design, construction and working.

2. Farmers should not be in a position to tamper with the outlets easily. If tampered, it should be easily detectable.

3. It should function efficiently with very small working head.

4. It should be cheap and easy in initial construction as well as in maintenance.

5. It should draw its fair share of sediments from the parent channel, which may be minor, distributary, or branch.

6. From farmers point of view, the outlet should give constant discharge, irrespective of change in discharge of parent channel.

7. From distribution point of view, the outlet should draw only the proportional discharge from the varying discharge of the parent channel.

8. It should not be easily clogged by floating or suspended matter, in the parent channel.

9. It should not weaken the bank of the canal.

22.3 CLASSIFICATION OR TYPES OF OUTLETS

The outlets may be divided into following three categories:

1. Non-modular outlet
2. Semi-modular outlet
3. Modular or rigid outlet.

1. Non-modular Outlet. It is such an outlet whose discharge is dependent upon the difference in levels of water in the distributary channel and the water course. In this, discharge varies with the water levels in the distributary as well as the water course. Submerged pipe outlet, masonry or orifice outlets, are the examples of non-modular outlets.

2. Semi-modular Outlet. It is also known as flexible outlet. The discharge of this outlet is dependent upon the water level in the supply channel only. Fluctuations in the water level and in the water course do not in any way affect the outlet discharge. Free discharging pipe outlet, Kennedy's gauge outlet, crumps open flume outlet, are the common examples of semi-modular outlets.

3. Rigid or Modular Outlet. It is such an outlet, which within minimum working limits of head, always gives constant discharge. Fluctuation in levels of water in distributing channel and water course do not affect the discharge whatsoever. Gibb's rigid module is the example of rigid outlet.

22.4 TYPES OF NON-MODULAR OUTLETS

This outlet consists of a pipe circular or rectangular in shape. It may be in form of an open sluice also. Mostly pipe outlet is used as non-modular outlet. Its face, discharging water into the water course remains submerged into water. Pipes are laid in concrete with a masonry wall at its *D/S* face. Sometimes there may be masonry wall at *U/S* face of the pipe also. Being embedded in concrete for full length and also because of masonry wall at its *D/S* face, the outlet cannot be tampered easily. This outlet does not distribute water equitably. Low lying fields lower the water level in the water course and thus discharge from the outlet is increased, more head being available. On the contrary higher fields do not get water of their share, as when water is issued to them the water level in water course increases and consequently head causing flow is reduced. Discharge from this outlet is worked out from formula $Q = CA\sqrt{H}$,

where C is constant whose value for average conditions is taken as 2.75, A is the sectional area of the pipe and H is the total loss of head, which is calculated by following formula

$$H = \left(1.5+\frac{4fl}{d}\right)\frac{V^2}{2g}$$

where f is coefficient of friction, l is the length of the pipe in metres, d is diameter of the pipe in metres and V is the velocity of flow in m/sec through the pipe. A table used by U.P. Irrigation Deptt. has been given in Table 22.1. In this table l has

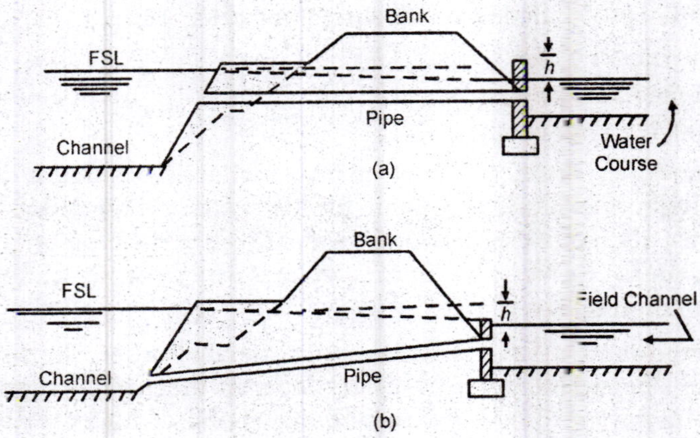

Fig. 22.1. *Pipe outlet.*

been considered 6 m. If length of pipe outlet exceeds 6 m next higher pipe diameter should be used. Silt discharge through the pipe outlet can be improved by keeping the *D/S* end at the same level but lowering the *U/S* end as shown in Fig. 22.1 (b).

Table 22.1. *Recommended size of outlets*

Pipe diameter	*Average discharge in cumecs*			
		Outlets with submerged outfall		
	Free overfall	*0 to 20% lift*	*21 to 50% lift*	*Over 50% lift*
(1)	(2)	(3)	(4)	(5)
2 × 15 cm	0.050	0.039	0.033	0.024
15 cm + 12.5 cm	0.044	0.035	—	—
15 cm + 10 cm	0.036	0.028	0.024	0.018
15 cm + 7.5 cm	0.031	0.024	—	—
15 cm	0.025	0.015	0.017	0.012
12.5 cm	0.019	0.010	—	—
10 cm	0.011	0.009	0.008	0.006
7.5 cm	0.006	0.0045	—	—

22.5 TYPES OF SEMI-MODULES

In semi-module outlets, the discharge varies with water level in distributing channel, but is independent from water level in the water course. The semi-modules are of following types:

1. Pipe outlet free fall
2. Kennedy's gauge outlet or venturi flume outlet
3. Open flume outlet
4. Orifice semi-modules.

1. Free Fall Pipe Outlet. It is one of the simplest types of semimodule outlets. The essential feature of this outlet is that it must discharge freely in air. It has high efficiency and its silt conducting power is good. Such an outlet can be provided only if there is sufficient difference between water levels in parent channel and water course. The best setting of this outlet is 0.3. But actually outlets are set lower than this and hence they are sub-proportional.

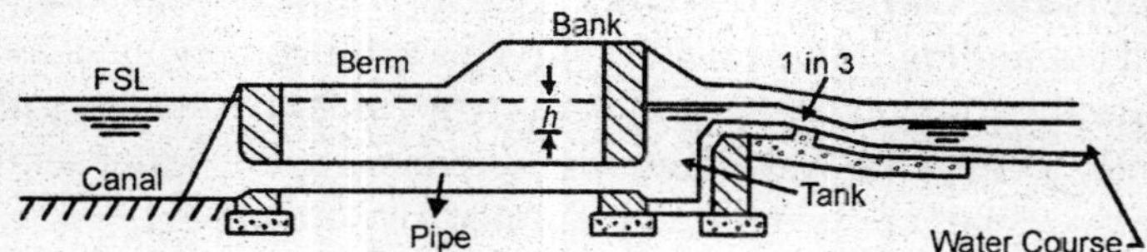

Fig. 22.2. *Semi-module pipe oulet.*

2. Venturi Flume Outlet. This was invented by Mr. R.G. Kennedy and its modified form is known as Kennedy's gauge outlet. It is made of cast iron and consists of an orifice with bell mouth entry, a long expanding delivery pipe, and an air vent pipe connected to the throat of the delivery pipe.

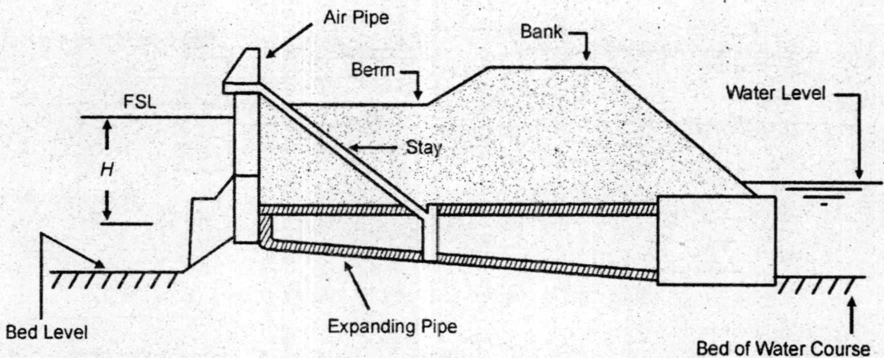

Fig. 22.3. *Kennedy's gauge outlet.*

Because of air vent pipe, the jet of water at throat remains at atmospheric pressure and as such the fluctuations in the water level in the water course do not affect the discharge being issued from the throat. The discharge of this outlet is found out by following formula.

$$Q = CA\sqrt{2gH}$$

Q = discharge of the outlet

C = constant of discharge whose value may be as high as 0.97

H = head measured from F.S.L. of the parent channel to the centre of the throat pipe.

A = area of cross-section of the pipe at throat.

In this particular outlet, the minimum head should be 0.22 H.

This outlet did not find much use as its discharge can be easily increased by the farmers by closing the airvent pipe. It is also a costly outlet, and easily tamperable by the users.

3. Open Flume Outlets. This outlet consists of a weir, with long constricted throat and expanding flume on *D/S* side. This arrangement ensures formation of the hydraulic jump on *D/S* side and thus relieves the dependence of discharge on the water level in the water course and rendering the outlet semi-modular. There are two types of open flume semi-modular outlets.

(i) Crump's open flume outlet, and

(ii) Punjab open flume outlet.

(i) Crump's open flume outlet. The main feature of this outlet is that wall lying on *D/S* of the flow in the parent channel is set projecting in the parent channel

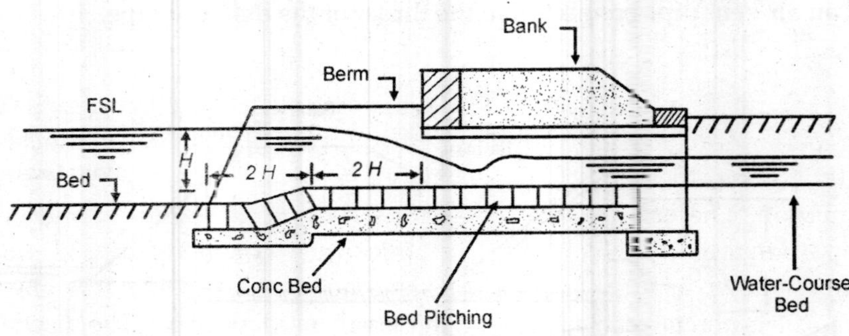

Fig. 22.4. *Open flume outlet (Punjab).*

by *d* distance. This measure accelerates the silt conduction capacity of the outlet. Length of the constricted throat is kept 2.5 H. Discharge of this outlet is found out by following formula. Sketch of this outlet is similar to Fig. 22.4 except that throat length is kept 2.5 H instead of 2 H.

$$Q = CB_t H^{3/2}$$

where B_t = width of throat in metres

C = Coefficient of discharge whose theoretical value is 51.71

H = Head in metres over the crest

Q = discharge in cumecs.

This outlet works as proportional oulet when it is set at $\frac{H}{D} = 0.9$. If set higher than this it works as hyper-proportional. It is a quite efficient outlet. Value of C varies from 1.60 for B varying from 6 cm to 9 cm width to 1.66 for B above 12 cm.

(ii) Figure 22.4 shows Punjab open flume outlet. It is slightly modified form the crump's open flume outlet. In this outlet *U/S* wall of the outlet is curved for greater length and length of throat is reduced to $2H$ instead of 2.5 H. Formula for discharge is same as given for crump's open flume outlet.

4. Orifice Semimodule Outlet. This outlet consists of an orifice, developed with the help of a roofing block. The main features of this outlet are similar to a flumed regulator. It consists of a horizontal crest and curved approach on *U/S*

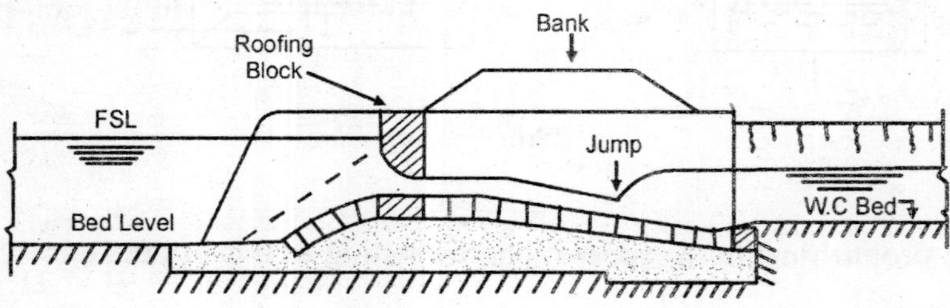

Fig. 22.5. *A.O.S.M.*

side. *D/S* side is gradually expanding flume type. The flow through the orifice is hyper-critical and thus develops a jump on *D/S* side. The water level in the water course does not affect the discharge of the outlet because of jump formation. The earliest form of this type of outlet is crump's adjustable proportional module (A.P.M.). This outlet was later modified and modified form is known as adjustable orifice semi-module (A.O.S.M.). It is one of the best outlets and is very much used in Punjab and Rajasthan. The discharge of the outlet is found from the following formula

$$Q = 4.04\, B_1\, V \sqrt{H_s}$$

where Q = discharge in cumecs

V = vertical height of the opening

B_1 = width of the throat

H_s = Head measured from water surface to the soffit of the opening.

N.D. Gulati gave following relation for minimum modular head (M.M.H.).

$$\text{M.M.H.} = 0.82\, H_s - 0.5\, B_1.$$

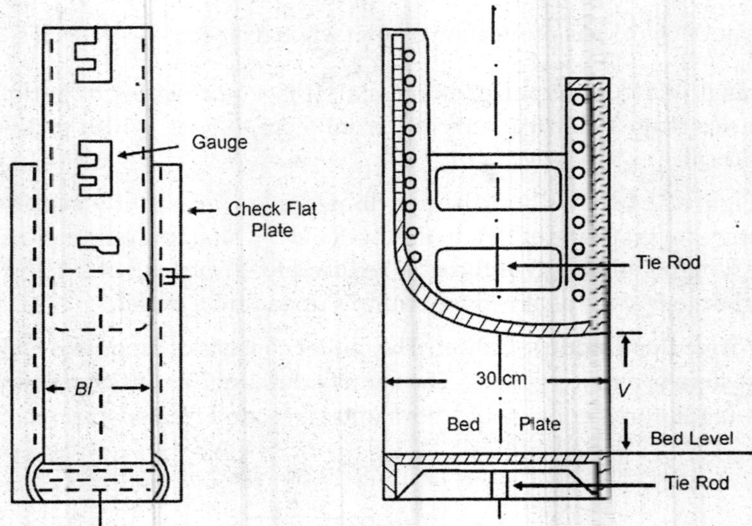

Fig. 22.6. *Roofing Block.*

For proportionality consideration, the setting of this outlet i.e. $\frac{H_s}{D}$, should be 0.3.

The roofing block of the outlet consists of a set of cast iron straps in which bolts are fitted. The opening of the outlet is adjusted with the help of C.I. straps. Because of straps, tampering of the roofing blocks becomes almost impossible and even if there is tampering it can be easily detected.

22.6 RIGID MODULES

There are several types of rigid modules but Gibb's rigid module is the most common type. Although Khanna also developed a rigid module but that did not work and hence not used.

1. Gibbs Module. The module consists of bell mouthed inlet pipe and a curved rising rectangular trough, known as Eddy chamber The rising circular trough is semi-circular in plan. A number of baffles are provided in the eddy chamber with their lower edges sloping at the required height above the bottom. The baffles are provided to prevent increased discharge passing through the module. The water, after entering the inlet pipe through bell mouth, is led to eddy chamber where it develops into a free vortex flow. The characteristic of free vortex flow is that the product of the velocity and radius remains constant at all the points in the circular motion of the eddy chamber. The water level at the outer surface of the eddy chamber remains higher and water surface thus remains sloping towards the inside. See Fig. 22.8.

When flow into the module is due to increased head, water banks up at the outer circumference of the eddy chamber, and strikes against the baffle walls. Thus excess energy due to increased head is dissipated and discharge through the module remains constant. The number of baffle walls coming into action depends upon the increase or decrease in head. The angle of eddy chamber varies from a semi-circle to $1\frac{1}{2}$ turns, depending upon the discharge and range of working required for the module. The discharge of the module is found from following formula :

$$Q = r_0\sqrt{2g}\,(d_1 + h_0)^{3/2}\left\{\frac{m^2-1}{m^3}\log e^m + \frac{1}{m}\log e^m - \frac{m^2-1}{m^3}\right\}$$

where Q = discharge

r_0 = outer radius of the eddy chamber

r_1 = inner radius of the eddy chamber

$m = \dfrac{r_0}{r_1}$.

d_1 = depth of water at inner circumference

h_0 = head loss in inlet pipe.

This formula is valid for the design given by Mr. Gibbs. He adopted $m = 2$ and $\dfrac{h_0}{D} = 7$.

D = difference in level measured from minimum water level in the distributing channel to the floor of eddy chamber.

2. Khanna's Rigid Module. This module is shown in Fig. 22.7. It is nothing but an orifice semi-module outlet. In addition, it consists of inclined shoots in

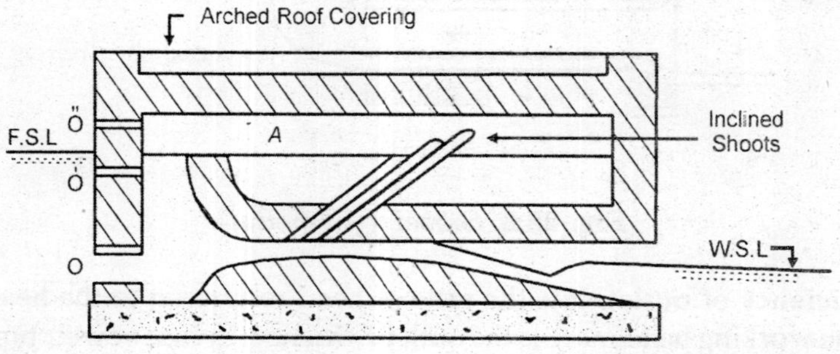

Fig. 22.7. *Khanna's rigid module.*

roof, which remain connected to the throat of the outlet. When head causing flow is small the outlet works as A.O.S.M. When level in the parent channel rises above predetermined level, the water level in the differential chamber also rises and water starts flowing in the inclined shoots. The water flowing through inclined shoots, strikes against the flow passing through the throat and dissipates the extra energy and thus discharge through the outlet is maintained constant. If more rise in water lever, in distributing channel takes place more inclined shoots come into action and excess energy due to increased head, dissipated. Inclined shoots are sloped at 1 in 2 slopes. This rigid module is not much in use as shoots can be easily closed and increased discharges can be obtained by the farmers. Inclined shoots may even get clogged.

22.7 SOME IMPORTANT DEFINITIONS

In order to understand the behaviour and performance of the outlets, following terms should be thoroughly understood.

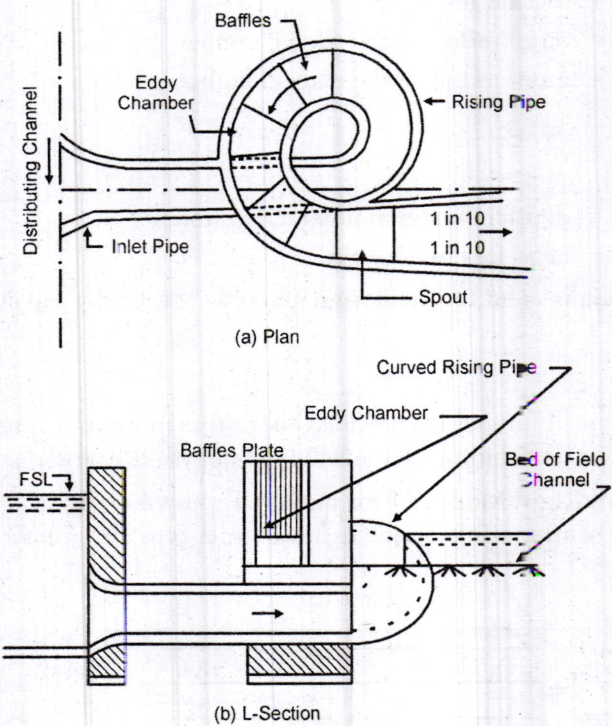

Fig. 22.8. *Gibbs' rigid module.*

1. Efficiency of outlet. It is the ratio of head recovered to the head put in. Lesser the working head more is the outlet efficiency. In case of weir type outlets the efficiency is same as drowning ratio.

2. Drowning ratio. The ratio between the depths of water over outlet crest, on *D/S* and *U/S* of module is known as *drowning ratio.*

3. Minimum modular head (M.M.H.). It is the minimum difference between the *U/S* and *D/S* water levels of the outlet which is required to enable the module to pass the designed discharge. In other words it is the minimum head which is essential so that module may do its intended work.

4. Modular range. It is the range of various factors within which a module or semi-module works as designed.

5. Modular limits. The upper and lower limits of any one or more factors beyond which a module ceases to function as desired by the design.

6. Flexibility. The ratio between the rate of change of discharge of an outlet to the ratio of change of discharge of the distributing channel is known as the *flexibility*. If q is the discharge of outlet and Q that of parent channel, the flexibility (F) by definition is given as follows:

$$F = \frac{\frac{dq}{q}}{\frac{dQ}{Q}} = \frac{Qdq}{qdQ}$$

If H is the working head of the outlet, then q may be expressed as $q = kH^m$ where k and m are dependent on the type of outlet.

$$q = kH^m$$

$$dq = mkH^{(m-1)}\, dH$$

$$\therefore \quad = \frac{mkH^{(m-1)}dH}{kH^m} = \frac{mdH}{H} \qquad (1)$$

Similarly discharge flowing in the parent channel can be expressed as

$$Q = CD^n.$$

$$dQ = nCD^{n-1}\, dD$$

$$\frac{dQ}{Q} = \frac{nCD^{n-1}dD}{CD^n} = n\frac{dD}{D} \qquad (2)$$

$$\therefore \text{ Flexibility} \quad F = \frac{dq}{q} \Big/ \frac{dQ}{Q} = \frac{mdH}{H} \times \frac{D}{ndD}$$

$$= \frac{m}{n}\,\frac{D}{H}\,\frac{dH}{dD} \qquad (3)$$

But change in head (dH) would be same as change in depth (dD) of the parent channel; hence $dH = dD$. Substituting this value in (3), we get

$$F = \frac{m}{n}\frac{D}{H} \tag{4}$$

7. Proportional Outlet. The outlet whose flexibility is one is known as proportional outlet. In other words rate of change of discharge of the outlet is same as that of parent channel.

$$\frac{m}{n} \times \frac{D}{H} = 1$$

or
$$\frac{H}{D} = \frac{m}{n}.$$

For a trapezoidal channel, value of n is $\frac{5}{3}$ and for outlet m is generally $\frac{1}{2}$.

$$\therefore \quad \frac{H}{D} = \frac{1}{2} \times \frac{3}{5} = 0.3$$

$$H = 0.3D$$

Hence for an outlet to be proportional it must be set at 0.3 D from the surface of water in the parent channel.

8. Hyper-proportional Outlet. The flexibility of this outlet is more than one. In other words the rate of change of discharge of the outlet is more than that of the parent channel.

$$\therefore \quad \frac{m}{n}\frac{D}{H} > 1$$

or
$$\frac{H}{D} < \frac{m}{n}.$$

This shows that the outlet is set at higher level than the setting, on the basis of proportional outlet

9. Sub-proportional Outlet. In this case, flexibility of the outlet is less than one.The rate of change of discharge of the outlet is less than that of the parent channel for varying head of water.

$$\frac{m}{n}\frac{D}{H} < 1$$

$$\frac{H}{D} > \frac{m}{n}.$$

The outlet will be sub-proportional if it is set lower than the setting that comes on the basis of proportional outlet.

10. Setting of the Outlet. It is ratio between head (*H*) of water causing flow through the outlet and depth (*D*) of water in the parent channel. Both *H* and *D* correspond to the F.S.L. of the parent channel.

$$\frac{H}{D} = \text{setting.}$$

11. Sensitivity. It is the ratio between the rate of change of discharge in the outlet and the rate of change of level of water surface in the parent channel, in reference to its normal depth of water. Sensitivity (*S*) can be expressed as follows.

$$S = \frac{dq}{q} \Big/ \frac{dG}{D}$$

where S = Sensitivity

G = Gauge reading

dG = dD because change in gauge reading is same as change in depth of channel.

$$\therefore \quad S = \frac{dq}{q} \Big/ \frac{dD}{D}$$

But $$F = \frac{dq}{q} \Big/ \frac{dQ}{Q} \text{ when } \frac{dQ}{Q} = n\frac{dD}{D}$$

$$\therefore \quad F = \frac{dq}{d} \Big/ n\frac{dD}{D} = \frac{S}{n}$$

or $$S = nF$$

Sensitivity of a rigid module is zero.

22.8 TAIL CLUSTER

As the canal goes on flowing from head towards tail its discharge goes on reducing, as outlets are continuously withdrawing water. At the tail of the canal when canal carries only discharge for two or three outlets, the discharge is distributed to the area with the help of open flumed type bunch of outlets. This bunch of outlets is known as tail cluster. Figure 22.9 shows a bunch of three open flume outlets.

The discharge in the canal at tail remains fluctuating. If some outlets in the head reach are taking excess discharge or some how irrigation water is being stolen by siphons or otherwise, the water reaching tail will be very small. On the contrary when head reach outlets are drawing less than the design discharge, excess amount of water reaches the tails. Hence uncertainty about the availability

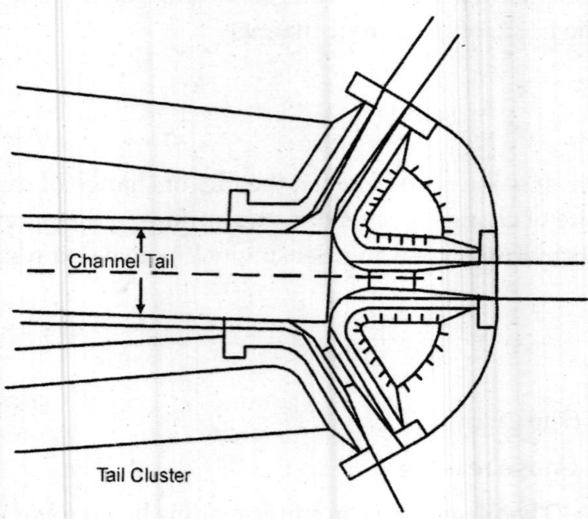

Fig. 22.9. *Tail cluster.*

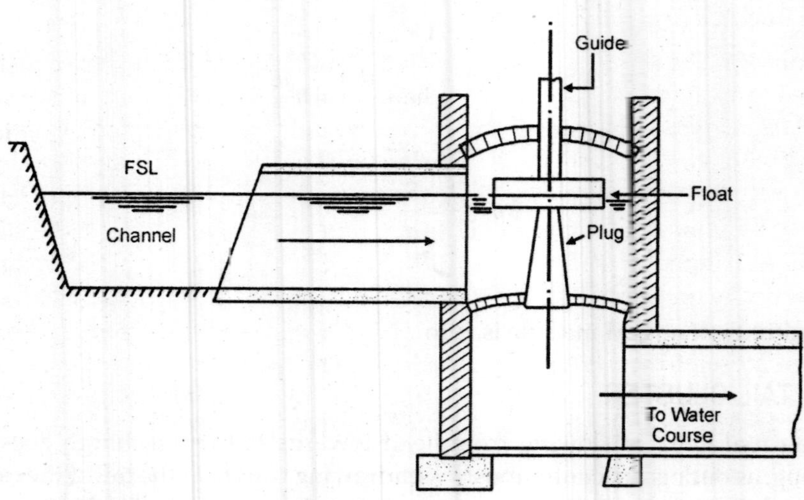

Fig. 22.10. *Spanish module.*

of irrigation water at tail always remains and hence no farmer is prepared to purchase lands at tails. In other words the cost of the land is generally very low at tails in relation to upper regions. To deal with excess water reaching tails a

Tail escape is also provided. Tail cluster takes its designed discharge or even more if required, and the remaining excess water is let loose in the open low lying land through Tail escape as shown in Fig. 22.11.

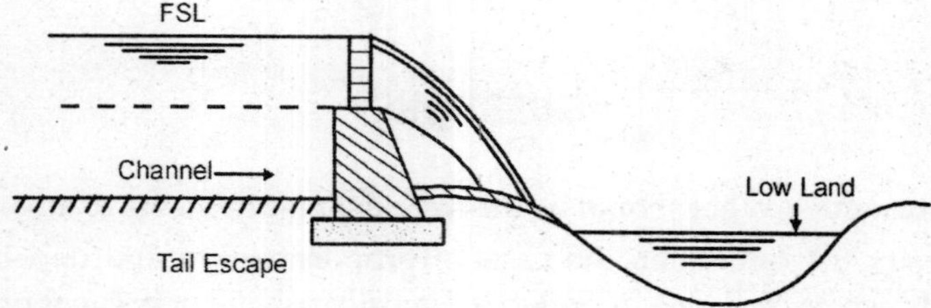

Fig. 22.11. *Tail reservoir.*

22.9 SELECTION OF OUTLET

While selecting the type of outlet the paramount consideration in the mind should be, that water is equitably distributed to the farmers. Besides this, assured suppliers of water should always be available. This develops confidence into the farmers and they can plan their crop patterns well in advance. From this point of view rigid modules are considered as the best choice. But this is not practicable. When canal has full supplies it can feed all the outlets with designed discharge, but when supplies in the canal are less than designed, only head reach outlets will receive water and all the outlets on tail side will go dry. This causes unequal distribution of water.

Secondly on all the new canals, position of outlets may have to be frequently changed to achieve better locations. But rigid modules once set cannot be easily shifted. Hence from this consideration non-modular outlets should be initially installed and after some years of run of the canal, when area gets stabilized and outlet positions are almost finally fixed, the non-modular outlets should be replaced by semi-modular outlets. Rigid modules also do not draw the silt of their share. It is one of the biggest drawbacks with rigid modules. Rigid modules are practically not used any where. Mostly non-modular outlets are installed in the beginning and after some time when conditions become stabilized they are replaced by semi-modules, predominantly A.O.S.M. type.

22.10 SPANISH MODULES

It is rigid module operated with the help of a float. The head of water over the throat or opening is maintained constant with the help of a float. The outlet consists of an entrance pipe and a float chamber. The discharge given by this outlet is $A \times V$. The float remains connected to a needle valve. See Fig. 22.10.

Tail Reservoir. Sometimes instead of directly tailing the canal by Tail cluster, a tail reservoir is constructed. Canal water and excess water reaching at the tail, both get collected in the reservoir. The outlets are given from this reservoir.

The reservoir eliminates the fluctuations in the discharge at tail cluster and excess stored water from the reservoir can be used to augment the discharge at the time of low discharges at tail. In this way, the water which would have escaped through tail escapes, and gone waste, is economically used.

QUESTIONS

22.1. Define outlet. What should be the essential requirements of an outlet?

22.2. Classify the outlets on the basis of proportionality and dependence of discharge. Which type of outlets are mostly used in new canal projects ?

22.3. Classify various types of semi-modules and given main characteristics of each.

22.4. Explain Gibbs' modular outlet, and non-modular outlet with the help of neat sketches.

22.5. Explain the terms : Flexibility, Tail cluster, setting of outlet efficiency, minimum modular head (MMH).

❑❑❑

23

Maintenance of Canals

23.1 INTRODUCTION

The irrigation projects have to be continuously maintained, so that they may perform their functions efficiently. If they are not maintained properly, the canals of the systems may get silted up. Weed growth may also occur. Capacity of the canal is reduced both by silting and weed growth and such a canal will not draw its full discharge and as such will not be in position to feed all its outlets. The canal may be scouring at some sections and if preventive measures are not taken in time, the canal may breach and cause lot of wastage of irrigation water. Hence canal system cannot work efficiently unless it is maintained properly.

23.2 BED BARS

Bed bars should be constructed at suitable intervals at the bed of the channels. The purpose of bed bars is to indicate the correct theoretical levels of the bed of the channel. They are constructed at an interval of say 200 m in case of small channels and 500 m in case of main and branch canals. In case of small channels the whole section of the channel is lined at intervals but in case of big canals, concrete blocks are used at mid-width of the canal bed. Bed bars indicate both silting as well as scouring of the canal. If channel section is scouring the bed bars will get exposed. The exposure of the bed bars indicates the extent of scouring. If the channel is silting the bed bars will get embedded in the silt. Height of silt above bed bars indicates the extent of silting.

When silt clearance is done, bed bars act as reference levels. The bed bar blocks should be sufficiently embedded in bed so that they are not unearthed by scouring. When complete profile is used as bed bar, the slope of the channel is also indicated by the slope of the profile. Bed bar profile should be properly anchored to the soil with the help of cut-offs as shown in Fig. 23.1 (b).

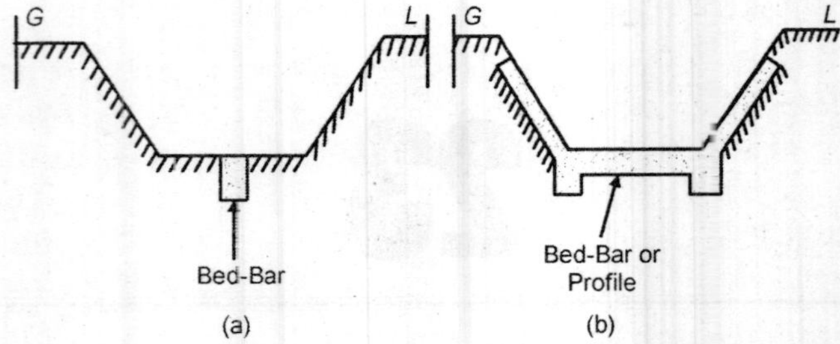

Fig. 23.1. *Bed bars.*

23.3 MAINTENANCE OF IRRIGATION CHANNELS

For the efficient working of an irrigation system, all the canals and connected works should be properly maintained. Following difficulties may hamper proper functioning of the irrigation canals.

1. Silting of the canals
2. Breaches in the canals
3. Weed growth in canals.

1. Silting of canals. The capacity of the canal is reduced if its bed and banks are silted up. The best method of preventing the silting, is to adopt the measures that would not allow excess silt to enter into the canal. Silt excluders, silt ejectors, raised crest, King's vanes etc. are the measures which can be adopted in this respect. But it is seen by experience that in spite of all these measures, some silting definitely takes place. Silting may not be taking place when canal is run at full capacity but will silt when it is run at less than full capacity. The following measures may be adopted to remove the deposited silt.

(i) By increasing the velocity of flow in the canal. For this either additional water will have to be admitted or canal escapes may be opened from time to time to develop increased velocity and scour out the deposited silt.

(ii) Loaded boats should be put in canal. It will decrease the area of cross-section of the canal and consequently velocity of flow will be increased and deposited silt scoured off.

(iii) *Dredging.* Dredger may be used to remove the deposited silt from the canal. Though it is a costly method but it can be employed where other methods fail to operate.

(iv) Silt is removed by manual labour. Silt clearance by manual labour is done during annual closer periods of the canal. If too much silting has taken place the canal may be closed for silt clearance.

(v) Grass which has grown into the canal at banks, should be cut from time to time.

(vi) The Bed and banks may be raked by iron bar rakers or thorny bushes. The raked silt is carried away by flowing water.

2. Breaches in the canal. The canals may breach in reaches of heavy filling. Breach in heavy filling reach is very serious, as breach develops into a very large opening at such places. At filling reaches, very large head of water is available and as soon as a certain breach occurs, it develops into a big gap, and lot of bank length is scoured due to water being passed through the breach with very large velocity. Canal breaches may occur due to following reasons.

(i) Rats may dig through holes in the canal banks and cause canal breach.

(ii) The seepage line at the outer toe of the bank may get exposed. This exposure will cause escape of seepage flow ultimately leading to canal breach.

(iii) If canal bank is weak or is defectively constructed, the breach may take place.

(iv) Piping at *D/S* toe may cause settlement of canal bank and water may flow over the bank of the canal.

(v) Canal may be breached by deliberate cut. This mischief is generally done in small distributaries in which discharge is not much. Farmers whose lands are in the vicinity of the canal, are generally responsible for such deliberate cuts. This way they soak their lands and sow crops. This tendency of farmers is predominantly noticed when canals run very irregularly.

(vi) If canal happens to pass through or by the side of the village or town, the canal banks are generally weakened by continuous use of the canal by the people. The canal may breach if banks are not continuously looked after.

(vii) If flow in the canal is some how obstructed, the water level in the *U/S* of the canal may over top the banks, and breach may occur.

(viii) Large deep erosion pits are developed at the canal banks during rains. If they are not immediately filled with soil, breach may occur.

Breaching of canals can be prevented by strengthening the canal banks. Following are the methods of strengthening the banks of canals.

(*a*) *External silting.* In this case additional banks are constructed parallel to the main banks, by leaving some open space between them. The intervening space is divided into silting compartments by constructing cross bunds. The canal water is allowed to enter these compartments from *U/S* side, held there for some time and then discharged back to the canal. The silt in the canal water settles in the compartment and in due course of time the compartments get silted up upto

F.S.L. of the canal and thus a wide bank is developed which cannot breach on its own. Additional banks are constructed outside of the main banks. See Fig. 23.2 (a).

(b) *Internal silting system*. In this case the original banks are constructed slightly away from their intended position. Thus section of the canal becomes larger than the designed section and consequently velocity of flow is reduced from

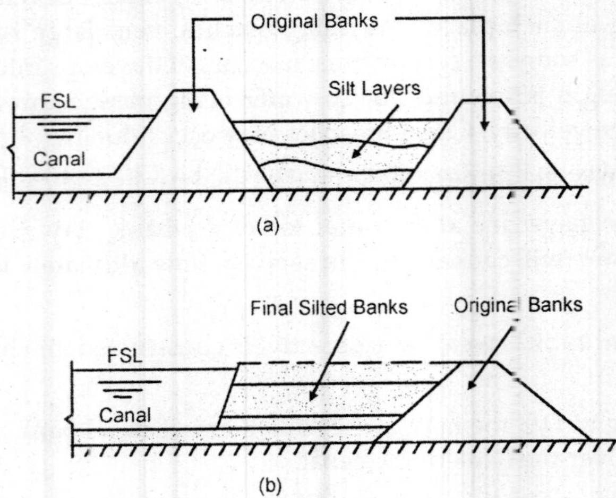

Fig. 23.2. *(a) External silting, (b) Internal silting.*

the designed velocity. Because of reduction in velocity, the banks of the canal get silted up. This process of silting continues till canal acquires the designed section. Thus by silting on the internal side of the banks, the banks become thick and thus strengthened. Sometimes, to accelerate the process of silting, low submerged spurs are constructed projecting towards the canal from banks. See Fig. 23.2 (b).

(c) *Internal berms*. Normally canal sections are excavated with 1 : 1 side slope. In due course of time, sides assume $\frac{1}{2}$: 1 slope due to silting. Such additional thickness of the berm is known as internal silted berms. This process also strengthens the canal banks.

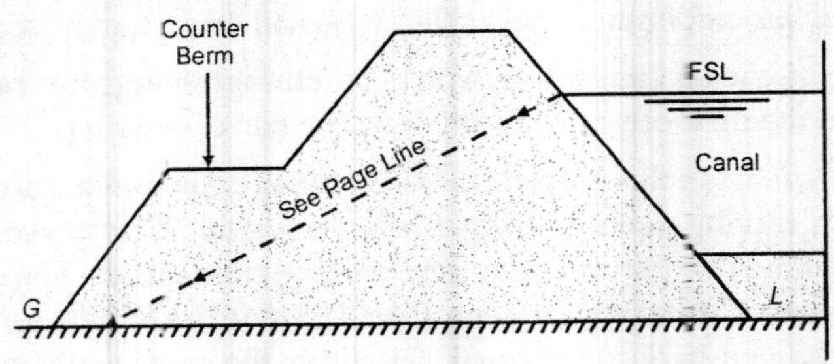

Fig. 23.3. *Counter-berm.*

(*d*) *Back berms.* If saturation line or seepage line gets exposed on the outer slope of the bank, additional soil may be deposited in form of a counter berm. Counter berms are known as back berms. See Fig. 23.3.

3. Weed growth in the canals. Aquatic weed growth hinders the flow in the canals. The result is, that adequate supplies of irrigation water cannot be delivered to the consumers. The growth of weed is very rapid. The problem of weed growth is more accuse in south India. Weeds may be annual, biennial, and perennial. Annual weeds live only for a year and then die automatically. Biennial weeds live for two years, but perennial weeds live for longer periods. Annual and Biennial weeds propagate by seeds alone and once the top is chopped, the roots have to power to regenerate. Perennial weeds are the most troublesome weeds. They can grow on canal bed, water surface and water marks. The methods of weed control can be summarized as follows :

1. Run the canals with relatively higher velocities of flow. The velocity of flow should never be allowed to fall down, 0.6 m/sec.

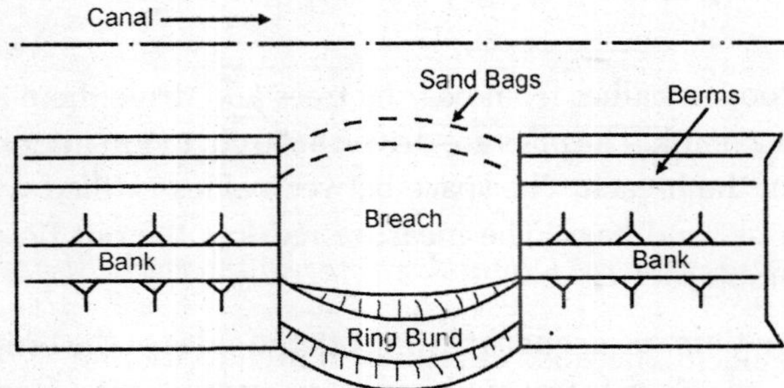

Fig. 23.4. *Closing of breach.*

2. The canal should be deep and run with full supply discharge. This will exclude most of the sun rays, and weed growth will be prevented.
3. Application of diesel oil, dilute solution of sodium chlorate, Ammate, and Benochor also destroy the vegetation.
4. During closure periods, the canals should be dewatered and dried. The dewatering of the canal can be done by providing drainage outlets.
5. The weed destructive insects if introduced in the canals will kill certain varieties of weeds. Cochined inset is one such insect.
6. The weed growth is cleared by manual labour during closure periods of canals. If weed growth is very heavy, continuous mechanical methods may be adopted to clear it.

23.4 CLOSING THE CANAL BREACH

Canal breach can be closed in following steps.

1. If the size of the channel is small, the breach can be closed without closing the discharge in the channel. For small breaches, the discharge escaping from

the breach is temporarily stopped by using wooden planks, and then breach is closed by taking soil either from outer face of the bank or from the top of the bank.

2. In case, size of the channel is moderate, the breach cannot be plugged without diverting the discharge.

3. As soon as information of the breach is received, the canal escape, *U/S* of the breach if any, is opened immediately. At the same time the discharge in the canal from head regulator is also reduced. Both these measures will cause considerable reduction in the discharge passing through the breach.

4. When discharge passing through the breach has considerably reduced the erosion of bank at breach would stop.

5. The loose and cracked soil of the breached bank is cut and solid bank is obtained at both the ends of the breach. Steps are cut in solid banks so that earth to be used for closing the gap develop good bond with old soil of the bank.

6. Cut the wooden ballies from near by trees and drive them as piles on the inside of the canal bank. The piles are driven at 50 cm to 1 m interval, depending upon the size of the breach. The space between allies is filled with either tree bushes, planks or sand bags. The measure reduces the out flow through the breach considerably.

7. In the mean time sufficient earth on both the sides of the breach is heaped. The earth may be taken from spoil banks, or borrow pits. Generally all the borrow pits are filled with water and bed of the canal heavily eroded and hence earth is not easily available. Earth is mostly taken from top of the banks and also from outer slope of the banks.

8. The closing of the gap should be started from both the sides by slipping earth from the heap in form of a ring bund. The last gap of about 1 m to 2 m width should be closed with a rush when enough earth has been collected on both the sides.

9. After having closed the gap by a ring bund which is on outer side of the bank, the open space between ring bund and pile line, is filled with earth in well compacted layers.

10. During filling grass clods should not be allowed to be embedded in the earth otherwise water will again seep along clods and cause breach again.

11. When breach is closed to sufficient height, the water in the canal is again left. By the time water level reaches the F.S.L., the breach is sufficiently strengthened.

12. Even after strengthening the breach, a continuous watch is maintained for about 10 days so that water may not seep through any defect and cause

breach again. During this period the gang keeping watch on the breach remains busy in further strengthening the breached section.

23.5 MAINTENANCE OF SERVICE ROADS

The service road is an unmetalled road constructed at left bank of the canal. The surface of this road has to be continuously maintained as it is frequently used by irrigation department to carry out inspections. This road is maintained by gangs of labour employed to maintain the canal. A particular length of the canal is entrusted to a gang of labour and all such gangs work under the guidance of a section officer. It is the duty of this gang, to maintain the canal and service road in proper order in the length allotted to it. Gang sprinkles water on the inspection road, fills pits developed by rains, removes grass, bushes and other unwanted vegetation from the road.

23.6 DUTIES OF MAINTENANCE GANG

As already stated that certain length of canal is put in the charge of certain gang. The duties of the gange, are the following :

(i) Gang daily takes a round of the length of canal in its charge and carries out all the functions that help in the efficient running of the canal.

(ii) Fills rain pits immediately after rains.

(iii) Strengthens the sections of the canal banks which has been rendered week due to certain reason.

(iv) Maintains service road in proper order by periodically levelling the roadsurface and sprinkling water.

(v) Removes unwanted vegetation.

(vi) Takes care that outlets are not tempered and siphons are not used by the farmers.

(vii) Arranges additional labour at the time of breaches.

(viii) Clears off an obstruction that might have developed in the fall or other regulation works, falling in its jurisdiction.

(ix) Daily report of the work is sent to the section officer incharge of the canal.

(x) Helps overseer in the levelling work, which is required for commanding the new areas or which may be required for fixing levels of water courses in villages.

23.7 REGULATION OF CANAL SYSTEM

The main canal draws its full supply from the river. The main canal feeds the branches, distributories and minors according to the demand of various off-taking channels and the total supplies available. The process of distribution of water into various sub-channels according to their demands and supplies available is termed as *Regulation of canal system.*

During keen demand the main canal has to run with full supplies. If demand is less than the keen demand it may be run with part supplies. Main canal is generally not closed, as it commands a very large area and some demand, in different parts of the area, always remains. When demand is limited to some particular distributory, it is run full but others which do not require water are closed. The difficulty in distribution arises when demand is more and supply available is less than the demand. In such circumstances, either all the distributories are run with reduced discharge or distributories are run with full discharge, in rotation. The later system is mostly adopted for regulation of supplies. In this method a group of channels is run with full discharge at a time and remaining are kept closed. Then after some days the group of channels which was running is closed and closed group of channels is run with full discharges. This process of regulation of supplies is known as *rostering* or *rotation*. Full control over regulation is exercised by the executive Engineer.

QUESTIONS

23.1. Enumerate the steps, for closing the breach in a canal of moderate size.

23.2. Enumerate the duties of a maintenance gang on an irrigation canal.

23.3. How breaches occur ? Give various methods of strengthening the canal bank.

23.4. How do you control weed growth in the canal ?

❐❐❐

24

Water Logging and Drainage

24.1 WATER LOGGING AND SALINITY

An agricultural land is said to be water logged when its productivity gets affected by the high ground water table. The productivity of the land is affected when the root zone of plant gets flooded with water and becomes ill-aerated. Because of the inadequate aeration of the root zone, oxygen required by the plant to maintain its growth is reduced and thus yield of the crop is reduced.

The depth of water table below ground level, which causes water logging is not same for all the conditions. It depends upon nature of soil and type of the crop. The effect of ground water, above ground water table depends upon the height. of capillary fringes which in turn depend upon the nature of the soil. The height of capillary fringes above water table, for normal agricultural soils, varies from 90 cm to as much as 1.50 m. The root zone depth, is considered about 0.6 m. Hence if at someplace, root zone depth is 0.6 m and height. of capillary fringes is say 1 m, the land will be called waterlogged if depth of table below ground level is 1 + 0.6 = 1.6 m or less. If depth of water table is more than 1.6 m, then it is not water logged. Depths of water table below ground level, for different crops are given below which will adversely affect their yield.

Crop	Wheat	Cotton	Sugarcane	Rice	Fodder
Depth of water table	1.2 m	1.5 to 1.8 m	0.3 m to 9.5 m	0.6 m	1.2 m

The ground water table does not remain constant. There shall be seasonal fluctuations, due to variation in rainfall.

24.2 CAUSES OF WATER LOGGING

Following may be the causes of rise in water table which ultimately leads to water logging of the area.

1. Over and intensive irrigation. When a policy of intensive irrigation is adopted, then, maximum irrigable water is supplied to certain region. This leads to heavy percolation and subsequent rise of water table. For this reason, to avoid water logging, a policy of extensive irrigation (on more area) should be preferred from the policy of intensive irrigation (more water on small area).

In short irrigation water applied to the fields is more than the requirements of crop and this results in deep percolation. Percolated water augments the ground water storage and causes rise in water table.

2. Seepage of water from adjoining high lands. The water from adjacent high lands may seep into the sub-soil of the affected land and may cause rise in water table.

3. Seepage of water through the canals. In nature the water table is always in a state of equilibrium. In other words amount of inflow is practically equal to the amount of outflow. Due to construction of canal, the seepage through bed and sides of the canal takes place. This increases the inflow into the ground water and thus rise in ground water table occurs. The rise in water table takes place upto the level when the increased inflow is again balanced by the increased outflow by way of increased soil evaporation, increased transpiration and increased discharge into seepage drains.

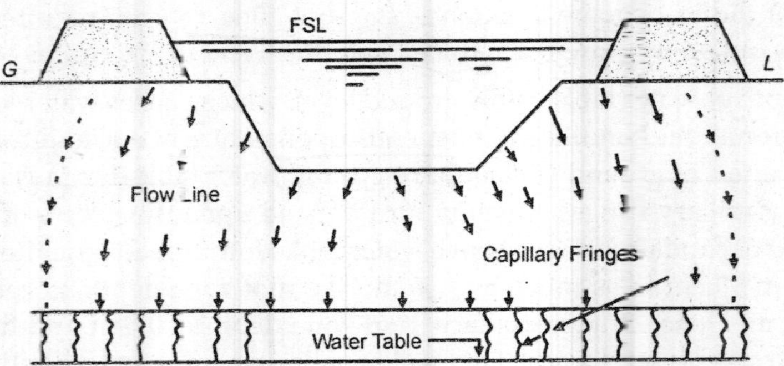

Fig. 24.1. *Loss of water by percolation and seepage.*

4. Impervious obstruction. Seeping water below the soil moves horizontally. If this flow finds an impervious obstruction, rise in water table occurs on the *U/S* side of the obstruction. On the same ground if an impervious stratum occurs below the top layer of pervious soil, seeping water will not be able to go deep and this also causes quick rise in water table.

5. Inadequate natural drainage. Soils having less permeable substrata (such a clay) below the top layers of pervious soils, will not be able to drain water deep into the ground. This will cause rise in water level.

6. Inadequate surface drainage. If the area is not properly drained, the storm water will accumulate into the depressions in the area. This accumulated water remains percolating and causes rise in the water table. If natural drains of the area are obstructed by railway or highway embankments, the water will get accumulated and thus causes water logging.

7. Excessive rains. Excessive rainfall may cause temporary water logged conditions. If the area is not properly drained the area may be subjected to continued water logging.

8. Submergence due to floods. If a land is continuously subjected to submergence by floods, water loving plants may grow in abundance. The weed growth will obstruct the natural surface drainage of the soil and chances of water logging get increased.

9. Irregular or flat topography. The water is immediately drained from regular steeply sloped area. If the area is flat and irregular, the drainage of the area is poor. This causes detention of water on the surface for longer periods longer water table may rise.

10. Inadequate capacity of arterial drains. If the capacity of the drains dealing with flood water, is smaller than the required size, the flood water of local drains will spread over country side for days. This will cause heavy percolation of water and may ultimately lead to water logging.

11. Construction of reservoirs. Seepage from reservoirs, canals, may cause water logging of the adjoining areas.

24.3 ILL-EFFECTS OF WATER LOGGING

The ill-effects of water logged soils are the following :

1. Water logged fields generally remain wet and their tillage becomes difficult.

2. Sowing of the crops is generally delayed because ploughing and mulching of fields is delayed in water logged areas.

3. The climate of the water logged area becomes damp. Standing water in pools become stagnant. The stagnant pools of water become good breeding places for mosquitoes. All these, effects ultimately lead to conditions detrimental to the health of the community.

4. **Growth of wild weeds.** Weed growth in water logged areas is generally very intensive. This reduces the supply of food to the crops, as weed shares the food. Farmers have to do lot of labour to clear the fields from weeds so that they may have good production of crops.

5. If water logged area has canal system of irrigation, the canal water will be used very little as such areas require very little irrigation. This aspect may lead to wastage of canal water, which remains filled up in low areas and may cause water logging of more areas.

6. Inadequate circulation of air in root zone. In water logged areas, air circulation in the root zone becomes inadequate, due to which activity of bacteria and consequent growth of the plants is affected.

7. Water logging causes salinity of the soil. The water that has risen up, continuously evaporates by capillary action. This establishes continuous upward flow of water from the water table to the land surface. With this upward flow, the salts which are present in water and also in lower layers of soil, rise towards the surface, resulting in the deposition of salts in the root zone of crops. The concentration of these salts (which are usually alikaline) has a corroding effect on the roots, which reduces the osmotic activity of the plants and reduces their growth. Such soils which contain large concentrations of harmfull salt in the root zone are known as *saline soils.* From this it can be easily said that wherever there is waterlogging, salinity is a must.

8. In waterlogged areas, temperature of the soil is generally low. Lower temperature does not allow bacteria to function properly and as such supply of food to the plants is reduced.

24.4 WATER LOGGING CONTROL OF REMEDIAL MEASURES

It is evident that water-logging can be controlled only if the quantity of water entering the soil is checked and reduced. To achieve this, the inflow of water into the ground water reservoir must be reduced and the outllow from this reservoir should be increased. Following are some of the measures which can be adopted for controlling water logging.

1. All the canals and water courses, used for irrigation and other purposes, should he lined. It is considered as one of the most effective measure of controlling waterlogging

2. Intensity of irrigation in the area likely to be water logged should be reduced. In areas where water tale is very high, irrigation during Kharif season may only be allowed. During Rabi season the cultivators may he asked to irrigate from wells.

3. Cultivators may be educated for economic use of water. They should be made to understand the importance of irrigation water They should be asked to divide fields into small Kiaries so that deep percolations do not occur.

4. In order to reduce seepage from Kucha canals, they should he designed with lesser depth of water. Lesser depth will induce lesser head for seepage losses.

5. Certain crops require more water than others. If a particular field is always sown with a crop requiring more water the chances of water-logging are more. In order to avoid this, crops requiring lesser, and more quantity of water, should be sown alternately in the fields.

6. Optimum use of water. We know that certain fixed quantity of water for a particular crop gives best results. Less than that and more than that reduces the yield. But most of our farmers are unaware of this fact. They always think that more the waterings you apply, to the field, the greater is going to be the yield and hence they try to use more and more water. This can be checked by educating the farmers.

7. Revenue should be charged on the basis of quantity of water utilized and not according to the area irrigated. This measure has its own practical drawbacks but theoretically the cultivator would be tempted to irrigate more area with same quantity of water.

8. **Providing intercepting drains.** Intercepting drains along the canals, particularly in high embankment reaches should be constructed. These drains intercept seepage and percolation from the canals. They should not be laid very close to canals, because then they would be drawing water from the canals directly.

9. **Introduction of lift irrigation (well irrigation).** Well irrigation utilises the underground water for irrigation. Hence the areas which are likely to be water logged in near future should be irrigated from wells and intensity of irrigation from canals should be reduced considerably. Well irrigation would lower the ground water-table and chances of land becoming water-logged are eliminated.

10. **Improving the natural drainage of the area.** The natural drainage system of the area should be such that no water is allowed to stand for longer period and as soon as rain water drops on the ground it should be immediately led to drains. For this natural drains should be maintained clear from weed growth and any obstructions should be removed. Natural slopes of drains should be maintained by clearing silt from time to time.

11. **Provision of an efficient drainage system.** An efficient drainage system should be provided to drain away the storm water. A good drainage system may consist of surface drains as well as sub-surface drains. If area is very intensively water- logged and soil is of poor drainability, under ground porous, or tile drains have to be laid to effectively lower down the waterable.

12. **Adoption of sprinkler method for irrigation.** This method almost completely eliminates the percolation losses from water courses. Maximum percentage of this water is utilized in form of consumptive use and losses are reduced to minimum.

24.5 DRAINAGE OF THE AREA

Surface irrigation is a blessing only if it is practised with great care. Only optimum amount of water should be supplied to crops, which depends upon the type of crop and properties of soil. Excess water, which the root zone of the soil fails to absorb, may percolate and help in raising the water table. Sometimes, this gravity water may encounter an impervious stratum and may not be drained up to tile water table. This excess water is not only waste but may be harmful to crop yield also. If such conditions are likely to occur, it becomes necessary that the excess water is removed and drained out from below the surface and discharged back either to river, canal or some where else. *Hence while designing a canal irrigation network,* it is sometimes desirable to provide a suitable drainage system, so as to remove the excess water. This may be essential in areas of high water table and in river deltas.

Drainage system is also required to drain out the storm water effectively and prevent its percolation.

Two types of drainages can be provided.

1. Surface drainage and
2. Sub-surface.

1. Surface drainage. Surface drainage consists of open ditches, field drains etc. and are used for the removal of excess water from the surface. The surface drainage is carried out with the help of (i) Shallow surface drains and (ii) deep surface drains.

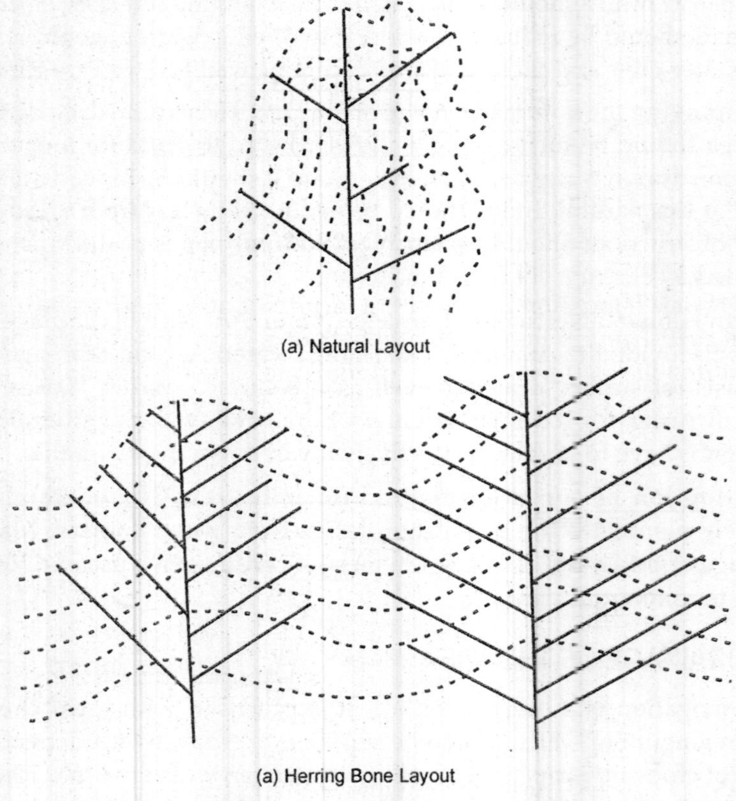

Fig. 24.2. *Layouts of underground pipe drains.*

(*i*) *Shallow surface drains.* These drains are constructed to remove the excess irrigation water, applied to the fields and also the storm water. They are wide but shallow in depth and trapezoidal in shape. These drains are designed to carry normal storm water plus the excess irrigation water if any.

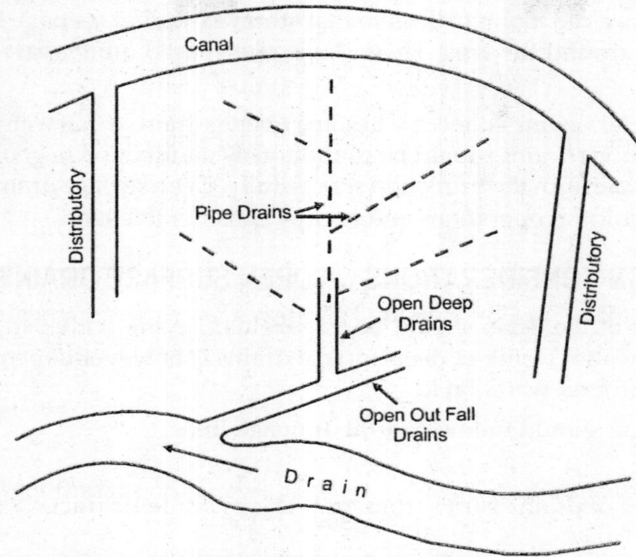

Fig. 24.3. *General layout of drains in an area.*

(*ii*) *Deep surface drains.* If open drains are dug upto the level below the ground water table, they are known as deep surface drains. Deep drains are also sometimes known as *outlet drains* as they accept discharge from closed drains also. These drains carry seepage water also along with storm water. They are therefore designed for combined discharge. Generally a cunnette is provided in

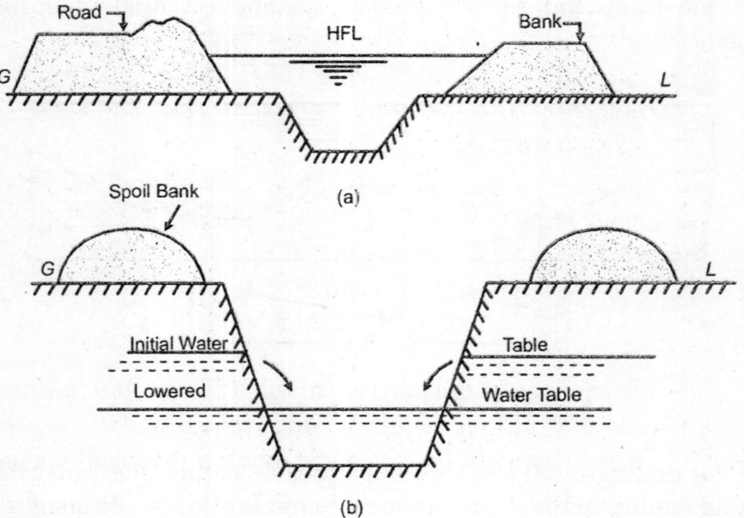

Fig. 24.4. *Surface drains.*

the centre of the drain-bed so as to carry the seepage water. A steeper slope is given to the cunnette and it is lined so as to withstand higher flow velocities. The full section of the drain would be coining into use only during the storms. These drains are run along valleys so that storm as well as seepage water may reach it from around the area. These drains are spaced quite apart from each other.

The surface drains are connected to some river or drain so that water is carried away. The surface drains should be continuously maintained in good shape so that they may perform their function successfully. Deep surface drains interfere with the agricultural operations and occupy considerable land.

24.6 DESIGN CONSIDERATIONS OF OPEN SURFACE DRAINS

1. The section of the drain should be trapezoidal having wide bed width but less depth of water. Depth of deep surface drains is however dependent upon the depth of around water table.

2. Alignment should follow natural drainage lines.

3. Bed slope of drains varies from $\frac{1}{300}$ to $\frac{1}{150}$ in deltic tract.

4. Capacity of drains should be according to normal storms and not according to heavy storms. Storm water may be allowed to remain accumulated for sometime before drainage and hence even smaller drains can perform the drainage of the area but area in that case may remain submerged for sometime. The maximum anticipated discharge for plains of Ganga is calculated at the rate of 0.0054 cumecs per hectare of catchment while for Punjab, the design figure is 0.043 cumecs per hectare.

5. **Sub-surface drainage.** Sub-surface drainage is installed in form of under-drains or tile drains. Sub-surface drainage becomes essential under following circumstances.

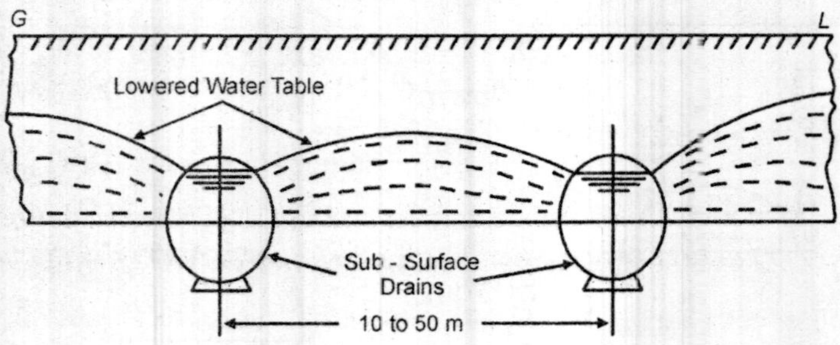

Fig. 24.5. *Effect of sub-surface drains.*

(i) When underground soil is of impermeable nature, the water field in it will not be readily drained, and as such, to cause effective drainages of such soils, tile drains are essential.

(ii) When surface open drains cover very large area of valuable agricultural land, it becomes economical to install underground tile drains.

(iii) Bridges will have to be constructed across the open drains for roads and railways, which are generally very costly. In order to avoid their construction under drains may be preferred.

Tile drains or closed drains are laid deep in the ground and covered so that they do not interfere with the agricultural operations. These drains are circular in section and generally made of porous earthen ware. Their diameter varies from 10 cm to 30 cm. To ensure effective drainage, tile drains should be placed in a permeable stratum or may be shrouded with filter material which is coarser towards the pipe and finer towards the trench. The average minimum depth of tile drain from ground surface is 1 m to 1.20 m.

24.7 DESIGN OF PIPE DRAINS

An approximate method of estimating the discharge and spacing of drains may be done as follows.

Let two adjacent closed drains be spaced L metres apart and located a metre above the impervious layer. Let b be the maximum height of water table above the impervious layer. Let at distance x from the centre of a drain, height of water table is y metre. See Fig. 24.6.

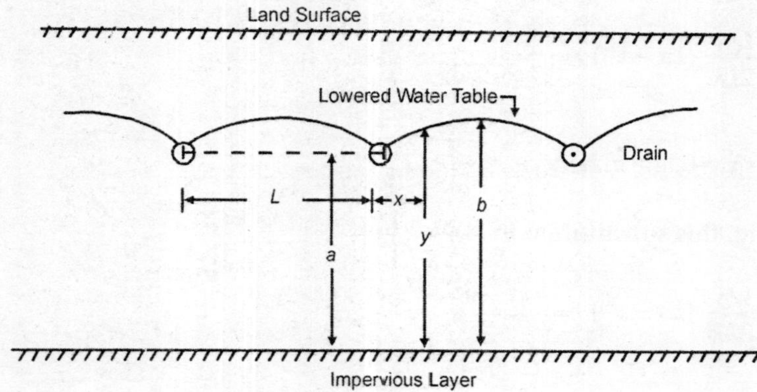

Fig. 24.6

By Darcy's law

$$q = ky \frac{dy}{dx} \tag{1}$$

where q = discharge passing the section at y per unit length of the drain

Q_D = design flow per metre of drain

At $x = 0 \quad q = \frac{Q_D}{2}$

At $x = \frac{L}{2} \quad q = 0$

$$\therefore \quad \frac{q}{\frac{Q_D}{2}} = \frac{\frac{L}{2} - x}{\frac{L}{2}}$$

$$q = \frac{Q_D}{2}\left[\frac{\frac{L}{2} - x}{\frac{L}{2}}\right] = \frac{Q_D}{2L}(L - 2x) \tag{2}$$

or

$$ky\frac{dy}{dx} = \frac{Q_D}{2L}(L - 2x)$$

$$ydy = \frac{Q_D}{2Lk}(L - 2x)\,dx \tag{3}$$

Integrating (3), we get

$$\frac{Q_D}{2Lk}(Lx - x^2) = \frac{y^2}{2} + c \tag{4}$$

Where $x = 0$, $y = a$, hence $c = -\frac{a^2}{c}$

Making, this substitution in (4), we get

$$\frac{Q_D}{2Lk}(Lx - x^2) = \frac{y^2 - a^2}{2}$$

or

$$Q_D = \frac{Lk\,(y^2 - a^2)}{Lx - x^2} \tag{5}$$

$$k = Q_D\frac{(Lx - x^2)}{L(y^2 - a^2)} \tag{6}$$

When $x = \frac{L}{2}$ $y = b$ and hence

$$Q_D = \frac{4k(b^2 - a^2)}{L} \tag{7}$$

or $$L = \frac{4k(b^2 - a^2)}{Q_D} \quad (8)$$

$$Q_D = VL$$

where V is the safe discharge per unit area of land surface.

Substituting the value of $Q_D = VL$ in (8), we get

$$L = \frac{4k(b^2 - a^2)}{L}$$

or $$L^2 = \frac{4K}{V}(b^2 - a^2)$$

$$L = 2\sqrt{\frac{k}{V}(b^2 - a^2)} \quad (9)$$

This equation is also known as Dupuits equation. Bureau of reclamation gave following formula for c/c distance between adjacent underground drains.

(i) $$L = \pi\sqrt{\frac{kD_a t}{f \log e^{\left(\frac{4}{x}\frac{y_0}{y}\right)}}}$$

where $$D_a = d - \frac{y_0}{2}$$

t = Time in which the water table is to be lowered for specified depth.

y = Maximum allowable height of water table midway between the drains.

y_0 = Height of water table before drainage.

f = drainage porosity, a function of k.

(ii) $$L = \pi\sqrt{\frac{kD_a t}{f.\log e^{\left(\frac{4}{x}\frac{y_0}{z}\right)}} - D_a \log e^{\frac{D_a}{4_r}}}$$

where $D_a = a + \frac{y_0}{2}$

r = radius of the drain or $\frac{d}{2}$.

24.8 LAYOUT OF TILE DRAINS

The layout of tile drains is dependent upon the topography of the area only. Following are some of the lay outs.

1. Natural layout. In this system of layout the drains arc aligned along the natural valleys or drains.

2. Grid iron layout.

3. Herring bone pattern.

4. Intercepting drains.

All these layouts are applicable to specific needs. The drains should not be laid near the trees as roots of trees may dislodge them.

24.9 CHANNEL LOSSES

Channel losses may be classified under two heads:

1. Losses due to seepage.
2. Losses due to evaporation.

1. Losses due to seepage are the following :

(i) Permeability of soil through which canal passes.

(ii) Level of ground water table.

(iii) Drainage conditions of the sub-soil.

(iv) Position of canal either in cutting or instilling.

(v) Amount of silt carried by the canal ; more the silt the lesser the losses.

(vi) Velocity of flow in canal.

(vii) Regulation of canal.

(viii) Depth of canal.

The losses due to evaporation are dependent upon climatic elements such as temperature. humidity, wind, velocity, and area of exposure. Evaporation losses are only 1.5 to 2% of the total losses in Kucha channels.

For design purpose evaporation and seepage losses are expressed in cumecs per million square metre of wetted perimeter. Harding and Etchevery gave following conveyance losses.

Nature of soil	*Losses in cumec/million sq-m wetted perimeter*
1. Gravelly sandy soil	7.0 to 8.0
2. Sandy soils	5.2 to 6.1
3. Sandy loam	3.6 to 5.2
4. Sandy clay loam	2.7 to 3.6
5. Silty soil or ordinary clay loam	1.2 to 2.7
6. Clay foams	0.9 to 1.2

In Punjab (India) combined losses are found out from following empirical formula.

Losses in cumec/million sq/m of wetted perimeter

$$= 1.9\, Q^{1/6}$$

where Q = discharge of canal in cumecs

For U.P. (India) losses in cumec/km length of channel are computed from following formula.

$$\text{Losses in cumec/km length} = \frac{(B+D)^{2/3}}{200}.$$

Where B and D are bed width and depth in metres.

24.10 FRENCH DRAINS

When amount of water to be removed from depressions is small, a blind inlet may be installed over the tile drains. The blind inlet is also called *French drains.*

24.11 RECLAMATION OF SALINE LANDS

Land reclamation is a process by which an unculturable land is made fit for cultivation. Saline and water-logged lands give very small yield unless they are reclaimed.

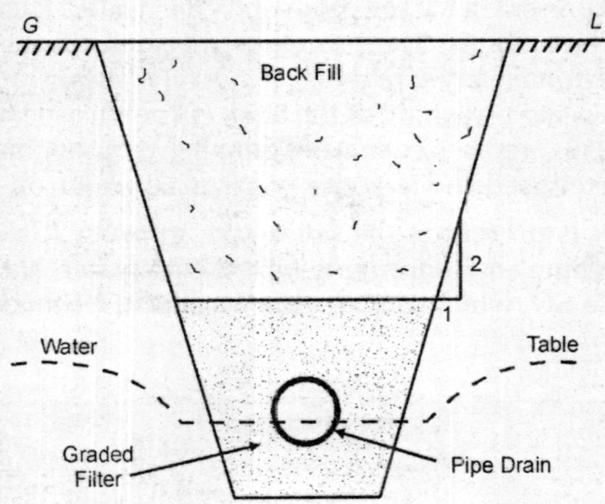

Fig. 24.7. *French drains.*

Every soil contains certain mineral salts, some of which are beneficial to plants and others harmful. Harmful salts are Alkaline salts and their common examples are NaCl, Na_2CO_4, Na_2CO_3. Out of these Na_2CO_3 is the most harmful whereas NaCl the least. All these salts are soluble in water. When water table rises up and capillary fringes reach the root zone, the water from water table starts flowing upwards. The soluble alkaline salts also move up with water and get deposited in the soil within the root zone, as well as on the surface of land. The phenomenon of salts coining up in solution and forming a thin (5 to 7.5 cm) crest on the surface after evaporation of water, is called efflorescence. Land affected by efflorescence is called *saline soil.* The saline water around the roots reduces the osmotic activity of plants. If the salt efflorescence continues for a longer period, a base exchange reaction sets up, particularly if the soil is clayey, thus sodiumising clay, making it impermeable and therefore, ill-aerated and highly unproductive. Such soils are called alkaline soils. The reclamation of *alkaline land* is more difficult than *saline soils.*

24.12 RECLAMATION OF SALT AFFECTED LANDS

It is evident from the above discussions that efflorescence can be avoided, by maintaining water table sufficiently below the root zone so that the capillary fringes are not able to reach the plant roots. Hence all those measures which were discussed to prevent water-logging hold good for preventing salinity of land also. An efficient and elaborate drainage system must be provided to lower the water table in saline lands. When water table has been sufficiently lowered, the soil is freed from the accumulated salts by the process called *leaching process.*

24.13 LEACHING PROCESS

In this process the field to be reclaimed is flooded with sufficient depth of water. The filled water is agitated either by driving animals in the field or by any other method. The salts present in the soil get dissolved in water. This water is drained away by sub-surface drains. The process is repeated till the salts in the top layer are reduced to such any extent that some salt resistant crop can be grown. This process of repeated washing of the land, is known as leaching. High salts resistant crops are rice and berseem. After growing, crops for one or two seasons, salinity is further reduced and now any crop can be grown on reclaimed fields.

When Na_2CO_2, is present in the saline soil, gypsum ($CaSO_4$) is generally added before leaching and thoroughly mixed with water, Na_2CO_3 reacts with $CaSO_4$ forming Na_2SO_4 which can be easily washed or leached out as explained earlier.

QUESTIONS

24.1. What do you understand by term water logging ? What are the ill-effects of water logging ?

24.2. Enumerate various causes of water logging with brief description of each.

24.3. Enumerate various measures of controlling water logging.

24.4. What do you understand by term drainage of the area ? Explain briefly surface drainage and sub-surface drainage.

24.5. Write short notes on :

(i) Land reclamation

(ii) Channel losses.

(iii) Layout of tile drains.

(iv) Saline land and alkaline land.

❑❑❑

25

Cross Drainage Works

25.1 INTRODUCTION

Whenever an irrigation channel intercepts a natural drainage, which may be small stream or river, a masonry work has to be constructed to pass one above the other or to pass at the same level. Such masonry works are known as *cross drainage works.* Canals are mostly run at ridges, and hence there should not be any necessity of such works. Following are some of the situations where necessity of cross drainage works occurs :

(i) In the head reaches where a main canal is led to the water shed, it has to cross a number of natural streams.

(ii) If, due to certain reason, the canal alignment leaves the watershed and catches it again after some distance, the area enclosed by the watershed and the alignment of canal, has to be drained. This requires construction of a cross drainage work.

(iii) When canal is aligned as contour canal it requires cross drainage works.

(iv) When various canal systems have to be linked with one another, necessity of cross drainage work may arise.

Cross drainage works are generally very costly works and should, as far as, possible be avoided. If they cannot be avoided at all, their number should at least be reduced by adopting following measures.

(i) The canal may be aligned in such a way that it does not cross large number of small drains but crosses smaller number of large size drains. Several tributaries after meeting each other form a large drain. The canal should be aligned to cross the large drain rather than smaller tributaries.

(ii) Several large independent drains can be connected together by link channels and canal should cross them at one point only.

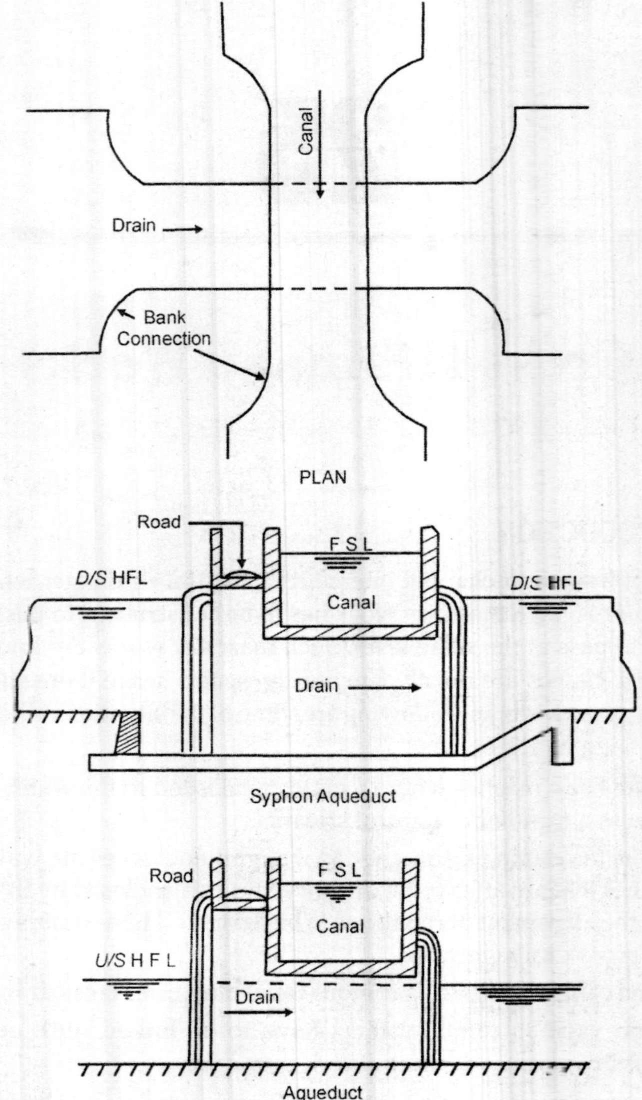

Fig. 25.1. *Aqueduct and syphon aqueduct.*

25.2 TYPES OF CROSS DRAINAGE WORKS

Cross drainage works can be grouped under following three heads:

1. Cross drainage works when canal crosses over the drainage.

2. Cross drainage works when drainage crosses over the canal.

3. Cross drainage works when drainage and canal waters are allowed to intermix.

25.3 CROSS DRAINAGE WORKS WHEN CANAL PASSES OVER THE DRAINAGE

In this case, irrigation canal passes over the drain. The cross drainage works connected with this particular condition are the following :

(i) Aqueduct.

(ii) Syphon aqueduct.

(*i*) *Aqueduct.* This is constructed when bed level of the canal is sufficiently higher than the H.F.L. of the drainage. This is the most commonly used cross-drainage work. The main advantages being that canal will not be led into cutting and canal is open for inspection. Moreover floods in drains do not cause any

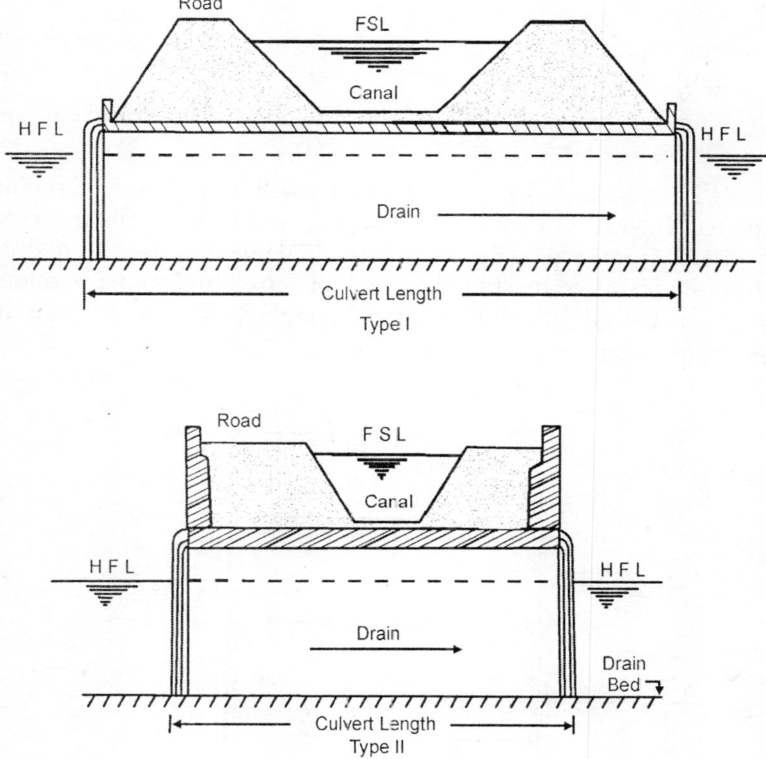

Fig. 25.2. *Aqueduct.*

damage to the canal as its bed is kept will above the H.F.L. of the drainage. Aqueduct structure may be arch culvert slab culvert, or even pipe culvert. If drainage and canal both are of substantial size a large masonry or concrete bridge like structure has to be constructed. Aqueduct is just a highway bridge, with the only difference that aqueduct carries canal above whereas highway bridge carries highway traffic. For channels of very small capacity, the pipe aqueducts may be constructed. The channel water is carried through pipes instead of water troughs. The pipes are supported on piers along the width of the drainage. The head wall and toe wall should be provided at both the ends. The pipes should be capable of discharging the full supply discharge of the channel.

(*ii*) *Siphon aqueduct.* If difference between bed level of canal and bed level of drain is not substantial, the drainage water passes under the canal by touching the under side of the canal. If H.F.L. of the drain is much higher than the bed level of the canal, the water passes under the canal, under syphonic action. In case of syphon aqueduct the bed of the canal is subjected to very large uplift pressure and thus have to be quite strong.

Sections of aqueduct of syphon aqueduct may be of three types as follows:

(i) Full canal section along with full side earthen banks are maintained at the aqueduct. Fig. 25.2 (i).

(ii) Full canal section is maintained as usual but outer slopes of both the banks are replaced by masonry walls. Fig. 25.2 (ii).

(iii) Canal section is flamed and confined into a trough and sides are made of concrete or masonry. Fig. 25.2 (iii).

In case I Fig. 25.2, width of the aqueduct will be quite large and hence cost per unit width will be maximum. In this case, bank connections are not required. Hence choice of this type will depend upon the relative cost of aqueduct proper and the cost of bank connections. The cost of bank connection is independent to the length of the aqueduct. Hence this type of construction is suitable when length of aqueduct is small.

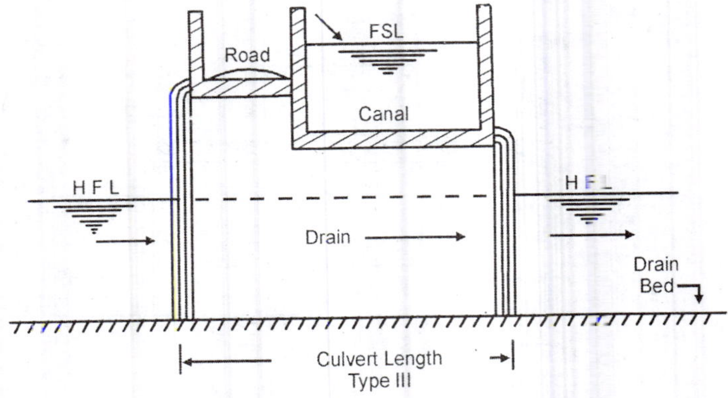

Fig. 25.3. *Aqueduct.*

For case III Fig. 25.3, the width of the aqueduct is minimum and thus cost of aqueduct per unit width is minimum. However the cost of bank connections will be additional. Hence this type is suitable when length of the culvert is very large (big drains). Type II Fig. 25.2 section is suitable for intermediate conditions. In short type III is suitable for very large drains while type I is suitable for very small drains. The correct way of choosing the type is to work out cost of all the three types and adopt the cheapest alternatively. Design of aqueduct is given ahead.

25.4 CROSS DRAINAGE WORKS WHEN DRAINAGE PASSES OVER THE CANAL

In this case, drainage is carried over the canal. This is found suitable when bed level of the canal lies below the bed level of the channel. Cross drainage works of this category are :

(i) Super passage and, (ii) Canal Syphon.

(*i*) *Super passage.* Where the bed level of the drainage is well above the F.S.L. of the canal passing below, the cross drainage work is known as super-passage. In this case canal water does not touch the bottom of the drainage. Super passage is possible when F.S.L. of canal is sufficiently below drain bed. Fig. 25.4.

(*ii*) *Canal syphon.* When the canal discharge is small in comparison to the drain and the canal bed level is lower than the F.S.L. of canal, syphon is the

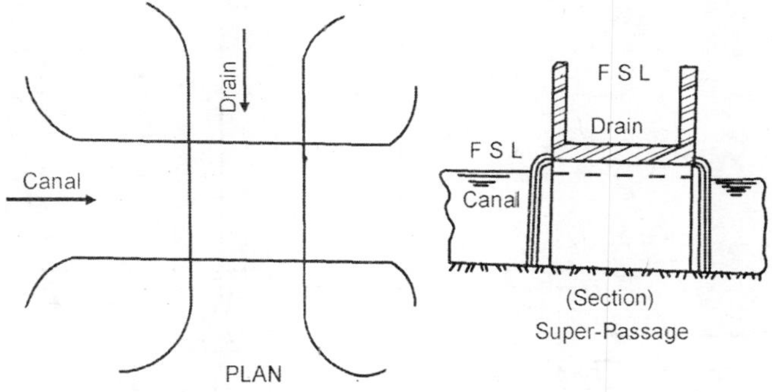

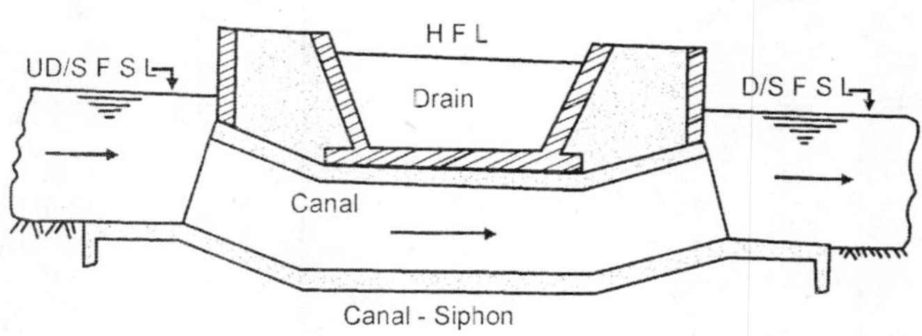

Fig. 25.4

preferred structure to be adopted. In this case canal is syphoned below the drain. See Fig. 25.4.

The siphon is disadvantageous in the respect that the canal bed is likely to silt up, and that the siphon barrel has to resist the undue uplift pressure on the top.

The siphon essentially consists of a tunnel or a barrel, laid under the drainage of for carrying canal water. The siphon may be designed with vertical drops at the ends and connected by a horizontal barrel in the centre. The vertical drop type is not suitable for canal water with heavy charge of silt.

The siphons may be constructed with a curved passage for water, so as to cause less disturbance to the canal flow and reduce the silting trouble. But constructing barrels of curved type is more difficult.

The siphons with gradually sloped inlet and exit are more suitable for canals having muddy flow.

Design of super passage and siphons are done similar to the design of aqueducts and siphon aqueducts.

25.5 CROSS DRAINAGE WORKS WHEN DRAINAGE AND CANAL WATERS ARE ALLOWED TO INTERMIX

The works connected with this situation are *level crossing*, *inlets* and *outlets*.

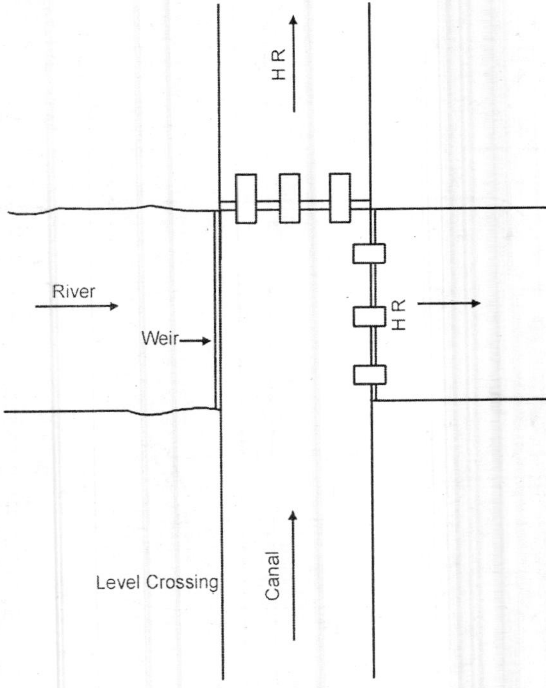

Fig. 25.5. *Level crossing.*

When bed level of canal and that of the drainage are at or about the same level, the canal water is carried across the drainage by mixing with it. The drainage and canal water mix with each other at the site of the cross drainage work. The discharge of the drainage decides whether level crossing, or inlet alone should be adopted.

When the drainage flow is small and drain is not perennial, it is generally admitted into the channel by the inlets. If the discharge in the channel is too much and admitting the drainage may raise the water level in the channel considerably, an outlet also becomes a necessity to drive out the excess excepted discharge.

If the drainage is small and runs throughout the year the water may be led to the channel by an inlet. This water helps to supplement the discharge of the channel and can be safely used for irrigation.

When the discharge of drainage is large and imermittent, the cross drainage work is provided by means of level crossing which consists of a cross-regulator across the channel and also cross the drainage. The work is similar to a diversion head work. After mixing of the two waters, the canal draws its designed discharge through a head regulator. The advantage of the regulator across the stream may be taken to raise the level of water so that the command of the channel *D/S* of crossing may be further increased. The canal regulator is generally kept closed during floods and flood discharge is passed through the regulator across the drainage. When floods recede, and the water is clear in the stream, the canal regulator is opened for drawing water for the channel.

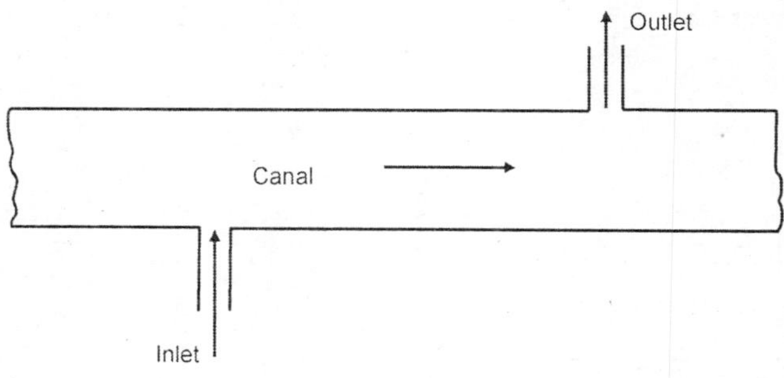

Fig. 25.6. *Inlet and outlet.*

When drainage is a big river in which the flow is perennial, permanent cross drainage works are necessary. Such a work consists of a dam or pick up weir across the river. The canal water is discharged into the river through the regulator gates at one side and at the other side water is drawn into the canal through another head regulator. Escapes may be provided to deal with the surplus water and also to remove silt from canals.

Inlet. The inlet is constructed to admit the flow of small perennial streams into the channel. The bank of channel is cut open and the sides and bed protected by pitching. If bed of the stream is slightly at higher level than the bed of canal, the stream may be admitted to the channel through inlets provided with drop. For very small streams inlet may be mere pipe laid in the bank of the canal.

25.6 OUTLET

Its job is reverse that of inlet. It is an arrangement in the bank of a channel to discharge out the surplus water from the channel. The outlet becomes necessary when discharge admitted into the channel by an inlet, raises the water level above the maximum water level of the channel. The outlet is similar in action to a surplus escape, and is provided at suitable site where surplus water can be easily discharged into a side valley.

25.7 PIPE AQUEDUCT

The inlets and outlets can be replaced by pipes in case of small drainages. The whole discharge of the drainage is carried across the canal by providing pipes extending from one bank to the other bank. The pipes are laid above the F.S.L. of the channel with a free board of 60 cm to 1 m. If the drainage bed width is more, the pipes are supported on piers. The bed width, in such cases, should be suitably widened so that the natural water way is not decreased by piers.

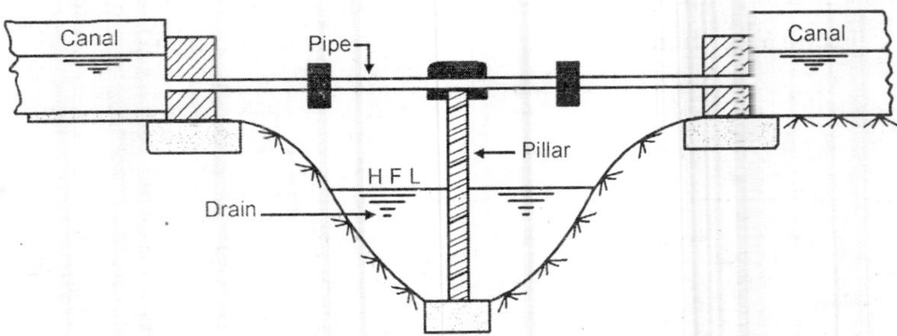

Fig. 25.7. *Pipe aqueduct.*

25.8 SELECTION OF SITE FOR CROSS DRAINAGE WORKS

While selecting site for a cross drainage work, the following points should be borne in mind:

1. Soil of good quality should be available so that safe foundations may be laid on it.
2. The natural banks should be high and stable, otherwise large cost is required for making good bank connections.
3. The canal and drainage, should cross each other at right ankles. There should be straight reaches in both on the *U/S* and *D/S* sides.
4. The section of the drainage should he uniform. The bed of the drainage should be even and the flow should be free from turbulances.

5. The type of the crossdrainage work should be carefully decided with reference to :
    (i) Bed levels.
    (ii) Discharge of the canal and drainage.
    (iii) Overall cheapness of construction.

25.9 SELECTION OF SUITABLE TYPE OF CROSS DRAINAGE WORK

While making selection of suitable type of cross drainage work following points should he carefully considered :

(i) If Bed level of the canal is sufficiently above the H.F.L. of the drainage, file construction of aqueduct is the obvious choice.

(ii) If bed level of drain is sufficiently above the F.S.L. of the canal super passage is the obvious choice.

(iii) If head way below the canal bed level and above H.F.L. of drainage is not adequate, it can be increased by shifting point of crossing *D/S* of the drainage. If however, shifting of canal alignment, *D/S* is not possible, syphon aqueduct should be provided. In this case H.F.L. of the drainage would be above the bed level of the canal.

(iv) If canal bed is lower than the H.F.L. of the drainage, canal syphon should be provided.

(v) If canal and drainage cross each other more or less at the same level, level crossing may be preferred. Level crossing should as far as possible be avoided.

If we want to change the type of cross drainage work it can be done by shifting the crossing point either *D/S* or *U/S* side. By doing so relative heights of each work can be changed and hence type of work can be changed.

25.10 DESIGN FEATURES OF CROSS DRAINAGE WORKS

Design of cross drainage work includes the design of following elements:

1. Determination of maximum flood discharge.
2. Determination of water way of the drain.
3. Contraction of canal water way if canal is to be flumed.
4. Head loss through syphon barrels.
5. Determination of uplift pressure on the bed of canal of drain whichever is above.
6. Determination of uplift pressure on the bed of canal or drain whichever is below.
7. Design of bank connections.

Besides the design of these elements, structural design of foundation, piers, abutment, syphon barrels, etc. also have to be done.

After having found out the above said hydraulic elements, the structural elements can be easily designed.

1. For the determination of maximum flood discharge, sufficient explanation has been given in Chapter on Hydrology.

2. Determination of water way of the drain. After having decided about the maximum flood discharge, water way for the drain can be easily fixed by following Lacey's equation

$$P = 4.75\sqrt{Q} \tag{25.1}$$

For large drains the perimeter (P) may be assumed equal to the width of the river. A contraction upto 20% of water way may be allowed in case of small drains. In large drains, no extra provision is made for the area covered by the piers.

While fixing the water way, it should ensured that a minimum velocity of flow from 2 to 3 m/sec may be maintained.

3. Contraction of canal water way. Contraction of canal is required only where type III Fig. 25.3 aqueduct is to be installed. Fluming of the canal requires the provision of extra transition wings for joining the flumed portion to the normal section. While fluming the canal following points should be taken care of:

(i) The velocity of flow through the flumed section of the canal is not more than 3 m/sec.

(ii) The flow should remain sub-critical without any formation of hydraulic jump.

(iii) The approach transition wings should not be steeper than 30° (splay of 2 : 1) and departure transition should not be steeper than $22\frac{1}{2}°$ (splay of 3 : 1). See Fig. 25.8.

(iv) The transitions are curved and flared so that there is minimum loss of head and the flow is streamlined.

The transitions can be designed for following two conditions:

(a) Depth of water remains constant.

(b) Depth of water varies.

(a) Depth of water remains constant. For this design, following two formula may be used.

(i) R.S. Chaturvedis's semi-cubical parabolic transition

$$x = \frac{L_f B_0^{3/2}}{B_0^{3/2} - Bf^{3/2}}\left\{1 - \left(\frac{B_f}{B_x}\right)^{3/2}\right\} \tag{25.2}$$

The values of x can be found out by choosing various convinient values of B_x.

(ii) *AC Mitra's hyperbolic transition*

$$B_x = \frac{B_0 B_f L_f}{L_f B_0 - x(B - B_f)} \quad (25.3)$$

The value of B_x can be found out by choosing various values of x.

In both the formulae, (Chaturvedis's and Mitras)

B_0 = normal width of the canal

B_f = Flumed width of the canal

B_x = Width at any distance x from the flumed section.

L_f = Total length of transition.

(b) *Depth of water varies.* When depth of water varies, the design of transition is done by Hind's method.

Consider Fig. 25.8. Let $A - A$ and $B - B$ be the sections of canal on the entrance side and $C - C$ and $D - D$ on the exit side. $A - A$ and $D - D$ are the normal sections of the canal and $B - B$ and $C - C$ are the limed sections of the canal.

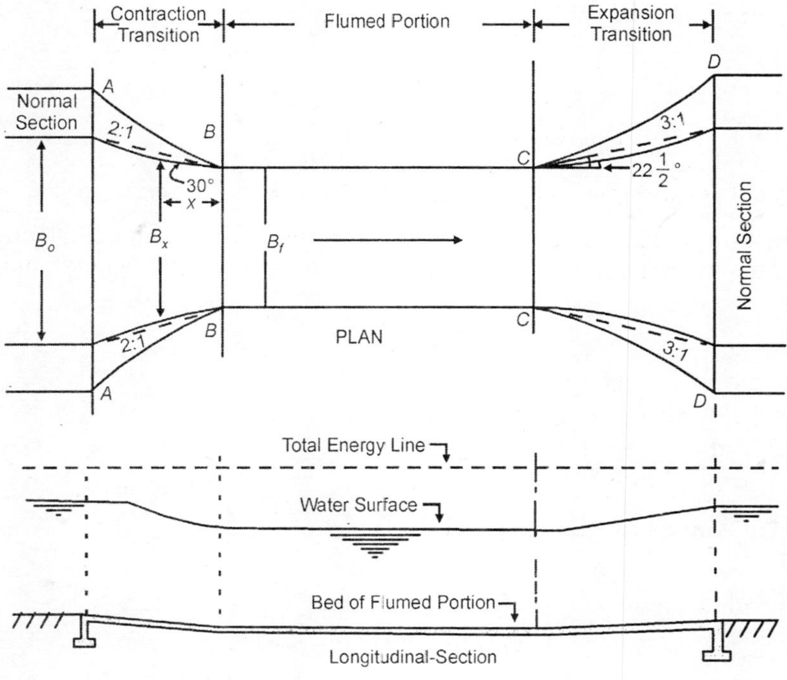

Fig. 25.8

Let D_A, D_B, D_C and D_D are the depths and V_a, V_b, V_c, V_d, the corresponding velocities of flow at sections $A - A$, $B - B$, $C - C$, and $D - D$ respecting. See Fig. 25.8.

Design is started from the D/S side i.e., from section $D - D$.

1. Let bed level and cross-section of the canal at section $D - D$ is completely known.

Level of water surface at $D - D$

$$= \text{Level of bed at } D - D + D_d$$

T.E.L. at section $D - D$

$$= \text{Level of bed at } D - D + D_d + \frac{V_d^2}{2g}$$

2. Loss of energy in expansion from section $C - C$ to $D - D$ is taken 0.3 $\left(\frac{V_c^2 - V_d^2}{2g}\right)$.

Neglecting friction losses between section $C - C$ and $D - D$.

$$\text{T.E.L. at } C - C = \text{T.E.L. at } D - D + 0.3\left(\frac{V_c^2 - V_d^2}{2g}\right)$$

$$\text{Level of water at } CC = \text{T.E.L. at } C - C - \frac{V_c^2}{2g}.$$

Bed level at CC = Level of water at $CC - D_c$.

3. The flumed channel section between $B - B$ and $C - C$ remains constant and no loss, other than friction occurs in it which can be worked out using Manning's formula

$$Q = \frac{A\,R^{2/3}\,S^{1/2}}{N}$$

$\therefore$ T.E.L. at $B - B$ = T.E.L. at CC + head loss.

$$\text{Level of water at } B - B = \text{T.E.L. at } B - B - \frac{V_b^2}{2g}.$$

Bed level of channel at $B - B$

$$= \text{Level of water at } B - B - D_b.$$

Since velocity and depth are constant in the flumed portion, T.E.L, level of water surface and level of bed are parallel to each other between sections $B - B$ and $C - C$.

4. Loss of energy or head, between section $B - B$ and $A - A$ due to contraction is taken $0.2\left(\frac{V_b^2 - V_a^2}{2g}\right)$.

Neglect friction losses.

$$\text{T.E.L. at } A - A = \text{T.E.L. at } B - B + 0.2\left(\frac{V_b^2 - V_a^2}{2g}\right).$$

$$\text{Level of water at } A - A = \text{T.E.L. at } A - A \ \frac{V_a^2}{2g}.$$

Level of bed at $A - A$ = Level of water at $AA - D_a$.

5. Now bed level, water surface level and T.E.L. for all the four sections are known. The T.E.L. may be drawn straight between adjacent sections. The bed levels are also joined by straight line. If there is any fall it is provided either by rounding off the corners or with smooth reverse curve tangential to the bed. Smooth reverse curve is provided where drop in the bed levels is appreciable.

Drop in water surface line depends upon following two aspects.

(*i*) Increased velocity head for contraction and decreased velocity head for expansion.

(*ii*) Fall in T.E.L. between two adjacent sections.

This drop is accomplished by drawing two parabolas opposite to each other and meeting at mid drop points tangentially. Equation for parabola is

$$y = \frac{y_1}{x_1^2}x^2.$$

Where y = ordinate at any distance x measured from the origin. Origin point of first parabola is section $A - A$ and of second parabola section $B - B$.

$x_1 = \frac{L_f}{2}$ and y_1 = half of the Total difference in water levels between $A - A$ and $B - B$.

6. After having plotted the water profile over the full length of the flumed canal, the velocity head (h_a) at any point can be found, by determining difference between T.E.L. and corresponding water surface level at that point. The velocity head can be converted into equivalent velocity by formula

$$V = \sqrt{2gh_a}.$$

Thus the velocity at any point can be computed.

After having computed value of V, the area of cross-section at any point is worked out as follows.

$$A = \frac{Q}{V}.$$

If $S : 1$ is the slope of banks and B the bed width

$$A = BD + SD^2$$

Value of side slope can be interpolated in proportion to the length of transition from the starting point of the slope. Now A, D and S are known and Value of B can be computed at any point by substituting these values is equation $A = BD + SD^2$. Thus all the elements of transition are fully known.

4. Head loss through syphon. In case of syphon aqueduct and canal syphon, the head loss (h) is found out by unvin's formula given below.

$$h = \left(1 + f_1 + f_2 \frac{L}{R}\right)\frac{V^2}{2g} - \frac{V_x^2}{2g} \tag{25.4}$$

where f_1 = Coefficient of head loss at entrance whose value is taken 0.505 for unshaped mouth and 0.08 for bell mouth.

V_a = Velocity of approach.

V = Velocity of flow in m/sec.

R = H.M.D. of barrel.

L = Length of barrel in metres.

$$f_2 = a\left(1 + \frac{b}{R}\right)$$

Values of a and b for different materials are given as follows :

Barrel Material	*a*	*b*
Smooth iron pipe	0.00497	0.025
Encrusted pipe	0.00996	0.025
Smooth cement plaster	0.00316	0.030
Brick work or Ashlar	0.00401	0.070
Rubble masonry as stone pitching	0.00507	0.250

The velocity of flow through syphon aqueduct is limited between 2 m to 3 m/sec. Knowing the velocity, the head required to generate the same, can be found out from Eq. 25.4. H.F.L. on the *D/S* side remains unchanged but H.F.L. on the *U/S* increases by the amount of afflux calculated by Eq. 25.4. The height of guide banks and marginal bonds is governed by the height of afflux.

5. Determination of uplift pressure on bed of the canal or drain whichever is above. Once the afflux is found by Unwin's formula the hydraulic gradient line can be drawn and uplift pressure at various points determined. Uplift pressure will be proportional to the height of hydraulic gradient line above the underside of the trough. The maximum uplift pressure will be at the *U/S* point near the entry point and minimum at the *D/S* point near the exit. The trough is designed for the following two conditions.

(i) There is H.F.L. in the drain but no water in the canal.

(ii) Canal running full but there is no uplift pressure.

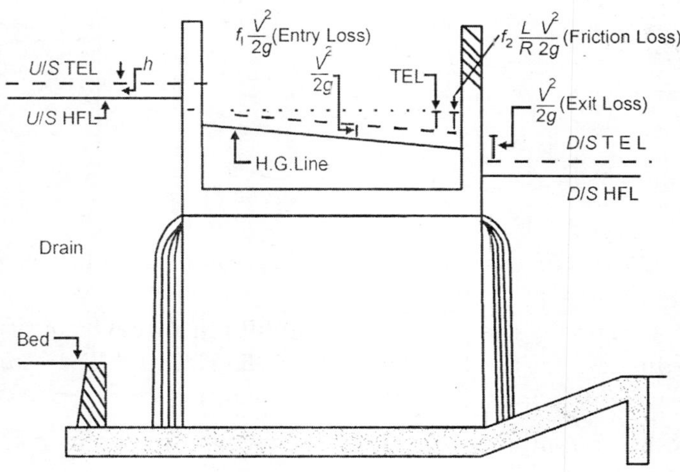

Fig. 25.9. *Pressures at various points of syphon aqueduct.*

6. Uplift pressure on bed of the canal or drain whichever lies at the lower level. The most critical condition for the uplift at the bottom of the drainage (lying below) occurs when there is no flow in drainage and canal is running at full supply level. In case of siphon aqueduct if the floor is depressed below the water table then most critical condition is considered when underground water Table has risen to drainage bed and there is no water in drainage but canal is running at F.S.L.

In order to find out the head causing seepage at point *C* at the *D/S* end of impervious floor of drainage, the seepage water is assumed as starting from point *A*, the *U/S* end point of the impervious floor of the canal. Seeping water is considered to follow the path *ABC*. Total length of creep path (*L*) is sum of length from *A* to *B* i.e. (L_1) and *B* to *C* (L_2). Total head causing seepage (H_s) is as follows. See Fig. 25.10.

H_s = F.S.L. of canal – D/S bed level of the drain Residual head (H_r) at point B is given as follows.

$$H_r = \frac{H_s}{L} \times L_2.$$

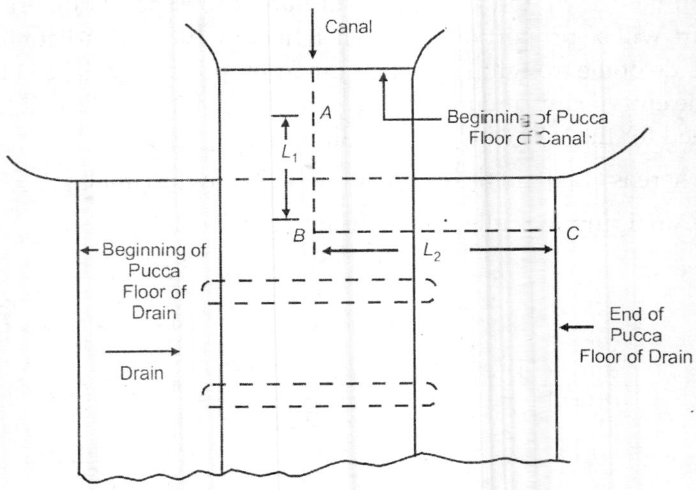

Fig. 25.10. *Seepage path of water from canal to drainage bed D/S.*

The floor is finally designed for total uplift pressure which is sum of uplift pressure due seepage head and static head. The Total uplift pressure is resisted partly by weight of floor and partly by the bending action.

The intensity of uplift pressure can bed reduced by following measures.

(i) Extend the impervious length of the canal bed *U/S*. This would increase the seepage length L_1.

(ii) By providing drainage holes in the floor of the drainage. To avoid the flow of soil through the drainage holes inverted filter is provided below them.

7. Design of Bank connections. The canal wings and drainage wings come under the category of Bank connections.

Drainage wings. They have to retain earth slope at their back. Their foundation should be deeper than the scour depth $\left\{R = 0.47\left(\frac{Q}{f}\right)^{1/3}\right\}$ which is taken as 2R. Entry and exit should be smooth. The length of the drainage wings should be large enough to accommodate the splay in the drainage transition.

Canal wings. They are usually extended upto the end of the splay. In order to provide easy transition from channel section to aqueduct side walls, they are sometimes warped. They should be designed as retaining walls. They should be taken deep into the soil to avoid any short cut by seeping water.

Example 25.1. *Design syphon aqueduct from following data.*

(i) *Canal discharge = 20 cumecs.*

(ii) *Width of canal = 17.5 m.*

(iii) *Depth of canal = 1.6 m.*

(iv) *Bed level of canal and general ground = 100.00.*

(v) *High Flood discharge of drainage 225 cumec.*

(vi) *H.F.L. of drainage on D/S side = 100.50 m.*

(vii) *Bed level of drainage = 98.00.*

Solution Canal section has been flumed as size of drainage is substantial.

(1) Design of drainage water way

$$P = 4.75\sqrt{Q} = 4.75\sqrt{225} = 71.25 \text{ metres.}$$

Provide 11 spans of 5 m each.

Width of one pier = 1.25 m.

Total length of water = 11 × 5 + 10 × 1.25 = 67.7 m.

Let velocity of flow through syphon = 2 m/sec.

$$\text{Height of barrels required} = \frac{225}{55 \times 2} = 2.045$$

Provide height of barrel = 2 m

$$\text{Actual velocity of flow} = \frac{225}{2 \times 5 \times 11} = 2.045 \text{ m/sec.}$$

2. Design of canal water way

Actual bed width = 17.5 m.

Let the width be reduced to 10 m.

Provide a splay of 2 : 1 in contraction.

The length of contraction transition

$$= \left(\frac{17.5 - 10}{2}\right) \times 2 = 7.5 \text{ m.}$$

Provide 3 : 1 splay in expansion.

$$\text{The length of expansion transition} = \left(\frac{17.5 - 10}{2}\right)^3$$

$$= 11.25 \text{ m.}$$

Length of Flumed portion of canal from abutment to abutment of the drainage = 6.75 m.

In the transitions the side slopes of the canal section will he warped from original slopes of 1.5 : 1 to vertical.

3. Levels at different sections

(*i*) *At section D – D*. (Fig. 25.8).

$A = BD + SD^2 = (17.5 + 1.5 \times 1.6) \times 1.6 = 31.84 \text{ m}^2$.

Velocity of flow $= \dfrac{Q}{A} = \dfrac{20}{31.84} = 0.628$ m/sec.

Velocity head $= \dfrac{V^2}{2g} = \dfrac{(0.628)^2}{2 \times 9.81} = 0.02$ m.

R.L. of Bed of canal = 100.00 (given)

∴ R.L.of water surface = 100.00 + 1.60 = 101.60

∴ R.L. of T.E.L. = 101.60 + 0.02 = 101.62 m.

(*ii*) *At section C – C*.

Area of Trough = 10 × 1.6 = 16 m².

∴ Velocity of flow $= \dfrac{20}{16} = 1.25$ m/sec

Velocity head $= \dfrac{V^2}{2g} = \dfrac{(1.25)^2}{2 \times 9.81} = 0.080$ m.

Loss of head in expansion from *C – C* to *D – D*.

$$= 0.3\left(\frac{(V_c)^2 - (V_d)^2}{2g}\right)$$

$$= 0.3\ (0.08 - 0.02) = 0.018 \text{ m.}$$

∴ Level of T.E.L at *C – C* = T.E.L. at *D – D* + head loss

= 101.62 + 0.018

= 101.638.

R.L. of surface of water at *C – C* = T.E.L. at *C – C* – velocity head

= 101.638 – 0.08

= 101.558 m.

R.L. of bed of maintain constant water depth

= R.L. if water surface – water depth

= 101.558 – 1.60 = 99.958 m.

At section B – B.

$$\text{H.M.D.} = R = \frac{A}{P} = \frac{10 \times 1.6}{10 + 2 \times 1.6}$$

$$= \frac{16}{13.2} = 1.212 \text{ m/sec}$$

From Mannings formula

$$V = \frac{R^{2/3} S^{1/2}}{N}$$

or $$V^2 = \frac{R^{4/3} S}{N^2}$$

$$S = \frac{V^2 N^2}{R^{4/3}} = \frac{(1.25)^2 \times (0.016)^2}{(1.212)^{4/3}}$$

$$= \frac{1.5625 \times 0.000256}{1.2922}$$

$$= \frac{1.5625 \times 2.56}{129.22} = \frac{1}{3230}$$

Length of flumed portion = 67.5 m.

Head loss in trough = $67.5 \times \frac{1}{3230} = 0.021$ m.

R.L. of T.E.L. at $B - B$ = R.L. of T.E.L. at CC + head loss in trough.

= 101.638 = 0.021 = 101.659 m.

R.L. of water surface at $B - B$ = 101.659 – 0.080 = 101.579 m.

R.L. of bed to maintain constant water depth

= R.L. of water surface at $B - B$ – Water depth.

= 101.579 – 1.600 = 99.979 m.

(*iv*) *At section A – A.*

Loss of head in contraction transition from section $A - A$ to $B - B$.

$$= 0.2\left(\frac{(V_b)^2 - (V_a)^2}{2g}\right)$$

Adopt $$V_a = V_d, V_b = V_c$$

$$= 0.2\left(\frac{(1.25)^2 - (0.628)^2}{2 \times 9.81}\right)$$

$$= 0.2\left(\frac{1.5625 - 0.3944}{19.62}\right)$$

$$= 0.2\left(\frac{1.1681}{19.62}\right) = 0.0119 \text{ m.}$$

$$= 0.012 \text{ m say.}$$

R.L. of T.E.L. at $A - A$ = R.L. of T.E.L. at $B - B$ + head loss

$$= 101.659 + 0.012$$

$$= 101.671 \text{ m.}$$

R.L. of water surface = R.L. of T.E.L. $- \dfrac{V^2}{2g}$

$$= 101.671 - 0.012$$

$$= 101.651 \text{ m.}$$

R.L. of bed to maintain constant water depth

$$= 101.651 - 1.60$$

$$= 100.051 \text{ m.}$$

4. Design of contraction transition

Use R.S. Chaturvedi's formula

$$x = \frac{L B_0^{3/2}}{B_0^{3/2} - B_f^{3/2}}\left[1 - \frac{B_f}{B_x^{3/2}}\right]$$

$$B_0 = 1.75 \text{ m}, B_f = 10 \text{ m}, L = 7.5 \text{ m.}$$

$$x = \frac{7.5 \times (17.5)^{3/2}}{(17.5)^{3/2} - (10)^{3/2}}\left\{1 - \frac{(10)^{3/2}}{B_x^{3/2}}\right\}$$

$$= \frac{549}{73.20 - 31.62}\left\{1 - \frac{31.12}{B_x^{3/2}}\right\}$$

$$= 13.20\left\{1 - \left(\frac{10}{B_x}\right)^{3/2}\right\}$$

Values of B_x for various values of x are given as follows.

x	0	2.5 m	5 m	7.5 m
B_x	10 m	11.50 m	13.74 m	17.50 m

Design of expansion transition

Again use R.S. Chaturvedi's formula

$$x = \frac{L B_0^{3/2}}{B_0^{3/2} - B_f^{3/2}}\left[1 - \left(\frac{B_f}{B_x}\right)^{3/2}\right]$$

$$L = 11.25 \text{ m}, B_f = 10 \text{ m}, B_0 = 17.5 \text{ m}.$$

$$x = \frac{11.25 \times (17.5)^{3/2}}{(17.5)^{3/2} - (10)^{3/2}}\left\{1 - \left(\frac{10}{B_x}\right)^{3/2}\right\}$$

$$= 19.8\left[1 - \left(\frac{10}{B_x}\right)^{3/2}\right]$$

For various values of x, values of B_x are given below.

x	0	2.5	5.0	7.5 m	10.0	11.25
B_x	10 m.	10.95	12.14	13.75	15.975	17.50

Design of Trough

Flumed width of canal = 10 m.

Provide it in two spans of 5 m width each ; with a partition wall 0.3 m thick. If side walls of the trough are 0.4 m thick, Total width of the Trough or in other words length of the syphon will be 10 + 0.3 + 2 × 0.4 = 11.1 m.

Height of Trough = Depth of water + free board. If thickness of bottom slab is also say 0.4 m the Total height of Trough including bottom slab will be 1.60 + 0.4 + 0.4 = 2.4 m.

The entire trough can be designed as a monolithic R.C.C. section.

Head loss through syphon barrels

$$h = \left(1 + f_1 + f_2 \frac{l}{R}\right)\frac{V^2}{2g}$$

V = Velocity through barrel = 2.045 m/sec.

f_1 = 0.505

$$f_2 = a\left(1+\frac{b}{R}\right)$$

a = 0.00316,

b = 0.03 for cement plaster, Length of barrel = 11.1 m.

$$R = \frac{A}{P} = \frac{5\times 2}{(5\times 2)^2} = \frac{10}{14} = 0.7143.$$

$$f_2 = a\left(1+\frac{b}{R}\right)$$

$$= 0.00316\left(1+\frac{0.03}{0.7143}\right)$$

$$= 0.00316\ (1 + 0.042)$$

$$= 0.00316 \times 1.042 = 0.00329$$

$$h = \left(1+f_1+f_2\frac{L}{R}\right)\frac{V^2}{2g}$$

$$= \left(1+0.505+0.00329\times\frac{11.1}{0.7143}\right)\frac{(2.045)^2}{2\times 9.81}$$

$$= (1 + 0.505 + 0.051)\ 0.21315$$

$$= 1.556 \times 0.21315$$

$$= 0.3317$$

D/S H.F.L. = 100.50

U/S H.F.L. = *D/S* H.F.L. + Loss of head

= 100.50 + 0.3317

= 100.8317 m.

Uplift pressure on the roof

R.L. of bottom of trough = R.L. of canal bed – Slab thickness

= 100.00 – 0.40 = 99.6 m.

Loss of head at entry of the barrel

$$= 0.505\frac{V^2}{2g} = 0.505\frac{(2.045)^2}{2\times 9.81}$$

= 0.1076 m.

∴ Uplift on the roof = 100 – 8817 – 0.1076 – 99.60

= 1.1241 m.

Uplift pressure = 1.1241 t/m^2.

Load of trough slab = 0.4 × 2.4 = 0.96 t/m^2

Unbalance upward pressure = 1.1241 – 0.96 = 01641 t/m^2

The un-balanced upward pressure has to be resisted by bending action of the slab.

Total load of Trough = (0.4 + 2 × 0.4 + 0.3 × 1) x 2.4

= (0.4 + 0.8 + 0.3) 2.4

= 1.5 × 2.4 = 3.6 t/m^2.

Since downward load of the trough (3.6 t/m^2) is more than the upward pressure (1.1241 t/m^2), hence no anchorage at piers is necessary.

When water level in the drain is low there would not be any uplift and trough slab will have to be designed for the full water load of the canal plus load of the canal itself. In this case, reinforcement bars shall be put at the bottom of the slab. For uplift, steel bars shall be provided near the top surface of the trough slab.

Uplift on the floor of the barrel and its design

(*a*) **Static head.** R.L. of barrel floor

= R.L. of bottom of trough slab height of the barrel

= 99.60 – 2.00 = 97.60 m

Assume thickness of the floor slab 60 cm tentatively.

∴ R.L. of bottom of floor = 97.60 – 0.60 = 97.00 m

Drainage bed level 98.00 given

Assume the ground water table upto the bed level of the drain.

The static lift on the floor

= 98.0 – 97.0 = 1 m

(*b*) **Seepage head.** Residual seepage head at point *B*, the centre of the first barrel. (Fig. 25.10).

Neglecting thickness of various floors, the total length of the seepage path will be sum of following lengths.

(i) *U/S* transition length = 7.5 m

(ii) Half of barrel width = 2.5 m

(iii) End of impervious floor = 10 m say

Total creep ABC = 20 cm

Creep length upto centre of the barrel 10 m. (Point B).

Total seepage head = canal F.S.L. – R.L. of drainage bed

$$= 101.60 - 98.00 = 3.60 \text{ m}$$

Residual head at B

$$= 3.60\left(1 - \frac{10}{20}\right)$$

$$= 1.80 \text{ m}$$

Total uplift = static head + seepage head

$$= 1 + 1.800 = 2.80 \text{ m}$$

$$= 2.80 \text{ t/m}^2$$

Weight of base slab of drain = 0.60×2.4 = 1.44 t/m^2

Net upward pressure to be resisted by the bending

$$= 2.800 - 1.440$$

$$= 1.36 \text{ t/m}^2$$

Suitable reinforcement shall have to provided at top of the floor to counteract the bending action.

Length of the floor of siphon has been adopted 20 m on following consideration.

1. Length of floor under barrel = 11.1 m

2. Extra length to accommodate pier projections on both the sides. Assume each projection 1 m = 2 m

3. Horizontal length of the D/S ramp joining to the bed level at slope of

$5 : 1 = 5\,(98 - 97.60) = 2$ m

Width of D/S cut off = 0.5 m

= 15.6 m

4.4 m length of floor has been provided on the U/S side of the piers of the canal.

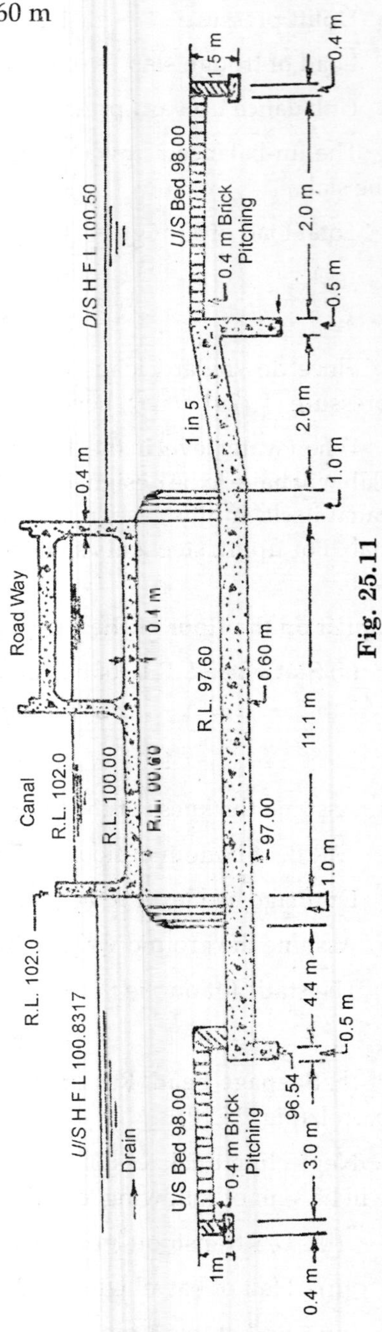

Fig. 25.11

Design of cut-off and protection works for the drainage floor.

$$\text{Scour depth } R = 0.47\left(\frac{225}{1}\right)^{1/3}$$

$$= 2.859 \text{ m}$$

Depth of *U/S* cut off below H.F.L.

$$= 1.5 \times R = 1.5 \times 2.859 = 4.29 \text{ m}$$

R.L. of bottom of *U/S* cut off

$$= U/S \text{ H.F.L.} - 4.29$$

$$= 100.8317 - 4.9$$

$$= 96.5417 \text{ m}$$

Depth of *D/S* cut off below H.F.L.

$$= 2R = 2.859 \times 2$$

$$= 5.718 \text{ m} = 5.72 \text{ m (say)}$$

∴ R.L. of bottom of *D/S* cut off

$$= D/S \text{ H.F.L.} - 572$$

$$= 100.50 - 5.72$$

$$= 94.78 \text{ m}$$

Length of *D/S* protection consisting of 40 cm brick pitching

$$= 2.5 \text{ (R.L. of } D/S \text{ bed} - \text{R.L. of bottom of } D/S \text{ cut off)}$$

$$= 2.5\,(98 - 94.78) = 2.5 \times 3.22$$

$$= 8.05 = \text{say } 8 \text{ m}$$

This pitching is supported by a toe wall 40 cm thick and 1.5 m deep at its *D/S* end.

Length of *D/S* protection consists of 40 cm brick pitching

$$= 2 \text{ (R.L. of } U/S \text{ bed} - \text{R.L. of bottom of } U/S \text{ cut off)}$$

$$= 2\,(98.00 - 96.5417)$$

$$= 2 \times 1.4583$$

$$= 2.9166 \text{ m}$$

$$= \text{say } 3 \text{ m}$$

This pitching may be supported by a toe wall 40 cm wide and 1 m deep at the *U/S* end.

For details see Fig. 25.11.

QUESTIONS

25.1. Enumerate the situations where cross drainage works are necessary to be provided.

25.2. How the cross-drainage works can be classified ? Draw sketches of syphon aqueduct, and super passage.

25.3. What do you understand by level crossing ? Under what circumstances level crossings are preferred ? What do you understand by inlets and outlets

25.4. (a) What points should be considered while choosing site for cross-drainage work ?

(b) Oil what grounds the selection of the type of cross-drainage works is made ?

26

River Training Works

26.1 INTRODUCTION

We have already discussed earlier in the book that river length may be divided in four lengths namely mountainous stage, sub-mountainous stage, alluvial stage, and deltic stage. Out of these stages the first two stages i.e. mountainous and sub-mountainous are also known as upper reaches of the river. In these two stages, the path of the river is well defined between solid high rocky banks and as such almost no river training works are required. These stages may be useful for storage works but not for irrigation works. The reasons are simple that these stages are quite undulating and hence construction of canal is very difficult. Secondly irrigable area is always available in alluvial stage and hence leading canal from sub-mountainous stage to alluvial plains increases length of the canal. Keeping above considerations in view, the most of the irrigation works are located in alluvial plains. In alluvial plains the area is plain and river slopes is very gentle. If weir or barrage for diverting river water is constructed across the river, the river water will spread on large areas *U/S* and cause lot of submergence. In order to contain the rivers in their specified route of flow, whatever works are constructed are known as *river training works.*

26.2 CLASSIFICATION OF RIVERS ON ALLUVIAL SOILS

The rivers on alluvial soils may be classified into three types:

1. The meandering type ;
2. The aggrading type ; and
3. The degrading type.

1. The meandering type river. A meandering type river has a series of consecutive curves. The curvatures of consecutive curves are opposite to each other. There may be short straight lengths between consecutive curves known as *crossing*. The axial distance between peaks of two curves of the same order is known as *length of meander* (M_L). The lateral or transverse distance between the peak point of one curve and peak point of reverse curve is known as *width of meandering belt* (M_b). The distance is measured along the width of the river. The ratio between the curved length and the straight length is known as *degree of sinuosity*. It is also sometimes known as *tortuosity*. The ratio $\frac{M_b}{M_L}$ is known as *meander ratio*. See Fig. 26.1.

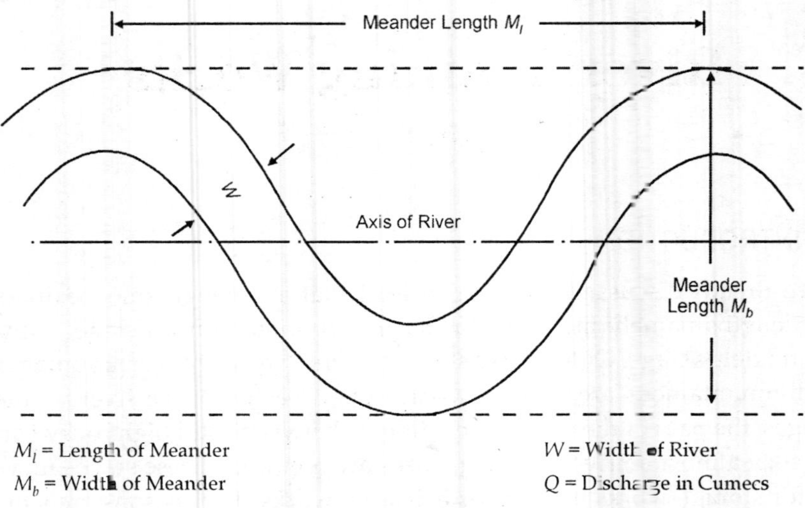

Fig. 26.1

Causes of meandering. According to J.S. Firedkin local bank erosion and subsequent local over-loading of silt causes meandering. According to Inglis if lower layers of water are charged with heavy bed load, the river cannot maintain symmetrical axial flow and results in an uneven bed. This causes cross flow in the river leading to scouring on one side and silting on the other side. This phenomenon continues till equilibrium is attained.

One more thought suggests the cause of meandering as the excess of silt charge during floods. Heavily charged silt water of the river tends to build steeper slopes, by depositing the silt charge on the bed. This increase in slope tends to widen the river, if banks are not resistant. At this stage a slight deviation from axial flow would cause more flow towards one bank than the other and this process once set, would develop curved flow and ultimately meandering flow.

Elements controlling the process of meandering

Following are the basic elements that control the process of meandering:

1. Discharge. The rate of bed-load movement is closely related to the rate of discharge and hence discharge in the river affects meandering.

2. Steam load. Composition of stream load as well as its rate of movement affect the meandering.

3. Valley slope. Change in bed slope produces changes in meandering.

4. Resistance of section. Composition of bed and sides of the river also have bearing on the meandering tendencies. Grain size, specific gravity, cohesion, and roughness are the important characteristic of the section.

General features of meandering. Various experiments reveal the following general features of meandering.

1. The width of river, width and length of meander vary roughly with square root of discharge.

2. A fully developed meander of homogeneous valley material with a constant discharge has a definite pattern of curvature, length, width and depth of the channel which can be reproduced in models.

3. Variation in discharge, slope of basin and slope of bed change the pattern of meander.

4. Increase in discharge or slope increases the meandering feature like size of bends, their length, width, and degree of sinuosity.

5. Increase in sediment charge increases the slope as well as width of the river while the depth of the river is reduced.

In general it can be appreciated that where a river has enough capacity to carry the incoming sediment *D/S* with forming large deposits, the whole or a part of it would be meandering type.

M_l, M_D, W and Q have been linked by Mr. Inglis after analysing large data on Indian rivers. The relation is given in Table 26.1.

Table 26.1. *Inter-relationship between* M_l, M_b, W *and* Q

M_l, M_b or W *in metres*	*River in flood plain*	*Incised rivers (rivers flowing in a cross-section cut below the natural ground surface)*
M_l	6.06 W	11.45 W
M_b	17.40 W	27.30 W
M_b	2.86 M_l	2.2 M_l
M_l	$53.25\sqrt{Q}$	$46\sqrt{Q}$
M_b	$153.75\sqrt{Q}$	$102.25\sqrt{Q}$
W	$8.87\sqrt{Q}$	$4.54\sqrt{Q}$

Q is discharge in cumecs.

The agrading type of river. This type of river is in the process of building up its bed to a certain slope. This may be due to :

(i) Sudden inflow of sediment from a tributary.

(ii) Excessive sediment entering the river with a sudden dimunition of slope on the plain.

(iii) Excessive sediment entering the river, but slope is flattened by works like weirs, barrages or dams.

(iv) Extension of delta at the river mouth.

An aggrading river section is usually wide straight reach with shoals in the middle.

The degrading type of river. This type of river is in process of loosing its bed i.e. its bed is scouring. When the sediment load of a river is held up by a dam *U/S*, a degrading reach may develop *D/S*. If sediment load is decreased suddenly at some point, it starts scouring *D/S* of that point to replenish the sediment load. The degrading rivers increase slope of the river.

A river in alluvial plain is seldom of a single type. All the three types may be found on the same river along its length. In rivers, re-adjustment of hydraulic dimensions is always occurring. The stream tries to adjust to the prevailing conditions. Variations in discharge and silt-charge get adjusted by corresponding changes in cross-section and slope. Hence, any particular section of a river may be aggrading, degrading or meandering at different times, depending upon the charge and size of sediment in that section.

The river portion on the *U/S* of a barrage of dam is of aggrading type. It is usually a straight and wide length with shoals in the middle and with a divided river flow.

The river portion on *D/S* of the barrage of dam is of *degrading type.*

The meandering type is the final stage of the river development, while other two types are only the interim stages.

26.3 RIVER TRAINING

This term includes various measures, that when adopted compel the river to the flow through the specified path only. Following are the usual objects of river training works.

(i) To achieve safe and expeditious passage of flood through the river. This is usually achieved by constructing marginal bunds or dikes and providing sufficient water way between them. Floods can also be controlled by constructing reservoirs on the river or sometimes by dredging certain sections or by inducing artificial cut-offs.

(ii) To achieve efficient transport of bed silt and suspended silt. This is achieved by keeping a channel in stable condition.

(iii) To achieve stable stream course with minimum bank erosion. This can be achieved by pitching the banks.

(iv) To achieve sufficient depth of flow during lean periods, so as to ensure navigation through the river.

(v) To achieve flow of river through defined width of river so as to ensure flow of river through the hydraulic structures like weirs, barrages, bridges etc. If flow through specified width is not ensured the river may change its course and all the structures constructed over it are out flanked.

Keeping in view the above objects, the river training works may be classified into following three categories:

(i) High water training ;

(ii) Low water training ; and

(iii) Mean water training.

(i) High water training. Under this, the emphasis is given to provide sufficient and efficient cross-section to pass down the maximum floods without any difficulty. The main concern is in regard to location and fixing height of marginal bunds and guide bunds. It is also sometimes known as training of the river for the discharge.

(ii) Low water training. In this case, river section is restricted, so that sufficient depth of water remains available in the river during dry months for navigation purposes. This training is sometimes known as training for the depth.

(iii) Mean water training. This training is done to correct the river bed configuration, so that transport of sediment load is effectively carried out. This training does not allow silting of channels. It is also called *training for the sediment*.

26.4 TYPES OF RIVER TRAINING WORKS

River training works may be directive or protective or both. Types of river training works depend on the purpose to be achieved. The following are the usual types of river training works :

1. Guide banks.
2. Marginal bunds.
3. Revetment of banks.
4. Artificial and natural cut-offs.
5. Groynes or spurs.
6. Pitched islands.

26.5 GUIDE BANKS

They are used to guide the river to pass through the constrained width of the river at the structure. Guide bank system of river training at bridges and diversion weirs, was first introduced by Mr. J.R. Bell. Hence guide banks are also known as Bell's Bunds in his honour.

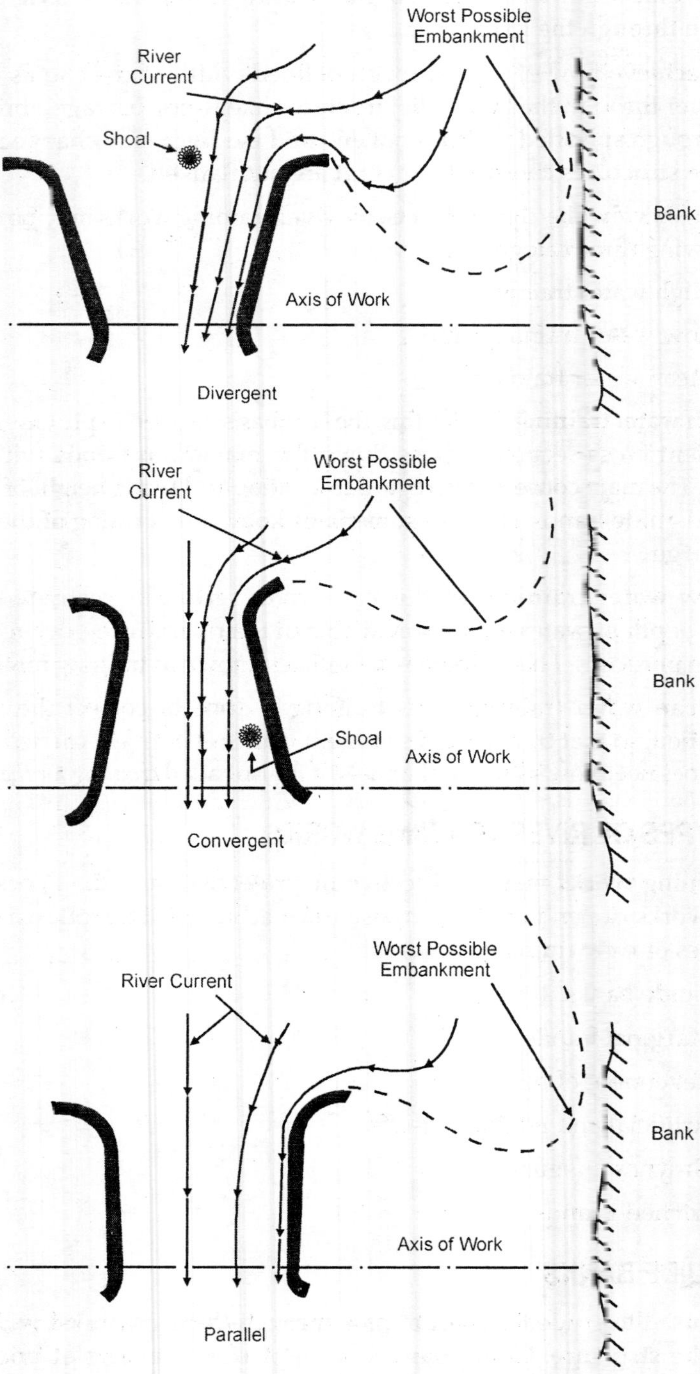
Worst Possible Embankment
River Current
Shoal
Bank
Axis of Work
Divergent
Worst Possible Embankment
River Current
Bank
Shoal
Axis of Work
Convergent
Worst Possible Embankment
River Current
Bank
Axis of Work
Parallel

Fig. 26.2

The guide banks are provided in pairs, symmetrical in plan. Both the guide banks may be kept parallel, slightly converging or slightly diverging. Mostly parallel system of guide banks is used. The length of the guide bank may be

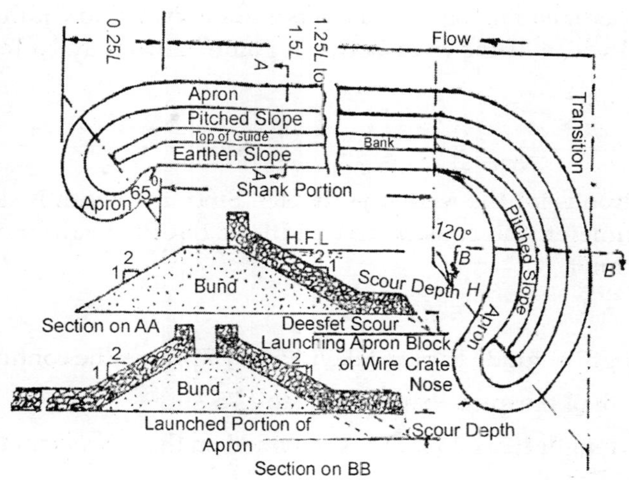

Fig. 26.3. *Bell's guide bank.*

equal to the lineal waterway of the weir or barrage. In plan guide banks have curved ends. The angle subtended at the centre, by the curved end at *U/S* side, ranges from 120° to 145° and that at *D/S* end ranges from 45° to 50°.

If one bank of the river is quite high and stable it can be used as one of the guide banks. In that case only one guide bank will be required to be constructed.

Purpose of Guide Banks

(i) In order to reduce the cost of construction of diversion works their length has to be reduced. The length of the diversion works is constrained or restructed between the guide banks.

(ii) They confine the river within a reasonable water way.

(iii) They direct the flow in such a manner that ensures safe and expeditious passage through the work.

(iv) They protect the works from being outflanked.

(v) They protect adjacent land from being flooded due to afflux caused by the construction of the work in the river.

The effects of confining the river between the guide banks are the following

(i) Water level on the *U/S* side is increased.

(ii) Slope of water surface *U/S* of the work is reduced.

(iii) Velocity of flow and scouring action between marginal bunds is increased.

(iv) Increase the rate at which the flood wave travels down the stream.

(v) Increase the maximum discharge at all point D/S.

Design Features of Guide Banks

1. Wetted perimeter in the case of rivers is almost equal to width of the river. The confined width of the river between guide banks may be found out by $4.75\sqrt{Q}$ formula

$$L = 4.75\sqrt{Q}.$$

This equation is for the wetted perimeter. Since the width is slightly more than the perimeter, the constrained width of river (L) can be found from following formula

$$L = 5\sqrt{Q}.$$

2. Total length of guide bank is taken 1.5 to 1.75 times the confined width L.

Total Length of the guide bank = 1.5 L to 1.75 L.

Out of this length 1.25 L to 1.50 L is provided on the U/S side of the weir and 0.25 L on the D/S side.

3. The shank or straight portion of the bond should have 6 m wide top. Side slopes vary from 2 : 1 to 3 : 1 depending upon the type of material.

4. 1 m to 1.50 free board should be provided.

5. The inside slope should be protected by stone pitching ; the usual thickness of pitching varies from 40 cm to 60 cm. The thickness of pitching as recommended by Inglis may be determined from following formula $t = 0.60\ Q^{1/3}$.

where t is thickness in cm and Q is discharge in cumecs.

6. The radius of the curve at the U/S end of the bank is 150 m to 300 m and central angle varies from 120° to 140°.

Mr. Spring suggested a value of R equal to 180 m to 250 m for rivers having velocities 2.4 m/sec to 3.1 m/sec gales on the other hand suggested a value of 250 m for rivers having flood discharge between 7000 to 20000 cumecs. Sharper curves are permissible for discharge less than 7000 cumecs. A value of 580 m is recommended for discharge varying from 40000 to 70000 cumecs. The radius may be obtained by interpolation for discharges between 20000 and 40000 cumecs. The value of R can be obtained by relation $R = 0.45\ L$.

Radius of curvature of D/S end varies from 90 m to 240 m and central angle from 45° to 60°.

7. Maximum scour depth (D) can be calculated from formula

$$D = 0.47\left(\frac{q}{f}\right)^{1/3}$$

Scour depth at the bond may be assumed 1.5 *D* to 1.75 *D*.

8. The toe of the slope should be protected by the Launching apron, the thickness of which is generally kept 1.25 times the thickness of pitching. At places where deep scour is expected, it may be kept 1.5 times the thickness of pitching. The thickness of apron is kept same as pitching at toe. Apron thickness on the *U/S* curved head is generally kept 1.25 times the normal apron thickness.

26.6 MARGINAL BUNDS

They are also called Levees. They are earthen bonds whose face towards the river is generally pitched. They are constructed parallel to river banks to protect the marginal land from inundation caused by floods. They should be located at the limiting line of meandering of river i.e. they should be beyond the meander belt of river. Effects of levels or marginal bunds are same as those enumerated in case of guide banks.

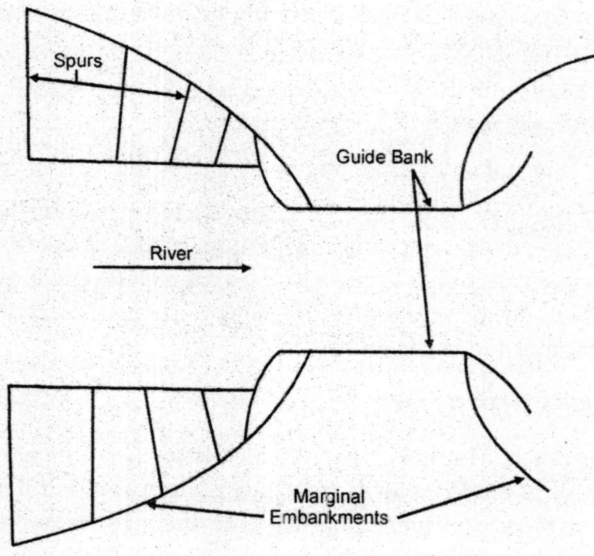

Fig. 26.4. *Marginal bunds.*

Advantages of Marginal Bunds

1. The restricted area of the river causes increased depth of water. This causes improvement in Navigation facilities.

2. It prevents submergence of the adjoining land during floods.

3. Since water is made to flow in between the dykes, the velocity of flow gets increased, causing the peak discharge to flow *D/S*.

4. They prevent course changing tendencies of the river.

Disadvantages of Marginal Bunds

1. There is rise in H.F.L.

2. They require heavy maintenance.

3. If fail, they inundate large low lying areas thus causing loss of property and life.

4. They can be easily eroded.

5. The rise in water level increases sub-soil water level also. This makes surrounding areas, water logged.

6. Spread of silt on the country is prevented.

Design Features of Marginal Bunds

Spacing and height. Both these elements are interdependent. If levees are located closer their height has to be more. If reverse is the case i.e., levees are retired their height will be less :

1 m to 1.25 m free board should invariably be provided over the maximum rise expected during floods.

Top width. Minimum Top width should be 3 m. Top width may be kept more than this if required.

Side slopes. Internal side slope may be 3 : 1 to 5 : 1.

Outer side slope may be 4 : 1 to 7 : 1 Banquette or counter-berms may be added on outer side, to prevent sloughing of high banks: Muck trenches should be constructed to act as cut-off to check the seepage along the foundation plan. In order to collect and dispose off any seepage that may be occurring, ditch is provided at the outer toe of the levee.

26.7 REVETMENT OF BANKS

It is a protection of boulders or concrete blocks to the river bank. A flexible apron from the toe of bank, towards river side is also provided Bank revetment also exerts an attracting influence by drawing the river water towards it on account of deep scours formed at the Toe of the bank. The river channel is, in this way held permanently at the pitched banks.

26.8 BANK PROTECTION

Any protective work whose purpose is to maintain the stability of the bank against erosive effect of water is known as *bank protection*. The purposes of bank protection are the following :

(i) To resist erosion of banks by water currents.

(ii) To prevent sliding of soil due to drawdown of flood.

(iii) To prevent piping of water through the banks.

(iv) To afford facilities for water transportation.

(v) To protect hydraulic structures.

(vi) To protect flood embankments.

If bank slopes are subjected to strong currents, the bank protection may be provided in form of vegetal cover, either by turfing or growing low shrubs. If the water currents are very strong, protection has to be provided by stone pitching or various types of mattresses such as willow, asphalt, or articulated concrete. Stone is the most commonly used material for protection of banks where available locally. The thickness of the pitching is governed by the velocity of the current near the bank. Mr. Spring gave following thicknesses of pitching for different types of soils and different bed slopes of the river.

River Material	*Thickness in metres for river bed slopes in cm per km*				
	5	*15*	*20*	*30*	*40*
Very coarse	0.40	0.475	0.55	0.575	0.70
Coarse	0.55	0.575	0.77	0.775	0.86
Medium	0.70	0.775	0.85	0.925	1.00
Fine	0.85	0.925	1.00	1.075	1.15
Very fine	1.00	1.705	1.15	1.225	1.30

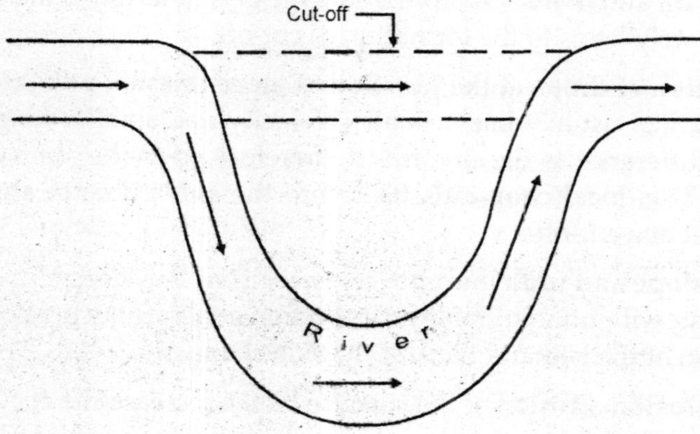

Fig. 26.5. *Cut-off.*

15 cm thick filter layer should be provided behind the pitching to prevent failure by sucking action of high velocity flow. Launching apron is a must at the Toe, to protect the bank against the effects of the flood scour.

26.9 ARTIFICIAL AND NATURAL CUT-OFFS

When meandering river develops very sharp horse-shoe bends, the land between successive peaks near the same bank of river is very much reduced. If small cut is given to connect the peaks, the water of the river will rush through the cut-off and develop into a full-fledged channel section. Rush of water takes place because more steep slope is available along the cut-off rather than the curved path. The cut-off may be artificial or natural, usually it is artificial. Cut-off can be achieved without giving any cut, by concentrating more of river water at the bottleneck by putting a spur or groyne in the natural flow at the curve. Cut-off induces following effects :

(i) Steep slope being available along the cut-off, river water flow is more through the cut-off rather than curved path.

(ii) Discharge in curved path, being considerably reduced, is silted up in due course of time and river flow becomes straight.

(iii) Water level in the river on *U/S* side is reduced.

(iv) Due to scouring action, the regime of *D/S* stream is disturbed.

(v) Channel taking off from just *D/S* of cut-off may get silted up.

Development of cut-off

Cut-off develop due to following reasons.

1. Development of bars at infliction. The flow through the main channel is reduced because of the growth of bars at inflections. This induces flow through already existing shallow side channels. Discharge is main channel being reduced continues to silt and more and more discharge gets diverted to the side channel which ultimately leads to the formation of cut-off.

2. Formation of drops at the junction of main channel with side channel. The side channel usually has a smaller velocity and smaller length with the same head difference as the channel. It, therefore, joins the main channel at a local drop. This local drop cuts back into the side channel and tends the development of a cut-off.

3. Steep slope and unfavourable cross-section. The steep slope of the side channel along with unfavourable cross-section generate such velocities which cause erosion of the bed and bank of the side channel.

4. Bend erosion. Erosion of the concave banks increase the curvature of the river. The erosion might occur to such an extent that the arms of a loop cut into one another and cut-off occurs.

5. Duration of flood. The duration of flood should be sufficient so that stream may erode its bed and banks.

26.10 SPURS OR GROYNES

Groynes are structures, built transverse to the river flow, extending from the bank towards the river. Groynes, or spurs are also sometimes known as transverse dykes. They perform the following functions :

(i) They contract the width of the river. Thus depth of water in the river increases, and navigation during slack periods becomes possible.

(ii) They protect the river banks by keeping the flow away from it.

(iii) They create slack flow in the vicinity and thus cause silting of the area.

(iv) They train the river to flow along a specified coarse.

Classification of Groynes. Following are the possible classifications of the groynes :

1. According to permeability.
2. According to height.
3. According to their actions.
4. Special groynes.

1. According to permeability. According to this classification the groynes may be :

(i) Permeable groynes.

(ii) Impermeable groynes.

Permeable groynes may be made of bushes, bamboos, piles, matresses. They are cheap and popularly used but they are temporary. The water can pass through them gradually the silt getting deposited in the permeable spurs make them almost impermeable.

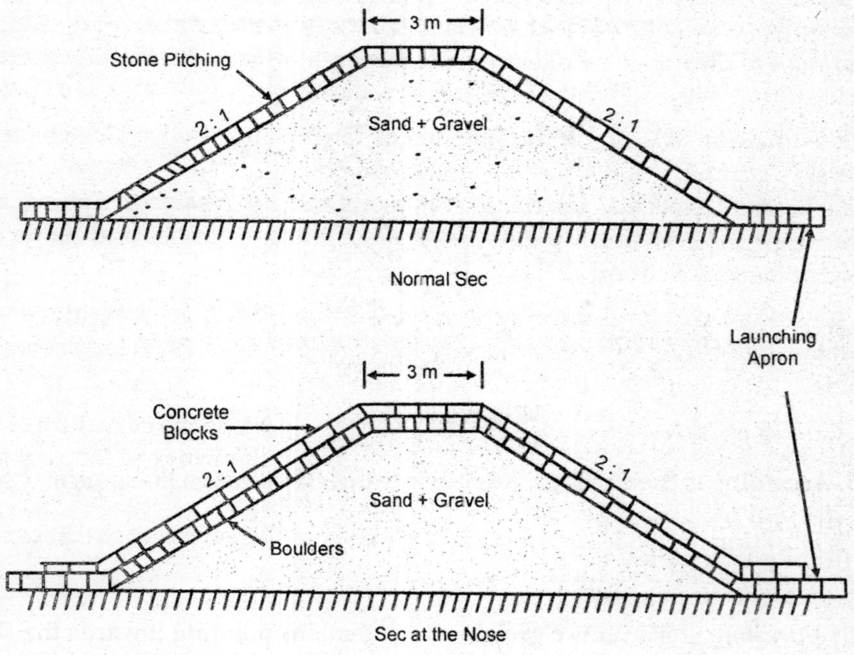

Fig. 26.6

Impermeable spurs may be made of earth embankments with necessary pitching and flexible apron. They do not allow water to pass through them. They are used at important places to protect the river banks.

2. According to height. The spurs may be submerged type or non-submerged type. If spurs remain projecting above water level during, floods they are known as non-submersible spurs. Those that get sub-merged are known as sub-merged spurs.

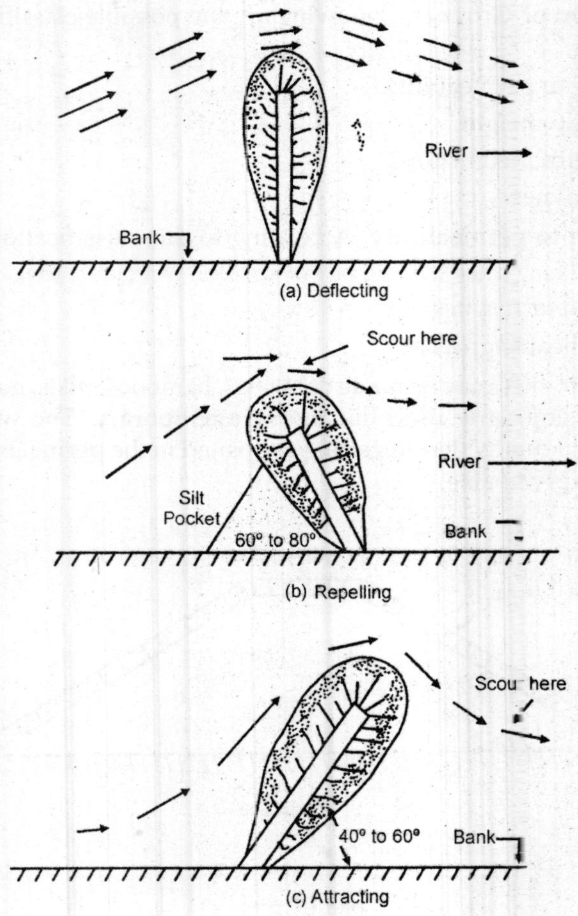

Fig. 26.7. *Spurs.*

3. According to their actions. According to this classification the spurs may be :

(i) Attracting groyne

(ii) Repelling groyne

(iii) Deflecting groyne.

(*i*) *Attracting groyne.* It is a groyne which remains pointing towards the *D/S* of the flow. Its main purpose is to attract river flow towards that bank from which the spur is originating. This spur has to be very strong as it has to bear the full frontal attack of the river on it *U/S* face.

(*ii*) *Repelling groyne.* This spur is constructed pointing towards *U/S*. The angle of inclination with the normal to the bank varies from 10° to 30°. The head of the spur should be strong to resist the swirling action of the currents. Its best location is just at the beginning of the damaging river loop. One spur is effective for a length of about 1 to 2 times its length. The distance between spur is kept more in case of convex banks and small for concave banks.

(*iii*) *Deflecting spur or groyne.* It is a spur that is short in length than repelling groyne. It is constructed perpendicular to the bank.

4. Special groynes. (i) *Hockey type groyne.* It is curved spur. It has attracting characteristic of flow. It is not very helpful for bank protection.

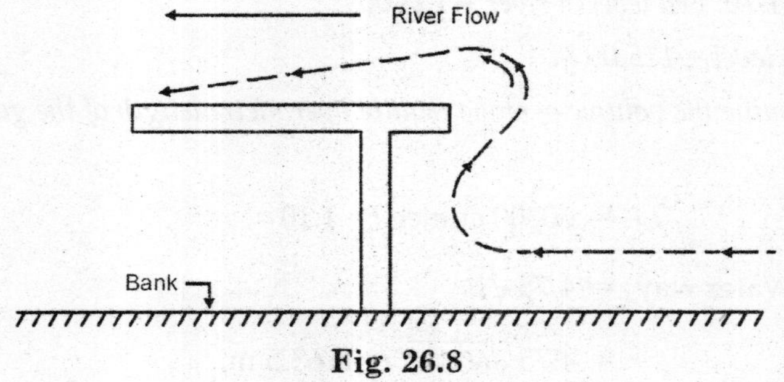

Fig. 26.8

(*ii*) *Denehy's T-headed groyne.* It is a *T*-shaped spur. The longer arm of the *T*-head remains in the *U/S* direction of the flow. The *T*-head is protected by stone pitching. Such groynes may be spaced at a distance of say 800 m or so.

26.11 PITCHED ISLANDS

It is an artificially created island in the river. It may be made of masonry or earth embankment, but pitched alround. On account of the disturbances created

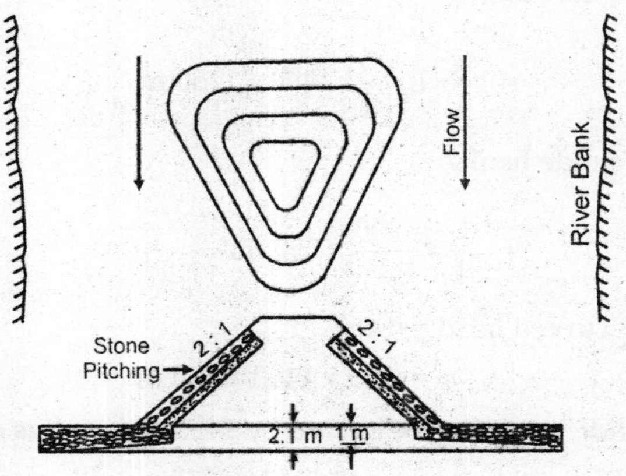

Fig. 26.9. *Pitched island.*

by the islands placed in series; a deep river channel is formed and flow is diverted away from the bank. The measure is not successful if river is shallow and wide, as scouring velocities cannot be developed by pitched islands.

26.12 BENGAL BANDHAL

It is permeable spur made of bamboos. A line of bamboos is prepared across the bed of the stream. The bamboos are tied together with ropes. The mats are fixed vertically to the bamboos to provide an obstruction or temporary fence.

Example 26.1. *Design a guide bank required for a bridge on a river having following particulars:*

Maximum flood discharge 60000 cumecs.

Silt factor 1.10, Bed level of river = 103.00

High flood level = 115.00 m

Also determine the volume of stone required per metre length of the guide bank.

Solution

$$Q = 60000 \text{ cumec } f = 1.10$$

$$\text{Water way} = 4.75\sqrt{P}$$

$$= 4.75\sqrt{60000} = 1163.5 \text{ m.}$$

Provide 20% more length to account for thickness of piers abutments, end contractions due to a butments and piers etc.

$$= 233 \text{ m (Approx.)}$$

Gross length between banks

$$= 1163.5 + 233 = 1396.5 \text{ m}$$

Adopt gross length of 1400 m

U/S length of guide banks

$$= \frac{5}{4} \times L = \frac{5}{4} \times 1400 = 1750 \text{ m}$$

D/S length of guide banks

$$= \frac{L}{4} = \frac{1400}{4} = 350 \text{ m.}$$

Radius of *U/S* curved head = 0.45 *L*

$$= 0.45 \times 1400 = 630 \text{ m}$$

U/S end of guide bank may be curved by 145° with radius of 630 m.

Radius of *D/S* curved head may be kept as 315 m with an angle of 60 at the centre.

Cross-section of guide bund

H.F.L. = 115.0 m . Assume free board of 1.5 m

Level of top of guide bund = 116.5 m.

To be make allowance for an settlement adopt top level of bund 117 m instead of 116.5 m.

Height of bund = 117 – 103.0 = 14 m.

Keep Top width of 4 m and side slope 2 : 1

The stone pitching and a launching apron must be provided on the slope on water side for the entire length of the bund. However the rear side also will be pitched for curved portions of the bund.

Design of stone pitching and apron

Thickness of stone pitching = $0.06\ Q^{1/3}$

$$= 0.06(60000)^{1/3} = 2.35 \text{ m}$$

This can be kept 1 m above H.F.L. i.e upto level 116 m

Depth of scour $R = 0.47 \left(\frac{Q}{f}\right)^{1/3}$

$$= 0.47\left(\frac{60000}{1.1}\right)^{1/3} = 17.82 \text{ m}$$

For straight reach of guide bund

Maximum scour = $1.25\ R = 1.25 \times 17.82 = 22.28$ m

R.L. at maximum anticipated scour

$$= 115 - 22.28 = 92.72 \text{ m}$$

Depth of maximum scour = 103 – 92.72 = 10.28 m.

Length of apron = $1.5\ D = 1.5 \times 10.28 = 15.42$ m

For curvilinear transition portion of guide bund, Maxi scour

$$= 1.5\ R = 1.5 \times 17.82 = 26.73 \text{ m}$$

R.L. of Maximum scour = 115 – 26.73 = 88.27 m

Depth of Maximum scour = D = 103 – 88.27 = 14.73

Length of apron = $1.5\ D = 1.5 \times 14.73 = 22.1$ m

Thickness of launching apron = $1.9 \times T$

$$= 1.9 \times 2.35 = 4.47 \text{ m}$$

Volume of stone

At shank – on slope = $\sqrt{5}\,(116 - 103)\ 2.35 \times 1 = 68.3 \text{ m}^3/\text{m}$

On apron with a slope 2:1

In a vertical length D and thickness $1.25\ T$.

$$\text{Volume} = \sqrt{5} \times D \times 1.25\, T \times 1 = \sqrt{5} \times 10.28 \times 1.25 \times 2.35$$
$$= 67.52 \text{ m}^2/\text{m}.$$

***U/S* and *D/S* curved portions**

$$\text{Volume of apron} = \sqrt{5} \times 14.73 \times 1.25 \times 2.35 \times 1$$
$$= 96.75 \text{ m}^3/\text{m}$$

$$\text{Volume of slope} = \text{same as for shank} = 68.3 \text{ m}^3/\text{m}$$

$$\text{Thickness of launching apron} = \frac{\text{Volume of apron}}{\text{Width}} = \frac{96.75}{1.5 \times 14.8}$$
$$= 4.38 \text{ m}$$

QUESTIONS

26.1. Explain the term meandering. What are its causes?

26.2. Give detailed classification of rivers on alluvial soils.

26.3. What do you understand by term river training? What are the works that come under this category ?

26.4. Draw a detailed sketch of a guide bank. Give its design features.

26.5. Write short notes on :

(i) Marginal bonds.

(ii) Pitched Islands.

(iii) Revetment.

(iv) Cut-off.

26.6. What are the functions of groynes ? Give their various classifications.

27

Introductory Water Power Engineering

27.1 SOURCES OF ENERGY

Following are the sources from which energy can be obtained to run our machines.

1. Coal. It is one of the oldest sources of energy. Even today it is one of the major sources. Wherever coal is available the concerned countries are trying to exploit it to the maximum extent so as to escape the effects of spiralling price of petrol and diesel.

2. Crude oil. Liquid fuels like petrol, diesel and kerosene, are obtained from crude oil. This source of energy is becoming very costly day by day. The demand of this source of energy is very large in relation to its formation by natural processes and still demand is going up day to day. This source is not unlimited and hence mankind is direct need of some alternative source of energy which may replace petroleum fuels.

3. Atomic power. This source of energy is being exploited on a very large scale throughout the world. But there is a risk attached with this source. The countries having atomic power sources, may develop atomic bombs.

4. Winds. In old days wind mills used to exploit this source. But this source is not of much use. The reason being that wind characteristics keep on changing.

5. Solar energy. Lot of research work is going on to exploit this limitless natural source of energy. We should hope that in near future this source would be supplying us ample of energy to run our machines.

6. Flowing water in rivers and canals. This is also a very large source of energy. But all the countries of the world may not be having rivers and conditions to exploit them. Fortunately India has a very large potential to develop power from this source. Since independence India is trying to make optimum use of its hydel rescurces.

7. Ocean waves and tides. Energy of ocean waves and tides can also be exploited. But its use will be limited and that too for the area in the vicinity of the sea coasts.

In this chapter we will be discussing the energy obtainable from flowing rivers and canals only.

27.2 HISTORY OF DEVELOPMENT OF HYDEL POWER IN INDIA

India has a very large potential of producing hydel power. It has been estimated that India's hydel power development potential is 40×10^6 kW; out of which only 0.6×10^6 kW was harnessed by 1951. In order to development more of hydel power, many multipurpose and hydel power projects had been started and completed since independence. The planned development of hydroelectric power was taken up from first five year plan in 1951. The total installed capacity in hydel installations increased from 0.6×10^6 kW in 1951 to about 0.94×10^6 kW by the end of 1955. In second five year plan hydel power was stepped up by about 2.1×10^6 kW. By the end of third five year plan, the hydel power capacity of the country was about 5×10^6 kW. By 1980 the target of electric energy is kept 60×10^6 kW out of which about 50% will be generated by hydro-plants and remaining 50% from thermal plants.

27.3 CHARACTERISTIC OF HYDEL POWER

1. No raw material is required. Natural falls occurring in the flow of big rivers are used scientifically and energy of falling water is used to generate electricity.

2. Hydro power plants are generally very large and generate comparatively very large amount of electricity.

3. Hydro-electric power is cheaper than the Thermal power.

4. There is no danger of pollution of atmosphere or of natural water resources.

5. Hydro-electric schemes are generally part of multi-purpose valley projects. Hence the fall of water is used to its optimum capacity.

27.4 REQUIREMENTS OF WATER POWER GENERATION

The following are the main requirements.

1. Adequate quantity of water should be available throughout the year to run the turbines.

2. Water should be available at sufficient head so that it may run the turbines.

The water may be available from natural perennial rivers or streams. It may also be made available artificially, as from an irrigation or navigation canal, or from a storage reservoir when the flow or river is seasonal.

The head at which this water is available depends on the Topography of the country. The head may be provided by a natural fall or by making dam or weir across the river or canal.

27.5 TYPES OF HYDEL PLANTS

In water power development, we essentially use the fall or head of water which is either naturally available or is artificially created by dams or weirs. The energy

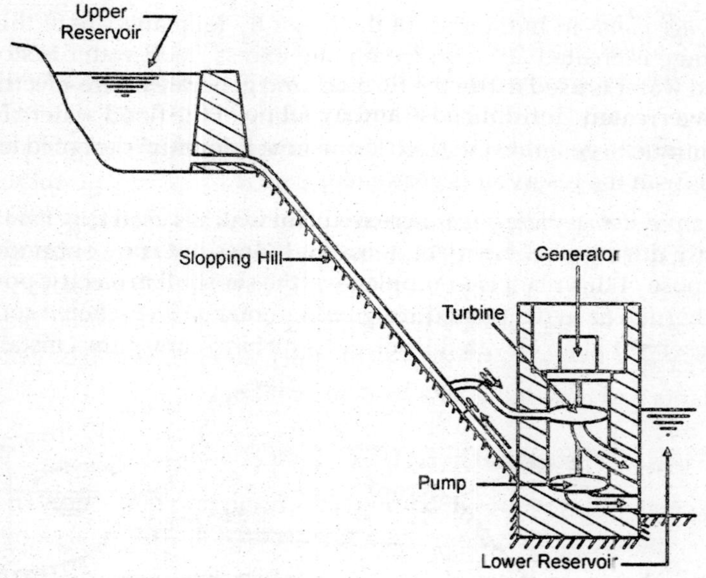

Fig. 27.1. *Pumped storage plant.*

of falling water runs turbines which in turn generate electric power. Hydel or hydro-electric schemes can be classified in following two ways :

1. Classification according to hydraulic features.
2. Classification according to head of water.

1. Classification according to hydraulic features

(i) Run-off river plants.

(ii) Storage plants.

(iii) Pumped storage plants.

(iv) Tidal plants.

(*i*) *Run-off river plants.* In such plants water storage is not done. A weir or barrage is constructed across the river, simply to raise the level of water slightly. Such a scheme can be adopted on such rivers which are perennial and maintain the minimum flow during dry weather, sufficient to run the turbines. This is essentially a low head scheme. Run-off river plants may be of two types.

(a) Those that operate under varying flow.

(b) Those that operate on the minimum available discharge.

In some cases small reservoir is provided so as to avoid frequent fluctuations of load. When the flow in the river is in excess of that required to run the turbines, it is stored *U/S* of the weir temporarily. This stored water can be utilized during peak demand.

(*ii*) *Storage plant.* In India most of the major hydel plants are of this type. A large storage is created *U/S* of the dam and excess flood water is stored in it. The stored water is used to run the turbines and generate hydro-electric power. The storage created *U/S* of dam helps for routing of the flood water also. These plants continue to generate the electric power at minimum designed level, even when inflow in the reservoir is very small.

(*iii*) *Pumped storage plants.* This is such a plant which is used to generate electric power only during peak hours of demand. It does not run continuously. The main purpose of this plant is to supplement the shortfall in electric power of an existing thermal or hydel plant during peak demand. The scheme consists of a storage reservoir at an elevated level and a turbine-cum-pump installation.

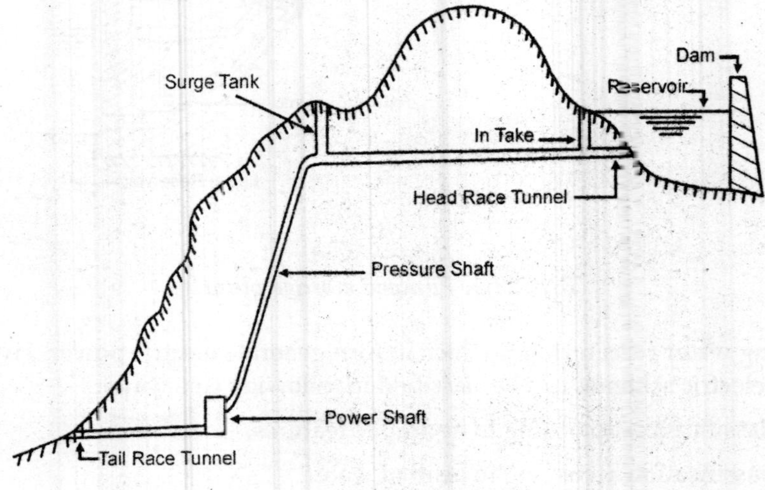

Fig. 27.2

During peak demand, the water from the reservoir is released to run the turbines and generate electric power. During minimum demand the flow of water from reservoir is stopped and their turbines stop generation of electricity. During this minimum demand period even power available from existing thermal or other hydroplant may be in excess. This excess power of thermal plant is utilized to pump back water into the elevated reservoir from the low reservoir. This again augments the capacity of power generation of the elevated reservoir. This stored water is used again during peak hours of demand. It is evident that no water is wasted in such schemes.

In order to reduce the cost of pumping, such machines should be used which work turbine as well as pump.

(*iv*) *Tidal plants.* In this scheme, ocean's tidal waves are used for the generation of hydro-electric power. A large basin is constructed and it is connected to the ocean by means of a connecting channel. Tidal plant is installed on this connecting channel. At the time of tide the ocean water goes to the basin through the channel and runs turbines and thus electricity is generated. At the time of fall in the level of ocean, water from the basin flows back to ocean through the channel and again runs turbines and thus electricity is generated. Hence electricity is generated by the plant both during rise as well as fall in the ocean level. A power house in France works on this principle.

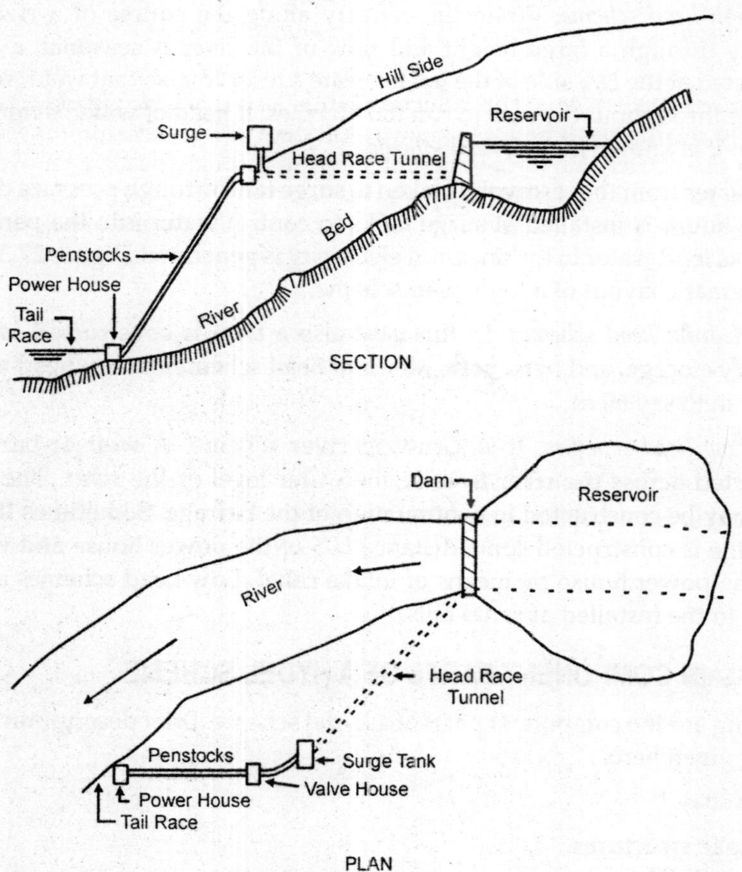

Fig. 27.3. *High head scheme.*

2. Classification according to head of water. The hydel schemes can be classified into following three categories, depending upon the available head.

(i) High head scheme.

(ii) Medium head scheme.

(iii) Low head scheme.

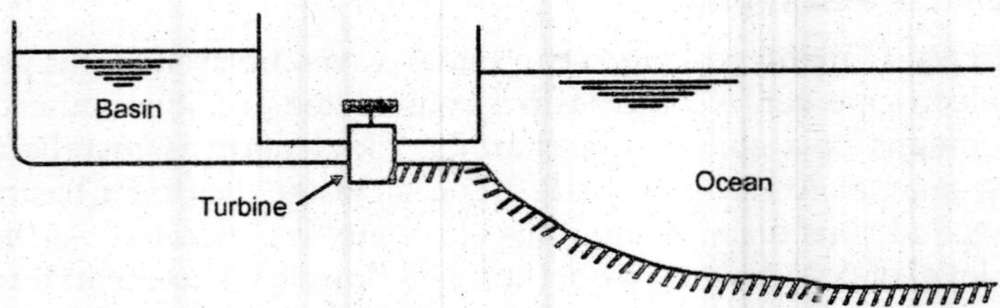

Fig. 27.4. *Tidal plant.*

(i) High head scheme. When the country along the course of a river falls naturally through a large height and flow of the river is seasonal, a dam is constructed at the *U/S* side of the fall to create a reservoir, so that water remains available throughout the year to run the turbines. If head of water is more than say 60 m it is known as *high head scheme.*

The water from the reservoir is taken to surge tank through pressure tunnels. A valve house is installed at surge tanks to control water into the penstocks. Penstocks lend water to turbines and electricity is generated. Figure 27.3 shows the systematic layout of a high head scheme.

(ii) Medium head schemes. In this case also a dam is constructed to create necessary storage, and head both. Medium head schemes have a head varying from 15 m to say 60 m.

(iii) Low head schemes. It is a run off river scheme. A weir or barrage is constructed across the river to raise the water level of the river. The power house may be constructed in continuation of the barrage. Sometimes the weir or barrage is constructed some distance *U/S* of the power house and water is led to the power house by means of intake canal. Low head schemes are also feasible to the installed at canal falls.

27.6 MAIN COMPONENT PARTS OF A HYDEL SCHEME

Following are the component parts of a hydel scheme. Brief description of each is being given here.

1. Forebay.
2. Intake structures.
3. Pen-stocks.
4. Surge tanks.

5. Power house.

6. Turbines and generators.

7. Transformers.

8. Transmission lines.

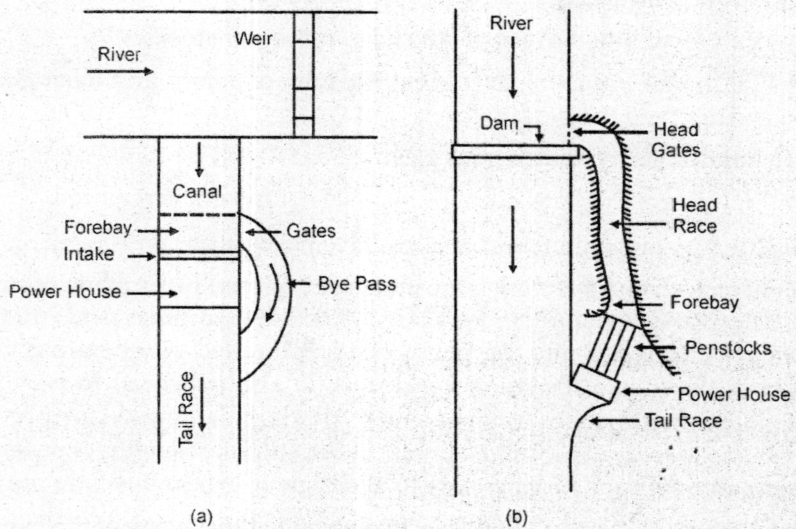

Fig. 27.5. *Component parts of a hydel plant.*

The last two elements are concerned with electricity distribution and hence not the subject of our concern.

1. Forebay. It is an enlarged body of water developed just in front of the intake. Its function is to temporarily store water, rejected by the power plant when generation of power is at reduced rate. This stored water is used when increased power generation is to be achieved. Thus the forebay acts as a buffer to absorb short period changes is intake water. If penstocks take water directly from reservoir, then reservoir itself acts as a forebay. If reservoir water is led to the turbines through canals, then canal acts as a forebay. Usually the mouth of the canal just *U/S* of the power house, in enlarged and acts as a forebay.

In some cases a bye pass is provided connecting forebay with the Tail race, *D/S* of power house. When power house is running at reduced load, the excess water reaching the forebay is lead to *D/S* side of the power house through bye pass channel. Bye-pass channel is provided with gates which operate automatically when level of water in the forebay reaches a predetermined level. When turbines run for reduced load, the turbine governors accept reduced amount of water with the result that level in Forebay increases. When level in Forebay reaches some specified level automatic gates operate and discharge excess water. Similarly when increased power is to be generated, more water is to be accepted by the turbines. To supply this increased quantity of water gates automatically move up and create sufficient extra-storage to be used for the generation of increased power.

2. Intake structure. Water from forebay in led to pen-stocks through intake structure. Intake structure consists of following fittings:

(i) Valves are fitted to control the flow in the pen-stocks.

(ii) Trash rack structure, to prevent the entry of debris, and other floating matter in to the pen-stocks.

(iii) Rake to keep the trash rack clear.

(iv) In very cold regions even ice removal equipment is required.

Intake structure is sometimes known as valve house also.

3. Penstocks. Pen-stocks are large diameter pressure pipes made of steel or R.C.C. They carry water under very large pressure to turbines, and run them. The entrance of water into the penstocks is controlled by operating valves situated in the intake structure. Penstocks are also subjected to very large additional pressures due to water hammer. Although surge tanks are provided on penstocks to reduce the water hammer effect, still short lengths of pen-stocks between surge tank and turbines have to be designed, to bear this extra pressure.

Sufficient head of water, should be maintained at the intake of the penstock especially at low water. It is very essential otherwise air may enter the pen-stocks and cause difficulty in running the turbines.

In addition to sufficient head during low water, the entrance to penstocks should be flared and sharp bends should be avoided.

Airvent pipe is provided near the top end of the pen-stock. This pipe permits air to enter the pen-stock when head gates are closed. If this pipe is not provided the pen-stocks may collapse. If penstock is empty, it should be filled with water with the help of a bypass before main head gate is opened. If main head gate is opened in an empty pen-stock, the turbines may be damaged.

4. Surge tanks. When flow in the turbine is suddenly reduced or closed, the pen-stock is subjected to water hammer. The surge tank is provided to reduce the effects of water hammer. A surge tank performs following functions.

(i) It provides a free reservoir near the discharge end of the pen-stock and thus dampens the pressure due to water hammer.

(ii) It stores water temporarily till the velocity in the pen-stock has steadied.

(iii) It supplies more water when the load is increased until such time that the increased velocity has again steadied.

A surge tank is an ordinary open topped storage tank connected by a branch pipe to the penstock. The surge tank should be located as near the discharge end of the penstock as possible so as to reduced the length of the penstock being subjected to water hammer. They should be built high enough so that water may not over flow. If topography permits the surge tank should be located at the point where slope of the penstock drops rapidly to the power house. The minimum height of a surge tank should be such that at no point the air should be drawn into the penstock.

Normal gradient exists under the normal conditions. When load is reduced the water level in surge tank rises and thus the hydraulic gradient is flattened. On the other hand, when load increases the level in surge tank drops and hydraulic gradient becomes steep.

5. Power house. The power house of a hydel scheme is the place where turbines and connected equipment are installed. The equipment that is installed at the power house are turbines, governors, gate valves relief valves for penstocks, generators, transformers, oil switches, oil circuit breakers etc. Fall of water sets turbines into motion. Turbines convert water energy into mechanical energy and A.C. generators convert the mechanical energy into electrical energy.

Power house is located either in continuation of the weir or barrage or at *D/S* foot of the dam. It may be located a few kilometres *D/S* but in that case water from the reservoir has to be carried to the power house in open hydel channels. Water of hydel channels again joins either river at *D/S* or runs into an irrigation channel after running the turbines. If fall of water is very large and topography available is stepped, more than one power houses may be run in series by the same water before it joins the river *D/S* or irrigation canal.

Turbines. The turbine is a machine which converts hydraulic energy of water into mechanical energy. The turbines are directly coupled to electric generators which generate electric power. The turbine consists of a wheel called runner. The runner further consists of specially designed blades or buckets. Water from pen-stocks strikes these blades and causes turbine runners to rotate. The turbine can be classified under two heads.

1. Velocity or impulse turbines, and

2. Pressure or reactions turbines.

1. Velocity or impulse turbine. In this case the whole of potential energy is converted into kinetic energy by passing water through a contracting nozzle or by guide vanes before water strikes the buckets of the turbine. The runner revolves free in air and water remains in contact with only part of the wheel at a time. A casing is provided, to prevent splashing and also to guide the water discharged from the buckets to the tail race.

The turbines of this type are low speed wheels and are useful for high heads exceeding 150 m; Pelton wheel, Girard turbine, Banki turbine etc. are examples of impulse turbines. Pelton wheel is the mostly used turbine these days.

2. Pressure or reactions turbines. In such turbines, only part of the available potential energy is converted to velocity head at the entrance to the runner. The balance potential energy remains as pressure head. The pressure at the inlet to the turbine is much higher than the pressure at the outlet and it varies throughout the passage of water through the turbine. A very large part of the power is derived from the difference in pressure acting on front and back of runner blades and very small part from the dynamic action of velocity. The entire flow from head race to tail race, takes place under pressure in closed conduit system. Atmospheric air cannot enter the turbine anywhere.

There are several types of turbines that work on this principle but Kaplan and Francis turbines are in most common use.

Turbines suitable for various heads may be classified under following three heads.

1. Low head turbines. Suitable for heads less than 30 m, Kaplan and propeller turbines are the examples of such turbines. These turbines require very large amount of water.

2. Medium head turbines. These turbines are capable to work under the heads ranging from 30 m to 150 m. Modern Francis turbine is the most suitable turbine.

3. High head turbines. Impulse turbines are suitable for very large heads. They require very small amount of water to run. They are suitable for heads varying from 150 m and above.

27.7 SCROLL CASING, DRAFT TUBE AND TAIL RACE

Scroll casing. The main purpose of the scroll casing is to maintain nearly uniform velocity at the entrance point to the guide vanes. The casing may be made of cast steel, plate steel, or concrete, depending upon the head of water. Concrete scroll casing is used for low heads and plate steel scroll for high heads of say 110 m. Scroll casing is a covering that covers the turbine runner and the guide mechanism. It is spiral shaped to avoid formation of eddies and also to distribute water evenly around the guide vanes.

Draft tube. It is a pipe of gradually increasing area. It connects runner exit with the tail race.

Tail race. It is a channel in which draft tube discharges water. Tail race may be river itself or an artificial channel may act as tail race.

27.8 ESTIMATION OF POWER AVAILABLE

Theoretical H.P. generated $= \dfrac{QwH}{75}$

where

Q = discharge in cumecs

H = Height of fall in metres

w = 1000 kg/m^3

1 metric H.P. = 75 kg m/sec.

$$\therefore \qquad \text{H.P.} = \frac{1000 \times Q \times H}{75} = 13.33\ QH$$

If η is the over all efficiency of the system H.P. available

$$= 13.33\ QH\eta,$$

One metric H.P. = 736 watts

$$\therefore \text{ Electrical energy} = 13.33\ QH\eta \times 746 \text{ watts}$$

$$= 13.33\ QH\eta \times 0.736 \text{ kilowatts}$$

$$= 9.8\ QH\eta \text{ kilo-watts}$$

For usual values of efficiencies, the power available may be estimated at the rate of 8 to 8.5 QH kilo-watts.

QUESTIONS

27.1. Enumerate the various sources of energy.

27.2. How hydel plants can be classified ?

27.3. What are the main component parts of a hydel scheme ? Give brief explanation of each.

27.4. What are different types of turbines ?

27.5. How does water power potential of a hydro-electric scheme is worked out.

27.6. Write short notes on :

(i) Surge tank,

(ii) Tail race,

(iii) Forebay

28

Water Resources Development

28.1 INTRODUCTION

Water Resource Development is a science which deals with conception, planning, design, construction and operation of facilities to control and utilize water. All these functions are basically functions of a civil engineer but services of specialists from other fields are also required. Further each water resources development project comes across a unique set of physical conditions to which it must conform and hence standard designs leading to simple solutions are seldom available .

Planning is the orderly consideration of a project right from its conceptions to the final decision. It includes all the work associated with the design of a project except the detailed engineering of the structure. Planning is the most important aspect as decision to proceed with or abandon a proposed project depends on it. Since each water resources development project is unique in its physical and economic setting, it is rather impossible to describe a simple process which leads to the best decision. Although nothing can take the position of engineering judgement during planning, each step towards the final decision should be supported by quantitive analysis rather than more judgement. This chapter deals with only some guideline for the planning of water resources development project.

28.2 OBJECTS OF WATER RESOURCES DEVELOPMENT PROJECTS

The water resources development projects are planned with the following objects in view.

1. Irrigation. The sole purpose of irrigation is to grow crops successfully to augment agricultural production. Dams, reservoirs, wells, canals, pumps, weed control, desilting works, distribution systems drainage works are the usual works concerning irrigation.

2. Hydro-electric power. The main objective in this case is generation of electrical power vitally necessary for the economic development of the region and also to achieve better living standard of the people of the region. Dams, reservoirs, penstocks, power channels power plants, transmission lines are the connected works with this objective.

3. Flood control. The various objectives to be achieved are the following.

I. Prevention or reduction of flood-damage.

II. Protection of economic development,

III. Conservation of storage.

IV. River training and regulation works.

The works are measured necessary to achieve the above objectives are dams, storage reservoirs, levees, flood watts, channel improvements, flood ways pumping stations, flood plane zoning and flood forecasting.

4. Navigation. Its objective is transportation of goods and passengers through water ways. The works connected with this objective are dams, reservoirs, canals, locks channel and harbour improvements, harbour facilities etc.

5. Water supply. Provisions of water for domestic, municipal, commercial and industrial purposes are the main objectives. Dams, reservoirs, wells, pumping stations, water treatment plants, distribution of water are the works involved.

6. Recreational use of water. Entertainment is the main objective in this case creation of reservoirs, development of recreational works, pollution control, selection and preservation of good scenes are the works involved for the purpose.

7. Water shed management. Conervation and improvement of soil, sediment abatement, run-off retardation, forest and grass land improvement are the objective which come under this heading.

8. Preservation or fish and wild life. Reduction of fish life losses and wild life losses, and provision for expansion of commercial fishing are the main objectives. The works connected with this objective are reservoir storage, regulation of flow, installation of fish ladders and screens, fish hatcheries, wild life refugies land management, pollution control.

9. Pollution control. The objective is protection or improvement of water supplies for Municipal, domestic industrial and agricultural use and for aquatic life. Treatment of water, reservoir storage, legal control measures etc. are the works involved in it.

10. Sediment control. The objective is to reduce the silt load in streams and projection of reservoirs. Soil conservation, desilting works, channel and revetment works, bank stabilization, special check dam construction and reservoir operation are the usual works involved.

11. Drainage. Agricultural production, urban development projection of public health are the objectives to be achieved. Construction of drainage ditches, tile drains, levees, pumping stations are usual works connected with this.

28.3 CLASSIFICATION OF WATER RESOURCE DEVELOPMENT PROJECTS

According to the purposes served the water resource development projects can be classified into following two categories.

1. Single purpose project
2. Multi-purpose project

The single purpose projects are designed and operated to serve only one purpose which may be irrigation, water supply, flood control or any other single purpose.

Multi-purpose projects are designed and operated to serve two or more than two objectives. Here it may be stated that a project which has been designed for single purpose but which is other benefits also for other purposes are not considered as multipurpose project.

Water resources are the most natural resources. They are available in abundance. However, since the world population is increasing at a very fast rate the food requirements are consequently increasing. It is imperative that food supplied can be increased only by making maximum use of available sources of water. Moreover multiple use of water increases benefits without much increase in costs and thus increase economic justification for the project. Hence most of the major water resources development projects are planned as multipurpose projects.

28.4 COMPATIBILITY OF MULTI-PURPOSE USES

The water required for irrigation, water supply and navigation cannot be used jointly. Hence a project being designed for these three functions should be provided with separate allocations of storage for since power generation is not a consumptive use of water, any water released for the other purposes can be used for power. If the power generation plant is operated as a base load plant is water requirement may fit in well with the relative uniform release of water for other purposes. However, if it is proposed to use the plant as peak load plant it may be necessary to construct an after bay or regulating dam D/S to smooth out the variations of the water released for power. The after bay storage capacity needs to be sufficient only to regulate flows for a few days at a time. Its function is same as that of a distribution reservoir which smooths out the load on a pumping plant is a water supply system. At the regulating dam a low head power plant may be established to produce a small amount of base load power.

Since the seasonal variations in power demand may not coincide with the requirements of other uses, it is usually necessary to allocate a curtain storage for power use.

Flood control requirement for empty storage space is the least compatible of all uses. The storage for flood control can be obtained either by permanent allocation of space exclusively for flood control or by seasonal allocation of space for flood, storage. Permanent allocated space is above the crest level of the spillway and is made available by closing the spillway crest gates. Any additional space available for flood control is used but it is not dependable. In the second case, the space for flood control is made available by operating the reservoir in such a way that the required space is available for flood storage.

Recreational benefits, fish production etc. are taken into consideration as opportunity permits and some portion of space may be allocated for this purpose.

28.5 STAGES OF WATER RESOURCES DEVELOPMENT PROJECT

The various stages in the planning of a water resources development project are the following

1. Objective or purpose of the project
2. Data collection
3. Lively future need projections
4. Project formulation and
5. Project evaluation.

1. Purpose of the project. Before any project for planning is taken up, the planner must be clear about the objectives to be achieved. The objective of a water resources development project may be single in form of irrigation, water supply, flood control etc. or multiple. The objectives however depend upon the level of planning also. For example the objective of planning a project by Central Govt. have a broad goal of achieving maximum economic and social benefits whereas objective of planning at state government level or district level may be single of reducing floods damage or irrigation or water supply. The objectives of the project also depend on the availability of funds. Generally funds are never available in abundance. In such cases a comprehensive plan for the project including all the possible objectives may be prepared but it may be implemented, in stages depending upon the availability of funds.

2. Data collection. For this data is collected in respects of hydrology, geology. economic and social states of the area. Except hydrologic data, all other data are current and may be collected by survey at the start of the planning. The hydrologic data are always historical and the data collected for the past several years are required. Data in respect of agriculture, flood control, hydro-electric power Municipal and industrial uses, drainage, recreational uses, navigation, fish and wild life are important.

3. Likely future need projections. All the projects are always planned keeping in view the present needs and also likely future needs. For this the projections of future needs are required. The historical and current data collected for or project planning are used to serve as the basis for projections of future needs. The projections for planning should be worked out with utmost care and caution, because unrealistic higher values of water needs may lead to over design and wastage of resources. Similarly too lower values may furl to accomplish the purpose of the project. Thus all attempts should be directed to achieve rational planning.

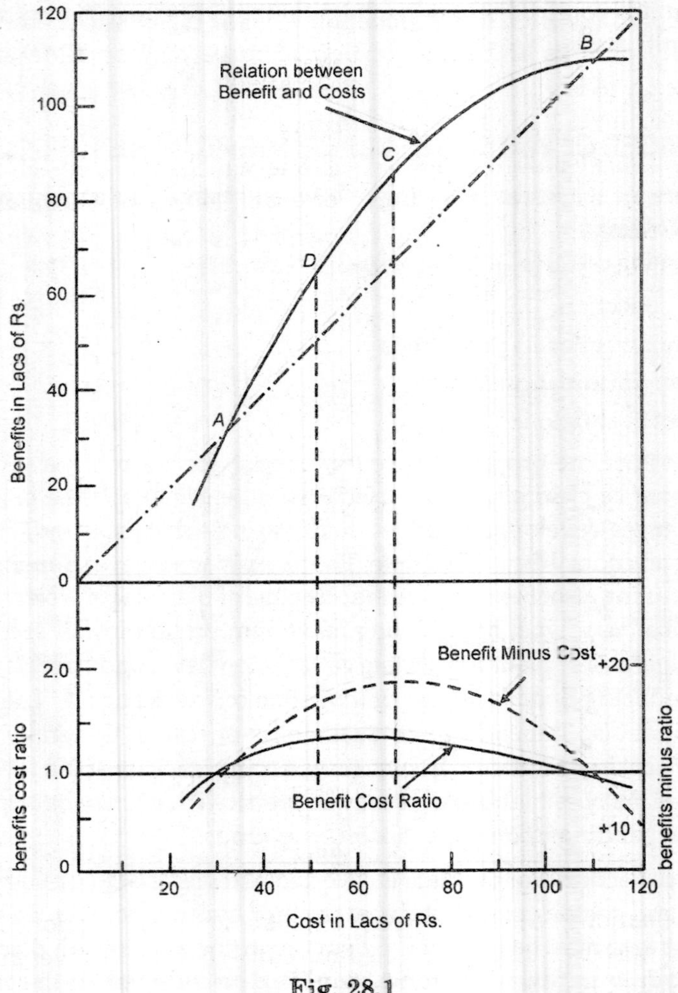

Fig. 28.1

4. Project formulation. Having obtained the data and future needs, the actual formulation of the project is started. First of all a very comprehensive list of alternatively is prepared and boundary conditions which restrict the project are defined. The restraints help eliminate sonic of the alternatives and thus simplify the process of project formulation. Lastly the cataloging of the possible project units is done. In this all the possible project units along with the alternative plans for each project unit is indicated. Preliminary cost estimates for each catalogued unit are prepared. At this stage detailed estimates are not required since some of them are definitely to be discarded.

5. Project evaluation. Having considered various alternatives proposals the next step is to select the most economically efficient proposal. The benefit cost ratio of the selected proposal should be more than one or any other specified minimum value.

If any unit is completely independent both physically and economically of all other units, it may be evaluated alone.

A physically independent unit is one which is not affected in respect of inflow by any other unit can contribute to the same.

An economical independent unit is one which is having no economic interconnection with any other unit. For example two power plants on two different streams are physically independent as far as flow is concerned but if they are serving the same power system they are not economically independent.

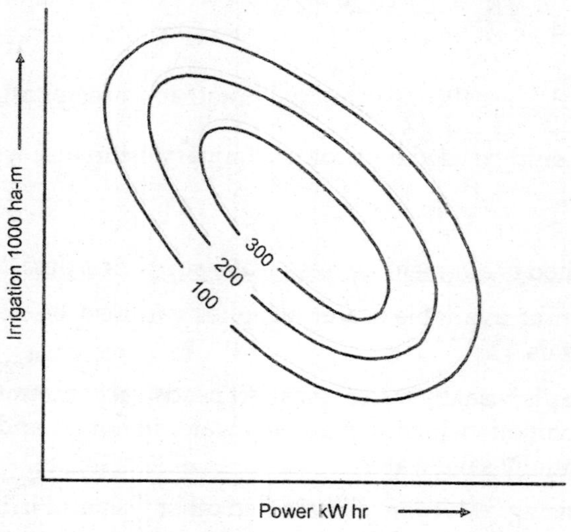

Fig. 28.2

The devaluation of a single purpose independent project is done by benefit cost ratio and also by rate of return method. Following three curves are plotted.

I. Benefit v/s cost

II. Benefit-cost ratio v/s costs

III. Benefits minus costs v/s costs

The three curves are shown in Fig. 28.1. From A to B benefits exceed costs. The curve of benefits minus costs shows a maximum at point C. The maximum benefit-cost ratio occurs at point D and thus should be the limit of project size.

Figure 28.2 shows a net-benefit surface for a multipurpose project serving two purposes. With the increase in the number of purposes served by a project the number of net-benefit surfaces required to be prepared will all increase, because a large number of combinations of the various purposes will have to be considered.

28.6 AUGMENTATION OF WATER SUPPLIES

Augmentation of water supply means increasing the utility of available water by conservation or regulation. It also includes methods for providing an entirely new supply of fresh water. Providing completely new supplies of water include weather modification. Presently the main effort has been to increase local precipitation by cloud seeding. Studies show that under favourable conditions on an average about 10% increase in precipitation can be expected by this technique. According to Mr. Lumb the increase is annual run-off ΔR resulting from a small increase is precipitation ΔP is given by

$$\Delta R = \Delta P\,(0.29 + 1.2)\frac{R_{avg}}{P_{avg}} \tag{28.1}$$

where R_{avg} and P_{avg} are the mean annual runoff and precipitation respectively. The equation should be used only for preliminary estimates, where $0.1 < \frac{R_{avg}}{P_{avg}} < 0.5$.

Another method of augmenting fresh water supplies is desalting of sea water.

Conservation of available water supplies can also be accomplished by following methods.

1. Sewage disposal by water transport needs vast amounts of pure water. Any technique which will use less water in homes and industry could substantially save water.
2. By reducing conveyance losses and other wasteful irrigation practices lot of useful water can be saved and used for other purposes.
3. By adopting measures which reduce evaporation of water from reservoirs.
4. Ground water is free from evapo-transpiration losses. Hence use of ground water storage together with surface storage will reduce large evaporation losses occurring from large reservoirs in arid regions.
5. Wild vegetation use large amounts of water by way of evapotranspiration. This water can be saved by eliminating wild

vegetation. Measures adopted to reduce evapo-transpiration by crops also help save useful water.

6. Reuse of water coming from industry and other uses also helps a great deal in saving precious water. If this water is to be used for higher uses it must be made safe both biologically and chemically.
7. Sending of soil surface using chemicals and other agent like asphalt greatly help augmentation of water yield in arid regions. Since, almost all the rain water falling oil the projected area is recovered the yield from the catchment stets increased. If this runoff water is lead to a covered tank evaporation loss is prevented.

□□□

Appendix I

Khosla Theory (Mathematical Solution)

Dr. A.N. Khosla, Dr. N.K. Bose and Dr. E.M. Taylor analysed the most common profile of composite floor shown in Fig. AI-1. They adopted the method of Schwarz – Christoffel transformation for the analysis. The flow lines and equipotential lines of the actual section are strictly neither ellipses nor hyperbolas. In the analysis of the complex profile the flow lines and equipotential lines are transformed into normal ellipses and hyperbola curves.

This Mathematical analysis is based on the fact that if a polygon is located in the Z-plane then the Schwarz – Christoffel transformation brings all the flow lines and equipotential lines take the shapes of ellipses and hyperbolas respectively. The Schwarz – Christoffel equation that maps the polygon conformally on to ξ-plane is given by

$$Z = A\int \frac{d\xi}{(\xi-\xi_1)^{\lambda_1}(\xi-\xi_2)^{\lambda_2}\ldots(\xi-\xi_n)^{\lambda_n}} \quad (1)$$

where

$\pi\lambda_1, \pi\lambda_2 \ldots\ldots \pi\lambda_n$ = change in angle of the polygon which are required to transform it into a straight line.

$\zeta_1, \zeta_2 \ldots \zeta_n$ = co-ordinates of its vertices, a complex constant

1. Transformation Equation (equation 1)

A complex profile of floor of most general case is shown in Fig. 15.12. This complex profile when transformed into straight line profile is shown in Fig. 15.13. The key points in actual profile are *A, E, D, C* and *B* and corresponding points on transformed straight line profile shown in Fig. 15.13 are A', E', D', C' and B' respectively. It can very easily be appreciated that the angles at verticaesl E, D and C are respectively $\frac{\pi}{2}$, 2π and $\frac{\pi}{2}$ and these have been changed to π each at E', D' and C' on transformation. The Schwarz–Christoffel formula is

$$Z = A\int \frac{d\xi}{(\xi-\xi_1)^{\lambda_1}(\xi-\xi_2)^{\lambda_2}(\xi-\xi_3)^{\lambda_3}} \quad (1)$$

where $\pi\lambda_1$ = change in angles at $C = \pi - \frac{\pi}{2} = +\frac{\pi}{2} = \frac{1}{2}\times\pi$

$\pi\lambda_2$ = change in angle at $D = \pi - 2\pi = -\pi = -1\,'\pi$

$\pi\lambda_3$ = change in angle at $E = \pi - \frac{\pi}{2} = +\frac{\pi}{2} = \frac{1}{2}\times\pi$

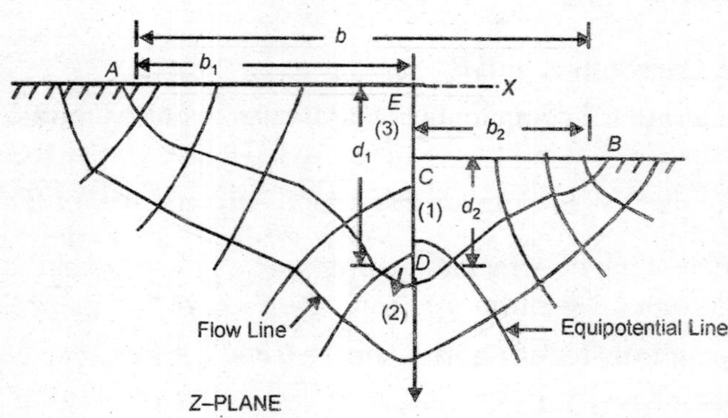

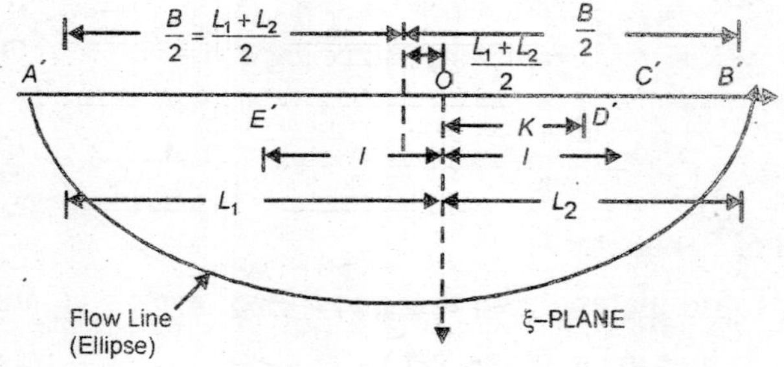

Fig. AI-1

Hence $\lambda_1 = \frac{1}{2}$, $\lambda_2 = -1$ and $\lambda_3 = \frac{1}{2}$

Select the orgin mid-way between C', and E' such that $OE' = OC'$

Let $OD' = K$ an absolute number

$$\xi_1 = \text{co-ordinate of } C = +1$$

$$\xi_2 = \text{co-ordinate of } D = K$$

$$\xi_3 = \text{co-ordinate of } E = -1$$

Substituting these values in Eq. (1) we get

$$z = \int \frac{d\xi}{(\xi-1)^{1/2}(\xi-k)^{-1}(\xi+1)^{1/2}}$$

$$= A\int \frac{(\xi-k)\,d\xi}{\sqrt{\xi^2-1}} \qquad (2)$$

Integrating Eq. (2) we get

$$z = A\sqrt{\xi^2-1} - AK\log e^{\left(\xi+\sqrt{\xi^2-1}\right)} + E \qquad (3)$$

Determining Constants *A* and *B*

In order to determine *B* compare the co-ordinates of point C and C′ in the two planes.

It is evident that the co-ordinate z of $c = i(d_1 - d_2)$ and that of C′ as $\xi = 1$

Putting these values in (2) we get

$$B = i(d_1 - d_2)$$

Similarly, co-ordinates of E and E' are $z = 0$ and $\xi = -1$

Putting these values is Eq. (3) we get

$$o = B - AK\log e^{(-1)}$$

or

$$AK\pi i = B + i(d_1 - d_2) \text{ (since } \log e^{(-1)} = \pi i)$$

$$AK = \frac{d_1 - d_2}{\pi} \qquad (4)$$

Determination of Parameter *K*

Comparing the co-ordinates of points D and D' we have $z = id_1$ and $\xi = K$.

Putting these values in Eq. (3) we get

$$id_1 = A\sqrt{k^2-1} - AK\log e\left(K+\sqrt{K^2-1}\right) - i(d_1-d_2)$$

or

$$id_1 = Ai\sqrt{1-k^2} - AK\log e\left(K+i\sqrt{1-K^2}\right) - i(d_1-d_2)$$

Putting $K = \cos\theta$ we get

$$id_1 = A\,i \sin\theta - AK \log e\,(\cos\theta + i\sin\theta) + i(d_1 - d_2)$$

or
$$id_1 = A\,i\sin\theta - AK\,i\theta + i(d_1 - d_2)$$

$\therefore$
$$d_2 = A\sin\theta - AK\,\theta$$

Putting the values of A and AK from Eq. (4) we get

$$d_2 = \frac{d_1 - d_2}{\pi k}\sin\theta - \frac{(d_1 - d_2)}{\pi}\theta$$

or
$$d_2 = \frac{d_1 - d_2}{\pi}\tan\theta - \frac{(d_1 - d_2)}{\pi}\theta$$

Hence
$$\frac{d_2\pi}{d_1 - d_2} = \tan\theta - \theta$$

or
$$\pi\delta = \tan\theta - \theta = \frac{\sqrt{1-k^2}}{k}\cos^{-1}k \tag{5}$$

where
$$\delta = \frac{d_2}{d_1 - d_2}$$

Thus from Eq. (5) the values of k for various values of δ can be determined.

Final Transformation Equation

Putting the values of A and AK is Eq. (3) final transformed equation reduces to following form

$$z = \frac{d_1 - d_2}{\pi k}\sqrt{\xi^2 - 1} - \frac{(d_1 - d_2)}{\pi}\log e^{\xi + \sqrt{\xi^2 - 1}} + i(d_1 - d_2) \tag{I}$$

where K is now a known parameter B

Determination of Co-ordinates of A' and N'

Equation (1) is applicable to all the points on the profile of the floor. Let the ξ co-ordinates of point A' and B' be $-L_1$ and $+L_2$ respectively. The z co-ordinates of the corresponding point A and B are $-b_1$ and $\{b_2 + i(d_1 - d_2)\}$ respectively. Substituting these co-ordinates separately in Eq. (I) we get

$$-b_1 = \frac{d_1 - d_2}{\pi k}\sqrt{L_1^2 - 1} - \frac{(d_1 - d_2)}{\pi}\log e\left(-L_1 + \sqrt{L_1^2 - 1} + i\,d_1 - d_2\right)$$

or
$$-b_1 = \frac{d_1 - d_2}{\pi k}\sqrt{L_1^2 - 1} - \frac{(d_1 - d_2)}{\pi}\log e\left(L_1\sqrt{L_1^2 - 1}\right) \tag{6}$$

and $b_2 + i(d_1 - d_2) = \frac{d_1 - d_2}{\pi k}\sqrt{L_2^2 - 1} - \frac{(d_1 - d_2)}{\pi}\log e\ L_2 + \sqrt{L_2^2 - 1} + i\ (d_1 - d_2)$

or $b_2 = \frac{d_1 - d_2}{\pi k}\sqrt{L_2^2 - 1} - \frac{(d_1 - d_2)}{\pi}\log e\left(L_2\sqrt{L_2^2 - 1}\right)$ (7)

Assuming $L_1 = \cosh Y_1$ and $L_2 = \cos h\ Y_2$ and putting these in the above we get

$$-b_1 = \frac{(d_1 - d_2)}{\pi k}\sinh Y_1 + \frac{(d_1 - d_2)}{\pi} Y_1$$

$$\sinh Y_1 + k\ Y_1 = \frac{\pi k b_1}{d_1 - d_2}$$

and $$\sinh Y_2 - k\ Y_2 = \frac{\pi k b_2}{d_1 - d_2}$$

Again let $\delta_1 = \frac{b_1}{d_1 - d_2}$ and $\delta_2 = \frac{b_2}{d_1 - d_2}$

Putting these values in the above we get

$\sinh Y_1 + k\ Y_1 = \pi\ k\ \delta_1$ (8)

$\sinh Y_2\ \ k\ Y_2 = \pi\ k\ \delta_1$ (9)

Thus from Eq. (8) and (9) values of Y_1 and Y_2 can be found out.

Hence $L_1 = \cosh Y_1$ and $L_2 = \cosh\ Y_2$ can be known.

Thus corresponding to given set of values of floor dimensions b_1, b_2, d_1 and d_2 we can first get k and there L_1 an L_2.

Pressure Distribution below the Floor

In Section 15.10 it has been indicated that the flow field (comprising stream lines and equipotential lines) for the case of a horizontal floor (Fig. 15.6) may be represented by the following equation

$$z = \frac{b}{2}\cosh w$$

where $z = x + iy$, $w = U + iV$, x and y are the co-ordinates of any point P in the flow field and U and V are the stream function and potential function (or pressure function) at P.

The above noted expression may be applied to the transformed floor of Fig. 15.13 in which case $\frac{b}{2}$ = half the width of floor $\frac{1}{2}$ $(L_1 + L_2)$ and $z = \xi + \frac{1}{2}$ $(L_1$

+ L_2).

Since the origin in Fig. 15.6 is at middle of the floor but in Fig. 15.13 it is at a distance of $\frac{1}{2}$ $(L_1 - L_2)$ from the middle point. The equation of this case shall thus become $\xi + \frac{1}{2}(L_1 - L_2) = \frac{1}{2}(L_1 - L_2)\cosh w$.

Putting $\frac{L_1 - L_2}{2} = \lambda$ and $\frac{L_1 - L_2}{2} = \lambda_1$ we get

$$\xi = \lambda \cosh w - \lambda_1 \qquad \text{(II)}$$

Putting this in Eq. (I) we get a general equation for the flow net diagram as under

$$z = \frac{(d_1 - d_2)}{\pi K}\sqrt{(\lambda \cosh w - \lambda_1)^2 - 1} - \frac{(d_1 - d_2)}{\pi}\log e\ \left\{(\lambda \cosh w - \lambda_1) + \sqrt{(\lambda \cosh w - \lambda_1)^2 - 1}\right\} + i(d_1 - d_2)$$

For the first stream line which hugs to the base we have $U = 0$ and hence $w = U + iV = iV$

Also $\qquad z = x + iy$

Hence

$$x + iy = \frac{d - d_2}{\pi k}(\lambda \cos v - \lambda_1)^2 - 1 - \frac{d_1 - d_2}{\pi}\log e\ \left\{(\lambda \cos V - \lambda_1) + \sqrt{(\lambda \cos V - \lambda_1)^2 - 1}\right\} + i(d_1 - d_2) \qquad \text{(III)}$$

From Eq. (III) we can get the pressure function V for any point whose co-ordinate is $x + iy$. We shall apply this equation to different regions of the floor of Fig. 15.12.

Pressure Distribution for Length *AE*

In this length $y = 0$

Also introducing a new variable β'_1 defined by equation

$$\lambda \cos v - \lambda_1 = -\cos h\beta'_1 \qquad (10)$$

We get from Eq. III

$$x = \frac{d_1 - d_2}{\pi k} \sinh \beta'_1 + \frac{d_1 - d_2}{\pi} \log e\left[-\cosh \beta'_1 + \sinh \beta'_1\right] + i(d_1 - d_2)$$

or $$x = \frac{d_1 - d_2}{\pi k} \sinh \beta'_1 + \frac{d_1 - d_2}{\pi} \beta'_1 - \frac{d_1 - d_2}{\pi} \log e(-1) + i(d_1 - d_2)$$

or $$x = \frac{d_1 - d_2}{\pi k} \sinh \beta'_1 + \frac{d_1 - d_2}{\pi} \beta'_1$$

Hence $$\frac{x}{d_1 - d_2} \pi K = \delta_n \ \pi k = \sinh \beta'_1 + K\beta'_1 \tag{11}$$

From this equation β'_1 can be calculated

Hence from Eq. (10), the pressure function V is found to be

$$V = \cos^{-1}\left(\frac{\lambda_1 - \cosh \beta'_1}{\lambda}\right) = \pi \frac{P}{H}$$

$$\therefore \quad P = \frac{H}{\pi} \cos^{-1}\left(\frac{\lambda_1 - \cosh \beta'_1}{\lambda}\right) \quad \text{[IV(a)]}$$

$$\phi = \frac{P}{H} = \frac{1}{\pi} \cos^{-1}\left(\frac{\lambda_1 - \cosh \beta'_1}{\lambda}\right) \quad \text{[IV(b)]}$$

This gives pressure distribution under AE.

Pressure Distribution for Length *CB*

Introducing a new variable β'_2 defined by equation $\lambda \cos V - \lambda_1 = \cosh \beta'_2$ we get from Eq. (III)

$$\sinh \beta'_2 - K\beta'_2 = \pi K \frac{x}{d_1 - d_2} = \delta_n \pi K \tag{12}$$

From this equation β'_2 can be known

Hence pressure distribution is given by

$$V = \cos^{-1} \frac{\lambda_1 - \cosh \beta'_2}{\lambda} = \pi \frac{P}{H}$$

Hence $$P = \frac{H}{\pi} \cos^{-1}\left(\frac{\lambda_1 - \cosh \beta'_2}{\lambda}\right) \quad \text{[V(a)]}$$

$$\phi = \frac{1}{\pi} \cos^{-1}\left(\frac{\lambda_1 - \cosh \beta'_2}{\lambda}\right) \quad \text{[V(b)]}$$

Combining Eq. IV (a) and V (a) in one common form we get

$$P = \frac{H}{\pi}\cos^{-1}\left(\frac{\lambda_1 \mp \cosh \beta'_{1,2}}{\lambda}\right) \quad \text{(VI(b)]}$$

– ive is used for *U/S* and + ive sign for the *D/S* part of the floor.

Pressure Distribution along *ED* and *DC*

For the pressure distribution along the *U/S* and *D/S* faces of the pile, we have $x = 0$. Hence from the Eq. (III)

$$iy = \frac{d_1 - d_2}{\pi k}\sqrt{(\lambda \cos v - \lambda_1)^2 - 1} - \frac{d_1 - d_2}{\pi}\log e\left[\lambda \cos V - \lambda_1 + \sqrt{(\lambda \cos V - \lambda_1)^2 - 1}\right] + i(d_1 - d_2)$$

Introducing a new variable β defined by the equation $\lambda \cos V - \lambda_1 = \cos \beta$ we get

$$iy = \frac{d_1 - d_2}{\pi k} i \sin \beta = \frac{d_1 - d_2}{\pi} i\beta + i(d_1 - d_2)$$

or

$$\left(\frac{y}{d_1 - d_2} - 1\right)\pi k = \sin \beta - k\beta \quad (13)$$

From this β can be calculate

Hence

$$V = \cos^{-1}\left[\frac{\lambda_1 + \cos \beta}{\lambda}\right] = \pi\frac{P}{H}$$

$$P = \frac{H}{\pi}\cos^{-1}\left(\frac{\lambda_1 + \cos \beta}{\lambda}\right) \quad \text{(VII)}$$

I Pressure at Key Points

Pressure at *E*

For $E'A'$ $\quad \xi = \lambda \cos V - \lambda_1$ $\quad$ (From Eq. II)

At E' $\quad \xi = -1$

Hence $\quad -1 = \lambda \cos V + \lambda_1$

$$V = \pi\frac{P}{H} = \cos^{-1}\frac{\lambda_1 - 1}{\lambda}$$

$$P_e = \frac{H}{\pi}\cos^{-1}\frac{\lambda_1 - 1}{\lambda} \qquad \text{[VIII (a)]}$$

II Pressure at *C*

For $C'B'$ $\quad \xi = \lambda \cos V - \lambda_1$ (From Eq. II)

At C' $\quad \xi = +1$

$$+1 = \lambda \cos V - \lambda_1$$

or

$$V = \pi\frac{P}{H} = \cos^{-1}\frac{\lambda_1 + 1}{\lambda}$$

$$P_C = \frac{H}{\pi}\cos^{-1}\frac{\lambda_1 + 1}{\lambda} \qquad \text{[VIII (b)]}$$

III Pressure at *D*

At points D' $\quad \xi = \lambda \cos V - \lambda_1 = K$

$$P_D = \frac{H}{\pi}\cos^{-1}\frac{\lambda_1 + k}{\lambda} \qquad \text{[VIII (c)]}$$

Special case

When $b_2 = 0$

Let $b_2 = 0$ and $b_1 = b$ as shown in Fig. AI-2.

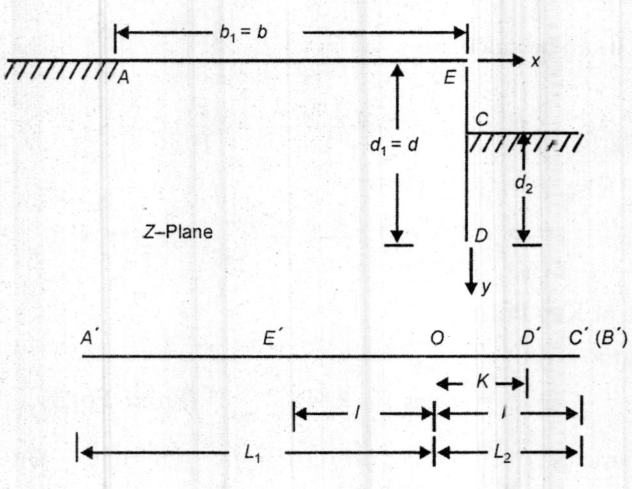

Fig. AI.2

$$\delta_2 = \frac{b_2}{d_1 - d_2} = 0$$

Hence from Eq. (9)

$$\sinh Y_2 - k\,Y_2 = \pi\, k\, \delta_2 = 0$$

or

$$Y_2 = 0$$

$$\therefore \quad L_2 = \cosh 0 = 1$$

This could also be found from Fig AI–2.

$$\lambda = \frac{L_1+1}{2}, \lambda_1 = \frac{L_1-1}{2} = \lambda - 1$$

$$\therefore \quad \lambda_1 - 1 = \frac{L_1-3}{2}$$

$$\lambda_1 + 1 = \frac{L_1+1}{2} = \lambda$$

Hence from Eq. (VIII)

$$P_E = \frac{H}{\pi}\cos^{-1}\left[\frac{\lambda_1-1}{\lambda}\right] = \frac{H}{\pi}\cos^{-1}\frac{L_1-3}{L_1+1} \qquad \text{[IX (a)]}$$

$$P_C = \frac{H}{\pi}\cos^{-1}\left[\frac{\lambda_1-1}{\lambda}\right] = \frac{H}{\pi}\cos^{-1}(1)$$

$$= 0 \qquad \text{[IX (b)]}$$

$$P_D = \frac{H}{\pi}\cos^{-1}\left[\frac{\lambda_1-k}{\lambda}\right] = \frac{H}{\pi}\cos^{-1}\left[\frac{2k+L_1-1}{L_1+1}\right] \qquad \text{[IX(c)]}$$

Exit Gradient

Along the sheet pile use Eq. (13)

$$\left[\frac{y}{(d_1-d_2)} - 1\right]\pi k = \sin\beta - k\beta$$

$$y = \frac{d_1-d_2}{\pi k}\sin\beta - \frac{(d_1-d_2)}{\pi}\beta + (d_1-d_2)$$

To determine the exit gradient at C, differentiate the above with respect of y

$$\frac{dy}{dy} = 1 = \frac{d_1-d_2}{\pi k}\cos\beta\frac{d\beta}{dy} - \frac{d_1-d_2}{\pi}\frac{d\beta}{dy}$$

But $\lambda \cos V - \lambda_1 = \cos\beta$

$$\therefore \qquad y \sin V \frac{dV}{dy} = \sin\beta \frac{d\beta}{dy}$$

or
$$\frac{d\beta}{dy} = \lambda \frac{\sin V}{\sin\beta} \frac{dv}{dy} \tag{14}$$

Substituting this we get

$$1 = \frac{d_1 - d_2}{\pi k} \cos\beta \frac{\lambda \sin V}{\sin\beta} \frac{dV}{dy} - \frac{d_1 - d_2}{\pi} \lambda \frac{\sin V}{\sin\beta} \frac{dV}{dy} \tag{15}$$

At exit end $V = 0$

$\cos\beta = \lambda \cos V - \lambda_1 = \lambda - \lambda_1 = 1, \beta = 0$

Hence when $V = 0,\ \beta = 0$ and $\frac{\sin V}{\sin\beta} = \frac{0}{0}$

It is therefore necessary to determine the limit of $\left(\frac{\sin V}{\sin\beta}\right)_V = 0, \beta = 0$

$$\text{Lt}\left(\frac{\sin V}{\sin\beta}\right)_V = \beta = 0 = \underset{\substack{V=0\\ \beta=0}}{\text{Lt}} \frac{\frac{d}{dV}(\sin V)}{\frac{d}{dy}(\sin\beta)} = \underset{\substack{V=0\\ \beta=0}}{\text{Lt}} \frac{\cos V \frac{dv}{dy}}{\cos\beta \frac{d\beta}{dy}}$$

$$= \underset{\substack{V=0\\ \beta=0}}{\text{Lt}} \left\{\frac{\cos V}{\cos\beta} \frac{\sin\beta}{\lambda \sin V}\right\} \qquad \text{From (14)}$$

$$= \underset{\substack{V=0\\ \beta=0}}{\text{Lt}} \left(\frac{\sin\beta}{\lambda \sin V}\right)$$

$$\therefore \qquad \left[\text{Lt}\left(\frac{\sin V}{\sin\beta}\right)\right]^{-2}_{\substack{V=0\\ \beta=0}} = \frac{1}{\lambda}$$

or
$$\underset{\substack{V=0\\ \beta=0}}{\text{Lt}}\left(\frac{\sin V}{\sin\beta}\right) = \frac{1}{\sqrt{\lambda}}$$

Putting this in Eq. (15) we get

$$1 = \frac{d_1 - d_2}{\pi k} \frac{\lambda}{\sqrt{\lambda}} \left[\frac{dV}{dy}\right]_{V=0} - \frac{d_1 - d_2}{\pi} \frac{\lambda}{\sqrt{\lambda}} \left(\frac{dV}{dy}\right)_{V=0}$$

$$\therefore \quad \left(\frac{dv}{dy}\right)_{V=0} = \frac{\pi k}{(d_1 - d_2)\sqrt{\lambda}} \times \frac{1}{1-k}$$

$$\text{Now } \left(\frac{dP}{dy}\right)_{\text{Exit}} = G_E = \frac{H}{\pi}\left(\frac{dV}{dy}\right)_{V=0}$$

$$= \frac{H}{\pi} \times \frac{\pi k}{(d_1 - d_2)\sqrt{\lambda}} \times \frac{1}{1-k}$$

$$G_E = \frac{H}{d_2} \times \frac{d_2}{d_1 - d_2} \times \frac{k}{1-k} \times \frac{1}{\sqrt{\lambda}} \qquad \text{[IX (I)]}$$

End Sheet Pile with $d_1 = d_2 = d$

For $d_1 = d_2$ we have $k = 0$

$$\therefore \quad \frac{k}{d_1 - d_2} = 0$$

Let us find the value of the above limit

From Eq. (V)

$$\tan\theta - \theta = \pi\delta = \frac{\pi d_2}{d_1 - d_2}$$

Also $\cos\theta = k$

Hence multiplying left side by $\cos\theta$ and right side by k we get

$$\sin\theta - \theta\cos\theta = \pi d_2 \frac{k}{d_1 - d_2}$$

For $k = 0$, $\quad \theta = \dfrac{\pi}{2}$

Hence the above equation becomes

$$1 - 0 = \pi d_2 \left[\frac{k}{d_1 - d_2}\right]_{4-d_1=d_2}$$

$$\therefore \text{Lt}\left[\frac{k}{d_1 - d_2}\right]_{d_1 = d_2 = d} = \frac{1}{\pi d_2} = \frac{12}{\pi d} \tag{16}$$

Hence from Eq. (X)

$$G_E = \left(\frac{H}{1-k}\right)\left(\frac{k}{d_1 - d_2}\right) \times \frac{1}{\sqrt{\lambda}} = \frac{H}{1} \times \frac{1}{\pi d} \times \frac{1}{\sqrt{\lambda}}$$

$$= \frac{H}{d}\frac{1}{\pi\sqrt{\lambda}} \tag{XI}$$

This is the important equation for exit gradient when floor is absent (i.e b = 0) $\lambda = 1$.

$$G_E = \frac{H}{\pi d} \tag{XII}$$

Eccentric Sheet Pile (Independent solution)

Refer Fig. AI- 3.

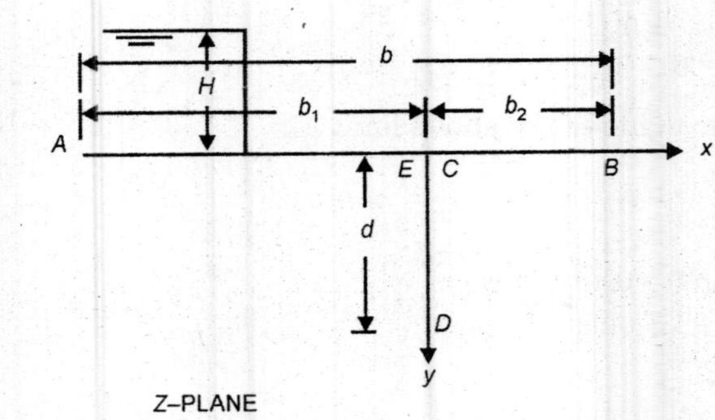

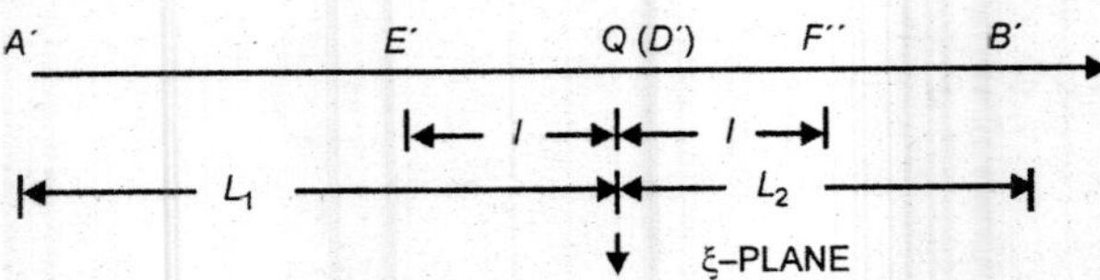

Fig. AI–3

The Schwarz – Christoffel equation is

$$z = A\int \frac{d\xi}{(\xi-\xi_1)^{\lambda_1}(\xi-\xi_2)^{\lambda_2}(\xi-\xi_n)^{\lambda_3}}$$

$$= A\int \frac{d\xi}{(\xi-1)^{1/2}(\xi-0)^{-1}(\xi+1)^{1/2}} = A\int \frac{\xi d\xi}{\sqrt{\xi^2-1}}$$

$$\xi = A\sqrt{\xi^2-1}+B \tag{17}$$

To determine constant B consider co-ordinates of points C and C′

$$z = 0 \text{ and } \xi = 1$$

Hence from Eq. (17), $B = 0$

To determine constant A, consider co-ordinates of D and D'

$$z = id \text{ and } \xi = 0$$

Hence from Eq. (17)

$$id = A\sqrt{-1} = iA$$

$$A = d.$$

Hence transformation equation becomes

$$z = d\sqrt{\xi^2-1} \tag{18}$$

Now, for the horizontal floor in ξ -plane we have

$$\xi+\frac{1}{2}(L_1-L_2) = \frac{1}{2}(L_1-L_2)\cosh w$$

Putting $\frac{L_1+L_2}{2} = \lambda$ and $\frac{L_1+L_2}{2} = \lambda_1$ we get

$$\xi = 1 \cosh w - \lambda_1$$

For the floor profile $U = 0$ hence $\cosh w = \cosh iv = \cos v$

$$\therefore \quad \xi = \lambda \cos V - \lambda_1 \tag{19}$$

Putting this in Eq. (18) we get

$$z = d\sqrt{(\lambda\cos V-\lambda_1)^2-1} \tag{20}$$

This equation is useful to find out the pressure distribution below the floor. For finding co-ordinates of A' and B' we get from Eq. (18)

$$-b_1 = d\sqrt{L_1^2-1} \quad \text{and} \quad b_2 = d\sqrt{L_2^2-1}$$

$$L_1 = \sqrt{\left(\frac{b_1}{d}\right)^2+1} = \sqrt{1+\alpha_1^2} \tag{21 (a)}$$

and $$L_2 = \sqrt{\left(\frac{b_2}{d}\right)^2 + 1} = \sqrt{1+\alpha_2^2} \tag{21 (b)}$$

When $$\alpha_1 = \frac{b_1}{d} \text{ and } \alpha_2 = \frac{b_2}{d}$$

Hence $$\lambda = \frac{L_1 + L_2}{2} = \frac{\sqrt{1+\alpha_1^2} + \sqrt{1+\alpha_2^2}}{2} \tag{22 (a)}$$

and $$\lambda_1 = \frac{L_1 - L_2}{2} = \frac{\sqrt{1+\alpha_1^2} - \pi + \alpha_2^2}{2} \tag{22 (b)}$$

For the floor $y = 0$.

Hence from Eq. (20)

$$z = x + iy = n = d\sqrt{(\lambda \cos v - \lambda_1)^2 - 1}$$

or $$\lambda \cos v - \lambda_1 = \pm\sqrt{\frac{d^2 + x^2}{d^2}}$$

or $$\cos V = \frac{\lambda_1 d \pm \sqrt{d^2 + x^2}}{\lambda d}$$

$$\therefore \quad V = \pi\frac{P}{H} = \frac{\cos^{-1} \lambda_1 d \pm \sqrt{d^2 + x^2}}{\lambda d}$$

or $$P = \frac{H}{\pi}\cos^{-1}\frac{\lambda_1 d \pm \sqrt{d^2 + x^2}}{\lambda d} \tag{23}$$

Use –ive sign for *U/S* floor and + ive sign for *D/S* floor

For point *E* $x = 0$

$$\therefore \quad P_E = \frac{H}{\pi}\cos^{-1}\frac{\lambda_1 - 1}{\lambda} \tag{24 (a)}$$

For point C $x = 0$

$$P_c = \frac{H}{\pi}\cos^{-1}\left[\frac{\lambda_1 + 1}{\lambda}\right] \tag{24 (b)}$$

For pressure along the sheet pile, put $x = 0$ in Eq. 20

$$\therefore \quad z = x + iy = iy = d\sqrt{(\lambda \cos V - \lambda_1)^2 - 1}$$

or
$$\lambda \cos V - \lambda_1 = \pm\sqrt{\frac{d^2 - y^2}{d^2}}$$

$$\cos V = \frac{\lambda_1 d \pm \sqrt{d^2 - y^2}}{\lambda d}$$

$$\therefore \quad V = \pi\frac{P}{H} = \cos^{-1}\left(\frac{\lambda_1 d \pm \sqrt{d^2 - y^2}}{\lambda d}\right)$$

or
$$P = \frac{H}{\pi}\cos^{-1}\left(\frac{\lambda_1 d \pm \sqrt{d^2 - y^2}}{\lambda d}\right)$$

For point D $\quad y = d$

$$P_D = \frac{H}{\pi}\cos^{-1}\left(\frac{\lambda_1}{\lambda}\right) \tag{24 (c)}$$

The curves of sheet No. 15.1 are based on Eq. 24 (a), (b), (c).

4. Floor with End Sheet Pile

Refer Fig. AI–4. both for z -plane and ξ - plane

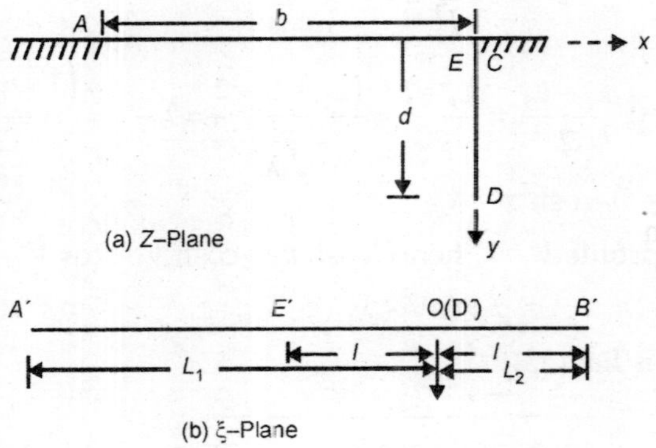

Fig. AI–4

Schwarz–Christoffel equation becomes

$$z = A\int\frac{d\xi}{(\xi-1)^{1/2}(\xi-0)^{-1}(\xi+1)^{1/2}} = A\int\frac{\xi d\xi}{\sqrt{\xi^2-1}}$$

or $\quad z = A\sqrt{\xi^2-1}+B$

For point C $\quad z = 0$ and $\xi = 1 \quad \therefore B = 0$

For point D $\quad z = id$ and $\xi = 0$

$\therefore \quad id = A\sqrt{-1} = iA \quad$ or $\quad A = d$

Thus the transformation equation becomes

$$z = d\sqrt{\xi^2-1} \tag{25}$$

From Fig. AI–4 $\quad L_2 = 1$. For L_1 consider co-ordinates of A and A'

$\therefore \quad -b_1 = d\sqrt{L_1^2-1} \quad$ (From Eq. 25)

$$L_1 = \sqrt{\left(\frac{b}{d}\right)^2+1} = \sqrt{1+\alpha^2} \text{ where } \alpha = \frac{b}{d}$$

Now for horizontal floor of z-plane

$$\xi+\frac{1}{2}(L_1-L_2) = \frac{1}{2}(L_1-L_2)\cosh w$$

$$\lambda = \frac{L_1-L_2}{2} = \frac{L_1+1}{2} = \frac{\sqrt{1+\alpha^2}+1}{2} \tag{26 (a)}$$

and $$\lambda_1 = \frac{L_1-L_2}{2} = \frac{L_1-1}{2} = \frac{(L_1+1)-2}{2} = \lambda-1 = \frac{(1+\alpha^2)-1}{2} \tag{26 (b)}$$

$$\xi = \lambda \cosh w - \lambda_1$$

For the floor profile $V = 0$, hence $\cosh w = \cosh V = \cos V$

$$\xi = \lambda \cos V - \lambda_1$$

Puttingh this is Eq. (25)

$$z = d\sqrt{(\lambda\cos V-\lambda_1)^2-1} \tag{27}$$

For the floor $y = 0$ and hence $z = n$

$$x = d\sqrt{(\lambda\cos v-\lambda_1)^2} \tag{27 (a)}$$

$$\therefore \qquad \lambda \cos V - \lambda_1 = \pm\sqrt{\frac{d^2 + x^2}{d^2}}$$

From this

$$V = \pi\frac{P}{H} = \cos^{-1}\frac{\lambda_1 d \pm \sqrt{d^2 + x^2}}{\lambda d} \text{ (Use – ive sign for } U/S \text{ floor)}$$

$$P = \frac{H}{\pi}\cos^{-1}\frac{\lambda_1 d \pm \sqrt{d^2 + x^2}}{\lambda d} \qquad (28)$$

For point E, $x = 0$

$$P_E = \frac{H}{\pi}\cos^{-1}\left(\frac{\lambda_1 d - d}{\lambda_1 d}\right) = \frac{H}{\pi}\cos^{-1}\frac{\lambda_1 - 1}{\lambda}$$

or

$$P_E = \frac{H}{\pi}\cos^{-1}\frac{\lambda - 2}{\lambda} \text{ or } (\sin a\, \lambda_1 = \lambda - 1) \qquad (29(a))$$

From C $\qquad x = 0$

$$\therefore \qquad P_C = \frac{H}{\pi}\cos^{-1}\left(\frac{\lambda_1 d + d}{\lambda_1 d}\right) = \frac{H}{\pi}\cos^{-1}\frac{\lambda_1 + 1}{\lambda}$$

$$= \frac{H}{\pi}\cos^{-1}\frac{\lambda}{\lambda} = 0 \qquad (29(b))$$

For the sheet pile, $x = 0$. Hence from Eq. (27)

$$Cy = d\sqrt{(\lambda \cos V - \lambda_1)^2 - 1}$$

or

$$\lambda \cos V - \lambda_1 = \pm\sqrt{\frac{d^2 - y^2}{d^2}}$$

$$\therefore \qquad V = \pi\frac{P}{H} = \cos^{-1}\frac{\lambda_1 d \pm \sqrt{d^2 - y^2}}{\lambda d}$$

or

$$P = \frac{H}{\pi}\cos^{-1}\frac{\lambda_1 d \pm \sqrt{d^2 - y^2}}{\lambda d}$$

From point $D \qquad y = d$

$$\therefore \quad P_D = \frac{H}{\pi}\cos^{-1}\frac{\lambda_1 d}{\lambda d} = \frac{H}{\pi}\cos^{-1}\left(\frac{\lambda - 1}{\lambda}\right) \qquad (29(c))$$

Curves of Plates 15.2 and 15.3 are based on Eq. 29 (a) (b) (c).

Exit Gradient

For finding exit gradient, consider Eq. 27 (b)

$$iy = d\sqrt{(\lambda \cos V - \lambda_1)^2 - 1} \qquad (30\ (a))$$

Let $\lambda \cos V - \lambda_1 = \cos \beta$

$$iy = d\sqrt{\cos^2 \beta - 1} = id \sin \beta$$

$$y = d \sin \beta$$

Differentiate it with respect to y

$$1 = d \cos\beta \frac{d\beta}{dy} \qquad (30\ (b))$$

But from Eq. 30(a) $\lambda \sin V \dfrac{dV}{dy} = \sin\beta \dfrac{d\beta}{dy}$ (30 (c))

Putting the value of $\dfrac{d\beta}{dy}$ in Eq. 30 (b)

$$1 = d\cos\beta \frac{\lambda \sin V}{\sin\beta}\frac{dv}{dy} \qquad (30\ (d))$$

At exit end, $V = 0$. Hence Eq. 30 (a) becomes

$\cos \beta = \lambda \cos V - \lambda_1 = \lambda \cos 0 - \lambda_1 = \lambda_1 - 1$

$\therefore \quad \beta = 0$ when $V = 0$. $\therefore \quad \dfrac{\sin V}{\sin\beta} = \dfrac{0}{0}$

Let us determine the limit of $\left(\dfrac{\sin V}{\sin\beta}\right)_{\substack{V=0\\ \beta=0}}$

$$\mathrm{Lt}\left(\frac{\sin V}{\sin\beta}\right)_{\substack{V=0\\ \beta=0}} = \beta = 0 = \underset{\substack{V=0\\ \beta=0}}{\mathrm{Lt}} \frac{\frac{d}{dv}(\sin v)}{\frac{d}{dy}(\sin\beta)} = \underset{\substack{V=0\\ \beta=0}}{\mathrm{Lt}} \frac{\cos V \frac{dV}{dy}}{\cos\beta\frac{d\beta}{dy}}$$

But
$$\frac{\frac{dV}{dy}}{\frac{d\beta}{dy}} = \frac{\sin\beta}{\lambda\sin V} \text{ from Eq. 30 (c)}$$

$$\therefore \quad \underset{\substack{V=0\\ \beta=0}}{\mathrm{Lt}}\left(\frac{\sin V}{\sin\beta}\right) = \underset{\substack{V=0\\ \beta=0}}{\mathrm{Lt}}\left(\frac{\sin\beta}{\lambda\sin V}\right)$$

Hence
$$\underset{\substack{V=0\\ \beta=0}}{\mathrm{Lt}}\left(\frac{\sin V}{\sin\beta}\right) = \frac{1}{\sqrt{\lambda}} \tag{31}$$

Putting the value in Eq. 30(d) we get, for $V = 0$ and $\beta = 0$

$$1 = d\frac{\lambda}{\sqrt{\lambda}}\left(\frac{dV}{dy}\right)_{\text{Exit}}$$

or
$$\left(\frac{dV}{dy}\right)_{\text{Exit}} = \frac{1}{d\sqrt{\lambda}} \tag{32}$$

Now
$$\left(\frac{dp}{dy}\right)_{\text{Exit}} = G_E = \frac{H}{\pi}\left(\frac{dV}{dy}\right)_{\text{Exit}} = \frac{H}{\pi}\frac{1}{d\sqrt{\lambda}}$$

or

$$G_E = \frac{H}{d}\frac{1}{\pi\sqrt{\lambda}} \tag{33}$$

This is the equation for exit gradient. In Plate 15.3 the exit gradient curve is based on this equation.

Example AI–1. A *horizontal impervious floor of length 25 m is put with a cut off of 5 m depth on its D/S end. Find out the uplift pressure at point E and D and also the exit gradient of the head causing seepage is 2.5 m.*

Solution. Pressure at point *E* $P_E = \frac{H}{\pi}\cos^{-1}\frac{\lambda-2}{\lambda}$

where $\lambda = \frac{1+\sqrt{1+\alpha^2}}{2}$ and $\alpha = \frac{b}{d} = \frac{25}{5} = 5$

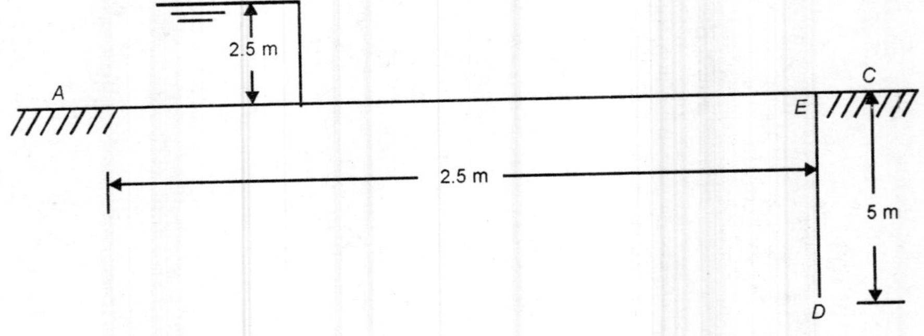

Fig. AI–5

$$\therefore \quad \lambda = \frac{1+\sqrt{1+\alpha^2}}{2} = \frac{1+\sqrt{1+25}}{2} = 3.05$$

$$P_E = \frac{2.5}{\pi}\cos^{-1}\left(\frac{3.05-2}{3.05}\right) = \frac{2.5}{\pi}(1.2195) = 0.97\text{m}$$

Also
$$P_D = \frac{H}{\pi}\times\cos^{-1}\frac{(\lambda-1)}{\lambda} - 1 = \frac{2.5}{\pi}\cos^{-1}\left(\frac{3.05-1}{3.05}\right)$$

$$= \frac{2.5}{\pi}\times 0.8337 = 0.663 \text{ m}$$

$$G_E = \frac{H}{d}\frac{1}{\pi\sqrt{\lambda}} = \frac{2.5}{5}\frac{1}{\pi\sqrt{3.05}} = \frac{1}{10.98} = \frac{1}{\pi}$$

Example AI–2. *A horizontal pucca floor of length 15 m is provided with cut-off of 4 m depth, placed eccentrically as shown in Fig. AI–6 compute the uplift pressures at point E, D and C.*

Given :

$b_1 = 10\ m$

$b = 5\ m$

$d = 4\ m$

$$b = 15\ m$$
$$H = 2\ m$$

Fig. AI–6

$$\alpha_1 = \frac{b_1}{d} = \frac{10}{4} = 2.5, \qquad \alpha_2 = \frac{b_2}{d} = \frac{5}{4} = 1.25$$

$$\lambda = \frac{\sqrt{1+\alpha_1^2} + \sqrt{1+\alpha_1^2}}{2} = \frac{\sqrt{1+(2.5)^2} + \sqrt{1+(1.25)^2}}{2} = 2.147$$

$$\lambda_1 = \frac{\sqrt{1+\alpha_1^2} - \sqrt{1+\alpha_1^2}}{2} = \frac{\sqrt{1+(2.5)^2} - \sqrt{1+(1.25)^2}}{2} = 0.546$$

$$P_E = \frac{H}{\pi}\cos^{-1}\frac{\lambda_1 - 1}{\lambda} = \frac{2}{\pi}\cos^{-1}\left(\frac{0.546-1}{2.147}\right) = \frac{2}{\pi}\times 1.784 = 1.135 \text{ m}$$

$$P_C = \frac{H}{\pi}\cos^{-1}\frac{\lambda_1 + 1}{\lambda} = \frac{2}{\pi}\cos^{-1}\left(\frac{0.546+1}{2.147}\right) = \frac{2}{\pi}\times 0.767 = 0.488 \text{ m}$$

$$P_D = \frac{H}{\pi}\cos^{-1}\frac{\lambda_1}{\lambda} = \frac{2}{\pi}\cos^{-1}\left(\frac{0.546}{2.147}\right) = \frac{2}{\pi}\times 1.3136 = 0.836 \text{ m}$$

Example AI–3. *For the profile shown in Fig. AI–7 compute the uplift pressure at points C_1 and E_2.*

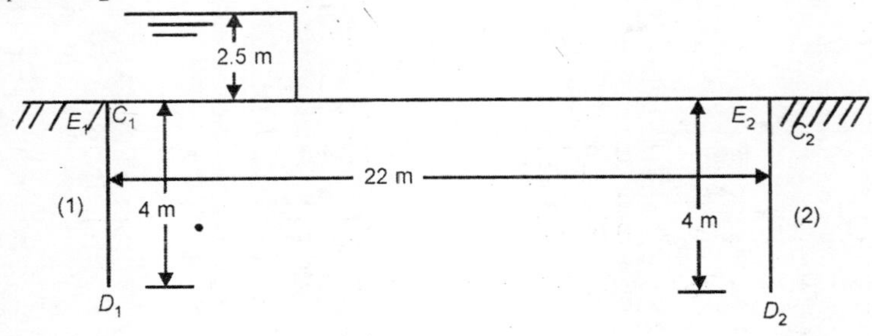

Fig. AI–7

Given $b = 22\ m$

$H = 2.5\ m$

$d = 4\ m$

Solution. Pile No. 2

$$\alpha = \frac{b}{d} = \frac{22}{4} = 5.5$$

Neglecting pressure of *U/S* pile

$$P_{E_2} = \frac{H}{\pi}\cos^{-1}\left(\frac{\lambda-2}{\lambda}\right)$$

$$\lambda = \frac{1+\sqrt{1+\alpha^2}}{2} = \frac{1+\sqrt{1+5.5^2}}{2} = 3.295$$

$$P_{E_2} = \frac{2.5}{\pi}\cos^{-1}\left(\frac{3.295-2}{3.295}\right) = \frac{2.5}{\pi}\times 1.1669 = 0.7425\ \text{m}$$

The correction due to interference of pile at *U/S*.

$$C = 19\sqrt{\frac{D}{b'}}\left(\frac{d+D}{b}\right)$$

where $D = b = 4$ m, $b = b' = 22$ m

$$C = 19\sqrt{\frac{4}{22}}\times\left(\frac{4+4}{22}\right) = 19\sqrt{\frac{1}{5.5}}\times\frac{8}{22} = 2.95\% \text{ (Substractive)}$$

Corrected $P_{E_2} = 0.7425 - 0.0295 \times 2.5 = 0.669$ m

Pile No. 1 Both piles are of same depth. By principle of reversibility of flow $P_{C_1} = H - P_{E_1} = 2.5 - 0.7425 = 1.7575$ m.

Correction due to interference of Pile No. 2 is

$$C = 19\sqrt{\frac{D}{b'}}\left(\frac{d+D}{b}\right)$$

$$= 19\sqrt{\frac{4}{22}}\left(\frac{4+4}{22}\right) = 2.95\% \text{ (Additive)}$$

$$= 0.0295 \times 2.5 = 0.07375 \text{ m}$$

Corrected $P_{C_1} = 1.7575 + 0.07375 = 1.8313$ m.

❑❑❑

Index

D

E

F

G

T

U

V

W

Y